Wine Chemistry and Biochemistry

M. Victoria Moreno-Arribas · M. Carmen Polo
Editors

Wine Chemistry and Biochemistry

Springer

Editors

M. Victoria Moreno-Arribas
Instituto de Fermentaciones
Industriales (CSIC), Madrid
Spain
mvmoreno@ifi.csic.es

M. Carmen Polo
Instituto de Fermentaciones
Industriales (CSIC), Madrid
Spain
mcpolo1@gmail.com

ISBN: 978-1-4419-2548-0 e-ISBN: 978-0-387-74118-5
DOI 10.1007/978-0-387-74118-5

Softcover reprint of the hardcover 1st edition 2009

Printed on acid-free paper

springer.com

Foreword

Winemaking is a most fascinating and complex transformation process of a raw plant material. It starts with the arrival of the harvest at the cellar and ends with the most active and decisive fermentation steps. After this, for some wines, comes the long aging period of the wine, during which the bouquet and taste of the wine is developed and refined. The transformation of grape must in wine is a priori a spontaneous phenomenon. The microbial complex present on the grape berry is exposed to a new ecosystem when the grapes are crushed and pressed. It then evolves spontaneously following the conditions dictated by both the nature of the microorganisms present and the composition of the community.

Without the skill and attention of the oenologist and winemaker, the system would evolve into a fermented product, the quality of which would have little chance of satisfying the consumer. This expertise is based on scientific knowledge of the phenomena that occur in this complex environment. After its beginnings mainly based on observation and empiricism, oenology now uses scientific data derived from research in chemistry, biochemistry and microbiology. Together with biochemical reactions catalyzed by enzymes of yeasts and bacteria, chemical reactions also occur between molecules already present in the must, those gradually extracted from the grape solids during fermentation, those derived from metabolisms and, possibly, also those released by the wood. For many of them the temperature and dissolved oxygen parameters related to technological operations of the winery can have dramatic effects and the quality of the final wine depends on the type and intensity of reactions taking place.

From the beginning of the twentieth century, chemistry and microbiology have been used in an attempt to interpret the observations used by winemakers. These constitute the foundations on which the basic rules for winemaking and aging were established. Hence, as producers' control of the events of winemaking and aging steadily increased, so did wine quality. First, defects and the most critical alterations have been avoided. After that, knowledge has become more accurate and reliable, and more technological tools have been developed, and now the winemaker can control the evolution of the system as a whole with great efficiency.

Continuously, researchers in oenology, both chemists and biologists, appropriate the most efficient analytical methods and data to conduct their research. New molecules of wine aroma, color and flavor have been identified. Sensory analysis, increasingly present in the laboratory alongside chemical analysis methods, reveals the importance of molecules present even at very low concentrations and the importance of interactions between them. Genomics is used in research on yeast and bacteria and reveals the extraordinary complexity of the microbial consortium, giving microbiologists keys for the optimal use of the natural biodiversity of species involved in fermentation.

The authors, invited by M.C. Polo and M.V. Moreno-Arribas to write this book, are recognized in their own field for their research and ability to transfer scientific results from the laboratory to the winemaking process and storage cellar, and here provide updates on the most recent advances in the field.

With this manual, oenologists will be able to update their knowledge and benefit from a deeper understanding of the phenomena they observe in practice. Moreover, researchers in oenology are now highly specialized, and must conduct their activities at the basic level, while finding in the cellars and caves the elements of their thinking. While in the laboratories, chemists specializing in macromolecules or volatile compounds and microbiologists specializing in yeasts or bacteria must continue their research into the interactions taking place. Working individually without knowledge of research in this field from other specialists their efforts lose all meaning and progress remains erratic or limited. Scientists will, therefore, benefit from this handbook that enables them to contemplate and understand the results and progress made in other specialities related to this area.

UMR oenologie. Aline Lonvaud
Université Victor Segalen Bordeaux 2
351 Cours de la Libération. 33405 Talence Cedex, France
e-mail: aline.lonvaud@oenologie.u-bordeaux2.fr

Contents

Part II Wine Chemical Compounds and Biochemical Processes

Contributors

André Barata Laboratório de Microbiologia, Departamento de Botânica e Engenharia Biológica, Instituto Superior de Agronomia, Technical University of Lisbon, Tapada da Ajuda, 1349-017 Lisbon, Portugal, abarata@isa.utl.pt

Begoña Bartolomé Instituto de Fermentaciones Industriales, (CSIC), C/Juan de la Cierva 3, 28006 Madrid, Spain, bartolome@ifi.csic.es

Raymond Baumes UMR Sciences pour l'Oenologie, 2 place Viala – 34060 Montpellier Cedex, France, baumes@supagro.inra.fr

Angel Anocibar Beloqui Enologist, Abadia Retuerta, E47340 Sardón de Duero, Spain

Alain Bertrand Regents professor, Université V. Segalen Bordeaux 2. Faculté d'œnologie, 351 Cours de la Libération, F33405 Talence, France, alain.bertrand@oenologie.u-bordeaux2.fr

Juan Cacho Laboratory for Flavor Analysis and Enology, Department of Analytical Chemistry, Faculty of Sciences, University of Zaragoza, 50009 Zaragoza, Spain, jcacho@unizar.es

Véronique Cheynier INRA, UMR Sciences pour l'Oenologie, 2 place Viala, F-34060 Montpellier, France, cheynier@supagro.inra.fr

Christopher Bruce Colby Arup, Level 2 Optus Centre, 431-439 King William Street, Adelaide, SA 5000, Australia, chris.colby@arup.com.au

Antonella Costantini CRA-Centro di Ricerca per l'Enologia, Via Pietro Micca 35 – 14100 Asti, Italy, micro@isenologia.it

Alberto Dávalos Servicio de Bioquímica-Investigación, Hospital Ramón y Cajal, Ctra. de Colmenar, km 9, E-28034, Instituto de Salud Carlos III, Madrid, Spain, adavalos.hrc@salud.madrid.org

Victor de Freitas Centro de Investigação em Química, Departamento de Química, Faculdade de Ciências, Universidade do Porto, Rua do Campo Alegre, 687, 4169-007 Porto, Portugal, vfreitas@fc.up.pt

M. Consuelo Díaz-Maroto Área de Tecnología de los Alimentos, Facultad de Ciencias Químicas, Campus Universitario 10, Universidad de Castilla-La Mancha, 13071 Ciudad Real, Spain, MariaConsuelo.Díaz@uclm.es

Marc Dubernet Laboratoires Dubernet, 9 quai d'Alsace, F-11100 Narbonne, France, marc.dubernet@dubernet.com

Denis Dubourdieu Université Victor Segalen Bordeaux 2, Faculté d' Œnologie, UMR, Œnologie, INRA, Institut des Sciences de la Vigne et du Vin, 351 cours de la Libération, 33405 Talence Cedex, France, denisdubourdieu@wanadoo.fr

Vicente Ferreira Laboratory for Flavor Analysis and Enology, Department of Analytical Chemistry, Faculty of Sciences, University of Zaragoza, 50009 Zaragoza, Spain, vferre@unizar.es

Emilia García-Moruno CRA-Centro di Ricerca per l'Enologia, Via Pietro Micca 35 – 14100 Asti, Italy, e.garciamoruno@isenologia.it

Paul A. Henschke The Australian Wine Research Institute, Urrbrae, Adelaide, South Australia, paul.henschke@awri.com.au

Philippe Jeandet Laboratory of Enology and Applied Chemistry, Research Unit N° 2069, Faculty of Science, University of Reims, PO Box 1039, 51687 Reims cedex 02, France, philippe.jeandet@univ-reims.fr

Miguel A. Lasunción Servicio de Bioquímica-Investigación, Hospital Ramón y Cajal, Madrid; Universidad de Alcalá, and CIBER Fisiopatología Obesidad y Nutrición (CB06/03), Instituto de Salud Carlos III, Spain, miguel.a.lasuncion@hrc.es

V. Loureiro Laboratório de Microbiologia, Departamento de Botânica e Engenharia Biológica, Instituto Superior de Agronomia, Technical University of Lisbon, Tapada da Ajuda, 1349-017 Lisbon, Portugal, vloureiro@isa.utl.pt

Manuel Malfeito-Ferreira Laboratório de Microbiologia, Departamento de Botânica e Engenharia Biológica, Instituto Superior de Agronomia, Technical University of Lisbon, Tapada da Ajuda, 1349-017 Lisbon, Portugal, mmalfeito@isa.utl.pt

Richard Marchal Laboratory of Enology and Applied Chemistry, Research Unit N° 2069, Faculty of Science, University of Reims, PO Box 1039, 51687 Reims cedex 02, France, richard.marchal@univ-reims.fr

Pedro J. Martín-Álvarez Instituto de Fermentaciones Industriales, Consejo Superior de Investigaciones Científicas (CSIC), C/Juan de la Cierva 3, 28006 Madrid, Spain, pjmartin@ifi.csic.es

Isabel Martínez-Castro Instituto de Química Orgánica General (CSIC), C/Juan de la Cierva 3, 28006 Madrid Spain, iqomc16@iqog.csic.es

Adolfo J. Martínez-Rodríguez Instituto de Fermentaciones Industriales (CSIC), C/Juan de la Cierva 3, 28006. Madrid, Spain, amartinez@ifi.csic.es

Juan C. Mauricio Professor of Microbiology, Departamento de Microbiología, Edificio Severo Ochoa, Campus Universitario de Rabanales, Universidad de Córdoba, 14071-Córdoba, Spain, mi1gamaj@uco.es

María Monagas Instituto de Fermentaciones Industriales, (CSIC), C/Juan de la Cierva 3, 28006 Madrid, Spain, monagas@ifi.csic.es

M. Victoria Moreno-Arribas Instituto de Fermentaciones Industriales (CSIC), C/Juan de la Cierva 3, 28006 Madrid, SPAIN, mvmoreno@ifi.csic.es

Rafael A. Peinado Assistant Professor of Agricultural Chemistry, Departamento de Química Agrícola y Edafología, Edificio Marie Curie, Campus Universitario de Rabanales, Universidad de Córdoba, 14071-Córdoba, Spain, qe1peamr@uco.es

M. Soledad Pérez-Coello Área de Tecnología de los Alimentos, Facultad de Ciencias Químicas, Campus Universitario 10, Universidad de Castilla-La Mancha, 13071 Ciudad Real, Spain, soledad.perez@uclm.es

M. Carmen Polo Instituto de Fermentaciones Industriales, (CSIC), C/Juan de la Cierva 3, 28006 Madrid, Spain, mcpolo1@gmail.com

Céline Poncet-Legrand INRA, UMR Sciences pour l'Oenologie, 2 place Viala, F-34060 Montpellier, France, andrea.wilkens@tu-braunschweig.de

María Ángeles Pozo-Bayón Instituto de Fermentaciones Industriales (CSIC), C/Juan de la Cierva 3, 28806 Madrid, Spain, mdelpozo@ifi.csic.es

Encarnación Pueyo Instituto de Fermentaciones Industriales (CSIC), C/Juan de la Cierva 3, 28006, Madrid, Spain, epueyo@ifi.csic.es

Gary Reineccius University of Minnesota, Department of Food Science and Nutrition, 1334, Eckles Avenue, St Paul, Minnesota, 55108, USA, greinecc@umn.edu

Michael Rentzsch Institut für Lebensmittelchemie, TU Braunschweig, Schleinitzstrasse 20, D-38106 Braunschweig, Germany, m.rentzsch@tu-braunschweig.de

Celestino Santos-Buelga Unidad de Nutrición y Bromatología, Facultad de Farmacia, Universidad de Salamanca, Campus Miguel de Unamuno s/n, E-37007 Salamanca, Spain, csb@usal.es

M. Luz Sanz Instituto de Química Orgánica General (CSIC), C/Juan de la Cierva 3, 28006 Madrid, Spain, mlsanz@iqog.csic.es

Nancy Terrier INRA, UMR Sciences pour l'Oenologie, 2 place Viala, F-34060 Montpellier, France, terrier@supagro.inra.fr

Takatoshi Tominaga Université Victor Segalen Bordeaux 2, Faculté d' Œnologie, UMR, Œnologie, INRA, Institut des Sciences de la Vigne et du Vin, 351 cours de la Libération, 33405 Talence Cedex, France, takatoshi.tominaga@oenologie.u-bordeaux2.fr

Maurizio Ugliano The Australian Wine Research Institute, PO Box 197, Glen Osmond (Adelaide), SA 5064, Australia, maurizio.ugliano@awri.com.au

Elizabeth Joy Waters The Australian Wine Research Institute, PO Box 197, Glen Osmond, SA 5064, Australia, Elizabeth.Waters@awri.com.au

Andrea Wilkens Institut für Lebensmittelchemie, TU Braunschweig, Schleinitzstrasse 20, D-38106 Braunschweig, Germany, andrea.wilkens@tu-braunschweig.de

Peter Winterhalter Institut für Lebensmittelchemie, TU Braunschweig, Schleinitzstrasse 20, D-38106 Braunschweig, Germany, p.winterhalter@tu-bs.de

Fernando Zamora Departament de Bioquímica i Biotecnologia, Facultat d'Enologia de Tarragona, Universitat Rovira i Virgili, Campus de Sescelades, C/Marcel.li Domingo s/n. 43007-Tarragona, Spain, fernando.zamora@urv.cat

Introduction

M. Carmen Polo and M. Victoria Moreno-Arribas

The aim of this book is to describe chemical and biochemical aspects of winemaking which are currently being researched. The areas of most interest at present and the subjects in which this interest is likely to continue or to increase in the following years have been selected.

The first part of the book concerns the most important aspects of winemaking technology and microbiology. The second part, the most extensive, deals with the different groups of compounds, how these are modified during the various steps of the production process, and how they influence the wine quality and its sensorial aspects and physiological activity. The third section describes undesirable alterations of wines, including those that affect quality and food safety. Finally, two aspects have been considered which have not yet been tackled in any other book on oenology – automatic analysers used in oenological laboratories for control and research purposes, and the statistical treatment of data. In this last subject, the author not only describes the tools available for analytical data processing but also indicates the most appropriate treatment to apply, depending on the information required. The chapter is illustrated throughout with examples from the oenological literature.

'Wine chemistry and biochemistry' is scientifically written including current trends but also in a style that is easy and clear to understand. It is hoped that it will serve as a most useful text and reference source for wine researchers and oenologists alike, as well as for winemakers and other professionals of the sector, and students of oenology, food technology and similar disciplines.

The editors would like to express their thanks to Springer and all the authors who contributed their expertise and know-how to the success of this book.

Part I
Chemical and Biochemical Aspects of Winemaking

Chapter 1
Biochemistry of Alcoholic Fermentation

Fernando Zamora

Contents

1.1 Introduction

Alcoholic fermentation is the anaerobic transformation of sugars, mainly glucose and fructose, into ethanol and carbon dioxide. This process, which is carried out by yeast and also by some bacteria such as *Zymomonas mobilis*, can be summarised by this overall reaction.

$$\underset{\text{Hexoses}}{C_6H_{12}O_6} \rightarrow 2\,\underset{\text{Ethanol}}{CH_3CH_2OH} + \underset{\text{Carbon dioxide}}{2CO_2}$$

However, alcoholic fermentation is fortunately a much more complex process. At the same time as this overall reaction proceeds, a lot of other biochemical, chemical and physicochemical processes take place, making it possible to turn the grape

F. Zamora (✉)
Departament de Bioquímica i Biotecnologia, Facultat d'Enologia de Tarragona, Universitat Rovira i Virgili, Campus de Sescelades, C/ Marcel.li Domingo s/n. 43007-Tarragona, Spain
e-mail: fernando.zamora@urv.cat

M.V. Moreno-Arribas, M.C. Polo (eds.), *Wine Chemistry and Biochemistry*,
DOI 10.1007/978-0-387-74118-5_1,

juice into wine. Besides ethanol, several other compounds are produced throughout alcoholic fermentation such as higher alcohols, esters, glycerol, succinic acid, diacetyl, acetoin and 2,3-butanediol. Simultaneously, some compounds of grape juice are also transformed by yeast metabolism. Without the production of these other substances, wine would have little organoleptic interest.

At the start of the winemaking process, several species of yeast may be present in the grape juice. This biodiversity depends on several factors such as grape variety, the ripening stage at harvest, the antifungal treatments, the climatic conditions of the year, the development of grey rot or other fungal plagues and the viticultural practices (Sapis-Domerq 1980; Pretorius et al. 1999). However, other factors are also important. All contact of grapes and must during harvest, transport and, in particular winery operations significantly influence the final distribution of yeasts at the beginning of alcoholic fermentation (Constantí et al. 1997; Mortimer and Polsinelli 1999).

Different yeast species participate in spontaneous alcoholic fermentation even when sulphur dioxide is present (Constantí et al. 1998; Beltran et al. 2002). Usually *Kloeckera*, *Hanseniaspora* and *Candida* predominate in the early stages of alcoholic fermentation. Later, *Pichia* and *Metschnikowia* prevail in the middle stages. Finally, during the latter stages of fermentation, *Saccharomyces cerevisiae* is the predominant yeast because of its greater resistance to high ethanol concentration (Fleet 1993; Fleet and Heard 1993). Some other yeast, such as *Torulaspora*, *Kluyveromyces*, *Schizosacchaomyces*, *Zygosaccharomyces* and *Brettanomyces* may also be present during alcoholic fermentation and even in the wine itself, which may cause some organoleptic defects (Peynaud and Domercq 1959; Ribéreau-Gayon et al. 2000a).

Evidently, the succession of these different yeast species throughout alcoholic fermentation influences the final composition of wine in a way that, depending on which yeasts have grown, may be positive in some cases or negative in others (Chatonnet et al. 1995; Ribéreau-Gayon et al. 2000a). To prevent undesirable yeasts developing, wineries add sulphur dioxide to the grape juice and inoculate selected strains of dry yeasts (*Saccharomyces cerevisiae*). Sulphur dioxide has a drastic selective effect on yeast development. As *Saccharomyces cerevisiae* is more resistant to sulphur dioxide than most other yeasts, using this additive favours its development (Lafon-Lafourcade and Peynaud 1974; Romano and Suzzi 1993).

On the other hand, the inoculation of selected dry yeasts greatly increases the initial population of *Saccharomyces cerevisiae*. Nowadays, most wineries inoculate selected dry yeast in order to guarantee alcoholic fermentation without any deviation. However, other wineries, especially traditional wine cellars, continue to use spontaneous alcoholic fermentation because they believe it gives their wines greater complexity.

1.2 Yeast Development During Alcoholic Fermentation

At the beginning of the winemaking process, the yeasts start to metabolize the sugars and other nutrients present in the grape juice. The yeasts use all these nutrients to obtain energy and increase their population (Boulton et al. 1996; Ribéreau-Gayon

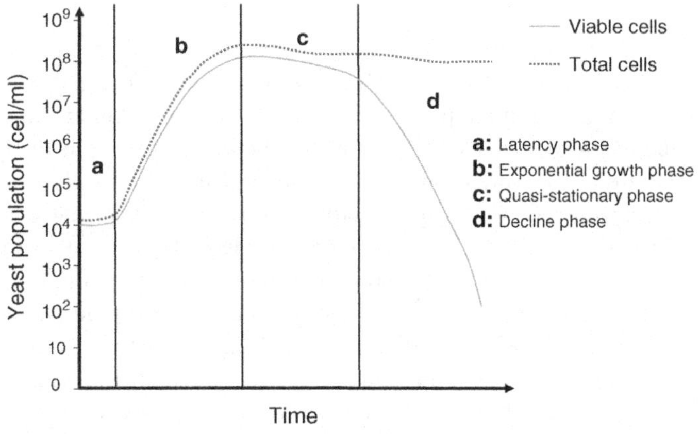

Fig. 1.1 Yeasts growth cycle

et al. 2000b). Figure 1.1 shows the classic yeast growth cycle under standard conditions (Fleet and Heard 1993; Del Nobile et al. 2003).

During the first hours the yeast population does not increase. During this period, also called the latency phase, it is necessary for the cell to adapt to the new environmental conditions. The initial population depends on several factors. If no yeasts are inoculated, the population is around 10^4 cells/ml. However, this population can be higher if the grapes have been attacked by grey rot or other fungal plagues. On the other hand, if selected dry yeasts were inoculated, the initial population would also be higher (around 5×10^6 cells/ml).

Once the yeasts have adapted to the environmental conditions, they begin to grow. This period, named the exponential growth phase, is highly influenced by temperature (Ough 1964), by the concentration of ammonia, amino acids and other nutrients (Lafon-Lafourcade 1983; Sablayrolles et al. 1996) and by the presence of oxygen (Sablayrolles and Barre 1986). During the exponential growth phase, the yeasts increase their population up to 10^7–10^8 cells/ml. This phase can last from 3 to 6 days. After that, yeast stops growing because some nutrients became deficient. During this new phase, called the quasi-stationary phase, the population of yeast remains nearly stable and can last from 2 to 10 days. Later, the decline phase begins and the population of yeast gradually decreases until it has almost completely disappeared. During this period yeasts die because of the lack of nutrients and also because ethanol and other substances produced during alcoholic fermentation are toxic to them (Lafon-Lafourcade et al. 1984).

The success of an alcoholic fermentation depends on maintaining the population of viable yeast at sufficient levels until all the fermentable sugars have been fully consumed (Bisson 1999; Zamora 2004). Otherwise, the winemaker is faced with the serious problem of stuck and sluggish fermentations. The causes and the ways to avoid stuck and sluggish fermentations are discussed later (Bisson and Butzke 2000).

1.3 Glycolysis

The word glycolysis comes from the Greek terms γλυκύς (glucus = sweet) and λύσις (lysis = rupture) and the process consists of the intracellular transformation of glucose (and fructose) into pyruvate. This biochemical pathway is the initial process of carbohydrate catabolism in most organisms and it takes place completely within the cytoplasm. This pathway was fully described in 1940 due, in great part, to the contributions of Gustav Embdem and Otto Meyenhorf. For that reason, it is also called the Embdem-Meyerhoff pathway in their honour although, regrettably, this name excludes other important contributors such as Gerti and Karl Cori, Carl Neuberg, Jacob Parnas, Hans von Euler and Otto Warburg (Kresge et al. 2005).

Yeasts use glycolysis as the main pathway for sugar catabolism (Gancedo 1988). The pentose pathway, which is used by some organisms such as acetic acid bacteria as the major pathway for sugar catabolism, is only used by yeast as a source of ribose and NADPH (Schaaf-Gersteenschaläger and Miosga 1996; Horecker 2002). Ribose is necessary for synthesizing nucleotides and nucleic acids whereas NADPH is required for some metabolic processes such as the lipid synthesis. Therefore yeasts use the pentose pathway not to obtain energy but rather to provide themselves with some of the substances indispensable for cell multiplication.

Glycolysis involves a sequence of 11 chemical reactions for breaking down hexoses and releasing energy in the chemical form of ATP (Barnett 2003). Figure 1.2 shows all the reactions in the glycolytic pathway.

Initially, hexoses are transported inside the cell by facilitated diffusion (Lagunas 1993). As the inner sugar concentration is lower than the external sugar concentration, no energy is necessary for this process.

The first step in glycolysis is the phosphorylation of glucose and fructose by a family of enzymes called hexokinases to form glucose 6-phosphate and fructose-6-phosphate (Gancedo 1988). This reaction consumes ATP, but it keeps the intracellular hexose concentration low and thus favours the continuous transport of sugars into the cell through the plasma membrane transporters. After this, phosphoglucose isomerase converts glucose-6-phosphate into fructose-6-phosphate.

Besides being intermediaries of glycolysis, glucose-6-phosphate and fructose-6-phosphate are also essential substrates for secondary metabolism. In fact, both hexose-phosphates are needed to synthesize the polysaccharides used to construct the cell wall (Cabib et al. 1982).

In the following stage, fructose-6-phosphate is phosphorylated again by the action of phosphofructokinase to form fructose-1,6-diphosphate. This reaction also consumes ATP. Later, the enzyme aldolase cleaves to fructose-6-phosphate. As a result of this reaction two triose phosphates are formed: dihydroxyacetone phosphate and glyceraldehyde-3-phosphate. This reaction produces a much greater proportion of dihydroxyacetone phosphate (96%), which is rapidly transformed into glyceraldehyde-3-phosphate by triose phosphate isomerase (Heinisch and Rodicio 1996).

Afterwards, the enzyme glyceraldehyde-3-phosphate dehydrogenase transforms glyceraldehyde-3-phosphate into 1,3-diphosphoglycerate. This reaction involves the oxidation of the molecule that is linked to reducing NAD^+ to NADH in order to redress the redox balance. Simultaneously, a substrate level phosphorylation takes

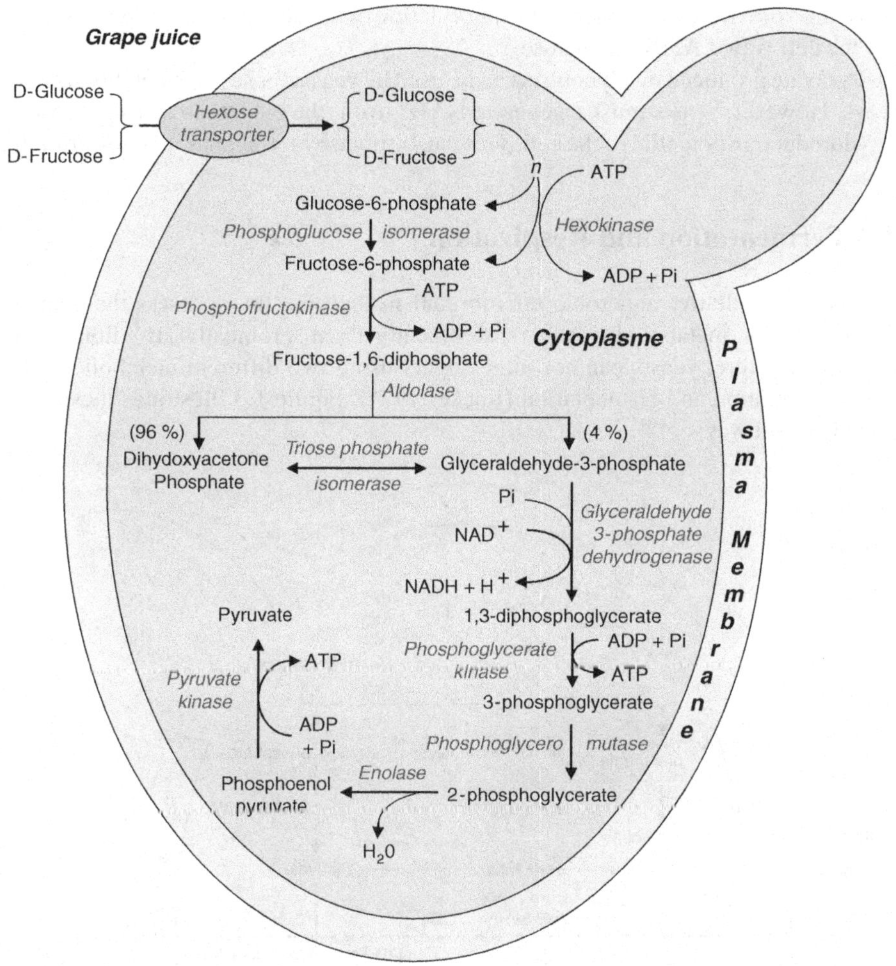

Fig. 1.2 Biochemical mechanism of glycolysis

place forming an energy rich bond between the oxidized carbon group and inorganic phosphate.

The next stage in glycolysis consists of transforming 1,3-diphosphoglycerate into 3-phosphoglycerate. This reaction, which is catalyzed by phosphoglycerate kinase, releases all the energy contained in the previously formed energy-rich bond, which the cell uses to phosphorylate one molecule of ADP into ATP.

After this, phosphoglycero mutase converts 3-phosphoglycerate into 2-phosphoglycerate, which is then dehydrated in phosphoenol pyruvate by the enzyme enolase. Phosphoenol pyruvate contains an energy-rich bond that is used by the enzyme pyruvate kinase to phosphorylate ADP into ATP. This reaction generates pyruvate, which is the final product of glycolysis.

As a consequence of glycolysis, each molecule of hexose generates two molecules of pyruvate, four of ATP and one of NADH. Since two molecules of ATP were consumed previously during the phosphorylation of the hexoses, the net energy gain for the cell is two ATPs per hexose.

Pyruvate produced by glycolysis can be used by yeasts for several metabolic pathways. However, yeasts must regenerate NAD^+ from the NADH to re-establish the oxydoreduction potential of the cell. This can be done by fermentation or respiration.

1.4 Fermentation and Respiration

Yeasts are facultative anaerobic microorganisms because they possess the genetic equipment for metabolizing sugars aerobically or anaerobically (Boulton et al. 1996). Therefore, yeasts can consume sugars using two different metabolic pathways: respiration and fermentation (Racker 1974). Figure 1.3 illustrates these biochemical pathways.

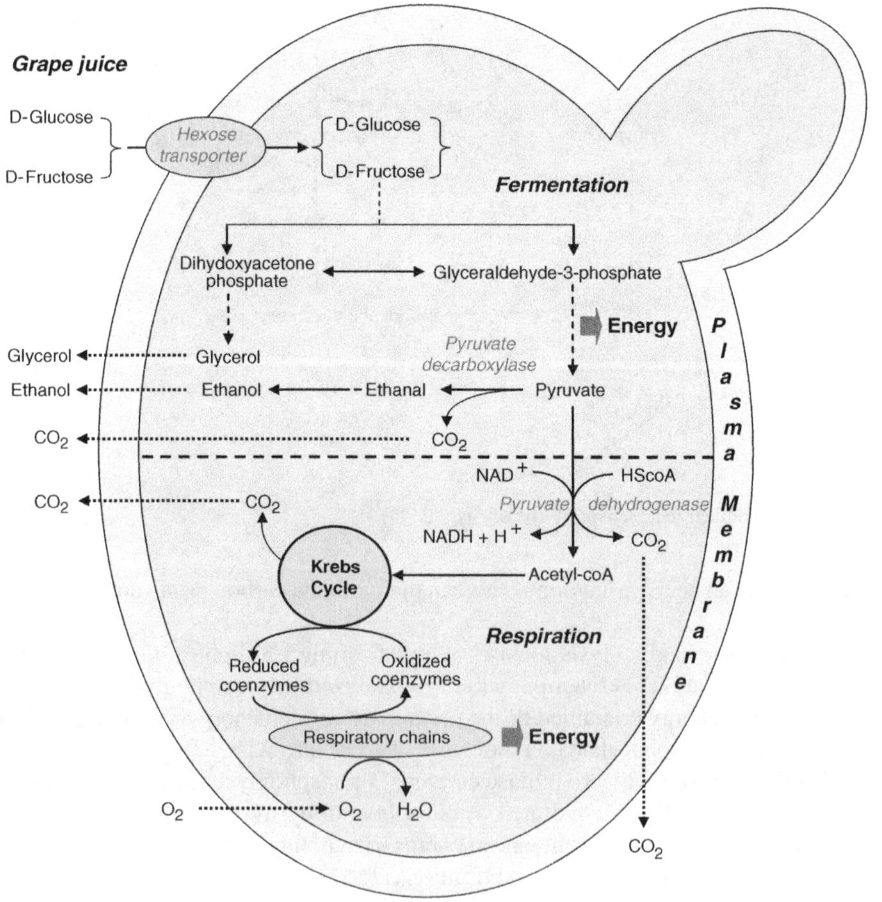

Fig. 1.3 Fermentation and respiration

Both pathways begin with glycolysis (as described above), which generates pyruvate as a final product. Pyruvate can be transformed into ethanal and carbon dioxide by the enzyme pyruvate decarboxylase and after ethanal can be reduced to ethanol. This process, named alcoholic fermentation, takes place within the cytoplasm. Alcoholic fermentation regenerates the NAD^+ consumed during glycolysis and gives yeast an energy gain of only two ATP molecules by metabolized hexose (Barnett and Entian 2005).

Nevertheless, pyruvate can also be transformed into acetyl-coA and carbon dioxide by pyruvate dehydrogenase. This reaction reduces NAD^+ to NADH and must incorporate the coenzyme A. Acetyl-coA can then be incorporated to the Krebs cycle, being transformed into carbon dioxide and producing several molecules of reduced coenzymes (NADH and $FADH_2$). The reduced coenzymes produced by the Krebs cycle, and also by glycolysis, are later reoxidized in the respiratory chains, reducing molecular oxygen to water (Barnett and Entian 2005). This process, known as respiration, yields an overall energy gain of 36–38 ATP molecules per metabolized hexose. Consequently, this process is much more beneficial to yeast than fermentation, in terms of energy. However, it needs oxygen as a substrate and it is inhibited by high sugar concentration (Crabtree 1929).

The transformation of pyruvate into ethanal or acetyl-coA is therefore a key point for regulating yeast metabolism (Ribéreau-Gayon et al. 2000c).

1.5 Regulation Between Respiration and Fermentation: Pasteur and Crabtree Effects

Louis Pasteur found that aeration increases biomass production and decreases the kinetics of sugar consumption and ethanol production (Pasteur 1861). He, therefore, concluded that aeration inhibits alcoholic fermentation (Racker 1974).

This phenomenon, which is known as the Pasteur effect, has been attributed to several mechanisms (Barnett and Entian 2005). Respiration needs very high amounts of ADP inside the mitochondria as a subtract for oxidative phosphorylation. Therefore, when respiration takes place, the cytoplasm lacks ADP and inorganic phosphate (Lagunas and Gancedo 1983), which in turn decreases the sugar transport inside the cell (Lagunas et al. 1982). These mechanisms explain how aeration inhibits the alcoholic fermentation.

Evidently, once the yeast starts to consume sugars, large quantities of carbon dioxide are produced. The release of carbon dioxide displaces the oxygen and creates semianaerobic conditions that favour fermentation. However, even in the presence of oxygen, *Saccharomyces cerevisiae* will not ferment if the sugar concentration is higher than 9 g/l. Crabtree first described this phenomenon in 1929 that is known by different names: the Crabtree effect, catabolic repression by glucose or the Pasteur contrary effect (Meijer et al. 1998; Ribéreau-Gayon et al. 2000c).

When *Saccharomyces cerevisiae* grow in a high sugar concentration, as is found in grape juice, their mitochondria degenerate. Simultaneously, the enzymes of the

Krebs cycle and the constituents of respiratory chains are repressed (Gancedo 1992; Polakis et al. 1965; Barnett and Entian 2005). Therefore, under wine fermentation conditions, *Saccharomyces cerevisiae* can only ferment sugars. *Saccharomyces cerevisiae* can only use respiration when the sugar concentration is really low and when oxygen is present in the medium. These conditions are used for the industrial production of selected dry yeast.

1.6 Alcoholic Fermentation

As was quoted above, when fermenting grape juice, *Saccharomyces cerevisiae* mainly directs the pyruvate to produce of ethanol in order to regenerate the NAD^+ consumed by glycolysis. This process, called alcoholic fermentation, is shown in Fig. 1.4.

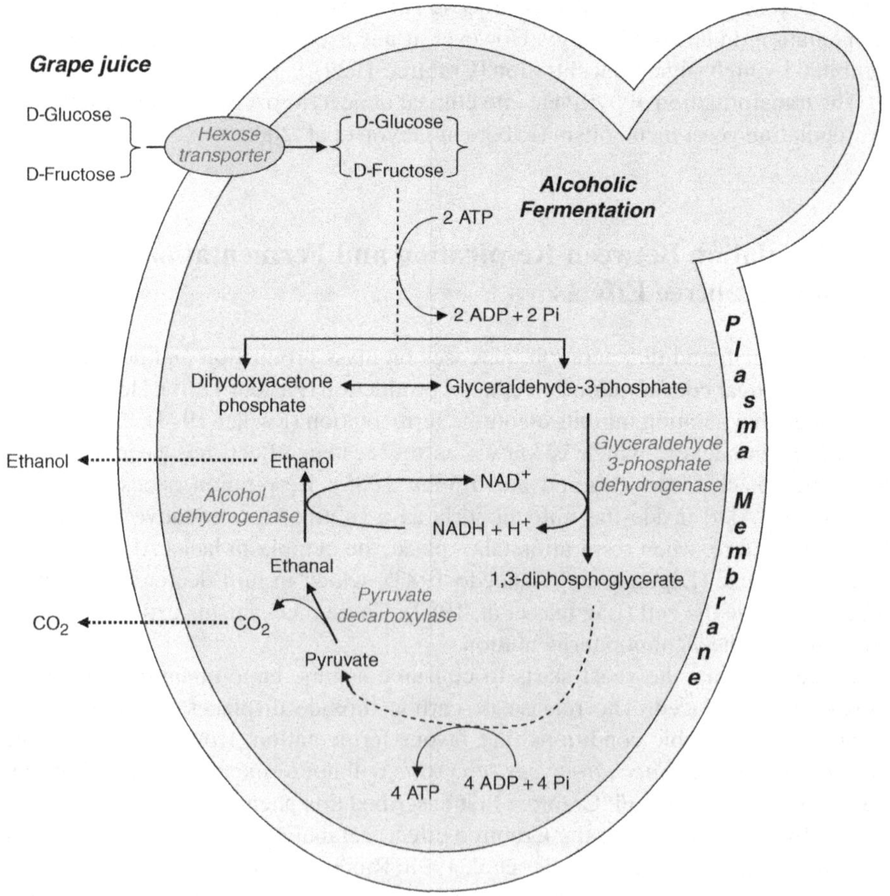

Fig. 1.4 Alcoholic fermentation

Pyruvate is initially decarboxylated into ethanal by pyruvate decarboxylase. This enzyme needs magnesium and thiamine pyrophosphate as cofactors (Hohmann 1996). Thereafter, alcohol dehydrogenase reduces ethanal to ethanol, recycling the NADH to NAD^+. There are three isoenzymes of alcohol dehydrogenase in *Saccharomyces cerevisiae*, but isoenzyme I is chiefly responsible for converting ethanal into ethanol (Gancedo 1988). Alcohol dehydrogenase uses zinc as cofactor (Ciriacy 1996).

Both final products of alcoholic fermentation, ethanol and carbon dioxide, are transported outside the cell by simple diffusion.

1.7 Glyceropyruvic Fermentation

Although the production of ethanol is the most important pathway to regenerate NAD^+, there is an alternative pathway for this purpose. This pathway, called glyceropyruvic fermentation, generates glycerol as its final product (Prior and Hohmann 1996). Figure 1.5 shows the biochemical mechanism of glyceropyruvic fermentation.

The first evidence of this pathway was found by Neuberg (1946). He discovered that the fermentation of glucose by yeast in the presence of sulphite produced a lot of glycerol. Sulphite combines with ethanal which then prevents NAD^+ from regenerating via alcohol dehydrogenase. Under these conditions, the yeasts need to oxide NADH through an alternative pathway in order to compensate for the NAD^+ deficit and the only way to do this is by producing glycerol.

Dyhroxyacetone phosphate, the main product of aldolase reaction, can be oxidized to glycerol-3-phosphate by the enzyme glycerol-3-phosphate dehydrogenase. This reaction is coupled to the oxidation of a molecule of NADH to NAD^+. Then, glycerol-3-phosphate phosphatase catalyzes the production of glycerol by dephosphorylating glycerol-3-phosphate. The production of glycerol consumes ATP but it is necessary to compensating for the redox imbalance in the cell (Barre et al. 1998).

Although glyceropyruvic fermentation was first described through the effect of sulphites, it can also be active in other situations. At the beginning of winemaking, yeasts need a lot of substrates to grow. Cell multiplication implies a very active biosynthesis of proteins, lipids, nucleotides, etc., and most of these biomolecules are synthesised using pyruvate as a substrate. Each time a molecule of pyruvate is used anabolically, a NAD+ deficit is produced which must be recovered through the glyceropyruvic pathway. For this reason, glycerol is mainly produced during the first steps of alcoholic fermentation, when yeasts are growing and they need a large proportion of pyruvate to increase their biomass (Ribéreau-Gayon et al. 2000c). Furthermore, yeasts produce glycerol as a protector against high osmotic pressures (Prior and Hohmann 1996).

For these reasons, glycerol is the third major component of dry wines (after water and ethanol). Its concentration is usually between 6 and 10 g/l and it improves wine quality because it confers sweet and mouthfeel sensations.

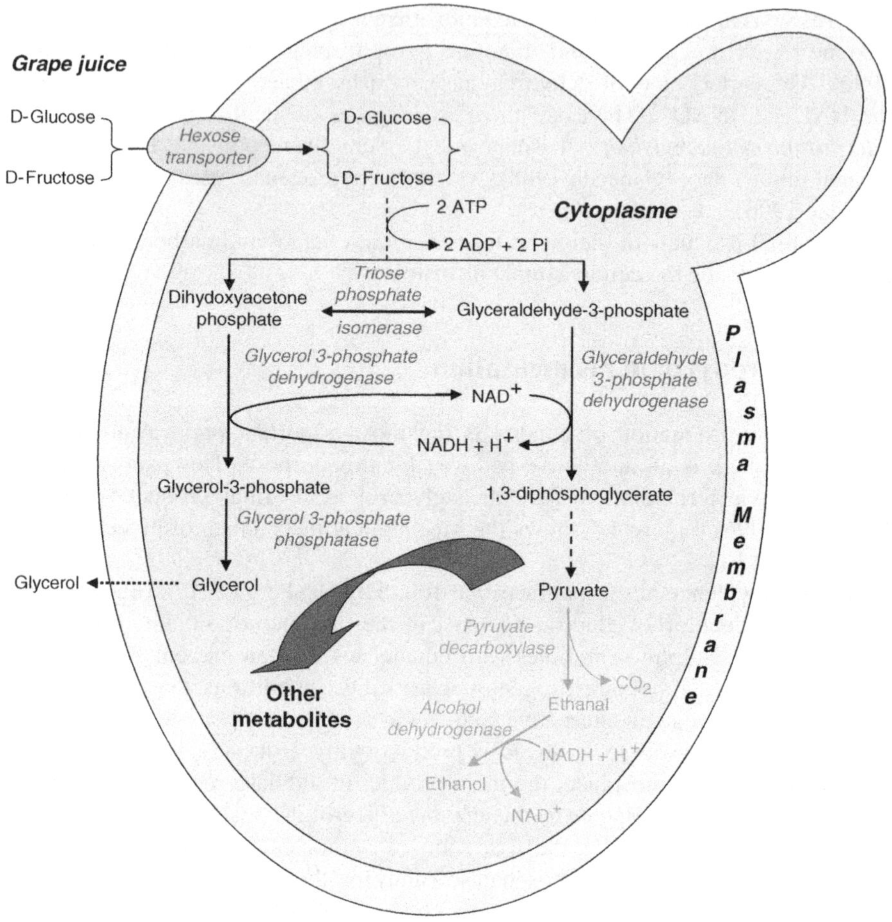

Fig. 1.5 Glyceropyruvic fermentation

1.8 Nitrogen Metabolism

When *Saccharomyces cerevisiae* grows in grape juice it needs significant amounts of assimilable nitrogen to synthesize biomass (Kunkee 1991). Grape juice contains a variety of nitrogen compounds such as ammonia, amino acids, peptides, proteins, etc., but only some of them can be assimilated by *Saccharomyces cerevisiae* (Hensche and Jiranek 1993). When fermenting grape juice fermentation, *Saccharomyces cerevisiae* can only use ammonia and amino acids, with the exception of proline, as an assimilable source of nitrogen (Barre et al. 1998). Proline can be assimilated by Saccharomyces cerevisiae but only under aerobic conditions (Boulton et al. 1996). For this reason, the term easily-assimilable nitrogen (EAN) has been proposed to describe collectively all the ammonia and amino acids, except proline. This EAN can simply be determined using the formol index (Taylor 1957; Aerny 1997).

Grape juice is relatively poor in ammonia and amino acids (Kunkee 1991). Consequently, the uptake of these compounds by the yeasts may be a critical stage in some cases. At a low concentration of EAN the risk of stuck and sluggish fermentations is really high (Bisson 1999; Zamora 2004). For that reason, winemakers used to supplement grape juice with ammonium salts (Barre et al. 1998). The EAN requirement for a complete alcoholic fermentation depends on the yeast strain (Manginot et al. 1998) and the potential alcoholic degree (Bisson and Butzke 2000). Generally, it is considered that an EAN lower than 130 mg/l can seriously affect the correct development of alcoholic fermentation.In contrast, excessive nitrogen can lead to the presence of non-assimilated residual nitrogen in the wine, which is a factor in microbiological instability and can even favour the production of ethyl carbamate and biogenic amines (Ribéreau-Gayon et al. 2000b). For that reason, nitrogen must be supplemented carefully and taking into account the initial EAN concentration of grape juice and its potential alcoholic degree (Bisson and Butzke 2000).

The first step in nitrogen assimilation is its transport inside the cell. Figure 1.6 illustrates nitrogen uptake in yeasts.

Ammonium ion is transported inside the cell by facilitated diffusion. However, the intracellular pH causes ammonium ion to release a proton, which must be sent outside the cell via H^+-ATPase. Once inside the cytoplasm, ammonia

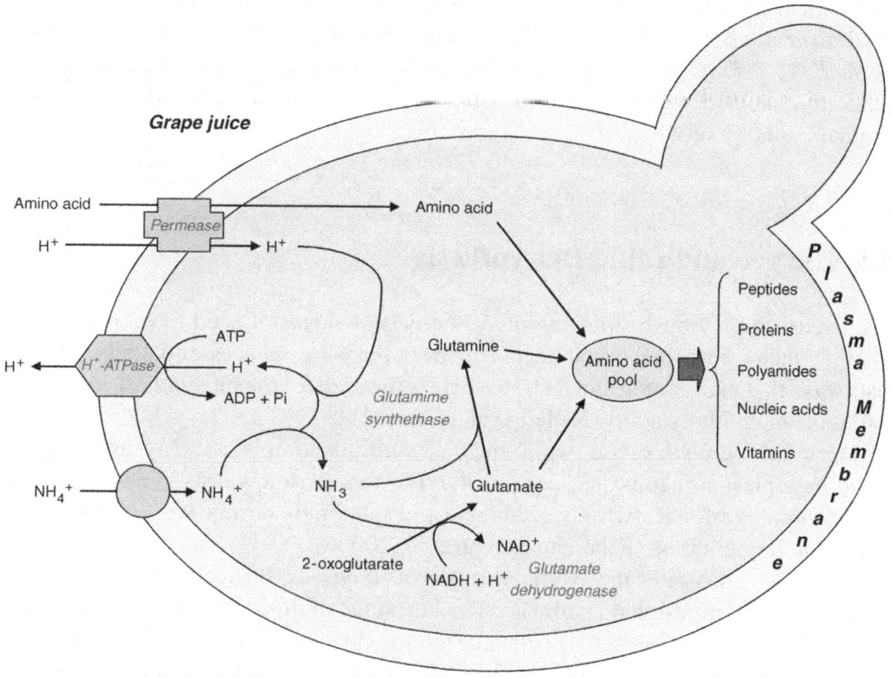

Fig. 1.6 Nitrogen metabolism

is incorporated to the amino acid pool via glutamate dehydrogenase or via glutamine synthetase, thus producing glutamate or glutamine respectively (Hensche and Jiranek 1993).

On the other hand, amino acids are transported inside the cell by different transporters. To date,15 transport systems have been identified for amino acids in *Saccharomyces cerevisiae* (Barre et al. 1998) and all of them are symport systems coupled to the entry of a proton. This proton must also be sent outside the cell in order to maintain the cellular homeostasis. Therefore, the uptake of ammonium and amino acids must be considered as active transport because it consumes ATP via H^+-ATPase.

All amino acids, except proline, may be used by *Saccharomyces cerevisiae* in grape juice fermentation. Amino acids can be directly used to synthesize proteins. However, the amino acid composition of the grape juice is not necessarily similar to the needs of the cell. For that reason, yeasts must use the remaining amino acids to synthesize those which it lacks (Hensche and Jiranek 1993; Ribéreau-Gayon et al. 2000b). In this case, ammonia is incorporated into other amino acids whereas the carbon skeleton is metabolized by the cell.

For this reason, the lack of enough EAN can make the yeast use sulphur amino acids (cysteine and methionine), thus releasing hydrogen sulphite and mercaptans. Thus, supplementing with ammonium salts is recommended not only to avoid stuck and sluggish fermentations but also to prevent reduction off-odours (Jiranek et al. 1995).

Finally, the relationship between the amino acid composition of grapes and the final aromatic composition of wine has been recently described (Hernández-Orte et al. 2002, 2006). Therefore, it is possible that in the near future grape juice will be complemented with specific mixtures of amino acids in order to improve the aromatic quality of wine.

1.9 Oxygen and Lipid Biosynthesis

As discussed previously,*Saccharomyces cerevisiae* does not need oxygen to obtain energy when fermenting grape juice. However, there are some essential biosynthetic pathways that use oxygen as substrate. This is the case for the biosynthesis of sterols and unsaturated fatty acids (Ratledge and Evans 1989).

During the growth phase, while the cell multiplication is active, yeast needs to build new plasma membranes continually. For that reason, yeasts must synthesize great amounts of sterols, fatty acids and phospholipids during the first stages of alcoholic fermentation (Ribéreau-Gayon et al. 2000b).

Figure 1.7 illustrates the synthesis of sterols in yeasts. Basically, sterols are synthesised by the mevalonate pathway. The key stage in this pathway is, without any doubt, the reaction catalysed by squalene monooxygenase. This reaction, which uses oxygen as substrate, transforms squalene into squalene 2,3, epoxide. Later, squalene epoxide lanosterol cyclase catalyses the synthesis of the first sterol of the pathway,

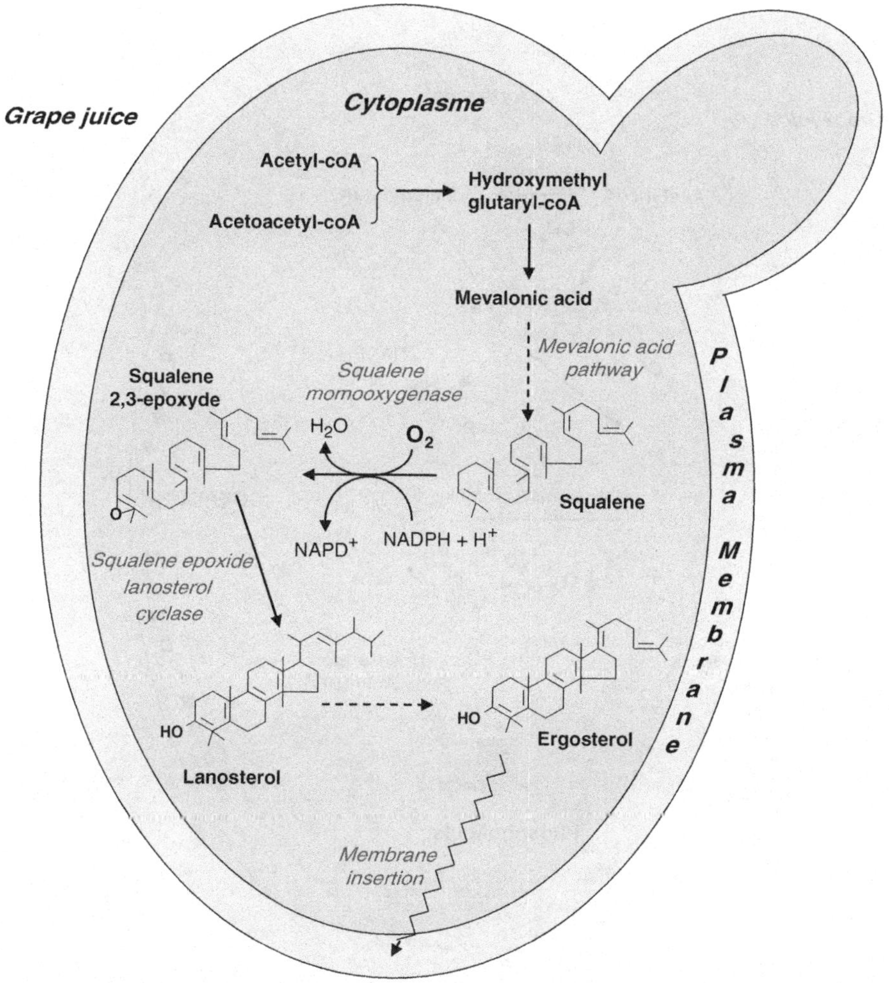

Fig. 1.7 Synthesis of ergosterol in yeasts

lanosterol. Afterwards, lanosterol is used to obtain ergosterol, which is the main sterol in *Saccharomyces cerevisiae* (Nes et al. 1993).

Therefore, in the absence of oxygen, this pathway is completely blocked and its final yeast product, ergosterol, cannot be produced.

Figure 1.8 shows the synthesis of fatty acids. This complex process is catalysed by the multienzymatic complex, fatty acid synthetase. This enzyme uses as substrates acetyl-coA and malonyl-coA to produce palmitic acid. Afterwards, palmitic acid, a saturated fatty acid of 16 carbon atoms, can be used to produce other fatty acids (Ratledge and Evans 1989). Fatty acids with more carbon units, such as estearic acid, are obtained by elongation of palmitic acid.

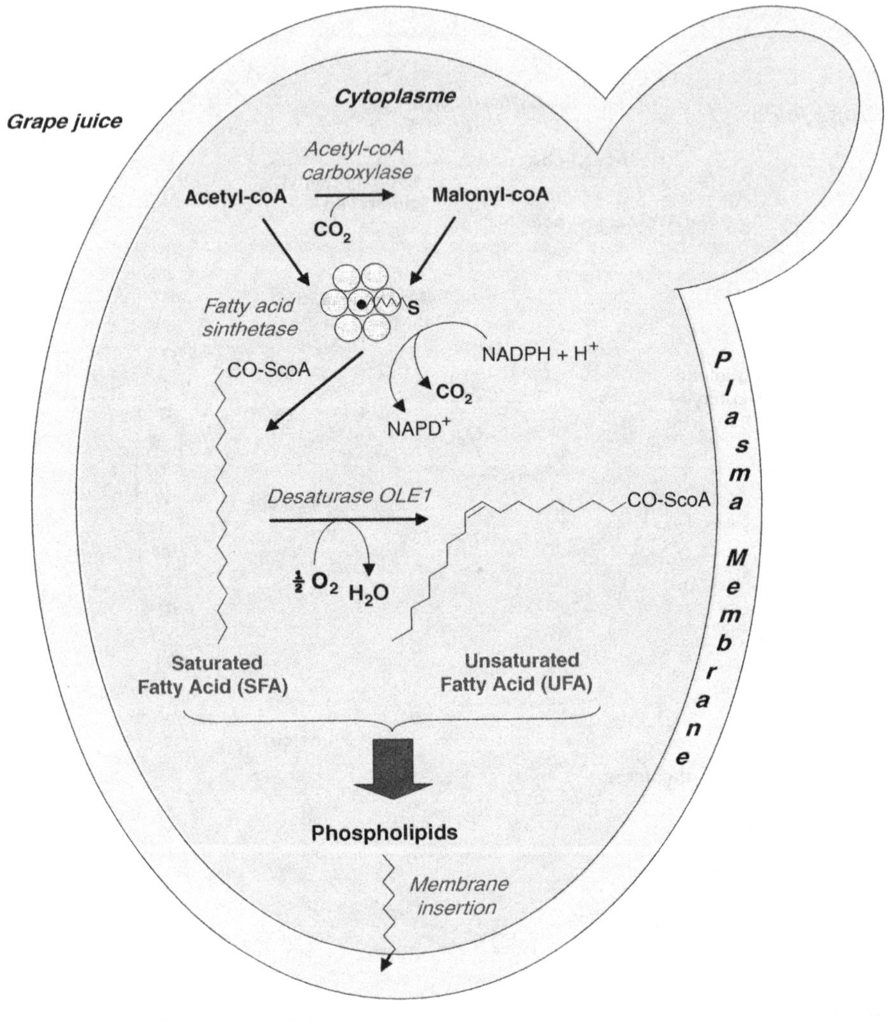

Fig. 1.8 Synthesis of fatty acids in yeasts

Nevertheless, unsaturated fatty acids (UFA) need oxygen to be synthesised (Alexandre et al. 1994). In *Saccharomyces cerevisiae*, UFA production is catalysed by the desaturase OLE1 (Kajiwara et al. 2000). This enzyme is activated by low temperatures and the presence of oxygen (Nakagawa et al. 2002). Afterwards, fatty acids are used to synthesize phospholipids, which are inserted into the plasmatic membranes.

Since oxygen is necessary to synthesize ergosterol and UFA, aeration during alcoholic fermentation limits the growth of *Saccharomyces cerevisiae* (Sablayrolles and Barre 1986; Ribéreau-Gayon et al. 2000b) and the lack of oxygen may sometimes cause stuck and sluggish fermentations (Sablayrolles et al. 1996; Bisson 1999;

Zamora 2004). For that reason, aerating the grape juice is recommended during the exponential growth phase of the fermentation in order to encourage yeast to build their membranes and avoid fermentation problems (Barre et al. 1998; Ribéreau-Gayon et al. 2000b).

Another aspect to take into account is that all microorganisms need to maintain adequate fluidity in their membranes (Rodríguez et al. 2007). Excessive rigidity can prevent cellular transport systems from functioning correctly (Los and Murata 2004).In contrast, excessive fluidity can alter the organization and the dynamic properties of phospholipidic bilayer (Laroche et al. 2001).

The fluidity of the plasmatic membrane is considerably affected by temperature (Rodríguez et al. 2007) and ethanol concentration (Jones and Greenfield 1987). Therefore, during alcoholic fermentation*Saccharomyces cerevisiae* must adapt the fluidity of the membrane to the changing environmental conditions. It should be emphasized that the temperature of fermentation and aeration depend on the type of winemaking. Usually, white wines are made at low temperatures (14–18 °C) and without aeration to conserve aromas whereas red wines are fermented at relatively high temperatures (28–30 °C) and are aerated in order to enhance colour extraction.

In white winemaking, *Saccharomyces cerevisiae* must develop at low temperature, which reduces membrane fluidity. To maintain adequate fluidity in their membranes, yeasts increase the proportion of UFA in the phospholipids (Thurston et al. 1981; Torrija et al. 2003). Phospholipids with unsaturated fatty acids have a lower melting point and more flexibility than phospholipids with saturated acyl chains (Rodríguez et al. 2007). Such adaptation involves inducing the fatty acid desaturase OLE1 which incorporate unsaturated bonds at defined positions in fatty acids (Nakagawa et al. 2002).

However, under normal white fermentation conditions, grape juice is very poor in fatty acids and fermentation is usually carried out under hypoxia conditions (Bertrand and Miele 1984). Under these conditions, yeasts cannot synthesize unsaturated fatty acids. Consequently, *Saccharomyces cerevisiae* need to use another strategy to fluidize their membranes and the only possibility is incorporating medium chain fatty acids (MCFA) within the phospholipids of the membrane (Rozès 1992). The effect of a short chain is similar to that of the double bond of a long chain (Quinn and Chapman 1980) and, therefore, the increased synthesis of MCFA could also modulate membrane fluidity.

Nevertheless, some of these MCFA can be released into the medium, reducing yeast viability and leading to stoppages in fermentation (Geneix et al. 1983). In standard red winemaking, this problem does not exist because fermentations are done at high temperatures and oxygen is introduced during the racking process.

Moreover, during alcoholic fermentation very important changes take place in the yeast's environment. Basically, the ethanol concentration increases progressively and yeasts need to adapt their plasmatic membranes to this aggressive new environment (Weber and Bont 1996). Apparently, the presence of ethanol in the medium alters drastically the fluidity of the membrane (Jones and Greenfield 1987). Under these conditions, *Saccharomyces cerevisiae* must increase its proportion of sterols and unsaturated fatty acids to compensate for this effect and consequently enhance its tolerance to ethanol (Alexandre et al. 1994). In the particular case of red winemaking, these changes can be done without problems because oxygen is introduced during the racking process. Nevertheless, as mentioned previously, white wines are usually made without aeration, and in that case the lack of oxygen may make it very difficult for yeast to adapt to ethanol.

1.10 Stuck and Sluggish Fermentations: Causes and Solutions

Sometimes alcoholic fermentation becomes too slow towards the end of the process. Yeasts drastically reduce their sugar consumption and fermentation may even stop before all the fermentable sugars have been completely metabolised. When this happens, oenologists are faced with two problems. First, the wine is not finished and something has to be done to finish it. Second, the risk of bacterial spoilage is very high. Heterolactic acid bacteria can metabolize the sugars and produce high amounts of acetic acid (Ribéreau-Gayon et al. 2000b). Winemakers are well aware of this problem, and for that reason they are continually concerned about the chances of having stuck and sluggish fermentation.

The causes of stuck and sluggish fermentations have been the subject of several studies (Larue et al. 1982; Ingledew and Kunkee 1985; Alexandre and Charpentier 1998; Bisson 1999). Some of them have already been described in previous points. The following list summarizes the possible causes and solutions of stuck and sluggish fermentations.

1. *Very high sugar concentration*: Excessive sugar concentration may be a factor which inhibits yeasts. Moreover, during the latter stages of fermentation, an excessive concentration of ethanol can seriously complicate the full consumption of the sugars. This is an important problem nowadays, especially with red wines, because winemakers look for complete phenolic maturity in grapes, which is usually attained at very high sugar concentration. In this case, using yeast with high ethanol resistance is recommended.
2. *Temperature extremes*: Yeasts have problems growing when the temperature is too low. Therefore, too low temperature at the beginning of the process can lead to a deficient yeast population. On the other hand, if the temperature is too high (more than 30 °C) the fermentation is at considerable risk of stopping. For that reason, thermic control of fermentations is nowadays indispensable. Another aspect to take into account is that sudden changes of temperature may provoke serious problems in fermentation. As previously mentioned, yeasts adapt their membrane composition to maintain the correct fluidity. A drastic temperature decrease can provoke excessive rigidity in the membranes before yeasts can adapt to it.
3. *Complete anaerobiosis*: Oxygen is necessary to synthesize ergosterol and MCFA. Without oxygen, yeast will struggle to grow and adapt their membranes to the environmental conditions. For that reason, aeration is recommended, at least during the exponential growth phase.

4. *Nutrient deficiencies*: The lack of some nutrients in the grape juice can cause serious problems during fermentation. Nitrogen, vitamins, minerals, etc. may be deficient in grape juice. For that reason yeast activators are usually added in wineries. The standard activators are made up of ammonium salts (phosphate and/or sulphate), thiamine and their application is certainly very useful. However, as previously mentioned, the nitrogen dose must be selected carefully, taking into account the initial concentration of easily assimilated nitrogen and the potential alcoholic content of the grape juice. Adding nitrogen will be more effective if it is done two or more times and is combined with aeration. The first dose should be added when fermentation starts, the second in the middle of quasi-stationary phase and finally the third at the end of this stage. During recent years a new generation of yeast activators has appeared on the market. These new activators are prepared from yeast and provide several other interesting substances such as sterols, UFA, minerals, panthothenic acid, etc. These activators are very useful during the latter stages of fermentation.
5. *Presence of anti-fungal substances*: Sometimes grape juice can contain pesticide residues which can seriously affect alcoholic fermentation. To avoid this, the vineyard's treatments must be closely inspected.
6. *Presence of medium chain fatty acids*: The presence of MCFA can decrease yeast viability and even stop alcoholic fermentation. This problem is more prevalent in white winemaking because fermentation is usually carried out at low temperatures and without any aeration. Yeast hulls have been very useful for avoiding this problem. Yeast hulls adsorb MCFA from the media and provide sterols and UFA to the yeasts. Yeast hulls can be used as preventives (20 g/hl) or as curatives (40–50 g/hl) of stuck and sluggish fermentations.
7. *Antagonism between microorganisms*: The different microorganisms present in grape juice compete for nutrients. *Saccharomyces cerevisiae* development is usually encouraged through LSA inoculation and by adding sulphur dioxide. However, sometimes autochthonous yeast or even bacteria can grow and cause organoleptic deviations and even stuck and sluggish fermentations (Edwards et al. 1999; Gerland 2000; Gao et al. 2002). This problem is greater when the pH of grape juice is high because the antiseptic effect of sulphur dioxide is less effective (Beech et al. 1979).

All these causes are possible and can prevent alcoholic fermentation from developing correctly. However, it is usually a synergistic combination of some of these causes which brings about stuck and sluggish fermentation. If these causes are eliminated, then stuck and sluggish fermentations are almost non-existent. Nevertheless, if a fermentation tank presents problems, winemakers must act as soon as possible. Abundant aeration and adding yeast hulls may solve the problem. Finally, if fermentation stops, the yeast must be reinoculated. The choice of the yeast and the way it is preadapted to ethanol are the key to the success of the inoculum. The use of commercial yeast with a high resistance to ethanol and the use of the classical protocols for adapting yeast to the champagnization process are highly recommended.

1.11 Other Subproducts of Alcoholic Fermentation

As mentioned previously, fermentation produces mainly ethanol and glycerol but also several other substances that contribute to their complexity (Lambrechts and Pretorius 2000). Figure 1.9 shows schematically the metabolic origin of some of the most important subproducts of alcoholic fermentation.

1. *Diacetyl, acetoin and 2,3-butanediol*: These compounds are produced by condensing of pyruvate with ethanal. This reaction produces acetolactate which is later decarboxylated. Diacetyl is produced if the decarboxylation is oxidative, whereas acetoin is produced if the decarboxylation is not oxidative. Acetoin can also be formed by directly reducing diacetyl. Finally, acetoin can be reduced to form 2,3-butanediol. This last reaction is reversible (Ribéreau-Gayon et al. 2000c). Acetoin and especially diacetyl give off a buttery smell that may

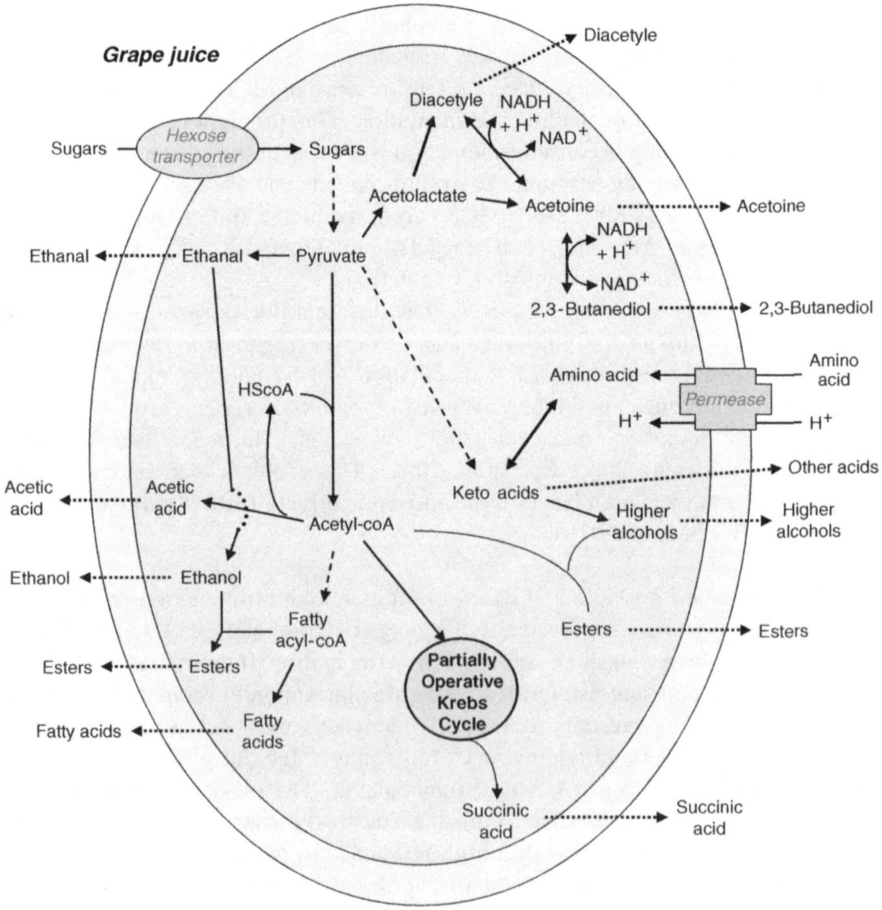

Fig. 1.9 Biosynthesis of the other subproducts

contribute to a wine's aroma. Although alcoholic fermentation does not produce enough acetoin and diacetyl to have a real impact on wine aroma, the subsequent development of lactic acid bacteria can considerably increase its concentration.

2. *Ethanal*: Ethanal, also called acetaldehyde, is an intermediary of alcoholic fermentation obtained by the decarboxylation of pyruvate. Later, ethanal is mainly reduced to ethanol, but little quantities of it may be released into the wine. Ethanal gives off a characteristic aroma that contributes to the perception that the wine is oxidized, although alcoholic fermentation usually produces only small amounts of this compound. However, ethanal can also be produced from ethanol by chemical or biological oxidation. Some wines, such as Fino and Manzanilla from Jerez, or the wine Jaune from Jura, give off a particular aroma which is characterized by a high concentration of ethanal. These wines are obtained by aging the wine under a film of *Saccharomyces cerevisiae*, which is responsible for producing ethanal from ethanol.
3. *Acetic acid*: Acetic acid is the main volatile acid of wine. Its presence at high concentrations gives off a vinegar odour and a disagreeable sensation in the mouth. For that reason, volatile acidity is one of the most important analytical parameters in oenology. Acetic acid may be produced by yeast, lactic acid bacteria and acetic acid bacteria. But normally *Saccharomyces cerevisiae* only produce small quantities of acetic acid if there are no problems during alcoholic fermentation (0.1–0.3 g/l). However, stuck and sluggish fermentations can generate large amounts of this acid. The high production of acetic acid may be due to the development of lactic disease or because yeasts produce more acetic acid than normal by hydrolysis of acetyl-coA.
4. *Higher alcohols*: Higher alcohols are produced as a deviation of the metabolism of amino acids. Higher alcohols are produced when keto acids corresponding to the carbon skeleton from the different amino acids are decarboxylated and reduced. Higher alcohols are normally below their limit of detection but they are the precursors of some esters, which have a large sensory impact.
5. *Esters*: Esters are synthesized from acyl-coA and alcohols by a group of enzymes, the alcohol-acylcoA transferases (Barre et al. 1998). Basically there are two types of esters in wine: the acetates of higher alcohols and the esters of fatty acids and ethanol. The first group are synthesised from acetyl-coA and the different higher alcohols. These esters give off different odours, such as glue (ethyl acetate), banana (isoamyl acetate) or rose (phenylethanol acetate). The other group of esters are synthesised from the different acyl-coA and ethanol. The different esters of fatty acids and ethanol give off a fruity aroma. All esters, with the exception of ethyl acetate, give off an agreeable smell and contribute positively to the wine aroma. Other esters such as ethyl lactate and diethyl succinate do not have any sensory impact at normal concentrations.
6. *Succinic acid*: Succinic acid is quantitatively the third product of alcoholic fermentation. Some authors suggest that succinic acid is synthesised via Krebs cycle although its functioning is seriously limited (Heerde and Radler 1978). However, other authors think that this cycle is not operative in conditions of grape juice fermentation (Salmon et al. 1987). In any case, succinic acid is present in wine

in concentrations between 0.6 and 1.2 g/l and it contributes significantly to the wine acidity. *Saccharomyces cerevisiae* also releases into the wine several other acids such as lactic acid, isovaleric and isobutyric acids, fatty acids, etc., but only in low concentrations.

It can be concluded that alcoholic fermentation is not only the simple transformation of sugars into ethanol. On the contrary, it is a very complex process that allows us to obtain a very pleasant beverage. It represents the transformation of sugars mainly into ethanol but also into other subproducts, which can contribute positively or negatively to sensory quality. A more complete knowledge about the regulation of all pathways implied that it is necessary to acquire a better understanding of the process and of designing strategies to enhance the quality of the product. Moreover, alcoholic fermentation also implies the transformation of other compounds present in the grape juice which have a high influence on wine quality. Considerable progress has been made during recent years and, without doubt, new solutions for new challenges will be developed in the immediate future (Pretorius 2000; Moreno-Arribas and Polo 2005).

Acknowledgments The author gratefully acknowledges CICYT (AGL2007-66338 and AGL 2004-02309) for financial support.

References

Aerny, J. (1997) Composés azotés des moûts et des vins. *Rev. Suisse Vitic. Hortic., 28*, 161–165.

Alexandre, H., & Charpentier, C. (1998) Biochemical aspects of stuck and sluggish fermentation in grape must. *J. Ind. Microbiol. Biotechnol., 20*, 20–27.

Alexandre, H., Rousseaux, I., & Charpentier, C. (1994) Relationship between ethanol tolerance, lipid composition and plasma membrane fluidity in Saccharomyces cerevisiae and Kloeckera apiculata. *FEMS Microbiol. Lett., 124*, 17–22

Barnett, J.A. (2003) A history of research on yeasts 5: the fermentation pathway. *Yeast, 20*, 509–543.

Barnett, J.A., & Entian, K.D. (2005) A history of research on yeasts 9: regulation of sugar metabolism. *Yeast, 22*, 835–894.

Barre, P., Blondin, B., Dequin, S., Feuillat, M., Sablayrolles, J.M., & Salmon, J.M. (1998). La levure de fermentation alcoolique. In C. Flanzy (Ed.), *Oenologie: fondements scientifiques et technologiques* (pp. 454–497). Paris: Tec Doc Lavoisier.

Beech, F.W., Burroughs, L.F., Timberlake, C.F., & Whiting, G.C. (1979) Progres recents sur l'aspect chimique et antimicrobienne de l'ahidride sulfureux. *Bull. OIV, 52*, 1001–1022.

Beltran, G., Torija, M.J., Novo, M., Ferrer, N., Poblet, M., Guillamón, J.M., Rozes, N., & Mas, A. (2002) Analysis of yeast populations during alcoholic fermentation: a six year follow-up study. *Syst. Appl. Microbiol., 25*, 287–293.

Bertrand, A., & Miele, A. (1984) Influence de la clarification du moût de raisin sur sa teneur en acides gras. *Conn. Vigne Vin, 48*, 293–297.

Bisson, L.F. (1999) Stuck and sluggish fermentations. *Am. J. Enol. Vitic., 50*, 107–119.

Bisson, L.F., & Butzke, C.E. (2000) Diagnosis and rectification of stuck and sluggish fermentations. *Am. J. Enol. Vitic., 51*, 168–177.

Boulton, R.B., Singleton, V.L., Bisson, L.F., & Kunkee, R.E. (1996). Yeast and biochemistry of ethanol fermentation. In R.B. Boulton (Ed.), *Principles and Practices of Winemaking* (pp. 139–172). New York: Chapman & Hall.

Cabib, E., Roberts, R., & Bowers, B. (1982) Synthesis of the Yeast Cell Wall and its Regulation. *Ann. Rev. Biochem., 51*, 763–793.

Chatonnet, P., Dubourdieu, D., & Boidron, J.N. (1995) The influence of *Brettanomyces/Dekkera sp.* yeasts and lactic acid bacteria on the ethylphenol content of red wines. *Am. J. Enol. Vitic., 46*, 463–468.

Ciriacy, M. (1996) Alcohol dehydrogenases. In F.K. Zimmerman & K.D. Entian (Eds.), *Yeast Sugar Metabolism: Biochemistry, Genetics, Biotechnology, and Applications* (pp. 213–224). Boca Raton: CRC Press.

Constantí, M., Poblet, M., Arola, L., Mas, A., & Guillamón, J.M. (1997). Analysis of yeasts populations during alcoholic fermentation in a newly established winery. *Am. J. Enol. Vitic., 48*, 339–344.

Constantí, M., Reguant, C., Poblet, M., Zamora, F., Mas, A., & Guillamón, J.M. (1998) Molecular analysis of yeast population dynamis: Effect of sulphur dioxide and the inoculum in must fermentation. *Int. J. Food Microbiol., 41*, 169–175.

Crabtree, H.D. (1929) Observations on the carbohydrate metabolism of tumours. *Biochem. J., 23*, 536–545.

Del Nobile, M.A., D'Amato, D., Altieri, C., Corbo, M.R., & Sinigaglia, M. (2003) Modeling the yeast Growth-Cycle in a model wine system. *J. Food Sci., 68*, 2080–2085.

Edwards, C.G., Reynolds, A.G., Rodríguez, A.V., Semon, M.J., & Mills, J.M. (1999) Implication of acetic acid in the induction of slow/stuck grape juice fermentation and inhibition of yeast by Lactobacillus sp. *Am. J. Enol. Vitic., 50*, 204–210.

Fleet, G.H. (1993) The microorganisms of winemaking – isolation, enumeration and identification. In G.H. Fleet (Ed.), *Wine Microbiology and Biotechnology* (pp. 1–25). Reading: Hrawood Academic.

Fleet, G.H., & Heard, G.M. (1993) Yeast-growth during fermentation. In G.H. Fleet (Ed.), *Wine Microbiology and Biotechnology* (pp. 27–54). Reading: Hrawood Academic.

Gancedo, J.M. (1988) La regulation du metabolisme des sucres chez la levure. In P. Bidan & J.R. Bonneviale (Eds.), *Application à l'œnologie des progrès récents en microbiologie et en fermentation* (pp. 133–143). Paris: OIV.

Gancedo, J.M. (1992) Carbon catabolite repression in yeast. *Eur. J. Biochem., 206*, 297–313.

Gao, Y.C., Zhang, G., Krentz, S., Darius, S., Power, J., & Lagarde, G. (2002) Inhibition of spoliage lactic acid bacteria by lysozyme during wine alcholic fermentation. *Aust. J. Grape Wine Res., 8*, 76–83.

Geneix, C., Lafon-Lafourcade, S., & Ribéreau-Gayon, P. (1983) Effet des acides gras sur la viabilité des populations de Saccharomyces cerevisiæ. *C.R. Acad Sci., 296*, 943–947.

Gerland, C. (2000) Gestion de la flore bactérienne lactique: enjeu important pour l'élaboration des vins de qualité. *Rev. Oenol., 96*, 31–36.

Heerde, E., & Radler, F. (1978) Metabolism of the anaerobic formation of succinic acid by *Saccharomyces cerevisiae. Arch. Microbiol., 117*, 269–276.

Heinisch, J.J., & Rodicio, R. (1996) Fructose-1,6 biphospohate aldolase, triose phosphate isomerase, glyceraldehide-3-phospkate deshidrogenases and phosphoglycerate mutase. In F.K. Zimmerman & K.D. Entian (Eds.), *Yeast Sugar Metabolism: Biochemistry, Genetics, Biotechnology, and Applications* (pp. 119–140). Boca Raton: CRC Press.

Hensche, P.A., & Jiranek, V. (1993) Yeast – metabolism of nitrogen compounds. In G.H. Fleet (Ed.), *Wine Microbiology and Biotechnology* (pp. 77–164). Reading: Hrawood Academic.

Hernández-Orte, P., Cacho, J., & Ferreira, V. (2002) Relationship between varietal amino acid profile of grapes and wine aromatic composition. Experiments with model solutions and chemometric study, *J. Agric. Food Chem., 50*, 2891–2899.

Hernández-Orte, P., Ibarz, M.J., Cacho, J., & Ferreira, V. (2006) Addition of amino acids to grape juice of the Merlot variety: Effect on amino acid uptake and aroma generation during alcoholic fermentation. *Food Chem., 98*, 300–310.

Hohmann, S. (1996) Pyruvate decarboxylases. In F.K. Zimmerman & K.D. Entian (Eds.), *Yeast Sugar Metabolism: Biochemistry, Genetics, Biotechnology, and Applications* (pp. 187–212). Boca Raton: CRC Press.

Horecker, B.L. (2002) The Pentose Phosphate Pathway. *J. Biol. Chem., 277*, 47965–47971.

Ingledew, W.M., & Kunkee, R.E. (1985) Factors influencing sluggish fermentations of grape juice. *Am. J. Enol. Vitic., 36*, 65–76.

Jiranek, V., Langridge, P., & Henschke, P.A. (1995) Regulation of hydrogen sulfite liberation in wine-producing *Saccharomyces cerevisiae* strains by assimilable nitrogen. *Appl. Environm. Microbiol., 61*, 461–467.

Jones, R.P., & Greenfield, P.F., (1987) Ethanol and the fluidity of the yeast plasma membrane. *Yeast, 3*, 223–232.

Kajiwara, S., Aritomi, T., Suga, K., Ohtaguchi, K., & Kobayashi, O. (2000) Overexpression of the OLE1 gene enhances ethanol fermentation by Saccharomyces cerevisiae. *Appl. Microbiol. Biotechnol., 53*, 568–574.

Kresge, N., Simoni, R.D., & Hill, R.L. (2005) Otto Fritz Meyerhof and the Elucidation of the Glycolytic Pathway. *J. Biol. Chem., 280*, 124–126.

Kunkee, R.E. (1991). Relationship between nitrogen content of must and sluggish fermentation. In *Proceedings of the International Symposium of Nitrogen in Grapes and Wine*, 18–19 de Juny de 1991, Seattle, Washington (pp. 148–155). Davis CA: American Society of Enology and Viticulture.

Lafon-Lafourcade, S. (1983). Wine and brandy. Biotechnology, In H.J. Rehm & G. Reed. (Eds.), *Food and Feed Production with Microorganisms*, Vol 5, (pp. 81–163). Weinheim: Verlag Chemie.

Lafon-Lafourcade, S., & Peynaud, E. (1974) Sur l'action antibacterienne de l'anhidride sulfureux sous forme libre et sous forme combinée. *Conn. Vigne Vin, 8*, 187–203.

Lafon-Lafourcade, S., Geneix, C., & Ribereau-Gayon, P. (1984). Inhibition of alcoholic fermentation of grape must by fatty acids produced by yeasts and their elimination by yeast ghosts. *Appl. Environm. Microbiol., 47*, 1246–1249.

Lagunas, R. (1993) Sugar transport in *Saccharomyces cerevisiae. FEMS Microbiol. Rev., 16*, 229–242.

Lagunas, R., & Gancedo, C. (1983) Role of phosphate in the regulation of the Pasteur effect in *Saccharomyces cerevisiae. Eur. J. Biochem., 137*, 479–483.

Lagunas, R., Dominguez, C., Busturia, A., & Sáez, M.J. (1982) Mechanisms of appearance of the Pasteur effect in *Saccharomyces cerevisiae*: inactivation of sugar transport systems. *J. Bacteriol., 152*, 19–25

Lambrechts, M.G., & Pretorius, S. (2000) Yeast and its importance to wine aroma – A Review. *South Afric. J. Enol. Vitic., 21*, 97–128.

Laroche, C., Beney, L., Marechal, P.A., & Gervais, P. (2001) The effect of osmotic pressure on the membrane fluidity of *Saccharomyces cerevisiae* at different physiological temperatures. *Appl. Microbiol. Biotechnol., 56*, 249–254.

Larue, F., Lafon-Lafourcade, S., & Ribéreau-Gayon, P. (1982). Inhibition de *Saccharomyces cerevisiae* dans le moût de raisin. *C.R. Acad. Sci., 294*, 587–590.

Los, D.A., & Murata, N. (2004) Membrane fluidity and its roles in the perception of environmental signals. *Biochim. Biophy. Acta, 1666*, 142–157.

Manginot, C., Roustan, J.L., & Sablayrolles, J.M. (1998) Nitrogen demand of different yeast strains during alcoholic fermentation. Importance of stationary phase. *Enz. Micro. Technol., 23*, 511–517.

Meijer, M.M.C., Boonstra, J., Verkleij, A.J., & Verrips, C.T. (1998) Glucose repression in *Saccharomyces cerevisiae* is related to the glucose concentration rather than the glucose flux. *J. Biol. Chem., 273*, 24102–24107.

Moreno-Arribas, M.V., & Polo, M.C. (2005) Winemaking biochemistry and microbiology: current knowledge and future trends. *Crit. Rev. Food Sci. Nutr., 45*, 265–286.

Mortimer, R., & Polsinelli, M. (1999). On the origin of wine yeast. *Res. Microbiol., 150*, 199–204.

Nakagawa, Y., Sakumoto, N., Kaneko, Y., & Harashima, S. (2002) Mga2p is a putative sensor for low temperature and oxygen to induce OLE1 transcription in *Saccharomyces cerevisiae. Biochem. Biophys. Res. Comm., 291*, 707–713.

Nes, D.W., Janssen, G.G., Crumley, F.G., Kalinowska, M., & Akihisa, T. (1993) The structural requirements of sterols for membrane function in *Saccharomyces cerevisiae. Arch. Biochem. Bioph., 300*, 724–733.

Neuberg, C. (1946). The Biochemistry of Yeast. *Ann. Rev. Biochem., 15*, 435–472.

Ough, C.S., (1964). Fermentation rates of juice.I. Effects of temperature and composition on white juice fermentation rates. *Am. J. Enol. Vitic., 15*, 167–177.

Pasteur, L. (1861) Influence de l'oxygène sur le développement de la levûre et la fermentation alcoolique. *Bulletin de la Société de Paris* (*Résumé de Séance du 28 juin 1861*), 79–80.

Peynaud, E., & Domercq, S. (1959) A review of microbiological problems in winemaking in France. *Am. J. Enol. Vitic., 1*, 69–77.

Polakis, E.S., Bartley, W., & Meek, G.A. (1965) Changes in the activities of respiratory enzymes during the aerobic growth of yeast on different carbon sources. *Biochem. J., 97*, 298–302.

Pretorius, I.S. (2000) Tailoring wine yeast for the new millennium: novel approcaches to the ancient art of winemaking. *Yeast, 16*, 675–729.

Pretorius, I.S., Van der Westhuizen, T.J., & Augustyn, O.P.H. (1999) Yeast biodiversity in vineyards and wineries and its importante to the South African wine industry. *S. Afr. J. Enol. Vitic., 20*, 61–74.

Prior, B.A., & Hohmann, S. (1996) Glycerol production and osmoregulation. In F.K. Zimmerman & K.D. Entian (Eds.), *Yeast Sugar Metabolism: Biochemistry, Genetics, Biotechnology, and Applications* (pp. 313–338). Boca Raton: CRC Press.

Quinn, A.J., & Chapman, D. (1980) The dynamics of membrane structure. *CRC Crit. Rev. Biochem., 8*, 1–117.

Racker, E. (1974) History of the Pasteur effect and its pathobiology. *Mol. Cell. Biochem., 5*, 17–23.

Ratledge, C., & Evans, C.T. (1989) Lipids and their metabolism. In A.H. Rose & J.S. Harrison (Eds.), *The Yeasts* (2nd ed.), Vol 3 (pp. 367–455). London: Academic Press.

Ribéreau-Gayon, P., Glories, Y., Maujean, A., & Dubourdieu, D. (2000a) Chemical nature, origins and consequences of the main organoleptic defects. In P. Ribéreau-Gayon (Ed.), *Handbook of Enology*, Vol 2, (pp. 209–253). Chichester: John Wiley & sons, Ltd.

Ribéreau-Gayon, P., Glories, Y., Maujean, A., & Dubourdieu, D. (2000b) Conditions of yeast development. In P. Ribéreau-Gayon (Ed.), *Handbook of Enology*, Vol 2, (pp. 75–107). Chichester: John Wiley & sons, Ltd.

Ribéreau-Gayon, P., Glories, Y., Maujean, A., & Dubourdieu, D. (2000c) Biochemistry of alcoholic fermentation and metabolic pathways of wine yeasts. In P. Ribéreau-Gayon (Ed.), *Handbook of Enology*, Vol 1, (pp. 51–74). Chichester: John Wiley & sons, Ltd.

Rodríguez, S., Sánchez, A., Martínez, J.M., Prieto, J.A., & Randez, F. (2007) Fluidization of Membrane Lipids Enhances the Tolerance of Saccharomyces cerevisiae to Freezing and Salt Stress. *Appl. Environm. Microbiol., 73*, 110–116.

Romano, P., & Suzzi, G. (1993) Sulfur dioxide and wine microorganisms. In G.H. Fleet (Ed.), *Wine Microbiology and Biotechnology* (pp. 373–393). Reading: Harwood Academic.

Rozès, N. (1992) Contribution à l'étude du métabolisme lipidique de *Saccharomyces cerevisiæ*. Application à la différenciation des levures de vin. Thèse de Doctorat de l'Université de Bordeaux II, Bordeaux, France.

Sablayrolles, J.M., & Barre, P. (1986) Evaluation des besoins en oxygen de fermentations alcooliques en conditions oenologiques silées. *Sci. Aliments, 6*, 373–383.

Sablayrolles, J.M., Dubois, C., Manginot, C., Roustan, J.L., & Barre, P. (1996). Efectiveness of combined ammoniacal nitrogen and oxygen additions for completion of sluggish and stuck fermentation. *J. Fermen. Bioeng., 82*, 377–381

Salmon, J.M., Vezinhet, F., & Barre, P. (1987) Anabolic role of L-malic acid in *Saccharomyces cerevisiae* in anaerobiosis during alcoholic fermentation. *FEMS Microbiol. Lett., 42*, 213–220.

Sapis-Domerq, S. (1980) Étude de l'influence des produits de traitement de la vigne sur la microflore des raisins et des vins. *Conn. Vigne Vin,* 14, 155–181.

Schaaf-Gersteenschaläger, I., & Miosga, T. (1996) The pentose phosphate pathway. In F.K. Zimmerman & K.D. Entian (Eds.), *Yeast Sugar Metabolism: Biochemistry, Genetics, Biotechnology, and Applications* (pp. 271–284). Boca Raton: CRC Press.

Taylor, W.H. (1957) Formol Titration: An evaluation of its various modifications. *Analyst, 82,* 488–498.

Thurston, P.A., Taylor, R., & Ahvenainen, J. (1981). Effects of linoleic acid supplements on the synthesis by yeast of lipids and acetate esters. *J. Inst. Brew., 87,* 92–95.

Torrija, M.J., Beltrán, G., Novo, M.T., Poblet, M., Guillamón, J.M., Mas, A., & Rozès, N. (2003) Effect of fermentation temperature and *Saccharomyces* species on the cell fatty acid composition and presence of volatile compounds in wine. *Int. J. Food Microbiol., 85,* 127–136.

Weber, F.J., & Bont, J.A.M., (1996) Adaptation mechanisms of microrganisms to the toxic effects of organic solvents on membranes. *Biochim. Biophys. Acta, 1286,* 225–245.

Zamora, F. (2004) Las paradas de fermentación. *Enólogos, 29,* 28–32.

Chapter 2
Biochemical Transformations Produced by Malolactic Fermentation

Antonella Costantini, Emilia García-Moruno, and M. Victoria Moreno-Arribas

Contents

E. García-Moruno (✉)
CRA-Centro di Ricerca per l'Enologia, Via Pietro Micca 35, 14100 Asti, Italy
e-mail: e.garciamoruno@isenologia.it

M.V. Moreno-Arribas
Instituto de Fermentaciones Industriales (CSIC), Juan de la Cierva, 3 - 28006 Madrid, Spain
e-mail: mvmoreno@ifi.csic.es

M.V. Moreno-Arribas, M.C. Polo (eds.), *Wine Chemistry and Biochemistry*,
DOI 10.1007/978-0-387-74118-5_2, © Springer Science+Business Media, LLC 2009

2.1 Introduction

Malolactic fermentation (MLF) in wine is by definition the enzymatic conversion of L-malic acid to L-lactic acid, a secondary process which usually follows primary (alcoholic) fermentation of wine but may also occur concurrently. This reduction of malic acid to lactic acid is not a true fermentation, but rather an enzymatic reaction performed by lactic acid bacteria (LAB) after their exponential growth phase. MLF is mainly performed by *Oenococcus oeni*, a species that can withstand the low pH (<3.5), high ethanol (>10 vol.%) and high SO_2 levels (50 mg/L) found in wine. More resistant strains of *Lactobacillus*, *Leuconostoc* and *Pediococcus* can also grow in wine and contribute to MLF; especially if the wine pH exceeds 3.5 (Davis et al. 1986; Wibowo et al. 1985). The most important benefits of MLF are the deacidification of high acid wines mainly produced in cool climates, LAB contribute to wine flavour and aroma complexity and improve microbial stability (Lonvaud-Funel 1999; Moreno-Arribas and Polo 2005).

Unfortunately, uncontrolled MLF also presents a risk of wine spoilage by compounds that can produce off-flavours (including acetic acid, volatile phenols and mousiness) or that may be hazardous to human health (such as ethyl carbamate and biogenic amines). The most important aspects of the development of LAB and MLF in wines are dealt with in this chapter.

2.2 Ecology and Development of Lactic Acid Bacteria During Vinification

2.2.1 Lactic Acid Bacteria in Wine

Winemaking is a complex microbial process involving yeasts and bacteria. They are both naturally present on grape skins (Renouf et al. 2005), but are also found in barrels, tanks and the equipment used during vinification. A large amount of research has focused on the description and ecology of LAB in wine; their involvement in winemaking, their distribution and their succession in musts, in wine and during fermentation have been extensively studied.

The LAB from grape, musts or wine belong to two families representing three genera. *Lactobacillaceae* are represented by the genus *Lactobacillus*, and *Streptococcaceae* are represented by *Oenococcus* and *Pediococcus*.

2.2.1.1 *Lactobacillus*

Lactobacillus represents a highly diverse group of Gram-positive, microaerophilic bacteria; its cells are non-mobile and they have long rod-like forms or short rods (Kandler and Weiss 1986) and can appear as single cells, in pairs or in chains of different sizes. Bacteria belonging to this genus are facultative anaerobes and require a rich medium containing fermentable sugar.

They are divided into two groups in relation to their hexose metabolism:

- Strict heterofermenters (*L. brevis, L. hilgardii*)
- Facultative heterofermenters (*L. casei, L. plantarum*)

In the heterofermentative metabolism, glucose is transformed into lactic acid and other compounds such as acetic acid, ethanol and carbon dioxide, as shown in Fig. 2.1.

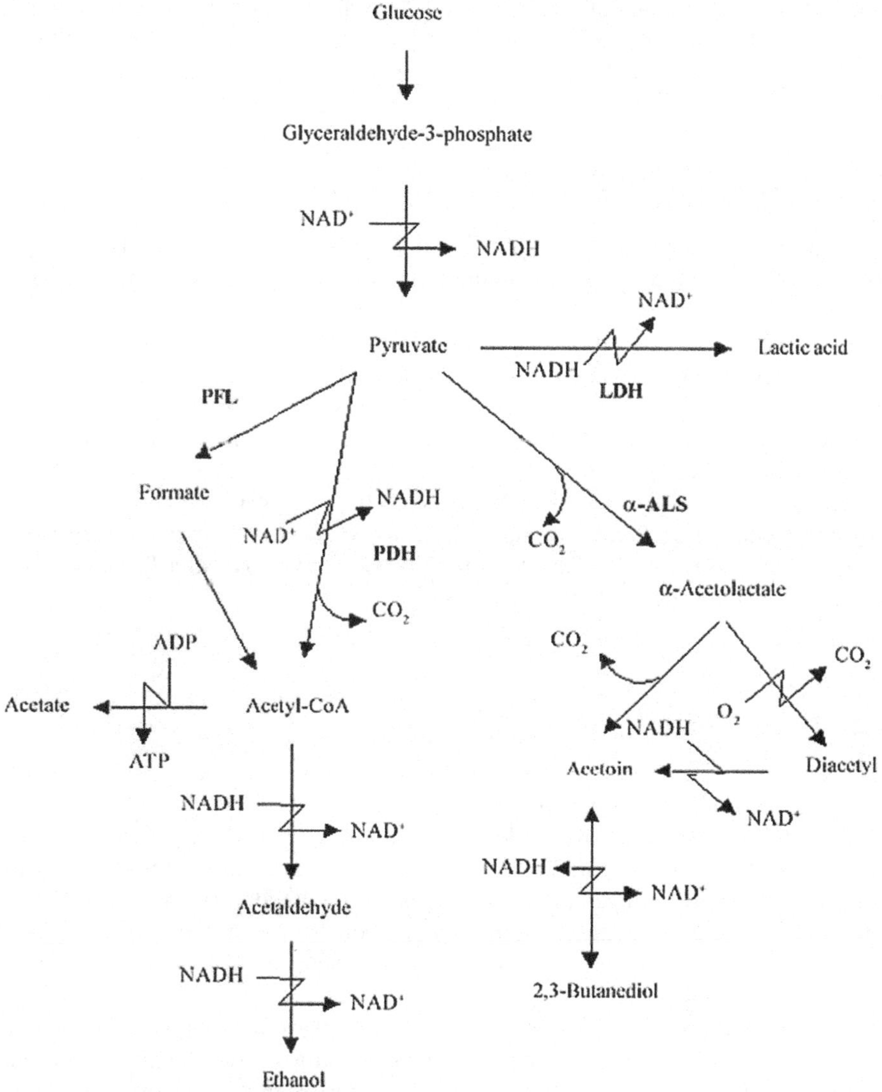

Fig. 2.1 Schematic pathway of heterofermentative metabolism. Intermediate and final glucose metabolism products are indicated by *arrows*. Catalytic enzymes are abbreviated in *bold* (LDH: lactate dehydrogenase; PDH: pyruvate dehydrogenase; PFL: pyruvate-formate lyase; α-ALS: acetolactate synthase) (Miyoshi et al. 2003)

There is also a third group comprised of strict homofermenters that has never been found in wine.

Several species of *Lactobacillus* have been isolated from grapes and wines worldwide, including *L. brevis*, *L. buchneri*, *L. casei*, *L. paracasei*, *L. cellobiosus*, *L. curvatus*, *L. delbrueckii*, *L. diolivorans*, *L. fructivorans*, *L. heterohiochii*, *L. hilgardii*, *L. jensenii*, *L. kunkeei*, *L. leichmanni*, *L. lindneri*, *L. mali*, *L. nagelli*, *L. paracasei*, *L. plantarum*, *L. trichodes*, *L. vermiforme*, *L. vini*, *L. yamanashiensis* and *L. zeae* (Douglas and Cruess 1936; Fornachon 1957; Costello et al. 1983; Lafon-Lafourcade et al. 1983; Davis et al. 1986; Sieiro et al. 1990; Edwards et al. 2000; Du Plessis et al. 2004; Beneduce et al. 2004; Moreno-Arribas and Polo 2008).

2.2.1.2 *Pediococcus*

Cells are non-mobile and have a spherical shape; these are the only LAB that separate into two planes, which results in the formation of pairs, tetrads or large clumps of spherical cells.

Bacteria belonging to this genera are facultative anaerobes and require a rich medium containing growth factor and fermentable sugar for their development. Their optimum temperature is 25–30 °C with a pH value of 6. They are homofermentative, which means that all the glucose is metabolized into lactic acid and they do not ferment pentose.

Among the approved species of *Pediococcus* (Garvie 1986), only four have been isolated from wines: *P. damnosus*, *P. parvulus*, *P. inopinatus* and *P. pentosaceus* (Davis et al. 1986; Edwards and Jensen 1992); *P. pentosaceus* and *P. parvulus* are the most common species in this medium.

2.2.1.3 *Oenococcus*

Oenococcus oeni is described as a Gram-positive non-mobile coccus and frequently occurs in pairs and chains of different sizes (Fig. 2.2).

Oenococcus is a facultative acidophilic anaerobe and grows at pH 4.8 with temperatures between 18 °C and 30 °C. It requires a rich medium supplemented with tomato juice or grape juice, and its growth is not inhibited in the presence of 10% ethanol. Glucose is fermented in lactic acid, carbon dioxide, acetic acid and ethanol (it is a heterofermenter). It converts malate into lactate and CO_2 in the presence of fermentable carbohydrate.

Wine bacteria belonging to the genus *Oenococcus* were previously classified as *Leuconostoc oenos* by Garvie (1967) and were the only acidophilic members of the genus *Leuconostoc*. Later, phylogenetic studies revealed that *L. oenos* represented a distinct subline separate from other *Leuconostoc* spp. (Martínez-Murcia et al. 1993), and this bacterium was, finally, assigned to a new genus: *Oenococcus* (Dicks et al. 1995).

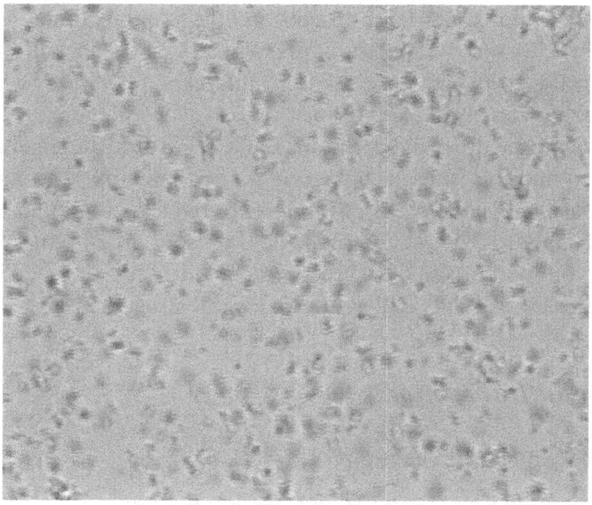

Fig. 2.2 *Oenococcus oeni* cells observed under optical microscope (CRA - Centro di ricerca per l'Enologia)

2.2.2 Development During Vinification

The growth of different microorganisms in wine tends to follow a specific order. During the harvest period, bacteria and yeasts colonize the winery. LAB are present on the grape surface and in must at very low levels, they are represented by *L. plantarum, L. casei, L. mesenteroides* and *O. oeni*. In the first few days of fermentation they multiply but their population is limited to levels of about 10^4 cells/mL. As alcoholic fermentation advances, these values decrease to 10^2 cells/mL: sensitivity to ethanol and low pH may explain the decline in cell population. After a lag phase, the surviving cells start multiplying and can reach populations of 10^6–10^8 cells/mL, during which stage MLF occurs (Fleet et al. 1984). This is completed when the bacteria reach a stationary phase.

O. oeni is the main species of LAB identified after primary fermentation and during MLF. Its development occurs naturally but it can be increased by raising the wine temperature to 20–25 °C and under conditions of low SO_2 (less than 15–20 mg/L "free"). After completion of MLF, other bacteria, such as *Lactobacillus* and *Pediococcus,* can take over. These stages overlap, giving rise to interactions between different types of bacteria, as well as between bacteria and yeasts.

Under standard conditions, LAB remain viable in wine during storage, exhibiting no tendency for further growth and showing only a slow progressive decline in viability over a long storage period. Carre (1982) observed a small decrease from 10^7 to 10^5cell/ml after 6 months of storage. Even if these cells do not multiply, they can metabolize some substances and produce unwanted compounds that can impair wine quality, especially due to the action of *Pediococcus* and *Lactobacillus* strains.

Wine is often a poor source of nutrients and these unfavourable conditions can make MLF very difficult. Temperature, pH, alcohol, SO_2 and nutrient availability all affect bacterial growth and activity. High and low temperatures will inhibit malolactic bacteria; high levels of alcohol and SO_2 can even kill them. Stuck or sluggish MLF may be caused by difficult conditions in the wine or by the malolactic bacteria not being able to multiply and reach the minimum population required for this process. In some cases, several weeks or months are required to obtain an appropriate number of cells able to degrade the malic acid present in red wines. Nowadays, it is becoming a common practice to directly inoculate a concentrated starter culture containing a selected malolactically-active bacterial strain in wine.

2.2.3 Microbial Interactions

2.2.3.1 Yeasts-Bacteria Interactions

The interrelationships between LAB and yeasts play an essential role during fermentation and in the final product. In complex ecosystems, the microorganisms may compete for the same substrates (Fleet 1990) or synergistically promote growth and wine is the product of these complex interactions between yeasts and bacteria. Results, however, are controversial. While some authors retain that these interactions are inhibitory, others consider them to be stimulatory.

Patynowski et al. (2002) showed that yeasts produce an unidentified inhibitory factor (maybe a toxic metabolite) that could be responsible for the inhibition of bacterial growth. These results could explain the antagonism between yeasts and malolactic bacteria, since yeasts are known to produce compounds during alcoholic fermentation such as ethanol, SO_2, medium-chain fatty acids and antibacterial proteins/peptides (Weeks et al. 1969; De Oliva et al. 2004; Comitini et al. 2005; Osborne and Edwards 2007). The nature and quantity of peptides and other molecules released by yeasts are different depending on winemaking techniques and the yeast strain.

In contrast to inhibition, in other studies these relationships have been shown to be positive for bacteria because yeasts may promote their growth and stimulate MLF. Challinor and Rose (1954) observed 13 interrelationships between yeasts and *Lactobacillus* spp. and in each of them the yeast appeared to be the active microorganism, synthesising the missing substances like vitamins, aminoacids or purine, essential for growth of the *Lactobacillus*. Kennes et al. (1991) showed that when *Lactobacillus plantarum* and *Saccharomyces cerevisiae* were grown in co-culture in a glucose-citrate medium under acid conditions, *S. cerevisiae* reduced the lactic acid produced by *lactobacillus* and thereby stabilized pH, encouraging the fermentation of citrate by *Lactobacillus*.

2.2.3.2 Bacteria-Bacteria Interactions

LAB can synthesise compounds with metabolic activity such as H_2O_2, organic acids and bacteriocins. Several studies have been conducted on bacteriocin production;

Lonvaud-Funel and Joyeux (1993) and Strasser de Saad and Manca de Nadra (1993) tackled this problem for wine LAB and two bacteriocins were discovered:

- Brevicin, produced by *Lactobacillus brevis,* has a broad range of action and can also inhibit *O. oeni, P. damnosus* and *L. brevis*; it is a small thermostable protein of 3 KDa and can act in a wide pH range.
- Caseicin, produced by *L. casei*, has a higher molecular weight, but is less stable.

Antibacterial activity has also been observed in *P. pentosaceus* and in one strain of *L. plantarum* that strongly inhibits the growth of *O. oeni, L. mesenteroides* and *L. hilgardii.* The discovery of these molecules gives only an indication of the true situation in wine. These could be species or strain-specific, so further studies are required to understand these relationships better. Fernandez and Manca de Nadra (2006) recently studied the interaction between a proteolitic strain of *O. oeni* and a non-proteolitic strain of *P. pentosaceus* and found a mutualism in the mixed culture, providing new knowledge about the metabolic interaction between LAB.

2.3 Isolation and Identification of Wine Lactic Acid Bacteria

Most bacteria growing in wine could be isolated by traditional microbiological techniques, such as plating them on a favourable nutritious medium. This involves serially diluting the wine sample in sterile physiological water (0.9% NaCl), then each solution is plated onto a specific medium. Usually, anaerobic Gram-positive bacteria, which comprise most LAB, are grown on MRS agar (de Man Rogosa and Sharpe) medium pH 4.8; and cyclohexamide 0.1% is added to inhibit yeast growth. Plates are incubated at 30 °C for 10–15 days. Wibowo et al. (1985) showed that the addition of tomato juice, grape juice, malic acid or different sugars to MRS medium increases bacterial growth. Usually, MRS supplemented with 10% tomato juice is the medium used to isolate and cultivate wine lactic acid bacteria. In order to obtain pure cultures, each colony is inoculated in liquid medium MRS and incubated at 30°C and the bacterial population obtained can be identified with traditional or molecular methods. Plating methods can yield ambiguous results, since many bacteria have similar nutritional needs and can grow under similar conditions.

2.3.1 Traditional Methods

Traditional methods used to identify LAB are based on phenotypic analysis: these methods study the morphological characteristics of the cells, the nature of their metabolic products and their ability to assimilate certain substrates.

Morphologic characteristic can be identified using microscopy, and depending on the shape of the cells it may be possible to establish which genus they belong to; this observation can be coupled with the Gram coloration test which verifies whether the cells studied are Gram-positive or not.

In the second step, the unidentified strain is grown in a medium containing only glucose as carbon source, after which the metabolic products are analysed:

- If the strain is a homofermenter, lactic acid will be the only metabolic product (*Pediococcus*)
- Of gas production is observed, the strain is a heterofermenter and this can be confirmed by analysing the presence of ethanol and acetic acid (*Oenococcus*, *Lactobacillus*)

The latest method that has been used to classify bacteria at species level, makes use of a system called API 50 CH (bio-Mérieux). This kit enables the genus *Lactobacillus* and related organisms to be identified. It is a ready-to-use medium which shows the fermentation profile of 49 carbohydrates (hexose, pentoses and others) on the API 50 CH strip of the microorganism to be studied. The bacterial suspension (made in a medium containing all the ingredients necessary for growth) is inoculated in each microtube of the strip. To assure anaerobiosis, the tubes are sealed with paraffin. During incubation, carbohydrates are fermented to acids, causing the pH to drop, detected by the colour change of the indicator: yellow indicating a positive character. The results make up the biochemical profile of the strain and are used in its identification or typing.

The fermentation profile is not well adapted to characterize LAB isolated from wine: bacteria are in optimal growth conditions and this does not give a true indication of the real metabolism in wine, which is influenced by environmental conditions. In general, the discriminating power is not high and several subcultures are required to obtain a stable profile. Therefore, a clear within-species identification by simple phenotypic tests may, sometimes, be difficult, and these tests are also labour-intensive and time-consuming.

2.3.2 Molecular Methods

The development of molecular techniques has opened up new perspectives for characterizing microorganisms from fermented foods and beverages. They provide outstanding tools for typing, taxonomy and evolution of bacteria in food processes (Giraffa and Neviani 2000; Germond et al. 2003).

2.3.2.1 16S rRNA Sequencing

The sequence of the 16S rRNA gene has been widely used as a molecular clock to estimate relationships among bacteria (phylogeny), but recently it has also become important as a means to identify an unknown bacterium to genus or species level. This gene is highly conserved, it is amplified with specific primers and the resulting sequence is inserted into the databases available online, and by the similarity obtained with other sequences it is possible to identify the unknown bacterium.

The sequence analysis method is very good to identify the organisms at genus and species level but it does not differentiate at the subspecies level.

2.3.2.2 G+C Content and DNA Hybridization

Estimation of the DNA nitrous base ratio (or G+C molar percent) is a classical genotypic method, constituting an integral part of a standard description of a bacterial taxon (Botina et al. 2006). These values vary from 24% to 76% among various bacteria (Schleifer and Kilpper-Balz 1987). It has been demonstrated, with high statistical significance, that among strains of a single species, the variation in the G+C ratio does not exceed 3%, compared with 10% in congeneric bacteria. LAB have a low (less than 50%) content of G+C pairs. In particular, *Oenococcus* has 38–44%; *Pediococcus* has 34–42% and *Lactobacillus* has 36–47%. This method does not allow the discrimination of species with similar GC values.

DNA-DNA hybridization is a method that provides more resolution than 16S rDNA sequencing, and has been used to describe bacterial species (Wayne et al. 1987). The 16S-23S rRNA spacer region has been suggested as a suitable region of the bacterial genome from which to derive useful taxonomic information, particularly with regard to identification at the species level (Whiley et al. 1995) and probes have been synthesized on its sequences to characterize bacterial species.

Lonvaud-Funel et al. (1989, 1991a) described the identification of LAB during vinification and wine storage by DNA-DNA hybridization. Genomic DNA of the strain to identify was hybridized with total genomic DNA probes extracted from reference strains. They found that this method was particularly efficient when used in colony hybridization to study mixed populations: at least five different species can be detected in a mixture with this system (Lonvaud-Funel et al. 1991b).

In spite of these values, the method is not popular. Major disadvantages include the laborious nature of pairwise cross-hybridizations and the impossibility of establishing a central database. Another disadvantage of the method is its high sensitivity to physiological parameters. Moreover, the data on DNA homology obtained in different laboratories are often discordant because of using different technical approaches or not complying with standard experimental conditions.

2.3.2.3 PCR-Based Methods

RAPD: This technique has been described as a useful technique for both identification and typing (Cocconcelli et al. 1995; Nigatu et al. 2001; Du Plessis and Dicks 1995; Sohier et al. 1999). Although variability has been observed in RAPD fingerprints, reproducibility can be achieved under carefully controlled conditions. The main advantage of the proposed system lies in the fact that, once a high reproducibility is reached, the method is fast, practical, easy to perform and inexpensive (Rossetti and Giraffa 2005). Figure 2.3 shows an example of RAPD analysis of different LAB with primer M13 (Rossetti and Giraffa 2005).

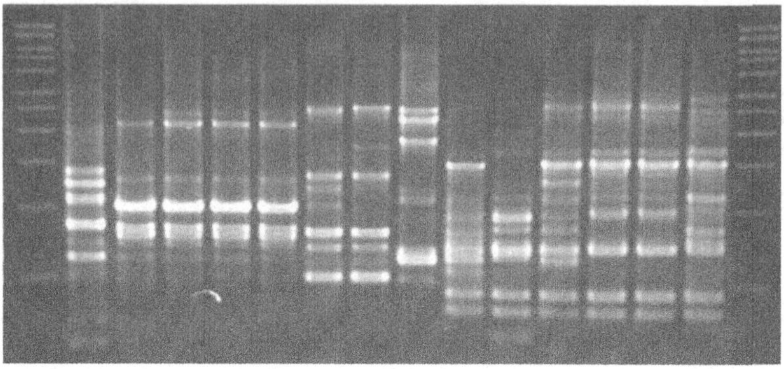

Fig. 2.3 RAPD-PCR fingerprinting of different wine lactic acid bacteria species (CRA-Istituto Sperimentale per l'Enologia)

Species-specific primer: Bartkowsky and Henschke (1999) designed specific primers to detect *O. oeni* in grape juice and wine samples. Recently, specific primers and fluorogenic probes, targeting the gene encoding malolactic enzyme of *O. oeni*, were developed and used in real time PCR assays (Pinzani et al. 2004). Real time PCR is an emerging technique that allows rapid quantification of microorganisms avoiding the plating step; this is a suitable method for monitoring fermentations and allows early and prompt corrective measures to regulate bacterial growth.

ARDRA: Restriction analysis of amplified rDNA (ARDRA) has been used to differentiate a variety of microorganism (Ventura et al. 2000; Rodas et al. 2003; Collado and Hernandez 2007). This technique is useful to simplify and clarify the identification of lactobacilli. 16S-ARDRA has advantages over RAPD: it is less dependent on reaction conditions and the interpretation of results is easier. 16S-ARDRA generates species-specific patterns in the majority of species studied, but is not useful for typing purposes because the 16S rRNA gene sequence is highly conserved at the species level (Rodas et al. 2005).

DGGE: Denaturing and temperature gradient gel electrophoresis (DGGE and TGGE) have been developed to analyze microbial communities rapidly by sequence-specific separation of PCR-amplified fragments (Fleske et al. 1998). This technique has been recently applied to evaluate the microbial diversity of several environments (Ampe et al. 1999; Gelsomino et al. 1999; Cocolin et al. 2000; Ercolini 2004) and to "profile" complex microbial communities (Heuer et al. 1997). It was also used to test the purity of bacterial strains, to monitor bacteria from environmental samples, and to study the dynamics of specific populations according to environmental variations (Tenske et al. 1996). This technique enables the separation of polymerase chain reaction amplicons of the same size but of different sequence; the amplicons in the gels are subjected to an increasingly denaturing environment; the migration is stopped when DNA fragments are completely denatured. Recently, DGGE has been applied to study wine microbial ecology giving an exhaustive profile of the species present in wine (Renouf et al. 2006, 2007). The results reported that this technique, based on *rpoB* gene as a molecular marker, is a reproducible and suitable tool and may be of great value for winemakers to monitor spoilage microorganism during wine fermentation (Spano et al. 2007).

2.3.2.4 Strain Identification

Pulsed field gel electrophoresis (PFGE) uses restriction enzymes to digest microbial DNA, which is then subjected to electrophoretic separation (Arbeit et al. 1990; Maslow et al. 1993). The restriction with endonuclease ApaI was shown to be an efficient method to reveal polymorphism between *O. oeni* strains (Zapparoli et al. 2000). DNA fragments after separation are then compared to evaluate the variability among strains belonging to the same species. The disadvantage is that this technique is laborious and time-consuming and requires special equipment. Figure 2.4 shows an example of PFGE profiles of *O. oeni*.

Zavaleta et al. (1997) and Reguant (2003) applied RAPD analysis, using different conditions, to evaluate intraspecific genetic diversity of *O. oeni*, and found that most strains showed unique RAPD patterns; they proposed this method as a good tool to study the population dynamics of bacteria during MLF.

Multilocus sequence typing (MLST) has emerged as a powerful new DNA-typing tool for the evaluation of intraspecies genetic relatedness. In MLST methods, several bacterial "housekeeping" genes are compared on the basis of the partial nucleotide sequence; all sequence types are represented by a single strain and all the strains can be distinguished from each other, because of a unique allele combination. This method has shown a high degree of intraspecies discriminatory power for bacterial and fungal pathogens; De las Rivas et al. (2003) applied this technique to discriminate *O. oeni* at the strain level; they determined the degree of allelic variation in

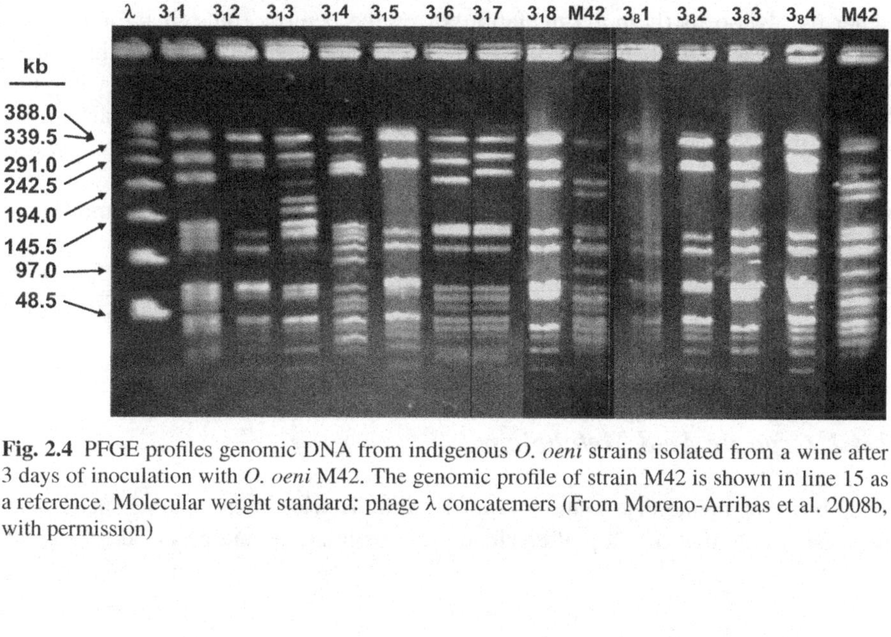

Fig. 2.4 PFGE profiles genomic DNA from indigenous *O. oeni* strains isolated from a wine after 3 days of inoculation with *O. oeni* M42. The genomic profile of strain M42 is shown in line 15 as a reference. Molecular weight standard: phage λ concatemers (From Moreno-Arribas et al. 2008b, with permission)

five genes of *O. oeni* and showed that the percentage of variable sites was high, indicating a considerably high degree of genetic diversity. Therefore, MLST was demonstrated to be a powerful method to discriminate *O. oeni* at the strain level and the data obtained could be applied to study the population structure and its evolutionary mechanism.

2.4 Relevant Aspects of Lactic Acid Bacteria Metabolisms in Wines

Of all the metabolic activities that lactic acid bacteria can carry out in wine, the most important, or desirable, in winemaking is the breakdown of malic acid, but only when it is intended for this to be removed completely from the wine by malolactic fermentation. Although the breakdown of malic and citric acids has considerable consequences from a winemaking perspective, it is also evident that lactic acid bacteria metabolise other wine substrates to ensure their multiplication, including sugars, tartaric acid, glycerine and also some amino acids. We will now describe some of the metabolic transformations that have received most attention in the literature, or which have important repercussions in winemaking.

2.4.1 Carbohydrate Metabolism

Sugars are the main energy sources for bacterial growth, which tend to prefer to use glucose and trehalose more than others. However, the metabolic routes of sugars have not yet been completely elucidated for enological lactic acid bacteria, especially for *O. oeni*. Depending on the species of lactobacilli and cocci, these ferment either by glycolysis (homofermentation) or by the pentose route (heterofermentation). However, only the latter process generates acetic acid that increases the wine's volatile acidity. Nonetheless, in normal vinification, without incidences, when the LAB multiply in the medium, only sugars not fermented by yeasts remain in the medium. In general, this corresponds to hundreds of mg/L of glucose and fructose and grape must pentoses (xylose and arabinose). The residual sugars are sufficient to supply the energy required for bacterial growth and to permit the formation of biomasses that will later carry out the MLF.

Wine lactic acid bacteria can degrade polysaccharides and *O. oeni* has been shown to have an extracellular β (1→3) glucanase activity (Guilloux-Benatier et al. 2000).

2.4.2 Organic Acid Metabolism

The ability to metabolise malic and citric acids is widespread among lactic acid bacteria strains that develop after alcoholic fermentation, and can lead to a great number and diversity of organoleptic changes. The one most studied is the breakdown of malic acid, in the phase known as "malolactic fermentation" and, more recently, citric acid breakdown, and its association with improved wine sensorial characteristics has also received attention.

2.4.2.1 Transformation of Malic Acid

This is the main reaction of MLF. Chemically it consists of a simple decarboxylation of the L-malic acid in wine into L-lactic acid. Biochemically, it is the result of activity of the malolactic enzyme, characteristic of lactic acid bacteria. This transformation has a dual effect. On the one hand, it deacidifies the wine, in other words, it raises the pH, an effect that is greater at higher initial quantities of malic acid. It also gives the wine a smoother taste, replacing the acidic and astringent flavour of the malic acid, by the smoother flavour of the lactic acid.

This is the main reaction by which MLF causes discrete changes in the organoleptic characteristics of a wine, and is why the second fermentation is especially recommendable for most red and many white wines. The duration of this transformation of malic acid depends on the initial amount of this acid present and the total population of bacteria that have multiplied in the wine. However, for the same biomass formed, this process can be slowed down as a consequence of certain inhibitors in the wine, which have not yet been identified.

The malolactic enzyme is dimeric and is comprised of two identical subunits of 60 kDa using NAD^+ and Mn^+ as cofactors. There are numerous studies into the biochemical characteristics of this enzyme in many bacteria species, such as *L. casei* (Battermann and Radler 1991) *L. plantarum* (Schüzt and Radler 1974), *L. mesenteroides* (Lonvaud-Funel and Strasser de Saad 1982) and *O. oeni* (Naouri et al. 1990). These studies have shown that this enzyme functions via a sequentially arranged mechanism, in which the cofactors of the reaction, Mn^{2+} and NAD^+, become fixed before l-malate. Moreover, the activity is induced by the reaction of the malic acid substrate. Also, the malolactic enzyme of *O. oeni* has been genetically characterised. In the *mle* locus of *O. oeni* the malolactic operon can be found, composed of three genes, gene *mleA* that encodes the malolactic enzyme, gene *mleP* that encodes the malate-permease and the *mleR* gene, which encodes the regulator that activates transcription of the malolactic operon (Labarre et al. 1996).

2.4.2.2 Breakdown of Citric Acid

While the wine contains several g/L of L-malic acid before MLF, it usually only contains between 200 mg/L and 300 mg/L of citric acid. Although the citric acid is only present in low concentrations, it is of considerable importance. On the one hand, its metabolic pathway leads to production of acetic acid, in other words, it increases the volatile acidity of the wine. However, the most important enological significance associated with fermentation of citrate is the production of diacetyl and other acetonic compounds, which affect the wine aroma.

At low levels (5 mg/L), diacetyl is considered to add complexity to wine aroma since it can impart positive *nutty* or *caramel* characteristics, although at levels above 5 mg/L it can result in spoilage, creating an intense *buttery* or *butterscotch* flavour, and is perceived as a flaw. Microbial formation of diacetyl is a dynamic process and its concentration in wine depends on several factors: bacterial strain, pH, wine contact with lees, SO_2 content (Martineau and Henick-Kling 1995; Nielsen and Richelieu 1999). The sensory threshold for the compound can vary depending on the levels of certain wine components, such as sulfur dioxide. It can also be produced as a metabolite of citric acid when all the malic acid has been used up. However, diacetyl rarely taints wine to levels where it becomes undrinkable.

2.4.3 Metabolism of Phenolic Compounds

To date, most studies on the interactions between phenolic compounds and LAB in wines refer to the metabolism of hydroxycinnamic acids (ferulic and coumaric acids), by different bacteria species, resulting in the formation of volatile phenols (4-ethylguaiacol and 4-ethylphenol) (Cavin et al. 1993; Gury et al. 2004). These derivatives can have a significant influence on wine aroma since they are regarded as sources of phenolic off-flavors in wine, due to their characteristic aroma and low detection threshold (Cavin et al. 1993). In wines, the amounts of these compounds are generally low and are, usually, limited by the concentrations of their precursors. Hernández et al. (2006) showed that *trans*-caftaric and *trans*-coutaric acids are substrates of LAB that can exhibit cinnamoyl esterase activities during MLF, increasing the concentration of hydroxycinnamic acids. An additional source of caffeic and *p*-coumaric acids may come from the hydrolysis of cinnamoyl-glucoside anthocyanins (Moreno-Arribas et al. 2008a), as well as from other hydroxycinnamic derivatives by LAB enzymatic activity. Furthermore, according to Hernández et al. (2007), it seems that among wine LAB, this activity could be strain-dependent and could also depend on the isomeric form of the above-mentioned esters, since only the *trans*-isomers were involved in the reaction. Besides wine LAB, free phenolic acids can also be metabolized by other wine microorganisms, mainly *Brettanomyces*/*Dekkera* (Chatonnet et al. 1995) to form 4-vinyl derivatives, which can be reduced to 4-ethyl derivatives in wine. Thus, on the basis of these observations, it can be deduced that LAB could contribute to the differences in vinylphenol levels found in wines.

2.4.4 Hydrolysis of Glycosides

The release of glycosidically-bound aroma compounds, such as monoterpenes, C_{13} norisoprenoids, and aliphatic alcohols, can be achieved by the action of glycosidase enzymes. β-Glycosidase activity has not been much studied in wine LAB. McMahon et al. (1999) detected a low β-glycosidase activity in *O. oeni* OSU and a low activity of α-L-ramnopyranosidase in *O. oeni* Viniflora oenos. However, other authors reported significant β-glycosidase activities in different *O. oeni* strains in model systems (Grimaldi et al. 2000; Ugliano et al. 2003; D'Inceddo et al. 2004) and during red wine production (Ugliano and Moio 2006). These results suggest that the LAB of wine have the potential to hydrolyse glycoconjugates that affect wine aroma.

2.4.5 *Metabolism of Amino Acids*

Cysteine and methionine are metabolised by bacteria that form diverse sulphated compounds, including hydrogen sulphide and methanethiol. The metabolism of methionine has been the most studied. The LAB isolated from wine are able to degrade methionine to form methanethiol, dimethyl disulphide, 3-(methyldisulphanyl) propan-1-ol and 3-(methyldisulphanyl) propionic acid. These compounds are formed in greater quantities by *O. oeni* than *Lactobacillus*. Methanethiol and 3-(methyldisulphanyl) propan-1-ol are characterized by putrid faecal like aromas and cooked cabbage descriptors, respectively (Pripis-Nicolau et al. 2004). A reaction in wine can occur between α- dicarbonyl compounds and aminoacids, in particular cysteine; several aromas can arise, including sulphury, floral, toasted and roasted aromas. Many of the compounds produced in this way have been identified in wine, and because of their low olfactory threshold could play an important role in wine aroma and flavour.

2.4.6 *Breakdown of Proteins and Peptides*

The peptide fraction of an industrially manufactured red wine has been studied during MLF, and it was found that wine LAB have the potential to hydrolyze wine proteins (Alcaide-Hidalgo et al. 2008), although some authors have consider that this activity is not widespread among oenococci strains (Leitao et al. 2000). However, the ability of *O. oeni* to exhibit extracellular protease activity able to release peptides and free amino acids during MLF in white (Manca de Nadra et al. 1997) and red wines has also been demonstrated (Manca de Nadra et al. 1999). The oligopeptide utilization of *O. oeni* was characterized only recently (Ritt et al. 2008) and *O. oeni* was found to be able to transport oligopeptides with two to five-amino acid residues and then to hydrolyse them further.

2.5 Contribution of Malolactic Fermentation to the Organoleptic Characteristics of Wine

Different studies have focused on the biosynthesis of aroma compounds during MLF and the concomitant organoleptic consequences (Laurent et al. 1994). Maicas et al. (1999) demonstrated that MLF noticeably changes major and minor volatile compounds which are beneficial to wine flavour. Pozo-Bayón et al. (2005) investigated the changes in volatile compounds before and after MLF, carried out by four different starter cultures of the species *Oenococcus oeni* and *Lactobacillus plantarum,* and found significant metabolic differences between both species. Aroma/ flavour attributes also seemed to vary according to the strain used for inducing MLF.

According to Henick-Kling (1993), MLF increases the fruity and buttery aromas but reduces vegetable or grassy aromas. Formation and hydrolysis of esters during MLF may also lead to an increase in the fruity aroma and it is, probably, due to the action of LAB esterases responsible for the synthesis and degradation of these compounds. However, to date there are no studies that demonstrate these changes. The reduction in vegetable or grassy aromas could be due to the catabolism of aldehydes by lactic acid bacteria. *O. oeni* can catabolise acetaldehyde, converting it into ethanol and acetate (Osborne et al. 2000).

As well as fruity and buttery aromas, MLF has also been associated with other characteristic aromas such as floral, roasted, vanilla, sweet, woody, smoked, bitter, honey, etc. (Henick-Kling 1993; Sauvageot and Vivier 1997). However, further studies are required to be able to relate the wine characteristics that are modified during malolactic fermentation with the production and/or degradation of a specific chemical compound by wine lactic acid bacteria. With this information, the winemaker can choose the best strain of lactic acid bacteria to obtain wine with a specific aroma or flavour.

In general, the change in colour of red wines after MLF corresponds to a reduced intensity with less blue tones, mainly due to the possible adsorption of anthocyanins, especially the methoxylated ones, by the bacterial cell walls, aided by the rise in pH which produces the transformation from malic to lactic acid and the decreased levels of free sulphurous anhydride (Suárez-Lepe and Íñigo-Leal 2003). Recently, new data were provided about the effect of MLF on the concentration of the phenolic compounds of red wines. The changes in four different groups of anthocyanins (simple glucosides, acetyl glucosides, cinnamoyl glucosides and pyroanthocyanins) were studied by HPLC-PAD-MS during MLF in barrel or in tank of an industrial red wine (Moreno-Arribas et al. 2008a). It was shown that the effect of the container used seems to be more important than the metabolic activity of the bacteria responsible for the process. Hydroxycinnamic acids (*trans*-caffeic and *trans-p*-coumaric) and their derivatives (*trans*-caftaric and *trans p*-coutaric acids) were the main compounds modified by MLF, independently of the use of stainless steel or barrel (Hernández et al. 2006). Taking into account that phenolic acids can act as anthocyanin copigments, stabilizing the colour of wine, higher contents of these compounds will have a positive effect on the colour.

The lactic acid bacteria may cause polysaccharides to be released in a wine (Dols-Lafalgue et al. 2007). These compounds can increase the sensation of volume or body of wines, and can also be polymerized with the grape or wood tannins, reducing sensations of roughness or astringency, and producing more complex flavours.

2.6 New Trends in the Performance of Malolactic Fermentation in Wineries

2.6.1 Use of Malolactic Starter Cultures

The induction of MLF through the use of selected starters gives some advantages: a better control of the start of fermentation, of its progress and of the strain that completes this process. In fact, the inoculum of selected bacteria, generally constituted by only one strain or mixtures of 2–3 strains of *O. Oeni,* prevents the development of bacteria belonging to the genera *Lactobacillus* and *Pediococcus*. These contaminating species can produce high concentrations of acetic acid that can impair the organoleptic quality of the wine and also substances undesirable from a health perspective such as the biogenic amines (Straub et al. 1995; Moreno-Arribas et al. 2003; Costantini et al. 2006).

In synthesis, the inoculum of selected bacteria permits:

- The beginning of MLF to be controlled: if the bacterial population has been correctly controlled at the end of alcoholic fermentation, the wine will contain few bacterial cells and spontaneous MLF could occur after weeks or even months
- Wine quality from being impaired by the development of contaminating bacteria
- The organoleptic characteristics of the wine to be selected; in fact, MLF not only represents a process of deacidification of the wine but, depending on the strain employed, it can also influence the organoleptic characteristics, preventing the production of negative secondary metabolites

Currently, different starters for MLF are commercialized as lyophilized preparations. Nevertheless, the vitality of these starters can decrease after they are inoculated in wine (Krieger et al. 1993). From the perspective of cell vitality it would, therefore, be preferable to use fresh or frozen preparations. However, these solutions are not feasible on a large scale: the fresh preparation must be produced in situ, and the frozen ones are difficult to keep, especially when they must be transported for long distances. The rehydration phase of the lyophilised cells is a delicate and important phase since it allows the cells to recover the viability required to survive in the wine (Nielsen et al. 1996). The use of a starter culture with a dilute microbial population renders the inoculum almost useless. On the other hand, although a low cell viability negatively influences the result of the inoculum, a high vitality does not always guarantee the success of MLF. A variable behaviour in wine has been shown for the different bacteria strains, reflecting their different ability to adapt and variable malolactic activity in wine (Malacrinò et al. 2003). One bacterial strain may be unable to adapt in wines with limiting chemical-physical characteristics, while another strain in the same conditions is able to adapt and to multiply.

Another important practical factor to consider in the use of starters for MLF is the possible sensitivity to bacteriophages. The sensitivity of bacteria to phages is very different, depending on the strain. Some authors (Sozzi et al. 1982) claimed that MLF may be interrupted, delayed, or completely inhibited by the action of phage active against lactic bacteria. Other authors (Poblet-Icart et al. 1998) studied the induction bacteriophage of *O. oeni* and showed that 45% of the strains studied proved to be lysogenic, suggesting that lysogeny is widespread among bacteria isolated from wines during MLF. Since the phages of *O. oeni* are lysogenic, they concluded that they would not represent a critical problem for MLF, even if they agree with Sozzi et al. (1982) on the need to inoculate a mixture of strains with different sensitivities to phages to avoid starter culture failure and problems during MLF.

2.6.2 Time of Inoculation/Co-inoculation

The most common decision is to inoculate selected bacteria at the end of alcoholic fermentation, to avoid an excess development of LAB, which can give high quantities of acetic acid. In recent years a co-inoculum of selected yeasts and bacteria has been proposed to induce simultaneous alcoholic fermentation and MLF to increase the adaptation of LAB to wine, particularly as this concerns adaptation to a high ethanol contents. Co-inoculation at different times has been studied by some authors (Henick-Kling and Park 1994; Rosi et al. 2003; Jussier et al. 2006). In the case of co-inoculum, but also when the selected bacteria are inoculated at the end of alcoholic fermentation, the phenomenon of yeast-bacteria interaction should be considered (King and Beelman 1986; Lemaresquier 1987; Delaquis et al. 2000; Larsen et al. 2003). In recent work, Alexandre et al. (2004) studied the interactions between *Saccharomyces cerevisiae* and *Oenococcus oeni* in wine and showed that yeasts can oppose or stimulate MLF. Recently, Osborne and Edwards (2007) found that a strain of *Saccharomyces cerevisiae* could produce a peptide responsible for inhibiting MLF. A successful co-inoculum will strongly depend on the selection of suitable yeast-bacterium combinations (Alexandre et al. 2004; Jussier et al. 2006).

2.6.3 Malolactic fermentation in Barrels/Microoxygenation

Microoxygenation is a technique that involves the addition of small and controlled quantities of oxygen to the wine. This technique is mainly used to stabilize the colour of red wines, since oxygen in small quantities favours polymerization reactions among anthocyanins and tannins (Atanasova et al. 2002). Globally, total anthocyanins decrease, but what is formed, combined with tannins, leads to a product which is more intensely coloured and more stable over time than the initial compounds.

An important reaction is that of acetaldehyde polymerization (Saucier et al. 1997; Romero and Baker 2000). This reaction not only increases the colouring intensity, but also intensifies the blue coloration (od 620 nm) that is responsible for the mauve tones in wine.

Because microoxygenation delays the beginning of MLF, this should be completed before inoculating the selected bacteria. Another reason to induce MLF after microoxygenation is because the lactic bacteria consume acetaldehyde, an essential intermediate in the polymerization reactions among phenolic compounds, as seen above.

Slow and controlled oxidation occurs in wooden containers. The use of wood in the refinement of wines furnishes volatile and non-volatile compounds, including polysaccharides and polyphenols that, together with a slow oxygenation process, help to stabilize the wine colour.

2.7 Wine Spoilage by Lactic Acid Bacteria

2.7.1 Aspects Related to the Organoleptic Quality of Wines

Besides carrying out MLF, under certain conditions LAB can also cause undesirable changes in wine flavour which render the wine undrinkable. Many species of LAB do not conduct MLF and their growth in wine can cause some serious wine spoilage. The nature and extent of this spoilage depend on several factors such as the type of bacteria, composition of the wine and vinification practices; specific types of spoilage are associated with a restricted number of lactobacilli.

These bacteria can partially metabolize more complex wine components such as phenolic compounds, aromatic compounds or aroma precursors present in small quantities and the resulting products can have important organoleptic repercussions on wine quality. Excess volatile acidity, mannitol taint, ropiness, mousiness, acrolein formation and bitterness, tartaric acid degradation, diacetyl overproduction and rancidness, as well as the very unpleasant geranium off-flavour are often the consequence of uncontrolled growth of some species of *Lactabacillus* (*e.g.*, *L. brevis*, *L. hilgardii*, *L. plantarum*), *Leuconostoc* (*e.g.*, *L. mesenteroides*), *Streptococcus* (*S. mucilaginosus*) and *Pediococcus* (*e.g.*, *P. cerevisiae*) (Pretorius, 2001).

When alcoholic fermentation is too slow or when it stops, conditions are favourable for bacterial development. LAB ferment different quantities of sugars that have not been totally fermented by yeasts and produce acetic acid and D-lactic acid. This alteration is called "*Lactic disease*" (piqûre lactique) and is characterised by a high volatile acidity that depreciates the wine. If this volatile acidity exceeds the limit of 1 g/L, the wine is unmarketable (Lonvaud-Funel 1999). This spoilage also occurs in fortified wine where *O. oeni, L. hilgardii, L. fructivorans* and L. *plantarum* are active in spite of very high ethanol contents.

The microbiological breakdown of glycerol forms acrolein, a product which causes *bitterness* in wine by binding with phenolic components (Singleton 1995). Ethanol increases the intensity of the bitter taste, as well as the duration of the bitter sensation (Noble 1994). An increased alcohol concentration resulted in an increase in the bitter sensation (Fischer and Noble 1994). *Lactobacillus brevis* and *L. buchneri*, isolated from spoiled wine, can metabolize glycerol in the presence of glucose or fructose, resulting in the formation of 3-hydroxy propanal (also known as 3-hydroxy propionaldehyde, 3-HPA), which is subsequently reduced to 1,3-propanediol (Schutz and Radler 1984a,b). 3-Hydroxy propionaldehyde is a precursor of acrolein. The conversion of glycerol to 3-HPA in co-metabolism with glucose or fructose is not restricted to wine lactobacilli. *L. collinoides,* isolated from spoiled cider and fermented apple juice, can also do this (Claisse and Lonvaud-Funel 2000). Physiologically, the co-metabolism of sugar and glycerol is important to these lactobacilli, since additional ATP is generated from acetyl phosphate (Veiga-da-Cunha and Foster 1992).

Some strains of *L. brevis* cause "*mannitol taint*" by enzymatic reduction of fructose to mannitol. Mannitol is a polyol produced in heterofermentative metabolism. Its perception is often complicated as it generally exists in wine alongside other defects, but it is usually described as viscous and ester-like, combined with a sweet and irritating finish (Du Toit and Pretorius 2000). Mannitol is usually produced in wines that undergo MLF with a high level of residual sugars still present.

Tartaric acid is relatively stable to bacterial activity and can only be metabolized by some *Lactobacillus* species with the production of acetic acid, lactic acid and succinic acid (Kandler 1983). When tartaric acid is metabolised, the volatile acidity increases and the wine acquires an acetic aroma and a disagreeable taste; this degradation can be total or partial depending on the bacteria population, but it always decreases wine quality. The tartaric acid degrading capacity is restricted to only a few species: Radler And Yannissis (1972) found it in four strains of *L. plantarum* and one strain of *L. brevis*.

Several strains of LAB isolated from wine were tested for their abilities to metabolize ferulic and *p*-coumaric acids. Cavin et al. (1993) showed that these acids were strongly decarboxylated by growing cultures of *Lactobacillus brevis, Lactobacillus plantarum*, and *Pediococcus;* when decarboxylation was observed, volatile phenols (4-ethylguaiacol and 4-ethylphenol) were detected, indicating the possibility of reduction of the side chain before or after decarboxylation. Couto et al. (2006) reported *L. collinoides* as a producer of volatile phenols, although strain specificity concerning this capacity was observed. *L. mali, L. sake, L. viridescens*, and *P. acidilactici* were also found to be able to produce volatile compounds but they only perform the decarboxylation step. Volatile phenols cause animal taints such as horse sweat, wet animal and urine that are usually attributed to *Brettanomyces* spoilage.

Wine with an increased viscosity and a slimy appearance is called "*ropy*". This aspect is due to the production of dextrane or glucane produced by *Leuconostoc* and *Pediococcus* (Fugelsang 1997; Lonvaud-Funel 1999). These polysaccharides are mainly produced by *P. damnosus* and their production is linked to a plasmid of approximately 5500 bp; the ropy phenotype disappears when the plasmid is lost. These ropy strains are much more tolerant to ethanol than others. Concentrations of glucane around 100 mg/L are high enough to give the wine this abnormal viscosity.

Mousiness is a wine fault most often attributed to *Brettanomyces* but can also originate from *L. brevis*, *L. fermentum*, and *L. hilgardii* (Du Toit and Pretorius 2000). The metabolism of ornithine and lysine is associated with the formation of N-heterocycles, 2-acetyl-1-pyrrolene, 2-acetyltetrahydropyridine and

2-ethyltetrahydropyridine. These compounds are not volatile at wine pH, but in the mouth where pH is neutral they become very apparent, giving a nauseating aroma of *mouse urine*.

2.7.2 Aspects Related to the Hygienic Quality of Wines

The metabolism of amino acids does not affect the taste, but is problematic at a toxicological level, because it increases the concentrations of biogenic amines and ethyl carbamate precursors in wine.

Biogenic amines are natural compounds present in different types of foods and beverages, such as cheese, fish, beer, and wine. Histamine and tyramine, when ingested, can have adverse reactions that affect the nervous and vascular systems (Silla 1996; Bover-Cid and Holzapfel 1999). Putrescine is also potentially dangerous, because it can react with nitrites to form carcinogenic nitrosamine (Halasz et al. 1994). Biogenic amines are mainly produced by decarboxylation of the precursor amino acid through the substrate-specific enzymes of microorganisms that can be present in food. The enzymes on which most research has focused are histidine decarboxylase (HDC), which catalyzes the formation of histamine (Coton et al. 1998); tyrosine decarboxylase (TDC), which is specific for tyramine formation (Moreno-Arribas et al. 2000); and ornithine decarboxylase (ODC), which catalyzes the formation of putrescine (Marcobal et al. 2004). The production of biogenic amines in wine should be considered an important criterion in the selection of starter cultures and in the study of the characteristics of the autochthonous microflora present in the wine environment. Several papers have reported conflicting results but, in general, the presence of biogenic amines in wine is correlated with wine spoilage and, especially due to the action of different *Lactobacillus* strains (Straub et al. 1995; Moreno-Arribas et al. 2003; Costantini et al. 2006). More information about the chemical and biochemical features of the production of biogenic amines in wines is found in Chapter 6A.

Ethyl Carbamate besides malic acid, some heterofermentative wine LAB are capable of forming small amounts of citrulline from degradation of the amino acid arginine. The excretion of citrulline is of toxicological concern, since it is a precursor in the formation of carcinogenic EC (ethyl carbamate) in wine (Zimmerli And Schlatter 1991). From the results obtained, Mira de Orduña et al. (2001) concluded that the risk of citrulline formation by malolactic bacteria in wines with high residual arginine concentrations can be reduced by carrying out MLF with pure oenococcal cultures and by precisely establishing complete malolactic conversion, which must be followed by inhibition of bacterial activity.

Also, in this case, research results indicate the need for caution in the selection of starter cultures for MLF in wine, since citrulline formation from arginine degradation could result in ethyl carbamate production, even at normal temperatures, during prolonged storage. In addition, spontaneous MLF by undefined strains should be avoided, as this may lead to formation of ethyl carbamate precursors (Liu et al. 1994).

2.8 Methods for Managing Lactic Acid Bacteria Growth

In winemaking it is especially important to control MLF effectively to avoid possible bacterial alterations. On the other hand, although MLF is sometimes difficult to induce in wineries, prevention or inhibition of the growth and development of LAB in wine is also a difficult task. In practise during vinification, by adding sulphur dioxide (SO_2) LAB are eliminated after all the wine's malic acid has been degraded. SO_2 has numerous properties as a preservative in wines; these include its antioxidant, antioxidasic and selective antimicrobial effect, especially against LAB. Today this is therefore considered to be an essential treatment in winemaking and preservative technology. However, the use of this additive is strictly controlled, since high doses can cause organoleptic alterations in the final product (undesirable aromas of the sulphurous gas, or when this is reduced to hydrosulphate and mercaptanes) and, especially, owing to the risks to human health of consuming this substance. In addition, a first move to increase food safety has been taken by the EU through a legislation that regulates the use of sulphites as preservatives. Henceforth directives 2000/13/EC, 2003/89/EC and 2007/68/EC request the systematic labelling of allergens or similar incorporated in food products, including wine. Since the 25th of November 2005, the mandatory and particular mention of the presence of sulphites in foodstuffs is also required as soon as the concentration exceeds 10 mg/L or 10 mg/kg. Because of these effects, in recent years there has been a growing tendency to reduce the maximum limits permitted in musts and wines. Although as yet there is no known compound that can replace SO_2 with all its enological properties, there is great interest in the search for other preservatives, harmless to health, that can replace or at least complement the action of SO_2, making it possible to reduce its levels in wines.

With regard to products with antimicrobial activity complementary to SO_2, recently dimethyldicarbonate (DMDC) has been described as being able to inhibit the development of yeasts and LAB, permitting the dose of SO_2 to be reduced in some types of wines (Renouf et al. 2008). Other alternatives have been introduced based on "natural antimicrobial agents", of which the use of lysozyme is especially important and some antimicrobial peptides or bacteriocins (Navarro et al. 2000; Du Toit et al. 2002). Since lysozyme can cause IgE-mediated immune reactions in some individuals (Mine and Zhang 2002), its presence in food products, including wine, can cause some concern. To date, nisin is the only bacteriocin that can be obtained commercially, and although this has been shown to be effective at inhibiting the growth of spoilage bacteria in wines (Radler et al. 1990a,b; Rojo-Bezares et al. 2007), it has not been authorized for use in enology. Other bacteriocins have been described to control the growth of LAB in wine (Bauer et al. 2005).

Recently, special attention has been paid to the effect of phenolic compounds on the growth and metabolism of LAB in wine in order to establish the extent to which these compounds are involved in malolactic fermentation during winemaking (García-Ruiz et al. 2008a). It has been suggested that phenolic compounds can behave as activators or inhibitors of bacterial growth depending on their chemical structure (substitutions in the phenolic ring) and concentration (Reguant et al. 2000;

Vivas et al. 1997). Recently, the evaluation of the dual antioxidant and antibacterial activity of 21 phenolic compounds mainly present in *Vitis Vinifera L.* belonging to different groups was examined (García-Ruiz et al. 2008b). Structure-activity relationships were probed for both antimicrobial and antioxidant properties of wine phenolics, confirming the potential of these compounds as an alternative to sulphites in winemaking.

2.9 Conclusions

In the last few years, the interest of the scientists in wine LAB has increased. Currently, the enologist has more and better ways to control the activity of lactic acid bacteria and to counter their effect on the quality of the wine through a multidisciplinary and more extensive vision. Of special importance, work on the natural diversity of the species *O. oeni*, the major control of its development during the MLF, and the contribution of precise aromatic notes depending on the type of wine is likely to continue. On the other hand, another line of prominent investigation will continue focusing on greater knowledge and control of the organoleptic impact and the security of the wine, and of the development and metabolism of LAB; the new tools involving advanced analytical techniques, as well as those of molecular biology, will enable continuous progress in this field.

Acknowledgments Authors are grateful to the AGL2006-04514, PET2007_0134 and CSD2007-00063 projects (Consolider Ingenio 2010 FUN-C-FOOD, Ministerio de Educación y Ciencia), the S-505/AGR-0153 ALIBIRD Project (Comunidad Autónoma de Madrid) and Regione Piemonte (regional resolution CIPE 2007) for financial support for this work.

References

Alcaide-Hidalgo, J., Moreno-Arribas, M.V., Polo, M.C., & Pueyo, E. (2008). Partial characterization of peptides from red wines. Changes during malolactic fermentation and aging with lees. *Food Chem., 107*, 622–630.

Alexandre, H., Costello, P.J., Remize, F., Guzzo, J., & Guilloux-Benatier, M. (2004). *Saccharomyces cerevisiae-Oenococcus oeni* interactions in wine: current knowledge and perspectives. *Int. J. Food. Microbiol., 93*, 141–154.

Ampe, F., ben Omar, N., Moizan, C., Wacher, C., & Guyot, J.P. (1999). Polyphasic study of the spatial distribution of microrganisms in Mexican pozol, a fermented maize dough, demonstrates the need for cultivation-independent methods to investigate traditional fermentations. *Appl. Environ. Microbiol., 65*, 5464–5473.

Arbeit, R.D., Arthur, M., Dunn, R., Kim, C., Selander, R.K., & Goldstein, R. (1990). Resolution of recent evolutionary divergence among *Escherichia coli* from related lineages: the application of pulsed-field electrophoresis to molecular epidemiology. *J. Infect. Dis., 161*, 230–235.

Atanasova, V., Fulcrand, H., Cheynier, V., & Moutounet, M. (2002). Effect of oxygenation on polyphenol changes occurring in the course of wine-making. *Anal. Chim. Acta, 458*, 15–27.

Bartkowsky, E.J., & Henschke, P.A. (1999). Use of polymerase chain reaction for specific detection of the MLF bacterium *Oenococcus oeni* (formerly *Leuconostoc oenos*) in grape juice and wine samples. *Aus. J. Grape Wine Res., 5*, 39–44.

Battermann, G, & Radler, F. (1991). A comparative study of malolactic enzyme and malic enzyme of different lactic acid bacteria. *Can. J. Microbiol., 37*, 211–217.

Bauer, R., Chikindas, M.L., & Dicks, L.M.T. (2005). Purification, partial amino acid sequence and mode of action of pediocin PD-1, a bactericin produced by *Pediococcus damnosus* NCFB 1832. *Int. J. Food Microbiol., 101*, 17–27.

Beneduce, L., Spano, G., Vernile, A., Tarantino, D., & Massa, S. (2004). Molecular characterization of lactic acid populations associated with wine spoilage. *J Basic Microbiol., 44*, 10–16.

Botina, S.G., Tsygankov, Yu.D, & Sukhodolets, V.V. (2006). Identification of industrial strains of lactic acid bacteria by methods of molecular genetic typing. *Russian Journal of Genetics, 42*, 1367–1379.

Bover-Cid, S., & Holzapfel, W.H. (1999). Improved screening procedure for biogenic amine production by lactic acid bacteria. *Int. J. Food Microbiol., 53*, 33–34.

Carre, E. (1982). Recherches sur la croissance des bacteries lactiques en vinification. Désacidification biologique des vins. Ph.D. thesis. Université de Bordeaux II, Bordeaux, France.

Cavin, J.-F., &ioc, V., Etievant, P. X., & Diviès, C. (1993). Ability of wine lactic acid bacteria to metabolize phenol carboxylic acids. *Am. J. Enol. Vitic., 44*, 76–80.

Challinor, S.W., & Rose, A.H. (1954). Interrelationships between a yeast and a bacterium when growing together in defined medium. *Nature, 174*, 877–878.

Chatonnet, P., Dubourdieu, D., & Boidron, J. N. (1995). The influence of *Brettanomyces/Dekkera* sp. yeasts and lactic acid bacteria on the ethylphenol content of red wine. *Am. J. Enol. Vitic., 46*, 463–468

Claisse, O., & Lonvaud-Funel, A. (2000). Assimilation of glycerol by a strain of Lactobacillus collinoides isolated from cider. *Food Microbiol., 17*, 513–519.

Cocconcelli, P.S., Porro, D., Galandini, S., & Senini, L. (1995). Development of RAPD protocols for typing of strains of lactic acid bacteria and enterococci. *Lett. Appl. Microbiol., 21*, 1–4.

Cocolin, L., Manzano, M., Cantoni, C., & Comi, G. (2000). Development of a rapid method for the identification of *Lactobacillus* spp. isolated from fermented Italian sausages using a polymerase chain reaction-temperature gradient gel electrophoresis. *Lett. Appl. Microbiol., 30*, 126–129.

Collado, C.M., & Hernandez, M. (2007). Identification and differentiation of *Lactobacillus*, *Streptococcus* and *Bifidobacterium* species in fermented milk products with bifidobacteria. *Microbiol. Res., 162*, 86–92.

Comitini, F., Ferretti, R., Clementi, F., Mannazzu, I., & Ciani, M. (2005). Interactions between *Saccharomyces cerevisiae* and malolactic bacteria: preliminary characterization of a yeast proteinaceous compound(s) active against *Oenococcus oeni*.*J. Appl. Microbiol., 99*, 105–111.

Costantini, A., Cersosimo, M., Del Prete, V., & Garcia-Moruno, E. (2006). Production of biogenic amines by lactic acid bacteria: screening by PCR, TLC and HPLC of strains isolated from wine and must. *J. Food Prot., 69*, 391–396.

Costello, P.J., Morrison, G.J., Lee, T.H., & Fleet, G.H. (1983). Numbers and species of lactic acid bacteria in wines during vinification. *Food Technol. Aust., 35*, 14–18.

Coton, E., Rollan, G.C., & Lonvaud-Funel, A. (1998). Histidine carboxylase of *Leuconostoc oenos* 9204: purification, kinetic properties, cloning and nucleotide sequence of the *hdc* gene. *J. Appl. Microbiol., 84*, 143–151.

Couto, J.A., Campos, F.M., Figueiredo, A.R., & Hogg, T.A. (2006). Ability of lactic acid bacteria to produce volatile phenols. *Am. J. Enol. Vitic., 57*, 166–171.

D'Incecco, N., Bartowsky, E,J., Kassara, S., Lante, A., Spettoli, P. & Henschke, P.A. (2004). Release of glycosidically bound flavour compounds of Chardonnay by *Oenococcus oeni* during malolactic fermentation. *Food Microbiol., 21*, 257–265.

Davis, C.R., Wibowo D.J., Lee, T.H., & Fleet, G.H. (1986). Growth and Metabolism of Lactic Acid Bacteria during and after MLF of Wines at Different pH. *Appl. Environ. Microbiol., 51*, 539–545.

De las Rivas, B., Marcobal, A., & Munoz, R. (2003). Allelic diversity and population structure in *Oenococcus oeni* as determined from sequence analysis of housekeeping genes. *Appl. Environ. Microbiol., 70*, 7210–7219.

De Oliva Neto, P., Ferreira, M.A., & Yokoya, F. (2004). Screening for yeast with antibacterial properties from an ethanol distillery. *Bioresour Technol., 92*, 1–6.

Delaquis, P., Cliff, M., King, M., Girare, B., Hall, J., & Reynolds, A. (2000). Effect of two commercial malolactic cultures on the chemical and sensory properties of Chancellor wines vinified with different yeasts and fermentation temperatures. *Am. J. Enol. Vitic., 51*, 42–48.

Dicks, L.M., Dell'aglio, F., & Collins, M.D. (1995). Proposal to reclassify *Leuconostoc oenos* as *Oenococcus oeni*. *Int. J. System Bacteriol., 45*, 395–397.

Dols-Lafalgue, M., Gindreau, E., Le Marrec, C., Chambat, G., Heyraud, A., & Lonvaud-Funel, A. (2007). Changes in red wine polysaccharide composition induced by malolactic fermentation. *J. Agric. Food Chem., 55*, 9592–9599.

Douglas, H.C., & Cruess, W.V. (1936). A *Lactobacillus* from California wine: *Lactobacillus hilgardii*. *Food Res., 1*, 113–119.

Du Plessis, E.M., & Dicks, L.M.T. (1995). Evaluation of random amplified polymorphic DNA (RAPD)-PCR as a method to differentiate *Lactobacillus acidophilus, Lactobacillus crispatus, Lactobacillus amylovorans, Lactobacillus gallinarum, Lactobacillus gasseri,* and *Lactobacillus johnsonii*. *Curr. Microbiol., 31*, 114–118.

Du Plessis, H.W., Dicks, L.M., Pretorius, I.S., Lambrechts, M.G., & du Toit, M. (2004). Identification of lactic acid bacteria isolated from South African brandy base wines. *Int. J. Food Microbiol., 91*, 19–29.

Du Toit, M., & Pretorius, I.S. (2000). Microbial spoilage and preservation of wine: using weapons from nature's own arsenal – A review. *S. Afr. J. Enol. Vitic., 21*, 74–96.

Du Toit, M., du Toit, C., Krieling, S.J., & Pretorius, I.S. (2002). Biopreservation of wine with antimicrobial peptides. *Bull OIV,* (855–856), 284–302.

Edwards, C.G., & Jensen, K.A. (1992). Occurrence and characterization of lactic acid bacteria from Washington state wine: *Pediococcus* spp. *Am. J. Vitic. Enol., 43*, 233–238.

Edwards, C.G., Collins, M.D., Lawson, P.A., & Rodriguez, A.V. (2000). *Lactobacillus nagelii* sp. nov., an organism isolated from a partially fermented wine. *Int. J. Syst. Evol. Microbiol., 50*, 699–702.

Ercolini, D. (2004). PCR-DGGE fingerprinting: novel strategies for detection of microbes in food. *J. Microbiol. Meth., 56*, 297–314.

Fernandez, P.A., & Manca de Nadra, M.C. (2006). Growth response and modifications of organic nitrogen compounds in pure and mixed cultures of lactic acid bacteria from wine. *Curr. Microbiol., 52*, 86–91.

Fischer, U., & Noble, A.C. (1994). The effect of ethanol, catechin concentration, and ph on sourness and bitterness of wine. *Am. J. Enol. Vitic., 45*, 6–10.

Fleet, G.H. (1990). Growth of yeasts during wine fermentations. *J. Wine Res., 1*, 211–223.

Fleet, G.H., Lafon-Lafourcade, S., & Ribereau-Gayon, P. (1984). Evolution of yeasts and lactic acid bacteria during fermentation and storage of Bordeaux wines. *Appl. Environ. Microbiol.,* 48, 1034–1038.

Fleske, A., Wolterink, A., van Lis, R., & Akkermans, A.D.L. (1998). Phylogeny of the main bacterial 16S rRNA sequences in Drentse a grassland soils. *Appl. Environ. Microbiol., 64*, 871–879.

Fornachon, J.C.M. (1957). The occurrence of malo-lactic fermentation in Australian wines. *Aust. J. Appl. Sci., 8*, 120–129.

Fugelsang, K.L. (1997) *Wine microbiology,* Chapman and Hall, London, pp. 117–42.

García-Ruiz, A., Bartolomé, B., Martínez-Rodríguez, A., Pueyo, E., Martín-Álvarez, P.J., & Moreno-Arribas, M.V. (2008a). Potential of phenolic compounds for controlling lactic acid bacteria growth in wine. *Food Control, 19*, 835–841.

García-Ruiz, A., López-Expósito, I., Díaz, S., Bartolomé, B., Pozo-Bayón, M.A., Martín-Álvarez, P.J., & Moreno-Arribas, M.V. (2008b). Evaluation of the dual antibacterial and antioxidant

activities of wine polyphenols. Proceeding of the 'WAC2008' International conference. David Chassagne (Eds.) Oenoplurimedia, Francia, pp. 36–38.
Garvie, E.I. (1967). *Leuconostoc oenos* sp. nov. *J. Gen. Microbiol., 48*, 431–438.
Garvie, E.I. (1986). Genus *Pediococcus* Claussen 1903, 68AL. In P. H. A. Sneath, N. S. Mair, M. E. Sharpe and J. G. Holt (Eds.), *Bergey's Manual of Systematic Bacteriology*, vol. 2 (pp. 1075–1079). Williams and Wilkins, Baltimore.
Gelsomino, A., Keijzer-Wolters, A.C., Cacco, G., & van Elsas, J.D. (1999). Assessment of bacterial community structure in soil by polymerase chain reaction and denaturing gradient gel electrophoresis. *J. Microbiol. Methods, 38*, 1–15.
Germond, J-E, Lapierre, L., Delley, M., Mollet, B., Felis, G.E., & Dell'aglio, F. (2003). Evolution of the bacterial species *lactobacillus delbrueckii*: a partial genomic study with reflections on prokaryotic species concept. *Mol. Biol. Evol., 20*, 93–104.
Giraffa, G., & Neviani, F. (2000). Molecular identification and characterization of food-associated lactobacilli. *Ital. J. Food. Sci., 12*, 403–423.
Grimaldi, A., McLean, H., & Jiranek, V. (2000). Identification and partial characterization of glycosidic activities of commercial strains of the lactic acid bacterium *Oenococcus oeni*. *Am. J. Enol. Vitic., 51*, 362–369.
Guilloux-Benatier, M., Pageault, O., Man, A., & Feuillat, M. (2000). Lysis of yeast cells by *Oenococcus oeni* enzymes. *J. Ind. Microbiol. Biotech., 25*, 193–197.
Gury, J., Barthelmebs, L., Tran, N. P., Diviès, C., & Cavin, J.-F. (2004). Cloning, deletion, and characterization of PadR, the transcriptional repressor of the phenolic acid decarboxylase-encoding *padA* gene of *Lactobacillus plantarum*. *Appl. Environ. Microbiol., 70*, 2146–2153.
Halasz, A., Barath, A, Simon-Sarkadi, L., & Holzapfel, W.H. (1994). Biogenic amines and their production by microorganisms in foods. *Trends Food Sci. Technol., 5*, 42–49.
Henick-Kling, T. (1993). Malolactic fermentation. En: Wine Microbiology and Biotechnology, ed. Fleet, G. H. pp. 286–326. Berlin, Springer-Verlag.
Henick-Kling, T., & Park, Y.H. (1994). Considerations fort he use of yeast and bacterial starter cultures: SO_2 and timing of inoculation. *Am. J. Enol. Vitic., 45*, 464–469.
Hernández, T., Estrella, I., Carlavilla, D., Martín-Álvarez, P.J., & Moreno-Arribas, M.V. (2006). Phenolic compounds in red wine subjected to industrial malolactic fermentation and ageing on lees. *Anal. Chim. Acta, 563*, 116–125.
Hernández, T., Estrella, I., Pérez-Gordo, M., Alegria, E-G., Tenorio, C., Ruiz-Larrea, F., & Moreno-Arribas, M.V. (2007). Contribution of *Oenococcus oeni* and *Lactobacillus plantarum* to the non anthocyanin phenolic composición of red wine during malolactic fermentation. *J. Agric. Food Chem., 55*, 5260–5266.
Heuer, H., Krsek, M., Baker, P., Smalla, K., & Wellington, E.M.H. (1997). Analysis of actinomycete communities by specific amplification of genes encoding 16S rRNA and gel-electrophoresis in denaturing gradients. *Appl. Environ. Microbiol., 63*, 3233–3241.
Jussier, D., Dubé Morneau, A., & Mira de Orduña, R. (2006). Effect of simultaneous inoculation with yeast and bacteria on fermentation kinetics and key wine parameters of cool-climate chardonnay. *App. Environ. Microbiol., 72*, 221–227.
Kandler, O. (1983). Carbohydrate metabolism in lactic acid bacteria. *Anton. Leeuw, 49*, 209–224.
Kandler, O., & Weiss N. (1986). Regular, non-sporing Gram-positive rods. In P. H. A. Sneath, N. Mair, M. E. Sharpe, and J. G. Holt (Eds.), *Bergey's manual of systematic bacteriology*, vol. 2. (pp. 1208–1234). William and Wilkins, Baltimore.
Kennes, C, Veiga, M.C., Dubourguier, H.C., Touzel, J.P., Albagnac, G., Naveau, H., & Nyns, E.J. (1991). Trophic relationships between *Saccharomyces cerevisiae* and *Lactobacillus plantarum* and their metabolism of glucose and citrate. *Appl. Environ. Microbiol., 57*, 1046–51.
King, S.W., & Beelman, R.B. (1986). Metabolic interactions between *Saccharomyces cerevisiae* and *Leuconostoc oenos* in a model juice/wine system. *Am. J. Enol. Vitic., 37*, 53–60.
Krieger, S.A., Hammes, W.P., & Henick-Kling, T. (1993). How to use malolactic starter cultures in the winery. *Austral. New Zeal. Wine Indust. J., 8*, 153–160.

Labarre, C., Guzzo, J., Cervin, J. F., & Divies, C. (1996). Cloning and characterization of the genes encoding the malolactic enzyme and the malate permease of *Leuconostoc oenos*. *Appl. Environ. Microbiol., 62*, 1274–1282.

Lafon-Lafourcade, S., Carre, E., & Ribereau-Gayon, P. (1983). Occurrence of Lactic Acid Bacteria During the Different Stages of Vinification and Conservation of Wines. *Appl Environ Microbiol., 46*, 874–880.

Larsen, J.T., Nielsen, J.C., Kramp, B., Richelieu, M., Bjerring, P., Riisager, M.J., Arneborg, N., & Edwards, C.G. (2003). Impact of different strains of *Saccharomyces cerevisiae* on MLF by *Oenococcus oeni*. *Am. J. Enol. Vitic., 54*, 246–251.

Laurent, M.H., Henick-Kling, T., & Acree, T.E. (1994). Changes in the aroma and odor of Chardonnay wine due to malolactic fermentation. *Wein-Wissenschaft, 49*, 3–10.

Leitao, M.C., Teixeira, H.C., Barreto Crespo, M.T., & San Romao, N.V. (2000). Biogenic amine ocurrence in wine. Amino acid decarboxylase and proteolytic activities expression by *Oenococcus oeni*. *J. Agric. Food Chem., 48*, 2780–2784.

Lemaresquier, H. (1987). Inter-relationship between strains of *Saccharomyces cerevisiae* from Champagne area and lactic acid bacteria. *Lett. App. Microbiol., 4*, 91–94.

Liu, S.-Q., Pritchard, G.G., Hardman, M.J., & Pilone, G.J. (1994). Citrulline production and ethyl carbamate (urethane) precursor formation from arginine degradation by wine lactic acid bacteria *Leuconostoc oenos* and *Lactobacillus buchneri*. *Am. J. Enol. Vitic., 45*, 235–242.

Lonvaud-Funel, A. (1999). Lactic acid bacteria in the quality improvement and depreciation of wine. *Anton. Leeuw, 76*, 317–331.

Lonvaud-Funel, A., & Joyeux, A. (1993) Antagonism between lactic acid bacteria of wines: inhibition of *Leuconostoc oenos* by *Lactobacillus plantarum* and *Pediococcus pentosaceus*. *Food Microbiol., 10*, 411–419.

Lonvaud-Funel, A., & Strasser de Saad, A. M. (1982). Purification and properties of a malolactic enzyme from a strain of *Leuconostoc mesenteroides* isolated from grapes. *Appl. Environ. Microbiol., 43*, 357–361.

Lonvaud-Funel, A., Fremaux, C., & Biteau, N. (1989). Identification de *L. oenos* per l'utilisation de sondes d'AND. *Sci Aliments, 8*, 33–49.

Lonvaud Funel, A., Fremaux, C., Biteau, N., & Joyeux A. (1991a). Speciation of lactic acid bacteria from wines by hybridisation with DNA probes. *Food Microbiol., 8*, 215–222.

Lonvaud-Funel, A., Joyeux, A., & Ledoux, O. (1991b). Specific enumeration of lactic acid bacteria in fermenting grape must and wine by colony hybridization with non isotopic DNA probes. *J. Appl. Bacteriol., 71*, 501–508.

Maicas, S., Gil, J.V., Pardo, I., & Ferrer, S. (1999). Improvement of volatile composition of wines by controlled addition of malolactic bacteria, *Food Res. Internat., 32*, 491–496.

Malacrinò, P., Zapparoli, G., Torrioni, S., & Dell'aglio, F. (2003). Adaptation in Amarone wine of indigenous Oenococcus oeni strains differentiated by pulsed-field gel electrophoresis. *Ann. Microbiol., 53*, 55–61.

Manca de Nadra, M.C., Farías, M., Moreno-Arribas, M.V., Pueyo, E., & Polo, M.C. (1997). Proteolytic activity of *Leuconostoc oenos*: Effect on proteins and polypeptides from white wine. *FEMS Microbiol. Lett., 150*,135–139.

Manca de Nadra, M.C., Farias, M., Moreno-Arribas, M.V., Pueyo, E., & Polo, M.C. (1999). A proteolytic effect of *Oenococcus oeni* on the nitrogenous macromolecular fraction of red wine. *FEMS Microbiol. Lett., 174*, 41–47.

Marcobal, A., De las Rivas, B., Moreno-Arribas, V., & Munoz, R. (2004). Identification of the ornithine decarboxylase gene in the putrescine producer *Oenococcus oeni*BIFI-83. *FEMS Microbiol. Lett., 239*, 213–220.

Martineau, B., & Henick-Kling, T. (1995). Performance and diacetyl production of commercial strains of malolactic bacteria in win. *J Appl. Bacteriol., 78*, 526–536.

Martínez-Murcia, A.J., Harland, N.M., & Collins, M.D. (1993). Phylogenetic analysis of some leuconostocs and related organism as determined from the large -subunit rRNA gene sequences:

assessment of congruence of small- and large- subunit r RNA derived trees.*J. Appl. Microbiol., 12*, 48–55.
Maslow, J. N., Mulligan, M. E., & Arbeit, R. D. (1993). Molecular epidemiology: application of contemporary techniques to the typing of microrganisms. *Clin. Infect. Dis., 17*,153–162.
McMahon, H., Zoecklein, B.W., Fugelsang, K., & Jasinski, Y. (1999). Quantification of glycosidase activities in selected yeasts and lactic acid bacteria. *J. Ind. Microbiol. Biotechnol., 23*, 198–203.
Mine, Y., & Zhang, J.W. (2002). Comparative studies on antogenity and allergenity of native and denatured egg white proteins. *J. Agric. Food Chem., 50*, 2679–2683.
Mira de Orduña, R., Patchett, M.L., Liu, S.-Q., & Pilone, G.J. (2001). Growth and arginine metabolism of the wine lactic acid bacteria *Lactobacillus buchneri* and *Oenococcus oeni* at different pH values and arginine concentrations. *Appl. Environ. Microbiol., 67*, 1657–1662.
Miyoshi, A., Rochat, R., Gratadoux, J-J., Le Loir, Y., Costa Oliveira, S., Langella, P., & Azevedo, V. (2003). Oxidative stress in *Lactococcus lactis*. *Genet. Mol. Res., 2*, 348–359.
Moreno-Arribas, M.V., & Polo, C. (2005). Winemaking microbiology and biochemistry: current knowledge and future trends. *Crit. Rev. Food Sci Nutr., 45*, 265–286.
Moreno-Arribas, M.V., & Polo, C. (2008). Occurrence of lactic acid bacteria and biogenic amines in biologically aged wines. *Food Microbiol., 25*, 875–881.
Moreno-Arribas, V., Torlois, S., Joyeux, A., Bertrand, A., & Lonvaud-Funel, A. (2000). Isolation, properties and behaviour of tyramine–producing lactic acid bacteria from wine. *J. Appl. Microbiol., 88*, 584–593.
Moreno-Arribas, V., Polo, M.C., Jorganes, F., & Muñoz, R. (2003). Screening of biogenic amine production by lactic acid bacteria isolated from grape must and wine. *Int. J. Food Microbiol., 84*, 117–123.
Moreno-Arribas, M.V., Gómez-Cordovés, C., & Martín-Álvarez, P.J. (2008a). Evolution of red wine anthocyanins during malolactic fermentation, postfermentative treatments and ageing with lees. *Food Chem., 109*, 149–158.
Moreno-Arribas, M.V., Ferrer, S., & Pardo, I. (2008b). Técnicas moleculares para la identificación y caracterización de bacterias lácticas de interés enológico. *Bull. OIV. (in press)*
Naouri, P., Chagnaud, P., Arnaud, A., & Galzy, P. (1990). Purification and properties of a malolactic enzyme from *Leuconostoc oenos* ATCC 23278. *J. Basic Microbiol., 30*, 577–585.
Navarro, L., Zarazaga, M., Sáenz, J., Ruiz-Larrea, F., & Torres, C. (2000). Bacteriocin production by lactic acid bacteria isolated from Rioja red wines. *J. Appl. Microbiol., 88*, 41–51.
Nielsen, J.C., & Richelieu, M. (1999). Control of flavor development in wine during and after MLF. *Appl. Environ. Microbiol., 65*, 740–745.
Nielsen, J.C., Prahl, C., & Lonvaud-Funel, A. (1996). MLF in wine by direct inoculation with freeze-dried Leuconostoc oenos cultures. *Am. J. Enol. Vitic., 47*, 42–48.
Nigatu, A, Ahrné, S, & Molin, G. (2001). Randomly amplified polymorphic DNA (RAPD) profiles for the distinction of *Lactobacillus* species. *Anton. Leeuw., 79*, 1–6.
Noble, A.C. (1994). Bitterness in wine. *Physiol. Behav., 56*, 1251–1255.
Osborne, J.P., & Edwards, C.G. (2007). Inhibition of MLF by a peptide produced by *Saccharomyces cerevisiae* during alcoholic fermentation. *Int. J. Food Microbiol., 118*, 27–24.
Osborne, J.P., Mira de Orduna, R., Pilone, G.J., & Liu, S.Q. (2000). Acetaldehyde metabolism by wine lactic acid bacteria. *FEMS Microbiol. Lett., 191*, 51–55.
Patynowski, R.J., Jiranek, V., & Markides, A.J. (2002). Yeast viability during fermentation and sur lie ageing of a defined medium and subsequent growth of *Oenococcus oeni*. *Aust J Grape Wine Res., 8*, 62–69.
Pinzani, P., Bonciani, L., Pazzagli, M., Orlando, C., Guerrini, S., & Granchi, L. (2004). Rapid detection of *Oenococcus oeni* in wine by real-time quantitative PCR. *Lett. Appl. Microbiol., 38*, 118–124.
Poblet-Icart, M., Bordons, A., & Lonvaud-Funel, A. (1998). Lysogeny of *Oenococcus oeni* (syn. *Leuconostoc oenos*) and study of their induced bacteriophages. *Curr. Microbiol., 36*, 365–369.

Pozo-Bayón, M.A., Alegría E.G., Polo, M.C., Tenorio, C., Martín-Álvarez, P.J., Calvo de la Banda, M.T., Ruiz-Larrrea, F., & Moreno-Arribas, M.V. (2005). Wine volatile and amino acid composición after malolactic fermentation: Effect of *Oenococcus oeni* and *Lactobacillus plantarum* starter cultures. *J. Agric. Food Chem., 53*, 8729–8735.

Pretorius, I. S. (2001). Gene technology in winemaking: New approaches to an ancient art. *Agric. Conspec. Sci., 66*, 27–47.

Pripis-Nicolau, L., de Revel, G., Bertrand, A., & Lonvaud-Funel, A. (2004). Methionine catabolism and production of volatile sulphur compounds by *Oenococcus oeni*. *J. Appl. Microbiol., 96*, 1176–1184.

Radler, F. (1990a). Possible use of nisin in winemaking. I. Action of nisin against lactic acid bacteria and wine yeasts in solid and liquid media. *Am. J. Enol. Vitic., 41*, 1–6.

Radler, F. (1990b). Possible use of nisin in winemaking. II. Experiments to control lactic acid bacteria in the production of wine. *Am. J. Enol. Vitic., 41*, 7–11.

Radler, F., & Yannissis, C. (1972). Weins`aureabbau bei Milchs´aurebakterien. *Arch. Mikrobiol., 82*, 219–239.

Reguant, C., & Bordons, A. (2003). Typification of *Oenococcus oeni* strains by multiplex RAPD-PCR and study of population dynamics during MLF. *J. Appl. Microbiol., 95*, 344–353.

Reguant, C., Bordons, A., Arola, L., & Rozès, N. (2000). Influence of phenolic compounds on the physiology of *Oenococcus oeni* from wine. *J. Appl. Microbiol., 88*, 1065–1071.

Renouf, V., Claisse, O., & Lonvaud-Funel, A. (2005). Numeration, identification and understanding of microbial biofilm on grape berry surface. *Aust J Grape Wine Res., 11*, 316–327.

Renouf, V., Claisse, O., Miot-Sertier, C, & Lonvaud-Funel, A. (2006). LAB evolution during winemaking: use of the rpoB gene as target for PCR-DGGE analysis. *Food Microbiol., 23*, 136–145.

Renouf, V., Claisse, O., & Lonvaud-Funel, A. (2007). Inventory and monitoring of wine microbial consortia. *Appl. Microbiol. Biotechnol., 75*, 149–164.

Ritt, J.F., Guilloux-Benatier, M., Guzzo, J., Alexandre, H., & Remize, F. (2008). Oligopeptides assimilation and transport by *Oenococcus oeni*. *J. Appl. Microbiol., 104*, 573–580.

Rodas A.M., Ferrer, S., & Pardo, I. (2003). 16S-ARDRA, a tool for identification of lactic acid bacteria isolated from grape must and wine. *Syst Appl Microbiol., 26*, 412–422.

Rodas A.M., Ferrer, S., & Pardo, I. (2005). Polyphasic study of wine Lactobacillus strains: taxonomic implications. *Int. J. Sys. Evol. Microbiol., 55*, 197–207.

Rojo-Bezares, B., Sáez, Y., Zarazaga, M., Torres, C., & Ruiz-Larrea, F. (2007). Antimicrobial activity of nisin against *Oenococcus oeni* and other wine bacteria. *Int. J. Food Microbiol., 116*, 32–36.

Romero, C., & Bakker, J. (2000). Effect of acetaldehyde and several acids on the formation of vitisin A in model wine anthocyanin and colour evolution. *Int. J. Food Sci. Technol., 35*, 129–140.

Rosi I., Fia, G., & Canuti, V. (2003). Influence of different pH values and inoculation time on the growth and malolactic activity of a strain of *Oenococcus oeni*. *Aust. J. Grape Wine Res., 9*, 194–199.

Rossetti, L., & Giraffa, G. (2005). Rapid identification of diary lactic acid bacteria by M-13 generated RAPD-PCR fingerprint database.*J. Microbiol. Methods, 63*, 135–144.

Saucier C., Little, D., & Glories, Y. (1997). First evidence of acetaldehyde-flavanol condensation products in red wine. *Am. J. Enol. Vitic., 48*, 370–372.

Sauvageot, F., & Vivier, P. (1997). Effects of malolactic fermentation on sensory properties of four Burgundy wines. *Am. J. Enol. Vitic., 48*, 187–192.

Schleifer, K.H., & Kilpper-Balz, R. (1987). Molecular and Chemotaxonomic Approaches to the Classification of Streptococci, Enterococci and Lactococci: A Review. *Syst. Appl. Microbiol., 10*, 1–19.

Schutz, H., & Radler, F. (1984a). Anaerobic reduction of glycerol to propanediol-1,3 by *Lactobacillus brevis* and *Lactobacillus buchneri*. *Syst. Appl. Microbiol., 5*, 169–178.

Schutz, H., & Radler, F. (1984b). Propanediol-1,2-dehydratase and metabolism of glycerol of *Lactobacillus brevis*. *Arch. Microbiol., 139*, 366–370.

Sch`utz, M., & Radler, F. (1974). Das vorkommen von malatenzym und malolactat-enzym bei verschiedenen milchsa´urenbakterien. *Arch. Mikrobiol., 96*, 329–339.

Shalaby, A. R. (1996). Significance of biogenic amines to food safety and human health. *Food Res. Int., 26*, 675–690.

Sieiro, C., Cansado, J., Agrelo, D., Velázquez, J.B., & Villa, T.G. (1990). Isolation and enological characterization of malolactic bacteria from the vineyards of North-western Spain. *Appl. Environ. Microbiol., 56*, 2936–2938.

Silla, M. H. (1996). Biogenic amines: their importance in foods. *Int. J. Food Microbiol., 29*, 213–231.

Singleton, V. L. (1995). Maturation of wines and spirits: comparisons, facts, and hypotheses. *Am. J. Enol. Vitic., 46*, 98–115.

Sohier, D., Coulon, J., & Lonvaud-Funel, A. (1999). Molecular identification of *Lactobacillus hilgardii* and genetic relatedness with *Lactobacillus brevis*. *Int. J. Syst. Bacteriol.*,49, 1075–1081.

Sozzi, T., Gnaegi, F., D'Amico, N., & Hose, H. (1982). Difficultes de fermentation malolactique du vin dues a des bacteriophages de *Leuconostoc oenos*. *Revue Suisse Vitic. Arboric. Hortic., 14*, 17–23.

Spano, G., Lonvaud Funel A., Claisse, O., & Massa, S. (2007). In vivo PCR-DGGE analysis of *Lactobacillus plantarum* and *Oenococcus oeni* populations in red wine. *Curr. Microbiol., 54*, 9–13.

Strasser de Saad, A.M., & Manca de Nadra, M.C. (1993). Characterization of bacteriocin produced by *Pediococcus pentosaceus* in wine.*J. Appl. Bacteriol., 74*, 406–410.

Straub, B. W., Kicherer, M., Schilcher, S.M., & Hammes W.P. (1995). The formation of biogenic amines by fermentation organisms. *Z. Lebensm.-Unters. -Forsch., 201*, 79–82.

Suárez-Lepe, J.A., & Íñigo-Leal, B. (2003). Desacidificación a cargo de bacterias. La fermentación maloláctica. En: Microbiología enológica. Fundamentos de vinificación. 3ª versión. Mundi-Prensa ed., Madrid, pp. 357–379.

Tenske, A, Sigalevich, P, Cohen, Y, & Muyzer, G. (1996). Molecular identification of bacteria from co-culture by denaturing gradient gel electrophoresis of 16S ribosomal DNA fragments as a tool of isolation in pure cultures. *Appl. Environ. Microbiol., 62*, 4210–4215.

Ugliano, M., & Moio, L. (2006). The influence of malolactic fermentation and *Oenococcus oeni* strain on glycosidic aroma precursors and related volatile compounds of red wine. *J. Sci. Food Agric., 86*, 2468–2476.

Ugliano M., Genovese, A., & Moio, L. (2003). Hydrolysis of wine aroma precursors during malolactic fermentation with four commercial starter cultures of *Oenococcus oeni*. *J. Agric. Food Chem., 51*, 5073–5078.

Veiga-da-Cunha, M., & Foster, M.A. (1992). Sugar-glycerol cofermentations in lactobacilli: The fate of lactate. *J. Bacteriaol., 174*, 1013–1019.

Veiga-da-Cunha, M., Santos, H., & van Schaftingen, E. (1993). Pathway and regulation of erythritol formation in *Leuconostoc oenos*. *J. Bacteriol., 175*, 3941–3948.

Ventura, M., Casas, I.A., Morelli, L., & Callegari, M.L. (2000). Rapid amplified ribosomal DNA restriction analysis (ARDRA) identification of Lactobacillus spp. isolated from faecal and vaginal samples. *Syst. Appl. Microbiol., 23*, 504–509.

Vivas, N., Lonvaud-Funel, A., & Glories, Y. (1997). Effect of phenolic acids and anthocyanins on growth, viability and malolactic activity of a lactic acid bacterium. *Food Microbiol., 14*, 291–300.

Wayne, L.G., Brenner, D.J, Colwell, R.R, Grimont, P.A., Kandler, O., Krichevsky, M.I., & Truper, H.G. (1987). Report of the ad hoc committee on reconciliation of approaches to bacterial systematic. *Int. J. Syst. Bacteriol., 37*, 463–464.

Weeks, C. (1969). Production of sulfur dioxide-binding compounds and of sulfur dioxide by two *Saccharomyces* yeasts. *Am. J. Enol. Vitic., 20*, 32–39.

Whiley, R.A., Duke, B., Hardie, J.M., & Hall L.M. (1995). Heterogeneity among 16S-23S rRNA intergenic spacers of species within the *Streptococcus milleri* group. *Microbiology, 141*, 1461–1467.

Wibowo, D., Eschenbruch, R., Davis, C.R., Fleet, G.H., & Lee, T.H. (1985). Occurrence and growth of lactic acid bacteria in wine: a review. *Am. J. Enol. Vitic., 4*, 302–313.

Zapparoli, G., Reguant, C., Bordons A., Torrioni, S., & Dellaglio, F. (2000). Genomic DNA fingerprinting of Oenococcus oeni strains by pulsed-field electrophoresis and randomly amplified polymorphic DNA-PCR. *Current Microbiol., 40*, 351–355.

Zavaleta, A.I., Martinez-Murcia, A.J., & Rodriguez-Valera, F. (1997). Intraspecific genetic diversity of *Oenococcus oeni* as derived from DNA fingerprinting and sequence analyses. *Appl. Environ. Microbiol., 63*, 1261–1267.

Zimmerli, B., & Schlatter, J. (1991). Ethyl carbamate: analytical methodology, occurrence, formation, biological activity and risk assessment. *Mutat. Res., 259*, 325–350.

Chapter 3
Special Wines Production

Chapter 3A
Sparkling Wines and Yeast Autolysis

Adolfo J. Martínez-Rodríguez and Encarnación Pueyo

Contents

3A.1 Introduction

Sparkling wines are among several wine types that require special production methods. These wines have an additional value and are highly esteemed worldwide. Their most important characteristic is that they effervesce over the entire tasting process. The sparkling properties of these wines, and other sensorial properties, largely depend on the process of autolysis occurring during their manufacture, which takes place because these wines age for a given time in the presence of yeast lees.

3A.2 Elaboration of Sparkling Wines

A sparkling wine is a wine obtained by secondary fermentation of a still wine, called a base wine. When the bottle is uncorked, the sparkling wine gives off carbon dioxide that is exclusively produced by this fermentation (endogenous).

A.J. Martínez-Rodríguez (✉)
Instituto de Fermentaciones Industriales (CSIC). C/Juan de la Cierva 3, 28006. Madrid, Spain

M.V. Moreno-Arribas, M.C. Polo (eds.), *Wine Chemistry and Biochemistry*,
DOI 10.1007/978-0-387-74118-5_3, © Springer Science+Business Media, LLC 2009

Depending on the production process, sparkling wines can be classified into sparkling wines fermented in the bottle and sparkling wines fermented in large cuves or "granvas". For the wines fermented in large containers, the second fermentation is carried out in hermetically sealed tanks and later transferred to bottles to be sold, after having spent a minimum of 20 days in the presence of yeasts. Many Italian sparkling wines are made by this method, including some very well-known types such as Lambrusco and Asti. For sparkling wines fermented in bottles, at the end of the production process the wine can be filtered and transferred to another bottle "transfer method", or the entire production and aging process can be carried out in the bottle itself. This latter technique is referred to as the classical or traditional method.

Wines produced by the traditional method include some of the Italian sparkling wines, such as the highest quality ones called "Talento", "Cava" wines made in Spain and Champagnes in France; in the latter case these wines are said to be made by the Champenoise method. The greatest differences between these types of wines are mainly the varieties of grape used to make the base wines, the production zones (geographical areas) and the length of time they are kept with the yeasts. Each of these is regulated by its own legislation, which governs these aspects and also certain physico-chemical parameters that must remain within certain specified intervals (CO_2 pressure, alcoholic degree, SO_2 concentration, etc.) (Ough 1992; Hidalgo 2003).

3A.2.1 Steps in the Elaboration of a Sparkling Wine by the Traditional Method

3A.2.1.1 Tirage

Tirage involves filling the bottle with the base wine (wine from the first fermentation, produced by the usual techniques of white wine production), monovarietal or made from a mixture of wines (in both cases the grape varieties used are appropriate for this type of production) and the addition of a solution called tirage liquor.

The base wine should present certain organoleptic (pale colour, fruity aroma, etc.) and analytical characteristics (sufficient concentration of oxygen for yeast growth, low residual sugar contents, moderate alcohol contents and low volatile acidity). It should have no residual yeasts, or bacteria, and will usually have been subjected to tartaric stabilization. After being introduced into the bottle it does not undergo any further treatment. The bottles are then closed with a temporary crown cap.

The tirage liquor is a solution formed by the yeasts responsible for the second fermentation, saccharose, grape must, corrected or not, or partially fermented grape must, in the correct proportions to produce the desired pressure of carbon dioxide. Moreover, a small amount of bentonite is usually added in order to facilitate flocculation, followed by removal of the lees. The amount of bentonite used is around 3g/hL (Martínez-Rodríguez and Polo 2003).

The yeasts used in the second fermentation should also present a series of characteristics such as fermentative activity at low temperature, resistance to ethanol and to CO_2 pressure, flocculation potential etc. (Bidan et al. 1986; Martínez-Rodríguez et al. 2001a).

3A.2.1.2 Secondary Fermentation and Aging in the Bottle

The bottles are kept in places specially furnished for the aging process. In this stage the second fermentationoccurs, the CO_2 is formed and stabilized and the aging with the yeast takes place.

3A.2.1.3 Riddling

In this step the bottles are subject to a stirring process so that the sediment moves up to the bottle neck. This process was originally carried out with the bottles on desks, turning them manually 1/8 of a turn for 15days, until they were finally perpendicular to the floor. Now this system has been replaced by automated systems that can move all the bottles simultaneously.

3A.2.1.4 Disgorging

Disgorging involves separating the wine from the lees accumulated in the previous step at the neck of the bottle. This deposit of lees is frozen by immersing the bottle in a cryogenic bath. The bottle is then placed upside down, the cap is removed and the pressure inside causes the frozen deposit to be expelled.

If incorporated or immobilised yeasts are used (Fumi et al. 1987; Godia et al. 1991; Busova et al. 1994; Colagrande et al. 1994; Divies et al. 1994; Yokotsuka et al. 1997; Martynenko and Gracheva 2003; Martynenko et al. 2004) or yeasts inside a cartridge designed for this purpose called Millispark system are placed in the neck of the bottle, the riddling is avoided and disgorging is facilitated.

3A.2.1.5 Addition of Dosage Liquor

During the disgorging step, some of the wine may be lost; this can be compensated for by the addition of the "liqueur expedition", which may correspond to the sparkling wine itself or to saccharose, grape must, partially fermented must, or must with or without corrected concentration, the base wine or a mixture of these products, with addition, where relevant, of distilled wines. In this way, the sparkling wine can be given the desired degree of sweetness and a standard product can be obtained, regardless of the vintage, standardizing each product. Finally, the bottle is closed with the final cork, which is secured to the neck with a wire cap.

3A.3 Foaming Properties

The most important sensorial characteristics of sparkling wines include, undoubtedly, the foaming properties perceived by the consumer when serving the sparkling wine and also when drinking it.

A quality foam can be defined as one that causes a slow release of CO_2, in ring shapes from the depths of the liquid, with small bubbles that contribute to the formation of a crown over the surface of the wine, covering it completely, with bubbles two or three rows deep. The permanence of the foam on the surface of the wine depends on a balance between the rate of foam formation and the rate at which the bubbles making up this layer are destroyed.

A tasting card has been designed that can be used to evaluate the foaming quality of a sparkling wine objectively (Obiols et al. 1998). However, it is also important to have instrumental techniques that can be used to obtain a quantifiable value for foam quality, to be able to compare sparkling wines and also to be able to correlate these foaming properties with the wines' physical and chemical characteristics, in order to establish which of these variables affect these properties. With this knowledge, different technological aspects of the production process of sparkling wines could be modified with the objective of improving their foaming qualities.

The methods reported in the literature and used to quantify the foaming properties of sparkling wines can be classified into methods based on measuring the kinetics of CO_2 discharging, gas sparging methods and image analysis methods.

The method of gas discharging kinetics (Maujean et al. 1988) involves studying the time course of the mass quantification of the CO_2 discharged spontaneously from the bottle. When all the free CO_2 has been discharged, the bottle is shaken to quantify the so-called provoked CO_2.

Gas sparging methods are those most used in the past and still used today, and are based on the procedure developed by Bikerman (1938). This essentially consists of making a jet of gas pass at a controlled speed through a volume of wine to generate a foam. Based on this method, experiments have been carried out with very simple equipment made in the laboratory, that can be used to measure the height of foam reached and the time it takes for this foam to collapse after the CO_2 jet has stopped (Edward et al. 1982; Pueyo et al. 1995). The advantage of this method over that described previously is that it can even be used to evaluate the foaming capacities of still wines. In sparkling wine production it is very important since it permits evaluation of the foaming characteristics of the base wines used to elaborate the sparkling wines.

Later, Maujean et al. (1990) designed automated equipment to measure the foaming properties of wines, based on Bikerman's principle, called the Mosalux. This equipment uses a photoelectric cell to record the height reached by the foam. It incorporates a personal computer that can be used to control the experiments and, also, for data acquisition. The foaming properties measured by Mosalux are HM or maximum height of the foam or foaming power, HS, or stable height of the foam with time and, TS, or stability time after stopping the gas flow. This equipment has been used in numerous works, from 1990 until today (Maujean et al. 1990; Brissonet and Maujean 1991, 1993; Robillard et al. 1993; Dussaud et al. 1994; Malvy et al. 1994; Andrés-Lacueva et al. 1996a, 1997; Gallart et al. 1997, 2004; López-Barajas et al. 1998, 1999; Senèe et al. 1998; Lao et al. 1999; Marchal et al. 2001, 2002, 2006; Girbau-Sòla et al. 2002b; Vanrell et al. 2002; Dambrouck et al. 2005; Cilindre et al. 2007; Vanrell et al. 2007).

The validity of the results obtained with the Mosalux has been verified, since there is a good correlation between these parameters and other foaming parameters such as foam height and stability, established by Robillard et al. (1993) and Bikerman's coefficient. The foam quality of a wine can be predicted by a mathematical model developed by combining some of these parameters (Gallart et al. 1997; López-Barajas et al. 1999). A good relationship has also been established between the foaming properties obtained with Mosalux and those obtained by evaluating the foam quality by sensorial analysis (Gallart et al. 2004).

In the Instituto de Fermentaciones Industriales of the Spanish National Research Council Research (CSIC), a piece of equipment has been designed to measure foam produced by gas sparging that uses an ultrasound emitter-detector and a waveguide to detect fluctuations in the foam (Moreno-Arribas et al. 2000; Martínez-Rodríguez and Polo 2003; Martínez-Rodríguez et al. 2001b; Hidalgo et al. 2004; Pozo-Bayón et al. 2004; Nunez et al. 2005, 2006). Moreover, the value of the results obtained with this technique and their relationship with the parameters established in the sensorial analysis has also been demonstrated.

The image analysis methods can be used to evaluate the height and collar produced in the glass by the foam by a system of artificial vision (Machet et al. 1993; Robillard et al. 1995). This method has the advantage that it can evaluate the quality of the foam in an automated manner by using video cameras but in real conditions in a glass. The conclusions reached in this work are that it is important to take into account the quality of the glass when evaluating foam quality produced in sparkling wines, since it significantly affects this property, as also do the environmental parameters during the analysis.

Another extensive group of works, described in the literature, focus on different physico-chemical aspects of the effervescence in sparkling wines to try to explain bubble formation (nucleation) and survival, both in the interior of the liquid and also at its surface. Some of these works even propose models constructed with fibers to simulate bubble formation and behaviour (Casey 1987, 1995, 2000; Jordan and Napper 1994; Liger-Belair et al. 1999, 2001, 2002, 2006; Liger-Belair 2005; Péron et al. 2000, 2001, 2004; Senèe et al. 1999; Uzel et al. 2006; Tufaile et al. 2007; Voisin et al. 2005).

The studies published by Senèe et al. (1999, 2001) show that the endogenous particles (bentonite and yeasts) that can exist in a wine also have an important effect on foam quality, simultaneously having both positive and negative effects, since they have different effects on the various foaming parameters.

Most of the studies published in the literature on sparkling wine quality are aimed at establishing the effect of the chemical composition of these wines and the different technologies used in their production on their foaming qualities.

One, almost unanimous, criterion is that wine proteins are the chemical compounds with the greatest positive influence on the foam quality of wines (Molan et al. 1982; Brissonnet et al. 1991, 1993; Malvy et al. 1994; Andrés-Lacueva et al. 1996a; López-Barajas et al. 1997; Pueyo et al. 1995; Moreno-Arribas et al. 2000; Vanrell et al. 2002). However, other compounds have also been described in wines that can improve their foaming qualities, such as polysaccharides (Brissonnet et al. 1991; Moreno-Arribas et al. 2000), although Girbau-Sòla et al. (2002b) found the opposite effect. Glucose has also been attributed beneficial effects on the foaming qualities of wines (Pueyo et al. 1995; López-Barajas et al. 1997).

It has been shown recently that mannoproteins from yeast cell walls make a significant contribution to improving foaming properties (Nunez et al. 2006). These authors also demonstrated the importance of the method of mannoprotein extraction from the yeast cell wall, since not all mannoproteins have the same effect on the foam of sparkling wines.

The compounds with negative effects on foaming qualities include the lipids. The first work to explain this effect was carried out on beer (Roberts et al. 1978), where it was shown that the harmful effect can be reduced with time, owing to the micellar configuration of the lipids. Therefore, this effect will depend more on the physical state of the lipids than on their concentration. This was later verified by Dussaud et al. (1994) in work carried out in wines. This group also reported that this effect would depend on the concentration of ethanol. At low ethanol concentrations they will have negative effects but at high concentrations their effects will be negligible relative to the negative effects that the ethanol has on the foam.

Gallart et al. (2002) have demonstrated that the fatty acids C_8, C_{10} and C_{12}, do indeed damage the foaming qualities of the wines, but some fatty acid esters actually cause an improvement. Pueyo et al. (1995) have reported previously that linolenic acid when not present in the free state (forming part of another type of lipid) is positively correlated with foam quality.

Several authors also agree that sulphur dioxide can negatively affect the foaming qualities of wines (maujean 1990; Pueyo et al. 1995; López-Barajas et al. 1997; Girbau-Sòla et al. 2002b), as can ethanol, as mentioned above (Molan et al. 1982; Dussaud et al. 1994; Girbau-Sòla et al. 2002b).

There is some controversy about the effect of total acidity on foaming quality, since some authors consider it to have beneficial effects (Andrés-Lacueva et al. 1996a; Girbau-Sòla et al. 2002b) while others assign it negative effects (Brissonnet et al. 1993; López-Barajas et al. 1997).

Studies carried out to establish the effects of different technological variables on wine foaming qualities include those on the effects of stabilization treatments, either using clarifiers or filtrations (Brissonnet et al. 1991; Robillard et al. 1993; Viaux et al. 1994; Andrés-Lacueva et al. 1996a; López-Barajas et al. 1998; Lao et al. 1999; Puig-Deu et al. 1999; Marchal et al. 2002; Vanrell et al. 2002, 2007; Martínez-Rodríguez and Polo 2003; Dambrouck et al. 2005). In all cases, the foams are negatively affected by these treatments, and this is directly correlated with the reduced protein concentration they produce.

Another of the variables studied was the effect of the second fermentation and also of aging time. The second fermentation is the main factor responsible for the reduced foaming properties of the base wine (Girbau-Solà et al. 2002; Hidalgo et al. 2004), although during aging there is a reduction in foamability and an increased persistence of the foam (Andrés-Lacueva et al. 1996b). Some authors have reported that a peak in the foaming properties is reached during aging of the sparkling wines in the presence of lees, but over prolonged time periods these properties can diminish (Maujean et al. 1990; Pueyo et al. 1995; Andrés-Lacueva et al. 1997).

Grape variety is another variable to be taken into consideration, since it can, also, influence wine foaming properties. The wines made with the Chardonnay variety have the best foamability, although worse stability times (Andrés-Lacueva et al. 1996b, 1997). Of the varieties Xarello, Macabeo and Parellada, grape varieties used to produce the Cava wines (sparkling Spanish wines made by the traditional method in specific regions), the Xarello variety is the one that gives the best foaming properties (Andrés-Lacueva et al. 1996a). Girbau-Sòla et al. (2002a, 2002b) have studied the effect of using red wine varieties (Trepat and Monastrell) to make base wines and sparkling wines, and obtained an improvement. Similar results were obtained by Hidalgo et al. (2004) when they made sparkling wines with the Garnacha grape variety.

The influence of harvest yield has also been studied and was found to have no effect on foaming qualities (Pozo-Bayón et al. 2004), in spite of the findings in this work that too large a harvest can significantly affect the sensorial quality of the sparkling wine, producing wines of a poorer quality.

The use of yeast strains with a greater autolytic capacity can help to achieve, in sparkling wines, better foaming properties in a shorter time period (Nunez et al. 2005). The effect of *Botrytis cinerea* on the foaming characteristics of sparkling wines has also been studied (Marchal et al. 2001, 2006; Cilindre et al. 2007). In these works it can be concluded that this infection can cause a drastic reduction in wine foaming properties, since it uses up the proteins in the medium.

3A.4 General Features of Autolysis

As described previously, during the elaboration of sparkling wines by the traditional method, an aging process takes place that is closely associated with the sensorial quality of the final wine. In fact, it is during this aging of sparkling wines that yeast autolysis takes place, by which the yeasts release intracellular compounds into the wine that can significantly alter its final composition (Charpentier and Feuillat 1993).

Yeast autolysis could be defined as the hydrolysis of biopolymers under the action of hydrolytic enzymes which releases cytoplasmic (peptides, amino acids, fatty acids and nucleotides) and cell wall (glucans, mannoproteins) compounds into the wine (Alexandre et al. 2006).Usually, autolysis takes place at the end of the stationary growth phase and is associated with cell death (Babayan and Bezrukov 1985; Charpentier and Feuillat 1993). When sugars and other nutrients are consumed, the yeast cells turn to their own internal energy reserves, composed of glycogen and other elements. Once these reserves become insufficient for the continued energy demands of the cell, cell degeneration begins and autolysis is triggered (Connew 1998).

Several authors have attempted to describe the mechanism to explain the process of yeast autolysis (Arnold 1980; Charpentier and Feuillat 1993; Charpentier and Freyssinet 1989). The general features of the mechanism proposed are as follows: hydrolytic enzymes are released into the intracellular space due to the degradation of cell endostructures. Initially, these enzymes are inhibited by specific cytoplasmic inhibitors which are later degraded provoking the proteolytic activation of these enzymes. Then, an accumulation of hydrolysis products is produced by the enzymatic degradation of intracellular macromolecules. When the cell wall pores are large enough, the autolytic products are released into the extracellular environment. Finally, a further autolytic degradation of more polymerized compounds into low molecular weight compounds occurs in the extracellular environment.

It is interesting to differentiate between the natural autolysis caused by cell aging and death, and the induced autolysis. Induced autolysis is widely used in different industrial applications, such as the production of yeast extracts or culture media. Autolysis in industrial processes can be induced by several inductors such as temperature, osmotic pressure, detergents, pH, etc. and the process is usually very fast, around 48–72 h, depending on the inducer used. In the manufacture of sparkling wines elaborated by the traditional method, yeast autolysis occurs under very specific conditions. The pH ranges from 3 to 3.5 and the temperature is generally low, between 10 °C and 15 °C, the ethanol concentration is around 10% and CO_2 pressure is high. These conditions are very different to the optimum conditions described for autolysis to occur (Alexandre and Guilloux-Benatier 2006; Fornairon-Bonnefond et al. 2002) and are responsible for the long duration of the process.

In the research carried out on autolysis in enology, two major approaches have been followed to date: first, studying the breakdown of the yeast cell wall, by analyzing the change in its components during autolysis and carrying out structural and ultrastructural studies, and second, analysing the different products released into the medium during autolysis, following the changes in nitrogen compounds, polysaccharides, glycoproteins, nucleic acids, lipids and other macromolecules.

3A.4.1 Breakdown of the Yeast Cell Wall During Autolysis

The yeast cell wall is a complex structure that accounts for 15–25% of the dry weight of the cell. The mechanical strength of the wall is mainly due to the inner layer, which consists of β-1,3-glucans and chitin. The outer layer is mainly formed by glycosylated mannoproteins emanating from the cell surface, which are involved, among others, in cell-cell recognition events (Klis et al. 2002). The cell wall is a multifunctional organelle, the primary function of which is to protect the protoplast.

Without its protection the protoplast could lyse, since the inner osmotic pressure is high compared to the environmental pressure.

During autolysis, the glycoproteins and polysaccharides in the cell wall are hydrolyzed. Lysis of the cell wall has been studied by microscopy and ultramiscroscopy methods, also studying the products obtained after cell wall autolysis. However, the enzymes involved in cell wall autolysis have been less investigated than other autolytic enzymes, such as proteases, and the kinetics of glucanase activity is unknown in sparkling wines (Alexandre and Guilloux-Benatier 2006). Charpentier and Freyssinet (1989) have suggested the following mechanism for lysis of the wall in *Saccharomyces cerevisiae*: in the early stage of the process, glucanases act on glucans, releasing mannoproteins inserted or covalently linked to glucans. Later, these enzymes release the glucans into the wine. Mannoproteins and other polymerized compounds are, finally, degraded by proteolytic enzymes.

Microscopic observation of the yeast cell under autolysis has revealed that although glucanases and proteases degrade the wall, there is no break down of the cell wall. The yeast cell wall retains its shape during autolysis, a so the variation in optical density of the medium cannot be related with the degree of autolysis, as in the case of bacteria. Although microscopy is used less to investigate yeast autolysis in sparkling wines than studies based on analysing the products released into the medium, several researchers have used different microscopic techniques to observe the changes taking place in the yeast cell wall (Fumi et al. 1987b; Martínez-Rodríguez et al. 2001b, 2004; Piton et al. 1988; Takeo et al. 1989). Since the natural autolysis that takes place in wines is a long-lasting process, model systems are commonly used to study this phenomenon (Feuillat and Charpentier 1982; Hernawan and Fleet 1995; Martínez-Rodríguez and Polo 2000b) in order to obtain results in shorter periods of time. Structural and ultrastructural changes occurring during autolysis have been compared in model wines and in sparkling wines. Structural observations have revealed that yeast cells in fermentation are elongate, ovoid and present a large vacuole containing a number of spherical bodies, located mainly on the edges of the vacuole. However, after 24 h of induced autolysis in a model wine, the volume of cells is much smaller, due to solubilization of the cytoplasmic content that takes place during induced autolysis (Martínez-Rodríguez et al. 2001b; Martínez-Rodríguez et al. 2004). The observation of cells which have been aged in wine for 12 months shows the presence of more spherical bodies than in cells isolated from a model wine. As the spherical bodies can be considered as intermediates in the authopagy process (Takeshige et al. 1992; Cebollero and González 2007), these findings indicated that, after 24 h induced autolysis in a model wine yeast, a higher degree of autolysis has been reached than after 12 months of aging in wine. Ultrastructural observations, using different scanning microscopy techniques, of the yeast cell under autolysis have revealed the presence of wrinkles or folds on the wall which are mostly longitudinal. These wrinkles are due to plasmolysis and do not appear in yeast isolated during fermentation. Low Temperature Scanning Electron Microscopy (LTSEM) has shown three-dimensional images of empty yeast cells that have lost most of their cytoplasmic contents as a result of autolysis after 24 h of incubation in a model wine (Fig. 3A.1) (Martínez-Rodríguez et al. 2001c).

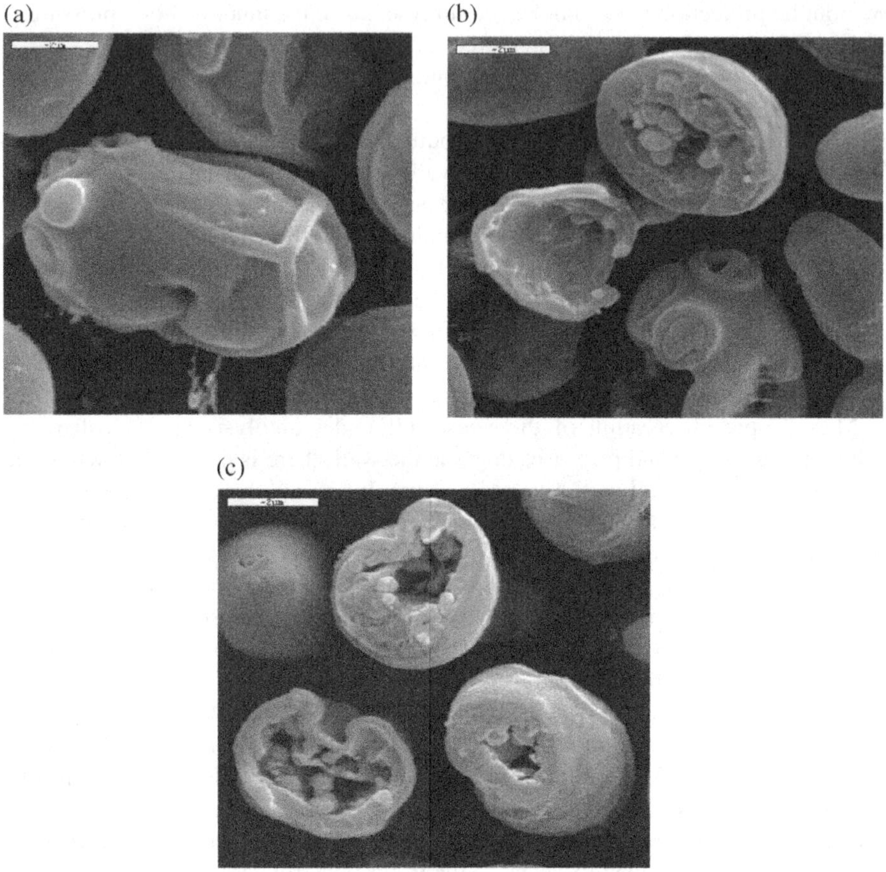

Fig. 3A.1 LTSEM images of yeast cells after 24 h of induced autolysis in a model wine system. **a** Superficial ultrastructure of a yeast cell. **b,c** Images of fractured empty yeast cells which have lost most of their cytoplasmic content during the autolysis (Martínez-Rodríguez AJ, Polo MC, Carrascosa AV (2001) Int J Food Microbiol 71:45–51. Copyright (2001). Elsevier)

However, a smaller amount of cytoplasmic contents were solubilized after 12 months of yeast aging in sparkling wines under autolysis in natural conditions (Fig. 3A.2). This shows that the autolysis conditions during aging of sparkling wines elaborated by the traditional method are not optimal and it may take several months or years before this is finished.

3A.4.2 Products Released by Yeast into Wine During Autolysis

Most studies carried out on the transformations taking place during the aging of sparkling wines elaborated by the traditional method have focused on analysing the compounds released by the yeast into the wine. Among them, nitrogen compounds

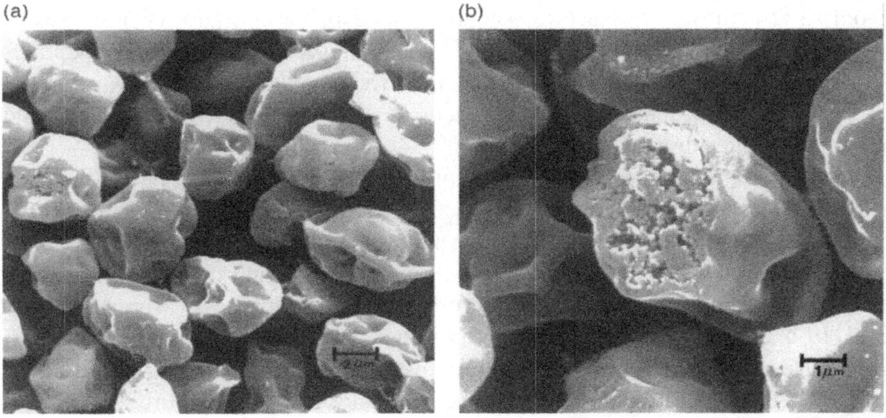

Fig. 3A.2 LTSEM images of yeast cells that have been in wine for 12 months. **a** Superficial ultrastructure of yeast cells. **b** Images of fractured yeast cells. (Martínez-Rodríguez AJ, Polo MC, Carrascosa AV (2001). Int J Food Microbiol 71:45–51. Copyright (2001). Elsevier)

have been considered to be the best markers of yeast proteolytic activity (Fornairon-Bonnefond et al. 2002; Lurton et al. 1989; Martínez-Rodríguez and Polo 2000a).

During the aging of wine in contact with yeast there is a steady drop in the protein content, since proteins are hydrolyzed to lower molecular weight compounds. For this reason, sparkling wines have lower values of proteins than those usually present in the base wine. Peptides and amino acids are, generally, considered the major compounds of those released into the wine during autolysis. For this reason, several authors have studied the role of protease activity in yeast autolysis and its relationship with foaming ability and other sensorial properties (Alexandre et al. 2001; Lurton et al. 1989). Several different types of proteases have been related with the proteolytic activity during autolysis in sparkling wines (Komano et al. 1990; Lurton et al. 1989; Olsen et al. 1999) and protease A is the main enzyme involved in the process. Protease A presents the greatest activity in a wine pH range of 3–3.5. Lurton et al. (1989) use pepstatine to inhibit protease A and obtain around 80% inhibition in the release of nitrogen compounds, demonstrating the key role of protease A in autolysis in an acid medium. More recently, Alexandre et al. (2001) found that protease A was responsible for 60% of the nitrogen released during autolysis in wine, suggesting that other acidic proteases may also be involved in the proteolytic process. This has been confirmed by other authors who found that yeast autolysis is mainly carried out by protease A, but with the help of other proteases.

Yeasts can release amino acids to the extracellular medium before autolysis starts. This release occurs as a cellular response to the absence of nutrients from the wine and was first defined as exsorption by Morfaux and Dupuy (1966). In contrast to autolysis, this has been described as a passive excretion of amino acids that occurs prior to the former. If we carry out a quantitative analysis of the increase in amino acid concentration during the process of autolysis, we observe that this is only a few milligrams per litre. Two possible explanations have been suggested for this observation: first, that protease A is an endoprotease that produce peptides rather than amino acids, and, second, that amino acids released are later transformed by decarboxylation and desamination reactions, which cause a reduction in the final amino acid concentration.

Peptides are the majority product of autolysis and although they have been used as markers of the process, they are less used than amino acids, mainly because of the complexity of the analytical techniques available for their analysis (Moreno-Arribas et al. 2002). It has been observed that high molecular weight peptides mainly of a hydrophobic nature are released in the first steps of the process (Moreno-Arribas et al. 1998a). These large peptides are hydrolyzed in the steps following autolysis, giving rise to less hydrophobic peptides of lower molecular weight and to free amino acids. The concentration of total wine amino acids increases before the concentration of free amino acids, demonstrating that first peptides are released and that later these are hydrolysed into amino acids (Moreno-Arribas et al. 1996). The final concentration of peptides in sparkling wines can be influenced by different variables such as temperature, wine-aging time, the yeast strain used in the second fermentation etc (Martínez-Rodríguez et al. 2002). By studying the amino acid composition of the peptides present in sparkling wines, it has been shown that both threonine and serine are present at the highest levels (Moreno-Arribas et al. 1998b), demonstrating that the peptides present in sparkling wines are mainly derived from yeast autolysis, since these two amino acids are involved in glycosidic bonds between the proteins and mannanes of the cell wall (Klis et al. 2002).

The polysaccharides present in sparkling wines can come from either the grape or from the yeasts and present constitutional differences. Arabinose is the sugar with the greatest presence in base wine polysaccharides, while in sparkling wines mannose is the majority sugar in the polysaccharides (43%), followed by glucose (31%) (Martínez-Rodríguez and Polo 2000b). During aging and yeast autolysis of the yeasts in sparkling wines, enzymatic activity (proteases and glucanases) causes the breakdown of glucanes and the release of cell wall mannoproteins (Feuillat et al 2003). Among the wine glycoproteins, the yeast mannoproteins, released during fermentation and autolysis, have been particularly studied in recent years for their ability, among others, to improve the tartaric stability (Moine-Ledoux and Dubourdieu 2000) and to increase wine stability against the protein haze (Dupin et al. 2000). Also, mannoproteins have been associated with improved foaming properties in sparkling wines (Nunez et al. 2006). Different authors have observed that the concentration of polysaccharides containing glucose and mannose in their structure increases three- or fourfold during the aging of sparkling wines in contact with yeast. The differences observed in the amount of mannoprotein and other polysaccharides released during autolysis into the wine by yeast depend on several variables such as yeast strain, temperature, and time of aging (Caridi 2006).

Other compounds released during autolysis are present in lower amounts, such as lipids and nucleic acids, but could play an important role in the sensorial character of the final wine. Lipids may affect wine flavour in that the fatty acids released could give rise to volatile components with low sensory thresholds, either directly or through derivatives such as esters, ketones and aldehydes (Charpentier and

Feuillat 1993). It has been seen that, after bottle aging in contact with yeast, the lipid content increases and there are qualitative and quantitative changes, depending on the time of aging (Piton et al. 1988). Pueyo et al. (2000) used an HPLC method with a light scattering detector to analyze the different families of lipids in induced autolysis assays in a model wine. They detected lipids in the autolysate after 5 h, and found these to decrease in the first two days, coinciding with the period of maximum loss of yeast viability. This release of lipids is attributed by the authors to the action of the hydrolytic enzymes on the cell wall, with the corresponding release of products to the surrounding medium. In the autolysates they detected triacylglycerols, free fatty acids, sterol esters and sterols. Triacylglycerols and sterol esters were the major lipids in the three strains studied.

The concentration of nucleic acids is also modified during autolysis (Alexandre and Guilloux-Benatier 2006). In general, the concentration of nucleic acids falls during aging and the extent of DNA degradation during autolysis depend on the yeast strain used (Hernawan and Fleet 1995). This could possibly explain the variability between the results reported. While some authors found that the DNA was almost completely degraded after autolysis (Houg and Maddox 1970), others found no decrease or only a partial degradation of DNA (Trevelyan 1978; Zhao and Fleet 2003). The tendency of DNA to form complexes with proteins could protect it, according to Hernawan and Fleet (1995), from the action of DNAses involved in the autolysis.

RNA is also degraded during autolysis, and hydrolysis of around 95% has been reported (Zhao and Fleet 2005). However, it is expected that in the conditions used to elaborate sparkling wines (high ethanol concentration and low temperature and pH), RNA degradation could be much lower.

RNA hydrolysis mainly produces 2'- ,3'- and 5'-ribonucleotides. Although monophosphate nucleotides are well recognized in the food industry as flavour compounds (Charpentier et al. 2005), further studies are required to know their impact in the sensorial properties of sparkling wines.

Finally, several volatile compounds are formed or released during autolysis (Hidalgo et al. 2004; Molnar et al. 1981), some with low perception levels. Esters are the major family of volatile compounds released during autolysis. Different authors have related the quality of the sparkling wines to the concentration of esters, such as isoamyl caproate, octyl acetate, phenylethyl acetate, phenylethyl caprate, ethyl linoleate and diethyl succinate (Loyaux et al. 1981; Pozo-Bayón et al. 2003; Pueyo et al. 1995). Although it is accepted that the secondary fermentation and lees aging significantly alter the aromatic profile of sparkling wines (De La Presa-Owens et al. 1998; Francioli et al. 2003), the impact of these compounds on the sensorial properties of sparkling wines is poorly understood, and will be the subject of further studies.

3A.4.3 Autolysis in the Winemaking Industry

Because of the slow nature of the natural process of autolysis, considerable research has been aimed at accelerating this process. Two methods are currently available to accelerate the autolytic process during sparkling wine production: adding yeast autolysates to the wine, and increasing the temperature during aging (Charpentier and Feuillat 1992). However, both techniques result in organoleptic defects in the final product, which are often described as toasty. It has recently been suggested that by using a combination of killer and sensitive yeasts, it should be possible to accelerate the onset of yeast autolysis during sparkling wine production (Todd et al. 2000). However, the effect of this strategy on the organoleptic properties of wine has not yet been studied. Tini et al. (1995) also found interesting results by using autolytic strains derived through meiosis from an industrial second fermentation yeast. In previous works, we suggested that a yeast strain with a good autolytic capacity would produce better quality sparkling wine than a yeast with a low autolytic capacity, and recommended that autolytic capacity be studied when selecting a yeast for sparkling wine production (Martínez-Rodríguez et al. 2001b). Also, González et al. (2003) found interesting results by using mutant yeast strains obtained by UV irradiation. Recently, we showed that one of these mutant yeasts, presenting accelerated autolysis, produced sparkling wines with improved foaming properties compared to the control strain (Nunez et al. 2005). In this experiment, aging time was reduced from nine to six months, which could also reduce production costs.

Cebollero et al. (2005) used a yeast mutant defective in the autophagic or the Cvt pathways to show that autophagy does take place under wine production conditions. Also, using the depletion of Ald6p as a marker, autophagy has been demonstrated to take place in commercial second fermentation yeasts during a real sparkling wine elaboration process (Cebollero and González 2006). As autophagy is a process that precedes autolysis and many of the genes involved in it are well known, genes related to autophagy could be good candidates to obtain genetic engineering wine yeast presenting accelerated autolysis. Yeast strains showing increased rates of autolysis could also be useful, not only for the elaboration of sparkling wines by the traditional method, but also for the production of still wines aged on lees, an enological practice that has been increasing in popularity over the past few years.

3A.5 Conclusions

Sparkling wines are typically classified in relation to the production method by which they are made. Of these, sparkling wines elaborated by the traditional method are usually considered to present the best sensory properties. During aging, numerous changes occur while the yeast remains in contact with the wine, modifying its sensory characteristics, and the foaming properties of these wines are highly appreciated. The process takes place slowly and continuously, and the yeast remains in contact with the wine for prolonged periods of time in order to give the wine its special characteristics. Morphological and analytical studies on yeast autolysis in sparkling wines have revealed the major changes produced during yeast autolysis and its implication in wine quality. Although the autolytic process has been extensively studied, there are still some aspects that remain unclear. This is the case for the release kinetics of some compounds, such as lipids or nucleotides, and some molecular mechanisms, such as those involved in the induction of autolysis are still unknown. This knowledge could help to reveal potential targets to accelerate the autolytic process. All the advantages of the molecular principles and basic aspects of autophagy and autolysis may be potentially useful for the future of yeast biotechnology.

References

Alexandre, H., & Guilloux-Benatier, M. (2006). Yeast autolysis in sparkling wine – a review. *Aust. J. Grape Wine Res., 12*, 119–127.

Alexandre, H., Heintz, D., Chassagne, D., Guilloux-Benatier, M., Charpentier, C., & Feuillat, M. (2001). Protease A activity and nitrogen fractions released during alcoholic fermentation and autolysis in enological conditions. *J. Ind. Microbiol. Biotechnol., 26*, 235–240.

Andrés-Lacueva, C., López-Tamames, E., Lamuela-Raventós, R.M., Buxaderas, S., & de la Torre-Boronat, M.C. (1996a). Characteristics of sparkling base wines affecting foam behavior. *J. Agric. Food Chem., 44*, 989–995.

Andrés-Lacueva, C., Gallart, M., López-Tamames, E., & Lamuela-Raventós, R.M. (1996b). Influence of variety and aging on foaming properties of sparkling wine (Cava). 1. *J. Agric. Food Chem., 44*, 3826–3829.

Andrés-Lacueva, C., Lamuela-Raventós, R.M., Buxaderas, S., and de la Torre-Boronat, M.C. (1997). Influence of variety and aging on foaming properties of Cava (sparkling wine). 2. *J. Agric. Food Chem., 45*, 2520–2525.

Arnold, W.N. (1980). Yeast Cell Envelopes. Biochemistry, Biophysics and Ultraestructure Volume 2. In W.N.Arnold (Ed), *Autolysis* (pp. 93–103). Boca Raton: CRC Press.

Babayan, T.L., & Bezrukov, M.G. (1985). Autolysis in yeast. *Acta Biotechnol., 5*, 129–136.

Bidan, P., Feuillat, M., & Moulin, J.P. (1986). Les vins mousseux. Rapport de la France. 65ème Asemblèe Génerale de L'OIV. *Bull. OIV, 59*, 563–626.

Bikerman, J.J. (1938). The unit of foaminess. *Trans. Faraday Soc., 34*, 634–638.

Brissonet, F., & Maujean, A. (1991). Identification of some foam-active compounds in Champagne base wines. *Am. J. Enol. Vitic., 42*, 97–102.

Brissonet, F., & Maujean, A. (1993). Characterization of foaming proteins in a champagne base wine. *Am. J. Enol. Vitic., 44*, 297–301.

Busova, K., Magyar, I., and Janky. F. (1994). Effect of immobilized yeasts on the quality of bottle-fermented sparkling wine. *Acta Aliment. Hung., 23*, 9–23.

Caridi, A. (2006). Enological functions of parietal yeast mannoproteins. *Antonie Van Leeuwenhoek, 89*, 417–422.

Casey, J. (1987). Effervescence in sparkling wines. *Austr. Grapegr. Winemaker, 280*, 55–62.

Casey, J. (1995). Effervescence in sparkling wines: the sequel. *Austr. Grapegr. Winemaker, n° 378a*, 37–47.

Casey, J. (2000). Méthode champenoise or mythes champenoises. *Austr. Grapegr. Winemaker, n° 438a*, 168–178.

Cebollero, E., & González, R. (2006). Induction of autophagy by second-fermentation yeasts during elaboration of sparkling wines. *Appl. Environ. Microbiol., 72*, 4121–4127.

Cebollero, E., & González, R. (2007). Autophagy: From basic research to its application in food biotechnology. *Biotechnol. Adv., 25*, 396–409.

Cebollero, E., Carrascosa, A.V., and González, R. (2005). Evidence for yeast autophagy during simulation of sparkling wine aging: A reappraisal of the mechanism of yeast autolysis in wine. *Biotechnol. Prog., 21*, 614–616.

Charpentier, C., & Feuillat, M. (1992). The Yeasts, vol. III. In A.H. Rose and J.S. Harrison (Eds), *Yeast autolysis* (pp. 315–386). New York: Academic Press.

Charpentier, C., & Feuillat, M. (1993). Wine Microbiology and Biotechnology. In G.H. Fleet (Ed), *Yeast autolysis* (pp. 225–242).Chur: Harwood Academic Publishers.
Charpentier, C., and Freyssinet, M. (1989). The mechanism of yeast autolysis in wine. *Yeast, 5,* 181–186.
Charpentier, C., Aussenac, J., Charpentier, M., Prome, J.C., Duteurtre, B., & Feuillat, M. (2005). Release of nucleotides and nucleosides during yeast autolysis: kinetics and potential impact on flavor. *J. Agric. Food Chem., 53*, 3000–3007.
Cilindre, C., Castro, A.J., Clément, C., Jeandet, P., & Marchal, R. (2007). Influence of Botrytis cinerea infection on Champagne wine proteins (characterized by two-dimensional electrophoresis/immunodetection) and wine foaming properties. *Food Chem., 103,* 139–149.
Colagrande, O., Silva, A., & Fumi, M.D. (1994). Recnt applications of biotechnology in wine production. *Biotecnol. Progr., 10*, 2–18.
Connew, S. (1998). Yeast autolysis. A review of current rersearch. *Aust. NZ Wine Ind. J., 13*, 61–64.
Dambrouck, T., Marchal, R., Cilindre, C., Parmentier, M., & Jeandet, P. (2005). Determination of the grape invertase content (using PTA-ELISA) following various fining treatments versus. Changes in the total protein content of wine. Relationship with wine foamability. *J. Agric. Food Chem., 53*, 8782–8789.
De La Presa-Owens, C., Schlich, P., Davies, H.D., & Noble, A.C. (1998). Effect of Methode Champenoise process on aroma flavour of four *V. vinifera* varieties. *Am. J. Enol. Vitic., 49*, 289–294.
Divies, C., Cachon, R., Cavin, J.F., & Prevost, H. (1994). Theme-4-immobilized cell technology in wine production. *Crit. Rev. Biotechnol., 14*, 135–153.
Dupin, I.V.S., McKinnon, B.M., Ryan, C., Boulay, M., Markides, A.J., Jones, G.P., Williams, P.J., & Waters, E.J. (2000). *Saccharomyces cerevisiae* mannoproteins that protect wine from protein haze: Their release during fermentation and lees contact and a proposal for their mechanism of action. *J. Agric. Food Chem., 48*, 3098–3105.
Dussaud, A., Robillard, B., Carles, B., Duteurtre, B., & Vignesadler, M. (1994). Exogenous lipids and ethanol influences on the foam behavior of sparkling base wines. *J. Food Sci., 59*, 148.
Edward, M., Eschenbruch, R., & Molan, P.C. (1982). Foaming in winemaking. *Eur. J. Appl. Microbiol. Biotechnol., 16*, 105–109.
Feuillat, M. (2003). Yeast macromolecules: Origin, composition, and enological interest. *Am. J. Enol. Vitic., 54*, 211–213.
Feuillat, M., & Charpentier, C. (1982). Autolysis of yeast in champagne.*Am. J. Enol. Vitic., 33*, 6–13.
Fornairon-Bonnefond, C., Camarasa, C., Moutounet, M., & Salmon, J.M. (2002). New trends on yeast autolysis and wine ageing on lees: A bibliographic review. *J. Int. Sci. Vigne Vin, 36*, 49–69.
Francioli, S., Torrens, J., Riu-Aumatell, M., López-Tamames, E., & Buxaderas, S. (2003). Volatile compounds by SPME-GC as age markers of sparkling wines. *Am. J. Enol. Vitic., 54*, 158–162.
Fumi, M.D., Trioli, G., & Colagrande, O. (1987a). Preliminary assessment on the use of immobilized yeast-cells in sodium alginate for sparkling wine processes. *Biotechnol. Lett., 9*, 339–342.
Fumi, M.D., Trioli, G., & Silva, A. (1987b). Autolisi di *Saccharomyces cerevisiae* in vini spumantizzati in bottiglia: uno studio biochimico ultraestructurale. *Ind. Bevande, 16*, 340–344.
Gallart, M., López-Tamames, E., & Buxaderas, S. (1997). Foam measurements in wines: comparasion of parameters obtained by gas sparging method. *J. Agric. Food Chem., 45*, 4687–4690.
Gallart, M., López-Tamames, E., Suberbiola, G., and Buxaderas, S. (2002). Influence of fatty acids on wine foaming. *J. Agric. Food Chem., 50*, 7042–7045.
Gallart, M., Tomás, X., Suberbiola, G., López-Tamames, E., & Buxaderas, S. (2004). Relationship between foam parameters obtained by gas-spaging method and sensory evaluation os sparkling wines. *J. Sci. Food Agric., 84*, 127–133.

Girbau-Sòla, T., López-Tamames, E., Buján, J., & Buxaderas, S. (2002a). From aptitude of Trepat and Monastrell red varieties in Cava elaboration. 1. Base wine characteristics. *J. Agric. Food Chem., 50*, 5596–5599.

Girbau-Sòla, T., López-Barajas, E., López-Tamames, E., & Buxaderas, S. (2002b). From aptitude of Trepat and Monastrell red varieties in Cava elaboration. 2. Second fermentation and aging.. *J. Agric. Food Chem., 50*, 5600–5604.

Godia, F., Casas, C., & Sola, C. (1991). Application of immobilized yeast-cells to sparkling wine fermentation. *Biotecnol. Progr., 7*, 468–470.

Gónzalez, R., Martínez-Rodríguez, A.J., & Carrascosa, A.V. (2003). Yeast autolytic mutants potentially useful for sparkling wine production. *Int. J. Food Microbiol., 84*, 21–26.

Hernawan, T., & Fleet, G.H. (1995). Chemical and cytological changes during the autolysis of yeast. *J. Ind. Microbiol., 14*, 440–450.

Hidalgo, J. (2003). *Tratado de enología*. Madrid: Mundi prensa.

Hidalgo, P., Pueyo, E., Pozo-Bayón, M.A., Martínez-Rodríguez, A.J., Martín-Álvarez, P., Polo, M.C. (2004). Sensory and analytical study of rose sparkling wines manufactured by second fermentation in the bottle. *J. Agric. Food Chem., 52*, 6640–6645.

Houg, M.D., & Maddox, I.S. (1970). Yeast autolysis. *Process Biochem., 5*, 50–52.

Jordan, A.D., & Napper, D.H. (1994). Some aspect of the physical chemistry of bubble and foam phenomena in sparkling wine. *Sixth Australian Wine Industry Conference*, 237–246.

Klis, F.M., Mol, P., Hellingwerf, K., & Brul, S. (2002). Dynamics of cell wall structure in *Saccharomyces cerevisiae*. *FEMS Microbiol. Rev., 26*, 239–256.

Komano, H., Rockwell, N., Wang, G.T., Krafft, G.A., & Fuller, R.S. (1999). Purification and characterization of the yeast glycosylphosphatidylinositol-anchored, monobasic-specific aspartyl protease yapsin 2 (Mkc7p). *J. Biol. Chem., 271*, 24421–24437.

Lao, C., Santamaría, A., López-Tamames, E., Bujan, J., Buxaderas, S., & de la Torre-Boronat, M.C. (1999). Effect of grape pectic enzyme treatment on foaming properties of white musts and wines. *Food Chem., 65*, 169–173.

Liger-Belair, G. (2005). The physics and chemistry behind the bubbling properties of Champagne and sparkling wines: a state-of-the-art Review. *J. Agric. Food Chem., 53*, 2788–2802.

Liger-Belair, G., Marchal, R., Robillard, B., Vignes-Adler, M., Maujean, A., & Jeandet, P. (1999). Study of effervescence in a glass of Champagne: frecuencies of bubble formation, growth rates and velocities of rising bubble. *Am. J. Enol. Vitic., 50*, 317–323.

Liger-Belair, G., Lemaresquier, H., Robillard, B., Duteurtre, B., & Jeandet, P. (2001). The secrets of fizz in Champagne wines: a phenomenological study. *Am. J. Enol. Vitic., 52*, 88–92.

Liger-Belair, G., Marchal, R., and Jeandet, P. (2002). Close-up on bubble nucleation in a glass of Champagne. *Am. J. Enol. Vitic., 53*, 151–153.

Liger-Belair, G., Tufalle, A., Jeandet, P., and Sartorelli, J.C. (2006). Champagne experiences various rhythmical bubbling regimes in a flute. *J. Agric. Food Chem., 54*, 6989–6994.

López-Barajas, M., Viu-Marco, A., López-Tamames, E., Buxaderas, S., and de la Torre-Boronat, M.C. (1997). Foaming in grape juices of white varieties. *J. Agric. Food. Chem., 45*, 2526–2529.

López-Barajas, M., López-Tamames, E., Buxaderas, S., and de la Torre-Boronat, M.C. (1998). Effect of vinification and variety on foam capacity of wine. *Am. J. Enol. Vitic., 49*, 397–402.

López-Barajas, M., López-Tamames, E., Buxaderas, S., Tomás, X., and de la Torre, M.C. (1999). Prediction of wine foaming. *J. Agric. Food Chem., 47*, 3743–3748.

Loyaux, D., Roger, S., and Adda, J. (1981). The evolution of champagne volatiles during ageing.*J. Sci. Food Agric., 32*, 1254–1258.

Lurton, L., Segain, J.P., and Feuillat, M. (1989). Etude de la protéolyse au cours de l'autolyse de levures en milieu acide. *Sci. Aliments, 9*, 111–124.

Machet., F., Robillard, B., and Duteurtre, B. (1993). Application of image analysis to foam stability of sparkling wines. *Sci. Aliments, 13*, 73–87.

Malvy, J., Robillard, B, and Duteurtre, B. (1994). Influence des protéines sur le comportement de la mousse des vins de Champagne. *Sci. Aliments, 14*, 87–98.

Marchal, R., Tabary, I., Valade, M., Moncomble, D., Viaux, L., Robillard, B., and Jeandet, P. (2001). Effects of *Botrytis cinerea* infection on Champagne wine foaming properties. *J. Sci. Food Agric., 81*, 1371–1378.

Marchal, R., Chaboche, D., Douillard, R., and Jeandet, P. (2002). Influence of lysozyme treatments on Champagne base wine foaming properties. *J. Agric. Food Chem., 50*, 1420–1428.

Marchal, R., Warchol, M., Cilindre, C., and Jeandet, P. (2006). Evidence for protein degradation by *Botrytis cinerea* and relationships with alteration of synthetic wine foaming properties. *J. Agric. Food Chem., 54*, 5157–5165.

Martínez-Rodríguez, A.J., and Polo, M.C. (2000a). Characterization of the nitrogen compounds released during yeast autolysis in a model wine system. *J. Agric. Food Chem., 48*, 1081–1085.

Martínez-Rodríguez, A.J., and Polo, M.C. (2000b). Recent Res. Devel. Microbiology Volume 4. In S.G. Pandalai (Ed), *Enological aspects of yeast autolysis.* (pp. 285–301). Trivandrum: Research Signpost.

Martínez-Rodríguez, A.J., and Polo, M.C. (2003). Effect of the addition of bentonite to the tirage solution on the nitrogen composition and sensory quality of sparkling wines. *Food Chem., 81*, 383–388.

Martínez-Rodríguez, A.J., Carrascosa, A.V., Barcenilla, J. M., Pozo-Bayón, M. A.,and Polo, M. C. (2001a). Autolytic capacity and foam analysis as additional criteria for the selection of yeast strains for sparkling wine production. *Food Microbiol., 18*, 183–191.

Martínez-Rodríguez, A.J., Polo, M.C., and Carrascosa, A.V. (2001b). Structural and ultrastructural changes in yeast cells during autolysis in a model wine system and in sparkling wines. *Int. J. Food Microbiol., 71*, 45–51.

Martínez-Rodríguez, A.J., Carrascosa, A.V., Martín-Álvarez, P.J., Moreno-Arribas, V., and Polo, M.C. (2002). Influence of the yeast strain on the changes of the amino acids, peptides and proteins during sparkling wine production by the traditional method.*J. Ind. Microbiol. Biot., 29*, 314–322.

Martínez-Rodríguez, A.J., González, R., and Carrascosa, A.V. (2004). Morphological changes in autolytic wine yeast during aging in two model systems. *J. Food Sci., 69*, 233–239.

Martynenko, N.N., and Gracheva, I.M. (2003). Physiological and biochemical characteristics of immobilized Champagne yeasts and their participation in champagnizing processes: A review. *Appl. Biochem. Microbiol., 39*, 439–445.

Martynenko, N.N., Gracheva, I.M., Sarishvili, N.G., Zubov, A.L., El'Registan, G.I., and Lozinsky, V.I. (2004). Immobilization of Champagne yeasts by inclusion into cryogels of polyvinil alcohol: Means of preventing cell release from the carrier matrix. *Appl. Biochem. Microbiol., 40*, 158–164.

Maujean, A., Gomerieux, T., and Garnier, J.M. (1988). A survey of the durability and quality of foam in sparkling wines. I. Adjustment of a technique to measure the spontaneous as well as provoked effervescence in fizzy drinks. *Bull. l'OIV, 61*, 25–35.

Maujean, A., Poinsaut, P., Dantan, H., Brissonnet, F., and Cossiez, E. (1990). Study of the performance and quality of the foam in sparkling wines. II. Perfecting of a measuring techniques for foaming abality, performance and stability of the foam in sparkling wines. *Bull. l'OIV, 63*, 405–427.

Moine-Ledoux, V., and Dubourdieu, D. (2000). Role yeast mannoproteins with regard to tartaric stabilisation of wines. *Bull. l'O.I.V, 75*, 471–482.

Molan, P.C., Edwards, M., and Eschenbruch, R. (1982). Foaming in winemaking. *Eur. Appl. Microbiol. Biotechnol., 16*, 110–113.

Molnar, I., Oura, E., and Suomalainen, H. (1981). Study of volatile substances produced during the autolysis of champagne yeast. *Acta Alimentaria, 10*, 27–36.

Moreno-Arribas, V., Pueyo, E., and Polo, M.C. (1996). Peptides in musts and wines. Changes during the manufacture of cavas (Sparkling wines). *J. Agric. Food Chem., 44*, 3783–3788.

Moreno-Arribas, V., Bartolomé, B., Pueyo, E., and Polo, M.C. (1998a). Isolation and characterization of individual peptides from wine. *J. Agric. Food Chem., 46*, 3422–3425.

Moreno-Arribas, V., Pueyo, E., Polo, M.C. and Martín-Álvarez, P.J. (1998b). Changes in the amino acid composition of the diffeent nitrogenous fractions during the aging of wine with yeast. *J. Agric. Food Chem., 70*, 309–317.

Moreno-Arribas, V., Pueyo, E., Nieto, F.J., Matín-Álvarez, P.J., and Polo, M.C. (2000). Influence of the polysaccharides and the nitrogen compounds on foaming properties of sparkling wines. *Food Chem., 70*, 309–317.

Moreno-Arribas, M.V., Pueyo, E., and Polo, M.C. (2002). Analytical methods for the characterization of proteins and peptides in wines. *Anal. Chim. Acta, 458*, 63–75.

Morfaux, J.N. and Dupuy, P. (1966). Comparaison de l exsorption des acides amines par une souche de *Saccharomyces cerevisiae* et un mutant resistant a la canavanine. *C. R. Biologies, 263*, 1224.

Nunez, Y.P., Carrascosa, A.V., González, R., Polo, M.C., and Martínez-Rodríguez, A.J. (2005). Effect of accelerated autolysis of yeast on the composition and foaming properties of sparkling wines elaborated by a champenoise method. *J. Agric. Food Chem., 53*, 7232–7237.

Nunez, Y.P., Carrascosa, A.V., González, R., Polo, M.C., and Martínez-Rodríguez, A. (2006). Isolation and characterization of a thermally extracted yeast cell wall fraction potentially useful for improving the foaming properties of sparkling wine. *J. Agric. Food Chem., 54*, 7898–7903.

Obiols, J.M., de la Presa-Owens, C., Buxaderas, S., Bori, J. L., and de la Torre-Boronat, M.C. (1998). Protocolo de evaluación de la formación de la efervescencia y espuma en un vino espumoso. *ACE, Revista d' Enología, 15*, 3–8.

Olsen, V., Cawley, N.X., Brandt, J., Egel-Mitani, M., and Loh, Y.P.(1999). Identification and characterization of *Saccharomyces cerevisiae* yapsin 3, a new member of the yapsin family of aspartic proteases encoded by the YPS3 gene. *Biochem. J., 339*, 407–411.

Ough, C.S. (1992). *Winemaking basics*. New York: Food Product Press.

Péron, N., Cagna, A., Valade, M., Marchal, R., Maujean, A., Robillard, B., Aguié-Béghin, V., and Douillard, R. (2000). Characterisation by drop tensiometry and by ellipsometry of the adsorption layer formed at the air Champagne wine interface. *Adv. Colloid Interfac. Sci., 88*, 19–36.

Péron, N., Cagna, A., Valade, M., Bliard, C., Aguié-Béghin, V., and Douillard, R. (2001). Layers of macromolecules at the Champagne/air interface and the stability of Champagne bubbles. *Langmuir, 17*, 791–797.

Péron, N., Meunier, J., Cagna, A., Valade, M., and Douillard, R. (2004). Phase separation in molecular layers of macromolecules at the Champagne-air interface. *J. Microsc., 214*, 89–98.

Piton, M., Charpentier, C., and Troton, D. (1988). Cell wall and lipid changes in *Saccharomyces cerevisiae* during aging of champagne wine. *Am. J. Enol. Vitic., 39*, 221–226.

Pozo-Bayón, M.A., Pueyo, E., Martín-Álvarez, P.J., Martínez-Rodríguez, A.J., and Polo, M.C. (2003). Influence of yeast strain, bentonite addition, and aging time on volatile compounds of sparkling wines. *Am. J. Enol. Vitic., 54*, 273–278.

Pozo-Bayón, M.A., Polo, M.C., Martín-Álvarez, P.J., and Pueyo, E. (2004). Effect of vineyard yield on the composition of sparkling wines produced from the grape cultivar Parellada. *Food Chem., 86*, 413–419.

Pueyo, E., Martín-Álvarez, P.J., and Polo, M.C. (1995). Relationship between foam characteristics and chemical composition in wines and cavas (sparkling wines). *Am. J. Enol. Vitic., 46*, 518–524.

Pueyo, E., Martínez-Rodríguez, A., Polo, M.C., Santa-Maria, G., and Bartolomé, B. (2000). Release of lipids during yeast autolysis in a model wine system. *J. Agric. Food Chem., 48*, 116–122.

Puig-Deu, M., López-Tamames, E., Buxaderas, S., and de la Torre-Boronat, M.C. (1999). *Food Chem., 66*, 35–42.

Roberts, R.T., Keeney, P.J., and Wainwright, T. (1978). Effects of lipids and related materials on beer foam. *J. Inst. Brewing, 84*, 9–12.

Robillard, B., Delpuech, E., Viaux, L., Malvy, J., Vignes-Adler, M. and Duteurtre, B. (1993). Improvements of methods for sparkling wine foam measurements and effect of wine filtration on foam behavior. *Am. J. Enol. Vitic., 44*, 387–392.

Robillard, B., Viaux, L., and Duteurtre, B. (1995). Medición de la espuma por visión artificial. *Sevi., (2567–68)*, 3656–3659.

Senèe, J., Viaux, L., Robillard, B., Duteurtre, B., and Vignes-Adler, M. (1998). The endogenous particles of a sparkling wine and their influence on the foaming behaviour. *Food Hidrocoll., 12*, 217–226.

Senèe, J., Robillard, B., and Vignes-Adler, M. (1999). Films and foams of Champagne wines. *Food Hydrocoll., 13*, 15–26.

Senèe, J., Robillard, B., and Vignes-Adler, M. (2001). The ζ-potential of the endogenous particles of a wine of Champagne in relation to the foam behaviour. *Colloid. Surf. B- Biointerfaces, 21*, 59–67.

Takeo, K., Yamamura, M., and Kamihara, T. (1989). Ultrastructural alterations in *Saccharomyces cerevisiae* cells in association with elevated temperature-induced autolysis.*FEMS Microbiol. Lett., 6*1, 297–300.

Takeshige, K., Baba, M., Tsuboi, S., Noda, T., and Ohsumi, Y. (1992). Autophagy in yeast demostrated with proteinase-deficient mutants and conditions for its induction. *J. Cell Biol., 19*, 301–311.

Tini, V., Zambonelli, C., Beneveli, M.,and Castellari, L. (1995). I ceppi autolisogeni di *Saccharomyces cerevisiae* in spumantizzazione. *Ind. Bevande XXIV*, 113–118.

Todd, B.E.N., Fleet, G.H., and Henschke, P.A. (2000). Promotion of autolysis through the interaction of killer and sensitive yeasts: potential application in sparkling wine production. *Am. J. Enol. Vitic., 51*, 65–72.

Trevelyan, W.E. (1978). Effect of procedures for the reduction of the nucleic acid content of SCP on the DNA content of Saccharomyces cerevisiae. *J. Sci. Food Agric., 29*, 903–908.

Tufaile, A., Sartorelli, J., Jeandet, P., and Liger-Belair, G. (2007). Chaotic bubbling and nonstagnant foams. *Phys. Rev. E, 75*, 066216.

Uzel, S., Chappell, M.A., and Payne, J. (2006). Modeling the cycles of growth and detachment of bubbles in carbonated beverages. *J. Phys. Chem., B, 110*, 7579–7586.

Vanrell, G., Cabanillas, P., Albet, S., Canals, J.M., Arola, L., and Zamora, F. (2002). Étude des composés influençant la mousse des cavas. *Rev. Franc. Oenol., n° 196*, 30–36.

Vanrell, G., Canals, R., Esteruelas, M., Fort, F., Canals, J.M., and Zamora, F. (2007). Influence of the use of bentonite as a riddling agent on foam quality and protein fraction of sparkling wines (Cava). *Food Chem., 104*, 148–155.

Viaux, L., Morard, C., Robillard, B., and Duteurtre, B. (1994). The impact of base wine filtration on Champagne foam behavior. *Am. J. Enol. Vitic., 45*, 407–409.

Voisin, C., Jeandet, P., and Liger-Belair, G. (2005). On the 3D-reconstruction of Taylor-like bubbles trapped inside hollow cellulose fibers acting as bubble nucleation sites in supersaturated liquids. *Colloid. Surf. A-Physiochem. Eng. Asp., 263*, 303–314.

Yokotsuka, K., Yajima, M., and Matsudo, T. (1997). Production of bottle-fermentated sparkling wine using yeast immobilized in double-layer gel beads or strands. *Am. J. Enol. Vitic., 48*, 471–481.

Zhao, J., and Fleet, G.H. (2003). Degradation of DNA during the autolysis of *Saccharomyces cerevisiae. J. Ind. Microbiol. Biotechnol., 30*, 175–182.

Zhao, J., and Fleet, G.H. (2005). Degradation of RNA during the autolysis of *Saccharomyces cerevisiae* produces predominantly ribonucleotides . *J. Ind. Microbiol. Biotechnol., 9*, 415–423.

Chapter 3B
Biologically Aged Wines

Rafael A. Peinado and Juan C. Mauricio

Contents

The biological aging of wines has aroused increasing interest in recent years, as reflected in the large number of papers on this topic over the last decade. Biological aging in wine is carried out by flor yeasts. Once alcoholic fermentation has finished, some *Saccharomyces cerevisiae* yeast races present in wine switch from a fermentative metabolism to an oxidative (respiratory metabolism) and spontaneously form a biofilm called "flor" on the wine surface. Wine under "flor" is subject to special conditions by effect of oxidative metabolism by yeasts and of the reductive medium established as they consume oxygen present in the wine. These conditions facilitate

J.C. Mauricio (✉)
Professor of Microbiology, Departamento de Microbiología, Edificio Severo Ochoa, Campus Universitario de Rabanales, Universidad de Córdoba, 14071-Córdoba, Spain
e-mail: mi1gamaj@uco.es

M.V. Moreno-Arribas, M.C. Polo (eds.), *Wine Chemistry and Biochemistry*,
DOI 10.1007/978-0-387-74118-5_4, © Springer Science+Business Media, LLC 2009

various transformations in wine components that lead to the acquisition of special sensory features.

Flor yeasts efficiently transform young wine obtained from sensory neutral grapes into genuine, internationally appreciated wine with a high aroma potential. In fact, wine aroma is dictated by the particular flor yeasts rather than by the grape variety used.

Fino wine, which is the best known type of biologically aged wine, is obtained by using the *criaderas and solera* system, which essentially involves the periodic homogenization of wines of different age. This process is complex and expensive, but produces wines of uniform quality over time. Also, it makes the "vintage" notion meaningless.

The need to store *fino* wine over long periods of time, the analysis and maintenance operations involved, and the need to obtain an effective yeast biofilm substantially increase its price. Hence the interest in shorting the aging time, whether by physical (e.g. periodic aeration) or biological means (e.g. by using specially efficient yeasts or genetically altering existing races).

3B.1 The Winemaking of Biologically Aged Wines

Although biological aging of wines under flor films is done in Italy (Sardinia and Sicily), France (Jura), Hungary (Tokay), USA (California) and various South African and Australian regions, the best-known biologically aged wines are produced in southern Spain (particularly in Jerez and Montilla-Moriles), using a traditional procedure aimed at ensuring uniform quality over time.

The complex homogenization system used for this purpose is depicted in Fig. 3B.1. A few months after alcoholic fermentation has completed, the wine is transferred and any lees removed. In Jerez, wines to be biologically aged are fortified

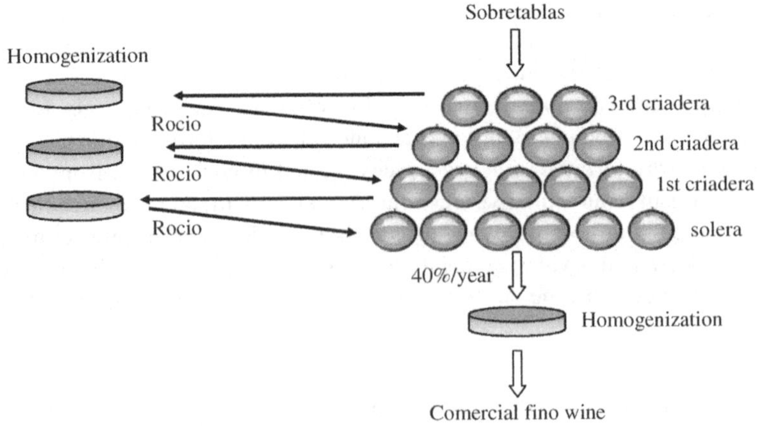

Fig. 3B.1 Scheme of biological aging system

with wine alcohol to an ethanol content of 15.0–15.5 vol.%; this operation is known as *encabezado*. In Montilla-Moriles, however, the favourable climatic conditions and the characteristics of Pedro Ximénez grapes, which constitute the dominant variety in the region, allow musts with alcohol contents in excess of 15 vol.% to be obtained in a natural manner, so no fortification is required.

The wine is stored prior to incorporation into the biological aging system. During the intervening time, it undergoes malolactic fermentation and spontaneously develops a yeast film, thereby incipiently acquiring the typical features of biologically aged wine. Wine stored in this way is referred to as *sobretablas* wine.

Subsequently, the wine is aged in American oak casks of variable capacity depending on their position in the aging system. The casks are filled to four-fifths in order to allow a biofilm of flor yeasts to develop on the wine surface. Biological aging proper is accomplished by using the *criaderas and solera* system, which involves stacking the casks in rows called *criaderas* (scales) in such a way that all casks in a row contain wine of the same type and aged for an identical length of time.

The row standing on the floor, which is called the *solera*, contains the oldest wine in the system. It is from this row that the commercial wine is withdrawn for bottling, extraction never exceeding 40% of the cask contents per year, and this process is carried out three or four times each year. The amount of wine extracted from the *solera* is replenished with an identical volume of wine from the upper row, which is called the *first criadera*. Likewise, the amount extracted from the *first criadera* is replenished with wine from the next row (the *second criadera*) and so on. Finally, the topmost *criadera*, which contains the youngest wine, is replenished with *sobretablas* wine. The number of scales typically ranges from four to six; usually, the greater the number, the higher the quality of the final wine.

The transfer of wine from one scale to the next is called *rocío* and is preceded by a series of operations intended to homogenize the wine in each scale and hence its degree of biological aging (Berlanga et al. 2004a). The wine extracted from the casks in a scale is homogenized in a tank and then transferred to the casks in the next (older) row. This operation must be performed with care in order not to disrupt the "flor" film present on the wine surface.

This dynamic process results in the *solera* casks containing a complex mixture of wine in terms of age. However, is very homogeneous, which allows wine of similar sensory features to be obtained year after year irrespective of the particular vintage. Also, the *rocío* operation brings older wine into contact with younger wine, the latter supplying the nutrients required for the yeast film to form and remain. Also, it provides aeration, which is highly beneficial for wine and flor yeasts (Berlanga et al. 2001, 2004a).

In other world regions, the biological aging process is static; such is the case with the production of "jaune" (yellow) wines in Jura (France). In other countries, USA (California) or South Africa, winemakers use a shorter dynamic process in order to cut costs. "Jaune" wines are obtained from Savagnin grapes (Traminer type) and possess an alcohol content of ca. 12 vol.%. Following malolactic fermentation, they are held in used casks of 228 L which are filled to 5–6 L less and tightly closed for storage in cellars at temperatures from 7 °C in winter to 17 °C in summer, and 60–80% relative humidity, for 6 years and 3 months. In this way, a yeast film develops on the surface of the wine that alters its sensory properties. Thus, the wine acquires a typical golden yellow colour and an acetaldehyde concentration of 600–700 mg/L (Pham et al. 1995).

3B.1.1 The Significance of Rocío in the Biological Aging Process

Recently, the effect of *rocío* in the different scales of a *criadera and solera* system and that on wine uniformity during the aging process have been examined (Berlanga et al. 2004a, 2004b).

The effect of *rocío* can be exposed by using a principal component analysis with the contents in compounds closely related to yeast metabolism (e.g. acetaldehyde, acetic acid, ethanol, glycerol, L-proline) as predictor variables. *Rocío* has significant effects on the scales containing the young and medium-aged wine (namely the fourth, third and second *criadera*). On the other hand, it has little effect on the first *criadera* or the *solera*, which contain the oldest wine (Fig. 3B.2). This can be ascribed to the biological activity of the yeasts at these two stages - the last in the process - being weaker than in the previous ones and physico-chemical processes prevailing over it as a result (Berlanga et al. 2004b).

A similar effect is observed when one compares the point immediately after *rocío* with that seven days later. Thus, blending has a significant effect on the third and second *criadera*, but not on the fourth - possibly because it receives younger wine each year and its contents are less uniform as a result. However, the analysis of each

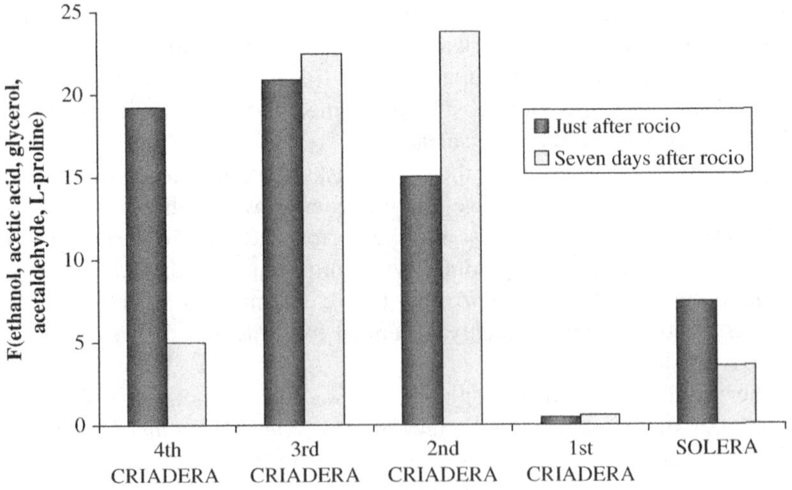

Fig. 3B.2 Effect of the *rocío* on the compounds highly related with yeast metabolism during the biological aging. **a** Just after *rocío*. **b** Seven days after *rocío*

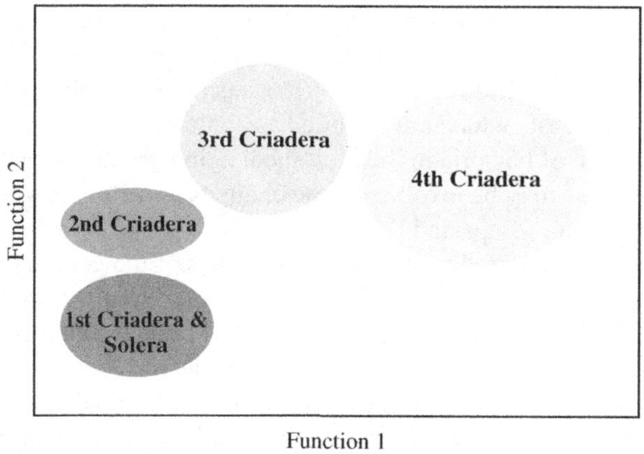

Fig. 3B.3 Heterogeneity of wine samples in the biological aging system

individual cask in the fourth *criadera* exposed the effect of blending and substantial metabolic activity in the yeasts, which is reflected in the differences observed between the young wine and that in the fourth *criadera* (Berlanga et al. 2004a).

There have been several attempts at distinguishing wine from different scales in a biological aging system. Thus, Pérez (1982) established a relationship between aging time and the different scales for an average permanence time of four years in a system consisting of four scales (three criaderas plus the solera). Moreno et al. (2001) applied a single regression model to some winemaking variables with a view to discriminating scales in a biological system, but chose the compounds to be monitored on an individual basis.

Recently, Berlanga et al. (2004b), using discriminant analysis, established differences among wine samples from the same *criaderas y solera* system.

Dispersion of wine samples was greater at an early stage (fourth and third criadera) by effect of the increased heterogeneity of the wine and the also increased physiological activity of the yeasts (Fig. 3B.3). Dispersion decreased from one scale to the next; also, the first criadera was indistinguishable from the solera. Because dispersion decreases from one scale to the next, the biological aging system efficiently converts heterogeneous wines from different vintages into homogeneous wine in the *solera*.

3B.2 Flor Film

3B.2.1 Microbiota in the Flor Film

Flor microbiota was first studied by Pasteur in 1875 (Charpentier et al. 2000), who called it *Mycoderma vini*, in Jura wines. Microbiological analyses of flor films in Jerez and Montilla-Moriles wines have revealed a high variability in microbial population between wines from different cellars. However, more than 95% of the film microbiota usually consists of film-forming *Saccharomyces cerevisiae* races (Martínez et al. 1997a). The remainders are occasional microbes such as bacteria and other, non-flor yeasts which are regarded as contaminants.

Presence and role of bacteria in the biological aging process have been not studied very much. These may be involved in the changes of different wine components as organic acids (Suárez-Lepe and Iñigo-Leal 2004).

The other yeasts found in flor films include species of the genera *Debaryomyces*, *Zygosaccharomyces*, *Pichia*, *Hansenula* and *Candida* (Benítez and Codón 2005; Suárez-Lepe and Iñigo-Leal 2004). Guijo et al. (1986) also isolated *Torulaspora delbrueckii* and *Zygosaccharomyces bailii*, which were deemed contaminants, from flor films in Montilla-Moriles wines. Some authors have additionally isolated species of the genera *Dekkera* and *Brettanomyces*, which are believed to cause an abnormal acidity increase in casks containing biologically aging wines (Ibeas et al. 1996).

Possibly, flor yeasts are present at very low concentrations during alcoholic fermentation; once this completes and the wine contains 15–15.5 vol.% ethanol, their presence increases as result of the disappearance of fermentation yeasts from the medium (Esteve-Zarzoso et al. 2001). This is consistent with the results of Guijo et al. (1986), who isolated flor yeasts during the fermentation of grape musts in the Montilla-Moriles region. Flor yeasts constitute a resident population in *criadera* and *solera* casks that vary among cellars. On the other hand, *sobretablas* wine contains a spontaneous population of yeasts that may come from the grapes.

Although flor yeasts possess good fermentation ability, they differ from typically fermentative yeasts in metabolic, physiological and genetic respects (Esteve-Zarzoso et al. 2001, 2004; Budroni et al. 2005).

The restrictive conditions of the biological aging process of wine (namely low pH, presence of sulphite, high ethanol and acetaldehyde concentrations, lack of sugars and low oxygen concentration) are compatible with only a few *S. cerevisiae* races. Also, the presence of specific flor races has been correlated with the aging stage of the wine and the sensory features of the end product (Mesa et al. 2000).

In Jerez, researchers have isolated four *S. cerevisiae* races or varieties: two major races (*beticus* and *montuliensis*) and two minor ones (*cheresiensis* and *rouxii*). The race *beticus* is faster in forming a flor film; also, it is more abundant in the scales holding the younger wines. On the other hand, the race *montuliensis* prevails in the scales containing the oldest wine by virtue of its increased tolerance and production of acetaldehyde (Martínez et al. 1997a).

In Montilla-Moriles, Guijo et al. (1986) isolated five film-forming *S. cerevisiae* races. Worthy of special note among them is one which forms a thick, rough film and exhibits decreased production of acetaldehyde, and another which forms a thin film and produces large amounts of this aldehyde (Mauricio et al. 1997).

Jura yellow wines have also been found to contain the races *beticus* (67%), *montuliensis* (26%) and *cheresiensis* (7%) (Charpentier et al. 2000).

Molecular techniques have enabled the identification of up to six types of flor yeasts according to electrophoretic karyotype and mitochondrial DNA restriction patterns (Ibeas et al. 1997). However, yeast types distribute in a non-uniform manner across scales. One type prevails throughout the system and coexists with other yeast types in the younger scales–with which it might be closely related judging by the similarity of their chromosomal and mitochondrial DNA patterns. As aging progresses, yeast variability decreases through selection of the specific strains best responding to the increasingly stringent conditions in the aged wine. The solera casks are those most clearly reflecting the evolution towards the best adapted strains (Infante 2002). This trend has also been observed in Jura yellow wines, where only one of the six karyotypes identified at the beginning of the aging process remained after five years (Charpentier et al. 2000).

3B.2.2 Factors Influencing the Formation of Flor Films

The way the flor film forms, and its thickness, appearance and colour, depend on a variety of factors, mainly for the race *S. cerevisiae*. Especially influential among the cellar environmental conditions are temperature and moisture. The acceptable temperature range for this yeast race is 15–20°C (Marcilla et al. 1936) and the optimum value 15–17 °C. Ibeas et al. (1997) found a temperature of 22.5 °C to be the threshold above which the frequency of respiratory mutants (rho^-), which are unable to form a flor film, rises to a substantial extent. The relative humidity should be higher than 70%.

One other major influence on film development is that of the oxygen concentration, which affects the rate with which the film is formed and its thickness (Mauricio et al. 2001).

Formation of the flor film is also affected by the ethanol content of the wine, the optimum value for which is 14.5–15.5 vol.%; in fact, the film rarely forms above a 16.5 vol.% content.

The influence of residual sugars on film formation is unclear. Thus, some authors have deemed their presence indispensable for the film to form and calculated that aging wine under flor yeasts should contain at least a 1–1.6 g/L concentration. Other authors, however, claim that these concentration levels have no effect on film formation (Suárez-Lepe and Iñigo-Leal 2004).

The nitrogen requirements of flor yeasts are supplied largely by L-proline, which they metabolize under aerobic conditions as the biological aging of wines. L-Proline is converted into glutamic acid, which favours growth and persistence of the flor film (Botella et al. 1990).

Other factors seemingly influencing the formation of a flor film include the presence of phenol compounds (Budroni et al. 2005), biotin (Bravo-Abad 1986) and pantothenate (Suárez-Lepe and Iñigo-Leal 2004).

3B.2.3 Formation of the Flor Film

The formation of a flor film is the adaptive response of the yeasts to the extreme conditions prevailing in the medium, which include high concentrations of ethanol and acetaldehyde, oxidative stress resulting from the metabolization of unfermentable carbon sources, and nitrogen sources of assimilation difficult as L-proline. The adaptive mechanism involves changes in cell size, shape and hydrophobicity, all of which can reduce the density of a flor yeast population below that of wine and float in it.

Some authors have found the transition from a fermentative status to a film forming status in the yeasts to occur simultaneously with an increased saturation in long-chained fatty acids and, especially, an increased C18:1/C18:0 ratio (Aguilera et al. 1997; Farris et al. 1993; Valero et al. 2002); this probably increases the tolerance to ethanol of the yeasts (Aguilera et al. 2006) and their hydrophobicity, thereby facilitating flotation of cells and the formation of a stable film.

The hydrophobicity of yeasts is a result of their producing hydrophobic proteins; in fact, treating yeast cells with proteinases K, I and VI inhibits formation of the biofilm (Martínez et al. 1997b).

The molecular base of the film formation has recently been examined by some authors (Fidalgo et al. 2006; Ishigami et al. 2004, 2006; Purevdorj-Gage et al. 2007; Verstrepen and Klis 2006). The molecular mechanism behind it remains obscure owing to its extreme complexity. The film is formed by effect of flor yeasts growing on ethanol as their aerobic carbon source and its development is repressed by glucose. The primary factor for the biofilm formation is *FLO11*, which encodes a hydrophobic glycoprotein in cell walls. *FLO11* is required for the air-liquid interfacial biofilm to form and also for the biofilm cells to reach a buoyant density greater than that of the suspending medium (Zara et al. 2005). Ishigami et al. (2006) have shown *NRG1* to be closely related with *FLO11* expression in flor yeasts and to facilitate formation of the biofilm. *NRG1* expression makes cell surfaces highly hydrophobic. There is thus abundant evidence that the formation of a biofilm by flor yeasts relies heavily on an increased hydrophobicity and that such hydrophobicity allows the yeasts to float by effect of surface tension; this provides an adaptive mechanism for gaining direct access to oxygen in oxygen-deficient liquid environments (Alexander et al. 1998; Fidalgo et al. 2006; Martínez et al. 1997c).

3B.2.4 Genetic Characteristics of Flor Yeasts

Flor yeasts exhibit high variability in terms of nuclear and mitochondrial genome (Ibeas and Jiménez 1996; Infante et al. 2003; Martínez et al. 1995; Mesa et al. 1999). Genetic studies have exposed many aneuploidies and a surprisingly large number of additional copies of some chromosomes, particular chromosome XIII, which contains the genes that code alcohol and aldehyde dehydrogenases, which are related with the metabolism of ethanol, acetaldehyde and acetic acid (Blandino et al. 1997; Fernández et al. 1972; Guijo et al. 1997; Mauricio et al. 1997).

Aneuploidies and chromosomal rearrangements allow flor yeasts to withstand the conditions prevailing in aging wines and play a central role in sexual isolation (Jiménez and Benítez 1987; Sancho et al. 1986;). Sexual isolation prevents the random distribution of favourable features; hence most flor strains sporulate very little, if any, and the resulting meiotic products are frequently inviable (Martínez et al. 1995). Intergenic region restriction analyses (ITS internal transcribed spacers) of 5.8S rRNA gene have revealed the presence of a 24 bp deletion in *S. cerevisiae* flor strains that may be useful with a view to their identification and authentication (Fernández-Espinar et al. 2000; Esteve-Zarzoso et al. 2004).

Aranda et al. (2002) examined the stress resistance of flor yeasts and found races isolated from wine in the solera row to be more resistant to acetaldehyde and ethanol than those in the other scales. Also, they found the resistance to stress from these compounds to be correlated with high expression levels for *HSP* genes.

3B.3 Chemistry and Biochemistry of Biological Aging

The biological aging process involves various changes in wine composition. Such changes are essentially the result of the metabolism of flor yeasts and, to a lesser extent, of other phenomena common to all types of aging processes including crystal precipitation, chemical reactions between wine components and extraction of substances from cask wood.

The formation of a yeast biofilm is largely associated with the presence of the nutrients required for the yeasts to grow. The major sources of carbon for this process are ethanol, glycerol and acetic acid, and those of nitrogen ammonium ion and amino acids (particularly L-proline). Oxygen is required for the synthesis of unsaturated fatty acids and sterols (Mauricio et al. 1991), and also for the assimilation of L-proline (Mauricio et al. 2001) (see Fig. 3B.4).

Flor yeasts can undergo autolysis during biological aging of wine (Charpentier et al. 2004). This biochemical process occurs under special conditions such as those in wines stored and aged in contact with yeasts for months or even years (e.g. in the aging of sparkling wines, white wines on lees and biological aging of wines). Yeast autolysis releases a variety of compounds into the wine (amino acids, peptides, nucleotides, manoproteins, esters, alcohols, aldehydes, acids and lactones) all of which influence its sensory properties (Charpentier and Feuillat 1993; Martínez-Rodríguez and Polo 2000).

Metabolite concentrations in wine depend on the particular aging conditions, number of rows in the *criaderas and solera* system, number of *rocios* and volume extracted for bottles per year, ratio area of the flor film to volume of wine, climatic conditions of the cellar (temperature and relative humidity) and alcoholic concentration, in addition to the particular flor yeasts present. Below are described the most common changes observed during the biological aging of wine, whether related to yeast metabolism or otherwise.

3B.3.1 Ethanol

Yeasts use ethanol as a source of carbon and energy. A fraction of the alcohol is converted into other compounds including acetaldehyde, acetic acid, butanediol,

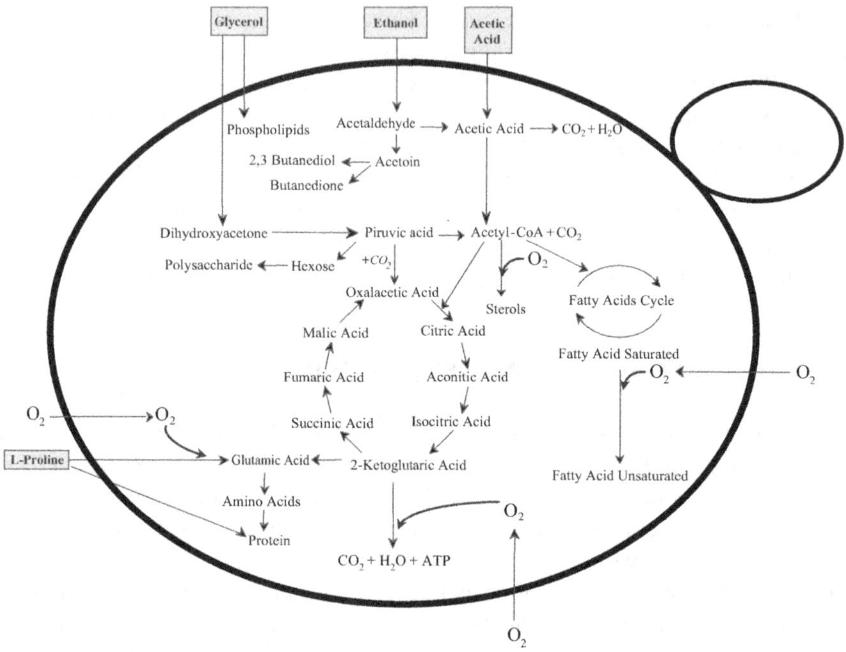

Fig. 3B.4 General scheme of the flor yeast metabolism

diacetyl and acetoin, and the remainder is metabolized via the tricarboxylic acid pathway for incorporation as carbohydrates, lipids and proteins into cellular material (Suárez-Lepe and Iñigo-Leal 2004). The ethanol uptake is governed by various factors such as the dominant yeast race in the flor film, film formation stage and aging stage. Thus, ethanol is consumed in increased amounts at the film formation stage, and also in the scales containing the youngest wine (Martínez et al. 1993, 1998). Usually, up to 1 vol.% ethanol can be used during the aging process - nearly 4 vol.% has been lost in some laboratory tests.

3B.3.2 Glycerol

Glycerol is the third most abundant component of wine after water and ethanol. Yeasts use it as a carbon source and reduce its concentration with time. This has prompted the use of the glycerol content of wine as a measure of its degree of aging (Cortés et al. 1998; Moreno et al. 2001).

3B.3.3 Acetaldehyde

Acetaldehyde is one of the most special compounds in the biological aging process and influences the sensory properties of wines by introducing a typical pungent aroma. This compound typically reaches concentrations from 350 to 450 mg/L and occasionally as a high as 1000 mg/L (Martínez et al. 1998). Acetaldehyde results mainly from the oxidation of ethanol by alcohol dehydrogenase II (ADH II), which produces NADH. This isoenzyme is repressed by glucose, so it plays an active role in the biological aging of wine. Flor yeasts races with a high specific activity of ADH II in pure cultures have been found to release large amounts of acetaldehyde into wine (Mauricio et al. 1997). The acetaldehyde content increases as the wine ages; however, the most marked changes occur in the scales containing the youngest wine, where yeast metabolism is more intense (Berlanga et al. 2004b; García-Maiquez 1988; Martínez de la Ossa et al. 1987).

Acetaldehyde takes part in a number of reactions during the biological aging of wine one of the most important of which is the formation of 1,1-diethoxyethane. This acetal results from the combination of acetaldehyde and ethanol. Its concentration, which is closely related to those of the parent compounds, can easily exceed 100 mg/L (Muñoz et al. 2005). Also, 1,1-diethoxyethane is the only acetal contributing to the aroma of wine (Etiévant 1991).

Acetaldehyde is also involved in the formation of acetoin and 2,3-butanediol. Although some authors have stressed the significance of a chemical pathway, these compounds are more likely to originate from yeast metabolism (Romano and Suzzi 1996).

Acetaldehyde also takes part in the synthesis of sotolon (Guichard et al. 1997; Pham et al. 1995).

Other reactions involving this aldehyde include the following: (1) combination with sulphite ion, which substantially increases the proportion of bound sulphite in wine; (2) formation of addition compounds with some polyphenols such as tannins and procyanidins, where it acts as a "bridging molecule" (Haslam and Lilley 1998); and (3) chemical oxidation to acetic acid, which only occurs to a small extent and has little influence on wine composition and quality.

3B.3.4 Nitrogen Compounds

Biological aging reduces the contents in amino acids, ammonium ion and urea of wine (Mauricio and Ortega 1997). Amino acids constitute the main source of nitrogen for yeasts; especially important in this respect is L-proline (Botella et al. 1990; Mauricio and Ortega 1997; Valero et al. 2003), which is not degraded by the yeasts during fermentation since this requires the presence of molecular oxygen (Ingledew et al. 1987). Its transformation into glutamic acid allows flor yeasts to synthesize all other amino acids they require to grow (Cooper 1982; Botella et al. 1990).

Flor yeast may be able to use amino acids not only as nitrogen source but also as redox agent to balance the oxidation-reduction potential under conditions of restricted oxygen. Thus, amino acids as threonine, methionine, cysteine, tryptophan, and proline can be released to the wine to restore the intracellular redox balance by means of the oxidation of NADH in excess (Berlanga et al. 2006; Mauricio et al. 2001; Moreno-Arribas and Polo 2005; Valero et al. 2003).

3B.3.5 Organic Acids

Malolactic fermentation usually occurs in *sobretablas* wine; as a result, the wine incorporated into the aging system contains no appreciable concentrations of malic acid. The decrease in tartaric acid contents during biological aging of the wine is a result of crystal precipitations.

Gluconic acid is widely used as a measure of rotting. Wines containing concentrations of this acid below 1 g/L are suitable for biological aging. Flor yeasts metabolize such acid during biological aging without altering the quality of the final wine (Peinado et al. 2003, 2006a).

Acetic acid is produced at concentrations rarely exceeding 0.7 g/L by yeasts during fermentation. The acid is metabolized by flor yeasts during the biological aging process, its concentration being reduced through consumption via acetyl-CoA for incorporation into the Krebs cycle or fatty acid synthesis.

Other fatty acids result from the metabolism of flor yeasts during the biological aging of wine. Butanoic, isobutanoic and 2- and 3-methylbutanoic acids increase its content with the aging time. Isobutanoic acid reaches the highest concentrations (8 mg/L), and those of the other acids range from 2 to 5 mg/L. Medium-chain fatty acids (hexanoic, octanoic and decanoic) exhibit the opposite behaviour and can even completely disappear from the medium after prolonged aging (Cortés 2002).

3B.3.6 Higher Alcohols and Ester

Higher alcohols are very important contributors to the aroma of fino wines. Although their overall content changes little during the biological aging, some individual alcohols exhibit marked changes. Thus, the contents in isobutanol, 2-phenylethanol and isoamyl alcohols (2-methyl-1-butanol and 3-methyl-butanol) increase only slightly with time; by contrast, that in propanol can be doubled during the process (Moreno 2005).

The production of higher alcohols from their corresponding amino acids is coupled to the oxidation of NADH; therefore, yeasts use them as alternative electron acceptors in the absence of oxygen. These alcohols are largely produced in the fourth, third and second *criadera*, coinciding with the periods of greatest activity of flor yeasts.

Yeast autolysis has a strong effect on the contents in higher alcohols, as shown by the presence of propanol, isobutanol and isoamyl alcohols in yeast extracts from wines under biological aging (Muñoz et al. 2002).

Ester concentrations change by effect of synthetic and hydrolysis reactions, and also as a result of enzymatic activity of the flor yeasts. The effect of the latter depends on the specific yeast race prevailing in the medium and its physiological status (Mauricio et al. 1993; Plata et al. 1998).

Broadly speaking, the contents in acetates of higher alcohols decreases through hydrolysis over the first few months of aging; by contrast, those in ethyl esters of organic acids (particularly lactic and succinic) increase with time (Usseglio-Tomasset 1983; Martínez de la Ossa et al. 1987).

3B.3.7 Polyphenols

Flor yeasts have the ability to protect the wine from atmospheric air and retain the compounds that lead to browning (Barón et al. 1997; Mérida et al. 2005). The total polyphenol value is low and decreases with time during biological aging.

3B.3.8 Lactones

Wines aged under a flor film typically contain various compounds especially important among which are lactones. Thus, solerone (4-acetyl-γ-butyrolactone) abounds in *fino* and Jura wines and has for years been deem one of the greatest contributors to their aroma profiles (Augustyn et al. 1971; Muller et al. 1973). However, subsequent studies have suggested that this lactone has no impact on wine aroma (Martin and Etievant 1991).

Sotolon (3-hydroxy-4,5-dimethylfuranone) possesses a strong, typical odour reminiscent of nut and curry with a very low perception threshold (10 μg/L) (Pham et al. 1995; Da Silva Ferreira et al. 2002). This lactone has been detected in Jura (Dubois et al. 1976) and *fino* wines (Moreno 2005), and also in some botrytized white wines (Guichard et al. 1997; Masuda et al. 1984). Sotolon is formed by aldolization of α-ketobutyric acid and acetaldehyde (Guichard et al. 1997; Pham et al. 1995). α-Ketobutyric acid results from the deamination of L-threonine by flor yeasts. Some authors have associated it with yeast autolysis (Pham et al. 1995). Its final content in wine depends on the aging time, but usually amounts to levels in the region of 200 μg/L (Dubois et al. 1976; Guichard et al. 1997; Moreno et al. 2004)

Other lactones detected include α-butyrolactone and pantonolactone (2,4-dihydroxy-3,3-dimethylbutyrolactone), which is deemed typical of sherry wines. Both have been associated with specific yeast races (Zea et al. 1995) and the aging time (Cortés 2002).

Wines are biologically aged in old casks which, unlike those used for vintage aging, are rarely emptied or scrubbed. As a result, biologically aged wines contain minimal amounts of lactones from wood such as β-methyl-γ-octalactone (Chatonnet et al. 1990) and only in the scales holding the oldest wine.

3B.4 Sensory Properties

In Montilla-Mories and Jerez winemaking regions there are fundamentally three different types of white wines, namely *fino*, *amontillado* and *oloroso*, which are obtained by using the *criaderas and solera* system.

Fino wines are only produced by biological aging. The aerobic metabolism of flor yeasts causes changes in the aroma fraction that endows them wine with their typical flavour. In addition, they protect these wines against browning and allow them to retain their pale colour for years.

Oloroso wines are only obtained by oxidative aging, which is accomplished by fortifying the initial wine to an ethanol content of about 18 vol.% in order to prevent growth of flor yeasts. Under such conditions, *oloroso* wine acquires a dark colour by effect of the oxidation of phenol compounds.

Amontillado wines are obtained in a two-step process involving biological aging under the same conditions to those of *fino* wines, followed by fortification and oxidative aging as in *oloroso* wines. *Amontillado* wines are thus the oldest and most valued of the three sherry types by virtue of their acquiring a more complex flavour than the other two.

Differences in aroma between biologically and oxidatively aged wines become even more marked when the wines are compared in terms of sensory properties. Thus, *fino* wines are clear, bright and light in the glass, and possess a straw yellow colour with olive-like hues. Also, they are pungent, almondy in the nose, and are easily recognized by their yeast or bread odour. *Oloroso* wines are dark mahogany in colour and possess an aroma of balsamic, freshly cracked nut with spicy notes. Because they are subjected to two types of aging (biological and oxidative), *amontillado* wines are very complex in terms of sensory properties. Thus, they possess and amber, yellowish brown colour, and an aroma of dried fruits such as hazel nut with strong balsamic and spicy notes.

Some authors (Peinado et al. 2004a, 2004b; Zea et al. 2007) have grouped aroma compounds in aromatic series based on their aroma descriptors. The overall value for each series is obtained by combining the odour activity value (OAV, which is defined as the ratio of the concentration of a compound to its perception thresholds) of the individual aroma compounds in the series. The combination of individual OAVs in order to calculate the overall value for a series cannot be interpreted as an arithmetic addition of odorant sensations. Some aspects of this classifying scheme can be subject to criticism, but in any case, the scheme is effective for comparing wines obtained with different aging methods inasmuch as the odour series always comprise the same compounds.

Figure 3B.5 shows the odour profiles of three wines as obtained using the OAV classification scheme. The example illustrates the differences between three types of wine in terms of volatile components, which have been classified into seven aroma series.

3B.5 Acceleration of Biological Aging

As noted earlier, biological aging is necessarily a slow process as it involves storing wine in vast cellars over long periods of time and performing many operations, such as several rocios per year to ensure homogeneity and quality of the wine, and control analyses. This substantially raises the production costs of sherry wines,

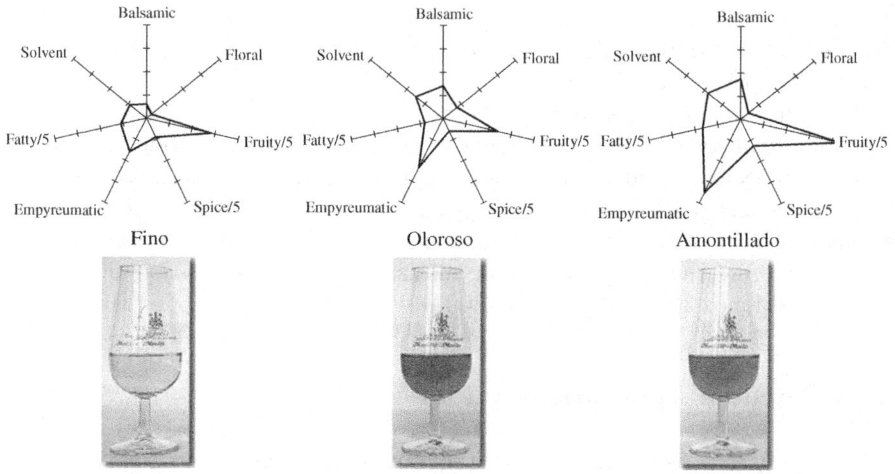

Fig. 3B.5 Odour profiles of fino, amontillado and oloroso wines. For a better visualization the values of the fatty, fruity and spice series are divided by 5

which has led some authors to seek effective ways of shortening the aging time without altering the quality of the resulting wines. Some procedures (Ough and Amerine 1958) involve aerating or agitating the wine in order to accelerate yeast aerobic metabolism without the need for yeasts to form a film on its surface (i.e. submerged cultures). The resulting wines have sensory properties that differ from those typical of *fino* wine and are used for blending with baked sherry. Other procedures use a tray system in order to increase the surface area/volume ratio of the wine in order to facilitate oxygenation and the development of flor yeasts; this procedure, however, entails individually processing each tray and produces large amounts of biomass which impair the quality of the end product. Rankine (1955) proposed the use of yeasts packed into oak chips in batch production processes. Ough and Amerine (1972) sought to make the aging process faster by pumping wine to the top of a tank in order to facilitate aeration and then dropping it on the surface. This breaks the flor film, which makes this method inadvisable in practice.

Recently, periodic microaeration of *fino* wines was found to shorten substantially their biological aging time (Muñoz et al. 2005). The procedure preserves the integrity of the flor film. Its effects were examined in a study involving the microaeration of wines aged for variable lengths of time in the presence of two different yeasts races (namely *S. cerevisiae* var. *capensis* G1 and *S. cerevisiae* F12). The best results were those for the wines which were previously subjected to biological aging. Also, *S. cerevisiae* G1 proved more effective than *S. cerevisiae* F12 for the intended purpose.

Aerating wine under biological aging on a monthly basis has been found to inhibit partially ADH I and ADH II enzyme activity immediately upon aeration. However, once all oxygen has been used by the yeasts and become a limiting factor again, ADH II activity rises, possibly as a result of the need to alleviate the excess of NAD^+ accumulating by effect of oxidative metabolism and maintain the redox balance. The increased ADH II activity reflects in increased production of acetaldehyde and its derivatives (Berlanga et al. 2001; Cortés et al. 1999; Mauricio and Ortega 1997; Mauricio et al. 2003).

The content in higher alcohols of wine is altered by periodic microaeration; however, their final contents have been found to be similar to those previously obtained in wines subjected to traditional biological aging procedures (Cortés 2002) The differences between traditionally produced wines and wines obtained by microaeration in stainless-steel vessels are mainly due to the compounds extracted from the cask wood. Therefore, the aging time can be substantially shortened by combining both processes (Muñoz 2003; Muñoz et al. 2007).

3B.6 Potential Applications of Flor Yeasts

Because they can grow in such a hostile environment, flor yeasts are highly suitable for stress-related physiological and genetic studies (Aranda and del Olmo 2003; Aranda et al. 2002). Flor yeasts are highly tolerant to ethanol and acetaldehyde; they also possess a high fermentation capacity, so they could be useful for producing bioethanol. A recent study on ethanol tolerance revealed that the increased tolerance of the flor yeast *S. cerevisiae* var. *capensis* G1 might be a result of an increased ergosterol content in its plasma membrane (Aguilera et al. 2006).

The usefulness of flor yeasts for the biological aging of red wines has also been examined (Suárez-Lepe and Iñigo-Leal 2004). The use of flor films on wine over short periods may provide a new red wine production technology facilitating the obtainment of a more uniform colour by formation of pyranoanthocyanin and polymeric pigments. Aging times under a flor film must necessarily be short in order to prevent excessive production of acetaldehyde and unduly altering the olfactive profile of the red wine as a result (Morata et al. 2007).

One other potential use of flor yeasts is in correcting and preventing browning in white wines or other types of drinks (Barón et al. 1997; Mérida et al. 2005).

Future insight into the role of gene *FLO11* in flor yeasts may be highly useful with a view to advancing cell immobilization technology (Purevdorj-Gage et al. 2007). So far, one *S. cerevisiae* var. *capensis* race has been successfully co-immobilized with *Penicillium chrysogenum* in order to obtain biocapsules for potential use in a number of fermentation processes (Peinado et al. 2006b).

Acknowledgments We would like to acknowledge the financial support from the Ministry of Science and Technology (grant AGL2005-01232/ALI) of the Spanish Government, cofinanced with the European Fund for the Regional Development (FEDER) of the European Union.

References

Aguilera, F., Valero, E., Millán, C., Mauricio, J.C., & Ortega, J.M. (1997). Cellular fatty acid composition of two physiological races of *Saccharomyces cerevisiae* during fermentation and flor veil formation in biological aging of fine wine. *Belgian J. Brew. Biotechnol.*, 2, 39–42.

Aguilera, F., Peinado, R.A., Millán, C., Ortega, J.M., & Mauricio, J.C. (2006). Relationship between ethanol tolerance, H^+-ATPase activity and the lipid composition of the plasma membrane in different wine yeast strains. *Int. J. Food Microbiol., 110*, 34–42.

Alexander, H., Bertrand, F., & Charpentier, C. (1998). Ethanol induced yeast film formation with cell surface hydrophobicity as a major determinant. *Food Technol. Biotechnol., 36*, 27–30.

Aranda, A., & del Olmo, M. (2003). Response to acetaldehyde stress in the yeast *Saccharomyces cerevisiae* involves a strain-dependent regulation of several *ALD* genes and is mediated by the general stress response pathway. *Yeast, 20*, 747–759.

Aranda, A., Querol, A., & del Olmo, M. (2002). Correlation between acetaldehyde and ethanol resistance and expression of *HSP* genes in yeast strains isolated during the biological aging of sherry wines. *Arch. Microbiol., 177*, 304–312.

Augustyn, O.P.H., Van Wyk, C.J., Muller, C.J., Kepner, R.E., & Webb, A.D. (1971). The structure of solerone (5-acetyldihydro-2-(3H)furanone) a substituted gamma-lactone involved in wine aroma. *J. Agric. Food Chem., 19*, 1128–1130.

Barón, R., Mayén, M., Mérida, J., & Medina, M. (1997) Changes in phenolic compounds and browning during biological aging of sherry-type wine. *J. Agric. Food Chem., 45*, 1682–1685.

Benítez, T., & Codón, A.C. (2005). Levaduras. *Saccharomyces cerevisiae* III. Levaduras de vinos de crianza biológica. In A.V. Carrascosa, R. Muñoz & R. González (Eds.), *Microbiología del vino* (pp. 78–113). Madrid, Spain: A.M.V. Ediciones.

Berlanga, T. M., Atanasio, C., Mauricio, J. C., & Ortega, J. M. (2001). Influence of aeration on the physiological activity of flor yeasts. *J. Agric. Food Chem., 49*, 3378–3384.

Berlanga, T., Peinado, R., Millán, C., Mauricio, J. C., & Ortega, J. M. (2004a). Influence of blending on the content of different compounds in the biological aging of sherry wines. *J. Agric. Food Chem., 52*, 2577–2581.

Berlanga, T., Peinado, R., Millán, C., & Ortega, J.M. (2004b). Discriminant analysis of sherry wine during biological aging. *Am. J. Enol. Vitic., 55*, 407–411.

Berlanga, T.M., Millán, C., Mauricio, J.C., & Ortega, J.M. (2006). Influence of nitrogen on biological aging of sherry wine. *J. Sci. Food Agric., 86*, 2113–2118.

Blandino, A., Caro, I., & Cantero, D. (1997). Comparative study of alcohol dehydrogenase activity in flor yeast extracts. *Biotechnol. Lett., 19*, 651–654.

Botella, M.A., Pérez-Rodriguez, L., Domecq, B., & Valpuesta, V. (1990). Amino acid content of fino and oloroso sherry wines. *Am. J. Enol. Vitic., 41*, 12–15.

Bravo-Abad, F. (1986). Crianza biológica de vino: procedimiento tradicional de vinos finos de D.O. Jerez y D.O. Montilla-Moriles. *Enología y Enotecnia, 1*, 15–19.

Budroni, M., Zara, S., Zara, G., Pirino, G., & Mannazzu, I. (2005). Peculiarities of flor strains adapted to Sardinian sherry-like wine ageing conditions. *FEMS Yeast Res., 5*, 951–958.

Charpentier, C., & Feuillat, M. (1993). Yeast autolysis. In G.H. Fleet (Ed.), *Wine Microbiology and Biotechnology* (pp. 225–242). Langhorne: Harwood Academic Publisher.

Charpentier, C., Etiévant, P., & Guichard, E. (2000). Vinificación de los vinos de velo: vino Jaune, Jerez y otros. In C. Flanzy (Ed.), *Enología: Fundamentos Científicos y Tecnológicos* (pp. 531–539). Madrid, Spain: A. Madrid Vicente, Ediciones, & Ediciones Mundi-Prensa.

Charpentier, C., Dos Santos, A.M., & Feuillat M. (2004). Release of macromolecules by *Saccharomyces cerevisiae* during ageing of French flor sherry wine "Vin jaune". *Int. J. Food Microbiol., 96*, 253–262.

Chatonnet, P., Boidron, J.N., & Pons, M. (1990). Élevage des vins rouges es fût de chene: évolution de certains composés volatils et de leur impact aromatique. *Sci. Aliment., 10*, 565–578.

Cooper, T.G. (1982). Nitrogen metabolism in *Saccharomyces cerevisiae*. In J.N. Strathern, E.W. Jones & J.B. Broach (Eds.), *The Molecular Biology of the Yeast Saccharomyces. Metabolism and Gene Expression* (pp. 39–99). New York: Cold Spring Harbor Laboratory.

Cortés, M.B. (2002) Aplicación de cultivos puros de Saccharomyces cerevisiae a la crianza biológica de vinos finos de la D.O. Montilla-Moriles. (Doctoral dissertation, University of Córdoba, Spain, 2002).

Cortés, M. B., Moreno, J., Zea, L., Moyano, L., & Medina, M. (1998). Changes in aroma compounds of sherry wines during their biological aging carried out by *Saccharomyces cerevisiae* races *bayanus* and *capensis*. *J. Agric. Food Chem., 46*, 2389–2394.

Cortés, M.B., Moreno, J., Zea, L., Moyano, L., & Medina, M. (1999) Response of the aroma fraction in sherry wines subjected to accelerated biological aging. *J. Agric. Food Chem., 47*, 3297–3302.

Da Silva Ferreira, A.C., Barbe, J.C., & Bertrand, A. (2002). Heterocyclic acetals from glycerol and acetaldehyde in port wine: Evolution with aging. *J. Agric. Food Chem., 50*, 2560–2564.

Dubois, P., Rigaud, J., & Dekimpe, J. (1976). Identification de la diméthyl-4,5-terahydrofuranedione-2,3-dans le vin jaune du Jura. *Lebens. Wiss. Technol., 9, 366–368.*

Esteve-Zarzoso, B., Peris-Torán, M.J., García-Maiquez, E., Uruburu, F., & Querol, A. (2001). Yeast population dynamics during the fermentation and biological aging of sherry wines. *App. Environ. Microbiol., 67*, 2056–2061.

Esteve-Zarzoso, B., Fernández-Espinar, M.T., & Querol, A. (2004). Authentication and identification of *Saccharomyces cerevisiae* "flor" yeast races involved in sherry ageing. *Antonie van Leeuwenhoek, 85*, 151–158.

Etiévant, P.X. (1991). Wine In H. Maarse (Ed.), *Volatile compound in food and beverages* (pp. 483–546). Zeist (The Netherlands): H.TNO-CIVO. Food Analysis Institute.

Farris, G.A., Sinigaglia, M., Budroni, M., & Guerzoni, M.E., (1993). Cellular fatty acid composition in film-forming strains of two physiological races of *Saccharomyces cerevisiae. Lett. App. Microbiol., 17*, 215–219.

Fernández, M.J., Gómez-Moreno, C., & Ruiz-Amil, M. (1972). Induction of isoenzymes of alcohol dehydrogenase in "flor" yeast. *Arch. Microbiol., 84*, 153–160.

Fernández-Espinar, M.T., Esteve-Zarzoso, B., Querol, A., & Barrio, E. (2000). RFLP analysis of the ribosomal internal transcribed spacers and the 5.8S rRNA gene region of the genus *Saccharomyces*: a fast method for species identification and the differentiation of flor yeasts. *Antonie van Leeuwenhoek, 78*, 87–97.

Fidalgo, M., Barrales, R.R., Ibeas, J.I., & Jimenez, J. (2006). Adaptive evolution by mutations in the *FLO11* gene. *PNAS, 103*, 11228–11233.

García-Maiquez, E. (1988). Les levures de voile dans l'elaboration des vins de xéres. In *Application à l'oenologie des progrés récents en microbiologie et en fermentation.* (341–372). Paris, France: Commission des Communautés européennes Programme COMET.

Guichard, E., Pham, T.T., & Charpentier, C. (1997). Le sotolon, marqueur de la typicité de l'arome des vins du Jura. *Rev. Oenol., 82*, 32–34.

Guijo, S., Millán, C., & Ortega, J.M. (1986). Fermentative features of vinification and maduration yeasts isolated in the Montilla-Moriles region of Southern Spain. *Food Microbiol., 3*, 133–142.

Guijo, S., Mauricio, J.C., Salmon, J.M., & Ortega, J.M. (1997). Determination of the relative ploidy in different *Saccharomyces cerevisiae* strains used for fermentation and "flor" film ageing of dry sherry-type wines. *Yeast, 13*, 101–117.

Haslam, E., & Lilley, T.H. (1988). Natural astringency in foodstuffs. A molecular interpretation. *CRC Crit. Rev. Food Sci. Nut., 27*, 1–40.

Ibeas, J.I., & Jiménez, J. (1996). Genomic complexity and chromosomal rearrangements in wine-laboratory yeast hybrids. *Curr. Gen., 30*, 410–416.

Ibeas, J.I., Lozano, I., Perdigones, F., & Jiménez, J. (1996). Detection of *Dekkera-Brettanomyces* strains in sherry by a nested PCR method. *App. Environ. Microbiol., 62*, 998–1003.

Ibeas, J.I., Lozano, I., Perdigones, F., & Jiménez, J. (1997). Dynamics of flor yeast populations during the biological aging of sherry wines. *Am. J. Enol. Vitic., 48*, 75–79.

Infante, J.J. (2002). Aplicación de métodos de Biología Molecular para la caracterización y mejora genética de levaduras de velo de flor responsables de la crianza biológica de los vinos de Jerez. (Doctoral dissertation, University of Cadiz, Spain, 2002).

Infante, J.J., Dombek, K.M., Rebordinos, L., Cantoral, J.M., & Young, E.T. (2003) Genome-wide amplifications caused by chromosomal rearrangements play a mojor role in the adaptive evolution of natural yeast. *Genetics, 165*, 1745–1759.

Ingledew, W.M., Magnus, C.A., & Sosulski, F.W. (1987). Influence of oxygen on proline utilization during the wine fermentation. *Am. J. Enol. Vitic., 38*, 246–248.

Ishigami, M., Nakagawa, Y., Hayakawa, M., & Iimura, Y. (2004). *FLO11* is essential for flor formation caused by the C-terminal deletion of *NRG1* in *Saccharomyces cerevisiae*. *FEMS Microbiol. Lett., 237*, 425–430.

Ishigami, M., Nakagawa, Y., Hayakawa, M., & Iimura, Y. (2006). *FLO11* is the primary factor in flor formation caused by cell surface hydrophobicity in wild-type flor yeast. *Biosci. Biotechnol. Biochem., 70*, 660–666.

Jiménez, J., & Benítez, T. (1987). Genetic analysis of highly ethanol-tolerant wine yeasts. *Curr. Gen., 12*, 421–428.

Marcilla, J, Feduchy, E., & Alas, G. (1936). Contribución al estudio de las levaduras que forman velo sobre ciertos vinos de elevado grado alcohólico. *Anales del Centro de Investigaciones Vinícolas de Madrid, 1*, 1–219.

Martin, B., & Etievant, P.X. (1991). Quantitative determination of solerone and sotolona in flor sherries by two dimensional-capillary GC. *HRC CC., 14*, 133–135.

Martínez, P., Valcárcel, M.J., González, P., Benítez, T., & Pérez, L. (1993). Consumo de etanol, glicerina y aminoácidos totales en vinos finos durante la crianza biológica bajo "velo de flor". *Aliment. Equipos Tecnol. 12*, 61–65.

Martínez, P., Codón, A.C., Pérez, L., & Benítez, T. (1995). Physiological and molecular characterization of flor yeasts: polymorphism of flor yeast populations. *Yeast, 11*, 1399–1411.

Martínez, P., Pérez Rodríguez, L., & Benítez, T. (1997a). Evolution of flor yeast population during the biological aging of fino sherry wine. *Am. J. Enol. Vitic., 48*, 160–168.

Martínez, P., Pérez Rodríguez, L., & Benítez, T. (1997b). Factors which affect velum formation by flor yeasts isolated from sherry wine. *Syst. App. Microbiol., 20*, 154–157.

Martínez, P., Pérez Rodríguez, L., & Benítez, T. (1997c). Velum formation by flor yeasts isolated from sherry wine. *Am. J. Enol. Vitic., 48*, 55–62.

Martínez, P., Valcárcel, M.J., Pérez, L., & Benítez, T. (1998). Metabolism of *Saccharomyces cerevisiae* flor yeasts during fermentation and biological aging of fino sherry: by-products and aroma compounds. *Am. J. Enol. Vitic., 49*, 240–250.

Martínez de la Ossa, E., Caro, I., Bonat, M., Pérez, L., & Domecq, B. (1987). Dry extract in sherry and its evolution in the aging of sherry. *Am. J. Enol. Vitic., 38*, 321–325.

Martínez-Rodríguez, A. J., & Polo, M. C. (2000). Enological aspects of yeast autolysis. In *Recent Res. Devel. Microbiology, 4*, (pp. 285–301). Trivandrum, India: Research Signpost.

Masuda, M., Okawa, E.C., Nishimura, K.I., & Yunome, H. (1984). Identification of 4,5-dimetyl-3-hydroxy-2(5H)-furanone (sotolona) and ethyl 9-hydroxy-nonanone in botrytised wine and evaluation of the roles of compounds characteristic of it. *J. Agric. Food Chem., 48*, 2702–2710.

Mauricio, J.C., & Ortega, J.M. (1997). Nitrogen compounds in wine during its biological aging by two flor film yeasts: an approach to accelerated biological aging of dry sherry–type wines. *Biotechnol. Bioeng., 53*, 159–167.

Mauricio, J.C., Guijo, S., & Ortega, J.M. (1991). Relationship between phospholipid and sterol contents in *Saccharomyces cerevisiae* and *Torulaspora delbrueckii* and their fermentation activity in grape musts. *Am. J. Enol. Vitic., 42*, 301–308.

Mauricio, J.C., Moreno, J.J., Valero, E.M., Zea, L., Medina, M., & Ortega, J.M. (1993). Ester formation and specific activities of *in vitro* alcohol acetyltransferase and esterase by *Saccharomyces cerevisiae* during grape must fermentation.*J. Agric. Food Chem., 41*, 2086–2091.

Mauricio, J.C., Moreno, J., & Ortega, J.M. (1997). *In vitro* specific activities of alcohol and aldehyde dehydrogenases from two flor yeasts during controlled wine aging. *J. Agric. Food Chem., 45*, 1967–1971.

Mauricio, J.C., Valero, E. Millán, C., & Ortega, J.M. (2001). Changes in nitrogen compounds in must and wine during fermentation and biological aging by flor yeasts. *J. Agric. Food Chem., 49*, 3310–3315.

Mauricio, J.C., Moreno, J., & Ortega, J.M. (2003). Aceleración de la crianza biológica mediante aireaciones periódicas. In *VII Jornadas Científicas de los Grupos de Investigación Enológica* (pp. 117–119). Logroño, Spain: Gobierno de la Rioja.

Mérida, J., Lopez-Toledano, A., Marquez, T., Millan, C., Ortega, J.M., & Medina, M. (2005). Retention of browning compounds by yeasts involved in the winemaking of sherry type wines. *Biotechnol. Lett., 27*, 1565–1570.

Mesa, J.J., Infante, J.J., Rebordinos, L., & Cantoral, J.M. (1999). Characterization of yeasts involved in the biological ageing of sherry wines. *Lebensm. Wiss. Technol., 32*, 114–120.

Mesa, J.J., Infante, J.J., Rebordinos, L., Sanchez, J.A., & Cantoral, J.M. (2000). Influence of the yeast genotypes on enological characteristics of sherry wines. *Am. J. Enol. Vitic., 51*, 15–21.

Morata, A., Calderón, F., Gómez-Cordovés, M.C., González, M.C., Colomo, B., & Suárez-Lepe, J.A. (2007). Evolución del contenido de antocianos en mostos de Cabernet Sauvignon (*Vitis vinifera* L.) fermentados y criados con levaduras de velo. In M. Ramirez (Ed.), *Avances en Ciencias y Técnicas Enológicas* (pp. 188–189). Badajoz, Spain: Junta de Extremadura.

Moreno, J.A. (2005). Influencia del tipo de envejecimiento sobre el perfil aromático de los vinos generosos andaluces. (Doctoral dissertation, University of Córdoba, Spain, 2005).

Moreno, J.A., Zea, L., Moyano, L., Moreno, J.J., & Medina, M. (2001). Aplicación de un modelo de regresión simple a la crianza biológica de vinos finos de la Denominación de Origen Montilla-Moriles. In *VII Congreso Nacional de Enólogos*. Pamplona, Spain.

Moreno, J.A., Zea, L., Moyano, L., & Medina, M. (2004). Aroma compounds as markers of the changes in sherry wine subjected to biological ageing. *Food Control, 16*, 333–338.

Moreno-Arribas, M.V., & Polo, M. C. (2005). Winemaking biochemistry and microbiology: Current knowledge and future trends. *Crit. Rev. Food Sci. Nut., 45*, 265–286.

Muller, C.J., Kepner, R.E., & Webb, A.D. (1973). Lactones in wines – a review. *Am. J. Enol. Vitic., 24*, 47–57.

Muñoz, D. (2003). Aceleración de la crianza biológica de vinos finos mediante oxigenaciones controladas. (Doctoral dissertation, University of Córdoba, Spain, 2003)

Muñoz, D., Valero, E., Moreno, J., Moyano, L., Zea, L., & Medina, M. (2002). Estudio del aprovechamiento de la biomasa procedente de la crianza biológica de vinos finos. In *XXII Jornadas de Viticultura y Enología Tierra de Barros* (pp. 243–252). Badajoz, Spain: Cultural Santa Ana.

Muñoz, D., Peinado, R.A., Medina, M., & Moreno, J. (2005). Biological aging of sherry wines using pure cultures of two flor yeast strain under controlled microaeration. *J. Agric. Food Chem., 53*, 5258–5264.

Muñoz, D., Peinado, R.A., Medina, M., & Moreno, J. (2007). Biological aging of sherry wines under periodic and controlled microaeration with *Saccharomyces cerevisiae* var. *capensis*: Effect on odorant series.*Food Chem., 100*, 1188–1195.

Ough, C. S., & Amerine, M. A. (1958). Studies on acetaldehyde production under pressure, oxygen and agitation. *Am. J. Enol. Vitic., 9*, 111–123.

Ough, C. S., & Amerine, M. A. (1972). Further studies with submerged flor sherry. *Am. J. Enol. Vitic., 23*, 128–131.

Peinado, R.A., Moreno, J.J., Ortega, J.M., & Mauricio, J.C. (2003). Effect of gluconic acid consumption during simulation of biological aging of sherry wines by a flor yeast strain on the final volatile compounds. *J. Agric. Food Chem., 51*, 6198–6203.

Peinado, R.A., Moreno, J., Bueno, J.E., Moreno, J.A., & Mauricio J.C. (2004a). Comparative study of aromatic compounds in two young white wines subjected to pre-fermentative cryomaceration. *Food Chem., 84*, 585–590.

Peinado, R.A., Mauricio, J.C., Medina, M., & Moreno, J. (2004b). Effect of *Schizosaccharomyces pombe* on aromatic compounds in dry sherry wines containing high level of gluconic acid. *J. Agric. Food Chem., 52*, 4529–4534.

Peinado, R.A., Mauricio, J.C., & Moreno, J. (2006a). Aromatic series in sherry wines with gluconic acid subjected to different biological aging conditions by *Saccharomyces cerevisiae* var. *capensis*. *Food Chem., 94*, 232–239.

Peinado, R.A., Moreno, J.J., Villalba, J.M., González-Reyes, J.A., Ortega, J.M., & Mauricio, J.C. (2006b). Yeast biocapsules: a new immobilization method and their applications. *Enz. Microb. Technol., 40*, 79–84.

Pérez, L. (1982). Consideraciones técnicas en la elaboración del Jerez. In *II Jornadas universitarias sobre el Jerez* (pp. 167–197). Cádiz, Spain: Universidad de Cádiz.
Pham, T.T., Guichard, E., Schlich, P., & Charpentier, C. (1995). Optimal conditions for the formation of sotolon from α-ketobutyric acid in the french "Vin Jaune". *J. Agric. Food Chem., 43*, 2616–2619.
Plata, M.C., Mauricio, J.C., Millán, C., & Ortega, J.M. (1998). *In vitro* specific activity of alcohol acetyl-transferase and esterase in two flor yeast strains during biological aging of sherry wines. *J. Ferment. Bioeng., 85*, 369–374.
Purevdorj-Gage, B., Orr, M.E., Stoodley, P., Sheehan, K.B., & Hyman, L.E. (2007). The role of *FLO11* in *Saccharomyces cerevisiae* biofilm development in a laboratory based flow-cell system. *FEMS Yeast Res., 7*, 372–379.
Rankine, B.C. (1955). Yeast cultures in Australian winemaking. *Am. J. Enol. Vitic., 6*, 11–15.
Romano, P., & Suzzi, G. (1996). Origin and production of acetoin during wine yeast fermentation. *App. Environ. Microbiol., 62*, 309–315.
Sancho, E.D., Hernandez, E., & Rodriguez-Navarro, A. (1986). Presumed sexual isolation in yeast populations during production of sherrylike wine. *App. Environ. Microbiol., 51*, 395–397.
Suárez-Lepe, J.A., & Iñigo-Leal, B. (2004). *Microbiología Enológica. Fundamentos de vinificación* (pp. 673–716). Madrid, Spain: Ediciones Mundi-Prensa.
Usseglio-Tomasset, L. (1983) Influence de la température de conservation sur les caractéristiques physico-chimiques et organoleptiques des vins (Vins aromatiques). *Bull. OIV, 623*, 19–34.
Valero, E., Millán, C., & Ortega, J.M. (2002). Changes in the lipid composition of *Saccharomyces cerevisiae* race *capensis* (G1) during alcoholic fermentation and flor film formation. *Lebensm. Wiss. Technol., 35*, 593–599.
Valero, E., Millán, C., Ortega, J.M., & Mauricio, J.C. (2003). Concentration of amino acids in wine after the end of fermentation by *Saccharomyces cerevisiae* strains. *J. Sci. Food Agric., 83*, 830–835.
Verstrepen, K.J., & Klis, F.M. (2006). Flocculation, adhesion and biofilm formation in yeasts. *Mol. Microbiol., 60*, 5–15.
Zara, S., Bakalinsky, A.T., Zara, G., Pirino, G. Demontis, M.A., & Budroni, M. (2005). *FLO11*-Based model for air-liquid interfacial biofilm formation by *Saccharomyces cerevisiae. App. Environ. Microbiol., 71*, 2934–2939.
Zea, L., Moreno, J., Ortega, J.M., Mauricio, J.C., & Medina, M. (1995) Comparative study of the γ-butyrolactone and pantolactone contents in cells and musts during vinification by three *Saccharomyces cerevisiae* races. *Biotechnol. Lett., 17*, 1351–1356.
Zea, L., Moyano, L., Moreno, J.A., & Medina, M. (2007). Aroma series as fingerprints for biological ageing in fino sherry-type wines. *J. Sci. Food Agric., 87*, 2319–2326.

Chapter 4
Enzymes in Winemaking

Maurizio Ugliano

Contents

4.1 Introduction

It is generally accepted that grape composition plays a primary role in determining the final quality of wine. However, many of the sensory characteristics that are normally used to assess the quality of a wine, including those that are considered typical of the grape variety, cannot be detected in the grapes. They develop in fact, in large part, through a complex array of biochemical reactions that take place during the winemaking process. The majority of these reactions are catalyzed by different enzymes coming form various sources, particularly grapes and microorganisms. Many sensorially-active constituents of wine are affected, at different stages of winemaking, by biochemical transformations that are catalyzed by specific enzymes. For example, enzymes are involved in the oxidation of grape phenolics, in the formation of volatile compounds during pre-fermentative operations, and in the transformation of odorless precursors into odor-active compounds during alcoholic

M. Ugliano (✉)
The Australian Wine Research Institute, PO Box 197 Glen Osmond (Adelaide), SA 5064, Australia
e-mail: maurizio.ugliano@awri.com.au

M.V. Moreno-Arribas, M.C. Polo (eds.), *Wine Chemistry and Biochemistry*,
DOI 10.1007/978-0-387-74118-5_5, © Springer Science+Business Media, LLC 2009

and malolactic fermentations. Also, the extent of certain enzymatic reactions can determine the efficiency of specific technological steps considered to be of primary importance in the modern wine industry, such as juice and wine clarification, color extraction, and protein stabilization. Therefore, understanding the role played by enzymes during winemaking can help in the development of rational and effective strategies for optimizing wine processing to modulate wine composition and sensory properties.

Grape berries and wine yeasts are the major sources of enzymes involved in the various biochemical transformations that take place during winemaking. However, typical winemaking conditions such as high sugars and ethanol concentrations, low pH and high concentrations of polyphenols, can potentially inhibit the activity of grape and microbial enzymes, often with synergistic interactions which result in enhanced inhibitory effects. For this reason, the reactions catalyzed by grape and microbial enzymes during winemaking are often incomplete, with a significant portion of the substrate remaining untransformed and therefore available for further reaction. Because many of these reactions are considered beneficial to wine quality or to the efficiency of specific technological operations, the addition of exogenous enzymes that exhibit higher efficacy under winemaking conditions is frequently carried out in the winery to obtain the desired level of substrate transformation. Exogenous enzymes are an important component of modern winemaking, and many industrial preparations are now commercially available, particularly to assist during juice and wine clarification and to increase liberation of aroma compounds from odorless precursors.

This chapter provides an updated overview of the current knowledge on the role played by enzymes in the process of transforming grapes into wine and on the possibility offered by new technologies to modify wine composition though a more effective control of enzymatic reactions.

4.2 Polyphenol Oxydases

Oxidative browning during the production of white wine is a well known phenomenon that can occur at various stages of the winemaking process. In sound grapes, enzymatic oxidation of phenolic compounds takes place in the presence of oxygen during pre-fermentative operations, due to the action of grape tyrosinase. Following grape crushing, tyrosinase is partially released from berry chloroplasts into the juice (Dubernet 1974), where it can react with cinnamic acids and their tartaric esters to form *o*-quinones (Fig. 4.1). These highly reactive species can then enter two different reactions. In one possible mechanism, quinones can condense with other phenolic compounds to form polymerized products that exhibit a more or less brown color, according to the degree of polymerization (Singleton 1987). Alternatively, quinones can react with reductive species like glutathione, a tripeptide commonly found in grapes and containing an –SH group. This reaction forms a colorless derivative that is stable to tyrosinase action, and therefore does not result in

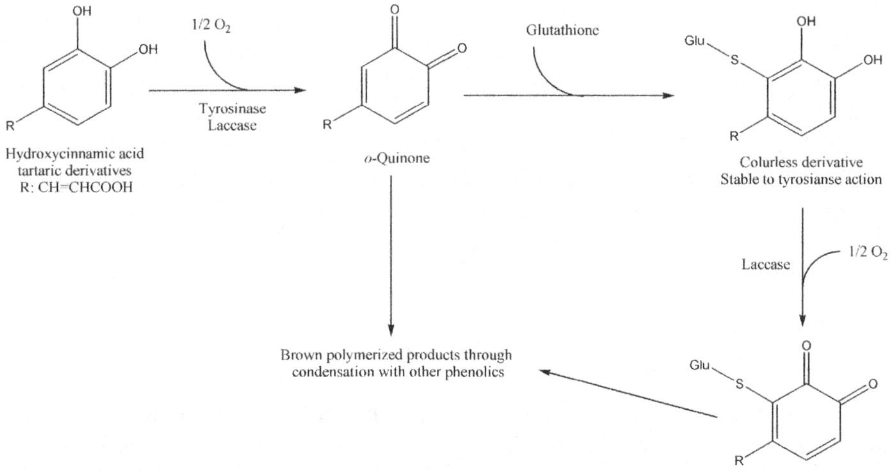

Fig. 4.1 Mechanism of formation of oxidized polymers responsible for must browning

significant modification to must color. The competitive nature of these two reaction mechanisms imply that, when a higher concentration of glutathione is present, less quinones will be transformed into brown, polymerized products. It has been shown that some grape varieties are typically more resistant to oxidative browning, due to a higher content of reductive species that can react with quinones, such as glutathione and ascorbic acid (Rigaud et al. 1990).

Oxidation of phenolic compounds becomes of greater concern when grapes are infested by *Botrytis*, due to the concomitant presence of a second powerful oxidase, namely laccase produced by *Botrytis* (Dubernet 1974). The activity of this enzyme, which can also be responsible for the loss of color in red wines, is about 30 times higher than that of tyrosinase. It is stable at juice pH, and is not inhibited by SO_2. Also, it can react with a variety of phenolic substrates, as well as with other classes of chemical compounds, including the quinone-glutathione complex (Salgues et al. 1986). Therefore, more severe browning is expected to occur in juices from botrytized grapes compared to juices from sound grapes (Fig. 4.1).

Various strategies can be adopted in the winery to reduce oxidation phenomena. Because oxygen is a cofactor of enzymatic browning reactions, careful protection of the must from oxygen and addition of ascorbic acid are often recommended. Sulfitation and cooling of grapes and juice during crushing/pressing are both very effective in controlling tyrosinase-driven juice oxidation, as tyrosinase is strongly inhibited by SO_2 and its activity is significantly reduced at low temperatures. Conversely, in the case of *Botrytis*-infected grapes, heating of the must at 50°C is the only treatment that can eliminate the risk of oxidation due to the activity of laccase.

Although it is generally accepted that protection of the must against oxidation can result in wines characterised by higher concentrations of volatile compounds that are responsible for fruity aromas (Moio et al. 2004), extreme antioxidant protection of the must can result later in wines that are more susceptible to oxidation. Based on this observation, a technique known as hyperoxygenation has been proposed in order to improve colour stability of white wines (Muller-Späth 1990; Artajona et al.1990; Dubourdieu and Lavigne1990; Cheynier et al. 1991). Accordingly, must phenolics are extensively oxidised by controlled supplementation of must with oxygen, and the oxidized forms are then removed during pre-fermentative clarification. Although hyperoxygenation has been shown to provide wines with increased colour stability (Schneider 1998), tentative evidence is available that it can be detrimental to aroma compounds that contribute to the varietal character of wine (Dubourdieu and Lavigne 1990; Ribereau-Gayon et al. 2006).

4.3 Lipoxygenase and Hydroxyperoxide Cleaving Enzymes

Six carbon atom (C_6) compounds are the major group of volatile compounds formed during the pre-fermentative stage of winemaking. This group includes volatiles such as hexanal, hexanol, *cis*-3-hexenol, *trans*-2-hexenol, and *cis*- and *trans*- isomers of 2- and 3-hexenal, which, depending on concentration, can have a detrimental effect on wine quality due to their grassy, herbaceous odors (Crouzet 1986).

This group of compounds is formed from the lipid fraction of the juice, through the sequential action of several enzymes. In Fig. 4.2. the mechanism described by Crouzet (1986) for the formation of *cis*-3- and *trans*-2-hexanal and their corresponding alcohol is shown. Linolenic acid is a common constituent of grape juice. Through the action of a lipoxygenase enzyme, linolenic acid is transformed

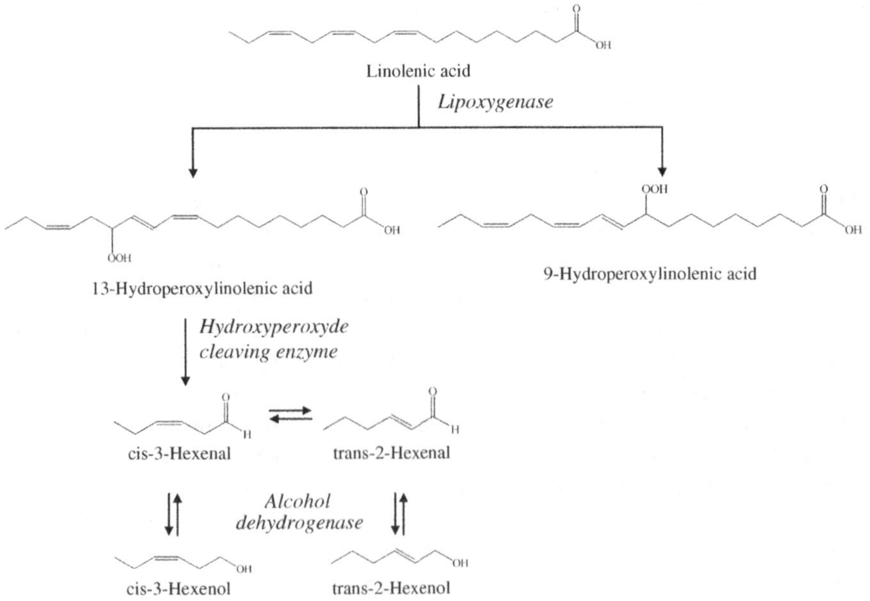

Fig. 4.2 Mechanism of formation of six carbon atom volatile compounds from long chain fatty acids during the pre-fermentative stages of vinification

into its corresponding 9- and 13-hydroperoxide. The subsequent action of a hydroxyperoxide-cleaving enzyme on the 13-hyroperoxide results in the formation of *cis*-3-hexanal, which is partially isomerized to *trans*-2-hexanal. The aldehydes are then transformed into their corresponding alcohol by alcohol dehydrogenase. A similar mechanism leads to the formation of hexanal and hexanol from linoleic acid (Crouzet 1986).

Control of C_6 compounds formation is normally achieved in the winery through careful clarification prior to alcoholic fermentation (Dubourdieu et al. 1986), as well as through removal of stalks and leaves, which have been shown to be a major source of these compounds (Cordonnier Bayonove 1981).

4.4 Wine Stabilization Enzymes

4.4.1 *Proteases*

Aggregation and subsequent precipitation of unstable proteins is the main cause of post-bottling haze formation in white wines (Ferreira et al. 2002). Although formation of haze is unlikely to affect wine taste and flavor, cloudy wines are normally rejected by consumers, resulting in significant economic losses for the wine industry. Wine proteins can originate from the grapes or from the metabolism of the various microorganisms involved in the winemaking process. Various studies have shown that the total protein content of a wine is a poor indicator of its haze instability, because only a fraction of the protein pool is responsible for the formation of haze (Paetzold et al. 1990; Waters 1991). Further investigations have indicated that this protein fraction is synthesized in grape berries, and appears to be associated with the occurrence of fungal diseases (Waters et al. 1996).

The occurrence of protease activity both in grape berries (Cordonnier and Dugal 1968) and wine yeast (Charoenchai et al. 1997; Dizy and Bisson2000) has been reported. Nevertheless, these proteases have low activity towards haze-forming proteins, which therefore survive the winemaking process. Pocock et al. (2003) have shown that heat treatment combined with the addition of proteolytic enzyme can reduce the incidence of haze formation, although the low specificity of commercially available proteases towards haze-forming proteins appeared to significantly reduce the possibilities offered by this strategy. At the moment, removal of wine proteins by means of bentonite fining remains the only effective method to control protein haze in white wine.

4.4.2 *Lysozyme*

Control of bacteria development is essential for the production of wines with consistent compositional and sensory characteristics. Although malolactic fermentation (MLF) is frequently carried out in the winery due to its beneficial effects on wine aroma and palate, for certain grape varieties and wine styles MLF is considered detrimental. Moreover, even in the cases where MLF is desirable, the wine might need further microbial stabilization against post-MLF bacterial spoilage. Sulfur dioxide is normally added to the wine to inhibit MLF as well as to prevent the development of other spoilage microorganisms. In an effort to reduce the concentrations of sulfur dioxide employed in winemaking, alternative strategies for wine stabilization have been explored. Lysozyme is an enzyme present in hen egg white, which can lyse Gram-positive bacteria, including wine lactic acid bacteria. Its application to stabilize the wine against bacteria, including those responsible for MLF, has been recently investigated. Although it has been proven that lysozyme can inhibit MLF and growth of other microbial species (Gerbaux et al. 2007), unwanted side effects have been observed in association with its use, namely loss of color in red wines and formation of haze in whites (Bartowsky et al. 2004). Furthermore, lysozyme has no inhibitory action against spoilage yeasts.

4.5 Pectolitic Enzymes

Grape berry skin and pulp contain significant amounts of pectic compounds that, along with other constituents such as cellulose, hemicellulose and lignin, contribute to the structure of cell walls (Vidal et al. 2003). Grape pectins are a heterogeneous group of compounds characterised by a structure consisting of linear chains of α-dgalacturonic acid joined through 1→4 linkages, partially esterified with methanol on the carboxylic function and with acetic acid in position O-2 and O-3. Insertion of rhamnose residues in alternate positions in this structure results in rhamnogalacturonanes, which also contain side chains of arabinose or galactopyranose (Doco et al. 1995).

During the production of white wines, part of pectic compounds from the berry is released into the juice with grape crushing and pressing, and forms a colloid that slows or prevents sedimentation of solid particles, particularly skin fragments. It has been reported that wines made from juice rich in suspended solids have increased levels of higher alcohols masking fruity aromas (Klingshirn et al. 1987) and higher concentration of sulphur compounds potentially responsible for off-flavors (Singleton et al. 1975; Lavigne-Cruège 1996). Therefore, elimination of solid particles is a key step in the production of quality white wines. Enzymatic hydrolysis of pectic structures is generally considered the most efficient way to break down their colloidal state and permit the separation of the entrapped solid particles. Due to the variety of possible structures, different enzymatic activities, particularly pectinesterases, polygalacturonases, pectin lyases, and pectate lyases can hydrolyse grape pectins. The occurrence of polygalacturonase and pectinesterase activities in grape berries has been reported (Usseglio-Tomasset 1978), which indicates that grape enzymes have the potential to promote juice clarification following crushing. Nevertheless, the activity of these enzymes is often insufficient, so that the time required to obtain an optimal degree of clarification thorough the action of grape pectinase can be incompatible with normal winery operations. To improve the efficiency of the clarification process, commercial preparations of pectinases can be added. These

Table 4.1 Commercial pectinase preparations commonly used in winemaking

Commercial pectinase preparations for clarification and filtration of juice and wine
Ultrazym 100 G (Novo Nordisk)
Novoclair speed (Novo Nordisk)
Rapidase Filtration (DMS)
Rapidase Vinosuper (DMS)
Vinoflow G (Novo Nordisk)
Rapidase CB (DMS)
Rapidase CR (DMS)
Lallzyme C (Lallemand)
White-style (Lallemand)
Zimopec PX1 (Perdomini)
Endozym Active (AEB)
Endozym Glucalyse (AEB)
Endozym Glucapec (AEB)
Endozym ICS 10 Éclair (AEB)
Endozym TMO (AEB)
Commercial preparation to increase extraction of colour and aroma compounds and pressing yields
Vinozym FCE G (Novo Nordisk)
Vinozym Vintage FCE (Novo Nordisk)
Rapidase Ex color (DMS)
Rapidase X Press (DMS)
Lallzyme EX (Lallemand)
Lallzyme EX-V (Lallemand)
Red style (Lallemand)
Endozym Rouge (AEB)
Endozym contact pelliculaire (AEB)

preparations possess a range of pectinase activities that can efficiently hydrolyse the pectic substances present in the juice, allowing sedimentation of the suspended solid particles. A large number of commercial preparations is now available on the market (Table 4.1), mainly obtained from the filamentous fungi *Aspergillus niger*.

Besides their primary application as clarification coadjutants, pectinase preparations can be used at various stages of the winemaking process. Due to their ability to degrade cell walls, pectinases are frequently used to improve the extraction of skin constituents. Particularly, during the production of white wine, pectinase can be added when pre-fermentative cold maceration is applied, in order to improve the extraction of aroma compounds and precursors located in the skins. In red winemaking, pectinase preparations, often in combination with cellulase and hemicellulase, are often used to increase the degradation of skin cell walls, in order to obtain increased pressing yields and improved extraction of color and aroma precursors during maceration (Ducruet et al. 2000; Gil and Valles 2001). Pectinases can also be used to prevent filter clogging prior to bottling.

The occurrence of unwanted side activities in commercial pectinase preparations has been reported, and represents one of the main problems associated with their use in winemaking. Notably, increased formation of methanol during fermentation has been observed in association with pectinase treatment of the juice, due to pectin-methyl esterase activity (Revilla and González-SanJosé 1998). Cinnamyl esterase activity responsible for increased production of volatile phenols can be also present in some pectinase preparation (Gerbaux et al. 2002). Great efforts are currently being made by enzyme producers to achieve a high degree of specificity for commercial pectinase preparations.

Because *S. cerevisiae* has limited pectinase and polysaccharide activities, expression of genes encoding for polysaccharide degrading activities have been investigated as an alternative strategy to improve skin cell degradation and colour extraction during maceration. Louw et al. (2006) reported that heterologous expression of endo-β-1,4-xylanase and endo-β-1,4-glucanase in a commercial *S. cerevisiae* strain resulted in increased pressing yield, lower hue, and increased color intensity, although the latter effect was only detectable after six months of bottle storage. The wines made with the recombinant strains were also preferred during sensory analysis. In theory, one attractive characteristic of recombinant strains compared to the addition of exogenous enzymes is that, since the strains will only produce the required enzymes, there is no risk of side activities which can be potentially detrimental to wine quality. Nevertheless, the strategies employed for the development of these recombinant strains appeared to affect other pathways related to the production of volatile compounds during fermentation (Perez-Gonzalez et al. 1993; Louw et al. 2006). Although major improvements are expected in this area in the next few years, the use of the recombinant yeast strains developed so far may result in unpredictable repercussions on wine flavor.

4.6 Aroma Enhancing Enzymes

4.6.1 Glycosidases

Since the occurrence of glycosylated precursors of volatile compounds in grapes was first reported (Cordonnier and Bayonove 1974), a considerable amount of research has been devoted to the understanding of the factors that determine their hydrolysis during winemaking. The ultimate goal has been to provide practical tools to modulate the release of the odor-active compounds contained in this type of precursor. Glycosides of volatile compounds are mainly *O*-β-D-glucosides or *O*-diglycosides, in which the aglycone moiety is always linked to the β-D-glucopyranose unit. In diglycoside structures, the glucose moiety is further substituted with a second monosacharide unit, which can either be α-L-arabinose, α-L-rhamnose, or β-D-apiose (Williams et al. 1982; Günata et al. 1988; Voirin et al. 1990, 1992). In grapes and wines, monoterpenes, C_{13} norisoprenoids, aliphatic alcohols, and benzene derivatives have all been identified as grape-derived volatile compounds that are largely present in the form of glycosidic aroma precursors (Fig. 4.3). Particularly, several powerful aroma compounds that have been shown to play an important role in the varietal aroma character of wine, such as linalool, geraniol, β-damascenone, 1,1,6-trimethyl-1,2-dihydronaphtalene (TDN), vinyl guaiacols, have been identified in the hydrolysates of precursors extracted from various

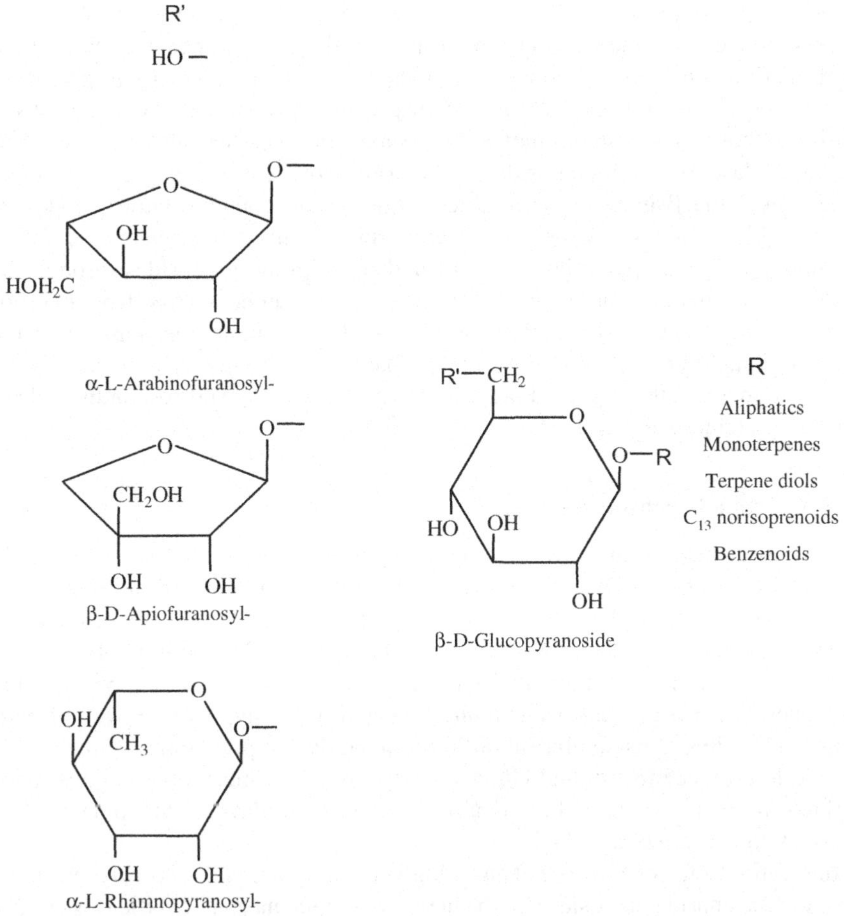

Fig. 4.3 Chemical structure of the various possible glycosides identified in grapes

grapes (Cordonnier and Bayonove 1974; Winterhalter et al. 1990, 1991; Winterhalter 1991; Sefton et al., 1993, 1994; Sefton 1998; Wirth et al. 2002). For this reason, it is generally accepted that hydrolysis of glycosides during winemaking is one of the main factors determining the formation of wine aroma characteristics (Abbott et al. 1991; Francis et al 1992, 1999). The liberation of glycosidically-bound aroma compounds can be achieved either via an acid-catalyzed process or by the action of glycosidase enzymes. Acid hydrolysis is however fairly slow under typical wine-making conditions, and is mainly regarded as a pathway for the formation of wine aging bouquet (Sefton 1998). Conversely, the action of glycosidase enzymes can rapidly hydrolyze the aroma precursors and release the bound volatile compound (Günata et al. 1993). This approach has therefore attracted considerable attention for its possible application as a tool to modulate wine aroma and enhance the expression of specific sensory characteristics.

Due to the occurrence of different chemical structures within the pool of grape glycosides, the complete enzymatic hydrolysis of these components requires in theory the action of multiple enzymatic activities. Studies with monoterpene glycosides have shown that liberation of sugar-bound aglycones involves, in the case of disaccharide glycosides, the preliminary action of an appropriate exo-glycosidase (arabinofuranosidase, rhamnopyranosidase, or apiofuranosidase) to release the terminal sugar before the β-glucosidase is able to release the bound volatile component, while for glucosidic precursors only β-glucosidase is needed (Günata et al. 1988). Enzymatic activities potentially able to hydrolyze grape glycosides through this sequential mechanism can be present in the grape or can be derived from microorganisms involved in the winemaking process, such as fermentation yeasts and lactic acid bacteria. Alternatively, commercial preparations with glycosidase activity can be used to increase the degree of precursor hydrolysis and the corresponding release of volatile compounds.

4.6.1.1 Grape Glycosidases

Grape berries have been shown to contain glycosidase activities which can, under optimal conditions, catalyze the release of volatile compounds from grape glycosides (Günata et al. 1990b). Studies on grape β-glucosidase have shown that this enzyme is strongly inhibited by glucose and ethanol, and has optimal pH around 5 (Aryan et al. 1987), indicating that its activity is largely limited during winemaking. Moreover, grape β-glucosidase is mainly located in the hypodermic wall of berry skins, and therefore its solubilisation in the juice during processing is likely to be of little importance (Sarry and Günata 2004). Based on these observations, grape β-glucosidase is considered to play a negligible role in flavor development during winemaking (Aryan et al. 1987).

Interestingly, the occurrence of an endoglycosidase capable of hydrolysing monoterpene disaccharide glycosides has been reported (Günata et al. 1988). This enzyme operates by liberating of the aglycone moiety of disaccharide with a single step mechanism instead of the sequential two step process proposed for other glycosidases. The involvement of this enzymatic activity in the process of flavor development during winemaking still needs to be clarified.

4.6.1.2 Microbial Glycosidases

The observation that a large part of the varietal character of wine develops during fermentation suggests that microbial species responsible for the fermentation process play a role in the release of varietal odor-active compounds from odorless precursors. This is particularly obvious in the case of wines obtained from non-aromatic grape varieties (e.g. Chardonnay, Semillon, Pinot Gris, Trebbiano, and most of the red varieties), in which specific varietal aroma attributes are often present at the end of fermentation, in spite of the fact that grapes of these varieties lack any characteristic aroma. Nevertheless, the ability of *S. cerevisiae*, the main agent of alcoholic fermentation, to hydrolyze glycosidic precursors during fermentation has been a topic of great controversy for many years. Early studies had indicated that *S. cerevisiae* possess β-glucosidase activity (Darriet et al. 1988), although in levels that were lower than found in other non-*Saccharomyces* yeast (Rosi et al. 1994). Typical winemaking conditions were shown to inhibit yeast β-glucosidase, α-arabinosidase α-rhamnosidase activities (Delcroix et al. 1994), suggesting that most of glycosides present in the juice remain virtually unchanged during fermentation. However, as the majority of these studies were carried out with non-fermenting yeast cells and using model substrates instead of native grape glycosides, the question remained open as to whether *S. cerevisiae* yeast were able to hydrolyze grape glycosides during fermentation. Also, in several studies where low stability of *S. cerevisiae* glycosidases were observed, quantification of yeast enzyme activities was carried out on cell enzyme extracts, regardless of the fact that glycosidases are located in the cell periplasmic space (Darriet et al. 1988; Mateo and Di Stefano 1997) and therefore not directly exposed to the action of potential inhibitors such as pH or ethanol. Recently, a series of studies have provided clear evidences for the major role played by yeast in the hydrolysis of glycosides during fermentation (Delfini et al. 2001; Ugliano 2006; Loscos et al. 2007). Under typical winemaking conditions, it was proven that the hydrolytic activity of yeast on various glycosides was responsible for the liberation of several bound volatile compounds mainly from glucosides, rhamnosides, and arabinosides, with concomitant acid hydrolysis playing a much less important role (Fig. 4.4). These experimental results are likely to renew the interest in yeast strain selection as a tool to modulate wine varietal character.

Although *S. cerevisiae* possesses enzymatic activities that can catalyze the hydrolysis of glycosides during fermentation, a large portion of the glycosides originally present in the must survive the fermentation process and are still present in the finished wine (Williams et al. 1996; Zoecklein et al. 1997a, b). This suggests that *S. cerevisiae* glycoside hydrolases, although capable of catalyzing the release of a significant portion of glycosidically-bound aroma compounds, are insufficient to induce complete hydrolysis of the whole glycosidic pool of grape juice. Based on this observation, other yeast species, particularly non-*Saccharomyces* yeast, have been largely studied as a possible source of additional glycosidase activities during fermentation (Günata et al. 1990a; McMahon et al. 1999; Arevalo Villena et al. 2006). Several species of non-*Saccharomyces* yeast, some of which are present during fermentation, were found to possess significant levels of glycosidase activities, suggesting that liberation of aroma compounds from glycosides during fermentation can partially be linked to the activity of non-*Saccharomyces* yeasts.

More recent research has focused on the construction of genetically modified organisms which over-express enzymatic activities that can catalyze the liberation of aroma compound from glycosidic precursors (Van Rensburg et al. 2005; Gil et al. 2005). However, the results obtained so far are still of limited practical interest compared to those achievable through the addition of exogenous enzymes, manly due the fact that the recombinant strains generated exhibit unpredicted patterns of production of volatile compounds, particularly fermentation esters. Interestingly, in the study of Gil et al. (2005), overexpression of the *S. cerevisiae* gene *EXG1* gene

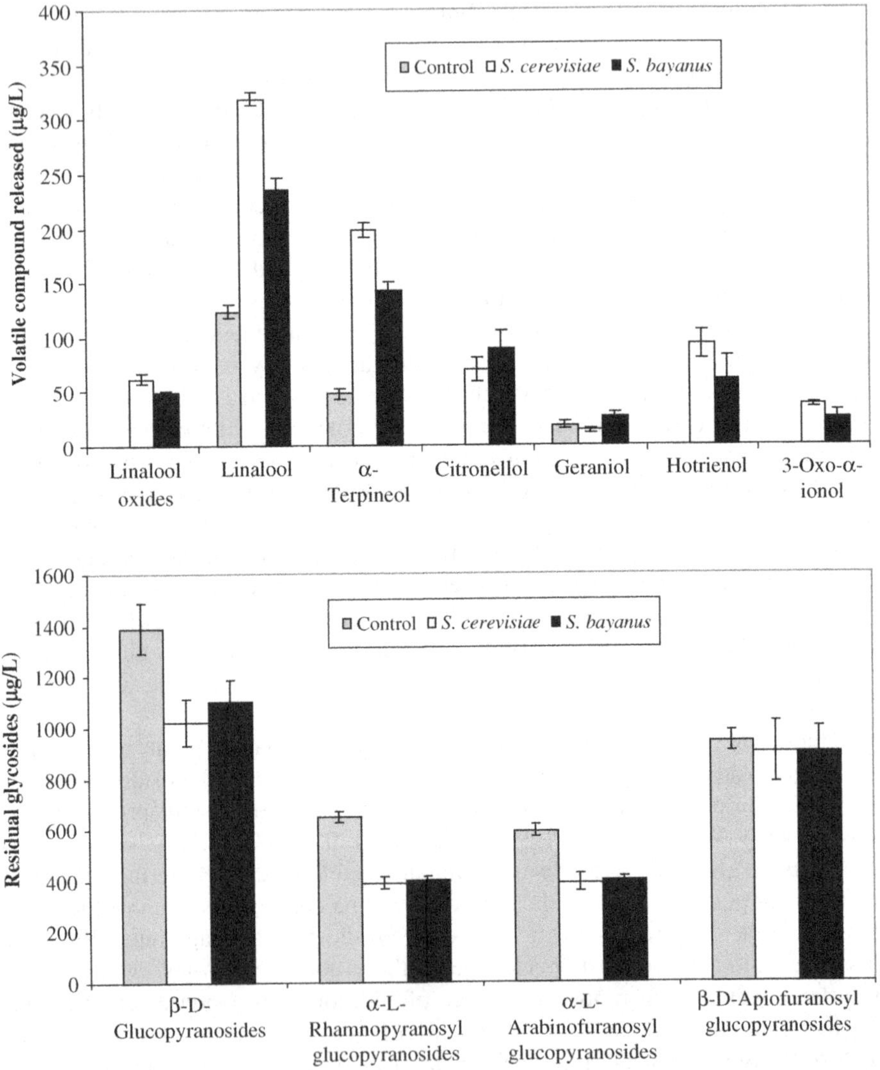

Fig. 4.4 Effect of alcoholic fermentation and yeast strain on the concentration of glycoside-derived volatile compounds and of different classes of glycosidic precursors during the fermentation of a model grape juice containing Muscat glycosides. The control refers to a non-fermented sample kept under the same conditions utilized for fermentation, which accounts for acid catalysed hydrolysis of glycosides (adapted from Ugliano 2006)

encoding for the production of an exoglucanase resulted in increased concentration of monoterpenes after fermentations, suggesting that enzymatic activities not strictly classified as glycosidases can play a role in the development of wine varietal flavor during fermentation. Similar conclusions were reached by Daenen et al. (2007) in a study on glucoside hydrolases in *Saccharomyces* and *Brettanomyces* brewing yeasts.

Although genetic engineering techniques have the potential to generate yeasts with enhanced ability to liberate glycosidically-bound aroma compounds, the use of genetically modified organisms in the wine industry is currently not allowed in the majority of wine-producing countries.

Lactic Acid Bacteria (LAB), particularly the species *Oenococcus oeni*, have also been studied for their ability to hydrolyze glycosidic aroma precursors in the course of malolactic fermentation. Studies in model systems have shown that different *O. oeni* strains possess enzymatic activities that can catalyze the hydrolysis of glycosides (Guilloux-Benatier et al. 1993; Grimaldi et al. 2000; Ugliano et al. 2003; D'Incecco et al. 2004). During the production of red wine, these activities have been shown to catalyze the hydrolysis of glycosides of different volatile compound (Boido et al. 2002; Ugliano and Moio 2006), the extent of the hydrolytic process mainly depending on the chemical structure of the aglycone moiety and on the bacterial strain employed for the induction of MLF (Fig. 4.5). Nevertheless, precursor hydrolysis during MLF does not result in a proportional increase in the concentration of the corresponding volatile compounds (Boido et al. 2002; Ugliano and Moio 2006), indicating the occurrence of other processes causing a partial loss of the liberated aglycons, such as adsorption on macromolecules or metabolisation by the bacteria (Boido et al. 2002; D'Incecco et al. 2004).

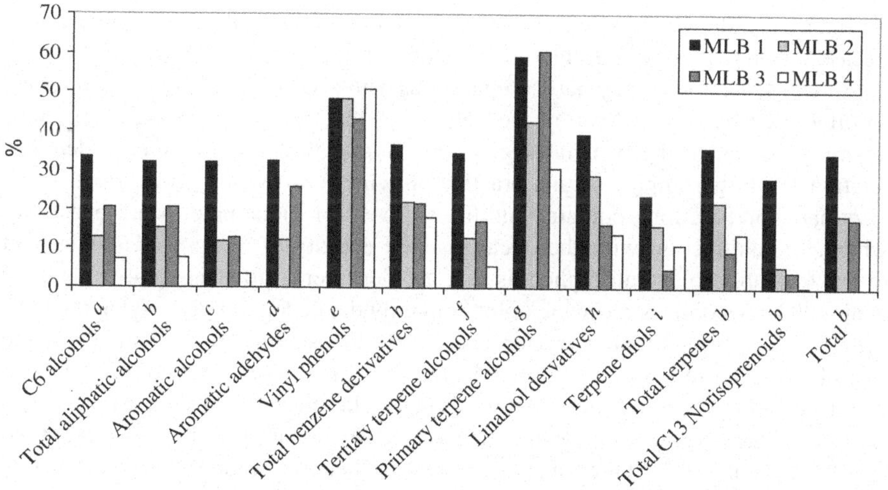

Fig. 4.5 Extent of hydrolysis of glycoside of different volatile compounds during MLF with four strains of malolactic bacteria (MLB). Values are calculated as a percentage ratio between the concentration of glycosides in MLF samples and in a non-MLF control. [a]Sum of 1-hexanol, *trans-* and *cis*-3-hexenol, *trans-* and *cis*-2-hexenol; [b]sum of isoamyl alcohols, heptanol, and 4-hydroxy-4-methyl-2-pentanol; [c]sum of benzyl alcohol and 2-phenylethanol; [d]sum of vanillin and benzaldehyde; [e]sum of 4-vinylphenol and 4-vinylguaiacol; [f]sum of linalool and α-terpineol; [g]sum of nerol and geraniol; [h]sum of *cis-* and *trans*-linalool oxides (pyranic and furanic); [i]sum of 3,7-dimethyl-1,5-octadien-3,7-diol and the two 2,7-dimethyl-2,7-octadien-1,6-diol isomers (from Ugliano and Moio 2006, reproduced with permission)

4.6.1.3 Exogenous Glycosidases

Due to the limited glycosidase activities associated with grape, yeast, and bacteria metabolism, exogenous enzymatic preparations with glycosidase activities can be added during winemaking to enhance the liberation of glycosidically-bound volatile compounds. Particularly, commercially available pectinase and hemicellulase preparations obtained from the fungus *Aspegillus niger* have side activities that can catalyze the release of volatile compounds from glycoconjugates. Addition of enzymatic preparation during winemaking generally results in higher concentrations of different classes of volatile compounds, particularly monoterpene alcohols, monoterpene polyols, norisoprenoids and benzenoids (Günata et al. 1990b, 1993). Experiments with monoterpene-dependant grape varieties have shown that the enzyme treated wines are characterised by increased floral attributes, due to the large release of monoterpene alcohols (Günata et al. 1993; Bayonove et al. 1996). In the case of non-terpene dependant grape varieties, the sensory impact of enzyme treatment on wine sensory characteristics is less predictable, and depends largely on the composition of the pool of glycosidically-bound volatiles of individual varieties. For example, some volatiles released with the enzymatic treatments, particularly those belonging to the class of norisoprenoids, are odorless, although they can later generate other powerful odorants during wine aging (Kotseridis et al. 1999). Also, when the enzyme treatment is performed before alcoholic fermentation on a grape juice rich in bound geraniol, increased concentrations of citronellol are then found in the final wine, due to the yeast-catalyzed formation of citronellol from the geraniol released from precursors (Ugliano et al. 2006).

Although the use of enzymatic preparations with glycosidase activities represents an effective way of accelerating the otherwise slow release of aroma compounds from glycosides during winemaking, several implications should be considered in relationship to the impact of this practice on wine sensory properties. First, some enzymatic preparations contain significant levels of cinnamate esterase activity which, in conjunction with the decarboxylase activity of *S. cerevisiae*, can lead to the formation of volatile phenols that are detrimental to wine quality (Günata et al. 1993). Second, as red wine color is mainly dependent on glycosylated anthocyanins, preparations containing non-specific β-glucosidase activity could cause the hydrolysis of these colored compounds, resulting in a general loss of wine colour (Le Traon-Masson and Pellerin 1998). Finally, the presence of esterase activities in some of the commercially available glycosidase preparations has been observed (Tamborra et al. 2004). These side activities have been shown to cause a decrease in the concentration of fermentation-derived esters such as isoamyl acetate and ethyl hexanoate, which are well known contributors to the fermentation bouquet of young wines. Careful attention has to be paid therefore to the selection of glycosidase preparations that are largely free of side activities detrimental to wine quality.

Moreover, altering the balance between free and bound forms of the pool of grape-derived volatile compounds of wine through addition of an exogenous glycosidase has to be considered in conjunction with the role played by both glycosides and aglycons as precursors to aroma compounds during wine aging. For example, the monoterpene alcohol geraniol and its glucoside, both identified in wine, can both act as precursor to various other terpene aroma compounds, including the monoterpene alcohol linalool. It has been shown that the rate of formation of linalool and other volatiles from geraniol during wine aging is much higher for free geraniol than for its glucoside (Fig. 4.6a) (Skouroumounis and Sefton 2000). Because linalool is a more powerful odorant than geraniol, this might have positive consequences for wine aroma. In contrast, in a study on the formation of the powerful odorant β-damascenone, it was found that acid-catalyzed hydrolysis of the glucoside of megastigm-5-en-7-yne-3,9-diol, one of the possible precursors to β-damascenone during wine aging (Sefton et al. 1989), yielded a higher proportion of β-damascenone compared to the aglycone, although the latter reacted more

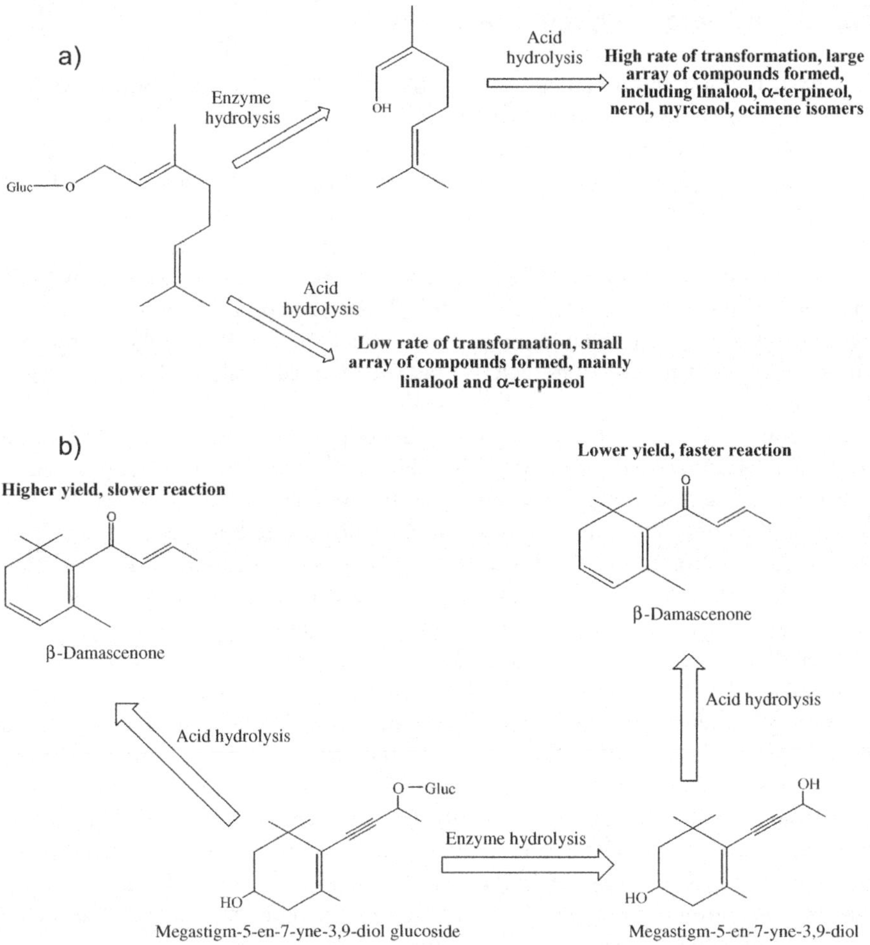

Fig. 4.6 Mechanisms of formation of: **a** terpenes from geraniol and geranyl glucoside; **b** β-damascenone from megastigm-5en-7yne-3,9-diol and its glucoside

quickly (Fig. 4.6b) (Skouroumounis et al. 1993). These observations were confirmed in a study on β-damascenone in Merlot wines submitted to accelerated aging, where it was found that glycosidase treated samples generated, during aging, about half the amount of β-damascenone observed in the non-glycosidase-treated control (Kotseridis et al. 1999). In this case, therefore, extensive hydrolysis of the pool of wine glycosidic precursors by exogenous enzymes appeared to reduce the potential for generating grape derived aroma compounds through acid-catalyzed reactions during aging. More research in required to understand the existence of this type of competing reactions for precursors of other key aroma compounds, in order to establish the potential benefits of enzyme treatments to the expression of wine varietal volatile character.

4.6.2 Cysteine-S-Conjugate β-Lyases

Volatile thiols are important aroma constituents of several food and beverages, including wines obtained from certain grape varieties such as Sauvignon Blanc, Scheurebe and Verdelho (Darriet et al. 1995; Guth 1997; Tominaga et al. 2000; Lopez et al. 2003). Three compounds in this group, namely 4-methy-4-mercaptopentan-2-one, 4-methyl-4-mercaptopentan-2-ol, and 3-mercaptohexanol, have been found in grapes only in the form of non-volatile odourless cysteine-conjugated precursors, and their concentration has been shown to increase with the fermentation process (Tominaga et al. 1998). It has been proposed that *S. cerevisiae* can hydrolyze the cysteine-conjugated precursor though a β-elimination reaction catalyzed by a cysteine-*S*-conjugate β-lyase (Tominaga et al. 1998; Howell et al. 2005).

Since a significant portion of the cysteine conjugates survives the fermentation process, as observed in the case of glycosylated aroma precursors, researchers have investigated possible strategies to enhance the cleavage of the conjugated thiols. Based on the observation that the concentration of volatile thiols in the wine at the end of fermentation is strictly linked to the *S. cerevisiae* strain used for fermentation, identification of strains with enhanced β-lyase activity has been indicated as a possible strategy to increase the concentrations of volatile thiols in Sauvignon blanc (Howell et al. 2005; Swiegers et al. 2006). A cysteine-*S*-conjugate β-lyase has also been isolated from various bacteria (Tominaga et al. 1998), but the fact that the optimal pH for these enzymes is around 8 prevents their possible utilization as exogenous enzymes during winemaking. Recently, the construction of a genetically modified *S. cerevisiae* commercial strain overexpressing an *E. coli* gene encoding for a tryptophanase with strong β-lyase activity has been reported (Swiegers et al. 2007). This strain was shown capable of releasing increased amounts of 4-mercapto-4-methylpentan-2-one and 3-mercaptohexanol during laboratory scale fermentations. Current legislation in the majority of wine producing countries does not allow the use of genetically modified organisms in the wine industry.

3-Mercaptohexyl acetate (3-MHA) is another powerful thiol that contributes the passion fruit-like character of Sauvignon Blanc wines (Tominaga et al. 2000). It is formed through a two step mechanism involving liberation of 3-MH from cystein-*S*-conjugates by the yeast and subsequent esterification with acetate (Dubourdieu et al. 2006). King et al. (2007) demonstrated that increases in the concentration of this powerful odorant can be achieved through the use of mixed yeast starter cultures containing a blend of two different strains characterised respectively by high S-cysteine lyase and high esterase activities.

4.6.3 Esterases

Volatile compounds synthesized by yeast during alcoholic fermentation play an important role in the aroma characteristics of wine. Among these, esters are considered the major contributors to the fruity character of wine. Two main groups of esters are synthetized by yeast, namely acetate esters and ethyl fatty acid esters. Acetate esters, often simply referred to as acetates, are obtained from condensation of yeast-derived higher alcohols with acetyl-CoA, catalysed in the cell by alcohol acyl transferase enzymes (Mason and Dufour 2000). It has been shown that the final concentration of these compounds is the result of the balance between alcohol acyl transferase enzymes promoting their synthesis, and esterase enzymes promoting their hydrolysis (Fukuda et al. 1998). The feasibility of improving acetates production by overexpressing one of the endogenous genes that control their production during fermentation has been recently explored in a number of studies. Experimental work on the alcohol acyl transferase gene *ATF1* has shown that overexpression of this gene can result in 2- to 12-fold increases in the concentration of different acetates (Lilly et al. 2000), while overexpression of a second alcohol acetyl transferase, encoded by the *ATF2* gene, affected acetate production to a lesser degree (Lilly et al. 2006a). Interestingly, manipulation of genes not strictly related to ester biosynthesis but involved in the production of ester precursors in the cell can also affect formation of acetates. Lilly et al. (2006b) demonstrated that overexpression of abranched-chain amino acid transferase gene *BAT1* enhanced the production of isoamyl acetate, which appeared to be linked to increased production of isoamyl alcohol, the precursor of isoamyl acetate.

Ethyl fatty acid esters are produced by fermentation yeasts through esterification of short and medium chain (C_4-C_{12}) fatty acids coming from lipid metabolism (**?** **?**). Only recently, researchers have started to investigate the mechanisms controlling formation of ethyl fatty acid esters during fermentation, in spite of the fact that compounds such as ethyl hexanoate and ethyl octanoate occur in young wines in concentration largely above their odour threshold, and are therefore likely to have a significant influence on the aroma characteristics of wine. A possible alcohol acyl transferase, designated Eht1p (ethanol hexanoyl transferase), has been suggested as a possible candidate for ethyl ester production (Mason and Dufour 2000), which was later confirmed by Saerens et al. (2006). Interestingly, overexpression of the *ETH1* gene did not result in increased concentration of ethyl fatty acid esters, possibly due to the bifunctional nature of this enzyme, being involved in both formation and hydrolysis of fatty acid esters (Saerens et al. 2006).

The studies on overexpression of ester-related genes, although carried out using GMO organisms, have indicated that optimized expression levels of genes encoding for ester synthase enzymes can assist in producing wines with specific ester profiles. This desirable feature is certainly going to become a major focus of yeast development research in the near future, either thorough the use of molecular biology techniques or by extensive screening of natural yeasts. Nevertheless, it has to be considered that the majority of the esters produced by yeast during fermentation are largely hydrolyzed during the first two years of wine storage, the hydrolysis rates being higher when the initial concentration is higher (Ramey and Ough 1980; Moio et al. 2004). This does not apply however to ethyl acetate, the most abundant ester produced during alcoholic fermentation, as this compound is generally stable during aging. Because ester profiles which are dominated by high concentrations of ethyl acetate can potentially result in wines with unwanted nail lacquer-like aromas, caution has to be used in choosing high ester producing strains.

Recent work has also indicated that commercially available strains of LAB are also able to synthesize acetates and ethyl fatty acid esters during winemaking (Maicas et al. 1999; Ugliano and Moio 2005), which appear to be in contradiction with early suggestions that the MLF could result in a decrease in wine esters (Davis et al. 1988; Du Plessis et al.2002). More research is required in this area before general conclusions can be drawn regarding the influence of MLF on the ester composition of wine.

4.7 Conclusion

The majority of wine sensorially-active compounds are formed or degraded during winemaking though pathways involving the intervention of enzymes of different origins, which underlines the importance of optimizing the extent of different enzymatic reactions to achieve wines with specific sensory attributes. Enzyme-driven transformations also play a primary role in determining the technological efficiency of various steps of the vinification process.

Addition of exogenous enzymes is currently the most commonly adopted practice in the wine industry to optimize the rate of enzymatic reactions taking place during winemaking. The large amount of research carried out on this topic has led to the presence of a considerable number of commercial preparations on the market, although in some areas, particularly protein stabilization of white wines, exogenous enzyme-based approaches have been unsuccessful so far.

In addition, promising results are being obtained through strain selection as well as by using molecular biology techniques to generate novel yeasts and bacteria with optimized enzyme activity, particularly for applications related to wine aroma enhancement.

Developments of successful tools for the control of enzymatic reactions during winemaking, regardless of the type of method employed to optimize enzyme activity (e.g. exogenous enzymes, selected yeast and bacteria, or genetically enhanced microorganisms), will benefit from further improvements in the understanding of wine flavor chemistry and sensory perception.

References

Abbott, N.A., Coombe, B.G. & Williams, P.J. (1991). The contribution of hydrolyzed flavour precursors to quality differences in Shiraz juice and wines: an investigation by sensory descriptive analysis. *Am. J. Enol. Vitic.*, *42*, 167–174.

Arevalo Villena, M., Ubeda Iranzo, J. & Briones Perez, A. (2006). Relationship between *Debaryomyces pseudopolymorphus* enzymatic extracts and release of terpenes in wine. *Biotechnol. Prog.*, *22*, 375–381.

Artajona, J., Bobet, R., Marco, J., Sabat, F. (1990). Expériences d'hyperoxygénation au Penedés. *Rev. Franc. Œnol.*, *124*, 65–67.

Aryan, A.P., Wilson, B., Strauss, C. & Williams, P.J. (1987). The properties of glycosidases of Vitis vinifera and a comparison of their β-glucosidase activity with that of exogenous enzymes. An assessment of possible application in enology. *Am. J. Enol. Vitic.*, 38, 182–188.

Bartowsky, E.J., Costello, P.J., Villa, A. & Henschke, P.A. (2004). The chemical and sensorial effects of lysozyme addition to red and white wines over six months cellar storage. *Aus. J. Grape Wine Res.*, *10*, 143–150.

Bayonove, C., Günata, Y., Sapis, J.C., Baumes, R., Dugelay, I. & Grassin, C. (1996). Enhancement of wine aromas by enzyme hydrolysis of glycosidic precursors from the grape. In C.S. Stockley, A.N. Sas, R.S. Johnstone & Lee, T.H., (Eds.), *Maintaining the competitive edge: proceedings of the ninth Australian wine industry technical conference. Adelaide, South Australia 16–19 July 1995.* Adelaide, South Australia.

Boido, E., Lloret, A., Medina, K., Carrau, F. & Della Cassa, E. (2002). Effect of β-glycosidase activity of Oenococcus oeni on the glycosylated flavour precursors of Tannat wine during malolactic fermentation. *J. Agric. Food Chem.*, *50*, 2344–2349.

Charoenchai, C., Fleet, G.H., Henschke, P.A. & Todd, B.E.N (1997). Screening of non-*Saccharomyces* wine yeast for the presence of extracellular hydrolytic enzymes. *Aus. J. Grape Wine Res.*, *6*, 190–196.

Cheynier, V., Souquet, J.M., Samson, A. & Moutounet, M. (1991). Hyperoxidation: influence of various oxygen supply levels on oxidation kinetics of phenolic compounds and wine quality. *Vitis, 30*, 107–115.

Cordonnier, R., & Bayonove, C. (1974). Mise en évidence dans la baie de raisin, variété Muscat d'Alexandrie, de monoterpènes liés révélables par une ou plusieurs enzymes du fruit. *Compte rendus Académie Science Paris, 278*, 3387–3390 Série D.

Cordonnier, R., & Bayonove, C. (1981). Etude de la phase préfermentaire de la vinification: extraction et formation de certain composes de l'arôme, cas des terpenols, des aldehydes et des alcools en C6. *Conn. Vigne Vin, 15*, 269–286.

Cordonnier, R. & Dugal, A. (1968). Les activites proteolytiques du raisin. *Ann. Technol. Agric.* 17, 189–206.

Crouzet, J. (1986). Les enzymes et l'arôme des vins. *Rev. Fran. Œnol., 102*, 42–49.

D'Incecco N., Bartowsky E.J., Kassara S., Lante A., Spettoli, P. & Henschke, P.A. (2004). Release of glycosidically bound flavour compounds of Chardonnay by *Oenococcus oeni* during malolactic fermentation. *Food Microbiol.*, 21, 257–265.

Daenen, L., Saison. D., Sterckx, F., Delvaux, F.R., Verachtert, H. & Derdelinckx, G. (2007). Scrrenening and evaluationof the glucoside hydorlase activity of Saccharomyces and Brettanomyces brewing yeasts. *J. Appl. Miocrobiol.* doi: 10.1111/j.1365-2672.2007.03566.x

Darriet, P., Boidron, J. N. & Dubourdieu, D. (1988). L'hydrolyse des hétérosides terpéniques du Muscat a petits grains par les enzymes périplasmatiques de *Saccharomyces cerevisiae. Conn. Vigne Vin, 22*, 89–195.

Darriet, P., Tominaga, T., lavigne, V., Boidron, J.-N. & Dubourdieu, D. (1995). Identification of a powerful aromatic compound if *Vitis vinifiera* L. var Sauvignon wines: 4-mercapto-4methypentan-2-one. *Flavour and Fragrance Journal, 10*, 385–392.

Davis, C.R., Wibowo, D., Fleet, G.H. & Lee, T.H. (1988). Properties of wine lactic acid bacteria: their potential oenological significance. *Am. J. Enol. Vitic., 39*, 137–142.

Delcroix, A., Günata, Z. Y., Sapis, J. C., Salmon, J. M. & Bayonove, C. (1994). Glycosidase activities of three enological yeast strains during winemaking: Effect on terpenols content of Muscat wine. *Am. J. Enol. Vitic., 45*, 291–296.

Delfini, C., Cocito, C., Bonino, M., Schellino, R., Gaia, P. & Baiocchi, C. (2001). Definitive evidence for the actual contribution of yeast in the transformation of neutral precursors of grape aromas. *J. Agric. Food Chem., 49*, 5397–5408.

Dizy, M., & Bisson, L. (2000). Proteolytic activity of yeast strains during grape juice fermentation. *Am. J. Enol. Vitic., 51*, 155–167.

Doco, T., Lecas, M., Pellerin, P., Brillouet, J.-M. & Moutounet, M. (1995) Les polysaccharides pectique de la pulpe et de la pellicule de raisin. Quel devenir pendant la phase pré-fermentaire? *Rev. Fran. Oenol.*, 153, 15–23.

Du Plessis, H.W., Steger, C.L.C., du Toit, M. & Lambrechts, M.G. (2002). The occurrence of malolactic fermentation in brandy base wine and its influence on brandy quality. *J. Appl. Microbiol., 92*, 1005–1013.

Dubernet, M. (1974). Recherches sur la tyrosinase de Vitis vinifera et la laccase de Botrytis cinerea. Applications technologiques. Thèse de doctorate, Université de Bordeaux II.

Dubourdieu, D. & Lavigne, V. (1990). Incidence de l'hyperoxygénation sur la composition chimique et les qualities organoleptiques des vins blancs secs du Bordelais. *Rev. Fran. Œnol., 124*, 58–61.

Dubourdieu, D., Ollivier, C., & Boidron, J.N. (1986). Incidences des opérations préfermentaires sure la composition chimique et les qualités organoleptiques des vins blancs. *Conn. Vigne Vin, 20*, 53–76.

Dubourdieu, D., Tominaga, T., Mnsneuf, I., Peyrot de Gachons, C. & Murat, M.L. (2006). The role of yeast in grape flavor development during fermentation: the example of Sauvignon blanc. *Am. J. Enol. Vitic., 57*, 81–88.

Ducruet, J., Glories, Y. & Canal, R. M. (2000). Mécanisme d'action et utilisation raisonnée d'une préparation enzymatique de maceration. *Rev. Oenol., 27*(96), 17–19.

Ferreira, R.B., Picarra-Pereira, M.A., Monteiro, S., Loureiro, V.B. & Teixeira, A.R. (2002). The wine proteins. *Trends Food Sci. Technol., 12*, 230–239.

Francis, I.L., Sefton, M.A., & Williams. P.J. (1992). Sensory descriptive analysis of the aroma of hydrolysed flavour precursor fractions from Semillon, Chardonnay, and Sauvignon blanc grape juices. *J. Sci Food Agric., 59*, 511–520.

Francis, I.L., Kassara, S., Noble, A.C. & Williams, P.J. (1999).The contribution of glycoside precursors to Cabernet Sauvignon and Merlot aroma: sensory and compositional studies. In A.L. Waterhouse & S.E Ebeler (Eds.), *Chemistry of wine flavour* (pp. 13–30) American Chemical Society: Washington, DC.

Fukuda, K., Yamamoto, N., Kiyokawa, Y., Yanagiuchi, T., Wakay, Y., Kitamoto, K., Inoue, Y. & Kimura, A. (1998). Balance of activities of alcohol acetyltransferase and esterase in *Saccharomyces cerevisiae* is important for production of isoamyl acetate. *App. Environ. Microbiol., 64*, 4076–4078.

Gerbaux, V., Villa, A., Monamy, C. & Bertrand, A. (1997). Use of lysozyme to inhibit malolactic fermentation and to stabilize wine after malolactic fermentation. *Am. J. Enol. Vitic.,48*, 49–54.

Gerbaux, V., Villa, A., Monamy, C. & Bertrand, A. (2002). Influence of maceration temperature and enzymes on content of volatile phenols in Pinot noir wines. *Am. J. Enol. Vitic., 53*, 131–137.

Gil, J.V. & Valles, S. (2001). Effect of macerating enzymes on red wine aroma at laboratory scale: addition or expression by transgenic wine yeast. *J. Agric. Food Chem., 49*, 5515–5523.

Gil, J.V., Manzanares, P. Genoves, S. Valles, S. & Gonzalez-Candelas, L. (2005). Over-production of the major exoglucanase of *Saccharomyces cerevisiae* led to an increase in the aroma of wine. *Int. J. Food Microbiol., 103*, 57–68.

Grimaldi, A., McLean, H. & Jiranek, V. (2000). Identification and partial characterization of glycosidic activity of commercial strains of the lactic acid bacterium, *Oenococcus oeni. Am. J. Enol. Vitic., 51*, 362–369.

Guilloux-Benatier, M., Son, H.S., Bouhier, M. Feuillat, M. (1993). Activités enzymatiques: glycosidases et peptidases chez Leuconostoc oenos au cours de la croissance bactérienne. Influence des macromolécules des levures. *Vitis, 32*, 51–57.

Günata, Y.Z., Bitteur, S., Brillouet, J.-M., Bayonove, C. L. & Cordonnier, R. E. (1988). Sequential enzymic hydrolysis of potentially aromatic glycosides from grapes. *Carbohydr. Research, 184*, 139–149.

Günata, Y.Z., Bayonove, C.L., Cordonnier, R.E., Arnaud, A. & Glazy, P. (1990a). Hydrolysis of grape monoterpenyl glycosides by *Candida molischiana* and *Candida wickerhamii* β-glucosidases. *J. Sci. Food Agric., 50*, 499–506.

Günata, Y.Z., Bayonove, C.L., Tapiero, C. & Cordonnier, R.E. (1990b). Hydrolysis of grape monoterpenyl β-D-glucosides by various β-glucosidases. *J. Agric. Food Chem., 38*, 1232–1236.

Günata, Y.Z., Dugelay, I., Sapis, J.C., Baumes, R., & Bayonove, C. (1993) Role of enzymes in the use of the flavour potential from grape glycosides in winemaking. In P. Schreier & P., Winterhalter, (Eds). *Progress in flavour precursor studies: analysis-generation-biotechnology*. Carol Stream Allured Publishing Corporation 219–234.

Günata, Y.Z., Dugelay, I., Sapis, J.-C., Baumes, R. & Bayonove, C. (1994). Role of enzymes in the use of flavour potential from grapes glycosides in winemaking. In P. Schreier & P. Winterhalter (Eds.), *Progress in flavour precursor studies. Proceedings of the international conference* (pp. 219–245), Allured Publishing Corporation: Carol Stream, IL.

Günata, Y.Z., Bondeel, C., Vallier, M.J., Lepoutre, J.P., Sapis, J.-C. & Watanabe, N. (1998). An endoglycosidase from grape berry skin of Cv. M. Alexandria hydrolyzing potentially aromatic disaccharides glycosides. *J. Agric. Food Chem., 46*, 2748–2753.

Guth, H. (1997). Quantitation and sensory studies of character impact odorants of different white wines varieties. *J. Agric. Food Chem., 45*, 3027–3032.

Howell, K.S., Klein, M., Swiegers, J.H., Hayasaka, Y., Elsey, G.M., Fleet, G.H., Høj, P.B., Pretorius, I.S. & de Barros Lopez, M.A. (2005). Genetic determinant of volatile-thiol release by *Saccharomyces cerevisiae* during wine fermentation. *App. Environ. Microbiol., 71*, 5420–54416.

King, E., Swiegers, J.H., Travis, B., Francis, I.L., Pretorius, I.S. & Connors, P. (2007). Modulating Sauvignon Blanc wine aroma through co-inoculated fermentations. *Australian & New Zealand Grapegrower & Winemaker, 525*, 102–108.

Klingshirn, L., Liu, M. & Gallander, J.F. (1987). Higher alcohols formation in wines as related to the particle size profiles of juice insoluble solids. *Am. J. Enol. Vitic., 38*, 207–209.

Kotseridis, Y., Baumes, R.L. & Skouroumounis, G. (1999). Quantitative determination of free and hydrolytically liberated β-damascenone in red grapes and wines using a stable isotope dilution assay. *J. Chromatog., 849*, 245–254.

Lavigne-Cruège, V. (1996). Recherches sur les composes volatils soufres formes par la levure au cours de la vinification et de l'élevage des vins blancs secs. Thèse doctorat, Université de Bordeaux II.

Le Traon-Masson, M-P., & Pellerin, P. (1998). Purification and characterization of two β-D-glucosidases from an *Aspergillus niger* enzyme preparation: affinity and specificity toward glucosylated compounds characteristic of the processing of fruits. *Enzyme Microb. Technol., 22*, 374–382.

Lilly, M., Lambrechts, M.G. & Pretorius, I.S. (2000). Effect of increased yeast alcohol acetyltransferase an activity on the flavour profile of wine and distillates. *App. Environ. Microbiol., 66*, 744–753.

Lilly, M., Bauer, F.F., Lambrechts, M.G., Swiegers, J.H., Cozzolino, D. & Pretorius, I.S. (2006a). The effect of increased yeast alcohol acetyltransferase and esterase activity on the flavour profile of wine and distillates. *Yeast, 23*, 641–659.

Lilly, M., Bauer, F.F., Styger, G., Lambrechts, M.G., & Pretorius, I.S. (2006b).The effect of increased branched-chain amino acid transaminase activity in yeast on the production of higher alcohols and on the flavour profiles of wine and distillates. *FEMS Yeast Res., 6*, 726–743.

Lopez, R., Ortin, N., Perez-Trujillo, J.P., Cacho J. & Ferreira, V. (2003) Impact odorants of different young white wines from the Canary Islands. *J. Agric. Food Chem., 51*, 3419–3425.

Loscos, N., Hernandez-Orte, P., Cacho, J., Ferreira, V. (2007) Release and Formation of Varietal Aroma Compounds during Alcoholic Fermentation from Nonfloral Grape Odorless Flavor Precursors Fractions. *J. Agric. Food Chem., 55*, 6674–6684.

Louw, C., La Grange, D., Pretorius I.S & van Rensburg P. (2006). The effect of polysaccharide-degrading wine yeast transformants on the efficiency of wine processing and wine flavour. *J. Biotechnol., 125*, 447–461.

Maicas, S., Gil, J.-V., Pardo, I. & Ferrer, S. (1999). Improvement of volatile composition of wines by controlled addition of malolactic bacteria. *Food Res. Inter., 32*, 491–496.

Mason, A.B. & Dufour, J-P. (2000). Alcohol acetyltranferase and the significance of ester synthesis in yeast. *Yeast, 16*, 1278–1298.

Mateo, J.J. & Di Stefano, R. (1997). Description of the β-glucosidase activity of wine yeasts. *Food Microbiol., 14*, 583–591.

McMahon, H., Zoecklein, B. W., Fugelsang, K. & Jasinsky, Y. (1999). Quantification of glycosidase activities in selected yeasts and lactic acid bacteria. *J. Ind. Microbiol. Biotechnol., 23*, 198–203.

Moio, L., M. Ugliano, A. Genovese, A. Gambuti, R. Pessina, & P Piombino (2004). Effect of antioxidant protection of must on volatile compounds and aroma shelf life of Falanghina (*Vitis vinifera* L.) wine. *J. Agric. Food Chem., 52*, 891–897.

Muller-Späth, H. (1990). Historique des expérimentations de vinification sans SO_2 et par oxygénation. *Rev. Fran. Œnol., 124*, 5–12.

Nagel, C.W. & Graber, W.R. (1988). Effect of must oxidation on quality of white wines. *Am. J. Enol. Vitic., 39*, 1–4.

Paetzold, M., Dulau, L. & Dubourdieu, D. (1990). Fractionnement et caractérisation des glycoproteines dans les moûts de raisins blancs. *J. Inter. Sci. Vigne Vin, 24*, 13–28.

Perez-Gonzalez, J.A., Gonzalez, R., Querol, A., Sendra, J. & Ramon, D. (1993). Constructionof a recombinant wine yeast strain expressing b-(1,4)-endoglucanase and its use in mictovinication process. *App. Environ. Microbiol., 59*, 2801–2806.

Pocock, K.F., Høj, P.B., Adams, K.S., Kwiatkowsky, M.J. & Waters, E.J. (2003). Combined heat and proteolytic enzyme treatment of white wines reduces haze forming protein content without detrimental effect. *Aus. J. Grape Wine Res., 9*, 56–63.

Ramey, D.D. & Ough, C.S. (1980). Volatile ester hydrolysis or formation during storage of model solutions and wines. *J. Agric. Food Chem., 28*, 928–934.

Revilla, I., & González-SanJosé, M.L. (1998). Methanol release during fermentation of red grapes treated with pectolytic enzymes. *Food Chem., 80*, 205–214.

Ribereau-Gayon, P., Dubourdieu, D., Donèche, B. & Lonvaud, A. (2006). Handobbok of Oenology, Vol I. John Wiley & Sons Ltd, Chichester, England.

Rigaud, J., Cheynier, V., Souquet, J.M. & Moutonet, M. (1990). Mécanismes d'oxydation des polyphenols dans les mouts blancs. *Rev. Fran. Oenol., 124*, 27–31.

Rosi, I., Vinnella, M. & Domizio, P. (1994). Characterization of β-glucosidase activity in yeast of enological origin. *J. App. Bacteriol., 77*, 519–527.

Saerens, S.M.G, Verstappen, K.J., Van Laere, S.D.M., Voet, A.R.D.Van Dijck, P., Delvaux, F.R. & Thevelein, J.M. (2006). The *Saccharomyces cerevisiae EHT1* and *EEB1* genes encode novel enzymes with medium-chain fatty acid ethyl ester synthesis and hydrolysis capacity. *J. Biol. Chem., 281*, 4446–4456.

Salgues, M., Cheynier, V., Günata, Z. & Wylde, R. (1986). Oxidation of grape juice 2-S-glutathionyl caffeoyl tartaric acid by Botrytis cinerea laccase and characterization of a new substance: 2,5-di-S-glutathionyl caffeoyl tartaric acid. *J. Food Sci., 51*, 1191–1194.

Sarry, J.-E. & Günata, Y.Z. (2004). Plant and microbial glycoside hydrolases: volatiles release from glycosidic aroma precursors. *Food Chem., 87*, 509–521.

Schneider, V. (1998). Must Hyperoxidation: A Review. *Am. J. Enol. Vitic., 49*, 65–73.

Sefton, M.A. (1998). Hydrolytically-released volatile secondary metabolites from a juice sample of Vitis vinifera cvs Merlot and Cabernet Sauvignon. *Aus. J. Grape Wine Res., 4*, 30–38.

Sefton, M.A., Skouroumounis, G.K., Massy-Westropp, R.A. & Williams, P.J. (1989). Norisoprenoids in Vitis vinifera white wine grapes and the identification of a precursor of β-damascenone in these fruits. *Aus. J. Chem., 42*, 2071–2084.

Sefton, M.A., Francis, I.L. & Williams, P.J. (1993). The volatile composition of Chardonnay juices. A study by flavor precursors analysis. *Am. J. Enol. Vitic., 44*, 359–370.

Sefton, M.A., Francis, I.L. & Williams, P.J. (1994). Free and bound secondary metabolites of *Vitis vinifera* grape cv. Sauvignon blanc. *J. Food Sci., 59*, 142–147.

Singleton, V.L. (1987). Oxygen with Phenols and Related Reactions in Musts, Wines, and Model Systems: Observations and Practical Implications. *Am. J. Enol. Vitic., 38*, 69–77.

Singleton, V.L., Sieberhagen, P.H.A., De Wet, P. & Van Wyk C.J. (1975). Composition and sensory qualities of wines prepared from white grapes by fermentation with and without grape solids. *Am. J. Enol. Vitic., 26*, 62–69.

Skouroumounis, G.K., Massy- Westropp, R.A., Sefton, M.A., Williams, P.J. (1993). ß-Damascenone Formation in juices and wines In P. Schreier & P. Winterhalter (Eds.), *Progress in Flavour Precursor Studies*, (pp. 275–278), Allured, Publishing Corp., Carol Stream, IL.

Skouroumounis, G.K. & Sefton, M.A. (2000). Acid-catalyzed hydrolysis of alcohols and their β-D-glucopyranosides. *J. Agric. Food Chem., 48*, 2033–2039.

Swiegers, J.H., Francis, I.L., Herderich, M. & Pretorius, I.S. (2006). Meeting consumer expectations through management in vineyard and winery, the choice of yeast for fermentation offers great potential to adjust the aroma of Sauvignon Blanc. *Wine Ind. J., 21*, 34–42.

Swiegers, J.H., Capone, D.L., Pardon, K.H., Elsey, G.M., Sefton, M.A., Francis, I.L. & Pretorius, I.S. (2007). Engineering volatile thiol release in *Saccharomyces cerevisiae* for improved wine aroma. *Yeast, 24*, 561–574.

Tamborra, P., Martino, N. & Esti, M. (2004). Laboratory test on glycosidase preparations in wine. *Anal. Chim. Acta, 513*, 299–303.

Tominaga, T., Peyrot de Gachons. C. & Dubourdieu, D. (1998). A new type of flavour precursors in *Vitis vinifera* L. cv. Sauvignon blanc: S-cysteine conjugates. *J. Agric. Food Chem., 46*, 5215–5219.

Tominaga, T., Baltenweck-Guyot, C., Peyrot de Gachons. C. & Dubourdieu, D. (2000). Contribution of volatile thiols to the aromas of white wines made from several *Vitis vinifera* grape varieties. *Am. J. Enol. Vitic., 51*, 178–181.

Ugliano, M. & Moio, L. (2005). Changes in the concentration of yeast-derived volatile compounds of red wine during malolactic fermentation with four commercial starter cultures of *Oenococcus oeni.J. Agric. Food Chem., 53*, 10134–10139.

Ugliano, M., & Moio, L. (2006). The influence of malolactic fermentation and *Oenococcus oeni* strain on glycosidic aroma precursors and related volatile compounds of red wine. *J. Sci. Food Agric., 86*, 2468–2476.

Ugliano, M., Genovese, A., & Moio, L. (2003). Hydrolysis of wine aroma precursors during malolactic fermentation with four commercial starter cultures of *Oenococcus oeni*. J*J. Agric. Food Chem., 51*, 5073–5078.

Ugliano, M., Bartowsky E.J., McCarthy, J., Moio, L. & Henschke P.A. (2006). Hydrolysis and transformation of grape glycosidically bound volatile compounds during fermentation with three *Saccharomyces* yeast strains. *J. Agric. Food Chem., 54*, 6322–6331.

Usseglio-Tomasset, L. (1978). Acquisitions récentes sur les phénomènes colloïdaux dans les mouts et les vins. *Ann. Technol. Agric.,* 27, 261–274.

Van Rensburg, P., T. Stidwell, M.G. Lambrechts, R.R. Cordero Otero & I.S. Pretorius. (2005). Development and assessment of a recombinant Saccharomyces cerevisiae wine yeast producing two aroma-enhancing β glucosidases encoded by the Saccharomycopsis fibuligera BGL1 and BGL2 genes. *Annal. of Microbiology 55*, 33–42.

Vidal, S., Williams, P., Doco, T., Moutounet, M. & Pellerin, P. (2003). The polysaccharides of red wine: Total fractionation and characterization. *Carbohydr. Polym., 54*, 439–447.

Voirin, S.G., Baumes, R., Bitteur, S.M., Günata, Z.Y. & Bayonove, C. (1990). Novel monoterpene disaccharide glycosides in *Vitis vinifera* grapes. *J. Agric. Food Chem., 83*, 1373–1378.

Voirin, S.G., Baumes, R., Sapis, C.L. & Bayonove, C. (1992). Analytical methods for monoterpene glycosides in grape and wine II. Qualitative and quantitative determination of monoterpene glycosides in grape. *J. Chromatogr., 595*, 269–281.

Waters, E.J. (1991). Heat unstable wine proteins and their interactions with wine polysaccharides. PhD thesis. University of Adelaide, Australia.

Waters, E.J., Shirley N.J. & Williams, P.J. (1996). Nuisance proteins of wine are grape pathogenesis-related proteins. *Aus. J. Grape Wine Res., 1*, 86–93.

Williams, P.J., Strauss, C.R., Wilson, B. & Massy-Westropp, R. (1982). Novel monoterpene disaccharide precursors of *Vitis vinifera* grapes and wine. *Phytochemistry, 12*, 2013–2020.

Williams, P. J., Francis, I.L. & Black. S. (1996). Changes in concentration of juice and must glycosides, including flavour precursors, during primary fermentation. In T. Henick-Kling & Wolf, T. (Eds.), *Proceedings of the IVth International Symposium on cool climate viticulture and enology* (pp. VI-5–VI-9). Rochester, NY.

Winterhalter, P. (1991). 1,1,6-trimethyl-1,2-dihydronaphtalene (TDN) formation in wine. 1. Studies on the hydrolysis of 2,6,10,10-tetramethyl-1-oxaspiro-[4,5] -dec-6-ene-2,8-diol rationalizing the origin of TDN and related C_{13} norisoprenoids in Riesling wine. *J. Agric. Food Chem., 39*, 1825–1829.

Winterhalter, P., Sefton, M.A., & Williams, P.J.. (1990). Two-dimensional GC-DCCC analysis of the glycoconjugates of monoterpenes, norisoprenoids, and shikimate-derived metabolites form Riesling wine. *J. Agric. Food Chem., 38*, 1041–1048.

Winterhalter, P., Sefton, M. A. & P.J. Williams. (1991). Volatile C_{13} norisoprenoids are generated from multiple precursors. *Am. J. Enol. Vitic., 41*, 277–283.

Wirth, J, Guo, W., Baumes, R. & Günata, Y. Z. (2002). Volatile compounds released by enzymatic hydrolysis of glycoconjugates of leaves and grape berries from *Vitis vinifera* Muscat of Alexandria and Shiraz cultivars. *J. Agric. Food Chem., 49*, 2917–2923.

Zoecklein, B.W., Marcy, J.E. & Jasinsky, Y. (1997a). Effect of fermentation, storage sur lie or post-fermentation thermal processing on White Riesling (*Vitis vinifera* L.) glycoconjugates. *Am. J. Enol. Vitic., 48*, 397–402.

Zoecklein, B.W., Marcy, J.E., Williams, J.M. & Jasinsky, Y. (1997b). Effect of native strain of Saccharomyces cerevisiae on glycosyl-glucose, potential volatile terpenes, and selected aglycones of White Riesling (*Vitis vinifera* L.) wines. *J. Food Comp. Anal., 10*, 55–65.

Chapter 5
Use of Enological Additives for Colloid and Tartrate Salt Stabilization in White Wines and for Improvement of Sparkling Wine Foaming Properties

Richard Marchal and Philippe Jeandet

Contents

5.1 Traditional White Wine Fining

After completion of the alcoholic fermentation in tanks or barrels, wine becomes a colloidal solution as well as a colloidal suspension. The particle density, which is close to that of the wine, electric repulsion forces and diffusion phenomena lead to very slow and insufficient spontaneous clarification. Moreover, natural

R. Marchal (✉)
Laboratory of Enology and Applied Chemistry, Research Unit N° 2069, Faculty of Science, University of Reims, PO Box 1039, 51687 Reims cedex 02, France
e-mail: richard.marchal@univ-reims.fr

Philippe Jeandet (✉)
Laboratory of Enology and Applied Chemistry, Research Unit N° 2069, Faculty of Science, University of Reims, PO Box 1039, 51687 Reims cedex 02, France
e-mail: philippe.jeandet@univ-reims.fr

M.V. Moreno-Arribas, M.C. Polo (eds.), *Wine Chemistry and Biochemistry*,
DOI 10.1007/978-0-387-74118-5_6, © Springer Science+Business Media, LLC 2009

sedimentation, centrifugation and clarification with continuous alluviation filters do not protect wines against colloidal haze. For these reasons, both organic and inorganic fining agents are still commonly used to clarify and stabilize white wines, thus avoiding natural haze. Fining involves the formation of a floccular precipitate in wine which will absorb the natural haze-forming constituents while settling. After a settling period, the supernatant can be withdrawn and receive a polishing filtration prior to sale. It is important that the settling process should be efficient in the removal of the natural haze. More rarely, juice particles are eliminated using a flotation technique, but the objective is the same. The fining process must also be reasonably rapid, and the loss of saleable products in the sediment or lees should be minimal. Finally, the clarified beverage should remain clear and neither the clarification nor the fining process should have any undesirable effects, like removal of desired flavorants or addition of undesired flavor components.

At this time, the chemistry of the fining process is poorly understood, and hence it is not surprising that this process is often considered as an art rather than a science. For this reason, the fining process sometimes remains inefficient. It is thus our intention here to review the chemistry of fining and present some results of our own investigations.

Fining is the addition to wine of a reactive or an adsorptive substance (in some cases, two and even three fining agents are used simultaneously) to remove or reduce the concentration of one or more undesirable constituents. Fining agents are added for the purposes of achieving clarity, color, flavor and/or stability modification in juices and wines. Fining agents are grouped according to their general nature (Zoecklein 1988a):

1. Earths: montmorillonite, bentonite, kaolinite (commonly named kaolin)
2. Animal proteins: gelatin, isinglass, caseins
3. Plant proteins: wheat gluten, soya, lupin, (garden) pea
4. Wood charcoal (carbons)
5. Synthetic polymers: PVPP
6. Silicon dioxide (kieselsols)
7. Metal chelators and enzymes (pectinases)

The most commonly used fining agents in the wine industry are "bentonites" (mainly containing montmorillonite) and proteins associated to tannins or a mineral agent. Gelatin has also been used in enology since the nineteenth century, but due to the crisis of the "mad cow" disease in 1999, plant proteins have widely been experimented and are now more and more used on an industrial scale. For this reason, this chapter will essentially focus on bentonite as well as plant proteins.

Many fining agents contain an electric charge. If this charge is the opposite of the particles in suspension, then neutralization and adsorption may occur. During fining, small particles of suspended solids coalesce so that they form larger particles which settle. In most cases, the fining agent adsorbs the suspended material and exerts some clarifying action by virtue of formation of particles of high density, thus increasing filterability.

Fining is one of the least expensive operations in wine production but one that can have the greatest impact on wine quality. Fining trials should always be done at several intervals in the winemaking process to ensure that the fining objective is achieved using the smallest possible amount of fining agent. Considering all the changes to color, flavors and aromas that occur during laboratory trials, fining is often a trade off between achieving a specific goal, such as protein stability, and producing a palatable wine. Many different fining agents can often be used to achieve the same goal. If a winemaker is not satisfied with the results of the fining trials, it is always possible to test another fining agent. To achieve consistent results, it is essential that the preparation methods, temperature, mixing and timing are the same between laboratory fining trials and winery applications. Bentonite, protein and polysaccharide fining agents must be prepared just prior to use. The efficacy of these fining agents can be reduced by 50% or more by improper preparation. Fining agents always have to be prepared in exactly the same way for both laboratory and winery uses. Preparation equipment, temperatures and timing are critical. The difference between a blender in the laboratory and a paddle mixer in the winery can be overfining.

Fining agents have to be removed from wine as they are not additives but technological "assistants" or adjuvants. Most fining agents react within seconds and the contact time between the fining agent and the wine should be as short as possible. Carbon and PVPP can be filtered out immediately or a few hours after fining. At the opposite extreme, formation of flocculates requires a few days when proteins are used (depending on wine temperature) and they require a week or two to settle.

5.1.1 Bentonite Fining of White Juices and Wines

Bentonite has been the most commonly used fining agent in the wine industry since the works of Saywell (1934) in the USA and Ribéreau-Gayon and Peynaud (1935) in Europe. Authorization for the use of bentonite followed the contribution of Milisavljevic (1963). The main bentonite uses are clarification and protein stability. A major problem encountered in juice and wine production (white and rosé wines) is protein stability, that is, removal of heat-unstable proteins (Hsu and Heatherbell 1987; Waters et al. 1992). This form of instability together with potassium bitartrate precipitation (see the section "Use of adjuvants for stabilizing wine with respect to tartrate salt crystallization") are the most common non-microbiological defects in commercial wines. Bentonite fining removes both stable and unstable proteins. The goal is to lower the unstable protein content to a level at which precipitation in the bottle will not occur, while using as little bentonite as possible (Zoecklein 1988c). The use of bentonite to obtain protein stability is a somewhat confusing issue due to the variation in bentonite characteristics, the nature of wine proteins, and the different phenomena by which protein stability is determined.

Bentonite is a volcanic material which was deposited millions of years ago in broad layers, which weathered and changed from a fragile glassy state into a mineral one. This mineral is classified as a montmorillonite, which referred to the small

French town (Montmorillon) where it was first discovered. In the USA, bentonite is principally mined in Wyoming – hence the term "Wyoming clay". The type of bentonite, the source, and its purity influence its properties (Marchal et al. 1995). Bentonite is a complex hydrated aluminum silicate with exchangeable cationic components: (Al, Fe, Mg) Si_4O_{10} $(OH)_2$ (Na^+, Ca^{++}). The most commonly used bentonite form in enology is the sodium bentonite. Sodium bentonite has enhanced protein binding capabilities over calcium bentonite.

Bentonite exists as small plates which, when hydrated, separate to form a colloidal suspension with enormous surface area (300–900 m^2/g). Its subsequent activity in solution is like that of a multiplated negatively charged structure able to exchange its cations with positively charged components of the juice or the wine (not only proteins). Bentonite absorption of uncharged molecules also occurs if they are polar. Additionally, due to the fact that the platelet edges are positively charged, some limited binding of negatively charged proteins may occur.

5.1.1.1 Mechanisms of Protein Removal

The mechanisms by which polypeptides and proteins adsorb on clay minerals have been studied using homopolyamino acids (polylysine and polyglutamic acid) and a synthetic montmorillonite (Gougeon et al. 2002, 2003). The use of complementary approaches (^{13}C NMR, specific surface area by BET measurements, X-ray diffraction, adsorption/desorption isotherms) have led to a detailed description of the interactions between "positively" charged proteins and "negatively" charged plate surfaces in a synthetic wine. ^{13}C NMR spectra showed that these polypeptides, which exhibit a mixture of helical and random coil conformations in the bulk, tend to unfold and adopt a more extended random coil structure upon adsorption on phyllosilicates. Values of the basal spacing measured by XRD on dehydrated samples clearly indicated that, in the presence of adsorbed polypeptides, the silicate layers do not collapse. Incorporation of polypeptide fragments within the interlayer spaces was also revealed by the decrease in the specific surface areas. Alogether, these results prove that polypeptides are adsorbed at the periphery of the montmorillonite particles through specific interactions between the protein side chains and the silicate sheets, whereas the polypeptide backbones do not enter the interlayer spaces (Gougeon et al. 2002, 2003).

Bentonite may indirectly adsorb some phenolic compounds via binding with proteins that have complexed with phenolics. However, the amount of phenols removed is usually not high. Bentonite is also known to enhance membrane filterability. Presumably, this is due to a reduction in the colloidal particle number in suspension.

Despite the vast literature on protein instability, the actual protein levels at which wines will remain protein-stable are unknown. Wine proteins are a mixture of probably more than 100 proteins derived from the grape, yeast (Dambrouck et al. 2003), autolyzed yeast (Charpentier et al. 1986) and sometimes *Botrytis cinerea* when grape berries are not sound (Cilindre et al. 2007). Variety, vintage, maturity of the fruit, pH, and processing methodology affect both the must and the wine protein contents. Yeast proteins, however, have not been shown to play a role in white wine protein clouding. Many wine proteins are not free but bounded to a minor quantity of grape phenolics (flavonoids).

White wines contain relatively large amounts of insoluble proteins that slowly precipitate from the solution (the initial protein precipitation begins during pressing). Most white wines are to deficient in phenolics, causing a quick and complete protein precipitation. Protein haze may be due to the fraction of residual wine proteins that have been rendered prone to precipitation by their interaction with reactive phenolics (from the grape, and also rarely from the cork). Bentonite removes different amounts of grape protein fractions (Moine-Ledoux and Dubourdieu 1999).

5.1.1.2 Electric Charges of Wine Proteins

At a certain pH, the positive and negative charges of each protein fraction are equal and the protein is less soluble. This pH value is known as the isoelectric point, or isoionic point, of the protein. The lower the difference between the juice or wine pH and the isoelectric point of the protein fraction, the lower the net charge on that protein fraction and the lower the solubility of that fraction. If the juice or wine pH values are substantially different from that of the protein isoelectric point, the protein charge is high and thus the higher will be the ability of that protein to bind electrostatically to fining agents. Therefore, the isoelectric properties of proteins influence not only their natural tendency to precipitate but also their affinity to be removed with various agents. The relationships between wine pH and isoelectric points can be illustrated in the following example. The Champagne base wine pH is 3.2. Then the protein fractions having a $pI > 3.2$ will be all positively charged and those with a $pI < 3.2$ will be negatively charged. The positively charged proteins will react with a fining agent of mainly the opposite charge (–) such as bentonite. In this case, there would remain in wine the protein fractions which, due to their negative charges, are not easily removed by the use of bentonite. Those protein fractions with isoelectric points closest to pH 3.2 have a limited charge and are not able to bind electrostatically to bentonite.

Protein clouding in white wines seems to be a greater problem when the wine pH is close to the isoelectric point of the various protein fractions. This is due to the fact that bentonite will remove, preferentially, the most positively charged proteins. The electrostatic charge of various protein fractions explains the observable phenomena of not being able to stabilize certain wines with the use of bentonite alone, or only with excessive amounts that can strip the wine character. But the pI of proteins only partially explains wine haze formation. It is also important to note that other factors, as yet not clearly identified, can intervene.

5.1.1.3 Methods for Bentonite Preparation

The method of preparation significantly affects the ability of bentonite to remove wine or juice proteins. Bentonite is made up of small platelets that are separated by a layer of water molecules. During hydration, the charged platelets repel each other and pop apart. As this occurs, swelling begins. Water molecules partially neutralize the exposed surfaces holding them apart, thus exposing the large reactive surfaces. In most wineries, bentonites are hydrated by addition of water. This allows the platelets to disperse and to form a gel. Most winemakers prepare bentonite by simply adding it slowly to water and letting the slurry stand for a day or two prior to use. When properly dispersed, bentonite sets up a network commonly known as the "house of cards". This network encases droplets of water, which protect the bentonite from coalescing or flocculating with itself. In order for bentonite to be effective in binding with proteins, the bentonite platelets must be separated into a homogeneous suspension. The bentonite/water ratio in the slurry is usually 5–10% w/v. The total quantity of water must not exceed 1% of the wine volume treated. Because bentonite's protein binding activity is due to its exposed surface area, slurries for laboratory trials must be prepared exactly in the same way as suspensions used for cellar fining.

5.1.1.4 Variability of Enological Bentonites

Bentonites of various types exist in different geographical locations, are mined from different depths, and obtained in different levels of purity, particle size, adsorption capacity, and swelling ability (Marchal et al. 1995). The type and source of the bentonite used can affect protein removal. This is generally the result of variations in the swelling capacity and cation exchange capacity of the bentonite. There can be slight differences in bentonite from one shipment to another (this makes it imperative that the same lot of bentonite should be employed for both laboratory trials and cellar activity). Sodium bentonite is generally employed in enology because it has a greater swelling power than calcium bentonite. Calcium bentonite platelets tend to clump together, thus reducing the exposed surface area, and therefore, protein binding. Calcium bentonite precipitates at a slower rate than sodium bentonite but produces more compact lees. Because of its compact lees, calcium bentonite is generally preferred vs sodium bentonite as a riddling agent in the *méthode traditionnelle*, the so-called *méthode champenoise*.

5.1.1.5 Bentonite Fining and Volume of Lees

The commonly expressed problem with sodium bentonite is indeed excessive lees production. Bentonite lees volumes often range from 5% to 10%. There are several methods employed to minimize this problem, such as centrifugation or filtration with a rotary vacuum filter. Bentonite needs only minutes to react with proteins and precipitate them. Therefore, the winemaker must not let his wine or juice settle following bentonite addition but may remove bentonite and proteins "in line" with the proper filtration or centrifugation equipment. Proteins react with bentonite within the first minute of contact (Blade and Boulton 1988). In wineries, bentonite often stays in contact with wine or juice for a prolonged period of time (depending on the capacity of the tank essentially). The possibility of leaching or "sloughing off" of proteins from the bentonite platelets perhaps may occur.

An additional method for avoiding excessive lees formation in wine is to hydrate the bentonite in the wine to be fined rather than in water (Marchal et al. 2002d).

Although this can significantly reduce the bentonite's binding ability (−50%), it often produces about one half of the normal lees volume. Bentonite may also be counterfined with kieselsol (aqueous silicon dioxide) to aid in lees compaction. Gelatin is a positively charged protein which will bind with negatively charged species such as tannins kieselsols and bentonite. Gelatin can be used to help bentonite flocculation and possibly aid in lees compaction. Additionally, fining juice or wine that is already relatively free from suspended solids will minimize lees formation and, consequently, the bentonite requirement. Some winemakers prefer multiple finings with bentonite rather than a single large addition. This approach may be successful in reducing the overall bentonite requirement, particularly if the wine to be fined is free from suspended solids.

A method used to solve the problems of excessive lees and flavor stripping caused by fining wine with bentonite is to ferment in contact with bentonite. Fermentation in the presence of bentonite is an old practice used in Europe for protein stabilization. Such a practice avoids or minimizes the need for subsequent bentonite addition into wine. Fermentation in contact with bentonite has several advantages: (1) only juice components are adsorbed onto bentonite and not the fermentation-by products or barrel-aging constituents and (2) fermentation lees have a lower monetary value than do finished wine lees. Thus, protein stabilization or partial stabilization during fermentation may be an important economic consideration.

Nevertheless, addition of bentonite at the beginning of the alcoholic fermentation may deplete the assimilable nitrogen content of the must due to electrostatic binding and adsorption. This may result in fermentation sticking and/or hydrogen sulfide production. Addition of an exogenous source of nitrogen eliminates these potential problems.

5.1.1.6 Dosage of Bentonite

The quantity of bentonite to be added to the juice to reach a protein-stable wine is generally determined using a heat test (Dubourdieu et al. 1988; Marchal et al. 2002d). In addition to protein stability, juice bentonite fining can help to enhance wine filterability via general removal of suspended solids. Protein removal is not proportional to the amount of bentonite added, but follows a power law (Marchal et al. 2002d). Although complete removal of wine proteins can generally be achieved by the use of bentonite (except for yeast proteins but the latter do not really participate to protein haze), it has been recognized that this may not be necessary to obtain protein stability and may have detrimental effects on the sensory qualities of wine (wine body, color, and possibly imparting an earthy, freshly "laundered" smell).

Care must be taken when fining a sparkling wine with bentonite in order to preserve its foaming properties. Excessive use of bentonite for the fining of sparkling wine *cuvées* can produce a finished product that has a large bubble size and a poor bubble stability as a result of a reduction in both protein and peptide contents. Cold stabilization procedures cause both a precipitation of potassium bitartrate crystals as well as proteins because of the downward shift in pH. This precipitation of proteins caused by cold stabilization is why some winemakers decide to fine with bentonite during or following potassium bitartrate stabilization. In certain wines, free tartaric acid can be complexed with proteins and polyphenolics, inhibiting potassium bitartrate crystal formation. Removal of a portion of these complexing compounds with bentonite can enhance potassium bitartrate stability. Additionally, bentonite fining of wines during cold stabilization allows potassium bitartrate crystals to compact the bentonite lees.

However, because the use of bentonite alters the organoleptic characteristics of wines (Lubbers et al. 1996; Main and Morris 1994; Miller et al. 1985), other authors have searched for alternative treatments for the use of bentonites, particularly mannoproteins (see Sect. 5.3).

5.1.2 Use of Gelatin in White Wine Fining

The primary reaction occurring with gelatin is a complex formation between polyphenols in the wine and the protein of gelatine to give the desired floccular precipitate. The second reaction, less well understood but equally important, is the complex formation between the natural wine proteins and the added protein, i.e. gelatin.

5.1.2.1 The Nature of Gelatin

Gelatin is prepared from derived sources of collagen. Gelatin is a protein, that is, a polymer of amino acids joined together by peptide bonds. Hence, proteins can be depicted as long molecules with many different side chains, which accounts for their varying properties. Proline, that is, a very important amino acid of gelatin, imparts a twist to the chain and affects the shape of the protein molecule and its rigidity. The protein chain is amphoteric and can carry either a positive or a negative charge depending on the pH of the medium. In wine and beverages at a pH of 2.9–3.6, most of the amino groups are positively charged and most of the acidic groups are uncharged. The molecule then behaves as a cation. It attracts and forms polar associates with anions in solution. In addition, proteins form associations due to hydrogen bonding using the negatively charged oxygen and nitrogen atoms in the molecules.

The isoelectric point (pI) of a protein is the pH at which the protein will not migrate in an electric field. At that pH, the molecule carries an equality of positive and negative charges: the molecule is isoionic, in the absence of added ions other than hydrogen and hydroxyl ions in solution. Gelatin, is rather unique in that it can have an isoelectric point anywhere between pH 9 and pH 5, depending upon the source and method of production. Gelatins used in enology are usually derived from acid pre-treated pigskin (Type A) and have isoelectric points between 6 and 9. Gelatins with a high gel strength (Bloom strength) have the higher pI and gelatins with a low Bloom strength have a pI closer to 6. Some enological gelatins sometimes derived from limed hide or limed ossein (Type B gelatins) and all of them have a pI close to 5. The significance of pI is, of course, that the higher the pI, the greater the cationic charge on the molecule at the wine pH. In other words, at pH 2.9–3.6, all gelatins would be positively charged, but the charge density would be far higher for high pI gelatins.

5.1.2.2 Polyphenol-Protein Reaction

Both tannins and anthocyanins in wines are molecules containing benzene rings with adjacent hydroxyl groups which are proposed as a major source of hydrogen bonds at the basis of complex formation between gelatin and tannins or anthocyanins in wines. Gelatin is held to be particularly suited to hydrogen bonding because one third of the amino acids are glycine, where R = H, and hence steric hindrance to hydrogen bonding would be far less than with proteins containing less glycine. However, the tannin/gelatin complex is also very pH dependent and disappears at approximately pH 8, which would be due to both molecules becoming negatively charged and hence mutually repulsive. Hence, the role of polar bonding between molecules of dissimilar charges must not be overlooked.

5.1.2.3 Protein-Protein Interactions

Beverage proteins essentially derived from the enzymes which are responsible for the diversity of the biological processes occurring during grape maturation (plant proteins) and during the conversion of the grape juice into a wine (protein secreted by yeasts). In wine, growth and ripening enzymes of the sound grape and the fermentation enzymes provide such proteins. With time, some of these proteins can also associate to form insoluble precipitates and participate to "protein instability". For protein-protein interactions, it is necessary that the two proteins should be of opposite charges at the beverage pH for polar associations to occur. These associations lead to a reduction of hydrophilic sites and hence precipitation. Also, further hydrophobic bonding due to association of hydrophobic sites in aqueous media can lead to an increase in effective molecular weight and precipitation.

Two other fining proteins must be mentioned, namely egg albumin and casein (more precisely egg proteins and cow milk proteins). However, in both cases the floc formation is due to the insolubility of the fining protein at pHs below their pIs, hence the fining action is not the same as in the case of gelatin which is soluble at all pHs, even at its isoelectric pH.

5.1.2.4 Gelatin Fining

In Europe, most of the available gelatins is Type A pigskin gelatin (ossein gelatin, having a higher viscosity, is mainly employed in film forming applications). There is a wealth of European data which show that low Bloom strength gelatin (Low Molecular Weight) is optimum for fining (Calderon et al. 1968). This "dictate of the art" is probably due to the facility of use of LMW gelatins when compared with high molecular weight (HMW) gelatins that need to be heated before incorporation in the wine. High Bloom Strength (HMW) gelatin used in fining have to be prepared as a dilute solution of gelatin; one must remember that, on cooling, the solution will gel. Hence it is of prime importance to ensure that when a warm gelatin solution is added, it is added at a point of very intense agitation such that the small amount of gelatin is intimately mixed into a large bulk of beverages before any gelling can occur. In fact, there is really no detectable difference in performance between the use of HMW and LMW gelatins (Marchal et al. 1993). This applied to all parameters tested, i.e. sediment volume, clarity of supernatant and protein stability. Hence, the gelatin to use is often (essentially!) determined by economics (the lowest Bloom strength gelatins commend the lowest price) and facility of use.

Remark 1: Gelatin solution gel is an excellent nutrient for most forms of microbiological life. Hence, a number of manufacturers produce a highly concentrated solution of non-gelling hydrolyzed gelatin preserved with SO_2.

Remark 2: The use of finings at the pressing stage has not received much acceptance, largely, it is said, because the vintner feels he should see what the grapes are providing before he modifies it in any way. Also, addition of gelatin to cold grapes would generally lead to waste due to gelling of the solution.

Remark 3: Gelatin is primarily used to soften red wines but it is also used to reduce the phenol level and brown color in white juices before fermentation. The potential for overfining with gelatin is great. Kieselsol, a negatively charged silica compound, is recommended for white wines. Kieselsol helps to moderate the effect of gelatin on wine flavor, and it reduces the amount of gelatin needed and the volume of lees produced. Kieselsol or tannins must be added to most white wines following gelatin fining or the gelatin will remain suspended in solution.

5.1.3 Fining with Plant Proteins: A Solution for the Future

Bentonite treatments are very efficient for protein removal and can also reduce browning. But this mineral fining agent is also responsible for loss of wine aromas, a negative sensorial perception that is a serious problem in enology. The influence of treatments with tannins complexed with gelatin on wine composition and sensory perception have largely been reported in scientific studies. These have often focused on the factors contributing to protein-polyphenolic interactions, which are responsible for the expected flocculation and clarification. The incidence of the bovine spongiform encephalopathy ("mad cow disease") led to grave concerns about the use of proteins derived from animal sources in the food supply, and winemakers are encouraged to discontinue use of gelatins (especially bovine gelatin) and more generally animal proteins. In Europe, the concern of transmitting this disease to humans led to a ban on the use of bovine plasma and blood cells, commonly but incorrectly called blood albumin and glue (Regulation CE 2087/97. Council, 20 October 1997). Some winemakers are also hesitant to use caseins because of their bovine origin. The problem is the same for isinglass because of its animal origin too. Given the above, it was important to develop treatments that could replace gelatin and bentonite finings. Since 1999, many investigations have been carried out with wheat prolamins, commonly called gluten, as white musts and wines clarifying agents.

Different experimental procedures were established to compare gluten efficaciousness with usual fining agents (Lefebvre et al. 2000; Marchal et al. 2002b, c).

5.1.3.1 Comparison of Wheat Gluten and Other Fining Agents for White Wine

The results presented here concern a Chardonnay base wine from the Champagne wine-growing area (Marchal et al. 2002c). As for bentonites or gelatins, preliminary laboratory tests showed some glutens were not able to clarify white wines. This is certainly related to the low dispersion of the viscoelastic particles, leading to a small contact surface. Also, glutens composed of prolamins with a low molecular weight cannot lead to the formation of floculates with wine particles. For all of the glutens tested, an addition of 5 g/hL led to an increase in the wine turbidity (Fig. 5.1). Yet, the clarifying effect (from 10 g/hL) increased proportionally to the gluten quantity added to the wine. Also, differences between glutens were observed. Gluten 3 at 20 g/hL (see Fig. 5.1) was nearly as efficient as gluten 4 at 40 g/hL. These variations can be explained by the different treatments applied to vital gluten, and to the protein composition which depends on the wheat species. As for classical fining treatments (bentonite, gelatin, casein....), the enology literature does not provide precise information (molecular mechanisms) able to explain the differences observed.

Turbidities obtained with gluten at 20 and 40 g/hL were between the minimal and maximal values obtained with the different associations of tannins-gelatins. The clarifying effect obtained with bentonites was greater with 60 g/hL than with 30 g/hl but the increase was quite small (Fig. 5.1). Wine fining with glutens (especially gluten 3 and gluten 4 at 40 g/hL) resulted in better clarification than with the two

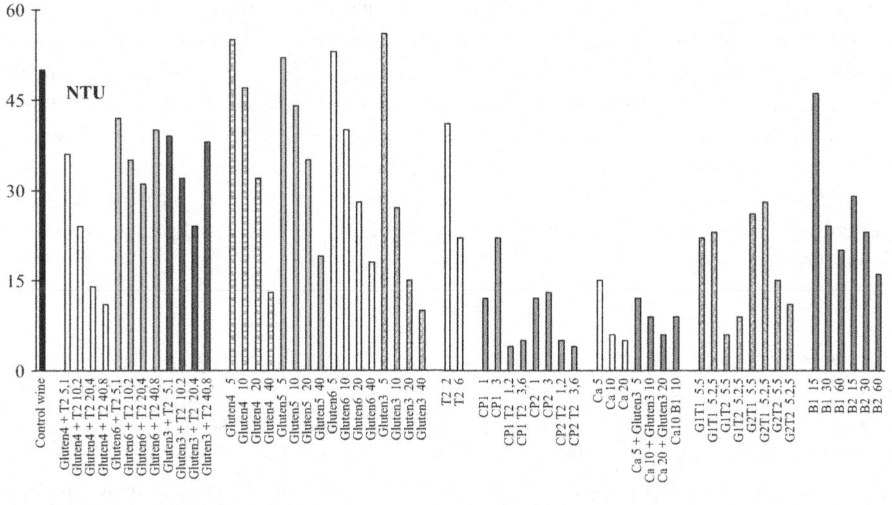

Fig. 5.1 Effect of different fining procedures on the turbidity of a Chardonnay wine. Measurements were made 28 h after treatments

bentonites tested. However, the best results were obtained with 1 g/hL fish glues (particularly when in combination with tannins) and with 10 and 20 g/hL (Fig. 5.1).

5.1.3.2 Combination of Gluten with Other Fining Agents for Chardonnay Wine Clarification

Clarifications obtained with the combination gluten 4-tannin 2 were always better than with gluten 4, only (Fig. 5.1). For example, the turbidity obtained with gluten 4 at 20 g/hL and tannin 2 at 4 g/hL was 60% lower than with gluten 4, only. One can also observe that flocculation/clarifying appeared when tannins were added in the wine without proteins (−56% for T_2 at 6 g/hL, compared to the control wine). In this experiment, there was a synergy between hydrolyzed gluten proteins and tannins. However, when the dose of gluten 4 increased (40 g/hL), the synergy between wheat proteins and tannins was less marked. The combination of gluten 3 or gluten 6 with tannins gave better results only for the dose of 5 g/hL, but the increase represented only 20% or 25% when compared to the untreated wine. With higher doses, the clarification was better with gluten 3 and 6 alone than with the combinations of gluten-tannins (except for gluten 6 at 10 g/hL and tannin 2 at 2 g/hL, but this treatment gave poor turbidity decrease and cannot be applied in the wine industry).

The combination of gluten-casein generates negative interactions. When compared to the untreated wine, gluten 3 alone or casein alone, both at 10 g/hL, led respectively to a decrease of 24 and 44 NTU (Fig. 5.1). But the combination of both in the same conditions made the turbidity fall by only 41 NTU. Therefore, the combination of the two is not advisable, nor is the combination caseins + bentonite. To sum up, these results show that gluten is really an efficient fining agent for clarifying white wines after alcoholic fermentation. Differences between the efficiencies at 28 and 48 h were quite small or even non-existent for numerous treatments. In conclusion, it seems that wine fining efficiency varies little during clarification. This makes it possible to accomplish small-scale fining tests with short reaction times. The influence of the fining type on lees volume has also to be considered as taken into account for the fining agent choice because it is related to the loss of wine. All the glutens generate volumes of lees that are similar to the values observed for casein, tannin-gelatin and fish glue. However, these volumes are smaller than those obtained with bentonite-casein or with bentonites at 30 g/hL. Globally, glutens generate volumes of lees comparable to those obtained with animal proteins used as fining agents.

5.1.4 Must Clarification Using the Flotation Technique

Some sparkling wines such as "Clairette de Die" (France), "Asti spumante" (Italy) and many others throughout the world (Spain, Portugal, Greece, Brazil for example) are sparkling wines with approximately 6–8% alcohol and 35–50 g/L residual sugars. Pectinolytic enzymes are often added to Muscat grapes in the crusher to increase the extraction of aroma compounds. The consequence of this treatment is that the must presents a high turbidity (1000–1500 NTU). During alcoholic fermentation, the must is bottled when the density is around 1040–1050 g/L; the fermentation continues in the bottle until the pressure reaches generally 6 bar, then stops spontaneously and the sediment is removed after 6–8 months bottle-aging. Riddling and disgorging are all the more easy that the turbidity of the must is low. Centrifugation after pressing or before bottling gives bad clarifications; filtration is very difficult to achieve and expensive. For these reasons, organic and mineral fining agents are commonly used to clarify white musts using the flotation technique (Davin and Sahraoui 1993; Déchaudat 1995; Ferrarini et al. 1992; Marchal et al. 2003; Sahraoui 1991).

5.1.4.1 Principle of the Flotation

Flotation is a solid-liquid separation process used when the density of the particles is lower than that of the liquid containing them. In a must, the greatest part of particles have a density close to that of the must. This density can be artificially reduced using gas bubbles that catch to insoluble particles from the grapeberry (Fig. 5.2a,b). When air bubbles are directly injected into the must to improve the separation of the particles, the process is called assisted flotation (Fig. 5.2a,ba). When air bubbles are injected into a must treated with a fining agent, the association between air bubbles and must floculates give insoluble complexes with very low density compared with that of the liquid. Then, these complexes reach easily the surface of the tank; the process is called induced flotation (Fig. 5.2a,bb). When the diameter of air bubbles is between 40 μm and 70 μm, the process is called dissolved-air flotation. The speed at which the gas-particle complexes rise through the liquid to reach the liquid/air surface largely depends on the diameter of microbubbles, the mass of particles, the

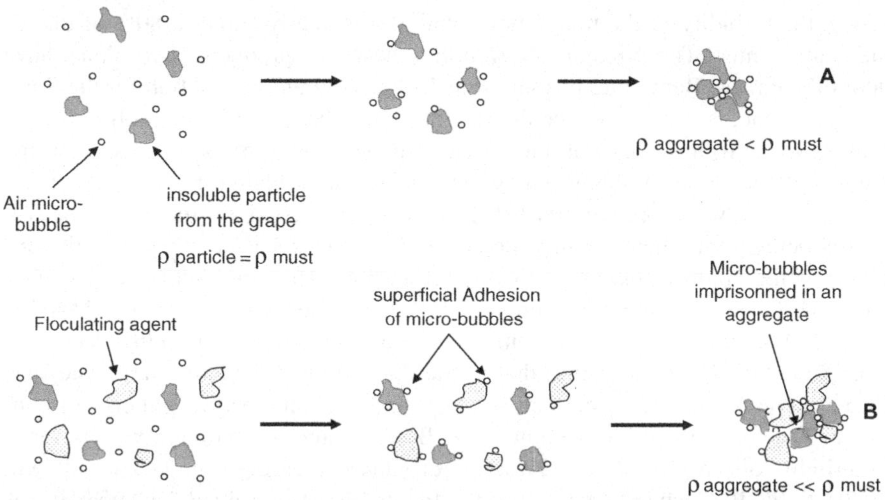

Fig. 5.2a,b Principle of the flotation

densities of the liquid, that of the gas and that of the particles. The temperature and the hydrophobicity of the particles are also important parameters for the must clarification using the flotation technique.

Density of must particles depends on the flocculation conditions following pressing. At first, mini-flocculates are formed. We observe after this an aggregation of these flocculates, leading to larger flocculates separable from the liquid. This second step is made easier when the must is treated with pectolytic enzymes, in the crusher or just after pressing. The vintage, the variety, and the maturity of grape berry also influence flocculation. In enology it is possible to use flocculating agents to bind particles. These agents are mineral adjuvants (bentonite, silica gel) or natural polymers extracted from animal substances (pork, bovine and fish gelatins). Proteins isolated from plants (wheat gluten, lupin, pea) are suitable alternatives.

5.1.4.2 Flotation Trials on a Laboratory Scale

Laboratory experiments were carried out with a miniflotator (Fig. 5.3). The capacity of the stainless steel tank was 1.5 l. The enzymed must (1 L) was poured into the tank. Fining agents were added with automatic pipettes. The miniflotator was then closed and turned over twice to homogenize fining agents. After this, the must was pressurized (6 bar) using the industrial pressurized air system. The miniflotator was then turned over 20 times to dissolve and saturate the liquid with the air. The must was poured into a graduated cylinder using a stainless pipe. After3 min a sample was taken as for static fining and the turbidity was measured. During the depressurization of the must (previously at 6 bar), micro-bubbles are formed because the liquid becomes supersaturated with air. This assisted flotation (Fig. 5.2a) gave a turbidity equal to 31% of that of the nontreated must (without any fining agent), but the turbidity still remained at 270 NTU, a value too high for this wine process (Fig. 5.4). When a mineral flocculating agent such as bentonite or silica gel was added, the turbidity of the must after flotation was nearly equal to 80% of that of the control must. These results showed that these two products, used alone, have poor efficiencies. Bentonite, in spite of its bad clarifying/fining action for the flotation technique, is used to reduce the risk of protein haze that is often observed for Muscat wines if musts are not fined. Bentonite was only (and is still) used for this main enological reason (that is a major and universal problem for white wines).

In contrast, wheat gluten and fish gelatin allowed good clarifications, the fish gelatin being a little better than gluten (−95.1% and −88.5% compared to the turbidity of the control must, respectively) (Fig. 5.4). The combination $B_{20} + FG_{10}$ gave an efficiency comparable with that of the must fined with FG_{10} alone (Marchal et al. 2003). When gluten was combined with bentonite, the turbidity decreased from 36% to 27%, compared to that of the flotated must without fining. The clarification was all the more marked if silica gel was combined to gluten and bentonite. The comparison $B_{20} + \text{Gluten}_{20}$ vs $B_{20} + \text{Gluten}_{20} + \text{Si}_{10}$ gave a difference in turbidity of DNTU = 23 (−32%); the turbidity decreased from 27% (−73%) to 20.7% (−79.3%), when compared to the flotated must (not fined); whereas it was −93.5% compared to the control must (no flotation). On the basis of these results,

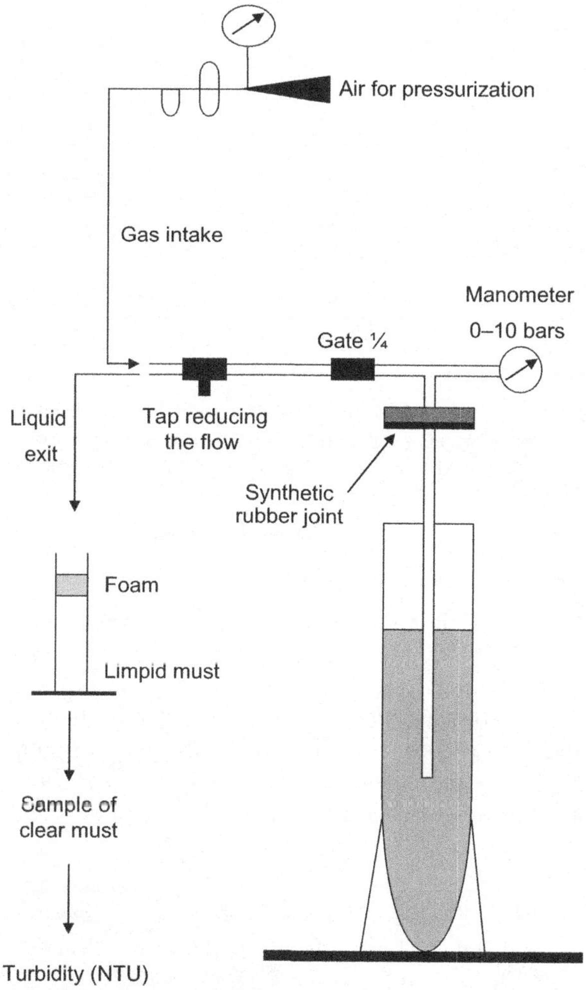

Fig. 5.3 Description of the laboratory flotation system

one can consider that the use of silica gel significantly improved the efficiency of the flotation technique when the protein used was the wheat gluten. The difference was less clear when the protein used was fish glue and it is not possible to write that $B_{20} + FG_{10} + Si_{10}$ (NTU = 35) was more efficient than FG_{10} (NTU = 42), probably because bentonite + fish glue gave together a turbidity already very low (Marchal et al. 2003).

Lastly, the combination $Gluten_{20} + B_{20} + Si_{10}$ was less efficient than the treatment $FG_{10} + B_{20} + Si_{10}$ (−87% and −95.9% in turbidity, compared to that of the flotated must and the non-treated must, respectively). Here again, results were confirmed with other experiments during the two previous harvests (Clairette and Muscat musts).

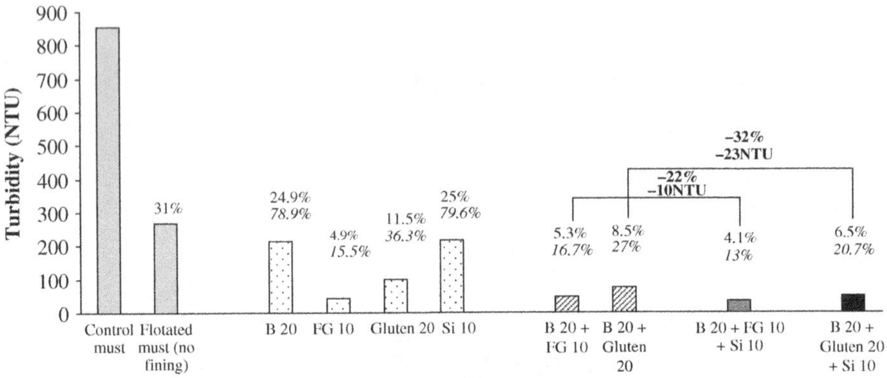

Fig. 5.4 Clarification of a Muscat must using the flotation technique (laboratory experiments). Influence of different fining agents (each value is the average of two experiments). *Upper values* (%): residual NTU compared to the control must; *lower values* (% in italics): residual NTU compared to the flotated must

5.1.4.3 Industrial Flotation Experiments

Trials were conducted on an industrial scale (Fig. 5.5). After fining, the must was pressurized with air in the saturation column (6 bar) and sent (140 hL/h) into the flotation tank (80 hL). An aspiration system collected the flotation foam (60 hL) on the top of the flotation tank. Particles of the foam were removed with a rotary filter and a clear must (35 hL) was obtained. In the following example, two identical tanks, containing exactly the same must, were treated with $B_{20} + Si_{10}$ + gluten 20 or with $B_{20} + Si_{10} + FG_{20}$. (FG, fish glue soluble gelatin; B, bentonite; Si, silica gel;

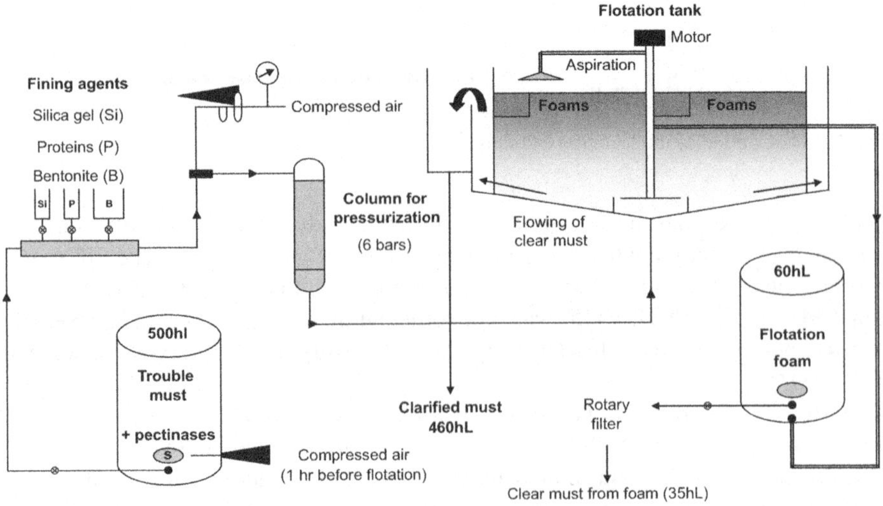

Fig. 5.5 Description of the industrial flotator

the number indicates the dosage in g/hL) (Fig. 5.5) (Marchal et al. 2003). The industrial experiment was made twice and the musts obtained with the same treatment were blended. So, 1000 hL of must clarified by flotation with gluten as protein fining agent were obtained, and also 1000 hL of must clarified by flotation using fish gelatin as protein fining agent.

Measurements were made just after flotation (between the flotation tank and the temporary storage tank). Turbidities decreased by 95.9% and 96.7% when the fining treatments were $B_{20} + Si_{10}$ + gluten 20 and $B_{20} + Si10 + FG_{10}$, respectively (Fig. 5.6a–c) when compared with the control must without flotation (1446 NTU). The turbidity of the must flotated with gluten as a fining protein was 20% higher than that of the must flotated with FG as a fining protein (Fig. 5.6a–cb). The same treatments were experimented with the laboratory flotator, with the same must. Results (Fig. 5.6a–ca) confirmed the good efficaciousness of gluten (−95.4% of the initial NTU) compared to that of FG (−97.5% of the initial NTU). If we now observe the residual turbidities, the must flotated with FG_{10} has a turbidity lower by 44% than that of the must flotated with gluten. Differences in turbidity were twice higher on a laboratory scale than at an industrial scale. Turbidities were also measured in temporary storage tanks (500-hL), 14 h after the industrial clarification by flotation. The turbidity of the must flotated with gluten was 14% lower than that of the must flotated with FG_{10} (Fig. 5.6a–cc). During these 14 h, a static settling occurred in the tank. Then, results of static settling (when the must was fined and clarified by sedimentation of flocculates) indicated a higher efficacy for fining with gluten, even if the kinetics was longer. One can hypothesize that small flocculates, present in the must after gluten fining, have higher density than flocculates formed with fish gelatin. According to this hypothesis: (1) during the flotation, the higher

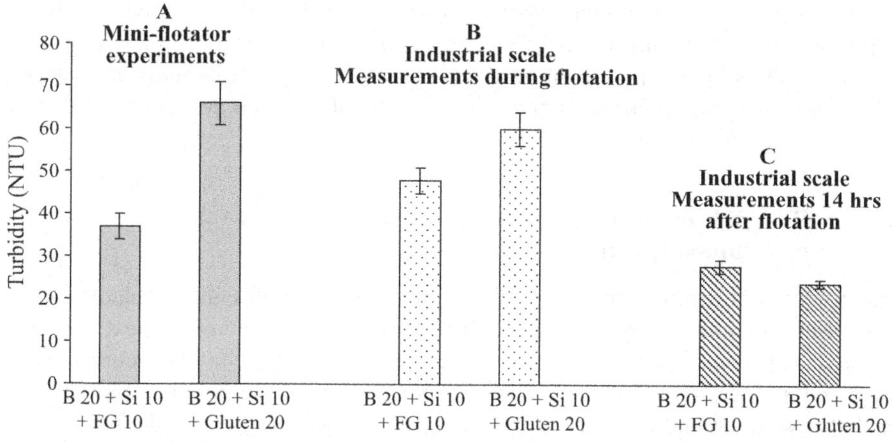

Fig. 5.6a–c Clarification of a Muscat must using the flotation technique. Comparison between industrial trials and laboratory experiments (three measurements), during and 14 h after flotation (average of three sampling). For industrial trials, measurements are the average of six measures made during the process (15 min between each sampling)

mass will reduce the velocity of the flocculates rising through the liquid and the flotation duration (approximately 15 min. at an industrial scale), which is not sufficient to allow a perfect clarification, and (2) during static settling, the sedimentation velocity will be higher for the same reason. On a production scale, clarification occurred in two steps: during flotation, FG_{10} was more efficient than gluten and gave lower turbidities; during the second step, the sedimentation of particles led to lower turbidities when gluten was added to the must before flotation (always combined with bentonite and silica gel in this study). To improve the efficacy of gluten fining, one can envisage reduction of the output of the pump that fills up the flotation tank; then flocculates will have more time to reach the surface. Theoretically, this should enable one to obtain clearer musts.

5.1.4.4 Flotation with Lupin Proteins

On the laboratory scale, turbidities of the musts fined with lupin or gluten were nearly equal after flotation, and higher than the turbidity of the must clarified with FG. However, industrial trials have shown that differences between plant and animal protein efficiencies were lower than those obtained using a laboratory flotator. More, the sedimentation step following the flotation reduces these differences.

On the basis of the molecular composition of wheat gluten (Babiker et al. 1996; Gupta and Shepherd 1990; Popineau and Denery-Papini 1996; Popineau and Pineau 1987; Shewry et al. 1994) and lupin proteins (Duranti et al. 1988, 1992), it is not possible at present to understand why this gluten is a little better than this lupin protein isolate. Glutens are composed of numerous gliadins and glutenins (Gupta and Shepherd 1990). Gliadins present differences in their amino acid composition, the consequence being a difference of wheat protein hydrophobicities (Popineau and Pineau 1987) and pH_i values. These biochemical characteristics are also induced by industrial processes (partial enzymatic hydrolysis for the gluten, no hydrolysis for lupin proteins). Now, these biochemical characteristics are responsible for protein and wine phenolic compound interactions, leading to flocculation and clarification. Moreover, the must is a very complex biochemical medium containing colloidal particles and soluble colloids.

5.1.4.5 Research of Residual Gluten Proteins in Muscat by Immunodetection

The research of residual gluten proteins, in Muscat musts and musts obtained from foam filtration, was assayed using an ELISA method. The gluten content was calculated with regard to two standard curves (10–200 μg/L). Standard curves were obtained with gliadins extracted from the gluten used for the flotation experiments and also with a blend of gliadins extracted from 30 wheat glutens. The two curves were superimposable. It is particularly important to insist on the detection of possible residual gluten, because it has clearly been shown that animal fining proteins are not completely removed by the settling and the filtration of treated wines (Douet et al. 1999). Coeliac disease (nontropical sprue) is an intolerance food and susceptible individuals show lesions of the intestinal mucosa after consumption of the gliadins of wheat. The drastic decrease in absorptive surface is the cause for the malabsorption accompanied by weight loss and disturbance of growth (Cornell et al. 1992; Sturgess et al. 1991). By using an ELISA immunotechnique, gluten-antibodies were not able to recognize their antigens in the Muscat must treated with 20 g/hL gluten, these values corresponding to the detection limits of the ELISA method used. This result clearly indicates the absence of residual gluten proteins in the must (less than 1 mg/L, that is the limit of the kit detection). For the Codex Alimentarius, a food is gluten-free when it contains less than 10 mg/L gluten. In this assay, the white must clarified with wheat gluten derivatives may thus be considered as gluten-free. These results seriously strengthen the interest one can take in the use of gluten to clarify musts.

This study shows that gluten proteins (combined with bentonite and silica gel) allowed very efficient clarification of the treated must when compared with the control must. More, the industrial scale trials showed that grape juice clarification using the flotation technique was a little better with bentonite-gluten than with the bentonite-fish gelatin combination which are the more commonly used fining agents. This confirms the possibility for plant proteins to replace animal proteins, as already observed for red and white wines (Marchal et al. 2002b,c).

5.1.5 Other Fining Agents

5.1.5.1 Polyvinyl Polypyrrolidone (PVPP)

PVPP is a high molecular weight fining agent made of crosslinked monomers of polyvinlypyrrolidone. It complexes with phenolic and polyphenolic components in wine by adsorption and attracts low molecular weight catechins. It removes bitter compounds and browning precursors in both red and white wines. PVPP is quick acting with no preparation required. Wines must be filtered to remove the PVPP and wines may seem more astringent when the bitter compounds are removed.

5.1.5.2 Kieselsol

Kieselsol is a generic name for aqueous suspensions of silicon dioxide. Kieselsol is a byproduct of the glass industry. Kieselsols are produced in Germany and are sold as 30% colloid solutions. The primary use of kieselsol is for clarification and as a replacement for tannins during gelatin fining of white wines. Kieselsols are negatively charged and electrostatically bind to and adsorb positively charged proteins and initiate flocculation and settling. Several different kieselsol formulations are available at a variety of pH levels. It is necessary to use a kieselsol that is recommended for wine. In general, kieselsol is used at a rate of seven times the amount of gelatin. Gelatin should be added first, and fining trials must be done to insure proper settling.

5.1.5.3 Isinglass

Isinglass is a positively charged fining agent derived from the air bladder of fish. It is available as sheet or flocculated isinglass. The flocculated form is the easiest form to work with, because it does not have to be rinsed to remove fishy odors. Isinglass is used principally in white still and sparkling wines to clean up the aromas, improve clarity and modify the finish without significantly modifying tannin levels. Usage levels are typically from 1 g/hL to 3 g/hL (Marchal et al. 2003).

5.1.5.4 Casein

Casein is nearly insoluble and must be dissolved at pH 11. Potassium caseinate is water-soluble and is preferred for this reason. Sodium caseinate is usually not used because it increases the sodium content of wine. Casein is a positively charged protein that flocculates in acidic media such as wine. When added to wine, casein adsorbs and mechanically removes suspended materials as it settles. In general, casein is used to remove undesirable odors, to bleach color and to clarify white wines. It is sometimes used as a substitute for carbon in color modification of juice and white wine. Usage levels are typically 10–30 g/hL.

5.1.5.5 Carbon

Activated carbons are nonspecific adsorptive agents made from wood. The sponge-like carbon binds with weakly polar molecules, especially those containing benzene rings. Carbon effectively removes phenolic compounds, especially small phenolic compounds. Compounds larger than dimers are too large to be adsorbed. Stripping of wine is often a problem with carbon because of its low selectivity and great care has to be taken with its use. Carbon also contains a large quantity of air, and oxidation sometimes follows carbon addition if the carbon is not quickly and thoroughly removed. The addition of carbon to juice rather than wine helps to diminish carbon-induced oxidation. Usage levels are typically 10–50 g/hL.

5.1.6 Equipment for the Addition of Fining Agents to Wine

The efficaciousness of fining is dependent upon the agent, the method of preparation and addition, the quantity employed, the pH, the metal content, the temperature, the age of the wine, and also previous treatments undergone by the wine. Fining is a surface action performed by the agent (adsorption); therefore, the method of hydration and addition of the agent is of extreme importance. Four common methods of adding fining agents are: (1) through a "Y" on the suction side of a positive displacement pump while transferring or mixing; (2) through an "in line" proportioning pump; (3) through a "T" into a Guth-type tank mixer; (4) added slowly in slurry form to a barrel using a dowel to stir in a figure-eight motion through the bung hole (Zoecklein 1988b).

5.1.7 Wine Fining: General Conclusion and Practical Recommendations

Carefully controlled laboratory fining trials must be performed before any agent is added to cellar wines. In evaluating fining trials, the winemaker must observe, note and record how each fining agent alters clarity, lees production, lees compaction, colloidal stability, color, body (front, middle, and finish), astringency, bitterness, the nose characteristics in general, the fruit, the finish, the aging potential, and overall wine palatability. To be able to duplicate laboratory trials in the cellar, the same lot of fining agent must be prepared and used in the same manner. A final analysis of protein stability should be performed just prior to bottling. Overall, winemakers have to remember than any changes in the wine pH and/or phenolic composition (oxidation) could modify protein stability.

5.2 Use of Adjuvants for Stabilizing Wine with Respect to Tartrate Salt Crystallization

The major physical instability in bottled wines remains precipitation of the tartaric salts, that is, potassium hydrogenotartrate (KHT) and calcium tartrate (TCa). Stabilization of these in bottled wines is desirable as consumers find them objectionable and an indication of poor quality control. The initial instability of KHT and TCa is caused by supersaturated levels in musts that are increased by a diminution in salt solubilities (due to the presence of ethanol) and the low temperatures used for wine storage. Thus, stabilization of wine with respect to potassium hydrogen tartrate and calcium tartrate crystallization is a critical point, especially in the winemaking process in Champagne or sparkling wine production. Before being stabilized, wines may be filtered on a simple continuous earth filter in order to remove macromolecules known as "protective colloids" which are able to inhibit crystallization of tartrate salts. The stability usually required, for instance, in Champagne wines corresponds to the temperature of −4°C. Briefly, potassium hydrogen tartrate stabilization is obtained by treating the wine with artificial cold using different technologies, namely, slow cold stabilization without KHT crystal seedling, rapid cold stabilization including KHT crystal seedling by the static contact process or by the dynamic continuous process.

Work has been done to find whether the very expensive treatment of wines with artificial cold could be advantageously replaced by the addition of inhibitors of the crystallization process of tartrate salts such as metatartaric acid, yeast mannoproteins or carboxymethylcellulose (CMC). Such inhibitors indeed increase the width of the supersaturation field of both KHT and TCa in the wine, thus delaying tartrate salt precipitation in the bottle. Metatartaric acid is currently the product most widely used for this purpose, though its efficacy is low as this compound does not remain stable over time. Yeast mannoproteins possess stabilizing properties, which result in the spontaneous improvement of protein and tartaric salt stability, as can be observed in white wine during its conservation on lees. Finally, experiments using carboxymethylcelluloses, polymers of β-D-glucose units whose primary and secondary alcohol groups are etherified by natrium carboxymethyl groups, are under investigation to evaluate their ability to stabilize wine as regard to tartrate salts precipitation.

5.2.1 Use of Metatartaric Acid for Stabilization of Tartrate Salt Precipitation

Metatartaric acid is a polydisperse polymer, that is, a mixture of polymers with different molecular weights as it consists of a polymerized compound formed by the intermolecular esterification between the carboxylic group of one tartaric acid unit and the secondary alcohol group of another molecule of tartaric acid (Maujean 2000). The legally imposed minimum esterification rate for metatartaric acid must be 40%. As the esterification rate than can be predicted theoretically is at equilibrium at 33%, tartaric acid solutions must be heated to 160°C in a partial vacuum to obtain higher esterification rates (Maujean 2000). Metatartaric acid contains some impurities such as oxaloacetic acid and, mainly, pyruvic acid (1—6 wt% of metatartaric acid) which are formed, respectively, by a loss of one molecule of water and the resulting decarboxylation of oxaloacetic acid. Metatartaric acid acts, as do other macromolecules known as "protective colloids", by opposing the growth of the submicroscopic nuclei around which crystals are formed. Metatartaric acid is thus able to inhibit the transfer of tartrate salt units from the bulk (the hydroalcoholic solution) to the growing crystal.

Though metatartaric acid is perfectly effective, this compound is not stable over time since a rapid hydrolysis of the ester group may take place, leading to an increase in acidity by release of tartaric acid units in wine. According to previous studies (Ribéreau-Gayon et al. 1977), a total hydrolysis of a 2% metatartaric acid solution can be observed within three months at a temperature of 23°C and ten months at 5°C followed by a decrease by 50% of the esterification rate. As a consequence, winemakers should prepare a concentrated solution of metatartaric acid, that is 200 g/L, in cold water just prior to use. Unfortunately, metatartaric acid was also proven to be unstable in wine, depending on the temperature used for wine storage. Accordingly, Ribéreau-Gayon et al. (1977) have shown that metatartaric acid is stable for several years at 0°C, over 2 years at 10–12°C, 1 year to 18 months at temperatures that are commonly found in a cellar, that is 10°C in winter and 18°C in summer, 3 months at 20°C, 1 month at 25°C, 1 week at 30°C and a few hours at temperatures ranging from 35°C to 40°C! This instability is very detrimental to the efficacy of treatments using metatartaric acid and some problems commonly occur with champagne wines that may have been stabilized with this compound. Since metatartaric acid can be removed through fining and since addition of this compound may induce an increase in wine turbidity, it is recommended to add it to wine, respectively, after the fining step and before the final clarification.

5.2.2 *Use of Yeast Mannoproteins for Stabilization of Tartrate Salt Precipitation*

Traditional practices, such as white wine aging on lees in barrels for several months, confer the wine's tartaric salt stability, which dispenses them from any cold stabilization treatment. It has indeed been observed (Moine-Ledoux and Dubourdieu 2007) that, in the Bordeaux wine-growing area, the majority of dry white wines aged on lees which are not stable in March after their first winter but become stable in June or July without any supplementary cold treatment (Table 5.1). In contrast, wines within the same crus which are not aged on lees systematically undergo cold treatment to obtain stability regarding tartaric salt crystallization.

It is well established that, during wine aging on lees, yeasts release mannoproteins that are able to prevent tartrate salt precipitation as was demonstrated first by Lubbers et al. (1993) in model wine solutions. Lubbers et al. (1993) used heat-extracted mannoproteins in alkaline buffers, that is, conditions very different from the spontaneous release of proteins during aging on lees. Nevertheless, Moine-Ledoux and Dubourdieu (1999) reported that heat-extracted mannoproteins added at a dose of 25 g/hL do not show any stabilizing effect as regard to tartrate salt precipitations in red, rosé and white wines.

Discovery of the potent crystallization inhibitor effect of mannoproteins extracted with enzymes (EEM) from yeast cell walls can be seen as an unforeseen consequence of the work of Moine-Ledoux and Dubourdieu on protein stabilization of wines. The mannoproteinfractions used for tartrate salt stabilization studies were indeed the same as those they used to reach protein stability: cell walls were treated at 40°C with an industrial preparation of the β-glucanase of *Trichoderma* sp (Glucanex-Novo), the use of which is authorized in enology for *Botrytis cinerea* β-d glucane hydrolysis (Dubourdieu et al. 1981). The resulting hydrolysate purified after ultra-filtration is a colorless and odorless solid, known as Mannostab^{TM}, perfectly soluble in water and wine. EEM used at a dose of 25 g/hL were shown to inhibit bitartrate salt precipitation in red, rosé and white wines even after having been kept at −4°C for six days (Moine-Ledoux and Dubourdieu 1999). Moreover, these authors observed that EEM are stable over time in contrast to metatartaric acid. Within 10 weeks at 30°C, wines treated with EEM remain stable as regard to

Table 5.1 Evolution of tartaric salt stability estimated by cold testing (six days at −4°C) of different white wines aged on lees throughout two vintages (Reprinted with authorization from Moine-Ledoux and Dubourdieu 2007)

Sample	March	June
Graves 1–94	Crystals present	Crystals absent
Graves 2–94	Crystals present	Crystals absent
Bordeaux 94	Crystals present	Crystals absent
Graves 1–95	Crystals present	Crystals absent
Graves 2–95	Crystals present	Crystals absent
Bordeaux 95	Crystals present	Crystals absent

Table 5.2 Tartrate stabilization of various white wines by adding Mannostab^{TM} as determined by visual observations of potassium bitartrate crystallization within six days at −4°C (redrawn with permission from Moine-Ledoux et al. 1997)

Wines		Mannostab^{TM}(g/hL)				
		0	15	20	25	30
1996 *Blanc de Blancs*	Visual test	+	0	0	0	0
	ΔK^+ (mg/L)	52	72	17	0	0
White *vin de table*	Visual test	+	0	0	0	0
	ΔK^+ (mg/L)	104	53	33	0	0
1996 white Bordeaux	Visual test	+	0	0	0	0
	ΔK^+ (mg/L)	62	21	0	0	21
1996 white Graves	Visual test	+	+	0	0	0
	ΔK^+ (mg/L)	155	52	0	0	62

bitartrate precipitations while the same wines treated with metatartaric acid were obviously proven to be unstable.

In other work, Moine-Ledoux et al. (1997) reported that the use of Mannostab^{TM} at doses ranging from 15 g/hL to 25 g/hL inhibit potassium bitartrate precipitation (Table 5.2) while excess amounts of this additive, that is 30 g/hL, are ineffective on potassium bitartrate crystallization (Table 5.2). Within the extracts, compounds responsible for the stabilizing effect observed were found to be highly glycosylated mannoproteins of molecular masses ranging from 30 kDa to 40 kDa possessing a glycosyl-phosphadityl-inositol anchor (GPI) (Moine-Ledoux and Dubourdieu 1999, 2002, 2007).

5.2.3 Use of Carboxymethylcelluloses for Stabilization of Tartrate Salt Precipitation

Natrium carboxymethylcelluloses (CMC) are polymers of β-D-glucose units whose primary or secondary alcohol groups are etherified by natrium acetate groups ($-CH_2-COONa$). This substitution is mostly 2-O- and 6-O-linked. Since CMC are polymeric substances, they may exhibit colloid protective effects against tartrate salt crystallization. CMC are a group of complex poorly-defined products, that is, dispersed polymers which are largely used in the food industry (E 468) as a thickener and stabilizing agent for instance in ice cream. CMC are defined by their polymerization and substitution (that is, the number of primary or secondary alcohol groups etherified) degrees. CMC may have different degrees of substitution, but it is generally in the range 0.6–0.95 derivatives per monomer unit. As thickeners, CMC are known to modify a wine's viscosity. By now, CMC have high purity (ca. 99.5%) with a natrium content ranging from 7% to 8.9% (Maujean 2000) and their low viscosity are no more detrimental to beverage's viscosity. CMC are sold as powders susceptible to absorb water in suspension in the air. For this reason, they must be kept in a dry place.

Use of CMC for enological applications is as yet not authorized in the European Community but, due to their capabilities (see below), a request for their use is in process. Through a collaborative study between the *Chambre d'Agriculture de la Gironde*and the Laboratory of Chemical Engineering of the University of Toulouse, it has been shown that a sodium CMC with both a low degree of substitution (between 0.65 and 0.90), a low polymerization degree and a 99.5% purity, at doses ranging from 2 g/hL to 4 g/hL, that is, doses 12- to 250-fold lower than those currently under use in the food industry, is able to stabilize various red and white wines as regard to tartrate salt crystallization (Crachereau et al. 2001). The protective effect observed was conserved even after heat treatment of the wines (55–60°C over a period of time ranging from 5 to 30 days), followed by a period at −4°C for 1 month, thus demonstrating the high thermostability of CMC compared to that of metatartaric acid (see above). Added at a dose of 4 g/hL, CMC shows a stabilizing effect similar to that obtained by using metatartaric acid at a dose of 10 g/hL (Crachereau et al. 2001). The stabilizing effect of CMC on tartrate salt precipitation both results from its capacity to reduce the transfer of bitartrate molecules from the bulk (the hydroalcoholic solution supersaturated with this salt) to growing crystals and the ability to decrease the speed with which some crystal faces grow. Gerbaux (1996) in his PhD dissertation reported that CMC specifically inhibits the growth of the face (010) of bitartrate crystals. In work developed in our laboratory (Achddou, unpublished results), CMC was shown to increase the supersaturation field of KHT in Champagne base wines, thus delaying tartrate salt precipitation.

5.3 Use of Mannoproteins to Stabilize Wines as Regard to Protein Haze

Protein haze remains as one of the key potential instabilities in white wine production that requires fining treatment with bentonites, animal proteins (casein and gelatin), or bentonites used together with casein (bentocasein), or tannins used together with gelatin (see Sect. 5.1). Haze formation is traditionally prevented by protein removal through adsorption onto bentonites, which have been used for more than 60 years as adsorbents in winemaking (Blade and Boulton 1988). This particular affinity of proteins for phyllosilicates is based on the commonly high specific-surface areas associated with swelling and cation-exchange properties of the latter. However, because this procedure alters the organoleptic characteristics of wines and since from 5% to 20% of the wine volume treated with bentonites may be lost as bentonite lees, there are numerous studies focusing on the mechanisms of protein haze and the compounds that are responsible for haze in white and rosé wines (Ferreira et al. 2004; Waters et al. 1996).

Other authors have searched for alternative treatments for the use of bentonites, particularly the yeast mannoproteins owing to the fact that a systematic improvement of protein stability can indeed be observed in white wines during aging on lees in barrels (Ledoux et al. 1992). Aged on lees, new wines become decreasingly cloudy under the effect of heat; correlatively, protein stability requires a lower dosage of bentonite. Thus, it has been observed that a racked Sauvignon wine with a turbidity of 38 NTU after completion of the alcoholic fermentation requires a bentonite treatment of 120g/hL to avoid protein haze. Within 10 months aging on lees, the same wine shows a turbidity of 12 NTU and remains stable following a bentonite treatment of only 30–40 g/hL. In further work, Moine-Ledoux and Dubourdieu (1999, 2007) demonstrated that proteins responsible for protein haze in a Sauvignon wine straight after fermentation can be separated by capillary electrophoresis into six main fractions already present in the must (Fig. 5.7), while, in the same wine kept on lees, an additional protein fraction (peak 7) was shown to appear 10 months after aging (Fig. 5.7). In contrast to proteins originating from grapes, this compound is thermostable and cannot be adsorbed by bentonites. Peak 7 which seems to play a key role in the improvement of protein stability in wines aged on lees was further purified and characterized as an invertase fragment of *Mr* 32 kDa, known as MP32 (Moine-Ledoux and Dubourdieu 1999). This protein fragment can be extracted from yeast cell walls by using Glucanex®, leading to various mannoproteins enriched in fraction 7.

Enzyme-extracted mannoproteins from the yeast cell wall added at a dose of 25 g/hL, can reduce by half the bentonite dosage necessary for protein stabilization of a very hazy wine (Table 5.3). During lees autolysis, MP32 is released from the

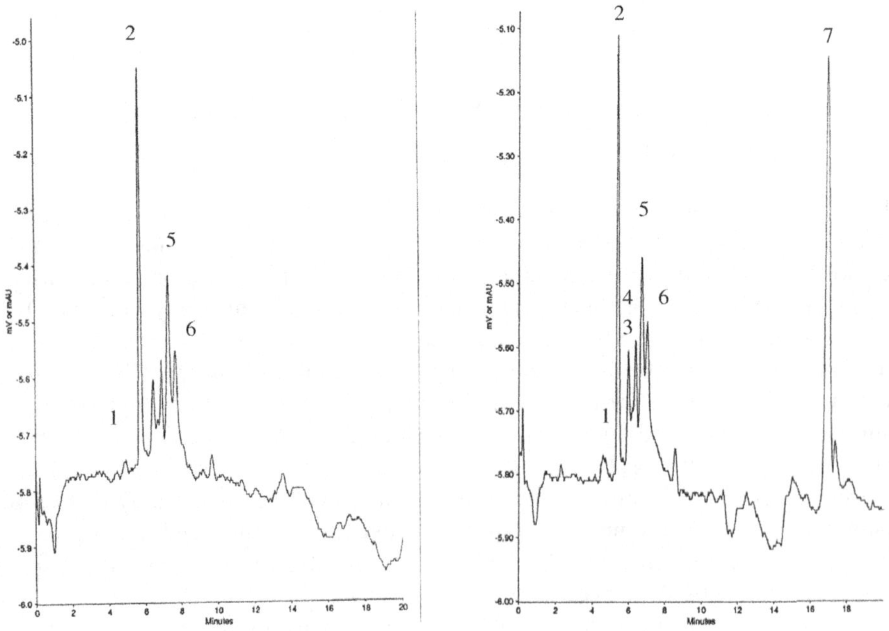

Fig. 5.7 Electrophoresis diagram of proteins in a Sauvignon wine after alcoholic fermentation (*left*) and after 10 months aging on lees (*right*) (redrawn with permission from Moine-Ledoux and Dubourdieu 2007)

Table 5.3 Protein stabilization obtained following addition of different invertase hydrolysates (redrawn with permission from Moine-Ledoux and Dubourdieu 2007)

	Turbidity (NTU)	Bentonite (g/hL)
Control wine	16	60
Wine + 25 g/hL EEM	10	30
Wine + 25 g/hL IPA	2	0
Wine + 25 g/hL IG	2	0

Table 5.4 MP32 content of EEM and of different invertase hydrolysates (redrawn with permission from Moine-Ledoux and Dubourdieu 2007)

	Percentage of MP32
Mannoproteins extracted by enzymes	18
Invertase hydrolysate by Glucanex	60
Invertase hydrolysate by protease A	65

yeast cell wall by the joint action of the parietal β glucanase and the vacuolar protease A from yeast. In vitro MP32 can be obtained by invertase hydrolysis using Glucanex. Hydrolysates obtained after hydrolysis of a commercial invertase preparation by the vacuolar protease A from yeast (IPA) and Glucanex (IG) were added to a given wine at a dose of 25 g/h. Their possible effects on protein stability and the bentonite dosage required to stabilize the wine were compared to that of yeast cell wall enzyme (Glucanex®)-extracted mannoproteins (EEM) (Table 5.3).

The invertase hydrolysate obtained by protease A (IPA) shows a remarkable heat stabilizing activity. The turbidity of the heated wine decreases from 16 to 2 NTU, an efficaciousness comparable to that of the invertase hydrolysate obtained with Glucanex® (IG) (Moine-Ledoux and Dubourdieu 1999, 2007). Likewise, analysis of the different invertase hydrolysates by capillary electrophoresis (Table 5.4) indicates that they all contain MP32 as do EEM. The hydrolysate obtained by protease A is slightly richer in MP32 than the hydrolysate obtained by Glucanex®. An extraction procedure for MP32 has been developed at an industrial scale, involving use of the same enzyme activities (glucanase and protease) (Dubourdieu and Moine 1994, 1996).

5.4 Use of Mannoprotein Fractions to Improve the Foaming Properties of Sparkling Wines

Mannoproteins have previously been shown to improve the quality of foam properties in sparkling wines (Nunez et al. 2005). We have seen that mannoproteins are currently obtained either by enzymatic or thermal procedures, or by a combination of them. The method chosen for protein extraction may greatly influence characteristics of the mannoprotein fraction obtained (Dupin et al. 2000).

It has been demonstrated that a model wine solution supplemented with a thermally-extracted mannoprotein fraction, that is, by heating 2 g of yeast cell walls at 80–85°C under stirring, shows better foaming properties than does the one supplemented with an enzymatically-extracted mannoprotein fraction, that is, 2 g of cell walls treated with Glucanex 200G (Martínez-Rodríguez et al. 2007; Nunez et al. 2006), the behavior of the last being very similar to that of the control model wine solution (Fig. 5.8). After supplementation of the model wine solution with varying, but low, concentrations of the thermally-extracted mannoproteins (0.05–0.4 mg/L), a positive correlation was observed between the quantity of mannoproteins added and the foaming parameters analyzed (Hpeak, Hplato and Sd300, that is, respectively, the maximum height reached by the foam after air injection through a glass frit, the foam height stability during air injection and the standard deviation of foam measurements in the last 300 points) (Nunez et al. 2006). Within the extracts, it was shown that the protein concentration is practically three times higher in the thermal extract than in the enzymatic one, confirming the prominent role played by proteins or glycoproteins in the foaming properties of champagnes and sparkling wines, which has previously been suggested by others (Brissonnet and Maujean 1993; Cilindre et al. 2008; Dambrouck et al. 2005; Marchal et al. 2002a; Moreno-Arribas et al. 2000).

Within the extracts (thermal or enzymatic), besides polysaccharides, mannoproteins were proven to be the main components, the molecular characterization of which has been carried out using SEC and SDS-PAGE (Martínez-Rodríguez et al. 2007; Nunez et al. 2006). Results showed in both cell wall extracts a protein band corresponding to a relative molecular mass of 30 kDa (Fig. 5.9). Moreover, three bands, which were absent from the extract obtained enzymatically, with relative molecular masses between 10 kDa and 21.5 kDa were observed in the thermal extract. Only glycoproteins with *Mr* between 10 kDa and 21.5 kDa, were proven to be foam-active though the protein at 30 kDa (also present in the enzymatic extract) was found to be inactive.

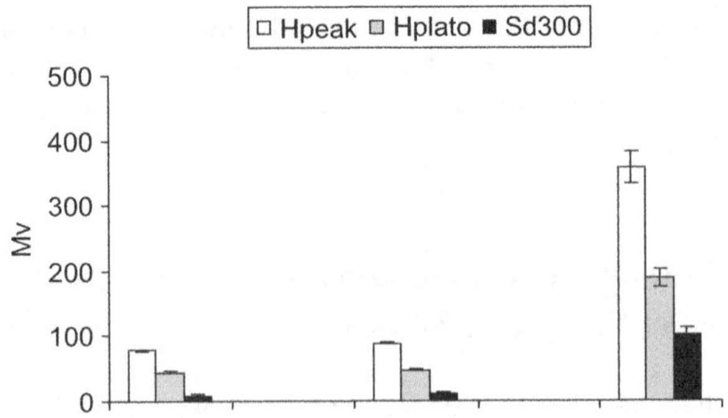

Fig. 5.8 Effect of two soluble yeast extracts (0.25 g/L) on the foaming properties of a model wine. Data presented are the mean of three different experiments. Model wine (control) (*left*); model wine supplemented with the enzymatic extract (*middle*) and model wine supplemented with the thermal extract (*right*) (reprinted with permission from Martínez-Rodríguez et al. 2007)

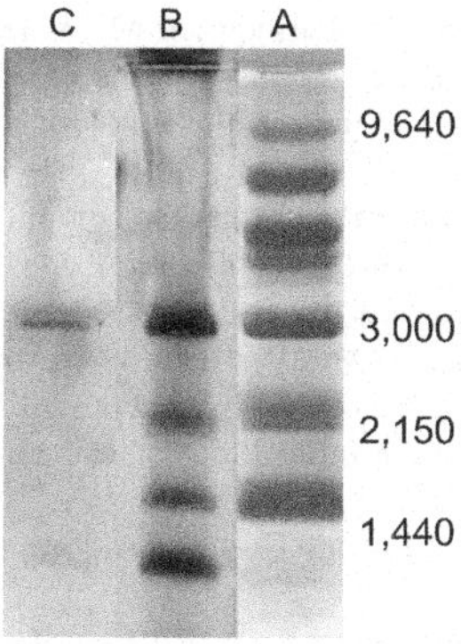

Fig. 5.9 SDS-PAGE analysis of the thermal and enzymatic yeast extracts. Line **A**: molecular weight markers; **B**: thermal extract; and **C**: enzymatic extract. Molecular weights of the standards are given on the *right* side of the gel (×10) (reprinted with permission from Martínez-Rodríguez et al. 2007)

Taken together, these results thus imply a more general relevance of the use of mannoproteins in enology and winemaking.

References

Babiker, E.F.E., Fujisawa, N., Matsudomi, N., & Kato, A. (1996). Improvement of the functional properties of gluten by protease digestion or acid hydrolysis followed by microbial transglutaminase treatment. *J. Agric. Food Chem., 44*, 3746–3750.

Blade, W.H., & Boulton, R. (1988). Adsorption of protein by bentonite in a model wine solution. *Am. J. Enol. Vitic., 39*, 193–199.

Brissonnet, F. & Maujean, A. (1993). Characterization of foaming proteins in a Champagne base wine. *Am. J. Enol. Vitic., 44*, 297–307.

Calderon, P., Van Buren, J.P., & Robinson, W.B. (1968). Factors influencing the formation of precipitates and hazes by gelatin and condensed hydrolysable tannins. *J. Agric. Food Chem., 16*, 479–482.

Charpentier, C., Nguyen Van Long, T., Bonaly, R., & Feuillat, M. (1986). Alteration of cell wall structure in *Saccharomyces cerevisiae* and *Saccharomyces bayanus* during autolysis. *Appl. Microbiol. Biotechnol., 24*, 405–413.

Cilindre, C., Castro, A.J., Clément, C., Jeandet, P., & Marchal, R. (2007). Influence of *Botrytis cinerea* infection on Champagne wine proteins (characterized by two-dimensional electrophoresis/immunodetection) and wine foaming properties. *Food Chem.*, 103, 139–149.

Cilindre, C., Jégou, S., Hovasse, A., Castro, A.J., Schaeffer, C., Clément, C., Van Dorsselaer, A., Jeandet, P., & Marchal, R. (2008). Proteomic approach to identify Champagne wine proteins as modified by *Botrytis cinerea* infection. *J. Proteome Res., 7*, 1199–1208.

Cornell, H., Wieser, H., & Belitz, H.D. (1992). Characterization of the gliadin-derived peptides which are biologically active in coeliac disease. *Clin. Chim. Acta, 213*, 37–50.

Crachereau, J.C., Gabas, N., Blouin, J., Hébrard, B., & Maujean, A. (2001). Stabilisation tartrique des vins par la carboxyméthylcellulose (C.M.C.). *Bull. l'OIV, 74*, 151–159.

Dambrouck, T., Marchal, R., Marchal-Delahaut, L., Parmentier, M., Maujean, A., & Jeandet, P. (2003). Immunodetection of proteins from grapes and yeast in a white wine. *J. Agric. Food Chem., 51*, 2727–2732.

Dambrouck, T., Marchal, R., Cilindre, C., Parmentier, M., & Jeandet, P. (2005). Determination of the grape invertase content (using PTA-ELISA) following various fining treatments *vs* changes in the total protein content of wine. Relationships with wine foamability. *J. Agric. Food Chem., 53*, 8782–8789.
Davin, A., & Sahraoui, A. (1993). Débourbage rapide des moûts de raisin par flottation à l'aide de bulles générées au sein du liquide par dépressurisation. *Rev. Fr. Oenol., 140*, 53–64.
Déchaudat, C. (1995). Clarifier des moûts avec des bulles de gaz. *La Vigne, 56*, 50–51.
Douet, J.P., Médina, B., & Castroviejo, M. (1999). Protéines résiduelles issues de colle de sang après traitement d'un vin blanc sec. *J. Inter. Sci. Vigne Vin, 33*, 177–185.
Dubourdieu, D., & Moine, V. (1994). Produit biologique pour la stabilisation physico-chimique d'un vin. French patent n° 94 13261.
Dubourdieu, D., & Moine, V. (1996). Produit de stabilisation protéique des vins. French patent n° 96 08187.
Dubourdieu, D., Villetaz, J.C., Desplanques, C., & Ribéreau-Gayon, P. (1981). Dégradation enzymatique du glucane de *Botrytis cinerea*. Application à l'amélioration de la clarification des vins issus de raisins pourris.*Conn. Vigne Vin, 15*, 101–107.
Dubourdieu, D., Serrano, M., Vannier, A.C., & Ribéreau-Gayon, P. (1988). Etude comparée des tests de stabilité protéique. *Conn. Vigne Vin, 22*, 261–273.
Dupin, I.V.S., Stockdale, V.J., Williams, P.J., Jones, G.P., Markides, A.J., & Waters, E.J. (2000). *Saccharomyces cerevisiae* mannoproteins that protect wine from protein haze: evaluation of extraction methods and immunolocalization.*J. Agric. Food Chem., 48*, 1086–1095.
Duranti, M., Guerrieri, N., Takahashi, T., & Cerletti, P. (1988). The legumin-like storage protein of *Lupinus albus* seeds. *Phytochemistry, 27*, 15–23.
Duranti, M., Guerrieri, N., & Vecchio, G. (1992). The legumin precursor from white lupin seed. Identity of subunits, assembly and proteolysis. *Eur. J. Biochem., 206*, 941–947.
Ferrarini, R., Zironi, R., Celotti, E., & Buiatti, S. (1992). Premiers résultats de l'application de la flottation dans la clarification des moûts de raisin. *Rev. Fr. Oenol., 138*, 29–42.
Ferreira, R.B., Monteiro, S., Piçarra-Pereira, M.A., & Teixeira, A.R. (2004). Engineering grapevine for increased resistance to fungal pathogens without compromising wine stability. *Trends Biotechnol., 22*, 168–173.
Gerbaux, V. (1996). Détermination de l'état de sursaturation et effet des polysaccharides sur la cristallisation du bitartrate de potassium dans les vins (Doctoral dissertation, University of Toulouse, 1996).
Gougeon, R.D., Reinholdt, M., Delmotte, L., Miéhé-Brendlé, J., Chézeau, J.M., Le Dred, R., Marchal, R., & Jeandet, P. (2002). Direct observation of polylysine side-chain interaction with smectites interlayer surfaces through ^{1}H-^{27}Al heteronuclear correlation NMR spectroscopy, *Langmuir, 18*, 3396–3398.
Gougeon, R.D., Soulard, M., Reinholdt, M., Miéhé-Brendlé, J., Chézeau, J.M., Le Dred, R., Marchal, R., & Jeandet, P. (2003). Polypeptide adsorption onto a synthetic montmorillonite: A combined solid-state NMR, X-ray diffraction, thermal analysis and N_2 adsorption study, *Eur. J. Inorg. Chem., 7*, 1366–1372
Gupta, R.B., & Shepherd, K.W. (1990). Two step one dimensional SDS-PAGE analysis of LGM subunits of glutenin. I- Variation and genetic control of the subunits in hexaploid wheats. *Theor. Appl. Genet., 80*, 65–74.
Hsu, J.C., & Heatherbell, D.A. (1987). Heat-unstable protein in wine. I- Characterization and removal by bentonite fining and heat-treatment. *Am. J. Enol. Vitic., 38*, 11–16.
Ledoux, V., Dulau, L., & Dubourdieu, D. (1992). Interprétation de l'amélioration de la stabilité protéique des vins au cours de l'élevage sur lies. *J. Inter. Sci. Vigne Vin, 26*, 239–251.
Lefebvre, S., Gerland, C., Maury, C., & Gazzola, M. (2000). Nouvelles colles végétales: origine, propriétés et performances. *Rev. Fr. Oenol., 184*, 28–32.
Lubbers, S., Leger, B., Charpentier, C., & Feuillat, M. (1993). Effet colloïde-protecteur d'extraits de parois de levures sur la stabilité tartrique d'une solution modèle. *J. Inter. Sci. Vigne Vin, 27*, 13–22.

Lubbers, S., Charpentier, C., & Feuillat, M. (1996). Etude de la rétention des composés d'arôme par les bentonites en moût, vin et milieux modèles. *Vitis, 35*, 59–62.

Main, G.L., & Morris, J.R. (1994). Color of Seyval blanc juice and wine as affected by juice fining and bentonite fining during fermentation. *Am. J. Enol. Vitic., 45*, 417–422.

Marchal, R., Sinet, C., & Maujean, A. (1993). Etude des gélatines œnologiques et du collage des vins de base champenois. *Bull. l'OIV, 751–752*, 691–725.

Marchal, R., Barret, J., & Maujean, A. (1995). Relation entre les caractéristiques physico-chimiques d'une bentonite et son pouvoir d'adsorption. *J. Inter. Sci. Vigne Vin, 29*, 27–42.

Marchal, R., Chaboche, D., Douillard, R., & Jeandet, P. (2002a). Influence of lysozyme on Champagne base wine foaming properties, *J. Agric. Food Chem., 50*, 1420–1428.

Marchal, R., Marchal-Delahaut, L., Lallement, A., & Jeandet, P. (2002b). Wheat gluten used as clarifying agent of red wines. *J. Agric. Food Chem., 50*, 177–184.

Marchal, R., Marchal-Delahaut, L., Michels, F., Parmentier, M., Lallement, A., & Jeandet, P. (2002c). Wheat gluten used as clarifying agent of musts and white wines. *Am. J. Enol. Vitic., 53*, 308–314.

Marchal, R., Weingartner, S., Voisin, C., Jeandet, P., & Chatelain, F. (2002d). Utilisation de modèles mathématiques pour optimiser les doses de bentonite gonflée et non gonflée lors du collage des vins blancs. Partie I: Clarification et stabilisation colloïdale. *J. Inter. Sci. Vigne Vin, 36*, 169–176.

Marchal, R., Lallement, A., Jeandet, P., & Establet, G. (2003). Clarification of Muscat musts using wheat proteins and the flotation technique. *J. Agric. Food Chem., 51*, 2040–2048.

Martínez-Rodríguez, A., Núñez, Y.P., Carrascosa, A.V., González, R., & Polo, M.C. (2007). A thermal-extracted fraction of mannoproteins potentially useful to improve the foaming properties of sparkling wines. In P. Jeandet, C. Clément & A. Conreux (Eds.), *Macromolecules and Secondary Metabolites of Grapevine and Wine* (pp. 371–375). Paris, London, New York: Intercept/Lavoisier.

Maujean, A. (2000). Organic acids in wine. In *Handbook of Enology vol. 2 The Chemistry of Wine, Stabilization and Treatments* (pp. 3–39). New York: John Wiley & Sons, LTD.

Milisavljevic, D. (1963). Prévention des troubles protéiques du vin par l'emploi des bentonites dans les moûts. *Ann. Technol. Agric., 12*, 315–330.

Miller, G.C., Amon, J.M., Gilson, R.L., & Simpson, R.F. (1985). Loss of wine aroma attributed to protein stabilization by bentonite or ultrafiltration. *Aust. NZ Grapegrower Winemaker, 256*, 49–50.

Moine-Ledoux, V., & Dubourdieu, D. (1999). An invertase fragment responsible for improving the protein stability of dry white wines. *J. Sci. Food Agric., 79*, 537–543.

Moine-Ledoux, V., & Dubourdieu, D. (2002). Role of yeast mannoproteins with regard to tartaric stabilization. *Bull. l'OIV, 75*, 471–482.

Moine-Ledoux, V., & Dubourdieu, D. (2007). Role of mannoproteins in wine stabilization. In P. Jeandet, C. Clément & A. Conreux (Eds.), *Macromolecules and Secondary Metabolites of Grapevine and Wine* (pp. 393–401). Paris, London, New York: Intercept/Lavoisier.

Moine-Ledoux, V., Perrin, A., Paladin, I., & Dubourdieu, D. (1997). Premiers résultats de stabilisation tartrique des vins par addition de mannoprotéines purifiées (MANNOSTAB). *J. Inter. Sci. Vigne Vin, 31*, 23–31.

Moreno-Arribas, V., Pueyo, E., Nieto, F.J., Martin-Alvarez, P.J., & Polo, M.C. (2000). Influence of polysaccharides and the nitrogen compounds on foaming properties of sparkling wines. *Food Chem., 70*, 309–317.

Nunez, Y.P., Carrascosa, A.V., Gonzalez, R., Polo, M.C., & Martínez-Rodríguez, A.J. (2005). Effect of accelerated autolysis of yeast on the composition and foaming properties of sparkling wines elaborated by a champenoise method. *J. Agric. Food Chem., 53*, 7232–7237.

Nunez, Y.P., Carrascosa, A.V., Gonzalez, R., Polo, M.C., & Martínez-Rodríguez, A. (2006). Isolation and characterization of a thermally extracted yeast cell wall fraction potentially useful for improving the foaming properties of sparkling wines. *J. Agric. Food Chem., 54*, 7898–7903.

Popineau, Y., & Denery-Papini, S. (1996). Les protéines de réserve du grain de blé. In B. Godon (Ed), *Protéines Végétales* (pp. 120–172). Paris: Lavoisier.
Popineau, Y., & Pineau, F. (1987). Investigation of surface hydrophobicities of purified gliadins by hydrophobic interaction chromatography, reverse-phase high performance liquid chromatography and apolar ligand binding. *J. Cereal Sci., 5*, 215–234.
Ribéreau-Gayon, J., & Peynaud, E. (1935). La clarification des vins. *Revue Viticole, 82*, 8–13.
Ribéreau-Gayon, J., Peynaud, E., Sudraud, P., & Ribéreau-Gayon, P. (1977). *Sciences et Techniques du Vin, Volume IV: Clarification et Stabilisation*. Paris: Dunod.
Sahraoui, A. (1991). Etude de la flottation par détente de liquide saturé sous pression. Application à la clarification des moûts (Doctoral dissertation, University of Toulouse, 1991).
Saywell, L.G. (1934). Clarification of wines. *Ind. Eng. Chem., 26*, 981–982.
Shewry, P.R., Miles, M.J., & Tatham, A.S. (1994). The prolamin storage proteins of wheat and related cereals. *Progress in Biophysics and Molecular Biology, 61*, 37–59.
Sturgess, R.P., Ellis, H.J., & Ciclitira, P.J. (1991). Cereal chemistry, molecular biology and toxicity in coeliac disease (Progress reports). *Gut, 32*, 1055–1060.
Waters, E.J., Wallace, W., & Williams, P.J. (1992). Identification of heat-unstable wine proteins and their resistance to peptidases. *J. Agric. Food Chem., 40*, 1514–1519.
Waters, E.J., Shirley, N.J., & Williams, P.J. (1996). Nuisance proteins of wine are grape pathogenesis-related proteins. *J. Agric. Food Chem., 44*, 3–5.
Zoecklein B.W. (1988a). Protein fining agents for wines and juices. Virginia Cooperative Extension, Publication No. 463–012.
Zoecklein B.W. (1988b). Bentonite fining of juice and wine. Department of Horticulture Virginia Polytechnic Institute & State University, Publication 463–014.
Zoecklein B.W. (1988c). Protein stability determination in juice and wine. Virginia Cooperative Extension, Publication No. 463–015.

Part II
Wine Chemical Compounds and Biochemical Processes

Chapter 6
Nitrogen Compounds

Chapter 6A
Amino Acids and Biogenic Amines

M. Victoria Moreno-Arribas and M. Carmen Polo

Contents

6A.1 Introduction

Together with proteins and peptides, amino acids constitute the main components of the nitrogenous fraction of musts and wines. They are also the most studied and best known nitrogenated components in wines. They serve as nutrients for yeasts in alcoholic fermentation and can also be metabolised by the lactic acid bacteria responsible for the process of malolactic fermentation. The concentration and composition of amino acids in wines and musts can also have an important effect on the aromatic complexity of wines (Bisson 1991; Rapp and Versini 1991). To obtain a good knowledge of the amino acid composition of a particular must or wine, depending on the grape variety, technological procedure or critical winemaking factors and steps, amino acid analysis is very important and has received special attention in the literature (Moreno-Arribas et al. 1998; Hernández-Orte et al. 2003).

Amines are widely occurring compounds in nature, containing one or more organic substituent bonded to a nitrogen atom. Biogenic amines are low molecular

M.V. Moreno-Arribas (✉)
Instituto de Fermentaciones Industriales, CSIC. Juan de la Cierva 3, 28006 Madrid, Spain
e-mail: mvmoreno@ifi.csic.es

M.V. Moreno-Arribas, M.C. Polo (eds.), *Wine Chemistry and Biochemistry*,
DOI 10.1007/978-0-387-74118-5_7,

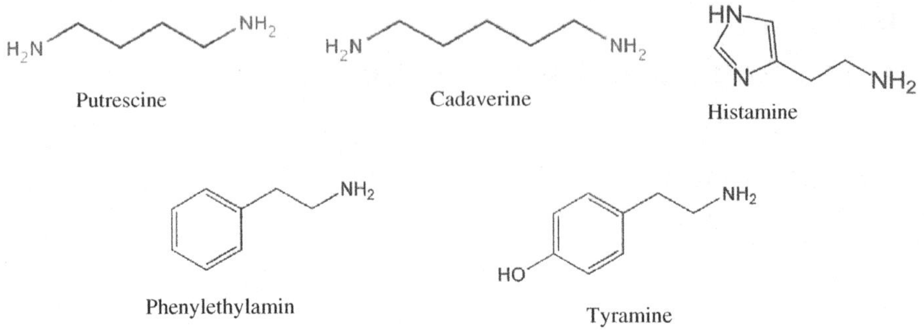

Fig. 6A.1 Chemical structure of the biogenic amines most frequently found in wines

weight compounds, derived from aromatic or cationic amino acids and all of them have one or more positive charge and a hydrophobic skeleton. The chemical structure of biogenic amines can be aliphatic (putrescine, cadaverine, spermine, spermidine), aromatic (tyramine, phenylethylamine) or heterocyclic (histamine, tryptamine). Figure 6A.1 shows the chemical structures of some biogenic amines. They are endogenous of plants but can also be found in fresh fruit and vegetables. However, amines are mainly formed in foods in fermentative processes and during aging and storage, produced by microbial decarboxylation of the corresponding amino acid precursors, which is why they are referred to as biogenic. The most studied of these biogenic amines is histamine.

Biogenic amines are essential at low concentrations for normal metabolic and physiological functions in animals, plants, and microorganisms. However, these biologically produced amines can have adverse effects at high concentrations and pose a health risk for sensitive individuals. Biogenic amines are likely to be found in foods and beverages that contain proteins or free amino acids, in conditions that favour microbial or biochemical activity (Hálasz et al. 1994; Silla Santos 1996).

In fermented foods, the non-volatile or biogenic amines (histamine, putrescine, cadaverine, spermine, spermidine, agmatine, tyramine tryptamine) and phenylethylamine (a volatile amine) are formed mainly by the microbial decarboxylation of the corresponding amino acids (Ten Brink et al. 1990). Volatile amines are believed to be formed by the reductive amination or transamination of the corresponding aldehyde or ketone.

This chapter tackles the most important aspects of amino acid and biogenic amine formation during wine manufacture and aging. The significance of biogenic amines to wine quality and safety will also be reported.

6A.2 Amino Acids

Around 20 free amino acids were detected in grape musts, representing on average 28–39% of the total nitrogen, depending on whether these corresponded to musts of white or red wine grapes, respectively (Rapp and Versini 1991). The amino acid content increases during grape maturation, and can reach 90% of the total nitrogen in musts from table wine grapes. At wine harvest, 70% of the organic nitrogen is constituted by amino acids, 3% by proteins and 2% by peptides.

Free amino acids in musts are of paramount importance, since they constitute a source of nitrogen for yeasts in alcoholic fermentation, for lactic acid bacteria in malolactic fermentation and can also be a source of aromatic compounds. In certain cases, some amino acids can produce undesirable compounds in wines, such as ethyl carbamate, biogenic amines, ochratoxin A (from 2-phenylalanine) and β-carbolines (from tryptophane) (Herraiz and Ough 1993; Herraiz et al. 1993).

6A.2.1 Evolution of Free Amino Acids During Alcoholic Fermentation

The free amino acids present in wine may be of different origins. Those from the grape may be partially or totally metabolised by the yeasts during fermentation. Other free amino acids are released by yeasts at the end of fermentation or by proteolysis processes during yeast autolysis. The two most abundant free amino acids are proline and arginine and both have glutamic acid as a precursor, which is also among the most abundant (Moreno-Arribas et al. 1998). Changes in free amino acids during alcoholic fermentation have been the focus of numerous and important studies. At the start of fermentation, yeasts use the nitrogen from ammonium salts for their development, followed by the nitrogen from free amino acids. Some of these, namely, arginine, glutamic acid, glutamine, aspartic acid, asparagine, threonine and serine, are preferentially assimilated. Simultaneously, by enzymatic processes, the yeasts degrade the proteins to peptides and to amino acids until fermentation has finished. The presence of exogenous proteases in yeasts has been demonstrated by Feuillat et al. (1980). Autolysis and the release of amino acids by yeasts also contribute to the increased concentration of free amino acids after fermentation (this subject is discussed in more detail in Chapter 3A. – Sparkling Wines and Yeast Autolysis)

Numerous factors are involved in nitrogen uptake by yeasts during alcohol fermentation. These include temperature (López et al. 1996), concentration of sugars – at higher sugar concentrations more nitrogen is required (Agenbach 1977) – and the concentration of oxygen (Ingledew and Kunkee 1985). Addition of weak concentrations of ammonia nitrogen (for example 0.23 mg/L of diammonium phosphate, a value close to the maximum permitted for nitrogen enrichment) has little effect an amino acid utilization (Monteiro and Bisson 1992). In contrast, the addition of high doses of this salt (2.0 g/L) greatly reduces arginine consumption and degradation, and the formation of polyamines, such as putrescine, spermine and spermidine.

In a study by Dizy and Polo (1996), the majority amino acids in grape must of the Malvar variety were glutamine, arginine, proline and γ-aminobutyric acid (GABA). The authors found that the concentration of proline rose during fermentation, while that of arginine diminished, because proline was derived from arginine metabolism. During the first step of fermentation, almost all the free amino acids were consumed. In the time between the end of fermentation and racking, there was an increase in free amino acids, which these authors associate with yeast autolysis.

According to Jiranek et al. (1995), all the amino acids, except glycine, can be removed from fermentation media to different degrees. Several authors (Cooper 1982; Henschke and Jiranek 1993, among others) have reported that the amino acids that constitute the best yeast nitrogen source are glutamic acid, glutamine, aspartic acid, asparagine, threonine, histidine, alanine, tyrosine and arginine. According to Fraile et al. (2000), the formation of different alcohols takes place at the end of the fermentation. Carrying out studies of partial least-squares regression models, Hernández-Orte et al. (2002) found that amino acid composition accounts for a high proportion of the variance in the volatile composition. Moreover, the composition of the amino acids remaining in the wine influences aromas during the maturing process (Escudero et al. 2000).

The importance of adding amino acids to the must and their effect on the biosynthesis of aromatic compounds by yeasts during alcoholic fermentation was evaluated by Hernández-Orte et al. (2005). These authors studied the effect of supplementing a must of Airen variety with ammonium (100, 300 mg/L) and amino acids (doubling the level of amino acids in that must) that was fermented with three different yeast strains. Statistical treatments showed that the yeast strain is the major factor affecting wine volatile composition, but must nitrogen supplementation also has an influence. From the sensory point of view, must supplementation brings about a decrease in sulphur notes and an increase in citric flavour. Also, the effect of adding selected amino acids (phenylalanine, alanine, aspartic acid and threonine) to grape juice on the generation of aroma compounds and on amino acid uptake were studied (Hernández-Orte et al. 2006a). The addition of amino acids produced differences in fermentation kinetics. Higher alcohols were generated at the same time as ethanol. From the sensorial point of view, the wines with more amino acids were better valued by the panel of tasters because the sulphured notes decreased significantly while the floral notes increased.

6A.2.2 Evolution of Free Amino Acids During Malolactic Fermentation

In the study of the changes in free amino acids during malolactic fermentation, the concentration of some of these (arginine, glycine, tyrosine, phenylalanine, histidine, α-alanine and serine) was found to decrease sharply, while the concentrations of the other amino acids tended to increase significantly (aspartic acid, glutamic acid, leucine, methionine, isoleucine and tryptophan) (Davis et al. 1986). Although there is only limited information about the presence of proteases in wine lactic acid bacteria, a higher concentration of amino acids was observed at the end of bacterial growth in the wine compared to the start. However, to date little attention has been paid to the influence of malolactic fermentation on the whole amino acid composition of wine (Soufleros et al. 1998), and in particular, the effect of the lactic acid bacteria strain involved in malolactic fermentation on this important wine nitrogenous fraction is still little known.

Pozo-Bayón et al. (2005) studied the evolution of free amino acids during industrial malolactic fermentation carried out by four different starter cultures of the species *Oenococcus oeni* and *Lactobacillus plantarum*. They showed a significantly different evolution of methionine in *O. oeni* and *L. plantarum* wines, probably due to the different capacities to catabolize methionine in these two lactic acid bacteria. In addition to methionine, significant differences were also found in *O. oeni* and *L. plantarum* wines for tryptophan and threonine. *L. plantarum* degraded these amino acids, while *O. oeni* did not. Moreover, concentration values of seven of the 21 amino acids determined in the wines studied by these authors, especially those of glutamine, glycine, β-alanine, α-alanine, GABA, valine and lysine, varied significantly depending on the bacterial strain performing the malolactic fermentation.

On the other hand, other studies focused on *O. oeni* amino acid requirements for growth and malolactic fermentation in several growth media (Tracey and Britz 1989). Remize et al. (2006) determined the essential amino acids for the growth of five different strains of *Oenococcus oeni*. These amino acids corresponded to glutamic acid, methionine, phenylalanine, serine and tyrosine for all the strains studied. They also found that the amino acids valine, leucine, tryptophan, isoleucine, histidine and arginine were essential or necessary for the strains studied, but that the amino acids alanine, glycine and proline were not essential.

In a recent study, Fernández and Manca de Nadra (2006) studied the changes in free amino acids produced by lactic acid bacteria. They identified the amino acids mainly consumed by a strain of *Pediococcus pentosaceus* (glutamic acid, arginine, phenylalanine, glycine, histidine, isoleucine, methionine, serine, tyrosine, threonine, tryptophan) and those released at the end of the growth process of *O. oeni* (aspartic acid, glutamic acid, alanine, arginine, glycine, isoleucine, leucine, lysine, serine, threonine and valine). When these two strains were grown together, a rise was observed in the amino acids glutamic acid, alanine, asparagine, phenylalanine, histidine, isoleucine, leucine, serine, tyrosine and valine, compared to the pure *O. oeni* strain. These results indicate that proteolytic stimulation of the system by the mixture of these two bacteria increases the release of essential amino acids.

6A.3 Biogenic Amines

Biogenic amines in wine can originate from the grape berries themselves or be produced during fermentation processes, aging or storage, when wine is exposed to the undesirable activity of decarboxylase-positive microorganisms. Contamination may occur from poor sanitary conditions of both grapes and processing equipment. Most biogenic amine contamination of wine is believed to take place during malolactic fermentation.

A number of studies have reported no remarkable rise in the concentration of biogenic amines during alcoholic fermentation, concluding that yeasts do not appear to be responsible for the production of most amines found in industrial commercial red wines (Herbert et al. 2005; Marcobal et al. 2006a). Figure 6A.2 illustrates the evolution of biogenic amines during industrial red wine winemaking. Most contamination of wine by biogenic amines takes place during natural or spontaneous malolactic fermentation (Landete et al. 2005; Marcobal et al. 2006a). Lactic acid bacteria are available in low populations in healthy grapes and are transferred to the cellar equipment where they develop rapidly. These indigenous bacteria are responsible for spontaneous malolactic fermentation. However, the metabolic characteristics of this microbial flora are not well known and in some strains enzymatic decarboxylase activities could be involved in biogenic amine production.

6A.3.1 Presence of Biogenic Amines in Wines: Toxicological Aspects

Problems related to biogenic amine formation affect numerous fermented food products consumed more frequently than wine, such as cheese, beer, some fermented sausages and meat products among others (Fernández-García et al. 1999; Izquierdo-Pulido et al. 2000; Kaniou et al. 2001), which have higher levels of these compounds. However, in alcoholic drinks, especially wine, biogenic amines have received more attention, because ethanol can increase the effects on health by directly or indirectly inhibiting the enzymes responsible for the detoxification of these compounds (Maynard and Schenker 1996). The human organism easily tolerates low concentrations of biogenic amines, since these are efficiently broken down by mono- and diaminoxidase enzymes in the intestinal tract. Although there are differences in individual susceptibility to intoxication by biogenic amines, several pharmacological reactions can take place after excess intake of these compounds. The best known reactions are those caused by histamine. More specifically, some histamine-induced symptoms include rash, edema, headaches, hypotension, vomiting, palpitations, diarrhoea and heart problems. Other amines, such as tyramine and phenylamine, can cause hypertension and other symptoms associated with vasoconstriction caused by the release of noradrenaline (especially cerebral hemorrhages and migraine). Although putrescine and cadaverine are not themselves toxic, they can increase the toxicity of histamine, tyramine and phenylethylamine, since they interfere in detoxification reactions. Moreover, putrescine and cadaverine can have negative effects on wine aroma, giving them flavors of putrefaction or rotting flesh, respectively.

Normally, if a low concentration of biogenic amines is ingested, these are quickly detoxified in the human body by amine oxidases or through conjugation. Amine

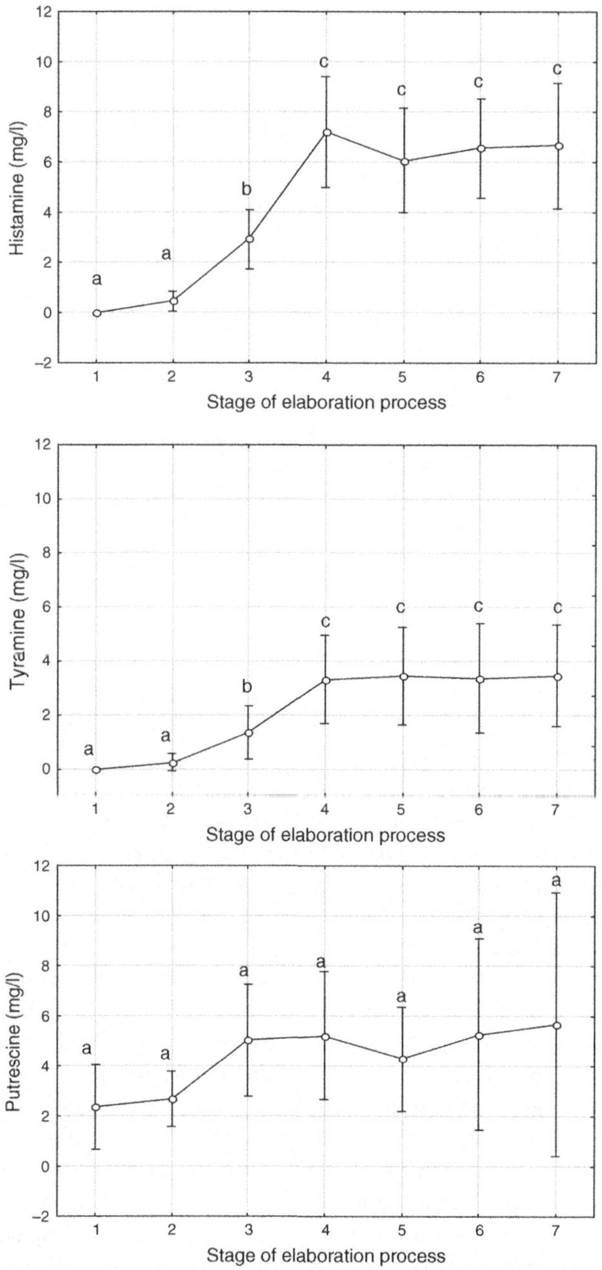

Fig. 6A.2 Means and 95% confidence intervals for histamine, tyramine, putrescine, methylamine, ethylamine, phenylethylamine and cadaverine in the seven stages of the elaboration process: (1) must; (2) after AF; (3) during MLF; (4) just after MLF; (5) after MLF and addition of SO_2; (6) after 6 months of aging, (7) after 12 months of aging. Means with the same letter are not significantly different by the Fisher LSD test (from Marcobal et al. 2006a, with permission)

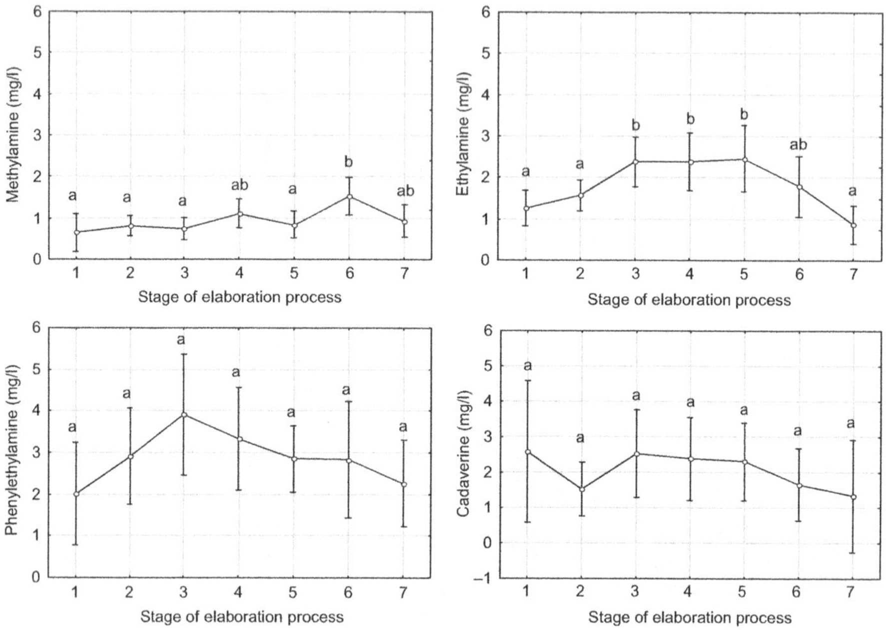

Fig. 6A.2 (continued)

oxidases catalyse deamination of biogenic amines to produce an aldehyde, hydrogen peroxide and ammonia (Gardini et al. 2005). However, if an excess amount of biogenic amines is ingested or if the normal catabolic routes are inhibited of genetically deficient, several disorders can occur (Ten Brink et al. 1990).

The toxic level of biogenic amines depends on the tolerance of the individual for the compound, the concentration of total biogenic amines and the consumption of ethanol and/or drugs. The toxicity of histamine and tyramine depends on the efficacy of the catabolic pathway which uses monoamine oxidase (MAO) and diamine oxidase (DAO) enzymes, which again varies among individuals (Ten Brink et al. 1990). Biogenic amines such as tyramine, putrescine and cadaverine that may also be present in the wine can inhibit the metabolism of histamine. These amines favour the passage of histamine from the intestines into the systemic circulation by competing for binding sites in the gastrointestinal tract or by interfering with the catabolism of histamine by saturating the activity of mono- or diamine oxidases. The amine oxidase enzymes are not very specific and alcohol, acetaldehyde and anti-depressant drugs can also cause interference (Ten Brink et al. 1990; Straub et al. 1995).

Generally, the toxic dose in alcoholic beverages is considered to be between 8 and 20 mg/L for histamine, 25 and 40 mg/L for tyramine, but as little as 3 mg/L phenylethylamine can cause negative physiological effects (Soufleros et al. 1998).

Kanny et al. (2001) report that a normal individual can tolerate 129 mg/L of histamine taken orally before symptoms occur, but only 7 μg administered intravenously. However, there are studies that conclude that oral ingestion of biogenic amines and wine tolerance are not related (Kanny et al. 2001; Jansen et al. 2003). Rather, these authors propose that wine may contain compounds (ethanol and acetaldehyde) that stimulate the release of histamine within the body. Amines are also important in wine from a financial point of view, since they could cause problems in commercial transactions (in their import and export). Although the legal limits have not yet been established for any biogenic amine, some countries have drawn up their own recommendations, especially for histamine. The exportation of wines to these counties could be paralyzed in the future, making biogenic amines a potential economic threat. In fact, Switzerland and Austria reject wines which contain more than 10 mg/L, and lower top limits have been recommended in Germany (2 mg/L), Holland (3 mg/L), Belgium (5–6 mg/L) and France (8 mg/L) (Lehtonen 1996).

Most wine-producing countries in the world, including Greece, Spain, Hungary, Oregon (EEUU), Canada, South Africa, Italy and France, have investigated the presence and concentration of biogenic amines in their wines (Glòria et al. 1998; Soufleros et al. 1998; Mafra et al. 1999; Gerbaux and Monamy 2000; Hajós et al. 2000; Herbert et al. 2005; Bover-Cid et al. 2006; mo Dugo et al. 2006; Soufleros et al. 2007; Smit 2007). In general, red wines have higher concentrations of biogenic amines than white wines. In the former, these higher values are attributed to the presence of lactic acid bacteria and malolactic fermentation. White wines, in general, contain lower concentrations of amino acids and lower pH, due to the absence of maceration during alcoholic fermentation. Evidently, other types of wines and winemaking processes, such as Cava wines and other sparkling wines made by the *traditional method*, and also biologically aged wines, are susceptible to the problem of biogenic amine formation, although there are very few studies of this in the literature (Bodmer et al. 1999; Moreno-Arribas and Polo 2008).

6A.3.2 Microorganisms Associated with Biogenic Amine Production in Winemaking

Potentially, all the microbial groups that participate in the winemaking process may be associated with biogenic amine production. However, there is general agreement that yeasts make a less significant contribution than lactic acid bacteria to the final concentration of biogenic amines in wine. It is also clear that yeasts form different biogenic amines to lactic acid bacteria. On the other hand, to date, there is much less information about the biochemistry, genetics and regulation of amine production by wine yeast, especially compared with the data available for lactic acid bacteria. The formation of biogenic amines by wine microorganisms is discussed below.

6A.3.2.1 Yeast

A large variety of indigenous yeast species can grow and perform alcoholic fermentation in wine, along with commercial *Saccharomyces cerevisiae* strains. Few studies have been conducted on the formation of biogenic amines by yeasts, and most of these only compared different yeast species and only quantified histamine (Torrea and Ancín 2002). Somavilla et al. (1986), using six yeast strains, demonstrated that small amounts of histamine are produced during alcoholic fermentation and that the association of yeasts and lactic acid bacteria can reduce the histamine level. The highest histamine values (from 3.7 to 8.3 mg/L) were obtained when histidine was added to the must (34 mg/L), in the other experiments histamine values were lower than 1.2 mg/L. Vidal-Carou et al. (1990a) did not detect histamine formation either during alcoholic fermentation although they did detect tyramine formation, but at very low levels (0.60 mg/L). In contrast, other authors disagree with the hypothesis that biogenic amines are formed by lactic acid bacteria during malolactic fermentation (Buteau et al. 1984). Torrea-Goñi and Ancín-Azpilicueta (2001) found a slight biogenic amine production by *Saccharomyces cerevisiae* depending on the strain. Recently, Landete et al. (2007b) screened 36 strains of yeast isolated from must and wines to produce biogenic amines. The yeast tested included strains belonging to the genera *Aureobasidum, Candida, Hanseniaspora, Hansenula, Kloeckera, Metschnikowia, Pichia, Rhodotorula* and strains of the species *Saccharomyces cerevisiae, S. cerevisiae* var.*bayanus, S. cerevisiae* var.*chevalieri and S. cerevisiae* var.*steiner*. No biogenic amines were produced by any of the yeast strains tested under the conditions present in this screening test, in synthetic medium, grape must or wine. These results are consistent with previous studies in which neither histamine, tyramine or putrescine production were detected in 50 yeast strains isolated from grapes and/or wine, including *Saccharomyces cerevisiae* and other non-*Saccharomyces* yeasts, while some of these yeasts varied in their capacity to form phenylethylamine (Caruso et al. 2002; Granchi et al. 2005). These findings, therefore, indicate that yeast does not appear to be at least the direct origin of most amines found in wines.

6A.3.2.2 Lactic Acid Bacteria

Extensive research has been done to correlate biogenic amine production in wine with species of lactic acid bacteria involved in the winemaking process. The first studies on histamine production by lactic acid bacteria indicate that *Pediococcus* were the main producers (Farias et al. 1993). Recent studies by Landete et al. (2007b) are in accordance since, although the percentage of *Pediococcus* species capable of producing histamine seems to be low, some strains are responsible for the highest concentrations. However, today it is well known that *Lactobacillus, Leuconostoc* and *Oenococcus* species are also implicated in biogenic amine production in wine. Different strains of *Lactobacillus hilgardii, L. brevis. L. buchneri* and *L. mali* have been found to be able to produce a variety of biogenic amines in wine (Moreno-Arribas and Lonvaud-Funel 1999; Moreno-Arribas et al. 2000, 2003; Constantini et al. 2006; Landete et al. 2007b).

Commercial *O. oeni* strains are selected for their oenological parameters, including the absence of amino acid decarboxylases. According to the in vitro studies done by Moreno-Arribas et al. (2003), none of the four commercial malolactic starter cultures tested could produce histamine, tyramine or putrescine. Martín-Álvarez et al. (2006) also compared inoculation with spontaneous malolactic fermentation in 224 samples of Spanish red wine. They found that inoculation with a commercial starter culture of lactic acid bacteria could reduce the incidence of biogenic amines compared to spontaneous malolactic fermentation in wines. Starter cultures could eliminate indigenous bacteria, or could possibly degrade the biogenic amines produced by the undesirable strains.

Other authors reported that no biogenic amine producers were present among naturally occurring *O. oeni* strains isolated from wine and must (Moreno-Arribas et al. 2003). Constantini et al. (2006) found no *O. oeni* among 92 strains able to produce biogenic amines in a broth medium. PCR screening was used by these authors to confirm the absence of the respective decarboxylase genes.

In contrast, other authors found that *O. oeni* is able to significantly contribute to the overall histamine content of wines and also that the ability of this species to produce biogenic amines varies among strains (Coton et al. 1998; Guerrini et al. 2002). Marcobal et al. (2004) isolated and identified a strain of the *O. oeni* species, a producer of putrescine and the capacity of another 42 strains of this species to produce this amine has also been studied at a molecular level and the gene that encodes biosynthesis of this amine was not present in any of them (Marcobal et al. 2004). These results indicate that although this species may be involved in the production of putrescine in wines, this is not a common property.

Amino acid decarboxylases are the enzymes responsible for the synthesis of biogenic amines. These proteins and the genes that encode them have been studied in several lactic acid bacteria in wine and are briefly discussed below.

Histidine decarboxylase (EC.4.1.1.22) is the enzyme that decarboxiles histidine to produce histamine. The first decarboxylase enzyme of a wine lactic acid bacterium was isolated from a histamine-producing strain by Lonvaud-Funel and Joyeux (1994). Coton et al. (1998) purified this enzyme to homogeneity and provided molecular data. The gene sequence aided researchers to develop rapid and specific detection systems based on polymerase chain reaction (PCR) to detect potential histamine-producing bacteria from wine (Le Jeune et al. 1995; Coton et al. 1998). Lucas et al. (2005) related histidine decarboxylase activity of *Lactobacillus hilgardii* 0006 isolated from wine, to the presence of an 80-kb plasmid on which the decarboxylase gene was located as part of a four-gene cluster. These authors suggest that a plasmid-encoded histidine decarboxylase could be transferred horizontally, and that the location of the gene on an unstable plasmid may explain the random distribution of histidine decarboxylase bacteria.

Tyrosine decarboxylase (EC.4.1.1.25), responsible for the production of tyramine from tyrosine, was first investigated in a wine lactic acid bacteria by

Moreno-Arribas and Lonvaud-Funel (1999). Moreno-Arribas et al. (2000) isolated and identified a number of tyramine-producing lactic acid bacteria in wine that had undergone malolactic fermentation; all belonging to the *lactobacilli*. Tyrosine decarboxylase was then purified (Moreno-Arribas and Lonvaud-Funel 2001) and the corresponding gene was purified and sequenced (Lucas and Lonvaud-Funel 2002; Lucas et al. 2003). As far as the literature suggests, no tyramine-producing *O. oeni* strain has yet been reported, with the exception of one strain (*O. oeni* DSM 2025) that was shown to be able to produce tyramine in a laboratory medium (Choudhury et al. 1990).

Ornithine decarboxylase (EC.4.1.1.17) is the enzyme that decarboxylates ornithine to produce putrescine. Marcobal et al. (2004) isolated an *O. oeni* putrescine-producing strain from fermentation lees of a Spanish red wine which showed a high concentration of putrescine. This led to the first report of the presence of the ornithine decarboxylase gene in the genome of *O. oeni* and detectable by PCR (Marcobal et al. 2004, 2005a). Later, the nucleotide sequence of a 17.2-kb chromosomal DNA fragment containing the ornithine decarboxylase gene was determined (Marcobal et al. 2006b). This DNA fragment contains 13 open reading frames, including genes coding for 5 transposases and 2 phage proteins. This could perhaps represent the first evidence of a horizontal gene transfer event as the origin of a biogenic amine biosynthetic locus. The structure of this gene is presented in Fig. 6A.3.

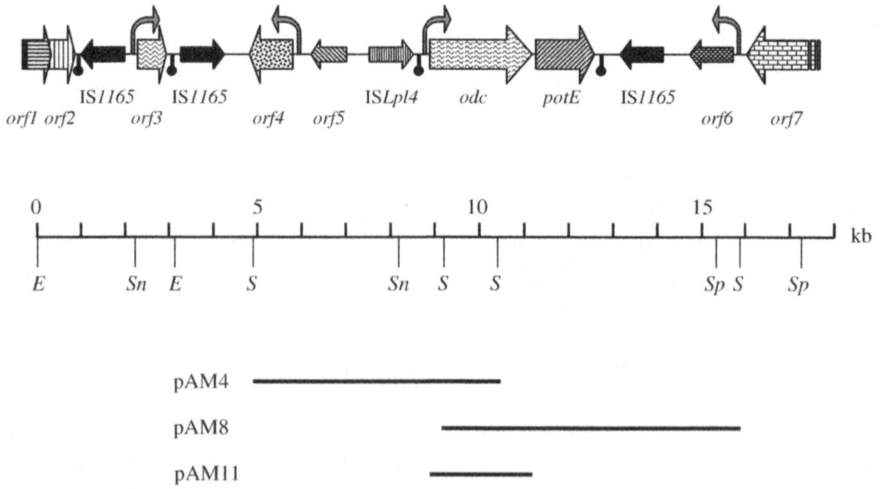

Fig. 6A.3 Genetic organization of the *O. oeni* RM83 17.2-kb DNA region containing the *odc* gene. *Thick and thin arrows* represent complete or interrupted ORFs, respectively. The localization of putative promoters (*vertical bent arrow*) and predicted transcriptional terminator regions (*ball* and *stick*) are indicated. Some of the plasmids used in this study are indicated, as are relevant restriction sites: *E*, EcoRV; *S*, Sau3AI; *Sp*, SpeI; *Sn*, SnaBI. Only some of the corresponding restriction sites present in this fragment are represented (from Marcobal et al. 2006b, with permission)

6A.3.2.3 Other Wine Microorganisms

The ability of 40 strains of acetic acid bacteria isolated from grape must and wine to produce biogenic amines was screened in synthetic medium and wine (Landete et al. 2007b), but no positive results were obtained. No further mention regarding the formation of biogenic amines by acetic acid bacteria was found in the literature.

When the grape vine is subject to biotic stresses, such as those caused by fungus *Botrytis cinerea*, this can lead to a rise in the amine content of the grape berries (Hajós et al. 2000). *B. cinerea* is responsible for the formation of Aszu grapes, which are characteristic of the famous Tokaj wines from Hungary. Amine concentrations in Aszu wines were found to be higher in three different cultivars studied when compared to corresponding normal wines from the same varieties. A positive correlation could be made between the biogenic amine content and the number of butts used for the Aszu wine (Sass-Kiss et al. 2000).

6A.3.3 Factors Affecting Biogenic Amine Formation During Wine Production

The levels of biogenic amines produced in wine largely depend on the abundance of amino acid precursors in the medium, since on the whole, biogenic amines increase with an increase in amino acids. Amino acid content may be influenced by vinification methods, grape variety, geographical region and vintage (Soufleros et al. 1998; Moreno-Arribas et al. 2000). While some factors increase the concentration of precursor amino acids, other factors influence the growth and enzyme activity of microorganisms that can form the biogenic amines.

Some amines, such as putrescine and other polyamines, may already be present in grape berries (Hálasz et al. 1994; Bover-Cid et al. 2006). It seems that putrescine concentration in wine may be influenced by the geographical region and grape variety (Glòria et al. 1998; Kiss et al. 2006). Potassium deficiencies in the soil have been linked to a rise in putrescine content in plants, while water deficiencies do not seem to influence the content of biogenic amines in grape berries and wines (Bover-Cid et al. 2006). Biogenic amines are also dependent on grape variety and wine nutrition, which determine the concentration and composition of precursor amino acids in grape must and wine (Herbert et al. 2005; Soufleros et al. 2007). Different studies have reported cultivar-related differences in the contents of biogenic amines (Glòria et al. 1998; Soleas et al. 1999; Soufleros et al. 2007), while the vintage and the region of production can also affect the free amino acid and amine content in must and wine (Herbert et al. 2005; Martín-Álvarez et al. 2006).

During alcoholic fermentation, the degree of maceration is the first factor that affects the extraction of some compounds present in the grape skin, especially phenolic compounds, which are responsible for the color of the wine. However, not only does the maceration affect the extraction of polyphenols but also of other grape components, such as proteins, polysaccharides and, also, amino acids, which are precursors of biogenic amines. In most red wines, alcoholic fermentation takes place in contact with the grape skin. Prolonged maceration after alcoholic fermentation at low temperatures is another frequently used option to prolong the extraction period. Commercial pectolytic enzymes are commonly used in winemaking to increase the extraction of compounds, facilitating the operations of pressing and clarification, increasing the clarity of musts and wines. In some cases, their addition is expected to improve the extraction of phenolic compounds and aroma. However, it is known that these commercial preparations can produce some secondary enzymatic activities, which are sometimes undesirable in wines. Among other activities, a significant proteolytic activity has also been mentioned, that would lead to an increase in amino acid concentration.

In order to gain a thorough understanding of the variables and technologies implicated in biogenic amine concentration, 55 different batches of red wine were produced by different technologies used in wine cellars. In addition to studying the changes in amine concentrations due to the producing cellar or the year of harvest, the following variables were also considered: the use of pectolytic enzymes, the type of container used for malolactic fermentation (stainless steel or oak barrel), inoculation with commercial preparations, permanence of the wine with lees, duration and intensity of the maceration (Table 6A.1). To study together the effects of the different technological processes used during winemaking on the concentration of biogenic amines in the wines studied, a multifactorial analysis of variance was used (Martín-Álvarez et al. 2006). Significant differences were found in the levels of biogenic amines in relation to the cellar producing the wine and also the year of harvest. Moreover, of the winemaking factors and practices studied, the use of pectolytic enzymes appeared to have no effect on the concentration of wine biogenic amines. Prolonged permanence of the wine with the lees affects the concentration of amino acids and favors amine production, while intense and prolonged macerations produce wines with higher levels of histamine, tyramine, putrescine and cadaverine. On the other hand, with maceration, and especially, extraction of polyphenols from the grape, these can react with amino acids giving rise to lower levels of wine biogenic amines in some cases.

Indirectly, yeasts may play an important role in the subsequent production of biogenic amines by lactic acid bacteria with decarboxylase activity, altering the composition of amino acids during alcoholic fermentation. Yeasts may also release amino acids during autolysis (Moreno-Arribas et al. 1998; Villamiel et al. 2008), which can act as precursors of biogenic amines during malolactic fermentation and during wine aging. Yeast autolysis favors the growth and activity of lactic acid bacteria due to the release of vitamins and large nitrogenated compounds. Lactic acid bacteria can hydrolyze and metabolize the proteins and peptides and use the amino acids as nutrients and sources of energy (Manca de Nadra et al. 1997, 1999; Leitao et al. 2000; Remize et al. 2006; Alcaide-Hidalgo et al. 2008). These amino acids may include precursors of biogenic amines. Yeast and bacterial lees may also be sources of bacteria with decarboxylase enzymes, potentially amine producers. In fact, the first gene of the enzyme ornithine decarboxylase, in a lactic acid bacteria of enological origin, was isolated from wine lees (Marcobal et al. 2004).

Table 6A.1 Average biogenic amine concentrations (mg/L) for different levels of vintage, pectolytic enzymes, aging on lees, maceration time and bacteria inoculation (from Martín-Álvarez et al. 2006, with permission)

	Number of samples	Histamine	Methylamine	Ethylamine	Tyramine	Phenylethylamine	Putrescine	Cadaverine
		**	**	**	**	*	**	
Vintage								
2001	117	4.87±0.67	1.36±0.12	3.07±0.20	2.12±0.55	3.36±0.50	9.69±1.24	1.16±0.44
2002	107	1.44±0.69	0.39±0.12	0.44±0.20	0.62±0.57	2.29±0.52	4.70±1.29	1.60±0.46
Pectolytic enzymes						**		*
No	162	2.76±0.59	1.01±0.11	1.66±0.17	0.96±0.49	3.86±0.44	8.08±1.10	1.93±0.39
Yes	62	3.55±0.81	0.74±0.14	1.85±0.24	1.77±0.67	1.79±0.61	6.31±1.51	0.83±0.53
Aging with lees			*		**		**	
No	159	3.82±0.63	0.71±0.11	1.92±0.19	2.30±0.52	2.68±0.47	3.93±1.18	1.21±0.42
Yes	65	2.49±0.78	1.05±0.14	1.58±0.23	0.43±0.64	2.97±0.58	10.46±1.45	1.55±0.51
Skin maceration (time)		**	*		**		**	
<10 days	123	1.83±0.58	0.71±0.10	1.83±0.17	0.51±0.47	2.70±0.43	3.32±1.07	1.02±0.38
>10 days	101	4.48±0.79	1.04±0.14	1.68±0.23	2.22±0.65	2.95±0.59	11.07±1.46	1.74±0.52
MLF inoculation		**			*			*
No	193	4.89±0.40	1.01±0.07	2.00±0.12	2.37±0.33	2.67±0.30	7.34±0.74	2.09±0.26
Yes	31	1.42±1.04	0.75±0.19	1.50±0.31	0.37±0.86	2.99±0.78	7.05±1.94	0.67±0.69
	224							

**The factor has statistically significant effect on the variable, $P < 0.01$; *the factor has a statistically significant effect on the variable, $P < 0.05$

It was suggested in 1971 that histamine production can be prevented by controlling the microbial flora of wines using inocula from malolactic fermentation (Smith 1980). Spontaneous malolactic fermentation is still extensively applied in cellars worldwide, although it is sometimes difficult to control the malolactic fermentations induced. Commercial strains of *O. oeni* have been selected for their enological properties and these must be analyzed to ensure the absence of amino acid decarboxylase enzymes. In vitro studies carried out in our laboratory studied the capacity of four commercial malolactic starters to form biogenic amines and found that none of them produced histamine, tyramine or putrescine (Moreno-Arribas et al. 2003). It has also been demonstrated that inoculation with commercial starter cultures that do not contain the genes that encode the enzymes involved in biogenic amine production, is a real and feasible option to control the generation of biogenic amines during wine manufacture in industrial conditions (Marcobal et al. 2006a). It also seems that the co-inoculation of starter cultures of *O. oeni* simultaneously to alcoholic fermentation, can potentially reduce amine formation even more than the conventional inoculation carried out when alcoholic fermentation has finished.

The effect of complex commercial fermentation activators on biogenic amine production was tested by Marques et al. (2008). On the whole, it does not seem that complex nutrient preparations produced for the use with fermentation starter cultures pose a serious threat to biogenic amine production in wines.

Currently, the most practical way to control the problem of biogenic amine production is based on inhibiting the growth of indigenous decarboxylase-positive bacteria and other microorganisms responsible for this alteration. Sulphur dioxide (SO_2) is the antimicrobial agent traditionally used in cellars after malolactic fermentation has finished, to remove yeasts and bacteria not wanted in the later stages. According to some authors, the addition of SO_2 does not affect the formation of biogenic amines during alcoholic fermentation (Garde-Cerdán et al. 2007).

Studies carried out throughout the process of industrial wine production indicate that addition of SO_2 to red wines, at concentrations known to reduce the bacterial population, prevents the formation of biogenic amines during wine aging and maturation (Marcobal et al. 2006a). Because of these conclusions, SO_2 is added immediately after malic acid breakdown to remove the lactic acid bacteria as quickly as possible so as to prevent biogenic amine formation. On the other hand, in several studies a rise in histamine after malolactic fermentation has finished and during wine aging has been observed, indicating that the addition of sulphur dioxide after malolactic fermentation has finished does not totally detain the enzymatic reactions carried out by the bacteria, including the decarboxylation of amino acids. Also, the use of SO_2 is less effective due to the high pH values of many wines and, often, the concentration of amines can rise in sulfited wines during aging. In fact, several studies have shown that red wines with high histamine contents (>10 mg/L) are characterized by pH values above 3.7 (Landete et al. 2005; Marcobal et al. 2006a).

Lysozyme is an enzyme that can cause lysis of the cell walls of Gram-positive bacteria, including lactic acid bacteria in wine (Delfini 2004). Lysozyme maintains its activity in wines with high pH values so it can be used successfully to delay or inhibit the growth of lactic acid bacteria, especially when used together with SO_2.

Therefore, in some wines it may be interesting to control the formation of amines by undesirable bacteria.

Generally, most studies in the literature agree that there are slight variations in biogenic amine concentrations, corresponding to a slight decrease or stabilization of these compounds during wine storage. Biogenic amines may be broken down to other compounds by oxidase type enzymes that may be present in some bacteria. Other factors of wine aging could also play an important role in the accumulation of biogenic amines. These include wine filtration using diatoms that can adsorb amino acids and cationic proteins at their surface, affecting changes in biogenic amine content during aging. On the other hand, the use of different clarifiers and enological coadjuvants, especially bentonite and polyvinylpirrolidone (PVPP), among others, at the dose used during winemaking, affects the final concentration of amines, since these treatments are able to adsorb certain biogenic amines.

It has also been shown that the type of oak used to make the barrel (American, French, etc.) used for wine aging does not affect the accumulation of biogenic amines in the final product (Jiménez-Moreno et al. 2003). However, the type of container used for malolactic fermentation seems to affect the final contents of biogenic amines. Significantly higher mean concentrations of biogenic amines were detected in wines undergoing malolactic fermentation in stainless steel tanks compared to those in which this process was carried out in oak barrels (Alcaide-Hidalgo et al. 2007), suggesting that components of the wood, mainly the phenolic compounds, may influence the production of biogenic amines by lactic acid bacteria (García-Ruiz et al. 2008a). Curiously, it has also been shown that some alternative aging techniques, based on the use of wood chips or enological tannins, may have an effect, depending on the product and the dose, by reducing the biogenic amine content of the wine (García-Ruiz et al. 2008b).

Another important aspect from a technological perspective was to study the presence of biogenic amines and how they are related to the ecology and diversity of lactic acid bacteria during the industrial manufacture of biologically aged wines from different Spanish wine cellars (Moreno-Arribas and Polo 2008). The concentrations of biogenic amines detected in the biologically aged wines analyzed were low, especially when compared with other types of wines, such as red wines, which is possibly due to the low levels of amino acids detected in these wines during the biological ageing. Although only a low incidence and population of lactic acid bacteria were detected, this made a clear metabolic contribution to wine quality; not only to degradation of the malic acid, but also to the production of biogenic amines.

6A.3.4 Analytical Tools to Control the Presence of Biogenic Amines in Wine

Analysis of biogenic amines, individually or simultaneously, is important because of their potential to be applied as indicators of food spoilage or authenticity. The two main problems encountered with amine determination in food products in general, and especially in wine, are the complexity of the matrix and their low concentrations. Several techniques have been developed for biogenic amine determination in foods, some to analyse one specific biogenic amine (histamine or tyramine), and others to detect more than one simultaneously. Initially, spectrofluorimetric techniques were used to determine biogenic amines (Vidal-Carou et al. 1990b). More recently, chromatographic methods are the techniques most commonly applied to analyse several biogenic amines simultaneously.

On the other hand, qualitative measurements have also been described to indicate the presence of amines in wine. The potential of biogenic amines to appear in wine can be determined by using molecular tools which detect the presence of decarboxylase-positive microorganisms.

6A.3.4.1 Quantification by Liquid Chromatography

Of the wide range of analytical techniques to detect and quantify biogenic amines in wines described in the literature, reverse phase high-power liquid chromatography (RP-HPLC) is the one most widely used in laboratories, since it can obtain the most information about all the amines identified in must and wine (Lehtonen 1996; Romero et al. 2002; Torrea and Ancín 2002; Vázquez-Lasa et al. 1998; Marcobal et al. 2005b; Hernández-Orte et al. 2006b; Gómez-Alonso et al. 2007). In fact, one RP-HPLC technique is currently being adopted as a reference technique by the International Organization of Vine and Wine (OIV).

Owing to the absence of chromophores from these compounds, it is necessary to form derivatives that are absorbed in ultraviolet or are fluorescent to facilitate detection. For this purpose, dansyl-chloride and *O*-phthaldialdehyde (OPA) are the reagents most used. In the case of dansyl-chloride, the reaction is carried out before chromatographic separation, while with OPA it can be done before the column or after (Vázquez-Lasa et al. 1998; Marcobal et al. 2005b). Other derivatizing reagents used, of more general application, are 9-fluorenylmethylchloroformate (FMOC), aminoquinolyl-*N*-hydroxysuccinimidyl caramate (AQC) (Hernández-Orte et al. 2006b), and diethyl ethoxymethylenemalonate (DEEMM) (Gómez-Alonso et al. 2007). Of all these derivatives, the most used is OPA, because of its rapid one-step derivatization reaction and the possibility of automating the reaction, which increases the reproducibility of the analytical method, among other reasons.

Besides HPLC methods, over the past few years other chromatographic techniques applicable to biogenic amine analysis have been performed. Among others, micellar liquid chromatography (MLC) (Gil-Agustí et al. 2007); micellar electrokinetic chromatography (MECC) separation with laser-induced fluorescence (LIF) detection (Nouadje et al. 1997), nanoliquid chromatography with UV detection (Hernández-Borges et al. 2007), ion-pair liquid chromatography (Hlabangana et al. 2006), and liquid chromatography-electronspray ionisation ion trap mass spectrometry (Millán et al. 2007) are used.

6A.3.4.2 Quantification by Capillary Electrophoretic Techniques

The attraction of using capillary electrophoresis (CE) methods mainly lies in the short analysis time, high separation efficiencies, and reduced consumption of reagents. CE also permits a wide variety of compounds to be analysed at low concentration levels with a minimum sample treatment. The analysis of different biogenic amines in foods by CE with diverse detection methods has been previously reported in the literature (Önal 2007) and applied to a wide variety of food matrices including wine (Kováks et al. 1999). A CE method using conductometric detection and which requires no derivatisation or sample cleaning steps was developed recently (Kvasnicka and Voldrich 2006). Also, a high-performance capillary electrophoresis (HPCE) method exists to determine biogenic amines in wine, amongst other foodstuffs (Kóvacs et al. 1999).

In recent years there has been a growing interest in the use of electrospray ionization-mass spectrometry (ESI-MS) either as a stand-alone technique, or following an analytical separation step like CE, to study and measure a wide variety of compounds in complex samples such us foods (Simó et al. 2005). ESI provides an effective means for ionising from large (e.g., proteins, peptides, carbohydrates) to small (e.g., amino acids, amines) analytes directly from solution prior to their MS analysis without a previous derivatization step. Santos et al. (2004) proposed the use of CE-ESI-MS for the separation and quantification of nine biogenic amines in white and red wines. More recently, the possibilities of two different CE-MS set-ups, namely, capillary electrophoresis-electrospray-ion trap mass spectrometry (CE-IT-MS) and capillary electrophoresis-electrospray-time of flight mass spectrometry (CE-TOF-MS) to analyze directly biogenic amines in wine samples without any previous treatment has been studied (Simó et al. 2008).

6A.3.4.3 Enzymatic Methods

Enzymatic methods to quantify histamine were first reported for use in fish. A direct enzyme-linked immunosorbent assay (ELISA) was applied to determine histamine in wine samples and the results were compared with those of an HPLC method of OPA-derivatives for this amine, obtaining concordant results (Marcobal et al. 2005a). Because of its simplicity, speed and low cost, the ELISA method can be considered as an alternative to HPLC and can, at least, be used as a screening tool in laboratories where chromatographs are not available or that have to analyse a large number of wines. The development of an enzymatic tool to determine the histamine contents in wines and its comparison with HPLC has also been published (Landete et al. 2004). Very good correlation was also established between biogenic amine quantification using this enzymatic method and HPLC analysis.

6A.3.4.4 Screening Methods Using Selective Media

Most studies to screen for biogenic amine-producing lactic acid bacteria use differential media that contain the precursor amino acid and a pH indicator. This indicator, usually purple bromocresol, will change colour when the medium is alkalinized and this colour change will be observed in the medium if the lactic acid bacteria produce amines (Bover-Cid and Holzapfel 1999; Choudhury et al. 1990; Maijala 1993).

However, some false negatives have been reported for lactic acid bacteria as a consequence of excess acid production or the generation of other alkaline compounds (Joosten and Northolt 1989) and also in some cases false-positives (Moreno-Arribas et al. 2003). Some loss of activity by these bacteria has also been described during successive steps and prolonged storage times (Izquierdo-Pulido et al. 1997), which could be disadvantageous when applying these methods in microbiological analysis.

6A.3.4.5 Thin-Layer Chromatography

Thin-Layer Chromatography (TLC) is a rapid and simple technique to detect several amines. This was one of the first techniques used to determine biogenic amines in foods (Hálasz et al. 1994). Recently, a simple and rapid qualitative TLC method to determine the ability of bacteria to produce biogenic amines in liquid culture media containing the amino acid precursor was described by García-Moruno et al. (2005).

6A.3.4.6 Polymerase Chain Reaction Methods

The molecular biology methods described to detect biogenic amine-producing bacteria are based on the polymerase chain reaction or PCR technique. With PCR, multiple copies of a specific DNA sequence (target) can be obtained in vitro. PCR techniques have been developed to detect bacterial amino acid decarboxylases in a rapid, sensitive and accurate way. They cannot determine quantitative (or qualitative) amounts of biogenic amines, but can only be used to estimate the potential risk of amine formation. Molecular methods for the detection of biogenic amine-producing bacteria have been reviewed recently by Landete et al. (2007a).

Since some DNA sequences of selected genes are highly conserved, PCR can be applied to detect specific genes in different organisms. Some conserved gene regions that code for histidine decarboxylase have been detected in different bacteria, such as *Lactobacillus* 30a, *Clostridium perfringes, Lactobacillus buchneri* and *Micrococcus* sp., and this has been used to design the oligonucleotides CL1, CL2, JV16HC and JV17HC (Le Jeune et al. 1995). By using these oligonucleotides and the PCR technique, lactic acid bacteria strains with the histidine decarboxylase gene can be used, therefore resulting in histamine production (Torres Alves and Teia dos Santos 2002). Another way to detect these strains is by hybridisation with DNA primers amplified with these oligonucleotides (Coton et al. 1998, Le Jeune et al. 1995). To sequence the gene that codes tyrosine decarboxylase of *Lactobacillus brevis* IOEB 9809 oligonucleotides P1 and P2 were designed from peptides obtained from the protein (Lucas and Lonvaud-Funel 2002). By a similar approach, universal primers were designed for the early detection of potential tyramine-producing bacteria in wine (Coton et al. 2004). Marcobal et al. (2005a) designed the first complete set of primers to amplify the gene coding for ornithine decarboxylase by aligning amino acid sequences of ornithine decarboxylases from Gram-negative bacteria.

Recently, *multiple* PCR systems have been published that use different pairs of primers to simultaneously detect lactic acid bacteria producers of histamine, tyramine and putrescine, the majority wine biogenic amines (Marcobal et al. 2005b; Constantini et al. 2006).

6A.4 Final Remarks

The characterization of amino acid composition in grape musts and wines will continue being of great interest because these compounds represent an important source of directly assimilable nitrogen and are also precursors for the synthesis of some compounds during the alcoholic fermentation. Other aspects, related to the modifications in the amino acid composition during malolactic fermentation will also be of great interest for the future. In a similar way, the occurrence of biogenic amines in fermented foods in general, and in wines in particular, has been extensively studied in the last few years. We have seen here that the formation of biogenic amines in wine is a complex process involving many different factors, mainly related to the winemaking conditions in the cellars and environmental factors. This information has allowed valuable data for control and prevention of biogenic amine accumulation in winemaking. Unfortunately, the concentration of biogenic amines in wines is still a problem for many wine cellars. It is expected, further investigation on the mechanisms and tools (both analytical and molecular) to avoid their presence in wines or for their successfully removing from wines. Oncoming research will reveal interesting findings within this field.

Acknowledgments The authors are grateful to AGL2006-04514, PET2007_0134 and CSD2007-00063 Consolider Ingenio 2010 FUN-C-FOOD Projects (Ministerio de Ciencia e Innovación) and S-505/AGR-0153 ALIBIRD Project (Comunidad Autónoma de Madrid) for their financial support for this work.

References

Agenbach W.A. (1977). A study of must nitrogen content in relation to incomplete fermentations, yeast production and fermentation activity. *South Afr. Society Enol. Vitic., 1*, 66–87.

Alcaide-Hidalgo J., Moreno-Arribas, M.V, Martín-Álvarez, P.J. & Polo, M.C. (2007). Influence of malolactic fermentation, postfermentative treatments and ageing with lees on nitrogen compounds of red wines. *Food Chem.,* 103, 572–581.

Alcaide-Hidalgo J., Moreno-Arribas, M.V., Polo, M.C. & Pueyo, E. (2008). Partial characterization of peptides from red wines. Changes during malolactic fermentation and aging with lees. *Food Chem., 107*, 622–630.

Bisson, L.F. (1991). Influence of nitrogen on yeast and fermentation of grapes. In: Proceeding of the International Symposium on Nitrogen in Grapes and Wine. The American Society for Enology and Viticulture (pp. 78–89). Seattle, Washington, USA.

Bodmer, S., Imark, C. & Kneubühl, M. (1999). Biogenic amines in foods: Histamine and food processing. *Inflamm. Res., 48*, 296–300.

Bover-Cid, S., Izquierdo-Pulido, M., Mariné-Font, A. & Vidal-Carou, M.C. (2006). Biogenic, mono-, di- and polyamine contents in Spanish wines and influence of a limited irrigation. *Food Chem., 96*, 43–47.

Buteau, C., Duitschaever, C.L., & Ashton, G.C. (1984). A study of the biogenesis of amines in a Villard Noir wine. *Am. J. Enol. Vitic., 35*, 228–236.

Caruso, M., Fiore, C., Contrusi, M., Salzano, G., Paparella, A. & Romano, P. (2002). Formation of biogenic amines as criteria for the selection of wine yeast. *World J. Microbiol. Biotech., 18*, 159–163.

Choudhury, N., Hansens, W., Engesser, D., Hammes, W.P. & Holzaptel, W.H. (1990). Formation of histamine and tyramine by lactic acid bacteria in decarboxylase assay medium. *Lett. Appl. Microbiol., 11*, 278–281.

Constantini, A., Cersosimo, M., Del Prete, V. & García-Moruno, E. (2006). Production of biogenic amines by lactic acid bacteria: screening by PCR, thin-layer chromatography, and high-performance liquid chromatography of strains isolated from wine and must. *J. Food Prot., 69*, 391–396.

Cooper, T.G. (1982). Nitrogen metabolism in *Saccharomyces cerevisiae*. In J.N. Strathern, E.W. Jones & J.R. Broach (Eds.), The molecular biology of the yeast *Saccharomyces*. Metabolism and gene expression (pp. 39–99). New York: Cold Spring Harbor.

Coton, E., Rollan, G., Bertrand, A. & Lonvaud-Funel, A. (1998). Histamine-producing lactic acid bacteria in wines; early detection, frequency and distribution. *Am. J. Enol. Vitic., 49*, 199–204.

Coton, M., Coton, E., Lucas, P. & Lonvaud, A. (2004). Identification of the gene encoding a putative tyrosine decarboxylase of *Carnobacterium divergens* 508. Development of molecular tools for the detection of tyramine-producing bacteria. *Food Microbiol., 21*, 125–130.

Davis, C.R., Wibowo, D.J., Lee, T.H. & Fleet, G.H. (1986). Growth and metabolism of lactic acid bacteria during and after malolactic fermentation of wines at different pH. *Appl. Environ. Microbiol., 51*, 539–545.

Delfini, C. (2004). Resistance screening assay of wine lactic acid bacteria on lysozyme: efficace of lysozyme in unclarified grape musts. *J. Agric. Food Chem., 52*, 1861–1866.

Dizy, M. & Polo, M.C. (1996). Changes in concentration of nitrogenous compounds during fermentation of white grape musts at pilot plant scale. *Food Sci. Technol. Int., 2*, 87–93.

Escudero, A., Hernández-Orte, P., Cacho, J. & Ferreira, V. (2000). Clues about the role of methional as character impact odorant of siome oxidiez wines. *J. Agric. Food Chem., 48*, 4268–4272.

Farias, M.E., Manca de Nadra, M.C., Rollan, G.C. & Strasser de Saad, A.M. (1993). Histidine decarboxylase activity in lactic acid bacteria from wine. *J. Int. Sci. Vigne Vin, 27*, 191–199.

Fernández, P.A.A. & Manca de Nadra, M.C. (2006). Growth response and modifications of organic nitrogen compounds in pure and mixed cultures of lactic acid bacteria from wine. *Curr. Microbiol., 52*, 86–91.

Fernández-García, E., Tomillo, J. & Nuñez, M. (1999). Effect of added proteinases and level of starter culture on the formation of biogenic amines in raw milk Manchego cheese. *Int. J. Food Microbiol., 52*, 189–196.

Feuillat, M., Brillant, G. & Rochard, J. (1980). Mise en Évidence d'une Production de Proteases Exocellulaires par les Levures au Cours de la Fermentation alcoolique du Mouêt de Raisin. *Conn. Vigne Vin, 14*, 37–52.

Fraile, P., Garrido, J. & Ancin, C. (2000). Influence of a *Saccharomyces cerevisiae* selected strain in the volatile composition of rose wines. Evolution during fermentation. *J. Agric. Food Chem., 48*, 1789–1798.

García-Moruno, E., Carrascosa, A & Muñoz, R. (2005). A rapid and inexpensive method for the determination of biogenic amines from bacterial cultures by thin-layer chromatography. *J. Food Protect., 68*, 625–629.

García-Ruiz, A., Bartolomé, B., Martínez-Rodríguez, A., Pueyo, E., Martín-Álvarez, P.J. & Moreno-Arribas, M.V. (2008a). Potential of phenolic compounds for controlling lactic acid bacteria growth in wine. *Food Control, 19*, 835–841.

García-Ruiz, A., Bartolomé, B. & Moreno-Arribas, M.V. (2008b). Understanding the effect of oak wood treatments and enological tannins in biogenic amine production by lactic acid bacteria in wines XXXI World Congress of Vine and Wine, Verone, 2008.

Garde-Cerdán, T., Arias-Gil, M. & Romano, P. (2007). Formation of biogenic amines throught spontaneous and inoculated wine alcoholic fermentations: effect of SO_2. *Food Control* (en prensa, doi:10.1016/j.foodcont.2006.07.003).

Gardini, F., Zaccarelli, A., Belleti, N., Faustini, F., Cavazza, A., Maruscelli, M., Mastrocola, D. & Suzzi, G. (2005). Factors influencing biogenic amine production by a strain of *Oenococcus oeni* in a model system. *Food Control, 16*, 609–618.

Gerbaux, V. & Monamy, C. (2000). Biogenic amines in Burgundy wines. Contents and origin in wines. *Rev. Fr. Oenol., 183*, 25–28.

Gil-Agustí, M., Carda-Brochm, S., Monferrer-Pons, L. & Esteve-Romero, J. (2007). Simultaneous determination of tyramine and tryptamine and their precursor amino acids by micellar liquid chromatography and pulsed amperometric detection in wines. *J. Chromatogr. A., 1156*, 288–295.

Glòria, M.B., Watson, B.T., Simon-Sarkadii, L. & Daeschel, M.A. (1998). A survey of biogenic amines in Oregon Pinot noir and Cabernet Sauvignon wines. *Am. J. Enol. Vitic., 49*, 279–282.

Gómez-Alonso, S., Hermosín-Gutiérrez, U. & García-Romero, E. (2007). Simultaneous HPLC analysis of biogenic amines, amino acids and ammonium ion as animoenone derivatives in wine and beer simples. *J. Agric. Food Chem., 55*, 608–613.

Granchi, L., Romano, P., Mangani, S., Guerrini, S. & Vincenzini, M. (2005). Production of biogenic amines by wine microorganisms. *Bull OIV*, 78, 595–609.

Guerrini, S., Mangani, S., Granchi, L. & Vincenzini, M. (2002). Biogenic amine production by *Oenococcus oeni*. *Curr. Microbiol., 44*, 374–378.

Hajós, G., Sass-Kiss, A., Szerdahelye, E. & Bardocz, S. (2000). Changes in biogenic amine content of Tokaj grapes, wines and Aszú-wines. *J. Food Sci., 65*, 1142–1144.

Hálasz, A., Baráth, Á., Simon-Sarkadi, L. & Holzapfel, W. (1994). Biogenic amines and their production by microorganisms in food. *Trends Food Sci. Tech., 5*, 42–49.

Henschke, P.A. & Jiranek, V. (1993). Yeasts – metabolism of nitrogen compounds. In G.H. Fleet (Ed.), Wine microbiology and biotechnology. (pp. 77–164). Chur, Switzerland: Harwood Academic Publishers.

Herbert, P., Cabrita, M.J., Ratola, N., Laureano, O. & Alves, A. (2005). Free amino acids and biogenic amines in wines and musts from the Alentejo region. Evolution of amines during alcoholic fermentation and relationship with variety, sub-region and vintage. *J. Food. Eng., 66*, 315–322.

Hernández-Borges, J., D'Orazio, G., Aturki, Z. & Fanali, S. (2007). Nano-liquid chromatography analysis of dansylated biogenic amines in wine. *J. Chromatogr. A, 1147*, 192–199.

Hernández-Orte, P., Cacho, J. & Ferreira, V. (2002). Relationship between varietal amino acid profile of grapes and wine aromatic composition. Experiments with model solutions and chemometric study. *J. Agric. Food Chem., 50*, 2891–2899.

Hernández-Orte, P., Ibarz, M.J., Cacho, J. & Ferreira, V. (2003). Amino acid determination in grape juices and wines by HPLC using a modification of the 6-aminoquinolyl-n-hydrosysuccinimidyl carbamate (AQC) method. *Chromatographia, 58*, 29–35.

Hernández-Orte, P., Ibarz, M.J., Cacho, J., & Ferreira, V. (2005). Effect of the addition of ammonium and amino acids to musts of Airen variety on aromatic composition and sensory properties of the obtained wine. *Food Chem., 89*, 163–174.

Hernández-Orte, P., Ibarz, M.J., Cacho, J. & Ferreira, V. (2006a). Addition of amino acids to grape juice of the Merlot variety: Effect on amino acid uptake and aroma generation during alcoholic fermentation. *Food Chem., 98*, 300–310.

Hernández-Orte, P., Peña-Gallego, A., Ibarz, M.J., Cacho, J., Ferreira, V. (2006b). Determination of the biogenic amines in musts and wines before and alter malolactic fermentation using 6-aminoquinolyl-N-hydroxysuccinimidyl carbamate as the derivatizing agent. *J. Chromatogr. A, 1129*, 160–164.

Herraiz, T. & Ough, C.S. (1993). Formation of ethyl-esters of amino acids by yeasts during the alcoholic fermentation of grape juice. *Am. J. Enol. Vitic., 44,* 41–48.
Herraiz, T., Huang, Z.X. & Ough, C.S. (1993). 1,2,3,4-Tetrahydro-beta-carboline-3-carboxylic acid and 1-methyl-1,2,3,4- tetrahydro-beta carboline-3-carboxylic acid in wines. *J. Agric. Food Chem., 41*, 455–459.
Hlabangana, L., Hernández-Cassou, S. & Saurine, L. (2006). Determination of biogenic amines in wines by ion-pair liquid chromatography and post-column derivatization with 1,2-naphthoquinine-4-sulfonate. *J. Chromatogr. A, 1130*, 130–136.
Ingledew, W.M. & Kunkee, R.E. (1985). Factors Influencing Sluggish Fermentations of Grape Juice. *Am. J. Enol. Vitic., 36*, 65–76.
Izquierdo-Pulido, M., Carceller-Rosa, J.M., Mariné-Font, A., & Vidal-Caron, M.C. (1997) Tyramine formation by *Pediococcus* spp. during beer fermentation. *J. Food Prot., 60*, 831–836.
Izquierdo-Pulido, M., Mariné-Font, A. & Vidal-Carou, M.C. (2000). Effect of tyrosine on tyramine formation during beer fermentation. *Food Chem., 70*, 329–332.
Jansen, S.C., van Dusseldorp, M., Bottema, K.C. & Dubois, A.E. (2003). Intolerance to dietary biogenic amines: A review. *Ann. Allerg. Asthama Im., 91*, 233–240.
Jiménez-Moreno, N., Goñi, D.T. & Ancín Azpilicueta, C. (2003). Changes in amine concentrations during aging of red wine in oak barrels. *J. Agric. Food Chem., 51*, 5732–5737.
Jiranek, V., Langridge, P. & Henschke, P.A. (1995). Amino acid and ammonium utilization by *Saccharomyces cerevisiae* wine yeasts from a chemically defined medium. *Am. J. Enol. Vitic., 46*, 75–83.
Joosten, H.M.L.J., & Northolt, M.D., (1989). Detection, growth, and amine-producing capacity of lactobacilli in cheese. *Appl. Environ. Microbiol., 55*, 2356–2359.
Kaniou, I., Samouris, G., Mouratidou, T., Eleftheriadou, A., & Zantopoulos, A. (2001). Determination of biogenic amines in fresh unpacked and vacuum-packed beef during storage at 4°C. *Food Chem., 74*, 515–519.
Kanny, G., Gerbaux, V., Olszewski, A., Frémont, S., Empereur, F., Nabet, F., Cabanis, J. & Moneret-Vautrin, D (2001). No correlation between wine intolerance and histamine content of wine. *J. Allergy Clin. Inmmun., 107*, 375–378.
Kiss, J., Korbász, M. & Sass-Kiss, A. (2006). Study of amine composition of botrytized grape berries. *J. Agric. Food Chem., 53*, 8909–8918.
Kovâks, A., Simon-Sarkadi, L. & Ganzler, K. (1999). Determination of biogenic amines by capillary electrophoresis. *J. Chromatogr. A, 836*, 305–313.
Kvasnicka, F. & Voldrich. M. (2006). Determination of biogenic amines by capillary zone electrophoresis with conductimetric detection. *J. Chromatogr. A, 1103*, 145–149.
Landete, J.M., Ferrer, S., Pardo, I. (2004) Improved enzymatic method for the rapid determination of histamine in wine. *Food Addit. Cont., 21*, 1149–1154.
Landete, J.M., Polo, L., Ferrer, S. & Pardo, I. (2005). Biogenic amines in wines from three Spanish Regions. *J. Agric. Food Chem., 53*, 1119–1124.
Landete, J.M., de las Rivas, B., Marcobal, A. & Muñoz, R. (2007a). Molecular methods for the detection of biogenic amine-producing bacteria on foods. *Int. J. Food Microbiol., 117*, 258–269.
Landete, J.M., Pardo, I. & Ferrer, S. (2007b). Biogenic amine production by lactic acid bacteria, acetic bacteria and yeast isolated from wine. *Food Control, 18*, 1569–1574.
Le Jeune, C., Lonvaud-Funel, A., Ten Brink, B., Hofstra, H. & van der Vossen, J.M.B.M. (1995). Development of detection system for histidine decarboxylating lactic acid bacteria based DNA probes, PCR and activity test. *J. Appl. Bacteriol., 78*, 316–326.
Lehtonen, P. (1996). Determination of amines and amino acids in wine – a review. *Am. J. Enol. Vitic., 47*, 127–133.
Leitao, M.C., Teixeira, H.C., Barreto Crespo, M.T. & San Romao, M.V. (2000). Biogenic amines occurrence in wine. Amino acid decarboxylase and proteolytic activities expression by *Oenococcus oeni*. *J. Agric. Food Chem., 487*, 2780–2784.
Lonvaud-Funel, A. & Joyeux, A. (1994). Histamine production by wine lactic acid bacteria: isolation of a histamine-producing strain of *Leuconostoc oenos*. *J. Appl. Bacteriol., 77*, 401–407.

López, R., Santamaría, P., Gutiérrez A.R. & Iñiguez, M. (1996). Changes in amino acids during the alcoholic fermentation of grape juice at different temperatures. *Sci. Aliment., 16*, 529–535.

Lucas, P. & Lonvaud-Funel, A. (2002). Purification and partial gene sequence of the tyrosine decarboxylase of *Lactobacillus brevis* IOEB 9809. *FEMS Microbiol. Lett., 211*, 85–89.

Lucas, P., Landete, J., Coton, M., Coton, E. & Lonvaud-Funel, A. (2003). The tyrosine decarboxylase operon of *Lactobacillus brevis* IOEB 9809: Characterization and conservation in tyramine-producing bacteria. *FEMS Microbiol. Lett., 229*, 65–71.

Lucas, P.M., Wolken, W.A.M., Claisse, O., Lolkema, J.S. & Lonvaud-Funel, A. (2005). Histamine-producing pathway encoded on an unstable plasmid in *Lactobacillus hilgardii* 0006. *Appl. Environ. Microb., 71*, 1417–1424.

Mafra, I., Herbert, P., Santos, L., Barros, P. & Alves, A. (1999). Evaluation of biogenic amines in some portuguese quality wines by HPLC fluorescence detection of OPA derivaties. *Am. J. Enol. Vitic., 50*, 128–132.

Maijala, R.L. (1993). Formation of histamine and tyramine by some lactic acid bacteria in MRS-broth and modified decarboxylation agar. *Lett. Appl. Microbiol., 17*, 40–43.

Manca de Nadra, M.C., Farías, M., Moreno-Arribas, M.V., Pueyo, E. & Polo, M.C. (1997). Proteolytic activity of *Leuconostoc oenos*: Effect on proteins and polypeptides from white wine. *FEMS Microbiol. Lett., 150*, 135–139.

Manca de Nadra, M.C., Farias, M., Moreno-Arribas, M.V., Pueyo, E. & Polo, M.C. (1999). A proteolytic effect of *Oenococcus oeni* on the nitrogenous macromolecular fraction of red wine. *FEMS Microbiol. Lett., 174*, 41–47.

Marcobal, A., de las Rivas, B., Moreno-Arribas, M.V. & Muñoz, R. (2004). Identification of the ornithine decarboxylase gene in the putrescine-producer *Oenococcus oeni* BIFI-83. *FEMS Microbiol. Lett., 239*, 213–220.

Marcobal, A., Polo, M.C., Martín-Álvarez, P.J. & Moreno-Arribas, M.V. (2005a). Biogenic amines content of red Spanish wines. Comparison of a Direct ELISA and an HPLC method for the determination of histamine in wines. *Food Res. Int., 38*, 387–394.

Marcobal, A., de las Rivas, B., Moreno-Arribas, M.V. & Muñoz, R. (2005b). Multiplex-PCR method for the Simultaneous Detection of Acid Lactic Bacteria Producing Histamine, Tyramine and Putrescine, Three Major Biogenic Amines. *J. Food Prot., 68*, 874–878.

Marcobal, A., Martín-Álvarez, P.J., Polo, M.C., Muñoz, R., & Moreno-Arribas, M.V. (2006a). Formation of biogenic amines throughout the industrial manufacture of red wine. *J. Food Prot., 69*, 391–396.

Marcobal, A., de las Rivas, B., Moreno-Arribas, M.V. & Muñoz, R. (2006b). Evidence for horizontal gene transfer as origin of putrescine-production in *Oenococcus oeni* RM83. *Appl. Envir. Microbiol, 72*, 7954–7958.

Marques, A.P., Leitao, M.C. & San Romao, M.V. (2008). Biogenic amines in wines: Influence of oenological factors. *Food Chem., 107*, 853–860.

Martín-Álvarez, P.J., Marcobal, A., Polo, C. & Moreno-Arribas, M.V. (2006). Technological factors influencing biogenic amine production during red wine manufacture. *Eur. Food Res. Technol., 222*, 420–424.

Maynard, L.S. & Schenker, V.J. (1996). Monoamine-oxidase inhibition by ethanol in vitro. *Nature, 196*, 575–576.

Millán, S., Sanpedro, M.C., Unceta, N., Goicolea, M.A. & Barrio, R.J. (2007). Simple and rapid determination of biogenic amines in wine by liquid chromatography-electronspray ionization ion trap mass spectrometry. *Anal. Chim. Acta, 584*, 145–152.

mo Dugo, G., Vilasi, F., La Torre, G.L. & Pellicanò, T.M. (2006). Reverse phase HPLC/DAD determination of biogenic amines as dansyl derivatives in experimental red wines. *Food Chem., 95*, 672–676.

Monteiro, F.F. & Bisson, L.F. (1992). Nitrogen supplementation of grape juice. II. Effect on amino-acid and urea release following fermentation. *Am. J. Enol. Vitic., 43*, 11–17.

Moreno-Arribas, M.V. & Polo, M.C. (2008). Occurrence of lactic acid bacteria and biogenic amines in biologically aged wines. *Food Microbiol., 25*, 875–881.

Moreno-Arribas, M. V., Polo, M.C., Jorganes, F. & Muñoz, R. (2003). Screening of biogenic amine production by lactic acid bacteria. Screening of biogenic amine production by lactic acid bacteria isolated from grape must and wine. *Int. J. Food Microbiol., 84*, 117–123.

Moreno-Arribas, V. & Lonvaud-Funel, A. (1999). Tyrosine decarboxylase activity of *Lactobacillus brevis* IOEB 9809 isolated from wine and *L. brevis ATCC 367. FEMS Microbiol. Lett., 180*, 55–60.

Moreno-Arribas, V. & Lonvaud-Funel, A. (2001). Purification and characterization of tyrosine decarboxylase of *Lactobacillus brevis* IOEB 9809 isolated from wine. *FEMS Microbiol. Lett., 195*, 103–107.

Moreno-Arribas, V., Pueyo, E., Polo, M.C. & Martín-Álvarez, P.J. (1998). Changes in the amino acid composition of the different nitrogenous fractions during the aging of wine with yeast. *J. Agric. Food Chem., 46*, 4042–4051.

Moreno-Arribas, V., Torlois, S., Joyeux, A., Bertrand, A. & Lonvaud-Funel, A. (2000). Isolation, properties and behaviour of tyramine-producing lactic acid bacteria from wine. *J. Appl. Microbiol., 88*, 584–593.

Nouadje, G., Simeón, N., Dedieu, F., Nertz, M., Puig, Ph. & Couderc, F. (1997). Determination of twenty eight biogenic amines and amino acids during wine ageing by micellar electrokinetic chromatography and laser-induced fluorescence detection. *J. Chromatogr. A, 465*, 337–343.

Önal, A. (2007). A review: current analytical methods for the determination of biogenic amines in foods. *Food. Chem., 103*, 1475–1486.

Pozo-Bayón, M.A., Alegría, E.G., Polo, M.C., Tenorio, C., Martín-Álvarez P.J., Calvo de la Banda M.T., Ruiz-Larrrea, F. & Moreno-Arribas, M.V. (2005). Wine Volatile and Amino acid Composition after Malolactic Fermentation: Effect of *Oenococcus oeni* and *Lactobacillus plantarum* Starter Cultures. *J. Agric. Food Chem., 53*, 8729–8735.

Rapp, A., & Versini, G. (1991). Influence of nitrogen compounds in grapes on aroma compounds of wines. In: Proceeding of the International Symposium on Nitrogen in Grapes and Wine. The American Society for Enology and Viticulture (pp. 156–164). Seattle, Washington, USA.

Remize, F., Gaudin, A., Kong, Y., Guzzo, J., Alexandre, H., Krieger, S. & Guilloux-Benatier, M. (2006). *Oenococcus oeni* preference for peptides: Qualitative and quantitative analysis of nitrogen assimilation. *Arch. Microbiol., 185*, 459–469.

Romero, R., Sánchez-Viñas, M., Vázquez, D. & Gracia Bagur, M. (2002). Characterization of selected Spanish table wine samples according to their biogenic amine content from Liquid Chromatographic determination. *J. Agric. Food Chem., 50*, 4713–4717.

Santos, B. Simonet, B.M., Rios, A. & Valcárcel, M. (2004). Direct automatic determination of biogenic amines in wine by flow injection-capillary electrophoresis-mass spectrometry. *Electrophoresis, 25*, 3427.

Sass-Kiss, A., Szerdahelyi, E. & Hajós, G. (2000). Study of biologically active amines in grapes and wines by HPLC. *Chromatographia, 52*, S316–S320.

Silla Santos, M. H. (1996). Biogenic amines: their importance in foods. *Int. J. Food Microbiol., 29*, 213–231.

Simó, C., Barbas, C. & Cifuentes, A. (2005). Capillary electrophoresis-mass spectrometry in food analysis. *Electrophoresis, 26*, 1306–1318.

Simó, C., Moreno-Arribas, M.V. & Cifuentes, A. (2008). Ion-trap vs. Time-of-flight mass spectrometry coupled to capillary electrophoresis to analyze biogenic amines in wine. *J. Chromatogr. A, 1195*, 150–156.

Smit, A.Y. (2007). Evaluating the influence of winemaking practices on biogenic amine production by wine microorganisms. Master Thesis. Stellenbosch University. Stellenbosch, South Africa.

Smith, T.A. (1980). Amines in food. *Food Chem., 6*, 169–200.

Soleas, G.L., Carey, M. & Goldberg, D.M. (1999). Method development and cultivar-related differences of nine biogenic amines in Ontario wines. *Food Chem., 64*, 49–58.

Somavilla, C., Bravo, F., Iñigo, B. & Burdaspal, P. (1986). Acumulación de histamina en medios naturales y semisintéticos. *Alimentaria, 86*, 37–42.

Soufleros, E., Barrios, M.L. & Bertrand, A. (1998). Correlation between the content of biogenic amines and other wine compounds. *Am. J. Enol. Vitic.*, *49*, 266–278.

Soufleros, E.H., Bouloumpasi, E., Zotou, A. & Lokou, Z. (2007). Determination of biogenic amines in Greek wines by HPLC and ultraviolet detection after dansylation and examination of factors affecting their presence and concentration. *Food Chem.*, *101*, 704–716.

Straub, B.W., Kicherer, M., Schilcher, S.M. & Hammes, W.P. (1995). The formation of biogenic amines by fermentation organisms. *Z. Lebensm. Unters. For.*, *201*, 79–82.

Ten Brink, B, Damink, C., Joosten, H.M.L.J. & Huis in't Veld, J.H.J. (1990). Occurrence and formation of biologically active amines in foods. *Int. J. Food Microbiol.*, *11*, 73–84.

Torrea, D. & Ancín, C. (2002). Content of biogenic amines in a Chardonnay wine obtained through spontaneous and inoculated fermentations. *J. Agric. Food Chem.*, *50*, 4895–4899.

Torrea-Goñi, D.T. & Ancín-Azpilicueta, C. (2001). Influence of yeast strain on biogenic amine content in wines: Relationship with the utilization of amino acids during fermentation. *Am. J. Enol. Vitic.*, *52*, 185–190.

Torres-Alves, R. & Teia dos Santos, A. (2002) Detection of histamine-producing bacteria using polymerase chain reaction techniques and DNA probes. *Eur. Food Res. Technol.*, *214*, 178–180.

Tracey, R.P. & Britz, T.J. (1989). The effect of amino acids on malolactic fermentation by *Leuconostoc oenos*. *J. Appl. Bacteriol.*, *67*, 589–596.

Vázquez-Lasa, M.B., Iñiguez-Crespo, M., González-Larraina, M. & González-Guerrero, A. (1998). Biogenic Amines in Rioja Wines. *Am. J. Enol. Vitic.*, *49*, 229.

Vidal-Carou, M.C., Codony-Salcedo, R. & Mariné-Font, A. (1990a). Histamine and tyramine in Spanish wines: Relationships with total sulfur dioxide level, volatile acidity and malolactic fermentation intensity. *Food Chem.*, *35*, 217–227.

Vidal-Carou, M.C., Veciana-Nogues, M.T., Marine-Font, A. (1990b). Spectrofluorometric determination of histamine in fish and meat products. *J. Assoc. Anal. Chem.*, *73*, 565 567.

Villamiel, M., Polo, M.C. & Moreno-Arribas, M.V. (2008). Nitrogen compounds and polysaccharides changes during the biological ageing of sherry wines. *LWT-Food Sci. Technol.*, *41*, 1842–1846.

Chapter 6B
Peptides

M. Victoria Moreno-Arribas, María Ángeles Pozo-Bayón, and M. Carmen Polo

Contents

6B.1 Introduction

Peptides form a heterogeneous group of compounds, due to the wide range of structures (amino acid composition and sequence) in which these compounds may appear. Generally speaking, the term oligopeptides, or low-molecular-weight peptides, refers to peptides with 10 or fewer residues of amino acids, while polypeptides is used for peptides with more amino acid residues. Although the point of transition from polypeptide to protein is not well defined, proteins are normally considered to have at least 100 residues (Mr $>$10,000) and this criterion will be followed within the chapter.

In wines, peptides are the least known nitrogen compounds, in spite of the fact that they are involved in diverse properties such as tensioactivity (González-Llano et al. 2004), sensorial activity (Desportes et al. 2001) and antihypertensive activity (Pozo-Bayón et al. 2005), among others. Also they can act as nutrients for yeasts

M.V. Moreno-Arribas (✉)
Instituto de Fermentaciones Industriales, CSIC. Juan de la Cierva 3, 28006 Madrid, Spain
e-mail: mvmoreno@ifi.csic.es

M.V. Moreno-Arribas, M.C. Polo (eds.), *Wine Chemistry and Biochemistry*,
DOI 10.1007/978-0-387-74118-5_8,

and bacteria (Feuillat et al. 1977; Alexandre et al. 2001; Manca de Nadra et al. 2005; Remize et al. 2006). The main reason for the dearth of studies on this fraction is due to the complexity and lack of specificity of the techniques used for its analysis.

6B.2 Analytical Methods

6B.2.1 Spectrophotometric Methods

The N-terminal α-amino groups on peptides react quantitatively with reagents such as ninhydrin (Doi et al. 1981) to form colored derivatives or with *o*-phthaldialdehyde (OPA) (Acedo et al. 1994) and fluorescamine to form fluorescent derivatives. These reactions are useful in making quantitative determinations of oligopeptides in wines and for their detection during chromatographic analysis. Polypeptide content can be determined by assaying the aromatic amino acids tyrosine and tryptophan, which is readily achieved by measuring absorbance at 280 nm or by measuring the color reactions of these amino acids with certain reagents like Folin-Ciocalteu. These methods are only applicable for the analysis of wine peptides after long stages of purification, owing to the presence of free amino acids, derivates in the case of OPA, and the presence of phenolic compounds, which also absorb at 280 nm and react with Folin-Ciocalteu reagent. These methods do not provide information regarding the quantity and the nature of peptides. For this purpose, chromatographic or electrophoretic techniques are required.

6B.2.2 Sample Preparation for Peptide Analysis

Analytical determination of peptides in wine requires sample preparation, involving their isolation from the remaining components, mainly high molecular weight nitrogen compounds, free amino acids and phenols. Table 6B.1 summarizes the procedures used in the literature for the extraction of wine peptides before their analysis by different analytical techniques and with different detection systems.

The traditional method used to remove proteins from a sample to be analyzed is to precipitate them using different precipitants obtaining a soluble fraction after centrifugation. The most common precipitants that have been used are 7% TCA (trichloroacetic acid) (Yokotsuka et al. 1975) and 95% ethanol (Moreno-Arribas et al. 1996, 1998a,b; Bartolomé et al. 1997; Martínez-Rodríguez et al. 2002). Ultrafiltration has been used as an initial step for isolation of the peptide fraction of wines by many authors (Usseglio-Tomasset and Di Stefano 1978; Colagrande and Silva 1981; Acedo et al. 1994; Martínez-Rodriguez and Polo 2000; Dos Santos et al. 2000; Desportes et al. 2000, 2001; Person et al. 2004; Pozo-Bayón et al. 2005, Alcaide-Hidalgo et al. 2007). The pore size of the membranes varies in the different works (from 10,000 to 200–300 Da). To obtain fractions of low relative molecular mass (less than 500Da) it is recommended to use membranes with an increase in

Table 6B.1 Procedure and material used for sample preparation and techniques for separation and detection of wine peptides

Sample preparation		Technique used		Reference
Procedure	Material	Peptides separation	Peptides detection	
Proteins precipitation ↓ Low pressure cationic exchange chromatography	TCA 7% ↓ Dowex 50-X2		*	Yokotsuka et al. (1975)
Low pressure cationic exchange chromatography	Amberlite IR-H 120			Usseglio-Tomasset and Bosia (1990)
Ultrafiltration ↓ Low pressure exclusion chromatography	Cut-off membrane 1000 Da ↓ Sephadex LH-20	RP-HPLC Ultrasphere ODS column	Absorbance at 214 nm and fluorescence of the OPA derivatives	Acedo et al. (1994)
Proteins precipitation ↓ Low pressure exclusion chromatography	Ethanol 95% ↓ Sephadex G-10	RP-HPLC Nova-Pak C18 column	Absorbance at 214 nm	Moreno-Arribas et al. 1996, (1998a,b); Martínez-Rodríguez et al. (2002)
Proteins precipitation ↓ Low pressure exclusion chromatography	Ethanol 95% ↓ Sephadex G-10	RP-HPLC Nova-Pak C18 column	Scanning from 190 to 280 nm and fluorescence of the OPA derivatives	Bartolomé et al. (1997)
Low pressure RP-chromatography ↓ GPC	Cosmosil 140 C_{18}-OPN ↓ Toyopearl HW-40	RP-HPLC μBondasphere C18 column	Absorbance at 220 nm	Takayanagi and Yokotsuka (1999)

Table 6B.1 (continued)

Sample preparation		Technique used		Reference
Procedure	Material	Peptides separation	Peptides detection	
Dialysis	Spectra POR3 membrane	PAGE-SDS	Silver staining	Martínez-Rodriguez and Polo (2000)
Ultrafiltration	Cut-off membrane 10,000 Da	RP-HPLC Nova-Pak C18 column	Absorbance at 214 nm	
↓	↓			
Low pression exclusion chromatography	Sephadex G-10			
Ultrafiltration	Cut-off membrane 5000 Da			Dos Santos et al. (2000)
	↓			
	Cut-off membrane 500 Da			
Ultrafiltration	Cut-off membrane 3000 Da	RP-HPLC Hypercarb column	Absorbance at 214 nm and fluorescence of the OPA derivatives	Desportes et al. (2000)
	↓			
	Cut-off membrane 200-300 Da	↓		
		CE	Absorbance at 214 nm	
↓	↓			
Low pressure exclusion chromatography	Sephadex LH-20			
Ultrafiltration	Cut-off membrane 3000 Da	RP-HPLC Hypercarb column	Absorbance at 214 nm and fluorescence of the OPA derivatives	Desportes et al. (2001)
	↓			
	Cut-off membrane 200-300 Da	↓		
↓		CE	Absorbance at 214 nm	
	↓			
Low pressure exclusion chromatography	Sephadex LH-20			
Ultrafiltration	Cut-off membrane 5000 Da			Alexandre et al. (2001)

Table 6B.1 (continued)

Sample preparation		Technique used		Reference
Procedure	Material	Peptides separation	Peptides detection	
Low pressure RP- chromatography ↓ GPC	Cosmosil 140 C_{18}-OPN ↓ Toyopearl HW-40 ↓ Superdex peptide HR 10/30	RP-HPLc Source 5 RPC ST		Yanai et al. (2003)
Ultrafiltration	Cut-off membrane 1000 Da	RP-HPLC Supercosil ABZ+Plus	MS/MS	Person et al. (2004)
Ultrafiltration ↓ Low-pressure exclusion chromatography	Cut-off membrane 10,000 Da ↓ Sephadex LH-20		Absorbance at 280 nm	Pozo-Bayón et al. (2005)
Low-pressure exclusion chromatography ↓ Low pressure RP-chromatography	Sephadex LH-20 ↓ Cosmosil 140 C_{18}-OPN	RP-HPLC Nova-Pak C18 column	Absorbance at 214 nm and scanning from 190 to 280 nm	Alcaide-Hidalgo et al. (2008)

the small pore size to favor the passage of wines (Dos Santos et al. 2000; Desportes et al. 2000, 2001).

After precipitation or ultracentrifugation of proteins, classical preparative chromatographic methods are generally used, such as molecular exclusion chromatography, ion exchange and, in a less extent, reversed phase chromatography. To separate wine peptides, low-pressure molecular exclusion chromatography on cross-like dextran (Sephadex G-10 and Sephadex LH-20) is widely employed. This type of chromatography allows the separation of peptides according to their molecular size, which is traditionally very useful as an initial step in the fractionation of peptides from foods. The choice of pore size is determined by the range of molecular weights of the peptides in the sample. At the same time, it may also be convenient, the clean up of the samples by removing interfering smaller or higher molecular components, such as salts, organic acids, carbohydrates, phenolic compounds and amino acids. Peptides can be eluted from the column with water, solutions of acetic acid or sodium chloride, or mixtures of water and methanol, or water and ethanol. This is a suitable strategy for analyzing low molecular weight peptides (<1000 Da). Since these compounds usually elute together with amino acids, it will be essential to separate them from the amino acids prior to analysis. Sephadex G-10 with acetate buffer as eluent has been used to separate peptides from amino acids in the ethanol-soluble fraction of wines (Moreno-Arribas et al. 1996, 1998a,b; Bartolomé et al. 1997; Martínez-Rodríguez et al. 2002) and Sephadex LH-20 has been used by Acedo et al. (1994), Desportes et al. (2000, 2001) and Pozo-Bayón et al. (2005) to eliminate the amino acids contained in the wine after ultrafiltration. Several successive chromatographic systems have been used by Takayanagi and Yokotsuka (1999), Yanai et al. (2003) and Alcaide-Hidalgo et al. (2008).

6B.2.3 Separation by Chromatographic Techniques

HPLC is the most widely used method for peptide analysis. It offers many advantages, such as versatility, short analysis times, high resolution, effective separations, and it is well suited to automation procedures, There are many mechanisms that could be employed in the chromatographic separation of peptides, e.g. those based on molecule size (GPC), charge (IEC), hydrophobicity (reversed-phase and interaction chromatography), and even on combinations of them. Nevertheless, only reversed-phase chromatography has been used to separate mixtures of peptides from wines (Table 6B.1). The mobile phase most frequently used is a mixture of water and acetonitrile, with trifluoroacetic acid as an ion pairing reagent, and in gradient conditions. Less employed are the mixtures of methanol, phosphate buffer, and tetrahydrofurane, also in gradient conditions.

In the first study appearing in the literature on wine peptides, Acedo et al. (1994) applied RP-HPLC to separate peptides from wine after the formation of peptide derivatives with *o*-phthaldialdehyde. Also an RP-HPLC procedure using a Nova-Pak C_{18} column under gradient conditions was applied to analyze peptides from wine (Moreno-Arribas et al. 1996 1998a,b; Martínez-Rodríguez et al. 2002). Figures 6B.1 and 6B.2 show the chromatograms obtained using this method to separate the peptides from the fraction of molecular weight higher (Fig. 6B.1) and lower (Fig. 6B.2) than 700 Da of one must, one base wine and four sparkling wines (9–12 months of aging) from the Parellada variety.

HPLC has also been used to isolate wine peptides for characterization by other analytical techniques. Gel filtration chromatography followed by RP-HPLC, using a μBondasphere C_{18} column, was used by Takayanagi and Yokotsuka (1999) to separate and fractionate several wine peptide fractions. They investigated the amino acid sequences of the isolated fractions and also their bioactive capacity using

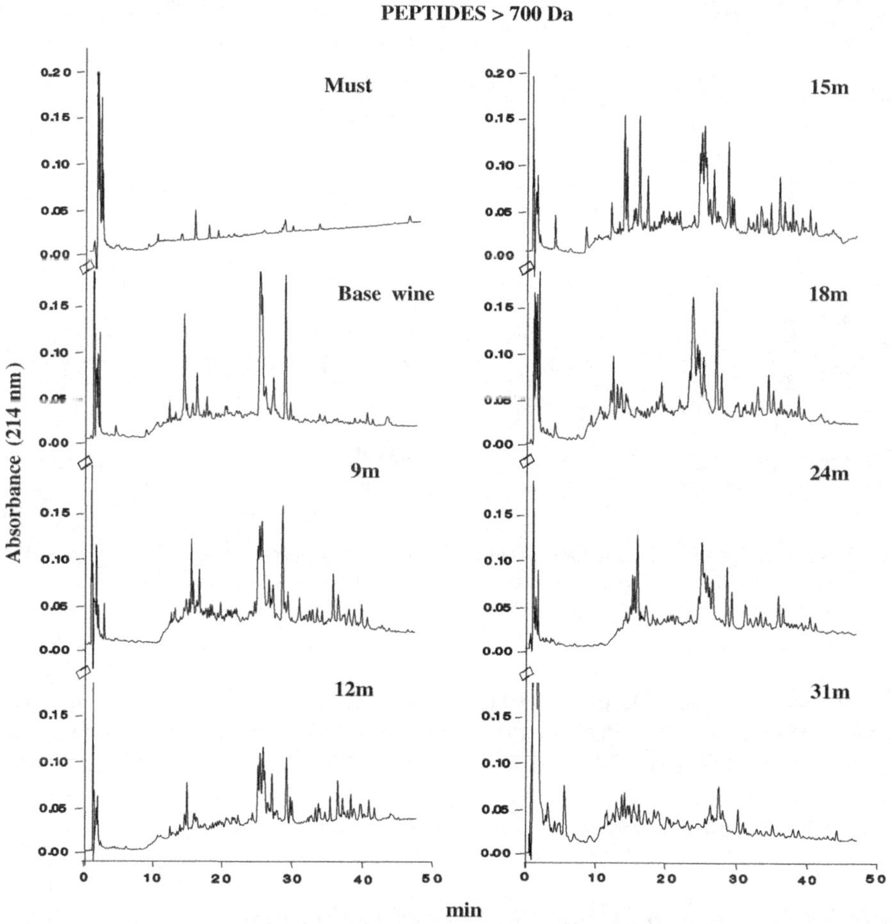

Fig. 6B.1 HPLC profiles of fraction >700 Da eluted from a Sephadex G-10 column of one must, one base wine and four sparkling wines of 9, 12, 15 and 18 months of aging with yeast (from Moreno-Arribas et al. 1996, with permission)

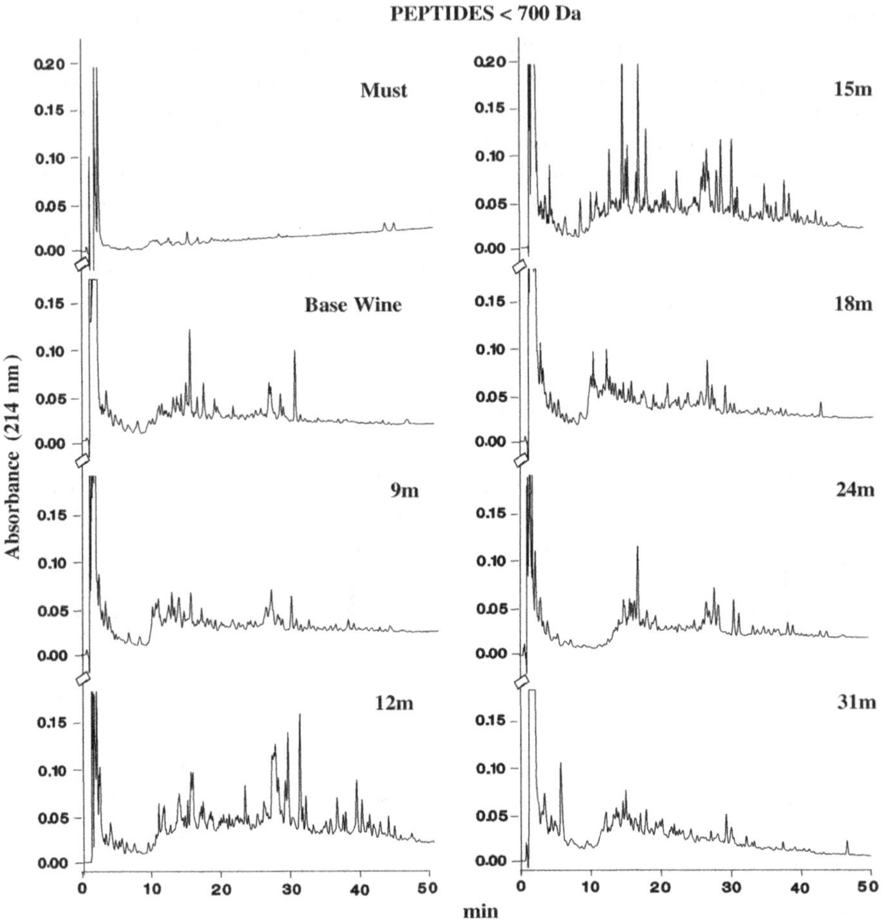

Fig. 6B.2 HPLC profiles of fraction <700 Da eluted from a Sephadex G-10 column of one must, one base wine and four sparkling wines of 9, 12, 15 and 18 months of aging with yeast (from Moreno-Arribas et al. 1996, with permission)

other methodologies. Desportes et al. (2000) isolated several small wine peptides (Mr <3000) using RP-HPLC on a porous graphite carbon Hypercarb column. The purity of the fractions collected was confirmed by free-solution capillary electrophoresis (FSCE). In a later study, Desportes et al. (2001) determined the peptide sequence of some isolated peptides by Edman degradation.

6B.2.4 Separation by Electrophoretic Techniques

Polyacrylamide gel electrophoresis with PAGE-SDS has been used by Martínez-Rodriguez and Polo (2000) to evaluate the release of polypeptides with molecular masses ranging from 16900 to 4400Da in a model wine during yeast autolysis. The authors used the Schägger and von Jagow's method (Schägger and von Jagow 1987) for discontinuous electrophoresis and silver staining.

Desportes et al. (2000, 2001) used capillary electrophoresis on fused-silica capillaries with 25 mM, pH 2.5 phosphoric acid as running buffer, to verify the purity of collected peaks after separation by HPLC. Both methods and some others described for the analysis of wine proteins, such as that described by Luguera et al. (1997), with slight modifications, could also be used for the analysis of medium molecular weight peptides.

6B.2.5 Detection Methods

Peptides are commonly detected by absorbance at 200–220 nm. However, most of the compounds present in wine may interfere in the ultraviolet detection of peptides when low wavelengths are used. Thus, for the analysis of these compounds it is useful to apply sensitive and selective detection methods. To this end, it is possible to form derivates of the peptides that can be detected at higher and more specific wavelengths. Detection by fluorescence can also be used to detect peptides containing fluorescence amino acids (tyrosine and tryptophan). For peptides without this property, the formation of derivates with derivatizing agents have been proved to be very useful (Moreno-Arribas et al. 1998a).

It is possible to use the formation of both pre- and post-column derivates. Post-column reaction with OPA has the advantage that enables the detection of natural peptides by ultraviolet and in a second step the on-line detection of the derivates by fluorescence). Bartolomé et al. (1997) reported the use of an on-line HPLC-photodiode array detection (HPLC-PDAD) and a post-column OPA derivatization system to characterize small peptides from wine. This methodology was initially applied to a wide range of standard peptides and, in further studies, it was successfully applied to identify the peptide fraction (<700 Da) of sparkling wines. Spectral parameters, such as wavelengths of the spectrum peaks, convexity interval, and wavelengths of the second-derivate spectrum peaks, obtained by PDAD, allow the identification of the aromatic amino acid residues within the peptides and also the identification of other compounds in wine that can coelute with peptides (i.e. phenolic compounds). Figure 6B.3 shows the original and the first and second derivative spectra of several peptides and Fig. 6B.4 the flowchart for the interpretation of both spectral data and OPA-fluorescence response proposed by Moreno-Arribas et al. (1998a) for the identification of peptides.

LC-ESI-MS/MS (liquid chromatography electrospray ionisation tandem mass spectrometry) has been used by Person et al. (2004) for identification and quantification of di- and tripeptides in champagne wine after filtration with a Mr 10,000 cut-off membrane. This procedure avoids the time-consuming prepurification and isolation steps.

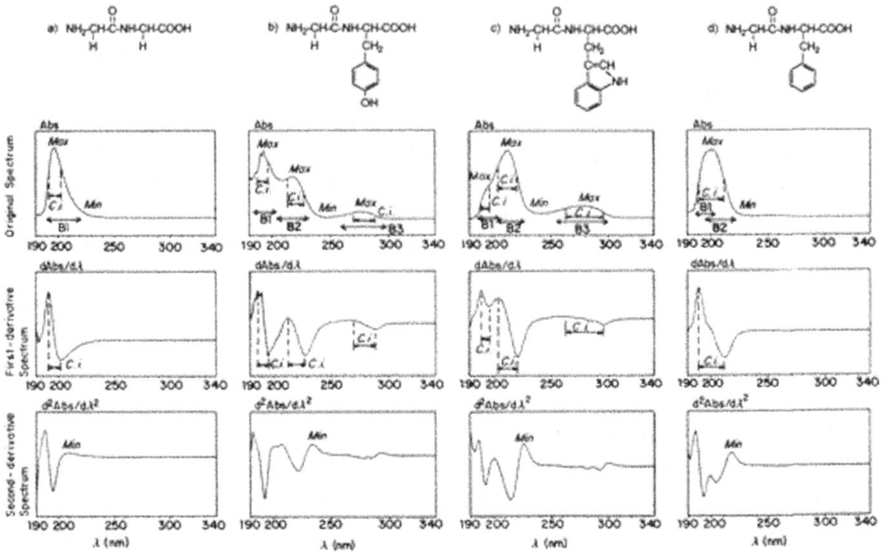

Fig. 6B.3 UV, original and first and second derivative spectra of four glycylpeptides (from Bartolomé et al. 1997, with permission)

6B.2.6 Determination of Amino Acid Composition and Sequence in Peptides

The amino acid composition of peptides is generally assayed by carrying out acid hydrolysis with 6 M hydrochloric acid, followed by determination of the individual amino acids by HPLC. Various types of HPLC have been employed to separate amino acids, but reversed-phase chromatography on C_{18}columns is the most commonly system used. Detection of amino acids normally involves derivatization, since their maximum absorbance is at 214 nm in which many other compounds also can absorb. The most frequently used derivatizing agents are dansyl chloride, phenyl isothiocyanate and *o*-phthaldialdehyde.

Peptide sequencing is normally performed by degradation of the N-terminal amino acid using phenyl isothiocyanate (Edman's reagent). The terminal amino groups react with the isothiocyanate, forming a phenylthiocarbamylic derivative. After its treatment with acid in an organic solvent, cyclization takes place forming the phenylthiohydantoinamino acid that can be separated from the rest of the chain, which remains intact, and the process can be repeated.

6B.3 Origin and Peptide Characterization

There are very few studies focused on determining the origin, concentration and composition of wine peptides. These studies have been carried out on peptide fractions of different molecular weights that have been obtained with different

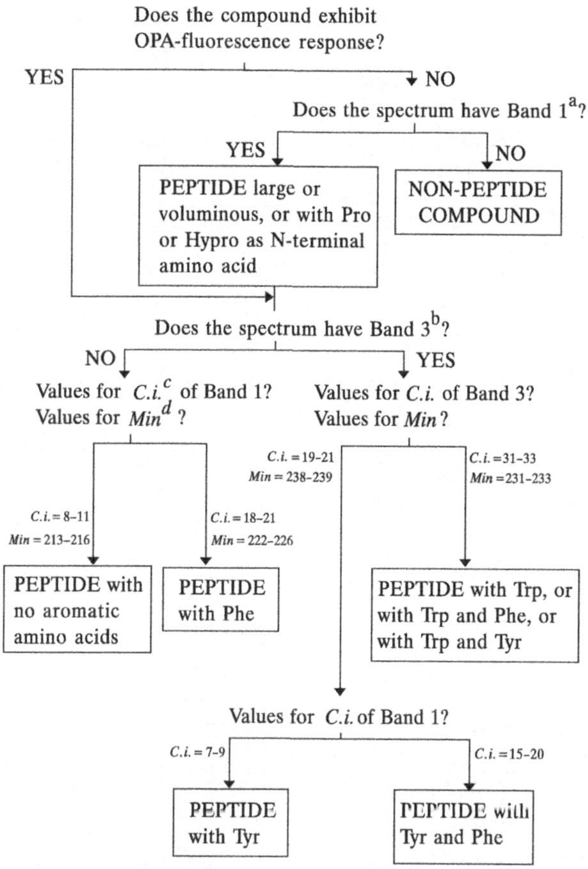

Fig. 6B.4 Flowchart for the interpretation of both spectral data and OPA-fluorescence response. [a]Band 1, spectral band (201–207 nm); [b]Band 3, spectral band (276–280 nm); [c]*C.i.* convexity interval (distance between the inflection points before and after the maximum in the original spectrum); [d]*Min*, minimum in the original spectrum (from Moreno-Arribas et al. 1998a, with permission)

procedures (Table 6B.1). Since the soluble fraction in TCA is not the same as the soluble fraction in ethanol or that obtained by ultrafiltration using membranes of different pore size, it is very difficult to compare the results reported in the literature. Moreover, in most of the cases, the authors express the results in different units. In this way, some results are expressed as mg/L of wine, percentage distribution of amino acids in the total fraction, ratio of peptide nitrogen/total nitrogen, number of residues of a peptide or a specific fraction, etc.

Table 6B.2 summarizes the data available in the literature on peptides contents of musts, wines and related substrates such as autolyzed yeasts obtained under winemaking conditions. This table also shows the fractions studied by the different authors, the type of product, the analytical methodology, the results obtained and the purpose of their research.

Table 6B.2 Resume of the information available in the literature on wine peptides

Studied fraction	Samples to which the study has been applied	Main information on peptides concentration and composition	Reference
Soluble in 7%TCA, eluted from a Dowex 50 column	3 musts and 3 wines	*Peptide nitrogen*: *Must* 18 mg/L *Wines* 22, 52, 144 mg/L N	Yokotsuka et al. (1975)
		Majority amino acids from peptides: *Must* Asx, Glx, Val, Pro *Wines* Asx, Glx, Gly, Pro, Ser, Ala	
Mr<5000 Da	3 must, 3 wines, and samples during elaboration	*Majority amino acids from peptides*: Glu, Asp, Ser, Gly, α-Ala, Lys.	Usseglio-Tomasset and Bosia (1990)
Mr<5000 Da	1 white wine	*Majority amino acids from peptides*: Glu, Asp, Ser, Gly, α-Ala, Lys	(Acedo et al. 1994)
Soluble in EtOH	4 musts, 4 base wines, 16 sparkling wines (9–18 months of aging with yeast)	*Peptide nitrogen*: Musts 108.9–162.2 mg/L	Moreno-Arribas et al. (1996)
Mr>700 Da and Mr Da		Base wines 14.5–48.5 mg/L Sparkling wines 16.9–90.2 mg/L HPLC relative areas at 214 nm of fractions	
Soluble in EtOH	4 sparkling wines (15 months of aging with yeast)	Tentative identification of peptides by spectral characteristics	Bartolomé et al. (1997)
Mr<700 Da			
Soluble in EtOH	1 sparkling wine (15 months of aging with yeast)		Moreno-Arribas et al. (1998a)
Mr>700 Da		*Majority amino acids in the fraction*: Asx, Glx, Ser, Thr, Ala, Gly *Amino acids in the isolated peaks*: Asx, Glx, Ser, Thr, α-Ala, Tyr (13 residues); Asx, Glx, Ser, Gly, Thr, α-Ala, Tyr, Val, Leu (21 residues); Asx, Glx, Ser, His, Thr, α-Ala, Leu (25 residues); Asx, Glx, Ser, Thr, α-Ala, Tyr, Leu (18 residues);	

Table 6B.2 (continued)

Studied fraction	Samples to which the study has been applied	Main information on peptides concentration and composition	Reference
Mr <700 Da		*Majority amino acids in the fraction*: Asx, Glx, Ser, Gly, Thr *Amino acids in the isolated peaks*: Asx, Ser, Gly, Thr, α-Ala, Tyr, Val (10 residues); Asx, Glx, Ser, Gly (6 residues); Asx, Glx, β-Ala, α-Ala, Ile, Leu (7 residues); Asx, Glx, Ser, Gly, Thr, α-Ala, Ile, Leu (9 residues) Glx, Ser, Gly, Thr, α-Ala, Ile, Leu (9 residues) Glu, Gly, Thr, α-Ala, Ile, Leu	
Soluble in EtOH	4 base wines, 24 sparkling wines (9-31 months of aging with yeast)		Moreno-Arribas et al. (1998b)
Mr >700 Da		*Majority amino acids in the fraction*: Asx, Glx, Ser, Gly, Thr, α.Ala, Pro	
Mr <700 Da		*Majority amino acids in the fraction*: Asx, Ser, Gly, α.Ala, Lys, Pro	
	1 white wine	*Sequences*[a]	Takayanagi and Yokotsuka (1999)
Mr<10,000 Da	Yeast autolysed in a model wine	Relative molecular masses HPLC profiles	Martínez-Rodriguez and Polo (2000)
Mr>700 Da Mr<700 Da			
500<Mr<5000 Da	"Flor" wines (0, 5, 8, 10 and 15 weeks of aging)	*Peptide nitrogen* Initial wine 124.4 mg/L; "Flor" wines 16.3, 24.5, 32.3 59.5 mg/L *Amino acids in the fraction*: Pro 12-23%; Gly 10.8–18.6%; Asp+Asn 7.5-9.7%; Glu+Gln 8.9–11.5%; α-Ala 5.4–8.5%; Lys 5.7–8.3%; Ser 7.2–10.9%; Thr 7.2–8.6%	Dos Santos et al. (2000)
200-300<Mr Da	1 white wine	*Majority amino acids in the fraction*: Asx, Glx, Gly, Pro, Thr, Ser, Lys α-Ala	Desportes et al. (2000)
200-300<Mr Da		*Sequences*[a]	Desportes et al. (2001)

Table 6B.2 (continued)

Studied fraction	Samples to which the study has been applied	Main information on peptides concentration and composition	Reference
Mr<5000 Da	One model wine after fermentation and yeast autolysis	*Majority amino acids*: *After fermentation* Asx, Glx, Thr, Arg, α-Ala, Val *After autolysis* Lys, Asx, Glx, His, Arg	Alexandre et al. (2001)
Mr>700 Da and Mr<700 Da	Sparkling wines	Relative areas of fractions > and <700 Da, respectively	Martínez-Rodríguez et al. (2002)
Fraction eluted from a RP-HPLC column	1 red wine	*Sequences*[a]	Yanai et al. (2003)
Mr<10,000 Da	1 white wine	*Sequences*[a]	Person et al. (2004)
Mr<5000 Da	1 white wine and 1 red wine	*White wine: Peak 1*: Asx, Glx *Peak 2*: Asx, Glx, ser, Gly, Thr, α-Ala, Val, Lys *Peak 3*:Asx, Glx, Ser, Gly, Thr, Arg, α-Ala, Val, Ile, Leu *Red wine*: *Peak 4*:Val *Peak5*:Asx, Glx, His, Thr, Arg, α-Ala, Met, Val, Ile, Leu *Peak 6*:Asx, Glx, Gly, Thr, Arg, α-Ala, Met, Val, Ile, Leu	Pozo-Bayón et al. (2005)
Mr<5000 Da	Red wines, changes during MLF and aging	*Peptide nitrogen:* 150–200 mg/L *Majority amino acids More polar peptides*: Asx, Glx, Ser, Gly, Ala *Less polar peptides*: Glx, Gly, Leu, Val	(Alcaide-Hidalgo et al. 2008)

aSee Table 6B.3

Yokotsuka et al. (1975) in the soluble fraction in TCA analysis, retained in a catioinic resin, found concentrations of 18 mg/L peptide nitrogen in a must and 22, 52 and 144 mg/L in two white wines and one red wine, respectively. These results correspond to 2.7% of the total nitrogen of the must, 5% and 9.5% of the total nitrogen of both white wines and 56% of the total nitrogen of the red wine. Moreno-Arribas et al. (1996), in their study of the nitrogenated fraction soluble in 95% ethanol of musts, base wines and sparkling wines obtained from the same musts of 9, 12, 15 and 18 months aging in bottle, estimated the nitrogen fraction as the difference between the total nitrogen and the sum of free amino nitrogen and proteic nitrogen. They obtained values ranging from 118.2 mg/L to 162.2 mg/L of peptidic nitrogen in three musts, from 14.5 mg/L to 57.6 mg/L peptidic nitrogen in the four white base wines studied and from 17.9 mg/L to 90.2 mg/L peptidic nitrogen in the 16 sparkling wines. These values constitute 47% of total must nitrogen, from 12% to 57% of total nitrogen of the base wines and from 21% to 70% of total nitrogen of the sparkling wines. Dos Santos et al. (2000) in the peptide fraction of relative molecular mass from 500 to 5000 Da found a concentration of 124 mg/L peptide nitrogen in the initial wine (base wine) and 16.3, 24.5, 32.3 and 59.5 mg/L peptide nitrogen in the same wine after 5, 8, 10 and 15 weeks of aging with "flor" yeasts. This nitrogen constitutes approximately 6%–21% of the total wine nitrogen.

Summarizing the data available in the literature, the concentration of peptide nitrogen in musts range from 18 mg/L to 162.2 mg/L, corresponding to 2.7%–47% when expressing such as total nitrogen; the concentration of peptidic nitrogen from table wines, range from 14.5 mg/L to 57.6 mg/L in white wines and 144 mg/L in red wines, equivalent to 12%–57% of the total nitrogen in white wines and 56% of red wine; in sherry type wines, the concentrations are between 16.3 mg/L and 59.5 mg/L of peptidic nitrogen and in sparkling wines from 17.9 mg/L to 90.2 mg/L of peptidic nitrogen (thus, between 21% and 70% of the total nitrogen).

In the case of peptide characterization, some of the studies of wine peptides describe the amino acid composition of the total peptide fraction obtained using different extraction methods (Table 6B.1) (Yokotsuka et al. 1975; Usseglio-Tomasset and Bosia 1990; Dos Santos et al. 2000; Alexandre et al. 2001). Other studies (Moreno-Arribas et al. 1998a,b; Pozo-Bayón et al. 2005) determine the amino acid composition of peptide fractions of different molecular sizes and the amino acid composition of some chromatographic peaks obtained by low pressure liquid chromatography (or open column chromatography) (Acedo et al. 1994) or HPLC (Moreno-Arribas et al. 1998a; Desportes et al. 2000).

It is interesting to note that aspartic acid and/or asparagine and glutamic acid and/or glutamine form part of the peptides of all the wine fractions of different studies, independently of the type of wine or the analytical methodology employed in the analysis. Serine, threonine, alanine and glycine appear in most of the fractions studied while, lysine, tyrosine, valine, leucine, histidine and isoleucine has been found in a minor extent in these fractions.

There are not many wine peptide sequences described in the literature. Table 6B.3 shows those that have been described. It can also be seen that in peptides with 2–10 amino acids, the amino acids that appear most frequently are tyrosine, proline, and isoleucine.

Table 6B.3 Sequences of peptides described in wines (range of concentration reported)

Wine of fraction studied	Sequence	Reference
White wine	LIPPGVPY	Takayanagi (1999)
	YYAPFDGIL	
	YYAPF	
	SWSF	
	WVPSVY	
	AWPF	
	YYYAPFDGIL	
200–300<Mr<3000	VGN o AGN o TGN	Desportes et al. (2001)
	KMN o AMN	
	FK	
	YK	
	FRR	
	SKTSPY	
	IV	
	VI	
	IR	
Red wine	VEIPE	Yanai et al. (2003)
	YPIPF	
Mr<1000	YQ (nd-0.295 mg/l)	Person et al. (2004)
	IV (0.615–1.627 mg/l)	
	VI (0.250–0.802 mg/l)	
	YK (0.311–3.067 mg/l)	
	IR (2.242–7.005 mg/l)	
	KY (nd-0.179 mg/l)	
	RI (1.042–2.138 mg/l)	
	KF (nd-0.325 mg/l)	

The wine peptide fraction is continuously changing both qualitatively and quantitatively. Some of the wine peptides come from the must, but most appear during the different stages of wine production. During alcoholic fermentation, especially if it takes place in nitrogen-deficient media, there is a reduction in the concentration of peptides. In the final stages of fermentation, peptides are again released into the medium, reaching the maximum release after cell death (Usseglio-Tomasset and Bosia 1990). This process is important in winemaking technologies based on aging the wines with yeasts for long periods of time, such as the biologically aged wines (Dos Santos et al. 2000) and the sparkling wines manufactured by the *traditional* method (Moreno-Arribas et al. 1996, 1998a,b). The existence of endo- and exocellular proteases have been described in winemaking conditions (Feuillat et al. 1980; Alexandre et al. 2001). The increase in wine peptides could be caused by their release from the yeast cells or due to the action of endo and exoproteases also released from the yeast acting on proteins and/or polypeptides present in the surrounding medium (Moreno-Arribas and Polo 2005). In wine model solutions employing a yeast autolysated, Martínez-Rodriguez et al. (2001) evidenced this phenomena. In a second study they also found that the extent of peptide release depends, at least in part, on the yeast strain (Martínez-Rodríguez et al. 2002).

The amino acid composition of sparkling wine peptides depends on the aging time (Moreno-Arribas et al. 1998b). The action of both types of proteases, but mainly of the exocellular ones, that are active in wines for months or even years, causes large variations in peptide contents as happen during the aging of wines with yeasts. Figures 6B.1 and 6B.2 illustrate these process showing the chromatograms corresponding to the peptide fractions larger and smaller of 700 Da molecular mass, respectively, of one must and the corresponding base wine and sparkling wines made by the *traditional* method. It can be observed that, although the must hardly contains any peptides in the fraction studied, their content increases during the production of the base wine. The variation in the peptide content observed in the sparkling wines is due to the successive and simultaneous degradation of some peptides and the release of others during the aging with yeast.

Some of the lactic acid bacteria strains of wines have also shown proteolytic activity in wine-making conditions (Feuillat et al. 1980; Manca de Nadra et al. 1997; Leitáo et al. 2000). This activity has been found even in the presence of ethanol and SO_2 (Manca de Nadra et al. 2005). It has been shown that this activity is greater in red than in white wines (Manca de Nadra et al. 1999). The proteolytic activity is a very important characteristic for some bacteria strains, allowing them to growth in nitrogen-deficient media, and also favouring them to carry out malolactic fermentation. According to Remize et al. (2005 and, 2006), peptides from 0.5 kDa to 10 kDa seem to be more favorable for the growth of wine lactic acid bacteria than other nitrogen sources (<0.5 kDa). Alcaide-Hidalgo et al. (2008) have observed a reduction in peptides during the malolactic fermentation of red wines followed by an increase during the aging in barrels that was more evident when the aging was in the presence of lees. The occurrence of proteolytic activity in all the stages of wine manufacture can explain the wide range of peptides that can coexist in a wine at any time and the discrepancies in most of the studies related to their identification.

Glutathione is a peptide of non-proteic origin, present in grapes and wines that merits a special mention. It was described for the first time in grapes by Cheynier et al. (1989). Because of its strong antioxidant properties, it has been recommended as an additive to prevent the enzymatic browning of white wines (Vaimakis and Roussis 1996). Nevertheless, neither its evolution during wine manufacture nor its precise role in wine manufacture are fully understood.

In an analysis of 29 varieties of grapes, Cheynier et al. (1989) detected glutathione, mostly in a reduced form, GSH, at levels of 56.3–371.8μmol/kg in grapes and of 41.9–332.7 mM in musts. According to Okuda and Yokotsuka (1999) and Park et al. (2000), the glutathione content decreases during the firsts days of fermentation, rising afterwards to reach values of 0.1–5.1 mg/L. According to Lavigne et al. (2007), the amount of glutathione present in a wine at the end of the alcoholic fermentation depends on the yeast strain, and decreases as the lees are removed and also during the aging in barrels. In yeast, GSH accounts for about 1% of dry weight of *Saccharomyces cerevisiae* and represents more than 95% of the low molecular thiol pool (Elskens et al. 1991); thereby some of this glutathione could possibly be released into the wine.

6B.4 Properties of Wine Peptides

Low molecular weight peptides have important functional properties, such as their tensioactive function and their bioactive properties (antioxidant, antimicrobial, antihypertensive). They are also responsible for sweet and bitter flavors (Polo et al. 2000). In spite of the importance of the content and structure of food peptides in food biological activities, in wines they have been very little evaluated apart from their antihypertensive effects. The increase in hypertension rates in recent years, which is being considered to be one of the commonest chronic diseases in developed countries, has favored these types of studies. Most of the antihypertensive peptides derived from foods act by inhibiting the angiotensin converting enzyme (ACE). This enzyme is responsible for the increase in blood pressure by converting angiotensin I into a strong vasoconstrictor, angiotensin II, and by degradation of bradykinin, a vasodilator. Therefore, ACE inhibition produces a hypotensive effect. More scarce, however, are studies dealing with the bioactivity of these compounds after ingestion, since they must be resistant to gastric enzymes and be absorbed in order to produce a biological effect. Thereby, in spite of the fact that some studies have demonstrated the in vitro inhibitory effect of some wine peptides on ACE, in vivo studies are required to demonstrate their true efficacy.

In spite of the lack of studies regarding the antihypertensive effect of wine peptides, in one of them, Takayanagi and Yokotsuka (1999) have determined the ACE inhibitory activity of two red and four white wines in grapes of the black Muscat Bailey A variety and in the fermented musts. They demonstrated that red wines had a higher ACE inhibitory activity than white wines. They also observed that this activity decreases during fermentation without any clear explanations for this reduction.

Pozo-Bayón et al. (2005) determined the amino acid composition of peptides from six fractions isolated from red and white wines that presented in vitro antihypertensive activity. They showed that the amino acids aspartic acid and/or asparagine, glutamic acid and/or glutamine valine, formed part of five of the six fractions studied, and threonine and alanine were present in four of these fractions.

In long-term studies with normotensive and spontaneously hypertensive rats, it has been shown (Perrot et al. 2003) that the extract from the low molecular weight fraction of a Champagne wine had antihypertensive activity in hypertensive rats but did not affect the normotensive ones. Owing to the complexity of this fraction, the authors claim that this reduction cannot be attributed only to the presence of a single compound. Since wine is rich in phenolic compounds and peptides, both families of compounds could include individual components with antihypertensive activity (Ling 2003; Zhang Choi et al. 2003). Pozo-Bayón et al. (2005) have demonstrated the contribution of peptides to this activity. However, they observed a greater ACE inhibitory activity in red wines, reflecting the possible contribution to the phenolic compounds. Ling (2003) has shown that phenols can be conjugated with proteins and produce non-competitive inhibition of the ACE. It is, therefore, necessary to carry out more studies to identify the compounds responsible for this activity in wines.

Recently ACE inhibitory activity has also been found in peptides released from *Saccharomyces cerevisiae* in a model wine (Alcaide-Hidalgo et al. 2007). In this assay there were no phenolic compounds, so this activity was exclusively attributed to yeast peptides. The same peptide-rich fractions also showed antioxidant activity, suggesting that peptides released by *S. cerevisiae* during autolysis in wine conditions could present multifunctional activity.

Also, prolylendopeptidase inhibitory peptides (PEP) have been found in wines (Yanai et al. 2003). PEP may have a role in the degradation of biologically active peptides containing proline, such as oxytocin, vasopressin, substance P, bradykinin, neurotensins and angiotensins. Two inhibitory peptides have been isolated and characterized, Pep A (Val-Glu-Ile-Pro-Glu) and Pep B (Tyr, Pro, Ile, Pro, Phe). Both of them showed PEP inhibition, thereby suppressing the degradation of neuropeptides, vasopressin, substance P and fragments 8–13 of neurotensin, which are involved in memory and neural communication.

6B.5 Conclusions

As a result of the extensive research carried out during the past 10 years, there is nowadays available a good number of methodologies for wine peptides purification and characterization. As a consequence, the knowledge of the origin, content and composition of wine's peptides is increasing. Despite this progress, the information available about the structure and amino acid sequence of peptides derived from wine is very limited. Further research should be carried out in this area. Such research may now benefit from the use of well-developed and sophisticated analytical techniques (e.g., MALDI-TOFMS, ES-MS/MS), which are available to assist with structural characterization of isolated peptides. While wine contains components with health-promoting effects, very few studies have addressed the impact of peptides on the biological properties of wines. On the other hand, it will necessary studies to explore the consequences of the interaction between peptides and other wine components during winemaking .The effect of wine micro-organisms and winemaking technologies on the peptide composition of wines and its functional properties also need to receive more attention.

Acknowledgments The authors are grateful to the AGL2006-04514, PET2007_0134 and CSD2007 -00063 projects (Consolider Ingenio 2010 FUN-C-FOOD, Ministerio de Educación y Ciencia) and the S-505/AGR-0153 ALIBIRD Project (Comunidad Autónoma de Madrid) for financial support of this work.

References

Acedo, M.I., Pueyo, E., & Polo, M.C. (1994). Preliminary studies on peptides in wine by HPLC. *Am. J. Enol. Vitic., 45*, 167–172.

Alcaide-Hidalgo, J.M., Pueyo, E., Polo, M.C., & Martínez-Rodriguez, A.J. (2007). Bioactive peptides released from *Saccharomyces cerevisiae* under accelerated autolysis in a model wine system. *J. Food Sci., 72*, M276–279.

Alcaide-Hidalgo, J.M., Moreno-Arribas, M.V., Polo, M.C., & Pueyo, E. (2008). Partial characterization of peptides from red wines. Changes during malolactic fermentation and ageing with lees. *Food Chem., 107*, 622–630.

Alexandre, H., Heintz, D., Chassagne, D., Guilloux-Benatier, M., Charpentier, C., & Feuillat, M. (2001). Protease A activity and nitrogen fractions released during alcoholic fermentation and autolysis in enological conditions. *J. Ind. Microbiol. Biotechnol., 26*, 235–240.

Bartolomé, B., Moreno-Arribas, V., Pueyo, E., & Polo, M.C. (1997). On-line photodiode array detection and OPA-derivatization for identification of small peptides from wine. *J. Agric. Food Chem., 45*, 3374–3381.

Cheynier, V., Souquet, J.M., & Moutounet, M. (1989). Glutathione content and glutathione to hydroxycinamic acid ratio in *Vitis vinifera* grapes and musts. *Am. J. Enol. Vitic., 40*, 320–324.

Colagrande, O., & Silva, A. (1981). Aspects qualitatives et technologiques de lapplication des methodes thermiques dans la production des vins mousseux. *Conn. Vigne Vin, 15*, 179–192.

Desportes, C., Charpentier, M., Duteurtre, B., Maujean, A., & Duchiron, F. (2000). Liquid chromatographic fractionation of small peptides from wine. *J. Chromatogr. A, 893*, 281–291.

Desportes, C., Charpentier, M., Duteurtre, B., Maujean, A., & Duchiron, F. (2001). Isolation, identification, and organoleptic characterization of low-molecular-weight peptides from white wine. *Am. J. Enol. Vitic., 52*, 376–380.

Doi, E., Shibata, D., & Matoba, T. (1981). Modified colorimetric ninhydrin methods for peptidase assay. *Anal. Biochem., 118*, 173–184.

Dos Santos, A.M., Feuillat, M., & Charpentier, C. (2000). Flor yeast metabolism in a model system similar to cellar ageing of the french Evolution of some by-products, nitrogen compounds and polysaccharides. *Vitis, 39*, 129–134.

Elskens, M.T., Jaspers, C.L., & Penninckx, M.J. (1991). Glutathione as an endogenous sulphur source in the yeast *Saccharomyces cerevisiae*. *J. Gen. Microbiol., 137*, 637–644.

Feuillat, M., Bidan, P., & Rosier, V. (1977). Croissance des bactéries lactiques à partir des principaux constituants azotés du vin. *Ann. Technol. Agric., 28*, 435–447.

Feuillat, M., Brillant, G., & Rochard, J. (1980). Mise en évidence d'une production de proteaes exocellulaires par les levures au cours de la fermentation alcoolique du moût de raisin. *Conn. Vigne Vin, 14*, 37–52.

Gómez-Ruiz J.A., Ramos, M., & Recio, I. (2004) Identification of angiotensin-converting enzyme inhibitory peptides in Manchego cheese by high-performance liquid chromatography-tamdem mass spectrometry. *J. Chromatog. A, 1054*, 269–277.

González-Llano, D., Herraiz, T., & Polo, M.C. (2004). Peptides. In: M.L. Leo Nollet (Ed.), *Handbook of Food Analysis*, Vol. 1: Physical Characterization and Nutrient Analysis. Chapter 6 (pp. 125–166). New York: Marcel Dekker.

Lavigne, V., Pons, A., & Dubourdieu, D. (2007). Assay of glutathion in must and wines using capillary electrophoresis and laser-induced fluorescence detection. Changes in concentration in dry white wines during alcoholic fermentation and aging. *J. Chromatog. A, 1139*, 130–135.

Leitáo, M.C., Teixeira, H.C., Barreto Crespo, M.T., & San Romáo, M.V. (2000). Biogenic amines occurence in wine. Amino acid decarboxylase and proteolytic activities expression by *Oenococcus oeni*. *J. Agric. Food Chem., 48*, 2780–2784.

Lin, J.Y. (2003). Antihypertensive effects of tannins isolated from traditional Chinese herbs as non-specific inhibitors of angiontensin converting enzyme. *Life Sci., 73*, 1543–1555.

Luguera, C., Moreno-Arribas, M.V., Pueyo, E., & Polo, M.C. (1997). Capillary electrophorestic analysis of wine proteins. Modifications during the manufacture of sparkling wines. *J. Agric. Food Chem., 45*, 3766–3770.

Manca de Nadra, M.C., Farias, M., Moreno-Arribas, M.V., Pueyo, E., & Polo, M.C. (1997). Proteolitic activity of *Leuconostoc oenos*: Effect on proteins and polypeptides from white wine. *FEMS Microbiol. Lett., 150*, 135–139.

Manca de Nadra, M.C., Farias, M.E., Moreno-Arribas, V., Pueyo, E., & Polo, M.C. (1999). A proteolytic effect of *Oenococcus oeni* on the nitrogenous macromolecular fraction of red wine. *FEMS Microbiol. Lett., 174*, 41–47.

Manca de Nadra, M.C., Farias, M.E., Pueyo, E., & Polo, M.C. (2005). Protease activity of *Oenococcus oeni* viable cells on red wine nitrogenous macromolecular fraction in presence of SO_2 and ethanol. *Food Control, 16*, 851–854.

Martínez-Rodriguez, A., & Polo, M.C. (2000). Characterization of the nitrogen compounds released during yeast autolysis in a model wine system. *J. Agric. Food Chem., 48*, 1081–1085.

Martínez-Rodriguez, A.J., Carrascosa, A.V., & Polo, M.C. (2001). Release of nitrogen compounds to the extracellular medium by three strains of *Saccharomyces cerevisaiae* during induced autolysis in a model wine system. *Int. J. Food Microbiol., 68*, 155–160.

Martínez-Rodríguez, A.J., Carrascosa, A.V., Martín-Álvarez, P.J., Moreno-Arribas, V., & Polo, M.C. (2002). Influence of the yeast strain on the changes of the amino acids, peptides and proteins during sparkling wines production by the traditional method. *J. Ind. Microbiol. Biotechnol., 29*, 314–322.

Moreno-Arribas, M.V., & Polo, M.C. (2005). Winemaking biochemistry and microbiology: Current knowledge and future trends. *Crit. Rev. Food Sci. Nut., 45*, 265–286.

Moreno-Arribas, V., Pueyo, E., & Polo, M.C. (1996). Peptides in musts and wines. Changes during the manufacture of cavas (sparkling wines). *J. Agric. Food Chem., 44*, 3783–3788.

Moreno-Arribas, M.V., Bartolomé, B., Pueyo, E., & Polo, M.C. (1998a). Isolation and characterization of individual peptides from wine. *J. Agric. Food Chem., 46*, 3422–3425.

Moreno-Arribas, V., Pueyo, E., Polo, M.C., & Martín-Alvarez, P.J. (1998b). Changes in the amino acid composition of the different nitrogenous fractions during the aging of wine with yeasts. *J. Agric. Food Chem., 46*, 4042–4051.

Okuda, T., & Yokotsuka, K. (1999). Levels of glutathione and activities of related enzymes during ripening of Koshu and Cabernet Sauvignon grapes and during winemaking. *Am. J. Enol. Vitic., 50*, 264–270.

Park, S.K., Boulton, R.B., & Noble A.C. (2000). Automated HPLC analysis of glutathione and thiol-containing compounds in grape juice and wine using pre-column derivatization with fluorescence detection. *Food Chem.*, 68, 475–480.

Perrot, L., Dukic, S., Charpentier, M., Duteurtre, B., Duchiron, F., & Kaltenbach, M.L. (2003). Antihypertensive effect of a low molecular weight fraction (1 kDa) of champagne wine in spontaneously hypertensive rats. In: A. Lonvaud-Funel, G. Revel, P. Darriet (Eds.), *Oenologie 2003* (pp. 688–691) Paris: TEC and DOC.

Person, M. de, Sevestre, A., Chaimbault, P., Perrot, L., Duchiron, F., & Elfakir, C. (2004). Characterization of low-molecular weight peptides in champagne wine by liquid chromatography/tandem mass spectrometry. *Anal. Chim. Acta, 520*, 149–158.

Polo, M.C., González-Llano, D., & Ramos, M. (2000). HPLC of peptides. In: Leo M.L. Nollet (Ed.), *Food Analysis by HPLC (Food Science and Technology Series)* (pp. 99–125). New York: Marcel Dekker.

Pozo-Bayón, M.A., Alcaíde, J.M., Polo, M.C., & Pueyo, E. (2005). Angiotensin I-converting enzyme inhibitory compounds in white and red wines. *Food Chem., 100*, 43–47.

Remize, F., Augagneur, Y., Guilloux-Benatier, M., & Guzzo, J. (2005). Effect on nitrogen limitation and nature of the feed upon *Oenococcus oeni* metabolism and extracellular protein production. *J. App. Microbiol., 98*, 652–661.

Remize, F., Gaudin, A., Kong, Y., Guzzo, J., Alexandre, H., Krieger, S., & Guilloux-Benatier, M. (2006). *Oenococcus oeni* preference for peptides: Qualitative assimilation and quantitative analysis of nitrogen assimilation. *Arch. Microbiol., 185*, 459–469.

Schägger, H., & von Jagow, G. (1987). Tricine-sodium dodecyl sulphate polyacrylamide gel electrophoresis for the separation of proteins in the range from 1 to 100 KDa. *Anal. Biochem., 79*, 544–552.

Takayanagi, T., & Yokotsuka, K. (1999). Angiotensin I converting enzyme-inhibitory peptides from wine. *Am. J. Enol. Vitic., 50*, 65–68.

Usseglio-Tomasset, L., & Bosia, P.D. (1990). Amino acids and oligopeptides development from the must to the wine. *Bull. lOIV, 63*, (707–708), 21–46.

Usseglio-Tomasset, L., & Di Stefano, R. (1978). Observazioni sulle sostanze azozate dei vini in realzione alle loro dimensioni molecolari Annali dell'Istituto sperimentale per l'enologia Asti 9, 101.

Vaimakis, V., & Roussis, I.G. (1996). Must oxigenation together with glutathione addition in the oxidation of white wine. *Food Chem., 57*, 419–422.

Yanai, T., Suzuki, I., & Sato, M. (2003). Prolyl endopeptidase inhibitory peptides in wine. *Biosc. Biotech. Biochem., 67*, 380–382.

Yokotsuka, K., Aihara, T., Umehara, Y., & Kushida, T. (1975). Free amino acids and peptides in must and wines from jspsnese grapes. *J. Fermen. Technol., 53*, 631–635.

Zhang, Y., Choi, H.J., Han, H.S., Park, J.H., Kim, S., Bae, J.H., Kim, H.K., & Choi, C. (2003). Polyphenolic compounds from korean pear and their biological activities. *Food Sci. Biotech., 12*, 262–267.

Chapter 6C
Proteins

Elizabeth Joy Waters and Christopher Bruce Colby

Contents

6C.1 Introduction

Grape and wine proteins have been of interest to winemakers for many years. This interest is primarily due to the ability of these proteins to aggregate together in finished white wines and form unattractive hazes and sediments. Proteins are significant in wine because they are a nuisance! Endogenous wine proteins, of course, also have other roles in wine and have been examined for their impact on the aroma and

E.J. Waters (✉)
The Australian Wine Research Institute, PO Box 197, Glen Osmond, SA 5064, Australia
e-mail: Elizabeth.Waters@awri.com.au

M.V. Moreno-Arribas, M.C. Polo (eds.), *Wine Chemistry and Biochemistry*, DOI 10.1007/978-0-387-74118-5_9, taste of wines (Jones et al. 2008; Peng et al. 1997), on bubble persistence and bead in sparkling wines (Girbau-Sola et al. 2002; Liger-Belair 2005; Senee et al. 1999), on lacquer-like bottle deposit in red wines (Peng et al. 1996a, b; Waters et al. 1994) and as allergens in rare cases of grape and wine allergy (Pastorello et al. 2003; Sbornik et al. 2007; Schad et al. 2005; Vassilopoulou et al. 2007). This chapter will focus on the role of endogenous wine proteins in white wine haze formation. It will not cover the other roles of protein listed above nor will it describe the roles and consequences of food proteins such as casein, egg white and isinglass, used in wine processing.

Published scientific studies of these proteins and of protein haze in wine began in earnest in the late 1950s and early 1960s with work by J. Koch in Geisenheim (Koch and Sajak 1959), H.W Berg, at the University of California, Davis (Bayly and Berg 1967; Berg and Akiyoshi 1961; Moretti and Berg 1965) and B. Rankine at the Australian Wine Research Institute in Adelaide (Rankine 1962).

6C.2 The Origin of Wine Proteins

Where do the proteins come from that subsequently form hazes and deposits in wine? Are they grape derived or do they come from the yeast? This is question that has occupied researchers since the 1950s. Bayly and Berg (1967) fermented a model juice solution and noted that protein levels contributed by the yeast were very low and probably negligible in most cases. Twenty years later, Hsu and Heatherbell (1987a) concluded the same using polyacrylamide gel electrophoresis. More recently, Ferreira et al. (2000) and Dambrouck et al. (2003) used modern immunological techniques to confirm that wine proteins originate predominantly from the grape. A contrasting possibility was raised by Yokotsuka et al. (1991), who analysed the protein profile of *Vitis vinifera* cv. Koshu grapes as well as the resulting wine made from the same grapes. In conflict with other authors, they found eight wine protein fractions not present in the juice and suggested they had come from yeast. Kwon (2004) demonstrated that there are multiple biological sources of wine proteins using nano-high-performance liquid chromatography/tandem mass spectrometry, although the relative levels of proteins from microbiological sources was not established by this methodology.

Waters et al. (1991, 1992) described two major wine protein fractions in *V. vinifera* cv. Muscat Gordo Blanc wine. By SDS PAGE, these proteins had molecular masses of 24 and 32kDa. By analysing the amino acid sequence of the proteins, Waters et al. (1996) showed that these 24-kDa and 32-kDa proteins shared homology with thaumatin and chitinases respectively and were highly similar to other plant pathogenesis-related (PR) proteins. Tattersall et al. (1997) further characterised the 24-kDa protein, *Vitis vinifera* thaumatin-like protein 1 (VvTL1). This protein is highly expressed in conjunction with the onset of sugar accumulation and softening in the grape berry. It has also been observed that there is relatively high expression after véraison of chitinase encoding genes in *V. vinifera* cv. Shiraz grape berries (Robinson et al. 1997).

6C.3 Characterization of Wine Proteins

6C.3.1 Size

In the 1960s, electrophoresis enabled separation of different sized wine proteins, with researchers at UC Davis (Bayly and Berg 1967; Moretti and Berg 1965), finding four different protein bands, varying in concentration within and among wines from different cultivars of *V. vinifera*. These researchers were the first to raise the possibility that certain protein fractions, rather than total protein, might be responsible for protein instability. Somers and Ziemelis (1973) used gel filtration chromatography to separate wine proteins from other constituents and, by using the exclusion limit of their gel, concluded that the wine protein size was between 10 and 50kDa. Hsu and Heatherbell (1987a) discovered many different protein fractions with a range of 11.2–65kDa. In a subsequent study (Hsu and Heatherbell 1987b), it was suggested that low molecular weight proteins (20–30kDa) were important to heat instability of wines, rather than those with high molecular weights. This was later confirmed (Waters et al. 1992). In an electrospray mass spectrometry study of the proteins in the juice of 19 cultivars of *Vitis vinifera* (Hayasaka et al. 2001), a range of masses (13–33kDa) was observed. The proteins were identified as mainly thaumatin-like proteins (range of 21,239–21,272 Da) and chitinases (range of 25,330–25,631 Da). Small variations in the masses of these protein due to small variation in their polypeptide sequence form the basis of a robust method of varietal identification based on mass spectrometry (Hayasaka et al. 2001).

6C.3.2 Isoelectric Point

Research into the isoelectric point of wine proteins has often been concurrent with studies of wine protein size. Proteins with low isoelectric points (pI) were found to be significant contributors to total wine protein (Moretti and Berg 1965) and to wine haze (Bayly and Berg 1967). Hsu and Heatherbell (1987a) confirmed this observation and suggested that the majority of wine proteins had a pI of 4.1–5.8, whilst Lee (1986) suggested the major protein fractions of wine had a pI of 4.8–5.7. Dawes et al. (1994) fractionated wine proteins on the basis of their pI and found that the five different fractions all produced haze after heat treatment. Haze particle formation was found to differ between the fractions however, leading to a statement that other wine components, such as phenolic compounds, need to be considered to understand fully protein haze.

6C.3.3 Stability

The unusual aspect of wine protein instability is that the proteins responsible for protein haze in the long term are, paradoxically, very stable themselves in the short term and survive the vinification process. It is, therefore, not surprising that wine proteins are resistant to low pH and enzymatic or non-enzymatic proteolysis (Waters et al. 1992). The mechanism for the resistance has not yet been fully elucidated, but resistance to low pH and proteolysis are characteristic of plant pathogenesis-related proteins in general. Experimental evidence suggests that it is also characteristic of the grape pathogenesis-related proteins involved, rather than phenolic association or glycosylation (Waters et al. 1995b). Limited proteolytic processing of the wine proteins can, however, occur during white table wine vinification (Waters et al. 1998) and during the Champagne winemaking process (Manteau et al. 2003). These data indicate that there could be scope to exploit this susceptibility to proteolysis under certain conditions and thus prevent wine protein haze.

6C.3.4 Levels in White Wines

Protein levels in white wine have been reported by several authors and have been shown to differ by variety. Lee (1986) reported a range of protein concentration from 18 to 81 mg/L in 14 wines from different Australian regions and made from different varieties. Some of these wines appeared to have been fined with bentonite prior to analysis. Pocock et al. (1998) reported concentrations in unfined Australian wines up to several hundred mg/L. Hsu and Heatherbell (1987b) found a range of 19–44 mg/L in four different unfined white wines from Oregon, while a very large variation (20—260 mg/L) was noted by Bayly and Berg (1967).

6C.4 Effect of Growing and Harvesting Conditions on Subsequent Wine Protein Levels

In grapevines, the synthesis of the PR proteins is regulated in a developmental and tissue specific manner. In *Vitis vinifera* cv. Muscat Gordo Blanco berries, both the expression of the VvTL1 gene, and the levels of the corresponding major thaumatin-like protein increased dramatically after the onset of berry softening (véraison) and continued throughout berry ripening (Tattersall et al. 1997). Similar developmental patterns were also found in the expression of genes encoding chitinases, some identical to those involved in wine protein haze (Derckel et al. 1996, 1998; Robinson et al. 1997). An immunological study of *V. labruscana* cv. Concord also showed that thaumatin-like proteins and chitinases accumulate during berry ripening (Salzman et al. 1998).

In all cultivars of *V. vinifera* studied so far, thaumatin-like proteins and chitinases are the major soluble components of grapes (Peng et al. 1997; Pocock et al. 1998, 2000; Tattersall et al. 1997). The predominance of these PR proteins was clear at all stages of berry development following véraison (Pocock et al. 2000). Importantly, as the concentration of these proteins in the berries continually increases during ripening, it can be presumed that the haze forming potential increases as ripening proceeds (Murphey et al. 1989; Pocock et al. 2000; Tattersall et al. 1997).

6C.4.1 Fungal Infection

While PR protein synthesis in healthy grapes appears to be triggered by véraison, this does not mean that the classical PR protein inducers – stress, wounding and pathogenic attack – cannot further modulate the levels of PR proteins in grapes. Grape PR proteins exhibit antifungal activity in vitro against common fungal pathogens of grapevines, including *Uncinula necator*, *Botrytis cinerea*, *Phomopsis viticola*, *Elsinoe ampelina* and *Trichoderma harzianum* (Giannakis et al. 1998; Jayasankar et al. 2003; Monteiro et al. 2003a; Salzman et al. 1998; Tattersall et al. 2001). It would be tempting to speculate that the antifungal activity observed in vitro reflects the main function of the PR proteins in vivo, and that their expression in fruit after véraison represents a preemptive defense mechanism for fruit. Jayasankar et al. (2003) added further credence to this hypothesis by demonstrating that grapevines regenerated after in vitro selection with *E. ampelina* culture filtrates had greater disease resistance and high constitutive expression of PR proteins, including VvTL1. Studies in which the synthesis of PR proteins is modified by gene technology would allow us to investigate this hypothesis further.

Increased expression of some PR genes and enhanced concentrations of some PR proteins have been observed in leaves and berries from grapevines infected with pathogens (Bézier et al. 2002; Derckel et al. 1998; Jacobs et al. 1999; Renault et al. 1996; Robert et al. 2002). In glasshouse experiments, Monteiro et al. (2003b) observed increased levels of thaumatin-like proteins in berries infected with *U. necator* compared to uninfected berries. Jacobs et al. (1999) demonstrated that chitinase and beta-1,3-glucanase activity increased in grape berries and leaves in response to powdery mildew infection, and that expression of genes, VvChi3, VvGlub, and VvTL2, coding for PR proteins, was also strongly induced. Of these three putative gene products, only VvTL2 has been detected as a soluble protein in grape juices and wines (Waters et al. 1996). Girbau et al. (2004) demonstrated that powdery mildew infection of *V. vinifera* cv. Chardonnay grape bunches resulted in increased levels of a grape minor thaumatin like protein, VvTL2, in wine. At high levels of infection (>30% of berries in a bunch infected), this had a significant impact on the level of haziness in the wine following a heat test.

Contrary to expectations that fungal diseases would lead to elevated levels of PR proteins in berries, Marchal et al. (1998) observed that juice from berries infected by *B. cinerea* showed reduced protein levels, and suggested that proteolytic enzymes from *B. cinerea* were responsible for this. Secretion of proteases by *B. cinerea* has been observed in culture media and on fruits such as apple (Zalewska-Sobczak et al. 1981) and tomato (Brown and Adikaram 1983). Girbau et al. (2004) also examined the impact of infection of grapes with *B. cinerea* in the vineyard and showed that infection resulted in marked decreases in the levels of PR proteins in the berries. Similar although less dramatic trends of reductions in protein levels were seen in laboratory experiments in which otherwise healthy berries were inoculated with *B. cinerea* (Girbau et al. 2004). Protein levels were also reduced in juice when *B. cinerea* was grown in this medium (Girbau et al. 2004).

The reduction in protein levels in the juice from Botrytis-infected grapes could, as suggested by Marchal et al. (1998), be due to proteolytic degradation of grape PR proteins by enzymes of *B. cinerea,* although there are other plausible explanations for these observations. If these effects are due to the activity of proteolytic enzymes from *B. cinerea,* these enzymes have the potential to replace bentonite fining for protein stabilization in oenology, a goal of many research efforts worldwide.

6C.4.2 Water Stress

The effect of water stress on the expression of PR proteins in grapes has been examined by analysing the PR protein content of *V. vinifera* cv. Shiraz berries from a replicated irrigation trial (Pocock et al. 2000). The lack of irrigation gave clear physiological signs of vine water stress but did not lead to elevated levels of PR proteins in the berries. At a fixed amount of protein per berry, however, the protein concentration in the juice from water stressed berries was higher than that from irrigated berries because berries from irrigated vines were larger and thus berry solutes were less concentrated. This effect of water stress on berry size is a general phenomenon (Smart and Coombe 1983) and it is likely that anecdotal reports that haze problems are greater in drought years are due to changes in berry sizes in these years rather than a direct physiological response of the berries to water stress in the form of enhanced PR protein production. Studies of gene expression under carefully controlled experimental conditions involving water stress would add further support to this hypothesis.

6C.4.3 Mechanical Harvesting

Paetzold et al. (1990) observed that mechanically harvested grapes produced must with increased protein content compared to that of hand harvested fruit pressed as whole bunches. They suggested that the lack of stems during crushing led to lower polyphenolic content in the must from mechanically harvested fruit compared to that from hand harvested fruit and therefore more protein was lost in complexes with phenolics from must from hand harvested fruit. Dubourdieu and Canal-Llaubères (1989) showed that wine made after maceration of destalked grapes for 18 h contained more protein than wine made after immediate pressing of whole bunches. Pocock and colleagues (Pocock and Waters 1998; Pocock et al. 1998) examined the impact of mechanical harvesting alone, not destemming, on the PR proteins in grapes and wine. Mechanical harvesting, with its associated prolonged transport of the fruit, resulted in higher PR protein levels in the juice and wine. Indeed mechanical harvesting of white grapes and subsequent transport was found to double the amount of bentonite required for prevention of protein haze when compared to fruit harvested by hand and transported from the same vineyard before destemming and crushing (Pocock and Waters 1998). This does not appear to be a result of increased protein synthesis in response to wounding as comparisons among hand harvested berries, mechanically harvested intact berries and the predominant form of mechanically harvested fruit: a mixture of broken fruit and juice, indicated little if any protein was produced as a result of stress caused by mechanical harvesting. Increases in protein content of juice from mechanically harvested fruit thus appear to be due to extraction or release of protein from or associated with pulp and skins rather than a physiological wounding response by the berry.

6C.5 Haze Formation in Wine

The mechanism of protein haze formation in wines is not fully understood. Slow denaturation of wine proteins is thought to lead to protein aggregation, flocculation into a hazy suspension and, finally, formation of visual precipitates. The importance of non-proteinaceous factors in white wine protein haze formation such as proanthocyanidins (Koch and Sajak 1959; Waters et al. 1995a; Yokotsuka et al. 1991) have been suspected for some time. Other factors such as polysaccharides, alcohol levels and pH have also been implicated (Mesquita et al. 2001; Siebert et al. 1996a). It has been observed that grape protein added to model wine does not precipitate or produce haze when heated, whereas visually obvious hazes occur when the same protein is added to a commercial wine (Pocock 2006).

The current theory of the mechanism of haze formation formulated by Pocock (2006) is as follows. Individual grape PR proteins probably exist as separate globular entities freely soluble in wine, tightly coiled and containing between six and eight disulfide bridges. The first step in the process of haze formation is to uncoil these proteins, or to denature them. This is accelerated by heating. The second step involves aggregation of the denatured proteins as haze particles.

The size and amount of protein haze formed in a wine is strongly influenced by other wine components. Pocock (2006) has demonstrated that one wine component, the sulfate anion, previously referred to as factor X, is essential for haze formation. If the sulfate anion is not present, heating does not result in sufficient denaturation of the proteins to lead to their aggregation, thus a haze will not form.

Sulfate is one of the Hofmeister series of anions, a ranking of the ability of various ions to precipitate proteins (Kunz et al. 2004). In simple terms, precipitation of proteins by kosmotropic anions occurs due to 'salting out' – a competition between the anion and the protein for water of solvation resulting in a loss of water from the protein surface. This process is classically applied in ammonium sulfate precipitation as the first step in many protein purification schemes, although the levels employed are several fold higher than those in wine. In the particular case of white wine, this loss of water of solvation, even by a relatively low amount of sulfate anion, by a protein in a solution containing a variety of cations and other anions and between 9 vol.% and 13 vol.% ethanol may be enough to affect the tertiary structure and/or aggregation of proteins.

Whilst sulfate appears to be fundamental to haze formation, other wine components such as phenolic compounds remain as candidate haze modulators. One possibility is that white wine phenolic compounds affect the particle size of denatured aggregated proteins, possibly through crosslinking. Several researchers (Oh et al. 1980; Siebert et al. 1996b) have suggested a hydrophobic mechanism for the interaction between phenolic compounds and proteins, in which the protein has a fixed number of phenolic binding sites. More of these sites are exposed when the protein is denatured.

Protein haze in white wine thus differs in several aspects from protein haze in beer. It is well established that beer protein haze is due to interactions between proteins, derived from the barley storage protein hordein and rich in proline, and hop polyphenolic compounds (Bamforth 1999; Miedl et al. 2005; Siebert 1999; Siebert and Lynn 2003). White wine proteins are not derived from storage proteins of grape seed nor are they as rich in proline as hordein. In addition, wine protein haze formation cannot be eliminated by removing polyphenolic compounds by PVPP (Pocock et al. 2006) while in beer this has been applied as a commercial strategy (Leiper et al. 2005; Madigan et al. 2000).

6C.6 Bentonite Fining

The major winemaking process to affect the levels of proteins in wine is bentonite fining. Bentonite, a montmorillonite clay, is used almost universally throughout the wine industry for the prevention of wine protein haze through removal of proteins before bottling (Blade and Boulton 1988; Ferreira et al. 2002; Høj et al. 2000). The adsorption of wine proteins onto bentonite is principally attributed to cationic exchange with the bentonite clay. Wine proteins are positively charged at wine pH, and thus can be exchanged onto bentonite, which carries a net negative charge (Blade and Boulton 1988; Ferreira et al. 2002; Høj et al. 2000).

A number of studies have indicated that different protein fractions require distinct bentonite concentrations for protein removal and consequent heat stabilisation (Duncan 1992; Ferreira et al. 2002). Bentonite fining has been shown to remove higher pI (5.8–8.0) and intermediate molecular weight (MW; 32–45 kDa) proteins first (Hsu and Heatherbell 1987b). However, these represent only a small portion of the soluble proteins. Proteins with a MW of 60–65 kDa, and with wide pI range (4.1–8.0) were highly resistant to removal by bentonite fining (i.e. required significant bentonite addition) and typically remained in protein-stabilised wine (Hsu and Heatherbell 1987b). Hsu and Heatherbell (1987b) concluded that it is necessary to remove lower pI (4.1–5.8), lower MW (12.6 kDa and 20–30 kDa) proteins, which represent a major component of proteins present, to protein stabilise wines. Contrary to these findings, a study by Dawes et al. (1994) found that there was no bentonite selectivity based on isoelectric point, and that bentonite fining resulted in the removal of all the different protein fractions. Further, the amount of protein depletion (across all protein fractions) observed in this study corresponded linearly with the level of bentonite addition (percentage reduction in protein concentration per g/L of bentonite added ranged from 70% to 89%). These different conclusions in the published literature might be attributed in part to the different methods used to fractionate proteins and assess their levels.

Several authors have investigated the extent of adsorption of standard and model proteins by bentonite. Gougeon et al. (2002, 2003) studied the absorption of two homopolypeptide preparations with average MW around 20 kDa onto a synthetic bentonite. Their data suggested that these polypeptides tended to unfold and take on a more random coil structure upon absorption. Using a range of physical measures, Gougeon et al. (2002, 2003) also hypothesized that the polypeptides were primarily absorbed near the edges of the bentonite sheets rather than within the interlayer spaces between the sheets. The adsorption of the standard protein, bovine serum albumin (BSA) by bentonite in model wine solutions was studied by Blade and Boulton (1988). Adsorption was shown to be independent of temperature, but varied slightly with protein content, pH and ethanol content. In another study (Achaerandio et al. 2001), bentonite adsorption was evaluated with three proteins (BSA, ovalbumin, lysozyme) in a model wine solution. The effect of ethanol content and protein molecular weight on the adsorption capacity of bentonite was also studied. Adsorption capacity tended to increase with increasing ethanol concentrations with regard to adsorption of BSA and lysozyme, however, no change was observed for ovalbumin. Blade and Boulton (1988) showed that maximal absorption was reached rapidly, and complete within 30 s of the addition. This is consistent with an earlier study (Lee 1986), in which Gewürztraminer wine fined with bentonite was rendered stable one minute after bentonite addition and with later studies of in-line dosing described below (Muhlack et al. 2006; Nordestgaard et al. 2006). In comparison, bentonite fining in a winery setting typically takes one to two weeks, depending on the tank size and rate of bentonite addition used.

Bentonite regeneration refers to the desorption of adsorbed wine protein from the bentonite surface, and would permit bentonite to be reused. However, a commercial process for bentonite regeneration does not currently exist, and thus bentonite is only used once before being discarded. An early study (Armstrong and Chesters 1964) investigated the effect of pH on the desorption of pepsin from bentonite. In this study, 62.8% of the pepsin (pI 2.8), which had been adsorbed onto the bentonite at pH 3.0, was desorbed by raising the pH to 5.2 using sodium hydroxide. In a more recent study (Churchman 1999), the batch treatment of bentonite-protein complexes with a range of bases at a variety of different concentrations, durations and agitation methods was examined. The greatest degree of desorption was achieved with sodium carbonate at a low solid: solution ratio and with agitation. Similar results were obtained with sodium hydroxide at pH 12 and 13 (protein released per gram of bentonite = 52 and 65 mg/g respectively). However, from the study it was concluded that the use of alkaline pH and salts was ineffective for substantial removal of protein from bentonites. Instead, the study suggested protein degradation followed by a treatment to displace the products of protein breakdown from the bentonite was required. For example, a batchwise treatment that employed hydrogen peroxide, sodium carbonate and photo-oxidation with UV radiation was found to be effective in removing residual protein from bentonite.

As discussed by Waters et al. (2005), in both these studies, protein was desorbed from bentonite by increasing the pH, but only in a batch system. As adsorption and desorption are equilibrium processes, it is hardly surprising that residual protein was retained on the bentonite surface. In contrast, in a continuous flow system, fresh reactant could be continually fed to the adsorption system, while ion exchange products could be continually transported away from the system, driving the exchange process to completion. Continuous flow contactors, such as continuous stirred tanks, packed bed columns and fluidised beds, are commonplace in many chemical industries, as is continuous catalyst regeneration (Fogler 1992). Therefore, it was postulated that effective regeneration of bentonite might be achieved by treatment with a base such as sodium hydroxide in a simple continuous flow system (Waters et al. 2005). However, the difficulty in regenerating bentonite by salts and alkalis could also suggest that protein adsorption on bentonite may involve other mechanisms than solely cation exchange; thus complete regeneration of bentonite by such a process may not be feasible.

6C.6.1 The Sensory Impact of Bentonite Fining

One of the perceived disadvantages of bentonite fining is that it is thought to reduce flavor and aroma of wines. This perception is not supported by strong evidence in the normal operational range of bentonite additions by winemakers. Sensory evaluation of wines treated with bentonite showed no significant differences between the control and the fined samples (Leske et al. 1995). Similarly, using difference testing, Pocock et al. (2003) reported that bentonite fining of a Chardonnay and Semillon wine had no effect on wine aroma and palate. This contrasts with previous findings (Miller et al. 1985) that demonstrated reduced concentration of aroma compounds after bentonite addition to juice, must or wine. More recently Pollnitz et al. (2003) elegantly confirmed that aroma compounds can be absorbed by bentonite, as did Cabaroglu et al. 2003, although the later study found no sensory effect of bentonite fining of *V. vinifera* cv. Muscat Ottonel or Gewurztraminer wine. Rankine (1989) stated that bentonite fining results in the loss of aroma and flavor and Martínez-Rodriguez and Polo (2003) extended this conclusion to sparkling wines when bentonite is added to the tirage solution. This has led to the widespread conclusion, throughout both literature and industry, that bentonite fining at typical addition rates has a detrimental effect on wine aroma and flavor, despite the fact that the conclusions of Miller et al. (1985) were influenced by many factors and "should not be considered general" (Miller et al. 1985). Similarly, Leske et al. (1995) concluded that "lack of significant differences observed in this trial suggests that the negative effect of fining may not be evident to a consumer in a commercial situation" and "the magnitude of any such [negative sensory] effect will presumably depend on the individual wine, the bentonite chosen, and the rate of addition chosen".

6C.7 Improving Bentonite Efficiency

As well as the perceived but questionable adverse sensory effects of bentonite fining, principal criticisms of the traditional method for bentonite fining are as follows:

1. Between 5 and 10% of the wine may be occluded in the lees (Tattersall et al. 2001). During recovery by RDV filtration, this occluded wine may be downgraded in quality (Waters et al. 2005).
2. Gravity settling of the bentonite normally requires several days to a week. Protein adsorption during bentonite fining is rapid – the treatment effect is completed within several minutes, as described above (Blade and Boulton 1988; Muhlack et al. 2006). Hence, the long processing time required for bentonite fining is constrained by gravity settling of the lees and to maximize wine recovery during racking. Furthermore, the long period of contact between bentonite and the wine during gravity settling might be a contributing factor to perceptions of adverse sensory outcomes.

Consequently, there is interest is developing more efficient ways of using bentonite.

6C.7.1 Predictive Assays for Heat Stability

The amount of bentonite required to prevent protein haze formation differs from wine to wine and is usually determined by a laboratory fining trial and assessed by a heat stability assay or test. The heat stability test most commonly employed by the Australian wine sector is to observe whether a wine sample produces excessive (e.g. >2 NTU) turbidity following heating at 80 °C for 6 h. Similar accelerated heating tests are used in other countries. Pocock and Waters (2006) have shown that this test does not always predict haze formation accurately during long-term storage trials of white wines. Furthermore, Pocock and Waters (2006) demonstrated that most of the current predictive assays used by winemakers could be biased towards over fining and it is also likely that technical errors performing both the heating and the haze measurement steps add to their inaccuracy. This could lead to fining with more bentonite than is needed under most conditions encountered commercially by bottled wines during transport and storage. Clearly, further trials with more wines and with storage under a wider range of conditions would be necessary to confirm these results.

Therefore, an alternative test method, such as heating at 80 °C for 2 h, an 'alkali' modified Coomassie-dye assay (e.g. Boyes et al. 1997), or the more recently available reagent based test kits such as Proteotest or Prostab, could present an opportunity to decrease bentonite dose and reduce volume of wine occluded in bentonite lees if confirmed to predict more accurately haze formation in wine. If these alternative test methods could replace the current heat test method of 80 °C for 6 h, this would also reduce the amount of time and/or effort presently required by a winery's laboratory to conduct heat stability testing.

6C.7.2 Selecting an Appropriate Bentonite Type

Sodium (Na) bentonites swell more than calcium (Ca) bentonites, and thus produce a greater volume of lees for an equivalent dose, but are more efficient at adsorbing proteins, and hence, require a lower dose than Ca bentonites to achieve heat stability (Zoecklin 1988). However, there is little quantitative data that shows what these relationships actually are. Analysis of a study of bentonites from more than a decade ago by AWRI (Morant 1990) suggests that Ca bentonites, whilst requiring (on average) a dose three times higher than Na bentonites to achieve heat stability, will produce one quarter of the lees. Furthermore, Na/Ca bentonites required a dose 50% greater than Na bentonites but generated one half the volume of lees.

The data presented indicates that using Na/Ca or Ca bentonites could significantly reduce values losses arising from bentonite fining. For example, estimates from the above data suggest that a Na bentonite dose of 0.4 g/L produces 6 vol.% lees whereas a Ca bentonite, by reducing lees volume, could regain up to 5% of the original volume of the wine that would ordinarily be occluded in Na bentonite lees.

However, if this is the case why are Na/Ca or Ca bentonites not more widely used? There are a number of possible reasons. First, higher doses of Ca bentonite require greater volume of slurry or more concentrated slurry. Calculations assuming a 2% constraint on water addition to the wine suggest that for a Na bentonite dose of 1 g/L, 5% w/v slurry would be required. At the higher doses of Na/Ca and Ca bentonite necessary to achieve the same treatment effect, the slurry concentrations are 7.5% and 15% w/v, respectively. It may not be possible or practical to achieve these higher slurry concentrations and thus Na bentonites are favored. However, Ca bentonites swell less and higher slurry concentrations may be possible when compared with Na bentonites. Investigations of the rheology of bentonite slurries would provide valuable data to answer this question. Second, Ca bentonites release Ca during bentonite fining and may contribute to tartrate instability. Third, Ca bentonites may contribute to heavy metal pick up (Boulton et al. 1996), and thus their use may be undesirable at elevated bentonite dose rates. However, this was not observed by Morant 1990) when she measured trace metal concentrations for arsenic and copper in wine following fining. In fact, no difference in heavy metal content was observed between the Na, Na/Ca and Ca bentonites.

6C.7.3 In-Line Dosing of Bentonite

The in-line dosing paradigm exploits the observation that protein adsorption by bentonite occurs rapidly – within several minutes (Blade and Boulton 1988; Muhlack et al. 2006). Furthermore, if combined with centrifugation, the bentonite and wine can be separated and the bentonite lees are simultaneously compacted to reduce value losses. In-line dosing methods for bentonite fining are already used by some wineries and anecdotal reports suggest that commercial applications are growing.

Results obtained from pilot scale field studies of in line dosing by Colby and colleagues (Muhlack et al. 2006; Nordestgaard et al. 2006) confirmed that a contact time of only a few minutes was sufficient for adsorption of the haze-forming proteins by bentonite. There was no detectable sensory impact on the wine of inline dosing compared to batch addition.

These studies also observed a carryover of bentonite from the centrifuge during in-line dosing of bentonite. This bentonite carryover could be reduced by decreasing the operating flow rate to the centrifuge. Furthermore, Na-Ca bentonite was separated more easily by centrifugation than Na bentonites, and thus produced less carryover at equivalent flowrates.

This carryover problem, if centrifugation is used, is a disadvantage of the in-line dosing method because it contributes to downstream wine losses from tanks containing settled lees and/or constrains the throughput achievable. Some potential solutions that could be used to overcome this problem include a more efficient centrifuge, multiple clarification steps, an alternative and more efficient clarification method, an alternative bentonite that is more easily clarified, and "counter-fining" with another fining agent to improve bentonite separation.

6C.8 Summary

Grape PR proteins are the main proteins that persist through white winemaking and remain in the wine. It is thought that their denaturation and subsequent aggregation leads to the formation of hazes and precipitates. The formation of wine protein haze is multifactorial and depends on the presence of components which are essential for protein denaturation, such as the sulfate anion, and potentially others as yet not formally identified. The main grape PR proteins in grape juice and wine are thaumatin-like proteins and chitinases; they range in mass between approximately 21 kDa to 26 kDa, have acidic isoelectric points and are relatively stable to proteolytic enzymes. The levels of these proteins in wines depend on grape variety, fungal infection, water stress and skin contact during winemaking. Bentonite fining is still the only commercially acceptable practical solution to avoid protein haze. The efficiency of this process can be increased by using appropriate test procedures to estimate the dose required, selecting the bentonite type best suited to the application and using inline dosing and centrifugation to add and remove the bentonite rather than the more time consuming process of batch addition and gravity settling.

Acknowledgments We thank Australia's grape growers and winemakers for their support through their investment body the Grape and Wine Research and Development Corporation, with matching funds from the Australian government. We also thank our past and current colleagues and students for their input into our understanding of this area.

References

Achaerandio, I., Pachova, V., Güell, C., & López, F. (2001). Protein adsorption by bentonite in a white wine model solution: Effect of protein molecular weight and ethanol concentration. *Am. J. Enol. Vitic., 52*, 122–126

Armstrong, D.E., & Chesters, G. (1964). Properties of protein-bentonite complexes as influenced by equilibration conditions. *Soil Sci., 98*, 39–52

Bamforth, C.W. (1999). Beer haze. *J. Am. Soc. Brew. Chem., 57*, 81–90

Bayly, F.C., & Berg, H.W. (1967). Grape and wine proteins of white wine varietals. *Am. J. Enol. Vitic., 18*, 18–32

Berg, H.W., & Akiyoshi, M. (1961). Determination of protein stability in wine. *Am. J. Enol. Vitic., 12*, 107–110

Bézier, A., Lambert, B., & Baillieul, F. (2002). Study of defense-related gene expression in grapevine leaves and berries infected with *Botrytis cinerea*. *Eur. J. Plant Pathol., 108*, 111–120

Blade, W.H., & Boulton, R. (1988). Adsorption of protein by bentonite in a model wine solution. *Am. J. Enol. Vitic., 39*, 193–199

Boulton, R., Singleton, V.L., L.F., Bisson & Kunkee, RE (1996). Principles and Practices of Winemaking. Gaithersburg, Maryland: Aspen Publishers, Inc.

Boyes, S., Strübi, P., & Dawes, H. (1997) Measurement of protein content in fruit juices, wine and plant extracts in the presence of endogenous organic compounds. *Lebensmittel-Wissenschaft und-Technologie, 30*, 778–785

Brown, A.E., & Adikaram, N.K.B. (1983). A role for pectinase and protease inhibitors in fungal rot development in tomato fruits. *Phytopathologische Zeitschrift, 106*, 239–251

Cabaroglu, T., Razungles, A., Baumes, R., & Gunata, Z. (2003). Effect of fining treatments on the aromatic potential of white wines from Muscat Ottonel and Gewurztraminer cultivars. *Sc. Aliment, 23*, 411–423

Churchman, G.J. (1999). Procedures for the efficient recycling of bentonite used for the removal of proteins from wine. Final Report to Grape and Wine Research and Development Corporation, Project CSS 98/1

Dambrouck, T., Marchal, R., Marchal-Delahaut, L., Parmentier, M., Maujean, A., & Jeandet, P. (2003). Immunodetection of proteins from grapes and yeast in a white wine. *J. Agric. Food Chem., 51*, 2727–2732

Dawes, H., Boyes, S., Keene, J., & Heatherbell, D. (1994). Protein instability of wines: Influence of protein isoelectric point. *Am. J. Enol. Vitic., 45*, 319–326

Derckel, J.P., Legendre, L., Audran, J.C., Haye, B., & Lambert, B. (1996). Chitinases of the grapevine (*Vitis vinifera* L.): five isoforms induced in leaves by salicylic acid are constitutively expressed in other tissues. *Plant Sci., 119*, 31–37

Derckel, J.P., Audran, J.C., Haye, B., Lambert, B., & Legendre, L. (1998). Characterisation, induction by wounding and salicylic acid, and activity against *Botrytis cinerea* of chitinases and β-1,3-glucanases of ripening grape berries. *Physiol. Plantarum, 104*, 56–64

Dubourdieu, D., & Canal-Llaubères, R.M. (1989). Influence of some colloids (polysaccharides and proteins) on the clarification and stabilization of wines. Seventh Australian Wine Industry Technical Conference, Adelaide, SA, Winetitles, Adelaide, SA

Duncan, B. (1992). Varietal differences in white grape protein: Implications for bentonite fining. *Aust. NZ Wine Ind. J., 7*, 189–193

Ferreira, R.B., Monteiro, S., Picarra-Pereira, M.A., Tanganho, M.C., Loureiro, V.B., & Teixeira, A.R. (2000). Characterization of the proteins from grapes and wines by immunological methods. *Am. J. Enol. Vitic., 51*, 22–28

Ferreira, R.B., Piçarra-Pereira, M.A., Monteiro, S., Loureiro, V.B., & Teixeira, A.R. (2002). The wine proteins. *Trends Food Sci. Tech., 12*, 230–239

Fogler, H.S. (1992). Elements of chemical reaction engineering. New Jersey: Prentice-Hall Inc.

Giannakis, C., Bucheli, C.S., Skene, K.G.M., Robinson, S.P., & Scott, N.S. (1998). Chitinase and ß-1,3-glucanase in grapevine leaves: a possible defence against powdery mildew infection. *Aust. J. Grape Wine Res., 4*, 14–22

Girbau, T., Stummer, B.E., Pocock, K.F., Baldock, G.A., Scott, E.S., & Waters, E.J. (2004). The effect of *Uncinula necator* (powdery mildew) and *Botrytis cinerea* infection of grapes on the levels of haze-forming pathogenesis-related proteins in grape juice and wine. *Aust. J. Grape Wine Res., 10*, 125–133

Girbau-Sola, T., Lopez-Tamames, E., Bujan, J., & Buxaderas, S. (2002). Foam aptitude of Trepat and Monastrell red varieties in Cava elaboration. 1 Base wine characteristics. *J. Agric. Food Chem., 50*, 5596–5599

Gougeon, R.D., Reinholdt, M., Delmotte, L., Miehe-Brendle, J., Chézeau, J.M., Le Dred, R., Marchal, R., & Jeandet, P. (2002). Direct observation of polylysine side-chain interaction with smectites interlayer surfaces through ^{1}H-^{27}Al heteronuclear correlation NMR spectroscopy. *Langmuir, the American Chemical Society Journal of Surfaces and Colloids, 18*, 3396–3398

Gougeon, R.D., Soulard, M., Reinholdt, M., Miehe-Brendle, J., Chézeau, J.M., Le Dred, R., Marchal, R., & Jeandet, P. (2003). Polypeptide adsorption onto a synthetic montmorillonite: A combined solid-state NMR, X-ray diffraction, thermal analysis and N_2 adsorption study. *Eur. J. Inorg. Chem., 2003*, 1366–1372

Hayasaka, Y., Adams, K.S., Pocock, K.F., Baldock, G.A., Waters, E.J., & Hoj, P.B. (2001). Use of electrospray mass spectrometry for mass determination of grape (*Vitis vinifera*) juice pathogenesis-related proteins: A potential tool for varietal differentiation. *J. Agric. Food Chem., 49*, 1830–1839

Høj, P.B., Tattersall, D.B., Adams, K., Pocock, K.F., Hayasaka, Y., van Heeswijck, R., & Waters, E.J. (2000). The 'haze protiens' of wine-a summary of properties, factors affecting their accumulation in grapes, and the amount of bentonite required for their removal from wine. Proceedings of the ASEV 50th Anniversary Annual Meeting; June 19–23, 2000; Davis, California, Davis, California, American Society for Enology and Viticulture

Hsu, J.C., & Heatherbell, D.A. (1987a). Isolation and characterization of soluble proteins in grapes, grape juice and wine. *Am. J. Enol. Vitic., 38*, 6–10

Hsu, J.C., & Heatherbell, D.A. (1987b). Heat-unstable proteins in wine. I. Characterization and removal by bentonite fining and heat treatment. *Am. J. Enol. Vitic., 38*, 11–16

Jacobs, A.K., Dry, I.B., & Robinson, S.P. (1999). Induction of different pathogenesis-related cDNAs in grapevine infected with powdery mildew and treated with ethephon. *Plant Pathol., 48*, 325–336

Jayasankar, S., Li, Z., & Gray, D.J. (2003). Constitutive expression of *Vitis vinifera* thaumatin-like protein after in vitro selection and its role in anthracnose resistance. *Funct. Plant Biol., 30*, 1105–1115

Jones, P.R., Gawel, R., Francis, I.L., & Waters, E.J. (2008). The influence of interactions between major white wine components on the aroma, flavour and texture of model white wine. *Food Qual. Prefer, 19*, 596–607

Koch, J., & Sajak, E. (1959). A review and some studies on grape protein. *Am. J. Enol. Vitic., 10*, 114–123

Kunz, W., Henle, J., & Ninham, B.W. (2004). 'Zur Lehre von der Wirkung der Salze' (about the science of the effect of salts): Franz Hofmeister's historical papers. *Curr. Opin. Coll. Interface Sci., 9*, 19–37

Kwon, S.W. (2004). Profiling of soluble proteins in wine by nano-high-performance liquid chromatography/tandem mass spectrometry. *J. Agric. Food Chem., 52*, 7258 – 7263

Lee, T. (1986). Protein instability: nature, characterization and removal by bentonite

Leiper, K.A., Stewart, G.G., McKeown, I.P., Nock, T., & Thompson, M.J. (2005). Optimising beer stabilisation by the selective removal of tannoids and sensitive proteins. *J. I. Brewing, 111*, 118–127

Leske, P.A., Bruer, N.G.C., & Capdeboscq, V. (1995). An evaluation of some characteristics of commercial bentonites. *Wine Industry J., 10*, 73–77

Liger-Belair, G. (2005). The physics and chemistry behind the bubbling properties of Champagne and sparkling wines: A state-of-the-art review. *J. Agric. Food Chem., 53*, 2788–2802

Madigan, D., Byrne, H., Matthews, S., Kelly, R., McEnroe, C., & Harmey, D. (2000). Studies on the effects of common process variables on the colloidal stability of beer. *J. Am. Soc. Brew. Chem., 58*, 160–164

Manteau, S., Lambert, B., Jeandet, P., & Legendre, L. (2003). Changes in chitinase and thaumatin-like pathogenesis-related proteins of grape berries during the Champagne winemaking process. *Am. J. Enol. Vitic., 54*, 267–272

Marchal, R., Berthier, L., Legendre, L., Marchal-Delahaut, P., Jeandet, P., & Maujean, A. (1998). Effects of *Botrytis cinerea* infection on the must protein electrophoretic characteristics. *J. Agric. Food Chem., 46*, 4945–4949

Martínez-Rodriguez, A.J., & Polo, M.C. (2003). Effect of the addition of bentonite to the tirage solution on the nitrogen composition and sensory quality of sparkling wines. *Food Chem., 81*, 383–388

Mesquita, P.R., Piçarra-Pereira, M.A., Monteiro, S., Loureiro, V.B., Teixeira, A.R., & Ferreira, R.B. (2001). Effect of wine composition on protein stability. *Am. J. Enol. Vitic., 52*, 324–330

Miedl, M., Garcia, M.A., & Bamforth, C.W. (2005). Haze formation in model beer systems. *J. Agric. Food Chem., 53*, 10161–10165

Miller, G.C., Amon, J.M., Gibson, R.L., & Simpson, R.F. (1985). Loss of wine aroma attributable to protein stabilization with bentonite or ultrafiltration. *Aust. J. Grape Wine Res., 256*, 46, 49–50

Monteiro, S., Picarra-Pereira, M.A., Teixeira, A.R., Loureiro, V.B., & Ferreira, R.B. (2003a). Environmental conditions during vegetative growth determine the major proteins that accumulate in mature grapes. *J. Agric. Food Chem., 51*, 4046–4053

Monteiro, S., Barakat, M., Picarra-Pereira, M.A., Teixeira, A.R., & Ferreira, R.B. (2003b). Osmotin and thaumatin from grape: A putative general defense mechanism against pathogenic fungi. *Phytopathology, 93*, 1505–1512

Morant, I. (1990). Evaluation of bentonites available on the Australian market. *Technical Review, 67*, 16–18, 31–33

Moretti, R.H., & Berg, H.W. (1965). Variability among wines to protein clouding. *Am. J. Enol. Vitic., 16*, 69–78

Muhlack, R., Nordestgaard, S., Waters, E.J., O'Neill, B.K., Lim, A., & Colby, C.B. (2006). In-line dosing for bentonite fining of wine or juice: contact time, clarification, product recovery & sensory effects. *Aust. J. Grape Wine Res., 12*, 221–234

Murphey, J.M., Spayd, S.E., & Powers, J.R. (1989). Effect of grape maturation on soluble protein characteristics of Gewürztraminer and White Riesling juice and wine. *Am. J. Enol. Vitic., 40*, 199–207

Nordestgaard, S., Chuan, Y.P., O'Neill, B., Waters, E., Deans, L., Policki, P., & Colby, C. (2006). In-line Dosing of White Wine for Bentonite Fining with Centrifugal Clarification. *Am. J. Enol. Vitic., 58*, 283–285

Oh, H.I., Hoff, J.E., Armstrong, G.S., & Haff, L.A. (1980). Hydrophobic interaction in tannin-protein complexes. *J. Agric. Food Chem., 28*, 394–398

Paetzold, M., Dulau, L., & Dubourdieu, D. (1990). Fractionnement et caractérisation des glycoprotéines dans les moûts de raisins blancs. *J. Int. Sci. Vigne Vin, 24*, 13–28

Pastorello, E.A., Farioli, L., Pravettoni, V., Ortolani, C., Fortunato, D., Giuffrida, M.G., Garoffo, L.P., Calamari, A.M., Brenna, O., & Conti, A. (2003). Identification of grape and wine allergens as an endochitinase 4, a lipid-transfer protein, and a thaumatin. *J. Allergy Clin. Immun., 111*, 350–359

Peng, Z., Waters, E.J., Pocock, K.F., & Williams, P.J. (1996a). Red wine bottle deposits, I: a predictive assay and an assessment of some factors affecting deposit formation. *Aust. J. Grape Wine Res., 2*, 25–29

Peng, Z., Waters, E.J., Pocock, K.F., & Williams, P.J. (1996b). Red wine bottle deposits, II: cold stabilisation is an effective procedure to prevent deposit formation. *Aust. J. Grape Wine Res., 2*, 30–34

Peng, Z., Pocock, K.F., Waters, E.J., Francis, I.L., & Williams, P.J. (1997). Taste properties of grape (*Vitis vinifera*) pathogenesis-related proteins isolated from wine. *J. Agric. Food Chem., 45*, 4639–4643

Pocock, K.F., & Waters, E.J. (1998). The effect of mechanical harvesting and transport of grapes, and juice oxidation, on the protein stability of wines. *Aust. J. Grape Wine Res., 4*, 136–139

Pocock, K.F., & Waters, E.J. (2006). Protein haze in bottled white wines: how well do stability tests and bentonite fining trials predict haze formation during storage and transport?. *Aust. J. Grape Wine Res., 12*, 212–220

Pocock, K.F., Hayasaka, Y., Peng, Z., Williams, P.J., & Waters, E.J. (1998). The effect of mechanical harvesting and long-distance transport on the concentration of haze-forming proteins in grape juice. *Aust. J. Grape Wine Res., 4*, 23–29

Pocock, K.F., Hayasaka, Y., McCarthy, M.G., & Waters, E.J. (2000). Thaumatin-like proteins and chitinases, the haze-forming proteins of wine, accumulate during ripening of grape (*Vitis vinifera*) berries and drought stress does not affect the final levels per berry at maturity. *J. Agric. Food Chem., 48*, 1637–1643

Pocock, K.F., Høj, P.B., Adams, K.S., Kwiatkowski, M.J., & Waters, E.J. (2003). Combined heat and proteolytic enzyme treatment of white wines reduces haze forming protein content without detrimental effect. *Aust. J. Grape Wine Res., 9*, 56–63

Pocock, K.F., Alexander, G.M., Hayasaka, Y., Jones, P.R., & Waters, E.J. (2006). Sulfate – a candidate for the missing essential factor that is required for the formation of protein haze in white wine. *J. Agric. Food Chem., 55*, 1799–1807

Pollnitz, A.P., Capone, D.L., Caldersmith, M.C., & Sefton, M.A. (2003). The effect of various wine bottle closures and fining agents on flavour and aroma compounds in wine. In: S.M. Bell, K.A. de Garis, C.G. Dundon, R.P. Hamilton, S.J. Partridge & G.S. Wall (Eds), Grapegrowing at the edge; managing the wine business; impacts on wine flavour, edn. (pp 59–63) Adelaide, South Australia: Australian Society of Viticulture and Oenology

Rankine, B.C. (1962). Bentonite and wine fining. *Austr. Wine Brew. Spirit Rev., 81*, 18–22

Rankine, B.C. (1989). Making good wine. Sydney: Pan Macmillan

Renault, A.S., Deloire, A., & Bierne, J. (1996). Pathogenesis-related proteins in grapevines induced by salicylic acid and *Botrytis cinerea*. *Vitis, 35*, 49–52

Robert, N., Roche, K., Lebeau, Y., Breda, C., Boulay, M., Esnault, R., & Buffard, D. (2002). Expression of grapevine chitinase genes in berries and leaves infected by fungal or bacterial pathogens. *Plant Sci., 162*, 389–400

Robinson, S.P., Jacobs, A.K., & Dry, I.B. (1997). A class IV chitinase is highly expressed in grape berries during ripening. *Plant Physiol., 114*, 771–8.

Salzman, R.A., Tikhonova, I., Bordelon, B.P., Hasegawa, P.M., & Bressan, R.A. (1998). Coordinate accumulation of antifungal proteins and hexoses constitutes a developmentally controlled defense response during fruit ripening in grape. *Plant Physiol., 117*, 465–472

Sbornik, M., Rakoski, J., Mempel, M., Ollert, M., & Ring, J. (2007). IgE-mediated type-I-allergy against red wine and grapes. *Allergy, 62*, 1339–1340

Schad, S.G., Trcka, J., Vieths, S., Scheurer, S., Conti, A., Brocker, E.B., & Trautmann, A. (2005). Wine anaphylaxis in a German patient: IgE-mediated allergy against a lipid transfer protein of grapes. *International Archives of Allergy and Immunology, 136*, 159–164

Senee, J., Robillard, B., & Vignes-Adler, M. (1999). Films and foams of Champagne wines. *Food Hydrocoll., 13*, 15–26

Siebert, K. (1999). Protein-polyphenol haze in beverages. *Food Technol., 53*, 54–57

Siebert, K.J., & Lynn, P.Y. (2003). Effects of alcohol and pH on protein-polyphenol haze intensity and particle size. *J. Am. Soc. Brew. Chem., 61*, 88–98

Siebert, K.J., Carrasco, A., & Lynn, P.Y. (1996a). Formation of protein-polyphenol haze in beverages. *J. Agric. Food Chem., 44*, 1997–2005

Siebert, K.J., Troukhanova, N.V., & Lynn, P.Y. (1996b). Nature of polyphenol-protein interactions. *J. Agric. Food Chem., 44*, 80–85

Smart, R.E., & Coombe, B.C. (1983). Water relations of grapevines. In: T.T. Kozlowski (Eds), Water deficits and plant growth, edn. (pp 137–196) New York: Academic Press

Somers, T.C., & Ziemelis, G. (1973). Direct determination of wine proteins. *Am. J. Enol. Vitic., 24*, 47–50

Tattersall, D.B., van Heeswijck, R., & Høj, P.B. (1997). Identification and characterisation of a fruit-specific, thaumatin-like protein that accumulates at very high levels in conjunction with the onset of sugar accumulation and berry softening in grapes. *Plant Physiol., 114*, 759–769

Tattersall, D.B., Pocock, K.F., Hayasaka, Y., Adams, K., van Heeswijck, R., Waters, E.J., & Høj, P.B. (2001). Pathogenesis related proteins – their accumulation in grapes during berry growth and their involvement in white wine heat instability. Current knowledge and future perspectives in relation to winemaking practices. In: K.A. Roubelakis-Angelakis (Eds), Molecular biology & biotechnology of the grapevine, edn. (pp 183–201) Dordrecht, The Netherlands: Kluwer Academic Publishers

Vassilopoulou, E., Zuidmeer, L., Akkerdaas, J., Tassios, I., Rigby, N.R., Mills, E.N.C., van Ree, R., Saxoni-Papageorgiou, P., & Papadopoulos, N.G. (2007). Severe immediate allergic reactions to grapes: Part of a lipid transfer protein-associated clinical syndrome. *International Archives of Allergy and Immunology, 143*, 92–102

Waters, E.J., Wallace, W., & Williams, P.J. (1991). Heat haze characteristics of fractionated wine proteins. *Am. J. Enol. Vitic., 42*, 123–127

Waters, E.J., Wallace, W., & Williams, P.J. (1992). Identification of heat-unstable wine proteins and their resistance to peptidases. *J. Agric. Food Chem., 40*, 1514–1519

Waters, E.J., Peng, Z., Pocock, K.F., Jones, G.P., Clarke, P., & Williams, P.J. (1994). Solid-state 13c nmr investigation into insoluble deposits adhering to the inner glass surface of bottled red wine. *J. Agric. Food Chem., 42*, 1761–1766

Waters, E.J., Peng, Z., Pocock, K.F., & Williams, P.J. (1995a). Proteins in white wine, I: Procyanidin occurrence in soluble proteins and insoluble protein hazes and its relationship to protein instability. *Aust. J. Grape Wine Res., 1*, 86–93

Waters, E.J., Peng, Z., Pocock, K.F., & Williams, P.J. (1995b). Proteins in white wine, II: Their resistance to proteolysis is not due to either phenolic association or glycosylation. *Aust. J. Grape Wine Res., 1*, 94–99

Waters, E.J., Shirley, N.J., & Williams, P.J. (1996). Nuisance proteins of wine are grape pathogenesis-related proteins. *J. Agric. Food Chem., 44*, 3–5

Waters, E.J., Hayasaka, Y., Tattersall, D.B., Adams, K.S., & Williams, P.J. (1998). Sequence analysis of grape (*Vitis vinifera*) berry chitinases that cause haze formation in wines. *J. Agric. Food Chem., 46*, 4950–4957

Waters, E.J., Alexander, G., Muhlack, R., Pocock, K.F., Colby, C., O'Neill, B.N., Høj, P.B., & Jones, P.R. (2005). Preventing protein haze in bottled wine. *Aust. J. Grape Wine Res., 11*, 215–225

Yokotsuka, K., Ebihara, T., & Sato, T. (1991). Comparison of soluble proteins in juice and wine from Koshu grapes. *J. Fermentation and Bioengineering, 71*, 248–253

Zalewska-Sobczak, J., Borecka, H., & Urbanek, H. (1981). Comparison of pectinase, xylanase and acid protease activities of virulent and less virulent isolates of *Botrytis cinerea. Phytopathologische Zeitschrift, 101*, 222–227

Zoecklin, B. (1988). Bentonite fining of juice and wine. VCESUSo America). Publication 463–014.

Chapter 7
Carbohydrates

M. Luz Sanz and Isabel Martínez-Castro

Contents

7.1 Introduction

Carbohydrates play an essential role in winery since they are the source of fermentation giving rise to the production of alcohol. They also contribute to the taste of wines conferring sweetness which varies from a background note in dry wines to a clear taste in sweet wines; they also generate derived-products as glycerol which gives body and roundness. Sugars also contribute to the aroma, through the formation of some volatile compounds.

The name "carbohydrates" is derived from their origin: they are formed through photosynthesis from CO_2 and water and their general formula is $C_n(H_2O)_n$ or $C_n(H_2O)_{n-x}$ when they are constituted by polymers. Together with lipids and proteins, carbohydrates are one of the main constituents of foods.

According to their polymerization degree, they can be classified as monosaccharides, disaccharides, trisaccharides, etc., and polysaccharides.

M. Luz Sanz (✉)
Instituto de Química Orgánica General (CSIC), Juan de la Cierva 3, 28006 Madrid, Spain
e-mail: mlsanz@iqog.csic.es

M.V. Moreno-Arribas, M.C. Polo (eds.), *Wine Chemistry and Biochemistry*,
DOI 10.1007/978-0-387-74118-5_10,

7.2 Chemical Structure

Monosaccharides are the basic pieces which constitute all higher carbohydrates. They are polyhydroxyaldehydes and polyhydroxyketones, with 3–8 carbon atoms. Sugars that contain four or more carbons can exist in a chain (open) form or a heterocyclic ring form. The carbonyl groups (aldehyde or ketone) react reversibly with an intramolecular hydroxyl to form hemiacetals, giving cyclic forms of five members (furanoses) or six members (pyranoses). When a reducing sugar is in solution, the different forms (α and β, pyranose and furanose) spontaneously interconvert through the open form until they reach equilibrium. This process can take several hours at room temperature. The most common monosaccharides are glucose and fructose, in aqueous solution the most abundant isomers being the pyranose forms of glucose and the β anomers (both pyranose and furanose) of fructose (Fig. 7.1).

Other functional groups are also found in natural sugars such as carboxylic acids (aldonic, uronic, aldaric) able to form lactones (reversibly), amines or amides (aminosugars) and both acid and amine groups (sialic acids).

7.3 Properties

7.3.1 Isomerism

Carbohydrates display a really high degree of isomerism. It is possible to find:

- *Anomers*, stereoisomers which differ in the position of the *substituent* of the hemiacetal or acetal linkage (*anomeric center*). They are named α when exocyclic O at the anomeric center is on the opposite face to the $-CH_2OH$ group, and β when the exocyclic O at the anomeric center is on the same face as the $-CH_2OH$ group (See Fig. 7.1).
- *Enantiomers*, stereoisomers that are mirror images of each other like L- and D-arabinose (Fig. 7.2).

Most natural sugars are D-isomers (glucose, fructose, sucrose, cellulose), but there is a number of sugars appearing in natural products which correspond to L-series (arabinose, sorbose, rhamnose):

- *Diastereomers*, stereoisomers that are not enantiomers (e.g. D-ribose and D-arabinose; Fig. 7.3)
- *Epimers*, diastereomers that differ only in a single chiral carbon that is not the reference carbon, such as D-glucose and D-galactose which differ only in -OH orientation at C-4. (e.g. D-glucose and D-galactose are epimers in C-4; Fig. 7.4)
- *Regioisomers,* which differ in the position of a substituent on the ring, as maltose and kojibiose (Fig. 7.5)

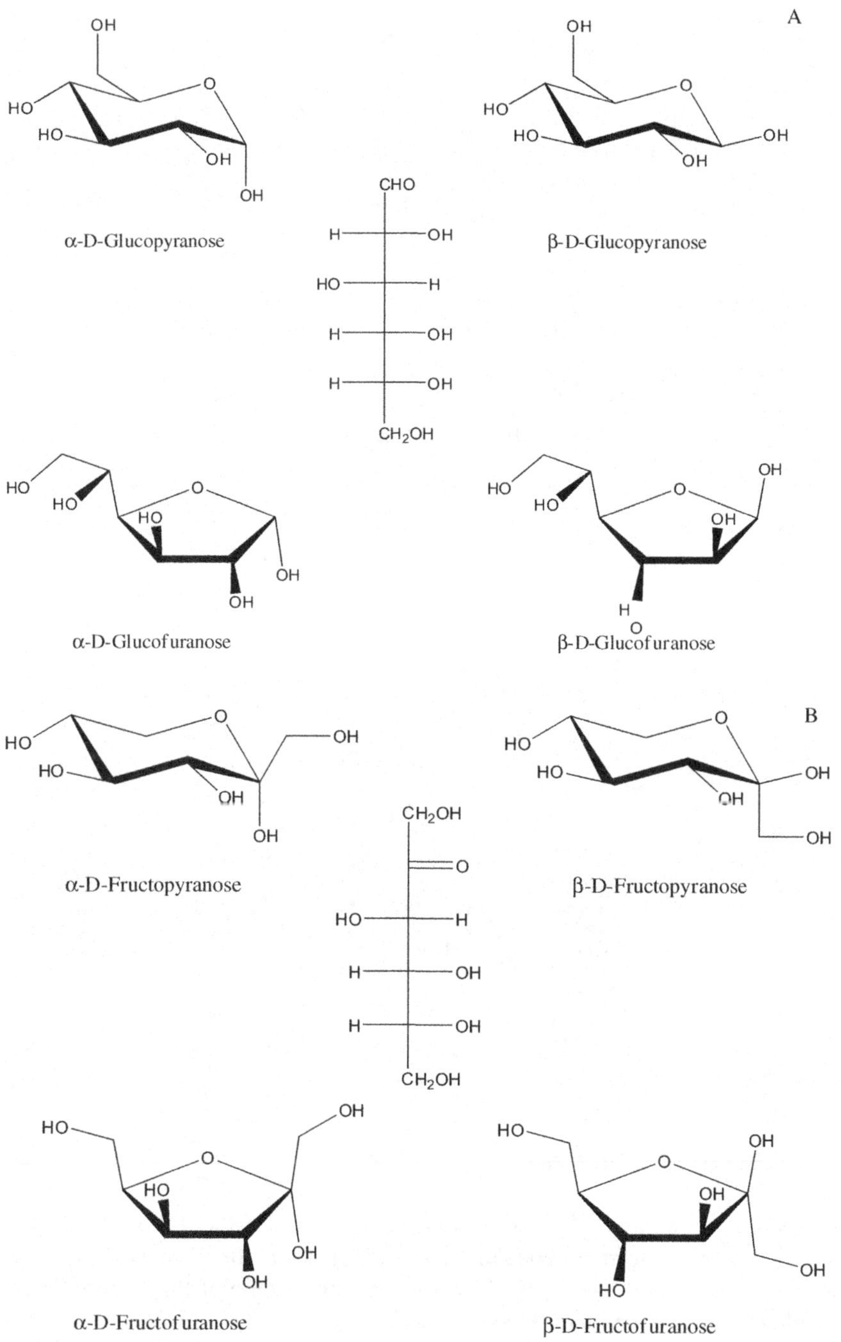

Fig. 7.1 Glucose (**A**) and fructose (**B**) tautomeric forms

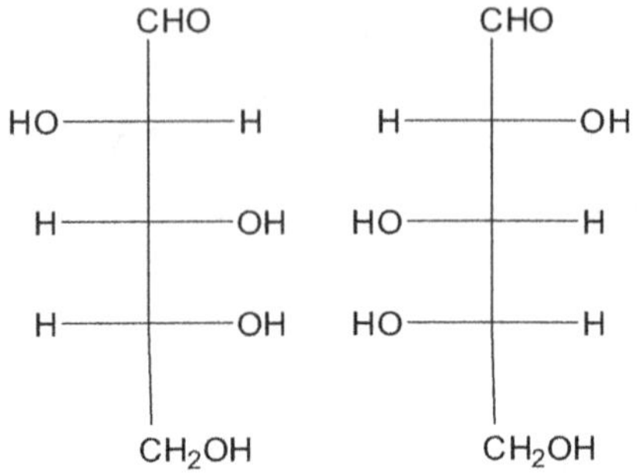

D-arabinose L-arabinose

Fig. 7.2 Enantiomeric forms of arabinose

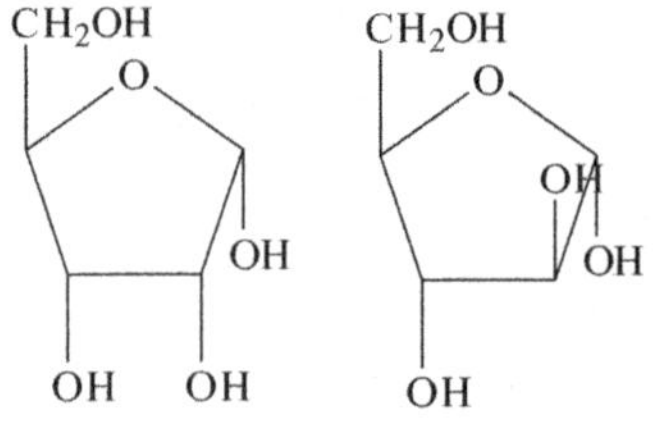

Ribose Arabinose

Fig. 7.3 Diastereomers: structures of ribose and arabinose

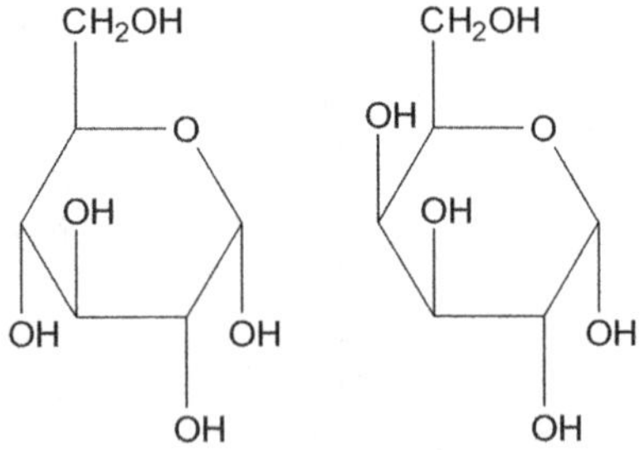

Glucose Galactose

Fig. 7.4 Epimers: structures of galactose and glucose

7.3.2 Chemical Reactivity

Carbohydrates are rather stable in solid state, at neutral pH and room temperature, but when they appear in aqueous matrices such as food, they can easily react with acids, bases and proteins to give different reactions. Some of them are relevant in wine chemistry and are mentioned below:

– *Reaction with acids:* in acid medium, disaccharides and polysaccharides may be hydrolyzed to monosaccharides, which in turn may be dehydrated and degraded

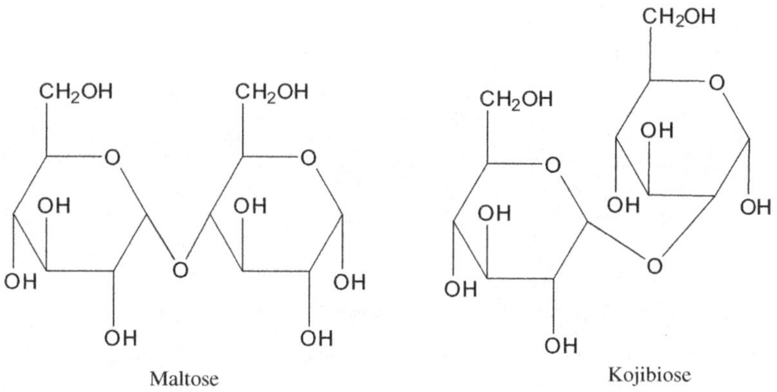

Fig. 7.5 Regioisomers: structures of maltose and kojibiose

to low-molecular weight compounds such as furanic compounds (furfural, HMF), cyclotene, etc., and also to coloured products. These reactions are accelerated by temperature giving rise to caramelization. In winery, caramelization can be found in some dessert wines (Kroh 1994) and in barrels where wood is toasted.

- *Reactions with bases:* in basic medium, reducing sugars undergo isomerization reactions through enediol intermediates (Lobry De Bruyn-Van Ekenstein reaction) so glucose is partially converted to fructose and mannose; a number of by-products is also produced.
- *Maillard reaction:* carbonyl compounds, mainly reducing sugars, can react with free amino groups (from peptides, proteins or amino acids) giving rise to Maillard reaction. This reaction takes place in a high number of foods and can produce not only the formation of desired colours and flavours but also some potentially toxic compounds. After alcoholic and malolactic fermentations in wine, dicarbonyl compounds are obtained, which are susceptible of participating in Maillard reaction (Pripis-Nicolau et al. 2000). The formation of some flavour components in different wine model systems (low pH, aqueous medium, and low temperatures) such as glucose with alanine, arginine and proline (Kroh 1994) or carbonyl (acetoin and acetol) and dicarbonyl (glyoxal, methylglyoxal, diacetyl and pentan-2,3-dione) compounds with different amino acids (Pripis-Nicolau et al. 2000; de Revel et al. 2004; Marchand et al. 2002) has been studied. Moreover, the influence of this reaction in the aroma of sweet fortified wines (Cutzach et al. 1999) and the production of Amadori compounds in Japanese white wines have been described (Hashiba 1978). Several by-products of Maillard reaction are identical or similar to some produced in degradation reactions catalysed by acids or bases.
- *Oxidation*: free carbonyl groups are able to reduce alkaline solutions of metal (mainly copper) salts to the free metal or to the oxide, to give aldonic acids. Thus glucose gives rise to gluconic acid (Fig. 7.6). When the hydroxyl group

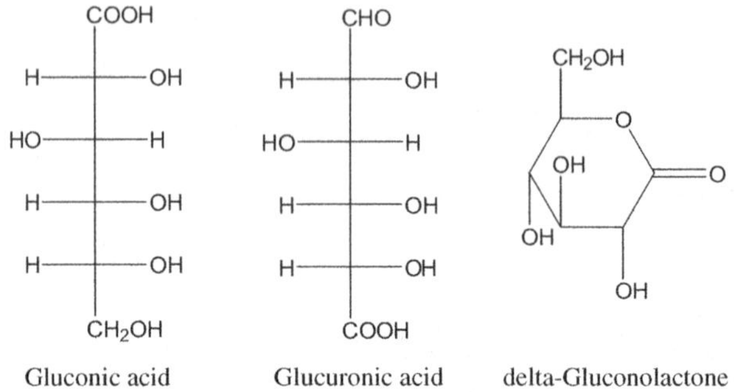

Fig. 7.6 Structures of gluconic acid, glucuronic acid and delta-gluconolactone

in position 6 of the chain is oxidized, the compound formed is a uronic acid (Fig. 7.6).

Aldonic acids are in equilibrium with their lactones, formed by reversible esterification of the carboxyl group with a hydroxyl in 4 or 5 position (gamma and delta lactones) (e.g. delta-gluconolactone; Fig. 7.6).

Both ends of an aldose chain can be simultaneously oxidized to carboxylic acids forming aldaric acids (e.g. galactose to galactaric acid; Fig. 7.7).

- *Reduction*: the carbonyl group of an aldose can be reduced with sodium borohydride to give a polyalcohol (alditol) with formula $HOCH_2(CHOH)_nCH_2OH$. When a keto group is reduced, two isomeric alditols are formed (e.g. whereas glucose is reduced to glucitol, fructose gives a mixture of glucitol and mannitol; Fig. 7.8).

Fig. 7.7 Structures of galactose and galactaric acid

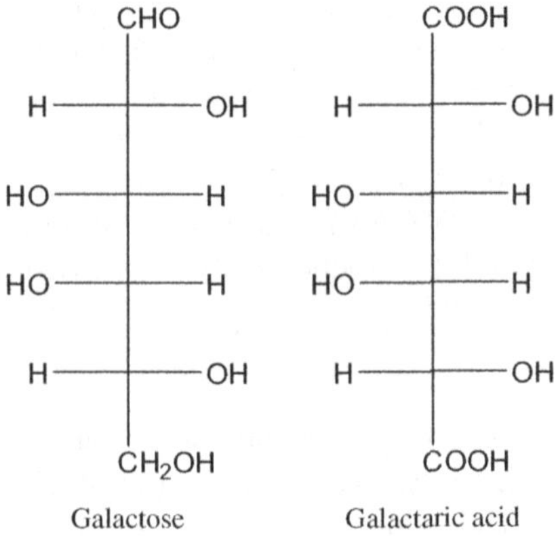

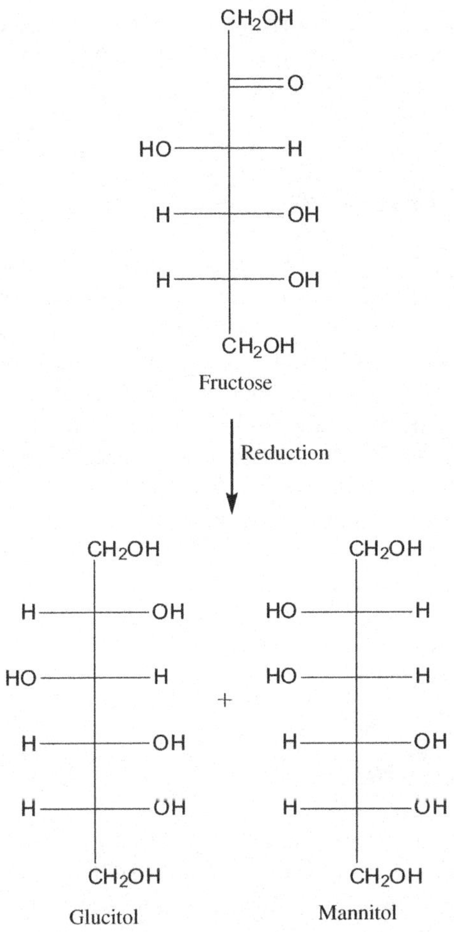

Fig. 7.8 Glucitol and mannitol produced after reduction of fructose

Reducing power of sugars is the basis of a number of classical analytical methods. Iodine, alkaline ferricyanide and other reagents have been used for analytical purposes.

- *Glycoside formation*: hemiacetals react with alcohols to form acetals, called glycosides. Acetal formation locks the ring structure, preventing oxidation-reduction and mutarotation reactions. When a carbonyl group reacts with a hydroxyl group from other sugar molecule, a disaccharide is created: the formed acetal is stable and it is called "glycosidic linkage". Monosaccharides can be polymerized through glycosidic linkages to form disaccharides, trisaccharides and polysaccharides.

– *Derivatization:* the active hydrogen of all hydroxyl groups can be substituted by non-polar groups such as methyl, acetyl or trimethylsilyl. The obtained derivatives are relevant for chromatographic analysis.

7.3.3 Biochemical Reactions

Sugars (mainly glucose and fructose) are converted to ethanol, carbon dioxide and several by-products by the action of yeast during alcoholic fermentation, the most important process in winemaking. Carbohydrates also participate in different biochemical reactions where the action of different enzymes (pectinases, cellulases, hemicellulases, glycosidases, etc.) is involved. As a result of enzymatic hydrolysis, different carbohydrate fractions are produced, whereas the formation of several by-products takes place. Polysaccharides are also affected by malolactic fermentation due to the high production of glycosidase activity by lactic acid bacteria which can hydrolyse their glycosidic bonds (Dols-Lafargue et al. 2007).

The addition of enzymes during winemaking is a common practice to improve the extraction of colour and aroma compounds, the clarification of the white wines and the filtration processes of musts and wines (Doco et al. 2007). More information about the behaviour of enzymes during winemaking is described in Chapter 4 (Enzymes in winemaking).

7.3.4 Physical Properties

Solubility: whereas monosaccharides are very soluble in water and in polar solvents such as alcohols, pyridine and dimethylsulfoxide, solubility decreases when molecular weight increases; thus some polysaccharides are totally insoluble. Formation of some polysaccharides in wines can produce defects (e.g. dextrans appearing in wines cause ropiness).

Taste: sweetness is the most characteristic property of soluble sugars. Sucrose is the reference substance for sweet taste; fructose is sweeter and glucose is less sweet than sucrose. There are some exceptions like gentiobiose which has bitter taste. Sweetness decreases when molecular weight increases, and the higher oligosaccharides are tasteless (Table 7.1). Polyalcohols also have sweet taste.

Table 7.1 Sweetness of some carbohydrates of wine (relative to that of sucrose)

Compound	Relative sweetness
D-Fructose	114
D-Glucose	69
Sucrose	100
α,α-Trehalose	45
Xylitol	100
Sorbitol	55
Mannitol	50

Table 7.2 Rotary power of some carbohydrates of wine

Compound	$[\alpha]_D$
D-Fructose	−92
D-Glucose	+52.7
α-D-glucose	+112
β-D-glucose	+18.7
Sucrose	+66.5
β-Cyclodextrin	+162

Rotary power: the specific rotation $[\alpha]_D$ is defined using the sodium D-line at 20–25 °C. It is not related to molecular weight, and is very characteristic of every sugar in solution (see Table 7.2). It depends on concentration, temperature and solvent. When a crystalline sugar is dissolved, the rotary power of the solution changes until the equilibrium is reached (mutarotation). Each anomeric form has their own rotary power, and the equilibrium value reflects the individual values of all present forms.

Crystallization: sugars may easily crystallize. The crystals are usually stable at room temperature, although in general are hygroscopic and they have to be held in a closed vessel. Sugars are present in wines at very low concentration; thus they remain soluble. Nevertheless some problems can arise when some insoluble crystals appear, such as mucic acid (galactaric).

7.4 Mono- and Disaccharides

Wine contains several sugars. Their structure and the reported concentration ranges are summarized in Table 7.3.

D-Glucose is the most abundant monosaccharide in nature, and it plays a central role in biochemistry, since it is the primary fuel for living cells. D-Fructose occurs in free form in many fruits and honey, as well as forming polysaccharides (inulin and fructans) in several plants. D-Glucose and D-fructose are the main sugars in must. Although both decrease during fermentation, the ratio fructose/glucose increases since glucose is the preferred substrate of many types of yeast.

D-Mannose is present in polysaccharides of plants (mannans) and also in combination with other sugars. Mannoproteins are considered one of the main polysaccharides of wines as it will be described later.

D-Galactose appears free in nature in low amounts, but it forms part of lactose in milk and appears in complex biomolecules as glycolipids and glycoproteins. Its content seems to be higher in wines aged in contact with lees (Doco et al. 2003).

L-Arabinose is found in free form in *Coniferae*; it appears in combined form in bacterial polysaccharides, pectic materials, hemicelluloses and plant glycosides.

Both galactose and arabinose contents were found to be higher in wines made from selected *Botrytis cinerea*-infected grapes than in those made from healthy grapes (Dittrich and Barth 1992).

D-Ribose is a constituent of nucleic acids, hence it is present in all plant cells; it is also found in several coenzymes.

Table 7.3 Neutral sugars content in different wines. Range in mg/L

	Sweet white (Liu and Davis 1994)	Medium-dry (Liu and Davis 1994)	Red (Liu and Davis 1994)	Red (del Alamo et al. 2000)	Red (Bernal et al. 1996)	Italian red (Cataldi and Nardiello 2003)	Italian white (Cataldi and Nardiello 2003)	Red (Usseglio-Tomasset and Amerio 1978)
Ribose	321	24	nd–71	15—21	nd–28	–	–	tr–31
Rhamnose	341	23	nd–76	10–15	–	–	–	2–15
Arabinose	2050	510	10–37	5–20	19–179	35–120	5–40	10–134
Fucose	313	13	nd	–	–	–	–	–
Xylose	1078	25	nd–11	23–33	11–58	5–45	5–27	6–24
Mannose	–	–	31–73	6–11	nd–11	–	–	–
Glucose	10170	9690	150–400	13–36	5–145	5–220	20–162	–
Galactose	80	–	140–940	28–41	25–79	22–85	10–65	tr–119
Fructose	–	–	–	16–50	nd–86	23–315	25–178	–
Sucrose	60	120	20–110	–	–	–	–	–
α,α-Trehalose	550	420	8–37	–	–	–	–	–

nd: not detected

tr: traces

D-Xylose is an important component of wood. It is only fermentable by certain microorganisms such as *Lactobacilli, Torula* and *Monilia*.

L-Rhamnose is a constituent of many glycosides and polysaccharides, whereas L-fucose appears in polysaccharides and glycoproteins.

Sugars in red wines are usually higher than in white wines (Ribereau-Gayon et al. 2006). This can be due to the contact with grape skins, to the longer aging which liberate sugars from polyphenol conjugates and also to the hydrolysis of hemicellulose when wine is aged in barrels. Monosaccharides, especially galactose, fructose and xylose vary during aging depending on the type of oak wood (del Alamo et al. 2000). This variation has been widely studied in brandies (Viriot et al. 1993).

The infection with *Botrytis cynerea* is responsible for certain changes in the carbohydrate composition of musts and wines, mainly by oxidation. Botrytized wines contain xylosone and 5-oxo-fructose (Ribereau-Gayon 1973), a hexodiulose which is formed by oxidation of fructose; it has been found in botrytized musts varying within 80–150 mg/L (Barbe et al. 2000, 2002). Acid sugars also increase (see below).

Aging in oak barrels changes the proportion of sugars in wine. During aging average levels of galactose, fructose and xylose were higher than other monosaccharides; type of oak and barrel manufacturer affected the monosaccharide composition of the aging wine, indicating structural differences between French and American oak hemicelluloses (del Alamo et al. 2000).

α,α-Trehalose is a non-reducing disaccharide formed by two units of glucose (1-*O*-α-D-glucopyranosyl-1-*O*-α-D-glucopyranoside). It is present at low concentration in fungi and plants. It is the main disaccharide in wines (Bertrand et al. 1975) and it is formed as a result of the metabolic activity of yeasts. Its level varies within 0–611 mg/L in wines (Bertrand et al. 1975) and within 0–53 mg/L in sherries (Santa-María et al. 1983)

Other disaccharides identified in wine (sucrose, isomaltose, lactose and turanose) are present in small amounts, seldom >50 mg/L and sometimes <5 mg/L (Bertrand et al. 1975). Sucrose appears in wines at very low levels. It is added occasionally to musts (chaptalization).

The presence of pentosylhexosides, trisaccharides and tetrasaccharides has been reported in different wines using positive- and negative-ion electrospray ionization Fourier transform ion cyclotron resonance mass spectrometry (Cooper and Marshall 2001). Nevertheless, at present there are still a number of minor carbohydrates in wines without a conclusive identification.

7.5 Polyalcohols

Wine contains different polyalcohols as shown in Table 7.4.

Tetritols and pentitols are not very frequent in nature, but some of them have been isolated from certain plants, algae and fungi. Hexitols appear more frequently in

Table 7.4 Polyalcohols in wine. Range in mg/L

	Different wines (De Smedt et al. 1981)	Sweet white (Liu and Davis 1994)	Medium-dry white (Liu and Davis 1994)	Red (Liu and Davis 1994)	Sherries (Estrella et al. 1986)
Erythritol	22–208	214	59	–	105–325
Threitol	–	25	7	17–19	–
Ribitol	–	23	3	106–140	–
Arabitol	0–266	18	8	83–108	17–42
Xylitol	–	37	nd	nd–15	22–149
Glucitol (Sorbitol)	0–163	237	8	4–138	tr–134
Mannitol	31–420	116	53	194–493	tr–731
Galactitol	nd–9	nd	nd	nd–9	–

nd: not detected

superior plants, especially mannitol. The alditols appearing in wine are considered products of fermentation.

Inositols (cyclohexanehexaols, cyclitols) are cyclic polyalcohols with general formula $C_6H_{12}O_6$ (as monosaccharides) but they are markedly stable to heat, acids and alkalis. They are also difficult to ferment.

Myo-inositol (formerly called *meso*-inositol) is present, free or combined in the tissues of all living species, excepting maybe a few bacteria. *Scyllo*-inositol has been reported in grape juice and wine (De Smedt et al. 1981; Versini et al. 1984); it has been proposed as genuinity control (Versini et al. 1984). *Chiro*-inositol has been reported in citric juices (Sanz et al. 2004) and in several wines (Carlavilla et al. 2006). Their (Table 7.5) structures appear in Fig. 7.9.

Quercitol is a deoxyinositol (1,3,4/2,5-cyclohexanepentol) which appears in certain Dicotyledons including all the *Quercus* genus (Plouvier 1963). This compound has been found in wines aged in contact with oak wood (barrels or chips) but not in wines aged in bottles (Carlavilla et al. 2006).

Composition in polyalcohols of oenological tannis (See Table 7.5) from different origins (oak, grape, quebracho, gall plant, etc.) has been recently determined (Sanz et al., 2008). These compounds appear to be useful markers to differentiate among these botanical origins.

Table 7.5 Cyclitols in wine. Range in mg/L

	Different wines (De Smedt et al. 1981)	White and sherries in oak (Carlavilla et al. 2006)	Red aged in oak (Carlavilla et al. 2006)	Wines without oak contact (Carlavilla et al. 2006)
Chiro-inositol	–	2–12	18–20	0–18
Scyllo-inositol	10–143	13–59	54–67	8–71
Myo-inositol	79–1044	348–478	306–412	199–427
Quercitol	–	15–39	19–69	nd

nd: not detected

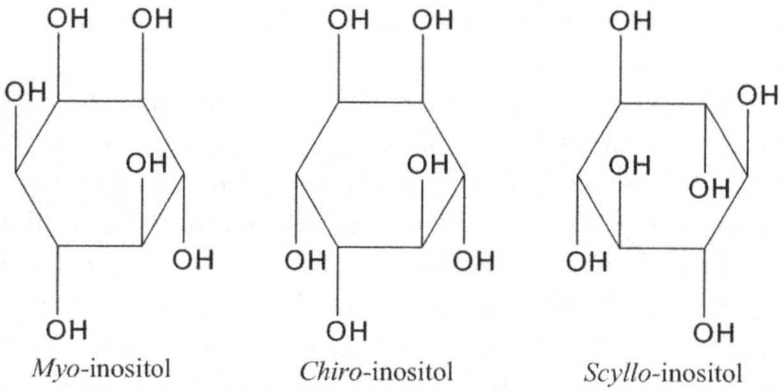

Fig. 7.9 Structures of the different inositols appearing in wines

2,3-Butanediol is a constituent of wine and an important neutral compound formed by yeast fermentation of carbohydrates. *Saccharomyces cerevisiae* produces mainly D(–)-2,3-butanediol and small amount of the *meso* form. Wines that had undergone malolactic fermentation contained traces of L(+)-2,3-butanediol which was absent from wines that had not undergone malolactic fermentation (Herold et al. 1995).

7.6 Sugar Acids

Sugar acids come from enzymatic oxidation of sugars, especially important in wines affected by *Botrytis cynerea* and by rot in general; acetic bacteria also produce these compounds. Some data about their level in wines are presented in Table 7.6.

Gluconic acid is produced by oxidation from glucose; gamma- and delta-gluconolactones (glucono-1,4-lactone and glucono-1,5-lactone, respectively) are formed by intramolecular gluconic acid esterification reactions. Both lactones are

Table 7.6 Acid sugars in wine. Range in mg/L

	Different wines (De Smedt et al. 1979)	White (Sponhloz and Dittrich 1985)	Red (Sponhloz and Dittrich 1985)	Sherries (Sponholz and Dittrich 1984)	White (Barbe et al. 2000)
Gluconic	0–2000 (including lactones)	–	–	–	–
Gluconolactone	–	–	–	–	155–270
Glucuronic	–	0–35	0–33	nd	–
Galacturonic	–	116–1048	381–1200	365–517	–
5-Oxogluconic	–	8–92	14–74	tr–5.3	–
2-Oxogluconic	–	0–122	0–17	nd–tr	–
2-Oxoglutaric	–	–	–	–	78–248

nd: not detected
tr: traces

in equilibrium with gluconic acid, and represent 5.8% and 4.1% of the acid level, respectively (Barbe et al. 2002).

The origin of glucuronic, galacturonic, 5-oxogluconic and 2-oxogluconic acids in musts and wines was showed by Sponholz and Dittrich (1985). They compared healthy and *Botrytis cynerea* infected products and found that the production of sugar acids was proportional to the number of infected berries. Application of pure cultures demonstrated that acetic bacteria (*Gluconobacter oxydans* and *Acetobacter xylinum*) produced most of the sugar acids found in botrytized wines.

Sugar acids have been related to the binding of SO_2 in wines (Ribereau-Gayon 1973); gluconic acid appeared to be indirectly responsible for about 8% of the bindable SO_2 in musts from botrytized grapes (Barbe et al. 2002).

The presence of mucic acid in wine causes calcium mucate crystals which are formed during prolonged storage when wine contains >0.1 g/L mucic acid. This level is reached when 10–25% of the grapes used are infected with *B. cinerea* (Wuerdig 1977).

7.7 Polysaccharides

Increased attention has been paid in the last few years to the study of polysaccharides in wine and great advances in the knowledge of their structures and intrinsic properties have been achieved (Vidal et al. 2003; Ayestarán et al. 2004).

Polysaccharides are considered those carbohydrates with a degree of polymerization higher than 20. Their presence in wines is due to the contribution of the cell walls of either microorganisms during alcoholic fermentation or grape berries after degradation by pectic enzymes during grape maturation or during winemaking. Polysaccharides constitute one of the main groups of macromolecules in wine and contribute to increase its viscosity and stability. Moreover, polysaccharides have been linked to the organoleptical qualities of wines (Vidal et al. 2003) because their interactions with other constituents such as polyphenols (Riou et al. 2002), aromatic compounds (Chalier et al. 2007), etc. The main polysaccharides coming from berries cell walls (pectic polysaccharides) are rhamnogalacturonans II (RG-II; Pellerin et al. 1996) and arabinogalactan-proteins (AGPs; Brilloue et al. 1990) whereas those from yeast cell walls are mainly mannoproteins (MPs) and mannans (Waters et al. 1994).

After their isolation by chromatographic techniques (anion-exchange chromatography, size exclusion, etc.), different analytical methodologies have been used to identify and quantify the polysaccharides in wine; the most commonly used being the traditional methylation analysis followed by GC-MS (Doco and Brillouet 1993). Polysaccharides have also been determined after solvolysis with anhydrous methanol containing HCl by GC-MS of their per-*O*-trimethylsilylated methyl glycosides (Vidal et al. 2003). Other techniques such as Fourier transform infrared spectroscopy (FTIR) have been more recently proposed (Coimbra et al. 2002, 2005; Boulet et al. 2007).

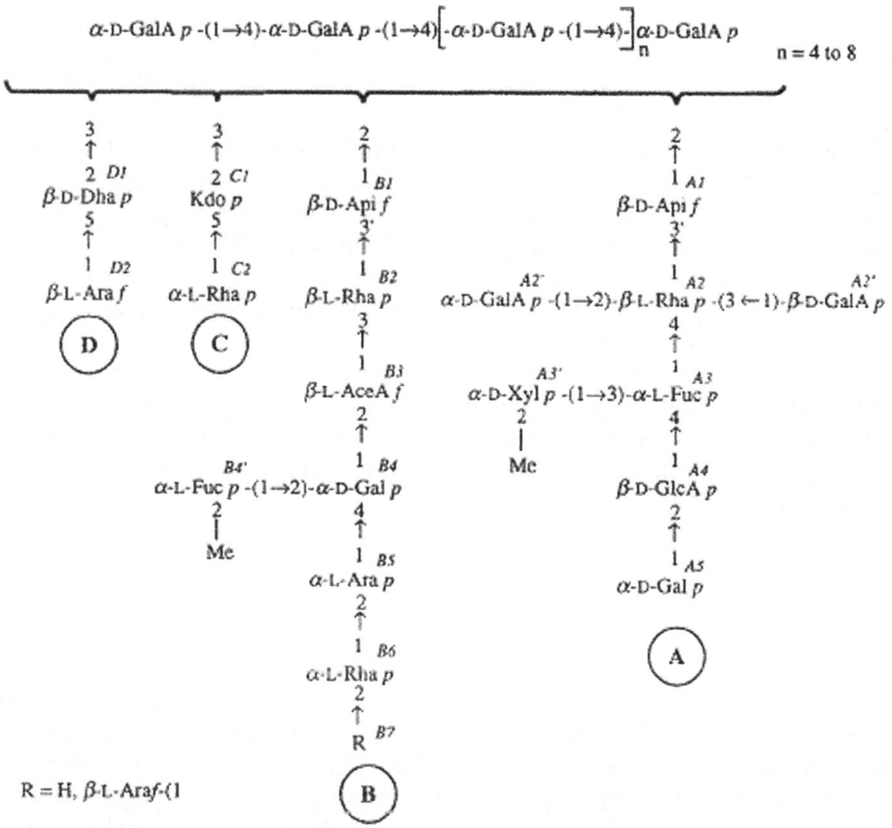

Fig. 7.10 Hypothetical structure of RG-II. The figure shows the four oligoglycosidic side chains A–D whose residues are numbered according to the model sequence. Reprinted from Pellerin et al. (1996), with permission from Elsevier

As shown in Fig. 7.10, RG-II contains 12 different glycosidic residues such as apiose, arabinose, galactose, 2-*O*-methyl-L-fucose, 2-*O*-methyl-D-xylose, 3-deoxy-D-*manno*-octulosic acid, 3-deoxy-D-*lyxo*-heptulosaric acid, and aceric acid among others, the main chain being formed by galacturonic acids linked at α-(1,4) positions (Boulet et al. 2007). It constitutes the 19% of total polysaccharides in wine (Vidal et al. 2003). This carbohydrate can lead to the formation of precipitates and haze due to their possible electrostatic and ionic interactions with other wine constituents (Pellerin et al. 1996). RG-I is also present in wines although it stands only for the 4% of total polysaccharides (Vidal et al. 2003). This carbohydrate is mainly constituted by rhamnose and galacturonic acid among others and it has been found to be linked with xyloglucans.

AGPs have been found to represent more than 40% of total red wine polysaccharides. They are mainly constituted by glycosidic residues of arabinose, galactose and glucuronic acid although other monosaccharides such as rhamnose, glucose and mannose are also present. Proteins represent less than 10% of AGPs (Pellerin et al. 1995).

Arabinans mainly constituted by short arabinose chains bonded in α-(1,5) can also be present (Belleville et al. 1993).

MPs are constituted by 90% of mannose, protein and phosphoric acids and represent the 35% of total polysaccharides in wine (Vidal et al. 2003). MPs combined with phenolic compounds have shown an indirect effect on astringency, although their stabilizing effect on protein precipitation in white wine and tartrate crystallization in both red and white wines are their main function (Ribereau-Gayon et al. 2006). *Glucomannoproteins* have been also detected in wines in lower amount than MPs (Ribereau-Gayon et al. 2006).

Glucans (Dubourdieu et al. 1981) from *Botrytis cinerea* have also been described in infected wines and can caused serious problems during their clarification. Botrytized wines also contain heteropolysaccharides with higher amounts of mannose. Other glucans produced by *Pediococcus* can cause ropiness in wines (Ribereau-Gayon et al. 2006).

7.8 Conclusions

Carbohydrates are minor components of wine which contribute to sensorial properties and play and important role in the different reactions occurring during fermentation and aging. Whereas monosaccharides and polyalcohols are rather well-known, more research is necessary on disaccharides, oligosaccharides and non-phenolic glycosides. The presence of cyclitols in wines and wine derivatives can constitute a useful tool to characterize their origin. Among all the studies related to carbohydrates in wine, polysaccharides have been the most important focus in the last years.

References

Ayestarán, B., Guadalupe, Z., & Leon, D. (2004). Quantification of major grape polysaccharides (*Tempranillo* v.) released by maceration enzymes during the fermentation process. *Anal. Chim. Acta, 513*, 29–39.

Barbe, J.C., de Revel, G., Joyeux, A., Lonvaud-Funel, A., & Bertrand, A. (2000). Role of carbonyl compounds in SO_2 binding phenomena in musts and wines from botrytized grapes. *J. Agric.Food Chem., 48*, 3413–3419.

Barbe, J.C., de Revel, G., & Bertrand, A. (2002). Gluconic acid, its lactones, and SO_2binding phenomena in musts from botrytized grapes. *J. Agric. Food Chem., 50*, 6408–6412.

Belleville, M.P., Williams, P., & Brillouet, J.M. (1993). A linear arabinan from a red wine. *Phytochem., 33*, 227–229.

Bernal, J.L., Del Nozal, M.J., Toribio, L., & Del Alamo, M. (1996). HPLC Analysis of carbohydrates in wines and instant coffees using anion exchange chromatography coupled to pulsed amperometric detection. *J. Agric. Food Chem.,44*, 507–511.

Bertrand, A., Dubernet, M.O., & Ribéreau-Gayon, P. (1975). Le tréhalose, principal diholoside des vins. *C. R. Acad. Sc. Paris series D*, 1907–1910.

Boulet, J.C., Williams, P., & Doco, T. (2007). A Fourier transform infrared spectroscopy study of wine polysaccharide. *Carbohydr. Polym., 69*, 79–85.

Brillouet, J.M., Bosso, C., & Moutounet, M. (1990). Isolation, purification, and characterization of an arabinogalactan from a red wine. *Am. J. Enol. Vitic., 41*, 29–36.

Carlavilla, D., Villamiel, M., Martínez-Castro, I., & Moreno-Arribas, M.V. (2006). Occurrence and significance of quercitol and other inositols in wine during oak wood aging. *Am. J. Enol. Vitic., 57*, 468–473.

Cataldi, T.R.I., & Nardiello, D. (2003). Determination of free proline and monosaccharides in wine samples by high-performance anion-exchange chromatography with pulsed amperometric detection (HPAEC-PAD). *J. Agric. Food Chem., 51*, 3737–3742.

Chalier, P., Angot, B., Delteil, D., Doco, T., & Gunata, Z. (2007). Interaction between aroma compounds and whole mannoprotein isolated from *Saccharomyces cerevisiae* strains. *Food Chem., 100*, 22–30.

Coimbra, M.A., Goncalves, F., Barros, A., & Delgadillo, I. (2002). FT-IR spectroscopy and chemometric analysis of white wine polysaccharide extracts. *J. Agric. Food Chem., 50*, 3405–3411.

Coimbra ,M.A., Barros A.S., Coelho E., Gonçalves F., Rocha S.M., & Delgadillo I. (2005). Quantification of polymeric mannose in wine extracts by FT-IR spectroscopy and OSC-PLS1 regression *Carbohydr. Polym., 61*, 434–440.

Cooper, H.J., & Marshall, A.G. (2001). Electrospray ionization fourier transform mass spectrometric analysis of wine. *J. Agric. Food Chem., 49*, 5710–5718.

Cutzach, I., Chatonnet, P., & Dubourdieu, D. (1999). Study of the formation mechanisms of some volatile compounds during the aging of sweet fortified wines. *J. Agric. Food Chem., 47*, 2837–2846.

De Revel, G., Marchand, S., & Bertrand, A. (2004). Identification of Maillard-type aroma compounds in winelike model systems of cysteine-carbonyls: Occurrence in wine *Nutraceutical Beverages: Chemistry, Nutrition, and Health Effects. ACS Symposium Series 871*, 353-364.

De Smedt, P., Liddle, P.A.P., Cresto, B., & Bossard, A. (1979). Application de la CPG sur colonne capillaire à l'analyse des composés fixes du vin. *Annal. Falsif. Expertise Chim. Toxicol., 72*, 633–642.

De Smedt, P., Liddle, P.A.P., Cresto, B., & Bossard, A. (1981). The analysis of non-volatile constituents of wine by glass capillary gas chromatography. *J. Inst. Brew., 87*, 349–351.

Del Alamo, M., Bernal, J.L., del Nozal, M.J., & Gómez-Cordovés, C. (2000). Red wine aging in oak barrels: evolution of the monosaccharides content. *Food Chem., 71*, 189–193.

Dittrich, H.H., & Barth, A. (1992). Galactose und Arabinose in Mosten und Weinen der Auslese-Gruppe. *Wein-Wissenschaft, 47*, 129–131.

Doco, T., & Brillouet, J.M. (1993). Isolation and characterisation of a rhamnogalacturonan II from red wine. *Carbohydr. Res., 243*, 333–343.

Doco, T., Vuchot, P., Cheynier, V., & Moutounet, M. (2003). Structural modification of wine arabinogalactans during aging on lees. *Am. J. Enol. Vitic., 54*, 150–157.

Doco, T., Williams, P., & Cheynier, V. (2007). Effect of flash release and pectinolytic enzyme treatments on wine polysaccharide composition. *J. Agric. Food Chem., 55*, 6643–6649.

Dols-Lafargue, M., Gindreau, E., Le Marrec, C., Chambat, G., Heyraud, A., & Lonvaud-Funel, A. (2007). Changes in red wine soluble polysaccharide composition induced by malolactic fermentation. *J. Agric. Food Chem., 55*, 9592–9599.

Dubourdieu, D., Ribéreau-Gayon, P., & Fournet, B. (1981). Structure of the extracellular β-D-glucan from *Botrytis cinerea. Carbohydr. Res., 93*, 294–299.

Estrella, I., Hernández, M.T., & Olano, A. (1986). Changes in polyalcohol and phenol compound contents in the ageing of sherry wines. *Food Chem, 20*, 137–152.

Hashiba, H. (1978). Isolation and identification of Amadori compounds from miso, white wine and sake. *Agric. Biol. Chem., 42*, 1727–1731.

Herold, B., Pfeiffer, P., & Radler, F. (1995) Determination of the three isomers of 2,3-butanediol formed by yeasts or lactic acid bacteria during fermentation. *Am. J. Enol Vitic., 46*, 134–137.

Kroh, L.W. (1994). Caramelisation in food and beverages. *Food Chem., 51*, 373–379.

Liu, S.Q., & Davis, R.C. (1994). Analysis of wine carbohydrates using capillary gas liquid chromatography. *Am. J. Enol. Vitic., 45*, 229–234.
Marchand, S., de Revel, G., Vercauteren, J., & Bertrand, A. (2002). Possible mechanism for involvement of cysteine in aroma production in wine. *J. Agric. Food Chem., 50*, 6160–6164.
Pellerin, P., Vidal, S., Williams, P., & Brillouet, J. M. (1995). Characterization of five type II arabinogalactan-protein fractions from red wine of increasing uronic acid content. *Carbohydr. Res., 277*, 135–143.
Pellerin, P., Doco, T., Vidal, S., Williams, P., Brillouet J.M., & O'Neill M.A. (1996). Structural characterization of red wine rhamnogalacturonan II. *Carbohydr. Res., 290*, 183–197.
Plouvier, V. (1963). Sur la recherche des itols a chaine droite et des cyclitols chez les végétaux. Relations entre leur répartition et la classification systématique. *Bull. Soc. Chim. Biol., 45*, 1079–1118.
Pripis-Nicolau, L., de Revel, G., Bertrand, A., & Maujean, A. (2000). Formation of flavor components by the reaction of amino acid and carbonyl compounds in mild conditions *J. Agric. Food Chem., 48*, 3761–3766.
Ribereau-Gayon, P. (1973). Reactions of sulphur dioxide in wine. *Bull. l'O.I.V., 46*, 406–416.
Ribéreau-Gayon, P., Glories, Y, Maujean, A., & Dubordieu, P. (2000). Glucids. In John Wiley and Sons Ltd. *Handbook of Enology* vol 2. (pp. 65–90). Baffins Lane, Chichester, West Sussex PO19 1UD, England.
Riou, V., Vernhet, A., Doco, T., & Moutounet, M. (2002). Aggregation of grape seed tannins in model wine-effect of wine polysaccharides. *Food Hydrocoll., 16*, 17–23.
Santa-María, G., Olano, A., & Tejedor, M. (1983). Quantitative determination of trehalose and inositol in white and red wines by gas liquid chromatography. *Chem. Mikrobiol. Technol. Lebensm., 9*, 123–126.
Sanz, M.L., Villamiel, M., & Martínez-Castro, I. (2004). Inositols and carbohydrates in different fresh fruit juices. *Food Chem., 87*, 325–328.
Sanz, M.L., Martínez-Castro, I., & Moreno-Arribas M.V. (2008) Identification of the origin of commercial enological tannins by the analysis of monosaccharides and polyalcohols. *Food Chem., 111*, 778–783.
Sponholz, W.R., & Dittrich, H.H. (1984). Galacturonic, glucuronic, 2-and 5-oxogluconic in wines, sherries, fruit and dessert wines. *Vitis, 23*, 214–224.
Sponholz, W.R., & Dittrich, H.H. (1985). Origin of gluconic acid, 2- and 5-oxogluconic, glucuronic and galacturonic acids in musts and wines. *Vitis, 24*, 51–58.
Usseglio-Tomasset, L., & Amerio, G .(1978). [Pentoses, rhamnose and galactose in dry wines.] *Vini d'Italia, 20*, (112), 27–33.
Versini, G., Dallaserra, A., & Margheri, G. (1984). [Polyalcohols and secondary sugars in concentrated rectified musts as genuineness parameters]. *Vignevini, 11*, 41–47.
Vidal, S., Williams, P., Doco, T., Moutounet, M., & Pellerin, P. (2003). The polysaccharides of red wine: total fractionation and characterization. *Carbohydr. Polym., 54*, 439–447.
Viriot, C., Scalbert, A., Lapierre, C., & Moutounet, M. (1993). Ellagitannins and lignins in aging of spirits in oak barrels. *J. Agric. Food Chem., 41*, 1872–1879.
Waters, E. J., Pellerin, P., & Brillouet, J. M. (1994). A *Saccharomyces* mannoprotein that protects wine from protein haze. *Carbohydr. Polym., 23*, 185–191.
Wuerdig, G. (1977). [Appearance of mucic acid in musts prepared from grapes contaminated with Botrytis.] *Bull. l'O.I.V., 50*, 50–56.

Chapter 8
Volatile and Aroma Compounds

Chapter 8A
Wine Aroma Precursors

Raymond Baumes

Contents

8A.1 Introduction

The aroma precursors herein described are non-volatile grape aroma precursors recognized to be related to the wine aroma varietal character. Indeed, grapes of most *Vitis vinifera* cultivars used in winemaking have no characteristic aroma, with a few exceptions, such as monoterpenoids in Muscat cultivars and some other aromatic cultivars, and 2-alkyl-3-methoxypyrazines in Cabernet-Sauvignon and related cultivars (Bayonove 1998). However, these grapes contain different groups of non volatile aroma precursors: unsaturated lipids, phenolic acids, carotenoids, *S*-cysteine conjugates, glycoconjugates and *S*-methylmethionine. These non-volatile, odorless constituents are susceptible to transformation into volatile varietal aroma compounds during the biotechnological sequence of wine, from the cellular disorganization of grape berries during harvest to the maturation of wine during its storage (Fig. 8A.1). They are generally secondary metabolites under genetic control, some of them associated with specific cultivars, but they are also dependent on the several factors which define the "terroir," e.g. situation, soil, climate, viticultural practices.

However, this chapter includes neither the main carbon, nitrogen and sulfur substrates of yeasts, giving rise to the volatile secondary products of alcoholic

R. Baumes (✉)
UMR Sciences pour l'Oenologie, 2 place Viala – 34060 Montpellier Cedex, France
e-mail: baumes@supagro.inra.fr

M.V. Moreno-Arribas, M.C. Polo (eds.), *Wine Chemistry and Biochemistry*,
DOI 10.1007/978-0-387-74118-5_11,

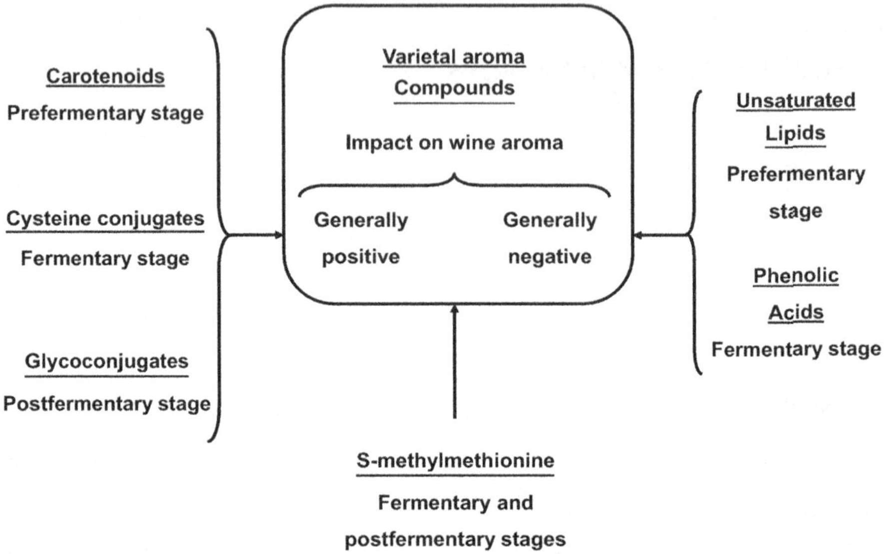

Fig. 8A.1 Wine aroma precursors, their main stages of degradation during the wine biotechnological sequence and general impact on wine aroma of the odorants generated

fermentation, constituents of the so-called fermentative aroma, nor those starting compounds, involved in more or less complex biological or chemical processes occurring under different wine aging conditions, such as Maillard reactions or biological and chemical oxidative aging. All these compounds are described elsewhere in this book.

8A.2 Unsaturated Fatty Acids

The majority of grape acyl lipids are esters of unsaturated fatty acids, and the major ones, linoleic and linolenic acid esters, are wine aroma precursors (Fig. 8A.2). However, their total contents in unsaturated fatty acids, about 350 mg/kg in grape berries, hardly depend on the cultivar and decrease with maturity (Bayonove 1998). These compounds, primarily located in the solid parts of the berry, are degraded by grape enzymes to the so-called C6-compounds, when grapes are crushed in air in the prefermentary stage; The powerful odorants, hexanal and 2-hexenal, are the major C6-compounds produced in grape must, and hexanol and hex-2- and 3-en-1-ols are minor products (Drawert et al. 1966; Roufet et al. 1986, 1987; Ferreira et al. 1995). The main enzymes involved in this degradation, are the grape lipoxygenase, which regiospecifically oxidizes only unsaturated lipids containing a 1-*cis*,4- *cis*-pentadiene system, to hydroperoxides with a *cis,trans*-diene system (linoleic and linolenic lipids), and grape hydroperoxide-lyase, which cleaves the hydroperoxides to the C6-aldehydes (Drawert et al. 1966). Thus, the potential in

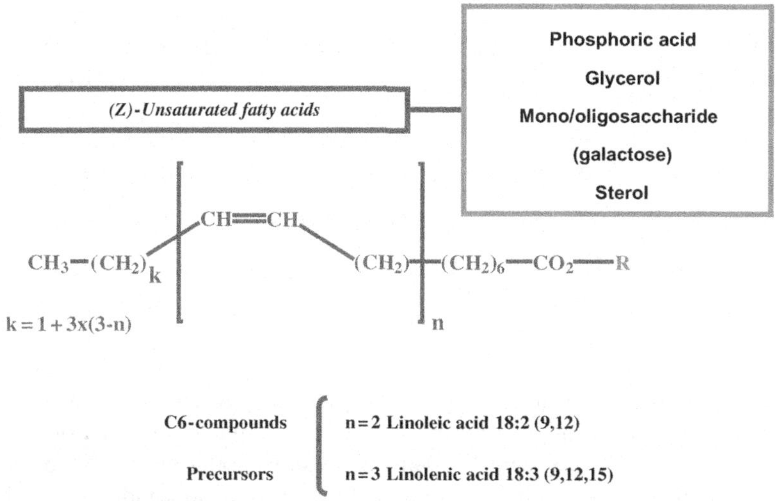

Fig. 8A.2 Major grape unsaturated acyl lipids, precursors of C6 compounds

these C6 compounds, odorants with green and grassy olfactive notes, depends not only on the contents of unsaturated lipids, but also on activities of the enzymes involved, and thus defined, some differences were observed between grape cultivars (Valentin 1993). However, these enzymatic activities are much more dependent on berry maturation than on grape cultivar. In addition, the C6-aldehydes are very powerful odorants in must, but they are almost totally reduced during the alcoholic fermentation by yeast to hexanol, which hardly exceeds its olfactive perception threshold in wine (Etiévant 1991). In contrast, the minor C6-compounds found in wine, hex-2- and 3-en-1- ols, are more powerful odorants than hexanol, but they usually occur in much lower levels and do not contribute to wine aroma, except for generating unpleasant herbaceous off-odors (Joslin and Ough 1978).

8A.3 Phenolic Acids

Another class of aroma precursors, phenolic acids, presents a similar case. These hydroxycinnamic acids occur mostly in grape as esters with tartaric acid (a hydroxyl group of tartaric acid is esterified by the phenolic acid), the E configuration of the cinnamic double bond being preferred (Fig. 8A.3) (Singleton et al. 1978). The most abundant tartaric esters of grape hydroxycinnamic acids are caftaric acid (<800 mg/kg), ester of caffeic acid, coutaric acid (<300 mg/kg), ester of coumaric acid, and fertaric acid (<60 mg/kg), ester of ferulic acid. These compounds are primarily located in the solid parts of the berry, and their contents decrease with maturation (Ribéreau-Gayon 1965; Romeyer et al. 1983). In contrast to lipids, their contents differentiate grape cultivars, and the percentages of their distribution have been proposed as taxonomic criterion (Boursiquot et al. 1986). However, their minor

Fig. 8A.3 Grape phenolic acid esters, and the chemical bonds cleaved to release odorous volatile phenols

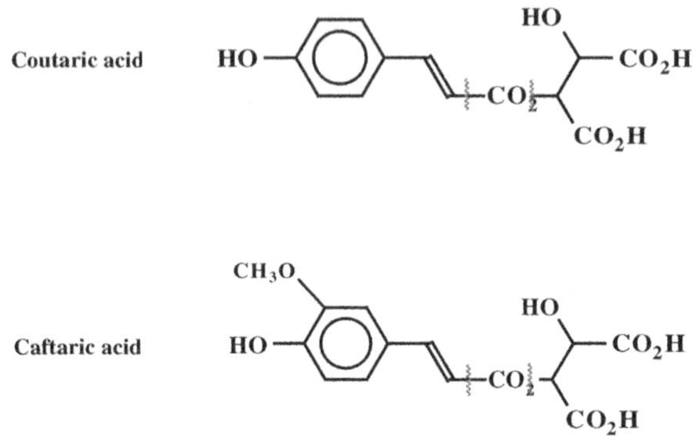

free acid forms only are precursors of wine volatiles, and little is known of the distribution and generation of these free acid forms from their tartaric esters. In this group, coumaric and ferulic acids only are precursors of odorous volatile phenols in wine, coumaric acid generating (4-vinyl- and 4-ethyl)-phenols and ferulic acid, (4-vinyl- and 4-ethyl)-guaiacols (Fig. 8A.3). The vinyl derivatives, 4-vinyl-phenol and 4-vinyl-guaiacol, are formed during the alcoholic fermentation by decarboxylation of the free cinnamic acids with a cinnamate decarboxylase of *Saccharomyces cerevisiae* yeasts (Albagnac 1975; Chatonnet et al. 1993a,b; Dugelay et al. 1992a, 1993). However, as this enzyme is inhibited by catechins and catechic tannins, abundant in red wines, the levels of volatile phenols formed in red wines are generally much lower than those in white and rosé wines, although the contents in hydroxycinnamic precursors in the corresponding red musts are higher (Chatonnet et al. 1993a,b).In contrast, the cinnamate decarboxylase of contaminant yeasts of the genus *Brettanomyces/Dekkera* is not inhibited by these polyphenolic compounds of red wines; furthermore, these yeasts can efficiently reduce the vinylphenols into ethylphenols with a vinylphenol reductase that *Saccharomyces cerevisiae* yeasts or lactic acid bacteria do not have, except for some strains of bacteria with very weak activity (Cavin et al. 1993; Chatonnet et al. 1992a,b, 1995; Herensztyn 1986). Therefore, red wines contain mostly very low levels of vinylphenols, and the formation of significant quantities of ethyl phenols, higher than their olfactive perception thresholds, occurs only in the presence of these contaminant yeasts, generally during aging prior to bottling of wine, which always causes the sensory depreciation of wine (phenolic off-flavor).

Thus, the vinylphenols may contribute to the aroma of white and rosé wines only. However, 4-vinylphenol seems to depreciate the aroma of white wine as soon as it is perceived by masking the fruity note, then at higher concentrations it is responsible for phenolic off-flavors (Chatonnet et al. 1993a,b). That was observed in winemaking, when enzymatic preparations containing high esterase activities, and *Saccharomyces cerevisiae* yeast strains with high cinnamate decarboxylase activities were used (Chatonnet et al. 1992c; Dugelay et al. 1992a, 1993). The influence of 4-vinylgaiacol on wine aroma would be less negative, as it was reported to play a role in the varietal aroma of Gewürztraminer from north of Italy (Versini 1985). However, its limit preference threshold is not much higher than its perception threshold, and as it always occurs in wine with 4-vinylphenol, it is rarely perceived positively (Chatonnet et al. 1993a,b). Furthermore, the levels of vinylphenols in wine decrease dramatically with aging, particularly by addition of ethanol on the vinyl double bond (Dugelay et al. 1995).

Therefore, volatile phenols play a minor role in the aroma of most wines, and when their influence is significant in certain wines, they have mostly a negative effect, which can definitely depreciate their aroma in limit cases (phenolic off-flavors). Thus, the corresponding precursors in grape, phenolic acids, as well as the above-mentioned unsaturated lipids, are hardly taken into account to capture an essential characteristic of the varietal aroma, but to avoid their transformation into off-flavors.

8A.4 Carotenoids

The carotenoids of mature grapes have all the corresponding bicyclic structures of carotenoids associated with PSI and PSII photosynthetic systems, the multiprotein complexes of plant chloroplast membranes (Britton 1982, 1995; Demmig-Adams et al. 1996; Goodwin 1980; Moneger 1968). Their 40 carbon skeleton is biosynthesized from eight isoprenic units: the two final 2,2,6-trimethylcyclohexylmethylene moieties (each made up of two isoprene units) occur at different oxidation levels and each is bound to both ends of a central linear conjugated system made up of four isoprene units, mostly of all *trans* configuration (Fig. 8A.4). The most common are β-carotene and lutein, which represent nearly 85% of the total, and the minor carotenoids are neoxanthin, violaxanthin, lutein-5,6-epoxyde, zeaxanthin, neochrome, flavoxanthin, luteoxanthin (Bureau 1998; Marais et al. 1989; Mendes Pinto et al. 2004; Razungles et al. 1987, 1996). The only carotenoid hydrocarbon, i.e. carotene, is β-carotene, as all others are oxidized carotenoids, i.e. xanthophylls. While a mature Syrah berry has similar proportions of β-carotene, luteine, violaxanthin and neoxanthin (36%, 50%, 2% and 9% of the total) as do leaves (27%, 51%, 10%, 11%), the quantities in the berry are a hundred times less than those in the leaves (Wirth et al. 2001). The levels of carotenoids in grape decrease from véraison to maturity, but depend also on climatic factors, agricultural practices, grape cultivar and clone (Bureau 1998; Marais et al. 1989; Razungles et al. 1987, 1993). Carotenoids are mostly found in the skin, at levels —two to three times higher than in the pulp, while they are absent from the juice (Razungles et al. 1988). In the same way, they are absent from wine, except in the case of fortified red wine, like Port wine, which contains more xanthophylls than carotene (Guedes de Pinho et al. 2001).

Carotenoids are regarded as part of the aroma potential of grape, as they are the biogenetic precursors of C13-norisoprenoids, a chemical family with many powerful odorants, which are mainly found as glycoconjugates in grape (Baumes et al. 2002; Enzell 1985; Mathieu et al. 2005; Winterhalter 1993). Sunshine favors the biosynthesis of carotenoids in the berry, from the first stage of fruit formation (berry set) until véraison, then it favors their degradation into C13-norisoprenoidic compounds, from véraison to maturity (Bureau 1998), when chlorophyllic photosynthetic pigments disappear (Hardie et al. 1996). This degradation of carotenoids in grape berry involves a carotenoid dioxygenase, VvCCD1 (*Vitis vinifera* Carotenoid Cleavage Dioxygenase), cleaving the 9,10- and 9′,10′-double bonds of carotenoids to release C13-norisoprenoidic carbonyl compounds, possessing the same oxidised backbone as their carotenoidic parents (Fig. 8A.4) (Mathieu et al. 2005). These primary products of cleavage are then transformed by oxidases and reductases to C13-norisoprenoidic compounds at different oxidation degrees, and finally glycosylated by grape glycosyltransferases (Ford and Hoj 1988; Wirth et al. 2003; Mathieu et al. 2005). The regiospecificity of the enzymatic cleavage of carotenoids in grape explains the predominance of the glycosidic norisoprenoids with 13 carbon atoms identified in this fruit, but their total contents are approximately 10 times lower than those of carotenoids (Baumes et al. 2002; Razungles et al. 1988).

However, in grape, VvCCD1 would not degrade carotenes into ionones, which would explain the absence in this fruit of monooxygenated C13-norisoprenoids , whereas β-carotene is a major carotenoid in grape (Mathieu et al. 2005). In addition, in contrast to other C13-norisoprenoidic precursors of odorants in wine, e.g. β-damascenone or 1,1,6-trimethyl-1,2-dihydronaphtalene (TDN), no glycosidic precursor of β-ionone was identified in grape or wine. Nevertheless, β-ionone is a constituent of wine, occurring in levels that do not exceed 1 μg/L in wine, but higher than those found in must from the same grapes (Kotseridis et al. 1998, 1999b). Thus, β-ionone would arise from the non enzymatic degradation of β-carotene during winemaking; it would be the same for α-ionone and β-cyclocitral, other norisoprenoidic odorants identified in Melon B. grape (Kotseridis et al. 1998; Schneider 2001; Silva Ferreira and Guedes de Pinho 2004). These carotenoid

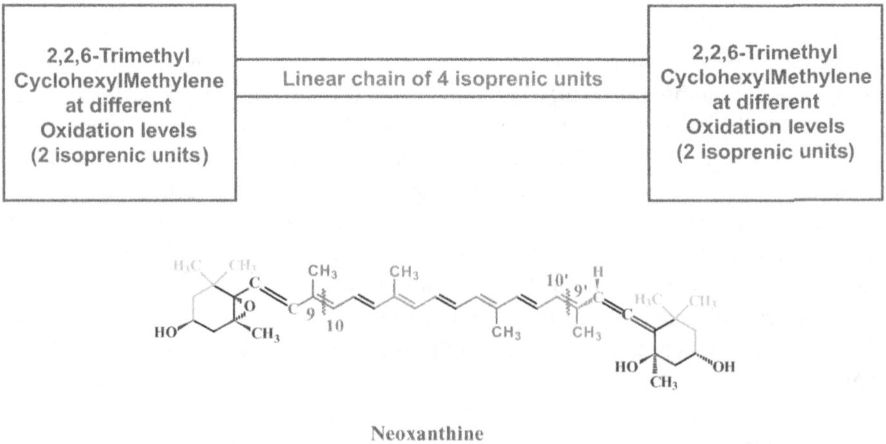

Fig. 8A.4 Grape carotenoid general structure, neoxanthine and the chemical bonds cleaved by *Vitis vinifera* Carotenoid Cleavage Dioxygenase to biosynthesize the corresponding primary products of cleavage (C13 norisoprenoids)

degradations by chemical, photochemical or coupled to oxidases (lipoxygenase, polyphenoloxidase, xanthine oxidase) reactions were known well before carotenoid cleavage dioxygenases in plants, as reported in reviews (Winterhalter and Rouseff 2000; Wirth et al. 2001), but they have not been unambiguously demonstrated in winemaking. If they occur, these reactions would be alternative pathways leading to norisoprenoidic compounds directly from xanthophylls, as hypothesized recently (Silva Ferreira and Guedes de Pinho 2004). However, they would be minor pathways indeed, as evidenced by the trace amounts of β-ionone found in most wines (about 1 μg/L), relative to that of β-carotene found in grapes (about 1 mg/L) (Bureau 1998; Kotseridis et al. 1998). Obviously, these pathways would be more important in fortified wines containing carotenoids, such as Port wines (Guedes de Pinho et al. 2001), than in other wines, for which the effective time for carotenoid degradation would be limited to the time of contact between the solid and liquid portions of the must.

8A.5 Glycoconjugates

The occurrence in *Vitis vinifera* grape of glycoconjugates of volatile compounds (glycosidically bound volatiles) were suggested first for monoterpenols by Cordonnier and Bayonove (1974). Afterwards, extensive research showed that this group of precursors consisted of a lot of derivatives, including four sub-groups according to the sugar moiety, β-D-glucopyranosides, 6-*O*-α-L-arabinofuranosyl-β-D-glucopyranosides, 6-*O*-α-L-rhamnopyranosyl-β-D-glucopyranosides and 6-*O*-β-D-apiofuranosyl-β-D-glucopyranosides, each with aglycon moiety belonging to different classes of volatiles, mostly monoterpenoids, C13-norisoprenoids, volatile phenols, C6 compounds, aliphatic, benzyl and 2-phenethyl alcohols and some other compounds (Fig. 8A.5) (Baumes et al. 1994; Cox et al. 2005; Günata 1984, 1985a,b; Sefton et al. 1989, 1993; Stahl-Biskup et al. 1993; Strauss et al. 1987a–c, 1988; Voirin 1990; Voirin et al. 1990, 1992; Williams et al. 1982a,b, 1983, 1984; Winterhalter and Skouroumounis 1997; Wirth et al. 2001). In grape berry, most volatile phenols, monoterpenoids and C13 norisoprenoids occur as glycoconjugates,

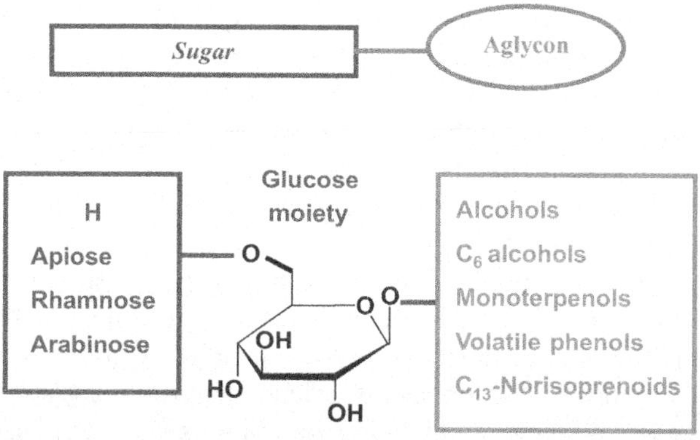

Fig. 8A.5 Glycoconjugate general structure, and the different groups of sugar and aglycon moieties occurring in grape

but monoterpenoids are also abundant as free compounds in Muscat and some other aromatic cultivars only. As a lot of these aglycons are odorants or precursors of odorants, their odorless glycosidic progenitors make up a reserve of grape flavor, which is generally more abundant than the free one (Günata et al. 1985a, b, 1986; Strauss et al. 1986; Günata et al. 1989a; Voirin et al. 1992). Indeed, these glycosides are susceptible to the release of their volatile aglycons in wine through hydrolysis. These aglycons are considered to be part of the varietal aroma only, except for the volatile vinylphenols, which are mostly generated from phenolic acids during winemaking. With regard to the other aglycons, C6 compounds, and fusel alcohols, they occur in grape in much lower amount than the corresponding volatiles formed during the prefermentary and fermentary steps of winemaking, so that their influence on wine aroma is not significant.

The total amount of glycoconjugates is not much different from a few mg/kg in grapes of neutral varieties (Kotseridis 1999; Schneider 2001; Schneider et al. 2002; Ségurel 2005; Dagan 2006), but it can be much higher in aromatic varieties, such as Muscats, in which it is up to ten times as much (Günata et al. 1985b; Voirin 1990; Voirin et al. 1992). The distribution of free and glycoconjugates among the solid and liquid fractions of different white and red grape varieties is quite variable and depends mostly on the aglycon structure, but generally the skin contains $\geq$50% of their total content, which is not much different from free volatiles (Gomez et al. 1994; Günata et al. 1985a; Park et al. 1991; Wilson et al. 1986). That is interesting on a technological basis, as this potential of aroma precursors, as well as the typical free aroma, can be enhanced by processes, such as skin contact or carbonic anaerobiosis (Baumes et al. 1989; Bitteur et al. 1992; Mc Mahon et al. 1999).

During berry maturation, the biosynthesis of glycoconjugates extends from the first stage of fruit formation (berry set), to the end of berry maturation, but the changes in glycoconjugate levels before and after véraison are dependent on the aglycon structure and on grape variety (Baumes et al. 2002; Günata 1984; Marais 1987; Park et al. 1991; Williams et al. 1984; Wilson et al. 1984). However, at the end of berry maturation, there is no simple relationship of glycoconjugate contents to the enological parameters used by winemakers to define maturity of grape (sugar and acid contents, pH) (Dagan 2006; Schneider 2001; Ségurel 2005). Therefore, more simple and less time-consuming methods than the reference analysis methods used in research laboratories to quantitate glycoconjugates should be valuable tools, providing winemakers with a useful index to evaluate the aroma glycosidic potential of grapes. Two methods were developed to reach that goal – the enzymatic red-free glycosyl-glucose (GG) method, which allows the total determination of glycoconjugates (Iland et al. 1996; Williams et al. 1995), and an FT-IR method, which allows glycoconjugate quantitation according to the different groups of aglycons (Schneider et al. 2005).

The formation of odorants from grape glycosides is generally a very slow chemical process, occurring mainly during wine aging, the kinetics of which depends on the structure of the aglycon and its site of glycosylation, on temperature and pH of the wine, but which seems independent on the structure of the sugar moiety (De La Presa-Owens and Noble 1997; Marais 1983; Puglisi et al. 2007; Skouroumounis et al. 1992; Voirin 1990; Winterhalter 1993). It is an acido-catalyzed process, which involves hydrolysis of the glycosides, and depending on the aglycon structure, its chemical transformation. The hydrolysis step release the volatile aglycon, i.e. a compound with an hydroxyl functional group (alcohol, phenol, acid), which can be odorous or not, but which can undergo a sequence of changes under wine aging conditions, susceptible to the generation of odorants from inodorous aglycons, and conversely (Cox et al. 2005; Francis et al. 1992, 1996; Rapp et al. 1985; Schneider 2001; Sefton 1998, 1993; Ségurel 2005; Versini et al. 1996; Voirin 1990; Williams et al. 1980, 1982b). Thus, in the first part of wine aging, the hydrolytic release of volatiles from glycosides and their possible chemical changes are concomitant. Then, when the corresponding glycosides are entirely hydrolyzed, the chemical changes of the volatiles only occur, so that their contents go down gradually, more or less sharply, depending on their chemical stability under wine aging conditions. More than a hundred volatiles generated from glycosides have so far been identified in wine, but only a few could reach levels higher than their olfactive perception thresholds. Such odorants mostly listed are linalool, geraniol, roseoxyde, 1,8-cineole and wine-lactone, as monoterpenoidic derivatives, eugenol, guaiacol, zingerone, methyl salicylate, as volatile phenolic derivatives, β-damascenone, 1,1,6-trimethyl-1,2-dihydronaphtalene (TDN) and (*E*)-1-(2,3,6-trimethylphenyl)buta-1,3-diene (TPB), as norisoprenoidic derivatives (Bonnländer et al. 1998; Etiévant 1991; Farina et al. 2005; Francis et al. 1992, 1996; Guth 1997; Janusz et al. 2003; Kotseridis 1999; Kotseridis et al. 1999a; Schneider 2001; Sefton 1998, 1993; Ségurel 2005; Simpson 1978; Skouroumounis et al. 1992; Winterhalter 1992; Winterhalter and Skouroumounis 1997). It must be noted that linalool, geraniol, eugenol, guaiacol, zingerone and methyl salicylate are genuine aglycons of grape glycosides, whereas the other odorants in this list arise from chemical changes of aglycons.

This general chemical reaction pathway can be modified when enzymatic preparation containing appropriate glycosidases are used in winemaking. The enzymatic hydrolysis of grape diglycosides is a sequential process. The first step consists in the hydrolysis of the terminal oses by the corresponding glycosidases, releasing β-D-glucosides similar to those occurring in grape. All these β-D-glucosides are hydrolyzed in the second step by a β-D-glucosidase to release the aglycons (Günata et al. 1989a). Filamentous fungi glycosidases are appropriate to hydrolyze grape glycoconjugates, and they are much more stable than those from grape or *Saccharomyces cerevisiae* yeast at wine acidic pH. Thus, despite the inhibition of their β-D-glucosidase by must glucose higher than that of grape or *Saccharomyces cerevisiae* yeast, they can be used during dry wine preparation to hydrolyze grape glycoconjugates, as they are still active at the end of fermentation, when the sugar content in must is residual (Günata et al. 1989a, 1990, 1993). Some non-*Saccharomyces* yeasts, with β-D-glucosidase weakly inhibited by glucose, were also used in winemaking (Gueguen et al. 1997; Belancic et al. 2003). However, gluconolactone was also reported as inhibitor of these β-D-glucosidases from grape, yeast and filamentous fungi, used in winemaking. This compound, produced in grapes infected by *Botrytis cinerea* fungi (Heyworth and Walker 1962; Günata et al. 1989b), may be present at high levels (5–10 mM) in the corresponding must and wine, which completely inhibits β-glucosidases, and prevents the enzymatic hydrolysis of grape glycoconjugates (Günata et al. 1993).

The increase in the rate of glycoconjugates hydrolysis by exogenous fungal enzymes in the preparation of dry wines, does not lead to a simple increase in the production rate of odorants from glycoconjugates. Indeed, it alters the whole dynamic release of odorants from glycoconjugates in wine, as well as the chemical changes that certain aglycons undergo (Skouroumounis et al. 1992; Winterhalter and Skouroumounis 1997). Furthermore, the structural change of aglycons released during the alcoholic fermentation could also be carried out by yeast, in much the same way as free volatiles, e.g. the reduction of geraniol to citronellol (Dugelay et al. 1992b).

As mentioned above, grape and *Saccharomyces cerevisiae* yeast have most glycosidases appropriate to hydrolyze grape glyconjugates, but they become effective only under conditions (pH, sugar content) quite different from those found in winemaking (Darriet et al. 1988; Delcroix et al. 1994; Günata et al. 1986, 1989a, 1993; Lecas et al. 1991; Sarry and Günata 2004). However, it was recently reported that some yeast strains hydrolyzed to the extent of 40%, glycoconjugates of Muscat grape during the fermentation of a model must at pH 3.2, and up to 70%, linalool β-D-glucoside, known to be not easily hydrolyzed by fungal β-D-glucosidases, but no data were reported for grape must (Ugliano et al. 2006). Finally, recent studies showed that enological strains of lactic bacteria have a slight influence only on the hydrolysis of grape glycoconjugates under the conditions of malo-lactic fermentation (D'Incecco et al. 2004; Grimaldi et al. 2005; Sarry and Günata 2004; Ugliano et al. 2003).

8A.6 *S*-Cysteine Conjugates

S-(L-cysteine) conjugates, odourless nonvolatile cysteinylated precursors of very odorous volatile thiols, were evidenced and identified in grape much later than the afore-mentioned classes of grape aroma precursors (Darriet 1993; Darriet et al. 1993; Tominaga 1998; Tominaga et al. 1995, 1998b). Chemically, they are S-substituted derivatives of L-cysteine, different by the nature of the moiety on the cysteine sulfur atom, and to date only three of these precursors were identified in grape: *S*-(1-hydroxyhex-3-yl)-L-cysteine (P3MH), *S*-(4-methyl-2-oxopent-4-yl)-L-cysteine (P4MMP) and *S*-(4-methyl-2-hydroxypent-4-yl)-L-cysteine (P4MMPOH (Fig. 8A.6) (Tominaga 1998; Tominaga et al. 1995, 1998b). However, an S-substituted tripeptide related to P3MH, *S*-(1-hydroxyhex-3-yl)-glutathion, was identified later in Sauvignon Blanc must, and probably in Gros Manseng must (Peyrot des Gachons et al. 2002b). This *S*-glutathione conjugate could be a biogenetic precursor of P3MH, as it could be cleaved by a γ-glutamyltransferase, which eliminates glutamic acid, and by a carboxypeptidase, which eliminates glycine, to give rise to P3MH (Peyrot des Gachons et al. 2002b). However, no other

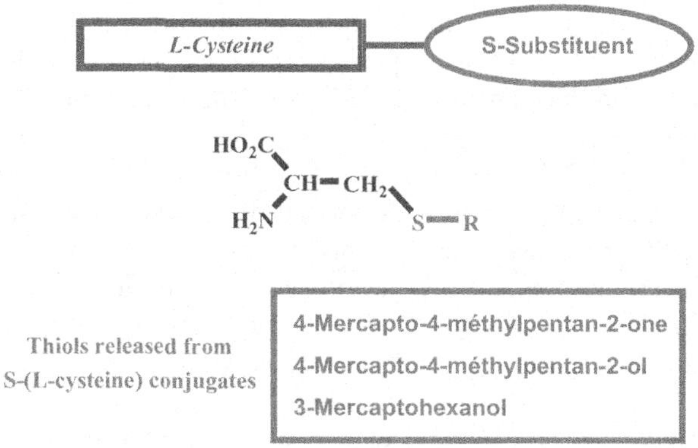

Fig. 8A.6 *S*-(L-Cysteine) conjugates occurring in grape, precursors of varietal thiols

S-glutathione conjugate was identified in grape, particularly no precursor corresponding to P4MMP and P4MMPOH.

As the analytical methods for assessing levels in grape *S*-cysteine conjugates were only recently available and are not easily implemented (Peyrot des Gachons et al. 2000; Dagan 2006), few quantitative data on these compounds were published. Their levels in must are low and do not exceed a 100 μg/L for the most abundant P3MH, in Sauvignon Blanc and Petit and Gros Manseng, and a few μg/L for P4MMP and P4MMPOH in Sauvignon Blanc, the latter being slightly more abundant (Peyrot des Gachons et al. 2000, 2002a, 2005; Dagan 2006). The changes in their levels during ripening were shown to be dependent on environmental conditions, soil and climate parameters and vineyard management techniques (Peyrot des Gachons et al. 2000, 2005; Chone et al. 2006; Dagan 2006). In Sauvignon Blanc grape berry, the levels of P4MMP and P4MMPOH are equivalent in the juice and skin, but the levels of P3MH are much higher in the skin (Peyrot des Gachons et al. 2002a). A lower enrichment in P3MH in Merlot and Cabernet Sauvignon grape skins was also reported (Murat et al. 2001a). Thus, skin contact in winemaking has a higher impact on the contents of P3MH in the must than on those of the other two precursors (Murat et al. 2001a; Peyrot des Gachons et al. 2002a).

Although they are few in number and not abundant, these odorless compounds are very important aroma precursors in grape. Indeed, they generate in wine, four very odorous volatile thiols, which make an important contribution to the aroma of certain wine varieties : 3-sulfanylhexan-1-ol (3MH), 3-sulfanylhexyl acetate (ac3MH), 4-methyl-4-sulfanylpentan-2-one (4MMP), and 4-methyl-4-sulfanylpentan-2-ol (4MMPOH), with very low olfactive detection thresholds, 0.8 ng/L, 60 ng/L, 4.2 ng/L and 55 ng/L, in hydroalcoholic model solution, respectively (Tominaga et al. 2000a). 3MH is the most abundant, and always present in wine, whatever be the grape cultivar, in levels generally higher than its olfactive perception threshold, up to a few μg/L. In contrast, 4MMPOH levels in wine are generally lower than its olfactive perception threshold. With regard to the other two thiols, their levels in wine are very dependent on the grape cultivar, but they are also very variable between different samples of the same cultivar. Thus, 4MMP and ac3MH levels in wines range from sub olfactive threshold values, sometimes below the analytical limit of quantification, to maxima of 100 ng/L for 4MMP and of a few hundreds ng/L for ac3MH (Bouchilloux et al. 1998; Darriet et al. 1993, 1995; Fretz et al. 2005; Guth 1997; Kotseridis and Baumes 2000; Lopez et al. 2003; Murat et al. 2003; Schneider et al. 2003; Tominaga et al. 1996, 1998a, 2000a).

During the alcoholic fermentation, a yeast carbon-sulfur lyase with β-elimination activity releases the odorous thiols from the *S*-cysteine conjugates by cleavage of the carbon-sulfur bond of the cysteine moiety (Tominaga et al. 1998b). Thus, this enzymatic activity of yeast is necessary to cleave the *S*-cysteine conjugates and it was shown that fermentation with different *Saccharomyces cerevisiae* yeast strains generate very different amounts of the thiol odorants (Murat et al. 2001b; Howell et al. 2004). However, although almost the whole amounts of precursors are degraded, the yields in thiol odorants from *S*-cysteine conjugates are always low, whatever be the precursor and the fermentation medium (model or natural), ranging from 0.06% to 1 0.6% for 4MMP (Murat et al. 2001b), and from 0.6% to 10.2 % for 3MH (Murat et al. 2001a). With regard to ac3MH, it has no direct precursor in grape, but the yeast acetylates partly the alcohol function of 3MH to give ac3MH, as it acetylates the fusel alcohols to generate their acetates. Thus, the biogeneration of these thiols is associated with the *S*-β-lyase and acetyl transferase activities of yeast, which depends greatly on the composition of the must, the conditions of fermentation and the yeast strain. Some strains of *Saccharomyces bayanus* var.*Uvarum* were reported to release efficiently these thiols (Murat et al. 2001b; Masneuf-Pomarede et al. 2002; Howell et al. 2004; Dubourdieu et al. 2006; Masneuf-Pomarede et al. 2006). However, yeast could hardly degrade directly the *S*-glutathione conjugates without hydrolyzing the tripeptide into the cysteine conjugate, as previous studies have shown that a bacteria *S*-β-lyase extract does not release any thiol from this kind of precursor (Tominaga et al. 1995). Regarding the stereoselectivity of 3MH release by certain *S*-β-lyases, recent studies showed that the degree of discrimination observed was low for *S*-β-lyases of enological yeasts, and moderate for *S*-β-lyases of other sources (Wakabayashi et al. 2003, 2004; Tominaga et al. 2006).

An alternative biogenetic pathway starting from (*E*)-2-hexenal in must and leading to 3-mercaptohexanol after yeast fermentation was recently evidenced (Schneider 2001; Schneider et al. 2006). Similarly, various yeasts were shown to produce 3-mercaptohexanol from both addition products of cysteine and hex-2-enal, *S*-(1-hydroxyhex-3-yl)-L-cysteine (P3MH) and 2-(2-*S*-L-cysteinylpentyl)-1, 3-thiazolidine-4-carboxylic acid (Wakabayashi et al. 2004). The mechanism of the pathways involved in the experiments reported by Schneider (2001) are still unknown, but they should involve the addition of the sulfhydryl group of a sulfur compound to the unsaturated hex-2-enal, leading to adducts which could be those reported by Wakabayashi et al. (2004). Then, the enzymatic reduction by yeast of the aldehydic group, and further reactions, such as a β-elimination by yeast *S*-β-lyase, if the sulfur compound is different from hydrogen sulfide, would yield the thiol. The formation of 4MMP from mesityl oxide during must fermentation by yeast was similarly observed, but this compound was not reported as a must constituent although its hydrate is a wine constituent (Schneider 2001; Escudero et al. 2002). Similar pathways could also explain the formation of other odorous volatile thiols identified in wine and also in beer, such as 3-methyl-3-sulfanylbutan-1-ol, 2-methyl-3-sulfanylbutan-1-ol, or 3-sulfanylpentan-1-ol, which have no *S*-cysteine or *S*-glutathione conjugates identified as yet (Bailly et al. 2006; Huyng-Ba et al. 2003; Sarrazin et al. 2007; Tominaga et al. 2000b, 2003a,b; Vermeulen et al. 2006).

During wine aging, the levels of all these thiols decrease, but this decrease is very dependent on the oxidative conditions of the aging. Thus, lees, sulfur dioxide, glutathione, and anthocyanins have protective effects on these volatile thiols, whereas increased contact with oxygen, particularly in presence of derivatives of catechin, favors their degradation (Murat et al. 2003; Blanchard et al. 2004). As for the evolution of these odorous thiols during the conservation of the wine, their contents generally decrease, but this phenomenon appears very dependent on the oxydative reactions related to this conservation. Thus, the factors which prevent the decrease of the reducing potential of the wine (contact limited with oxygen, dioxide of sulphur, lees, glutathion, anthocyanes) limit the degradation of these odorous thiol (Murat et al. 2003; Blanchard et al. 2004).

8A.7 Precursors of Dimethylsulfide

Dimethylsulfide (DMS) has been known as a wine constituent for a long time, as well as its formation during yeast fermentation and wine aging (Anness and Bamforth 1982; Etiévant 1991), but the ability of DMS precursors, differing from DMSO, to release this compound through a chemical process during wine aging, and their occurrence in grape, were reported recently (Loscos Deodad et al. 2007; Ségurel et al. 2004, 2005). DMS is a powerful odorant, which has an olfactive perception threshold of 27 μg/L in red wine (Anocibar Beloqui et al. 1996) and 25 μg/L in white wine (Spedding and Raut 1982). It is the key-aroma compound in truffle (Talou et al. 1987), an aroma descriptor often attributed to bottled bouquet in aged premium red wines and late harvest wines (Anocibar-Beloqui et al. 1996; Anocibar-Beloqui et al. 1998; Dagan 2006; De Mora et al. 1987; Du Plessis and Loubser 1974; Spedding and Raut 1982). However, its influence on wine aroma was perceived sometimes negatively (Spedding and Raut 1982; Goniak and Noble 1987).

During the alcoholic fermentation, DMS is biogenerated by yeast from DMSO and some sulfur amino acids, such as cysteine, cystine or glutathione (Anocibar Beloqui 1998; Bamforth and Anness 1981; De Mora et al. 1986; Schreier et al. 1976), but DMSO levels in must are generally very low (Ségurel 2005). In addition, lactic acid bacteria and yeast cannot generate DMS from methionine (Schreier et al. 1976; Pripis-Nicolau et al. 2004). Anyway, the amount produced during yeast fermentation is low, and as DMS is quite volatile (vapour pressure 53 KPa at 20 °C) (Lestremau et al. 2003), it is mainly stripped off by CO_2. Therefore, its levels in young wine after fermentation are generally lower than its olfactive preception threshold, but they can be higher in wines thought to have sulfide-related off-odors, in which DMS is generally associated with other volatile sulfur compounds eliciting such off-odors (Park et al. 1994).

In contrast, levels of DMS much higher than its perception threshold were reported to be specifically produced without generating off-odors in many bottled wines during aging, and levels increased with time and temperature (Marais 1979; De Mora et al. 1986, 1993; Anocibar Beloqui 1998; Goto and Takamuto 1987; Ségurel et al. 2004; Dagan 2006). Thus, high levels of DMS would contribute positively to the bottled bouquet in aged premium red wines and late harvest wines (Anocibar-Beloqui et al. 1996; Anocibar-Beloqui et al. 1998; Dagan 2006; Du Plessis and Loubser 1974; Spedding and Raut 1982). Furthermore, De Mora et al. (1987) reported the enhancement of fruity notes of Cabernet Sauvignon red wines by DMS, which was later confirmed in other samples (Ségurel et al. 2004; Escudero et al. 2007). However, sensorial data are not available for wines from a lot of cultivars, and the sensorial interactions of DMS with other aroma compounds are still unknown.

The potential DMS in wine, susceptible to be released during wine aging, was evaluated by an indirect method using a chemical treatment to degrade wine DMS precursors into DMS (Ségurel et al. 2004, 2005). That method showed that such DMS potential (PDMS) occurred in grape, and that DMS released during wine aging could not be explained by DMSO, which has long been regarded as the only wine DMS precursor (De Mora et al. 1986, 1993; Ségurel et al. 2004, 2005). Then, among the other possible DMS precursors known in plants (Bezzubov and Gessler 1992; Howard and Russell 1997), SMM was identified as the major DMS precursor in grape using MALDI-TOF MS (Fig. 8A.7) (Loscos Deosdad et al. 2007). During the first steps of winemaking, SMM gets from grape into wine, and it is only when wine is in stoppered bottle during its storage that DMS become accumulated, as it is released from SMM by an elimination reaction (Hoffmann degradation), a slow chemical process in usual wine storage conditions (Ségurel et al. 2004, 2005). Thus, the differences in DMS levels observed between wines depend primarily on the PDMS levels at bottling, then on the conditions of storage, particularly on temperature (Marais 1979).

PDMS measured in musts from grapes of different cultivars showed that all the samples analyzed contained DMS precursors, with high variation between cultivars, from levels lower than the perception threshold in some Grenache samples up to 4.5 mg/L in some Petit Manseng samples, and high variation between samples of the same cultivar (Ségurel 2005; Dagan 2006). Thus, grape PDMS is not only under genetic control, but it is also dependent on the several factors which define the "terroir", and it increases dramatically with over-ripening (Dagan 2006).

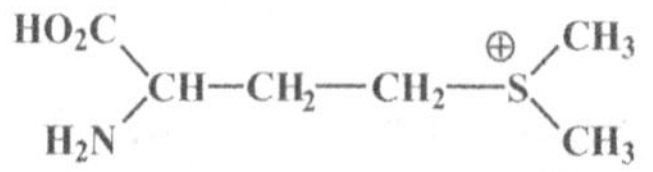

Fig. 8A.7 *S*-Methylmethionine, precursor of DMS

However, the PDMS levels of the young wines were always lower than those of the corresponding musts, and no clear relationship was found between them. The simple chemical degradation of SMM could hardly explain the high losses observed in some samples from must to wine, which could be due to must enzymatic activities, or to the microorganisms of the fermentation steps. Indeed, the degradation of SMM by enzymes of the metabolism of sulfur compounds or ethylene, such as SMM-homocysteine *S*-methyltransferase or *S*-methyl-L-methionine hydrolase or 1-aminocyclopropane-1-carboxylate synthase (ACC synthase), were previously reported in higher plants (Kiddle et al. 1999; Ko et al. 2004). Furthermore, although SMM fate in *Saccharomyces cerevisiae* yeast has not been studied, a SMM-homocysteine *S*-methyltransferase activity was evidenced in yeast lysates (Shapiro et al. 1964) and a recent work showed that SMM is transported in the yeast cells by two permeases (Rouillon et al. 1999). Thus, the partial degradation of SMM by some yeast strains seems conclusive, as this work brought evidence that yeast is able to use SMM as an efficient sulfur source, which explain PDMS levels in young wines differing from those in the corresponding musts. It must be noted that a long time ago, Schreier et al. (1976) hypothesized that DMS could be generated by yeast from SMM during fermentation, as it was shown during cheese ripening (Spinnler et al. 2001). Anyway, DMS produced from SMM during the first steps of winemaking should be mainly stripped off by CO2 during yeast fermentation, as mentioned above for DMS produced from other sulfur sources. Thus, the knowledge of the factors governing the recovery of SMM from grape and its degradation during the fermentation step, should allow its levels to be controlled in wine at bottling, and therefore, the levels of DMS generated in bottled wine during aging.

References

Albagnac, G. (1975). La décarboxylation des acides cinnamiques substitués par les levures. *Ann. Technol. Agric., 24*, 133–141.

Anness, B. J., & Bamforth, C. W. (1982). Dimethyl sulphide – a review. *J. Inst.Brew., 88*, 244–252.

Anocibar Beloqui, A. (1998). Les composés soufrés volatils des vins rouges.*Doctoral dissertation, University Victor Segalen Bordeaux II*.

Anocibar Beloqui, A., Kotseridis, Y., & Bertrand, A. (1996). Détermination de la teneur en sulphure de diméthyle dans quelques vins rouges. *J. Int. Sci. Vigne Vin, 30*, 167–170.

Bailly, S., Jerkovic, V., Marchand-Bryaert, J., & Collin, S. (2006). Aroma extraction dilution analysis of Sauternes wines. Key role of polyfunctional thiols. *J. Agric. Food Chem., 54*, 7227–7234.

Bamforth, C. W., & Anness, B. J. (1981). The role of dimethyl sulphoxide reductase in the formation of dimethyl sulphide during fermentation. *J. Inst.Brew., 87*, 30–34.

Baumes, R., Bayonove, C., Cordonnier, R., Torres, P., & Seguin, A. (1989). Incidence de la macération pelliculaire sur la composante aromatique des vins doux naturels de muscat. *Rev. Fr. Oenol., 116*, 5–11.

Baumes, R., Aubert, C., Günata, Z., De Moor, W., & Bayonove, C. (1994). Structure of two C13 norisoprenoid glucosidic precursors of wine flavor. *J. Essent. Oil. Res., 6*, 587–599.

Baumes, R., Wirth, J., Bureau, S., Günata, Z., & Razungles, A. (2002). Biogeneration of C_{13}-norisoprenoid compounds: Experiments supportive for an apo-carotenoid pathway in grapevines. *Anal. Chim. Acta, 458*, 3–14.

Bayonove, C. (1998). L'arôme variétal: le potentiel aromatique du raisin. In: C. Flanzy (Ed.), *Oenologie: fondements scientifiques et technologiques* (pp. 165–181). Paris : Lavoisier Tec et Doc.

Belancic, A., Günata, Z., Vallier, M.J., & Agosin, E. (2003). β-glucosidase from the grape native yeast *Debaryomyces vanriji* : Purification, characterization and its effect on monoterpene concentration of a Muscat grape juice. *J. Agric. Food Chem., 51*, 3083–3091.

Bezzubov, A. A., & Gessler, N. N. (1992). Plant sources of S-methylmethionine. *App. Biochem. Microbiol., 28*, 317–322.

Bitteur, S., Tesniere, C., Sarris, J., Baumes, R., Bayonove, C., & Flanzy, C. (1992). Carbonic anaerobiosis of muscat grape. I. Changes in the profiles ol free and bound volatiles. *Am. J. Enol. Vitic., 43*, 41–48.

Blanchard, L., Darriet, P., & Dubourdieu, D. (2004). Reactivity of 3-mercaptohexanol in red wine: Impact of oxygen, phenolic fractions, and sulfur dioxide. *Am. J. Enol. Vitic., 55*, 115–120.

Bonnländer, B., Baderschneider, B., Monika Messerer, M., & Winterhalter, P. (1998). Isolation of Two Novel Terpenoid Glucose Esters from Riesling Wine. *J. Agric. Food Chem., 46*, 1474–1478.

Bouchilloux, P., Darriet, P., Henry, R., Lavigne-Cruège, V., & Dubourdieu, D. (1998). Identification of volatile and powerful odorous thiols in Bordeaux red wine varieties. *J. Agric. Food Chem., 46*, 3095–3099.

Boursiquot, J. M., Sapis, J.C., & Macheix, J.J. (1986). Les esters hydroxycinnamiques chez le genre *Vitis*. Essai d'application taxonomique : premiers résultats.*C. R. Acad. Sci., 302*, 177.

Britton, G. (1982). Carotenoid biosynthesis in higher plants. *Physiol. Veg., 20*, 735–755.

Britton, G. (1995). Structure and properties of carotenoids in relation to function. *FASEB J., 9*, 1551–1558.

Bureau, S. (1998). Modification de l'environnement limineux des grappes et des ceps de vigne: effets sur le potentiel aromatique des baies de Syrah et de Muscat de Frontignan. *Doctoral dissertation*, University of Montpellier II.

Cavin, J.F., Andioc, V., Etievant, P., & Divies, C. (1993). Ability of wine lactic acid bacteria to metabolize phenol carboxylic acids. *Am. J. Enol. Vitic., 44*, 76–80.

Chatonnet, P., Dubourdieu, D., Boidron, J. N., & Pons, M. (1992a). The origin of ethylphenols in wines. *J. Sci. Food Agric., 60*, 165–178.

Chatonnet, P., Dubourdieu, D., & Boidron, J. N. (1992b). Le caractère phénolé des vins rouges: caractérisation, origine et moyens de lutte. *Rev. Fr. Oenol., 138*, 21–24.

Chatonnet, P., Barbe, C., Canal-Llauberes, R.M., Dubourdieu, D., & Boidron, J. N. (1992c). Incidences de certaines préparations pectolytiques sur la teneur en phénols volatils des vins blancs. *J. Int. Sci. Vigne Vin, 26*, 253–269.

Chatonnet, P., Dubourdieu, D., Boidron, J. N., & Lavigne, V. (1993a). Synthesis of volatile phenols by *Saccharomyces cerevisiae* in wines. *J. Sci. Food Agric., 62*, 191–202.

Chatonnet, P., Barbe, C., Boidron, J.N., & Dubourdieu, D. (1993b). Origines et incidences organoleptiques de phénols volatils dans les vins. Application à la maîtrise de la vinification et de l'élevage. In: J. Crouzet, C. Flanzy, C. Martin and J. C. Sapis (Eds.), *Connaissance aromatique des cépages et qualité des vins* (pp.279–287). Montpellier: Revue française d'oenologie.

Chatonnet, P., Dubourdieu, D., & Boidron, J. N. (1995). The influence of *Brettanomyces dekkera* sp. yeasts and lactic acid bacteria on the ethylphenol content of red wine. *Am. J. Enol. Vitic., 46*, 463–468.

Chone, X., Lavigne-Cruege, V., Tominaga, T., Van Leeuwen, C., Castagned, C., Saucier, C., & Dubourdieu, D. (2006). Effect of vine status on grape aromatic potential: flavor precursors (S-cysteine conjugates), glutathione and phenolic contents in *Vitis vinifera* L. cv. Sauvignon Blanc grape juice. *J. Int. Sc. Vigne Vin, 40*, 1–6.

Cordonnier, R., & Bayonove, C. (1974). Mise en évidence dans la baie de raisin, variété Muscat d'Alexandrie, de monoterpènes liés révélables par une ou plusieurs enzymes de fruits. *C. R. Acad. Sci. Paris, 278*, 3387–3390.

Cox, A., Skouroumounis, G.K., Elsey, G.M., Perkins, M. V., & Sefton, M. A. (2005). Generation of (*E*)-1-(2,3,6-Trimethylphenyl)buta-1,3-diene from C13-norisoprenoid precursors. *J. Agric. Food Chem., 53*, 6777–6783.

Dagan, L. (2006). Potentiel aromatique des raisins de *Vitis Vinifera* L. cv. Petit Manseng et Gros Manseng. Contribution à l'arôme des vins de pays Côtes de Gascogne. *Doctoral dissertation*, University of Montpellier II.

Darriet, P. (1993). Recherche sur l'arôme et les précurseurs d'arôme du Sauvignon. *Doctoral dissertation, University Victor Segalen Bordeaux II.*

Darriet, P., Boidron, J.N., & Dubourdieu, D. (1988). L'hydrolyse des hétérosides terpéniques du Muscat à petits grains par les enzymes périplasmatiques de *Saccharomyces cerevisiae*. *Conn. Vigne Vin, 22*, 89–195.

Darriet, P., Tominaga, T., Demole, E., & Dubourdieu, D. (1993). Mise en évidence dans le raisin de Vitis Vinifera L. (var. Sauvignon) d'un précurseur de la 4-méthyl-4-mercaptopentan-2-one. *C. R. Acad. Sci. Paris, 316*, 1332–1335.

Darriet, P., Tominaga, T., Lavigne, V., Boidron, J.N., & Dubourdieu, D. (1995). Identification of a powerful aromatic component of *Vitis vinifera* L. var. Sauvignon wines: 4-mercapto-4-methylpentan-2-one. *Flav. Frag. J., 10*, 385–392.

De La Presa-Owens, C., & Noble, A.C. (1997). Effect of storage at elevated temperatures on aroma of Chardonnay wines. *Am. J. Enol. Vitic., 48*, 310–316.

Delcroix, A., Günata, Z., Sapis, J.C., Salmon, J.M., & Bayonove, C. (1994). Glycosidase activities of three enological yeast strains during winemaking: effect on terpenols content of Muscat wine. *Am. J. Enol. Vitic., 37*, 112–114.

Demmig-Adams, B., Gilmore, A. M., & Adams, W. W. III (1996). In vivo functions of carotenoids in higher plants. *FASEB J., 10* , 403–412.

De Mora, S.J., Eschenbruch, R., Knowles, S.J., & Spedding, D.J. (1986). The formation of dimethyl sulphide during fermentation using a wine yeast. *Food Microbiol., 3*, 27–32.

De Mora, S.J., Knowles, S.J., Eschenbruch, R., & Torrey, W.J. (1987). Dimethyl sulphide in some australian red wines. *Vitis, 26*, 79–84.

De Mora, S.J., Lee, P., Shooter, D., & Eschenbruch, R. (1993). The analysis and importance of dimethylsulfoxide in wine. *Am. J. Enol. Vitic., 44*, 327–332.

D'Incecco N., Bartowsky, E., Kassara, S., Lante, A., Spettoli, P., & Henschke, P. (2004). Release of glycosidically bound flavour compounds of Chardonnay by *Oenococcus oeni* during malolactic fermentation. *Food Microbiol., 21*, 257–265.

Drawert, F., Heimann, W., Emberger, R., & Tressl, R. (1966). Uber die Biogenese von Aromastoffen bei Pflanzen und Früchten. II. Enzymatische Bildung von Hex-2-en-1-al, Hexanal und deren Vorstufen. *Ann. Chem., 694*, 200–208.

Dubourdieu, D., Tominaga, T., Masneuf, I., Des Gachons, C., & Murat, M.L. (2006). The role of yeasts in grape flavor development during fermentation: The example of Sauvignon blanc. *Am. J. Enol. Vitic., 57*, 81–88.

Dugelay, I., Günata, Z., Bitteur, S., Sapis, J.C., Baumes, R., & Bayonove, C. (1992a). Formation of volatile phenols from cinnamic precursors during wine making : the role of cinnamoyl esterase from commercial enzymic preparations. In: P. Schreier & P. Winterhalter (Eds.), *Progress in flavour studies* (pp. 189–193). Carol Stream Ils.: Allured Publishing Co.

Dugelay, I., Günata, Z., Sapis, J.C., Baumes, R., & Bayonove, C. (1992b). Etude de l'origine du citronellol dans les vins. *J. Inter. Sci. Vigne Vin, 26*, 117–184.

Dugelay, I., Günata, Z., Sapis, J.C., Baumes, R., & Bayonove, C. (1993). Role of cinnamoyl esterase activities from enzyme preparations on the formation of volatile phenols during winemaking . *J. Agric. Food Chem., 41*, 2092–2096

Dugelay, I., Baumes, R., Gûnata, Z., Razungles, A., & Bayonove, C. (1995). Evolution de l'arôme au cours de la conservation du vin: formation de 4-(1-éthoxy éthyl)-phénol et 4-(1-éthoxyéthyl)-gaïacol. *Sci. Alim., 15*, 423–433.

Du Plessis, C.S., & Loubser, G.J. (1974). The bouquet of late harvest wine. *Agrochemophysica, 6*, 49–52.

Enzell, C. (1985). Biodegradation of carotenoids – an important route to aroma compounds. *Pure Appl. Chem., 57*, 693–700.

Escudero, A., Asencio, E., Cacho, J., & Ferreira, V. (2002). Sensory and chemical changes of young white wines stored under oxygen. An assessment of the role played by aldehydes and some other important odorants. *Food Chem., 77*, 325–331.

Escudero, A., Campo, E., Farina, L., Cacho, J., & Ferreira, V. (2007). Analytical characterization of the aroma of five premium red wines. Insights into the role of odor families and the concept of fruitiness of wines. *J. Agric. Food Chem., 55*, 4501–4510.

Etiévant, P.X. (1991). Wine. In: H. Maarse (Ed.),*Volatile compounds in food and beverages* (pp.483–546). New york: Dekker Inc.

Farina, L., Boido, E., Carrau, F., Versini, G., & Dellacassa, E. (2005). Terpene compounds as possible precursors of 1,8-cineole in red grapes and wines. *J. Agric. Food Chem., 53*, 1633–1636.

Ferreira, B., Hory, C., Bard, M. H., Taissant, C., Olsson, A., & Le Fur, Y. (1995). Effects of skin contact and settling on the level of C18:2, C18:3 and C6 compounds on Burgundy Chardonnay musts and wines. *Food Qual. Prefer.,* 6, 35–41.

Ford, C.M., & Hoj, P.B. (1998). Multiple glucosyltransferas activities in the grapevine *Vitis vinifera* L.. *Aust. J. Grape Wine Res., 4*, 48–58.

Francis, I. L., Sefton, M. A., & Williams, P. J. (1992). Sensory descriptive analysis of the aroma of hydrolysed precursor fractions from Semillon, Chardonnay and Sauvignon blanc grape juices. *J. Sci. Food Agric., 59*, 511–520.

Francis, I. L., Tate, M. E., & Williams, P. J. (1996). The effect of hydrolysis conditions on the aroma released from Semillon grape glycosides. *Aus. J. Grape Wine Res., 2*, 70–76.

Fretz, C., Känel, S., Luisier, J. L., & Amado, R. (2005). Analysis of volatile components of Petite Arvine wine. *Eur. Food Res. Technol., 221*, 3–4.

Gomez, E., Martinez, A., & Laencina, J. (1994). Localization and bound aromatic compounds among skin, juice and pulp fractions of some grape varieties. *Vitis, 33*, 1–4.

Goodwin, T. W. (1980). *The biochemistry of the carotenoids. Vol. I. Plants, Second edition*. London: Chapman and Hall.

Goto, S., & Takamuto, R. (1987). Concentration of dimethyl sulfide in wine after different storage times. *Hakkokogaku, 65*, 53–57.

Grimaldi, A., Bartowsky, E., & Jiranek, V. (2005). A survey of glycosidase activities of commercial wine strains of *Oenococcus oeni. Int. J. Food Microbiol., 105*, 233–244.

Guedes de Pinho, P., Silva Ferreira, A., Mendes Pinto, M., Gomez Benitez, J., & Hogg, T. (2001). Determination of carotenoid profiles in grapes, musts and fortified wines from Douro varieties of *Vitis vinifera. J. Agric. Food Chem., 49*, 5484–5488.

Gueguen, Y., Chemardin, P., Pien, S., Arnaud, A., & Galzy, P. (1997). Enhancement of aromatic quality of Muscat wine by the use of immobilized β-glucosidase. *J. Biotechnol., 55*, 151–156.

Günata, Z. (1984). Recherche sur la fraction liée de nature glycosidique de l'arôme du raisin : importance des terpénylglycosides, action des glycosidases. *Doctoral dissertation, University Montpellier II.*

Günata, Z., Bayonove, C., Baumes, R., & Cordonnier, R. (1985a). The aroma of grapes. Localisation and evolution of free and bound fractions of some grape aroma components cv. Muscat during first devbelopment and maturation. *J. Sci. Food Agric., 36*, 857–862.

Günata, Z., Bayonove, C., Baumes, R., & Cordonnier, R. (1985b). The aroma of grapes. Extraction and determination of free and glycosidically bound fractions of some grape aroma components. *J. Chromatogr., 9*, 83–90.

Günata, Y.Z., Bayonove, C., Baumes, R., & Cordonnier, C. (1986). Stability of free and bound fractions of some aroma components of grape cv. Muscat during the wine processing: preliminary results. *Am. J. Enol. Vitic., 37*, 112–114.

Günata, Z., Bitteur, S., Baumes, R., Brillouet, J.M., Tapiero, C., & Bayonove, C., Cordonnier R. (1989a). Procédé d'obtention de composants d'arômes et d'arômes à partir de leurs précurseurs glycosidiques et composants d'arômes et arômes ainsi obtenus. French patent 88 02961 (INRA/Gist Brocades).

Günata, Z., Biron, C., Sapis, J.C., & Bayonove, C. (1989b). Glycosidase activities in sound and rotten grapes in relation to hydrolysis of grape monoterpenyl glycosides. *Vitis, 28*, 191–197.

Günata, Z., Dugelay, I., Sapis, J.C., Baumes, R., & Bayonove, C. (1990). Action des glycosidases exogènes au cours de la vinification : libération de l'arôme à partir des précurseurs glycosidiques. *J. Inter. Sci. Vigne Vin, 24*, 133–143.

Günata, Z., Dugelay, I., Sapis, J.C., Baumes, R., & Bayonove, C. (1993). Role of the enzymes in the use of the flavour potential from grape glycosides in winemaking. In: P. Schreier & P. Winterhalter (Eds.), *Progress in flavor precursors studies* (pp. 219–234). Carol Stream Ils.: Allured Publishing.

Guth, H. (1997). Quantitation and sensory studies of character impact odorants of different white wine varieties. *J. Agric. Food Chem., 45*, 3027–3032.

Hardie, W., Aggenbach, S., & Jaudzems, V. (1996). The plastids of the grape pericarp and their significance in isoprenoid synthesis. *Aust. J. Grape Wine Res., 2*, 144–154.

Heresztyn, T. (1986). Metabolism of phenolic compounds from hydroxycinnamic acids by *Brettanomyces yeasts. Arch. Microbiol., 146*, 96–98.

Heyworth, R., & Walker, P.O. (1962). Almond-emulsin β-D-glucosidase and β-D-galactosidase. *Biochem. J., 83*, 331–335.

Howard, A.G., & Russell, D.W. (1997). Borohydride-coupled HPLC-FPD instrumentation and its use in the determination of dimethylsulfonium compounds. *Anal. Chem., 69*, 2882–2887.

Howell, K., Swiegers, J., Elsey, G., Siebert, T., Bartowski, E., Flett, G., Pretorius, I., & de Barros Lopes, M. (2004). Variation in 4-mercapto-4-methylpentan-2-one release by *Saccharomyces cerevisiae* commercial wine strains. *FEMS Microbiol. Letters, 240*, 125–129.

Huyng-Ba, T., Matthey-Doret, W., Fay, L. B., & Bel-Rhlid, R. (2003). Generation of thiols by biotransformation of cystein-aldehyde conjugates with baker yeast. *J. Agric. Food Chem., 51*, 3629–3635.

Iland, P., Cynkar, W., Francis, I., Williams, P., & Coombe, B. (1996). Optimisation of methods for the determination of total and red-free glycosyl glucose in black grape berries of *Vitis vinifera. Aust. J. Grape Wine Res., 2*, 171–178.

Janusz, A., Capone, D. L., Puglisi, C.J., Perkins, M. V., Elsey, G.M., & Sefton, M. A. (2003). (*E*)-1-(2,3,6-Trimethylphenyl)buta-1,3-diene: A Potent Grape-Derived Odorant in Wine.*J. Agric. Food Chem., 51*, 7759–7763.

Joslin, W. S., & Ough, C. S. (1978). Cause and fate of certain C6 compounds formed enzymatically in macerated grape leaves during harvest and wine fermentation. *Am. J. Enol. Vitic., 29*, 11–17.

Kiddle, G., Bennett, R., Hick, A., & Wallsgrove, R. (1999). C-S lyase activities in leaves of crucifers and non-crucifers and the characterization of 3 classes of C-S lyase activities from oilseed rape (*Brassica rapus* L.). *Plant Cell Environ., 22*, 433–445.

Ko, S., Eliot, A., & Kirsch, J. (2004). S-methylmethionine is both a substrate and an activator of 1-aminocyclopropane-1-carboxylate synthase. *Arch. Biochem. Biophys., 421*, 85–90.

Kotseridis, Y. (1999). Etude de l'arôme des vins de Merlot et Cabernet Sauvignon de la région bordelaise. *Doctoral dissertation, University Victor Segalen Bordeaux II.*

Kotseridis, Y., & Baumes, R. (2000). Identification of impact odorants in Bordeaux red grape juice, in the commercial yeast used for its fermentation, and in the produced wine. *J. Agric. Food Chem., 48*, 400–406.

Kotseridis, Y., Baumes, R., & Skouroumounis, G. K. (1998). Synthesis of labelled [$^2H_4$]-β-damascenone, [$^2H_2$]-2-methoxy-3-isobutylpyrazine, [$^2H_2$]-α-ionone, and [$^2H_3$]-β-ionone, for quantification in grapes, juices and wines. *J. Chromatogr. A, 824*, 71–78.

Kotseridis, Y., Baumes, R., & Skouroumounis, G. K. (1999a). Quantitative determination of free and hydrolytically liberated β-damascenone in red grapes and wines using a stable isotope dilution assay. *J. Chromatogr. A, 849*, 245–254.

Kotseridis, Y., Baumes, R., Bertrand, A., & Skouroumounis, G. (1999b). Quantitative determination of β-ionone in red wines and grapes of Bordeaux using a stable isotope dilution assay. *J. Chromatogr. A, 848*, 317–325.

Lecas, M., Günata, Z., Sapis, J.C., & Bayonove, C. (1991). Purification and partial characterization of β-glucosidase from grape. *Phytochem., 30*, 451–454.

Lestremau, F., Andersson, F.A., Desauziers, V., & Fanlo, J.L. (2003). Evaluation of Solid-Phase Microextraction for Time-Weighted Average Sampling of Volatile Sulfur Compounds at ppb Concentrations. *Anal. Chem., 75*, 2626–2632.

Lopez, R., Orti, N., Perez-Trujillo, J. P., Cacho, J., & Ferreira, V. (2003). Impact odorants of different young white wines from Canary Islands. *J. Agric. Food Chem., 51*, 3419–3425.

Loscos Deodad, N., Segurel, M., Dagan, L., Sommerer, N., Marlin, T., & Baumes, R. (2007). Identification of S-methylmethionine in Petit Manseng grapes as DMS precursor in wine. *Anal. Chim. Acta, submitted.*

Loubser, G., & Du Plessis, C. S. (1976). The quantitative determination and some values of dimethyl sulfide in white table wines. *Vitis, 15*, 248–252.

Marais, J. (1979). Effect of storage time and temperature on the formation of dimethyl sulphide and on white wine quality. *Vitis, 18*, 254–260.

Marais, J. (1983). Terpenes in the aroma on grape and wine: a review. *South Afric. J. Enol. Vitic., 4*, 49–58.

Marais, J. (1987). Terpene concentration and wine quality of *Vitis vinifera* L. cv Gewürtraminer as affected by grape maturity and cellar practices. *Vitis, 26*, 231–245.

Marais, J., Van Wyk, C., & Rapp, A. (1989). Carotenoids in grapes. In: G. Charalambous (Ed.), *Flavors and off-Flavors* (pp. 71–85). Amsterdam: Elsevier Science.

Masneuf, I., Murat, M.L., Naumov, G.I., Tominaga, T., & Dubourdieu, D. (2002). Hybrids *Saccharomyces cerevisiae* x *Saccharomyces bayanus* var. *uvarum* having a high liberating ability of some sulfur varietal aromas of *Vitis vinifera* Sauvignon Blanc. *J. Int. Sc. Vigne Vin, 36*, 205–212.

Masneuf-Pomarede, I., Mansour, C., Murat, M.L., Tominaga, T., & Dubourdieu, D. (2006). Influence of fermentation temperature on volatile thiols concentrations in Sauvignon Blanc. *Int. J. Food Microbiol., 108*, 385–390.

Mathieu, S., Terrier, N., Procureur, J., Bigey, F., & Günata, Z. (2005). A Carotenoid Cleavage Dioxygenase from *Vitis vinifera* L.: functional characterization and expression during grape berry development in relation to C13-norisoprenoid accumulation. *J. Exp. Bot., 56*, 2721–2731.

Mc Mahon, H. M., Zoecklein, B. W., & Jasinski, (1999). The effects of prefermentation maceration temperature and percent alcohol (v/v) at press on the concentration of Cabernet Sauvignon grape glycosides and glycoside fractions. *Am. J. Enol. Vitic., 50*, 385–390.

Mendes Pinto, M., Silva Ferreira, A., Oliveira, M.B., & Guedes de Pinho, P. (2004). Evaluation of some carotenoids in grapesby reversed- and normal-phase liquid chromatography: a qualitative analysis. *J. Agric. Food Chem., 52*, 3182–3188.

Moneger, R. (1968). Contribution à l'étude de l'influence exercée par la lumière sur la biosynthèse des caroténoïdes chez *Spirodela polyrrhiza* L. Schleiden. *Physiol. Vég., 6*, 165–202.

Murat, M. L., Tominaga, T., & Dubourdieu, D. (2001a). Assessing, the aromatic potential of Cabernet Sauvignon and Merlot musts used to produce rose wine by assaying the cysteinylated precursor of 3-mercaptohexan-1-ol. *J. Agric. Food Chem., 49*, 5412–5417.

Murat, M. L., Masneuf, I., Darriet, P., Lavigne, V., Tominaga, T., & Dubourdieu, D. (2001b). Effect of *Saccharomyces cerevisiae* yeast strains on the liberation of volatile thiols in Sauvignon blanc wine. *Am. J. Enol. Vitic., 52*, 136–139.

Murat, M. L., Tominaga, T., Saucier, C., Glories, Y., & Dubourdieu, D. (2003). Effect of anthocyanins on stability of a key odorous compounds, 3-mercaptohexan-1-ol, in Bordeaux rose wines. *Am. J. Enol. Vitic., 54*, 135–138.

Park, S. K., Morrison, J. C., Adams, D. O., & Noble, A. C. (1991). Distribution of free and glycosidicaly bound monoterpenes in the skin and mesocarp of Muscat of Alexandria grapes during development. *J. Agric. Food Chem., 39*, 514–518.

Park, S. K., Boulton, R. B., Bartra, E., & Noble, A. C. (1994). Incidence of volatile sulfur compounds in California wines. A preliminary survey. *Am. J. Enol. Vitic., 45*, 342–344.

Peyrot des Gachons, C., Tominaga, T., & Dubourdieu, D. (2000). Measuring the aromatic potential of *Vitis vinifera* L. cv. Sauvignon Blanc grapes by assaying S-cystein conjugates, precursors of the volatile thiols responsible for their varietal aroma. *J. Agric. Food Chem., 48*, 3387–3391.

Peyrot des Gachons, C., Tominaga, T., & Dubourdieu, D. (2002a). Localization of S-cysteine conjugates in the berry: Effect of skin contact on aromatic potential of *Vitis vinifera* L. cv. Sauvignon Blanc must. *Am. J. Enol. Vitic., 53*, 144–146.

Peyrot des Gachons, C., Tominaga, T., & Dubourdieu, D. (2002b). Sulfur aroma precusor present in S-glutathione conjugate form: Identification of S-3-(hexan-1-ol)-glutathione in must from *Vitis vinifera* L. cv. Sauvignon Blanc. *J. Agric. Food Chem., 50*, 4076–4079.

Peyrot des Gachons, C., Van Leeuwen, C., Tominaga, T., Soyer, J. P., Gaudillère, J. P., & Dubourdieu, D. (2005). Influence of water and nitrogen deficit on fruit ripening and aroma potential of *Vitis vinifera* L. cv. Sauvignon blanc in field conditions. *J. Sci. Food Agric., 85*, 73–85.

Pripis-Nicolau, L., De Revel, G., Bertrand, A., & Lonvaud-Funel, A. (2004). Methionine catabolism and production of volatile sulphur compounds by *Oenococcus oeni*. *J. Appl. Microbiol.*, 96, 1176–1184.

Puglisi, C.J., Daniel, M.A., Capone, D.L., Elsey, G.M., Prager, R.H., & Sefton, M.A. (2005). Precursors to damascenone: synthesis and hydrolysis of isomeric 3,9-dihydroxymegastigma-4,6,7-trienes. *J. Agric. Food Chem., 53*, 4895–4900.

Rapp, A., Günther, M., & Ullemeyer, H. (1985). Uber Veränderung der Aromastoffe während der Flaschenlagerung von Weissweinen der Rebsorten Riesling. *Z. Lebensm. Untersuch. Forsch., 180*, 109–116.

Razungles, A., Bayonove, C., Cordonnier, R., & Baumes, R. (1987). Etude des caroténoïdes du raisin à maturité. *Vitis, 26*, 183–191.

Razungles, A., Bayonove, C., Cordonnier R., & Sapis, J.C. (1988). Grape carotenoids: changes during the maturation period and localization in mature berries. *Am. J. Enol. Vitic., 39*, 44–48.

Razungles, A., Günata, Y., Pinatel, S., Baumes, R., & Bayonove, C. (1993). Etude quantitative de composés terpéniques, norisoprénoïdes et de leurs précurseurs dans diverses variétés de raisin. *Z. Lebensm. Untersuch. Forsch., 13*, 59–72.

Razungles, A. J., Babic, I., Sapis, J. C., & Bayonove, C. L. (1996). Particular behavior of epoxyxanthophylls during veraison and maturation of grape. *J. Agric. Food Chem., 44*, 3821–3825.

Ribéreau-Gayon, P. (1965). Identification d'esters des acides cinnamiques et de l'acide tartrique dans les limbes et les baies de *Vitis Vinifera* L. *C. R. Acad. Sci., 260*, 341.

Romeyer, F., Macheix, J., Goiffon, J., Reminiac, C., & Sapis, J. C. (1983). The browning capacity of grapes. 3. Changes and importance of hydroxycinnamic acid-tartaric acid esters during development and maturation of the fruit. *J. Agric. Food Chem., 31*, 346–349.

Roufet, M., Bayonove, C., & Cordonnier, R. (1986). Changes in fatty acids from grape lipidic fractions during crushing exposed to air. *Am. J. Enol. Vitic., 37*, 59–72.

Roufet, M., Bayonove, C., & Cordonnier, R. (1987). Etude de la composition lipidique du raisin *Vitis Vinifera* L. Evolution au cours de la maturation et localisation dans la baie. *Vitis, 26*, 85–97.

Rouillon, A., Surdin-Kerjan, Y., & Thomas, D. (1999). Transport of sulfonium compounds. Characterization of the S adenosylmethionine and S-methylmethionine permeases from the yeast *Saccharomyces cerevisiae*. *J. Biol. Chem., 274*, 28096–28105.

Sarrazin, E., Shinkaruk, S., Tominaga, T., Bennetau, B., Frerot, E., & Dubourdieu, D. (2007). Odorous impact of volatile thiols on the aroma of young botrytized sweet wines: identification and quantification of new sulfanyl alcohols. *J. Agric. Food Chem., 55*, 1437–1444.

Sarry, E., & Günata, Z. (2004). Plant and microbial glycoside hydrolases: volatile release from glycosidic aroma precursors. *Food Chem., 87*, 509–521.

Schneider, R. (2001). Contribution à la connaissance du l'arôme et du potentiel aromatique du Melon B. (*Vitis vinifera* L.) et des vins de Muscadet. *Doctoral dissertation, University Montpellier II.*

Schneider, R., Razungles, A., Charrier, F., & Baumes, R. (2002). Effet du site, de la maturité et de l'éclairement des grappes sur la composition aromatique des baies de *Vitis vinifera* L. cv. Melon B. dans le vignoble du Muscadet. *Bull. l'OIV, 75*, 269–282.

Schneider, R., Koseridis, Y., Ray, J. L., Augier, C., & Baumes, R. (2003). Quantitative determination of sulfur-containing wine odorants at sub parts per billion levels. 2. Development and application of a stable isotope dilution assay. *J. Agric.Food Chem., 51*, 3243–3248.

Schneider, R., Charrier, F., Moutounet, M., & Baumes, R. (2005). Rapid analysis of grape aroma glycoconjugates using Fourier-transform infrared spectrometry and chemometric techniques. *Anal. Chim. Acta, 513*, 91–96.

Schneider, R., Charrier, F., Razungles, A., & Baumes, R. (2006). Evidence for an alternative biogenetic pathway leading to 3-mercaptohexanol and 4-mercaptopentan-2-one in wines. *Anal. Chim. Acta, 563*, 58–64.

Schreier, P., Drawert, F., Junker, A., Barton, H., & Leupold, G. (1976). Über die Biosynthese von Aromastoffen durch Mikroorganismen. *Z. Lebensm. Untersuch. Forsch., 162*, 279–284.

Sefton, M.A. (1998). Hydrolytically-released volatile secondary metabolites from a juice sample of *Vitis vinifera* grape cv. Merlot and Cabernet Sauvignon. *Aust. J. Grape Wine Res., 4*, 30–38.

Sefton, M.A., Skouroumounis, G.K., Massy-Westropp, R.A., & Williams, P.J. (1989). Norisoprenoids in *Vitis vinifera* white wine grapes and the identification of a precursor of damascenone in these fruits. *Aust. J. Chem., 42*, 2071–2084.

Sefton, M.A., Francis, I.L., & Williams, P.J. (1993). The volatile composition of Chardonnay juice: A study by flavor precursor analysis. *Am. J. Enol. Vitic., 44*, 358–370.

Ségurel, M. (2005). Contribution des précurseurs glycosidiques et du sulfure de diméthyle des baies de *Vitis vinifera* L. cv. Grenache noir et Syrah à l'arôme des vins de la vallée du Rhône. *Doctoral dissertation, University Montpellier II.*

Ségurel, M., Razungles, A., Riou, C., Salles, M., & Baumes, R. (2004). Contribution of dimethyl sulphide to the aroma of syrah and grenache noir wines and estimation of its potential in grapes of these varieties. *J. Agric. Food Chem., 52*, 7084–7093.

Ségurel, M., Razungles, A., Riou, C., Trigueiro, M.G.L., & Baumes, R. (2005). Ability of possible precursors to release DMS during wine aging and in the conditions of heat alkaline treatment. *J. Agric. Food Chem., 53*, 2637–2645.

Shapiro, S.K., Yphantis, D.A., & Almenas, A., (1964). Biosynthesis of Methionine in *Saccharomyces cerevisiae*. Partial purification and properties of S-adenosylmethionine:homocysteine methyltransferase. *J. Biol. Chem., 239*, 1551–1556.

Silva Ferreira, A., & Guedes de Pinho, P. (2004). Nor-isoprenoids profile during Port wine ageing-Influence of some technological parameters. *Anal. Chim. Acta, 513*, 169–176.

Simpson, R. (1978). 1,1,6-trimethyl-1,2-dihydronaphtalene: an important contributor to the bottle aged bouquet of wine. *Chem. Ind. Lond., 37*, 163.

Singleton, V., Timberlake, C., & Lea, A. (1978). The phenolic cinnamates of white grapes and wines. *J. Sci. Food Agric., 29*, 403–410.

Skouroumounis, G.K., Massy-Westropp, A., & Sefton, M.A. (1992). Precursors of β-damascenone in fruit juices. *Tetrahedron Lett., 33*, 3533–3536.

Spedding, D. J., & Raut, P. (1982). The influence of dimethyl sulphide and carbon disulphide in the bouquet of wines. *Vitis, 21*, 240–246.

Spinnler, H., Berger, C., Lapadatescu, C., & Bonnarme, P. (2001). Production of sulfur compounds by several yeasts of technological interest for cheese ripening. *Int. Dairy J., 11*, 245–252.

Stahl-Biskup, E., Intert, F., Holthuijzen, J., Stengele, M., & Schulz, G. (1993). Glycosidically bound volatiles – A review 1986–1991. *Flav. Frag. J., 8*, 61–80.

Strauss, C.R., Wilson, B., Gooley, P.R., & Williams, P.J. (1986). Role of monoterpenes in grape and wine flavor. In: T.H. Parliment & R. Croteau (Eds.), *Biogeneration of aromas*, (pp. 222–242). Washington: ACS Symposium Series 317.

Strauss, C.R., Wilson, B., Anderson R., & Williams, P.J. (1987a). Development of precursors of C_{13}-norisoprenoids. Flavorant in Riesling grapes. *Am. J. Enol. Vitic., 38*, 23–27.

Strauss, C.R., Wilson, B., & Williams, P.J. (1987b). 3-Oxo-α-ionol, vomifoliol and roseoside in *Vitis vinifera* fruit. *Phytochem., 26*, 1995–1997.

Strauss, C.R., Gooley, P.R., Wilson, B., & Williams, P.J. (1987c). Application of droplet countercurrent chromatography to the analysis of conjugated forms of terpenoids, phenols and other constituents of grape juice. *J. Agric. Food Chem., 35*, 519–524.

Strauss, C.R., Wilson, B., & Williams, P.J. (1988). Novel monoterpene diols and diol glycosides in *Vitis vinifera* grapes. *J. Agric. Food Chem., 36*, 569–573.

Talou, T., Delmas, M., & Gaset, A. (1987). Principal constituents of black truffle (*Tuber melanosporum*) aroma. *J. Agric. Food Chem., 35*, 774–777.

Tominaga, T. (1998). Recherches sur l'arôme variétal des vins de *Vitis vinifera* L. cv. sauvignon blanc et sa génèse à partir de précurseurs inodores du raisin. *Doctoral dissertation, University Victor Segalen Bordeaux II.*

Tominaga, T., Masneuf, I., & Dubourdieu, D. (1995). Mise en évidence d'un S-conjugué de la cystéine, précurseur d'arôme du Sauvignon. *J. Inter. Sci. Vigne Vin, 29*, 227–232.

Tominaga, T., Darriet, P., & Dubourdieu, D. (1996). Identification de l'acétate de 3-mercaptohexanol, composé à forte odeur de buis, intervenant, dans l'arôme des vins de Sauvignon. *Vitis, 35*, 207–210.

Tominaga, T., Murat, M.L., & Dubourdieu, D. (1998a). Development of a method for analyzing the volatile thiols involved in the characteristic aroma of wines made from *Vitis vinifera* L. Cv. Sauvignon blanc. *J. Agric. Food Chem., 46*, 1044–1048.

Tominaga, T., Peyrot des Gachons, C., & Dubourdieu, D. (1998b). A new type of flavor precursors in *Vitis vinifera* L. cv. Sauvignon Blanc: S-cystein conjugates. *J. Agric. Food Chem., 46*, 5215–5219.

Tominaga, T., Baltenweck-Guyot, R., Peyrot des Gachons, C., & Dubourdieu, D. (2000a). Contribution of volatile thiols to the aromas of white wines made from several *Vitis vinifera* grape varieties. *Am. J. Enol. Vitic., 51*, 178–181.

Tominaga, T., Blanchard, L., Darriet, P., & Dubourdieu, D. (2000b). A powerful aromatic volatile thiol, 2-furanemethanethiol, exhibiting roast coffee aroma in wine made from several *Vitis vinifera* grape varieties. *J. Agric. Food Chem., 48*, 1799–1802.

Tominaga, T., Guimberteau, G., & Dubourdieu, D. (2003a). Contribution of benzenemethanethiol to smoky aroma of certain *Vitis vinifera* L. wines. *J. Agric. Food Chem., 51*, 1373–1376.

Tominaga, T., Guimberteau, G., & Dubourdieu, D. (2003b). Role of certain volatile thiols in the bouquet of aged champagne wine. *J. Agric. Food Chem., 51*, 1016–1020.

Tominaga, T., Niclass, Y., Frérot, E., & Dubourdieu, D. (2006). Stereoisomeric distribution of 3-mercaptohexan-1-ol and 3-mercaptohexyl acetate in dry and sweet white wines made from *Vitis vinifera* (Var. Sauvignon Blanc and Semillon). *J. Agric. Food Chem., 54*, 7251–7255.

Ugliano, M., Genovese, A., & Moio, L. (2003). Hydrolysis of wine aroma precursors during malolactic fermentation with four commercial starter cultures of *Oenococcuc oeni*. *J. Agric. Food Chem., 51*, 5073–5078.

Ugliano, M., Bartowsky, E., McCarthy, J., Moio, L., & Henschke, P. (2006). Hydrolysis and transformation of grape glycosidically bound volatile compounds during fermentation with three *Saccharomyces* yeast strains. *J. Agric. Food Chem., 54*, 6322–6331.

Valentin, G. (1993). Etude des enzymes impliquées dans la formation des aldéhydes à six atomes de carbone chez le raisin. *Doctoral dissertation*, University of Montpellier II.

Vermeulen, C., Lejeune, I., Tran, T., & Collin, S. (2006). Occurrence of polyfunctional thiols in fresh lager beers. *J. Agric. Food Chem., 54*, 5061–5068.

Versini, G. (1985). On the aroma of ≪ Traminer aromatico ≫ or ≪ Gewuerztraminer ≫ wine. *Vignevini, 12*, 57–65.

Versini, G., Rapp, A., Marais, J., Mattivi, F., & Spraul, M. (1996). A new 1,1,6-trimethyl-1, 2-dihydronaphtalene (TDN) precursor isolated from Riesling grape products: Partial structure elucidation and possible reaction mechanism. *Vitis, 35*, 15–21.

Voirin, S. (1990). Connaissance de l'arôme du raisin: analyse et synthèse de précurseurs hétérosidiques.*Doctoral dissertation, University Montpellier II.*

Voirin, S., Baumes, R., Bitteur, S., Günata, Z., & Bayonove, C. (1990). Novel monoterpene dissacharide glycosides of *Vitis vinifera* grapes. *J. Agric. Food Chem., 38*, 1373–1378.

Voirin, S. G., Baumes, R., Sapis, J.C., & Bayonove, C.L. (1992). Analytical methods for monoterpene glycosides in grape and wine. II. Qualitative and quantitative determination of monoterpene glycosides in grape. *J. Chromatogr., 595*, 269–281.

Wakabayashi, H., Wakabayashi, M., & Engel, K.H. (2003). β-lyase-catalysed-transformations of sulphur-containing flavour precursors. In: J.L. Lequéré & P. Etievant (Eds.), *Flavour Research at the Dawn of the Twenty First Century* (pp. 350–355). Paris: Tec & Doc.

Wakabayashi, H., Wakabayashi, M., Eisenreich, W., & Engel, K.H. (2004). Stereochemical course of the generation of 3-mercaptohexanal and 3-mercaptohexanol by β-lyase-catalysed cleavage of cystein conjugates. *J. Agric. Food Chem., 52*, 110–116.

Williams, P.J., Strauss, C.R., & Wilson, B. (1980). Hydroxylated linalool derivatives as precursors of volatile monoterpenes of Muscat grapes. *J. Agric. Food Chem., 28*, 766–771.

Williams, P.J., Strauss, C.R., & Wilson, B. (1982a). Novel monoterpene dissacharide glycoside of *Vitis vinifera* grapes and wines. *Phytochem., 21*, 2013–2020.

Williams, P.J., Strauss, C.R., Wilson, B., & Massy-Westropp, R.A. (1982b). Studies on the hydrolysis of *Vitis vinifera* monoterpene precursor compounds and model β-D-glucosides rationalizing the monoterpene composition of grape. *J. Agric. Food Chem., 30*, 1219–1223.

Williams, P.J., Strauss, C.R., Wilson, B., & Massy-Westropp, R.A. (1983). Glycosides of 2-phenylethanol and benzyl alcohol in *Vitis vinifera* grapes. *Phytochem., 22*, 2039–2041.

Williams, P.J., Strauss, C.R., Wilson, B., & Dimitriadis, E. (1984). Recent studies into grape terpene glycosides. In: J. Adda (Ed.),*Progress in Flavour Research* (pp. 349–357). Amsterdam: Eselvier Science.

Williams, P.J., Cynkar, W., Francis, I.L., Gray, J.D., Hand, P.G., & Coombe, B.G. (1995). Quantification of glycosides in grape juices and wines through a determination of glycosyl glucose. *J. Agric. Food Chem., 43*, 121–128.

Wilson, B., Strauss, C.R., & Williams, P.J. (1984). Changes in free and glycosidically bound monoterpenes in developping Muscat grapes. *J. Agric. Food Chem., 32*, 919–924.

Wilson, B., Strauss, C.R., & Williams, P.J. (1986). The distribution of free and glycosidally-bound monoterpene among skin, juice and pulp fractions of some white grape varieties. *Am. J. Enol. Vitic., 37*, 107–111.

Winterhalter, P. (1992). Oxygenated C_{13}-norisoprenoids. Important flavor precursors. In: R. Teranishi, G. Takeoka, & M. Günther (Eds.), (pp. 98–115). *Thermal and Enzymatic Conversions of Precursors to Flavor Compounds*. Washington: ACS Symposium Series 490.

Winterhalter, P. (1993). The generation of C_{13}-norisoprenoid volatiles in Riesling wine. In: J. Crouzet, C. Flanzy, C. Martin, & J.C. Sapis (Eds.), *Connaissance aromatique des cépages et qualité des vins* (pp.65–73). Montpellier: Revue française d'oenologie.

Winterhalter, P., & Rouseff, R.(2000). Carotenoid-derived aroma compounds. In P. Winterhalter & R. Rouseff (Eds.),*Carotenoid-derived aroma compounds: an introduction* (pp. 1–17). Washington, DC: ACS Symposium Series 802.

Winterhalter, P., & Skouroumounis, G.K. (1997). Glycoconjugated aroma compounds: occurence, role and biotechnological transformation. *Adv. Biochem. Eng./Biotechnol., 55*, 73–105.

Wirth, J. (2001). Etude de la filiation entre les norisoprénoïdes et les caroténoïdes chez la vigne. *Doctoral dissertation*, University of Montpellier II.

Wirth, J., Guo, W., Baumes, R., & Günata, Z. (2001). Volatile compounds released by enzymatic hydrolysis of glycoconjugates of leaves and grape berries from *Vitis vinifera* Muscat of Alexandria and Shiraz cultivars. *J. Agric. Food Chem., 49*, 2917–2923.

Wirth, J., Sauvage, F.X., Baumes, R., & Günata, Z. (2003). Biogenesis of C_{13}-norisoprenoids in grape: biotransformation of $C_1$3-norisoprenoids by a cell suspension culture of cv. Gamay. In: J.L. Le Quéré & P.X. Etiévant (Eds.), *Flavour research at the dawn of the twenty-first century* (pp 393–396). Cachan, France: Tec. et Doc. Lavoisier.

Chapter 8B
Polyfunctional Thiol Compounds

Denis Dubourdieu and Takatoshi Tominaga

Contents

8B.1 Introduction

Sulfur compounds, and especially volatile thiols, are extremely odoriferous molecules, which contribute to the aromas of many fruits, plants, and foods (Shankaranarayana et al. 1982). For example, certain thiols contribute to the characteristic aroma of blackcurrant (Rigaud et al. 1986), grapefruit (Demole et al. 1982; Buettner and Schieberle 1999), passion fruit (Engel and Tressl 1991), guava (Idstein and Schreier 1985), and fringed rue (*Ruta chalepensis*) (Escher et al. 2006). The key role played by thiols in the odor of roasted coffee (Tressl and Silwar 1981), popcorn, and grilled meat is also well known (Gasser and Grosch 1988; Guth and Grosch 1994; Schieberle 1991). Finally, the contribution of compounds in this family to the aroma of beer was recently reported (Vermeulen et al. 2006; Takoi et al. 2007).

In wine, sulfur compounds (including thiols) were initially exclusively associated with fetid, nauseating odors, mainly due to molecules such as hydrogen sulfide, methanethiol, ethanethiol, and methionol. The involvement of sulfur compounds in the pleasant herbaceous, fruity, mineral, smoky, and toasty aromas of wine was only detected at a later stage, thanks to a specific method for purifying volatile thiols, using *p*-hydroxymercuribenzoic acid (*p*-HMB) (Tominaga et al. 1998a; Tominaga and Dubourdieu 2006).

D. Dubourdieu (✉)
Université Victor Segalen Bordeaux 2, Faculté d' Œnologie, UMR, Œnologie, INRA, Institut des Sciences de la Vigne et du Vin, 351 cours de la Libération, 33405 Talence Cedex, France
e-mail: denisdubourdieu@wanadoo.fr

M.V. Moreno-Arribas, M.C. Polo (eds.), *Wine Chemistry and Biochemistry*,
DOI 10.1007/978-0-387-74118-5_12,

8B.2 Volatile Thiols Involved in the Varietal Aromas of Wine

Since the early 1990s, several highly odoriferous volatile thiols have been identified in wines made from *Vitis vinifera* L. cv. Sauvignon blanc. These cover a broad aromatic palette, including such key descriptors as: green pepper, boxwood, broom, eucalyptus, blackcurrant buds, rhubarb, tomato leaves, nettles, grapefruit, passion fruit, white peaches, gooseberries, and asparagus broth, as well as acacia flowers and wood. After several years' bottle-aging, some Sauvignon blanc wines also have hints of smoke, roast meat, or even truffles. For many years, besides methoxypyrazines (Augustyn et al. 1982; Allen et al. 1991) with their green pepper aroma, the compounds responsible for the other odoriferous characteristics of Sauvignon blanc wines had not been identified.

The first molecule found to be a characteristic component of Sauvignon blanc wine aromas was 4-methy-4-sulfanylpentan-2-one (4MSP: **III**) (Darriet 1993; Darriet et al. 1995) (Table 8B.1). Assays of this mercaptoketone revealed concentrations considerably higher than the perception threshold in certain Sauvignon blanc wines (Tominaga et al. 1998a), confirming its decisive organoleptic role in wines characteristic of this grape variety. This compound, with its marked aromas of "boxwood" and "broom", has an extremely low perception threshold (0.8 ng/L in model solution) and concentrations in Sauvignon blanc wines where these aromas are very strong may be as high as 40 ng/L (Tables 8B.2 and 8B.3). This compound is found at concentrations ranging from a few to around a 100 ng/L in fresh box leaves and leafy broom branches (Tominaga and Dubourdieu 1997). Consequently, the "boxwood" and "broom" descriptors, used for many years to describe Sauvignon blanc aromas, actually reflect a chemical reality.

Several other odoriferous volatile thiols were later identified in Sauvignon blanc wines: 3-sulfanylhexyl acetate (3SHA: **XVI**) (Tominaga et al. 1996), 4-methyl-4-sulfanylpentan-2-ol (4MSPOH: **VI**), 3-sulfanylhexan-1-ol (3SH: **XVII**), and 3-methyl-3-sulfanylbutan-1-ol (3MSBOH: **VIX**) (Tominaga et al. 1998b) (Table 8B.1).

The complex aroma of 3-sulfanylhexyl acetate (3SHA) is reminiscent of boxwood, but also grapefruit zest and passion fruit. It had already been identified in that fruit by Engel and Tressl (1991), who described it as: "extremely fruity, reminiscent of passion fruit, with hints of Riesling". The concentrations of 3SHA in Sauvignon blanc wines analyzed ranged from 0 to several hundred ng/L. In view of its perception threshold, 4 ng/L, its olfactory impact is highly variable, but may be considerable in some young wines. During aging, 3SHA hydrolyzes to form 3SH.

3-Sulfanylhexan-1-ol (3SH) is always present in Sauvignon blanc, at concentrations of several hundred ng/L, ranging up to a few µg/L in some wines. As a result, this compound, which smells of grapefruit and has a perception threshold of 60 ng/L, also has considerable olfactory impact. Wines containing the most 3SH also have the highest concentrations of its acetate. The citrus zest aroma of 4-methyl-4-sulfanylpentan-2-ol (4MSOH) plays a more limited organoleptic role. Concentrations rarely exceed the perception threshold (55ng/L), although it may reach this value in a few wines (Tables 8B.2 and 8B.3).

Table 8B.1 Volatile thiols identified in *Vitis vinifera* wines

Compound	LRI (BP20)	Descriptor	Formula
I: Ethyl-2-sulfanylpropionate	n.d.	fruity	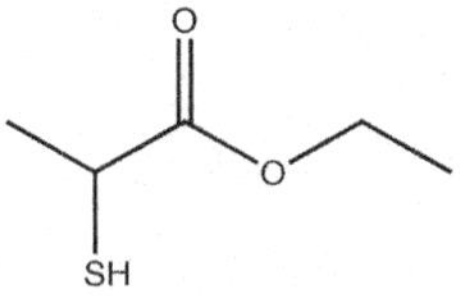
II: 2-Methyl-3-furanthiol	1341	meaty	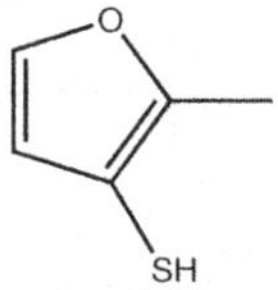
III: 4-Methyl-4-sulfanylpentan-2-one	1389	box-tree	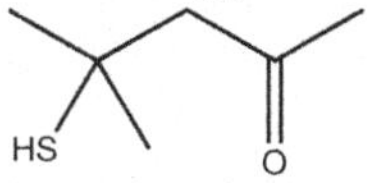
IV: 2-Furanmethanethiol	1420	Coffee	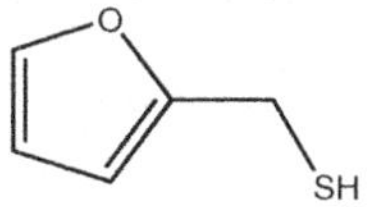
V: 2-Sulfanylethyl acetate	1420	meaty	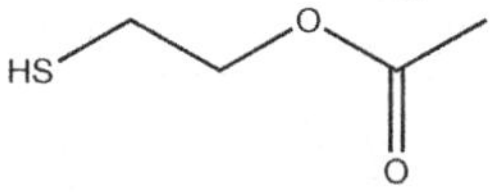
VI: 4-Methyl-4-sulfanylpentan-2-ol	1568	Citrus zest	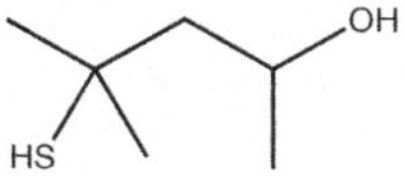
VII: Ethyl-3-sulfanylpropionate	n.d.	meaty	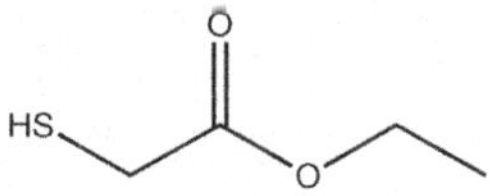
VIII: 3-Sulfanylbutan-1-ol	1670	onion, leek	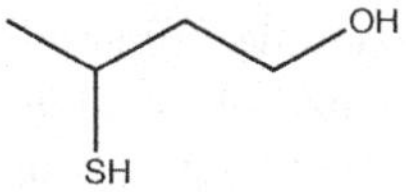
IX: 3-Methyl-3-sulfanylbutan-1-ol	1677	Cooked leek	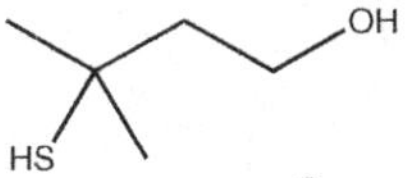
X: Sulfanylpropyl acetate	n.d.	Meaty	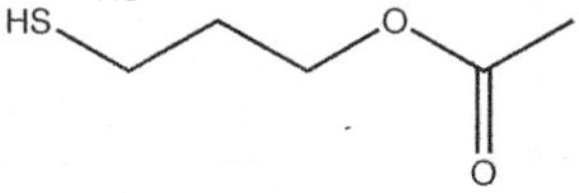
XI: 2-Methyl-3-sulfanylpropan–1-ol	1745	Broth, sweat	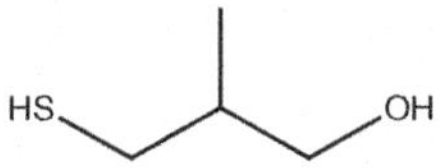

Table 8B.1 (continued)

XII: 2-Methyl-3-sulfanylbutan-1-ol	1745	Raw onion	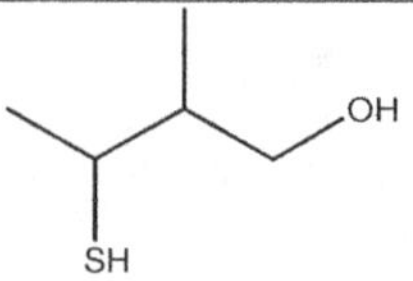
XIII: Benzenemethanethiol	n.d.	Smoky	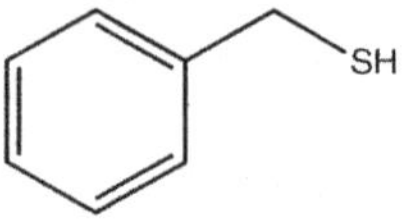
XIV: 3-Sulfanylpentan-1-ol	1772	Grapefruit	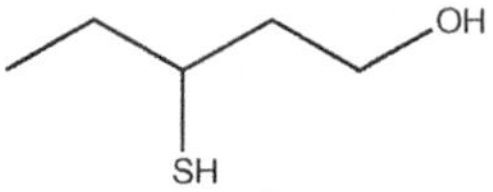
XV: 2-Methyl-3-sulfanylpentan-1-ol	1833	Raw onion	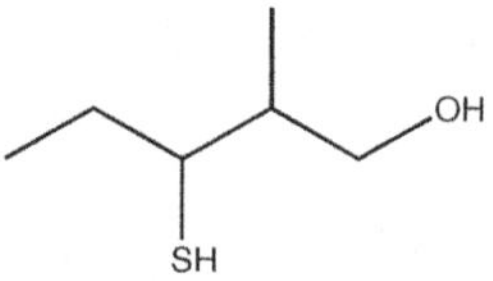
XVI: 3-Sulfanyl-hexyl acetate	n.d.	Box-tree	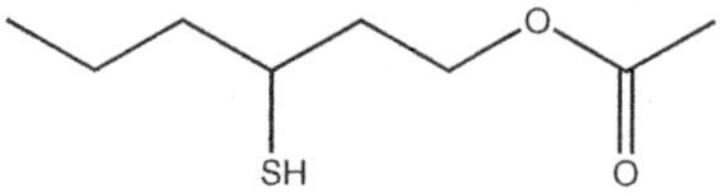
XVII: 3-Sulfanylhexan-1-ol	1857	Passion fruit Grapefruit	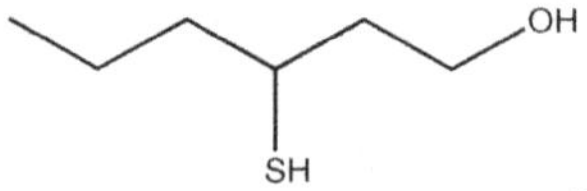
XVIII: 3-Sulfanylheptan-1-ol	1956	Grapefruit	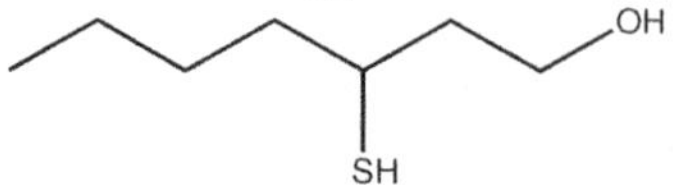

These volatile thiols are also present in wines made from other *Vitis vinifera* grape varieties (Tominaga et al. 2000a). Concentrations of 4MSPOH and 3SH are particularly high in white Muscat d'Alsace and Gew`urztraminer from Alsace, respectively (Fig. 8B.1) (Tominaga et al. 2001). Thus, 4MSPOH gives Muscat wines their "Sauvignon blanc" aroma, while 3SH, with its hints of grapefruit and tropical fruit, makes a significant contribution to Gewürztraminer aromas.

Table 8B.2 Assay of volatile thiols (ng/L) in four white Bordeaux wines made from Sauvignon blanc 1996 vintage (Tominaga et al. 1998a)

Compounds	Samples			
	1	2	3	4
4-Methyl-4-sulfanylpentan-2-one	5(6)	4(5)	10(13)	4(5)
4-Methyl-4-sulfanylpentan-2-ol	18(0.3)	20(0.4)	22(0.4)	20(0.4)
3-Sulfanylhexan-1-ol	8402(140)	12822(214)	7465(123)	3736(63)
3-Sulfanylhexyl acetate	724(181)	451(113)	451(113)	275(69)

Entries in parentheses are the aromatic index of each compound: concentration/perception threshold

Table 8B.3 Assay of volatile thiols (ng/L) in Sauvignon blanc wines from the same Bordeaux estate in several vintages (Tominaga et al. 1998a)

Compounds Compounds	Samples			
	1992	1993	1994	1995
4-Methyl-4-sulfanylpentan-2-one	7(9)	45(50)	10(13)	44(55)
4-Methyl-4-sulfanylpentan-2-ol	46(0.8)	111(2)	25(0.5)	33(0.6)
3-Sulfanylhexan-1-ol	871(15)	1178(20)	600(10)	1686(28)
3-Sulfanylhexyl acetate	0(0)	0(0)	0.4(0.1)	2.8(0.7)

Entries in parentheses are the aromatic index of each compound: concentration/perception threshold

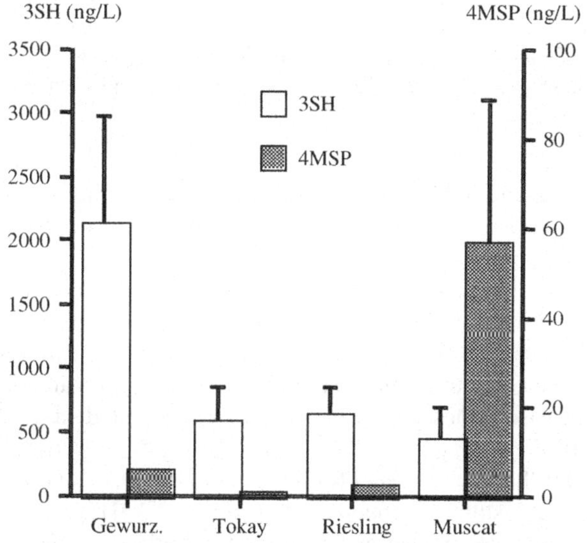

Fig. 8B.1 3SH and 4MSP concentrations in wines made from Alsace grape varieties (n = 5) (Tominaga et al. 2001)

Guth (1997) demonstrated the major olfactory impact of 4MSP in Scheurebe wines, which may contain as much as 400 ng/L.

3SH is also present at concentrations well above its perception threshold in wines made from Petit Manseng, Gros Manseng, Colombard, and botrytized Sémillon (Tominaga et al. 2000a) (Table 8B.4). This volatile thiol is present in small quantities, but still above the perception threshold, in Melon de Bourgogne (Schneider 2001) and Chardonnay wines (Tominaga, unpublished results). More recently, these volatile thiols were identified in wines from the Canary Islands (Lopez et al. 2003) and those made from Petite Arvine, a Swiss grape variety (Fretz et al. 2005). Similarly, 3SHA is found in young dry white wines made from Petit Manseng, Gros Manseng, and Colombard (Table 8B.4) (Tominaga et al. 2001).

The presence of 3SH was reported in red Cabernet and Merlot wines (Bouchilloux et al. 1998a; Kotseridis et al. 2000) and it is probably present in wine made from other red grape varieties. However, after alcoholic fermentation, especially

Table 8B.4 Volatile thiol concentrations (ng/L) in wines made from several Vitis vinifera grape varieties (Tominaga et al. 2000)

	4MSP	4MSOH	3SH	3SHA
Colombard				
Plaimont-98	n.d.	n.d.	423(7)	21(5)
Cave de Panjas-98	n.d.	n.d.	1053(18)	63(15)
Petit manseng				
Domaine Cauhapé-96	n.d.	n.d.	4468(74)	101(25)
Domaine Cauhapé-94			3747(62)	
Semillon (botrytized)				
Doisy-Daëne-95	40(50)	6(–)	4048(67)	n.d.
Doisy-Daëne-94	15(19)	13(–)	5969(99)	n.d.
Doisy-Daëne-90	8.5(10)	n.d.	5040(84)	n.d.

The number after the sample name signifies the vintage n.d.: not detected
Entries in parentheses are the aromatic index of each compound: concentration/perception threshold

during barrel-aging, the 3SH content of red wines decreases considerably, from a few µg/L to 300–600 ng/L after 12 months in barrel, as it oxidizes easily and is highly reactive with quinones.

3SH and 3SHA also contribute to the fruity aroma of some rosé wines made from Cabernet and Merlot (Murat et al. 2001a). Protecting rosé wines from oxidation (by aging in vat on fine lees and keeping racking to a minimum) preserves their high thiol content. Furthermore, the protective effect of anthocyanins on the volatile thiols in rosé wines has been clearly demonstrated in a model medium (Murat et al. 2003).

Certain volatile thiols, e.g. 3SH (Tominaga et al. 2000a), 3-sulfanylpentan-1-ol (3SPOH: **XIV**), and 3-sulfanylheptan-1-ol (3SHpOH: **XVIII**) (Sarrazin et al. 2007), may significantly enhance the varietal aromas of botrytized sweet white wines. The last two have pronounced aromas of citrus zest, while 3SH smells fruity. Table 8B.5 shows the concentrations of these three volatile thiols in botrytized white wines from various Bordeaux appellations. 3SH is remarkably stable: certain great Sauternes may still contain 5000 ng/L after around ten years' bottle-aging. In view of its perception threshold (60 ng/L in a model medium), this compound has an undeniable impact on the fruity aroma of botrytized wines. Concentrations of 3SPOH in

Table 8B.5 Assay of volatile thiols (ng/L) in young botrytized wines from various appellations of the Bordeaux region (Sarrazin et al. 2007)

Estate	Appellation	Vintage	3SH	3SPOH	3SHpOH
Doisy Daëne	Sauternes	2001	7033	299	63
Tour Blanche	Sauternes	2001	5606	217	59
Cantegrille	Barsac	2001	5034	223	44
Doisy Daëne	Sauternes	2002	4765	203	48
Dauphiné Rondillon	Loupiac	2002	4749	235	72
Doisy Daëne	Sauternes	2003	5386	199	44
Dauphiné Rondillon	Loupiac	2003	2450	91	26

Table 8B.6 Assay of volatile thiols (ng/L) at different stages in botrytizationyoung botrytized wines from various appellations of the Bordeaux region. Comparison with decrease in mean grape volume (mL/grape) (Sarrazin et al. 2007)

Variety	Botrytis stage	Mean grape volume[a]	Variation of mean grape volume (%)	3SH		3SPOH		3SHpOH	
Semillon	*Healthy*[b]	0.85	100	195[c]	±58[d]	tr[e]		tr	
	Pourri plein	0.68	80	2326	±419	93	±14	34	±5
	Pourri rôti	0.37	44	3678	±1765	124	±54	50	±26
	Late pourri rôti	0.38	45	6334	±1267	291	±128	118	±13
Sauvignon	*Healthy*	0.78	100	161	±27		tr	tr	
	Pourri plein	0.52	67	3003	±300	141	±8	95	±39
	Pourri rôti	0.21	27	9648	±1544	348	±42	263	±92
	Late pourri rôti	0.29	37	9319	±2050	375	±71	258	±44

[a]Based on the volume obtained from crushing 1000 grapes (mL/grape)
[b]*Healthy* (grapes not infected by *B. cinerea*), *pourri plein* (grapes entirely botrytized but not desiccated, picked two weeks after healthy grapes), *pourri rôti* (grapes botrytized and desiccated, picked two weeks after *pourri plein* grapes), and *late pourri rôti* (shriveled grapes left a further 10 days before picking)
[c]Mean value (n=3)
[d]Standard deviation s (n=3)
[e]tr: traces

botrytized wines are always well below the perception threshold (900 ng/L), while 3SHpOH rarely exceeds its perception threshold (60 ng/L). However an additive effect of these volatile thiols combined with 3SH has been clearly demonstrated (Sarrazin et al. 2007). Therefore, 3SPOH and 3SHpOH have a considerable impact on overall wine aroma in the presence of other volatile thiols, such as 3SH, although their concentrations in wines remain below the perception threshold.

The concentrations of 3SH, 3SPOH, and 3SHpOH in botrytized wines are strongly affected by the development of *Botrytis cinerea* (Table 8B.6). Wines made from healthy grapes contain 3SH but only traces of the other two thiols. Wines produced from grapes affected by noble rot, once they reach the *"pourri plein"* stage (entirely botrytized but not desiccated), have much higher thiol concentrations. The surprisingly high concentrations of volatile thiols in botrytized wines is due to their high cysteinylated precursor content (Thibon et al. 2008a).

8B.3 Stereoisomeric Distribution of 3-Sulfanylhexan-1-ol and its Acetate

In dry Sauvignon and Sémillon wines made from healthy grape must, totally unaffected by *Botrytis cinerea*, the R and S enantiomers of 3SH are relatively uniformly distributed (approximately 50:50), with slight variations from one vintage to another; e.g. there were slightly higher quantities of the R form of 3SH in wines from

Table 8B.7 Distribution of the two enantiomers of 3SH and 3SHA in dry white and sweet wines made from different grape varieties in different vintages. (Tominaga et al., 2006)

Vintages	Wine type	Varieties	3SH (R:S)	3SHA (R:S)
	Dry	Sauvignon 1	45:55	30:70
	Dry	Sauvignon 2	44:56	32:68
2004	Dry	Semillon 1	49:52	28:72
	Sweet	Semillon 1	24:76	n.d.
	Sweet	Semillon 2	34:66	n.d.
	Dry	Sauvignon 1	51:49	n.d.
	Dry	Sauvignon 2	57:43	n.d.
	Dry	Sauvignon 3	55:45	27:73
	Dry	Semillon 1	52:48	n.d.
2005	Dry	Semillon 2	51:49	n.d.
	Dry	Semillon 3	52:48	n.d.
	Sweet	Sauvignon	34:66	n.d.
	Sweet	Semillon 1	33:67	n.d.
	Sweet	Semillon 2	32:68	n.d.

n.d.: not detected

the 2005 vintage (Table 8B.7). However, in sweet Sémillon wines made from botrytized grapes, the ratio of the two enantiomers of 3SH (30:70) is very different from that in dry wines, irrespective of the vintage (Table 8B.7) (Tominaga et al. 2006).

The ratio of the two enantiomers of 3SHA was only determined in dry wines as this compound is never present in sweet, botrytized wines. The distribution of the two 3SHA enantiomers (R:S) in dry wines was approximately 30:70 for both grape varieties studied (Table 8B.7).

The yeast strain used for fermentation had no impact on the enantiomer distribution of these volatile thiols. 3SHA is generally considered to be formed by esterification of 3SH by yeast during alcoholic fermentation. The esterase or lipase involved probably acetylates 3SH with a certain enantioselectivity. In contrast, the enantiomer distribution of 3SH in wine made from botrytized grapes *(Botrytis cinerea)* is 25:75 in favor of the S form, which has also been found in botrytized must (Thibon et al. 2007, 2008a).

The perception thresholds of the R and S forms of 3SH are very similar. However these two enantiomers have different aromas that were significantly distinguished in the triangular tasting test. The R form's aroma is reminiscent of grapefruit, while the S form smells of passion fruit. The perception threshold of the S form of 3SHA is approximately four times lower than that of the R form. Furthermore, the S enantiomer is three times more abundant in wine than the R form. These two enantiomers have different aromas: the R form is more herbaceous, evocative of boxwood, while the S form is fruitier, reminiscent of passion fruit. These differences in aroma are in agreement with the findings of Weber et al. (1992). Significantly, these experiments demonstrated that the enantiomer distribution of thiols such as 3SH and 3SHA must be taken into account in assessing the olfactory impact of these compounds in wine (Table 8B.8).

Table 8B.8 Perception threshold and olfactory descriptors of the two enantiomers of 3SH and 3SHA (Tominaga et al. 2006)

Compounds	Enantiomeric forms	Threshold (ng/L)	Olfactory descriptors
	R	50	Grapefruits
	S	60	Passion fruit
3SH	Racemic mixture	60	Grapefruits
	R	9	Passion fruit
3SHA	*S*	2.5	Boxwood
	Racemic mixture	4	Boxwood
			Passion fruit

8B.4 Other Volatile Thiols also Contribute a Variety of Aromatic Nuances to Wines

Reports have also described volatile thiols with various aromatic nuances, such as cooked onions, soup broth, and grilled or smoked meat (Table 8B.1). 3-Methyl-3-sulfanylbutan-1-ol (**VIX**), which smells of cooked leeks, was initially identified in Sauvignon blanc wines (Tominaga et al. 1998b). Concentrations of this volatile thiol are always well below the perception threshold (1500 ng/L), so it is unlikely to make a significant contribution to Sauvignon blanc varietal aroma. The presence of 3-sulfanylbutan-1-ol (**VIII**) in Sauvignon blanc wines (Tominaga 1998) and 2-methyl-3-sulfanylpropan-1-ol (**XI**) (Bouchilloux et al. 1998a) in red Bordeaux wines has also been reported. The latter, with an asymmetrical carbon, has both R and S stereoisomers, but only the R form is found in wine (Bouchilloux et al. 2000). More recently, 2-methyl-3-sulfanylbutan-1-ol (**XII**) and 2-methyl-3-sulfanylpentan-1-ol (**XV**), which smell strongly of raw onions, were identified in Sauternes wines (Sarrazin et al. 2007). The latter was previously detected in raw onions (Widder et al. 2000).

Empyreumatic odors, evocative of smoke, burnt wood, and cooked meat, are common descriptors used by wine tasters. Until now, a few sulfur compounds with these odors had been identified and assayed in wine at concentrations above their perception thresholds: methylthiopropionic acid (Schreier and Drawert 1974), 2-sulfanylethyl acetate (**V**), and 3-sulfanylptopropyl acetate (**X**) (Lavigne et al. 1998). Two mercaptans, ethyl 3-sulfanylpropionate (**VII**) and ethyl 2-sulfanylpropionate (**I**), have been identified as constituents in the aroma of *Vitis labrusca*, Concord grapes (Kolor 1983) and both have also been found in *Vitis vinifera* wine (Tominaga et al. 2003a; Blanchard 2000).

The presence of 2-methy-3-furanthiol (**II**), an odoriferous compound evocative of cooked meat, has been reported in red Bordeaux (Bouchilloux et al. 1998b; Kotseridis et al. 2000) and de Rioja wines (Aznar et al. 2001). Concentrations of this compound in wine vary from 25 to 140ng/L, with the highest values in Champagnes. As the perception threshold of 2-methyl-3-furanethiol in model dilute alcohol solution is 4 ng/L, this thiol certainly contributes to the "toasty" aroma in wines.

Sauvignon blanc wine aromas often have both mineral and smoky nuances, described as "flinty", which is why this grape variety is sometimes known as "Fumé

Blanc" (Galet 1968). Benzenemethanethiol (**XIII**), a volatile thiol with a strong empyreumatic odor, described as "flint", was identified in boxwood, as well as red and white *Vitis vinifera* L. wines. Its perception threshold is on the order of 0.3 ng/L (Tominaga et al. 2003b). This volatile thiol has been assayed in several white Sauvignon blanc, Sémillon, and Chardonnay wines from various French appellations, as well as in a few red Bordeaux wines made from Cabernet Sauvignon and Merlot.

The benzenemethanethiol content of the Chardonnay wines, irrespective of their provenance, was two- or threefold higher than that of wines made from the other grape varieties studied (Fig. 8B.2). Sauvignon blanc wines from the Loire Valley and Bordeaux, as well as Sémillon from Bordeaux, had comparable benzenemethanethiol concentrations (10–15 ng/L). The red Bordeaux wines analyzed, from various different vintages and appellations, had benzenemethanethiol concentrations around 10 ng/L.

The composition of the odoriferous compounds responsible for hints of roasting coffee in wines fermented in new barrels was unknown until quite recently. Furan aldehydes, particularly furfural, smell toasty, but their perception threshold is much higher than the concentrations found in wine. Their olfactory impact is, therefore, negligible. Research by Tominaga et al. (2000b) demonstrated that the key compound in the roasting coffee aroma of barrel-aged wines was 2-furanemethanethiol (2FM) (**IV**), an extremely odoriferous volatile thiol (perception threshold: 0.4 ng/L). This molecule was first identified in roasted coffee, as early as 1926 (Reichstein and Staudinger 1926)

The mechanisms responsible for 2FM formation in wine have not yet been completely elucidated. During alcoholic fermentation of white wines in barrel, it is produced by bio-conversion of furfural released by the oak, via the yeasts' sulfur metabolism (Blanchard et al. 2001). Only traces of this volatile thiol are found in white wines fermented in stainless-steel vats. In red Bordeaux wines,

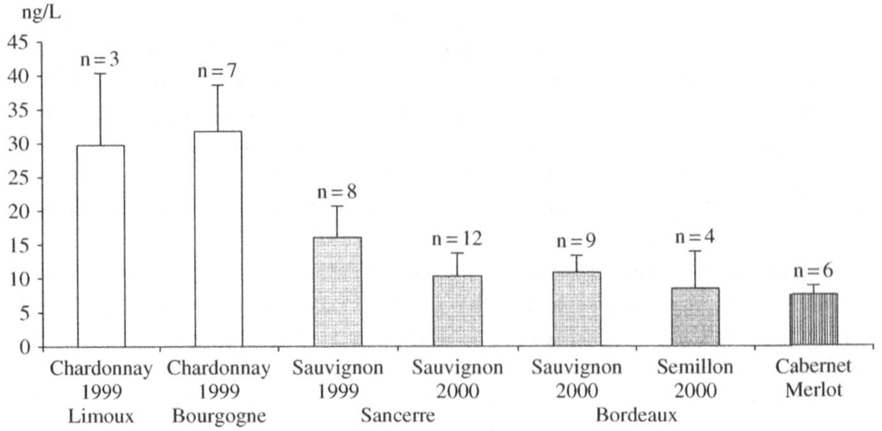

Fig. 8B.2 Concentrations (ng/L) of benzenemethanethiol in some white and red wines made from different *Vitis vinifera* grape varieties (Tominaga et al. 2003)

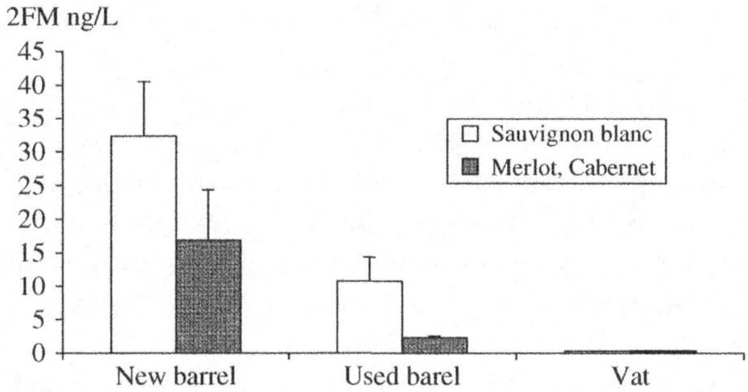

Fig. 8B.3 Contribution of oak barrels to 2FM formation in Sauvignon blanc and red Bordeaux wines (n = 4) (Tominaga et al. 2001)

2FM and its roasting coffee aroma develop at varying rates after the wine is put into new oak barrels (Fig. 8B.3) (Tominaga et al. 2001). Concentrations of this volatile thiol are always lower in young red wines than white wines. However the concentration/perception threshold ratio of 2FM in some red wines of varying ages remains high, so it very probably contributes to the toasty nuances in their aroma.

Furthermore, the contribution of 2FM to the roasting coffee bouquet characteristic of certain old champagnes was also demonstrated by Tominaga et al. (2003a). The 2FM content in the Champagnes analyzed increased in proportion to bottle aging time, both before and after disgorging (Fig. 8B.4).

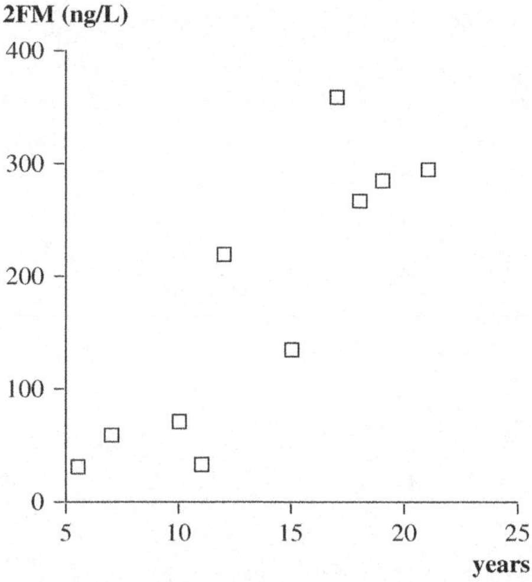

Fig. 8B.4 Development of 2FM content during bottle aging in champagne wines

8B.5 Precursors of Volatile Thiols Derived from Cysteine

Sauvignon blanc musts, like those of many grape varieties except Muscats, are not highly odoriferous. The characteristic aroma of the grape variety appears during alcoholic fermentation. Peynaud (1980) had a remarkable intuition of the existence of aroma precursors in Sauvignon blanc must, which he described as follows: 'When you taste a thick-skinned, golden Sauvignon blanc grape, you can detect its characteristic flavor, although it is not very intense. In the same way, freshly pressed juice is not highly odoriferous, and the initial flavor is quite discreet. Between 20 and 30 s after you have swallowed it, an intense aromatic Sauvignon blanc aftertaste suddenly appears in the rear nasal cavity. Fermentation brings out the primary aroma hidden in the fruit. Wine has more fruit aroma than grapes, etc. Fermentation reveals the aroma and releases the odoriferous substances from the grapes. This "aroma recurrence" or "aftertaste" could be explained by the action of the mouth enzymes on the aroma precursors.

The existence of a non-glycosilated precursor of 4MSP in Sauvignon blanc grapes and must was first demonstrated by Darriet (1993). The structure of this compound was elucidated by Tominaga et al. (1995, 1998c). A β-lyase specific to *S*-cysteine conjugates was used to release 3-SH, 4-MSP and 4-MSPOH from a non-volatile extract of Sauvignon blanc aroma precursors, suggesting that these three thiols were present in grapes in cysteinylated form: *S*-3-(hexan-1-ol)-L-cysteine, *S*-4-(4-methylpentan-2-one)-L-cysteine, and *S*-4-(4-methylpentan-2-ol)-L-cysteine (Fig. 8B.5). This was formally established by gas-phase chromatography/mass spectrometry of the precursors in trimethylsilylated form. *S*-3-(Hexan-1-ol)-L-cysteine has also been identified in passion fruit juice (Tominaga and Dubourdieu 2000) (Fig. 8B.6).

Methods to measure cysteinylated aroma precursors in must were developed by Peyrot des Gachons et al. (2000) and Murat et al. (2001b). They allowed the location of thiol precursors to be determined in Sauvignon blanc grapes (Fig. 8B.7). Around 80% of P-4MSP is in the juice and 50% of P-3SH in the skin. The skin contact improves the must aromatic potential, but its main effect is on the P-3SH content (Fig. 8B.8). Similarly, in Cabernet Sauvignon and Merlot grapes, 60% of the P-3SH is located in the skins This method has been also applied to study the influence of

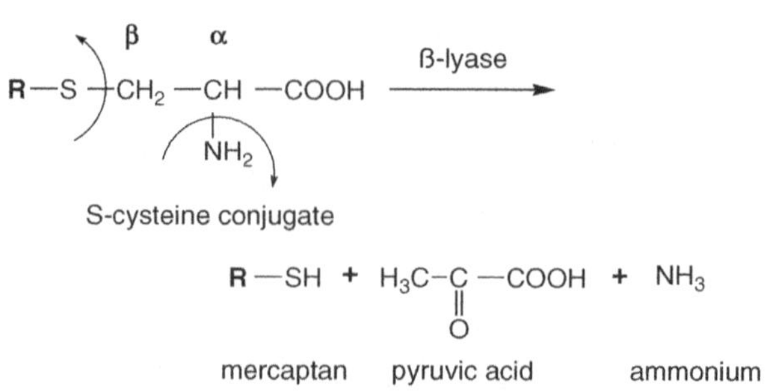

Fig. 8B.5 α,β-Elimination reaction of *S*-cysteine conjugate β-lyase (EC 4.4.1.13)

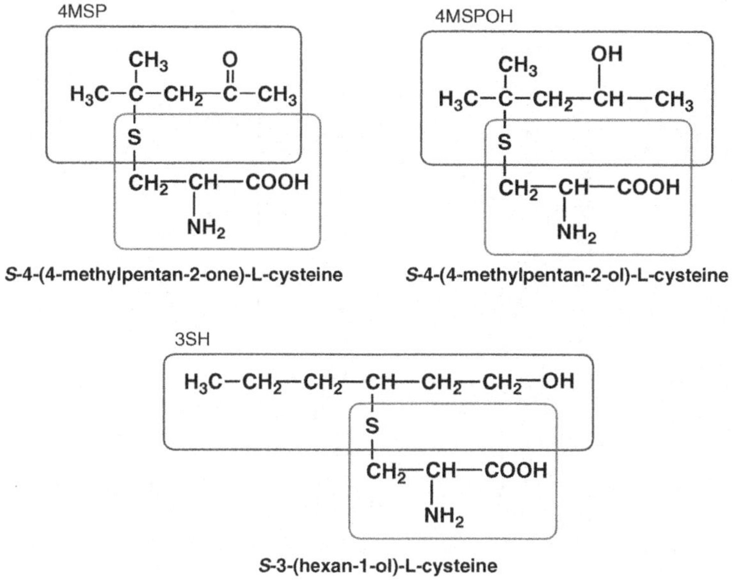

Fig. 8B.6 Structure of three aroma precursors of sauvignon: *S*-cyteine conjugates

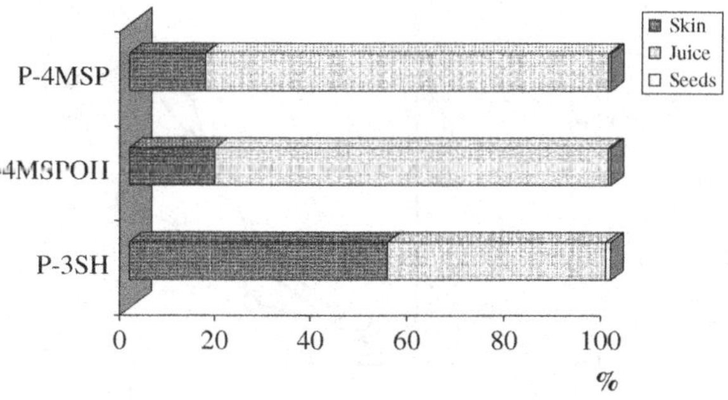

Fig. 8B.7 Distribution of cysteinylated precursors in a ripe berry of sauvignon

viticulture conditions on Sauvignon blanc aromatic maturity (Peyrot des Gachons et al. 2002a; Murat et al. 2001b).

More recently, a new method was developed to determine the diastereoisomeric distribution of P-3SH in grape juice. P-3SH assays in Bordeaux white grape juice affected by *Botrytis cinerea* showed an unusually increased proportion of the *RS* form of the precursor (approximately *RR*:*RS*=30:70) as compared to a diastereoisomer ratio (in the vicinity of 50:50) in healthy grape juice (Thibon et al. 2007, 2008a)

Sauvignon aroma is enhanced during fermentation (Fig. 8B.9) because of the conversion of cysteinylated aroma precursors. 3SH is released during the fermentation of a model medium added with *S*-3-(hexan-1-ol)-L-cysteine. When fermentation is inhibited by pimaricin, the development of aroma and the degradation of

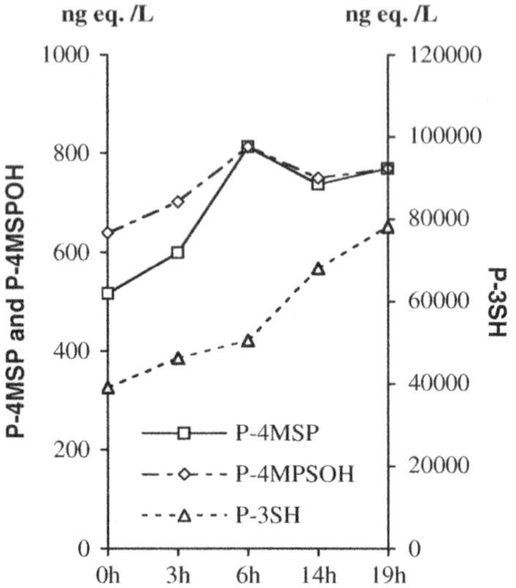

Fig. 8B.8 Influence of skin contact on the cysteinylated precursors of the must (Peyrot des Gachons et al. 2002a)

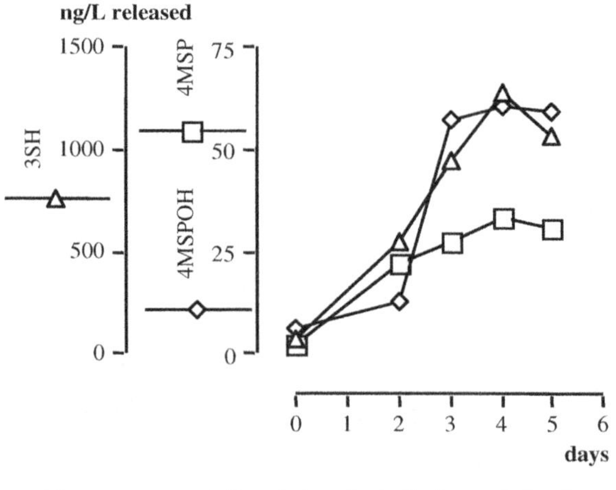

Fig. 8B.9 Sauvignon blanc aroma revelation during alcoholic fermentation in practical conditions: yeast strain VL3c (Tominaga et al. 1998c)

the precursor are very limited (Fig. 8B.10). The molar concentrations of 4-MSP 4-MSPOH, and 3-SH formed in Sauvignon blanc must (or a model medium supplemented with precursors) during alcoholic fermentation only account for less than 10% of the precursors degraded (Fig. 8B.11).

To be transformed in aroma, the precursor must enter into the yeast, then be cleaved by α–β elimination of the cysteinylated precursor and finally excreted in the medium where it is more or less stable according its reactivity with other compounds. Obviously, the yield of this transformation can be influenced by several

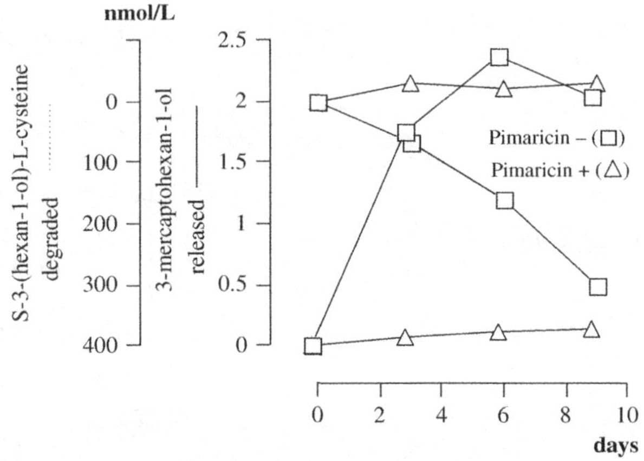

Fig. 8B.10 Release of 3SH during alcoholic fermentation from a model medium to which synthesized *S*-3-(hexan-1-ol)-L-cysteine was added in the absence (□) and presence (Δ) of pimaricin (Tominaga et al. 1998c)

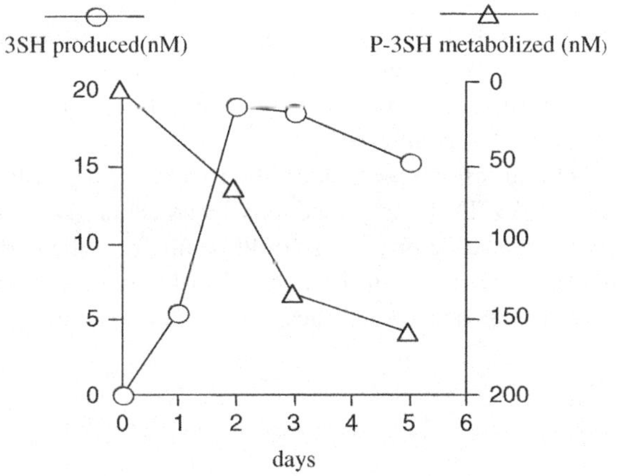

Fig. 8B.11 Production of 3SH and degradation of its precursor (P-3MH) during alcoholic fermentation of a Sauvignon blanc juice in practical conditions

factors such as yeast strains and species must composition and fermentation conditions. The yeast genes and enzymes involved in these reactions are to day better known (Howell et al. 2005; Thibon et al. 2008b).

The ability of *Saccharomyces cerevisiae* strains to enhance Sauvignon blanc aroma is variable (Murat et al. 2001c). Some strains produce Sauvignon blanc wines with higher quantities of volatile thiols and more pronounced varietal character. Some *Saccharomyces bayanus* var. *uvarum* yeast strains can also present a high aptitude to reveal the volatile thiols from their natural precursor comparing with *Saccharomyces cerevisiae* strains. Those cryophilic *S. bayanus* var. *uvarum* strains can be besides implicated in the spontaneous fermentation of various Sauvignon

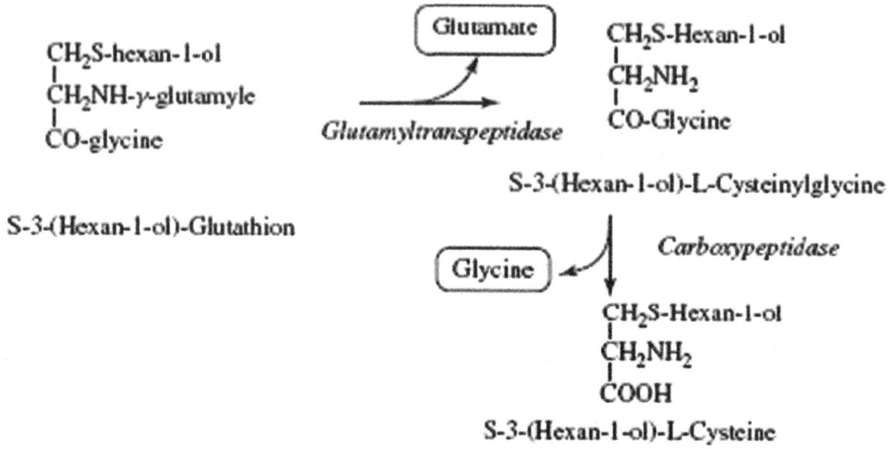

Fig. 8B.12 Proposed pathway for the conversion of glutathionylated "pro-precursor" into the cysteinylated precursor of 3SH

blanc musts from the Val de Loire area. *Saccharomyces uvarum* and their hybrids with *S. cerevisiae* are also commonly in the natural microflora of cool Structure of three aroma precursors of sauvignon: *S*-cysteine conjugates climate vineyards, such as Hungary, Alsace, etc. (Masneuf et al. 2002).

S-3-(Hexan-ol)-glutathion has been identified in Sauvignon blanc must(Peyrot des Gachons et al. 2002b). This compound may be considered a "pro-precursor" of 3SH. *S*-3-(hexan-ol)-L-cysteine in grapes could result from the catabolism of *S*-3-(hexan-ol)-glutathion, as proposed in Fig. 8B.12. *S*-Glutathion conjugates may be involved in detoxification processes in vines, as is the case in other plant and animal organisms.

References

Allen, M.S., Lacey M.J., Harris, R.L.N., & Brown, W.V. (1991). Contribution of methoxypyrazines to Sauvignon blanc wine aroma. *Am. J. Enol. Vitic., 42*, 109–112.

Augustyn, O.P.H., Rapp, A., & van Wyck, C.J. (1982). Some volatile aroma components of *Vitis vinifera* L. cv. Sauvignon blanc. *S. Afr. J. Enol. Viticult., 3*, 53–60.

Aznar, M., López, R., Cacho, J.F., & Ferreira, V. (2001). Identification and quantification of impact odorants of aged red wines from Rioja. GC-olfactometry, quantitative GC-MS, and odor evaluation of HPLC fractions. *J. Agric. Food Chem., 49*, 2924–2929.

Blanchard, L. (2000). Recherche sur la contribution de certains thiols volatils à l'arôme des vins rouges. Etude de leur genèse et de leur stabilité. Thèse de Doctorat de l'Université Victor Segalen Bordeaux2

Blanchard, L., Tominaga, T., & Dubourdieu, D. (2001). Formation of furfurylthiol exhibiting a strong coffee aroma during oak barrel fermentation from furfural released by toasted staves. *J. Agric. Food Chem., 49*, 4833–4835.

Bouchilloux, P., Darriet, P., Henry, R., Lavigne, V., & Dubourdieu D. (1998a). Identification of volatile and powerful odorous thiols in Bordeaux red wine varieties. *J. Agric. Food Chem., 46*, 3095–3099.

Bouchilloux P., Darriet P., & Dubourdieu D. (1998b). Identification d'un thiol fortement odorant, le 2-méthyl-3-furanthiol, dans les vins. *Vitis, 37*, 177–180.

Bouchilloux, P., Darriet, Ph., Dubourdieu, D., Henry, R., Reichter, A., & Mosandl, A. (2000). Stereodifferentiation of 3-mercapto-2-methylpropanol in wines. *Eur. Food Res. Technol., 210*, 349–352.

Buettner, A., & Schieberle, P. (1999). Characterization of the most odor-active volatiles in fresh, hand-squeezed juice of grapefruit (*Citrus paradisi* Macfayden). *J. Agric. Food Chem., 47*, 5189–5193.

Darriet, Ph. (1993). Recherches sur l'arôme et les précurseurs d'arôme du Sauvignon. Thèse de doctorat de l'université de Bordeaux II.

Darriet, Ph., Tominaga, T., Demole, E., & Dubourdieu, D. (1993). Mise en évidence dans le raisin de *Vitis vinifera* (var. Sauvignon) d'un précurseur de la 4-mercapto-4-méthylpentan-2-one. *C.R. Acad. Sci.* Paris, Biologie et Pathologie végétale, *316*, 1332–1335.

Darriet, P., Tominaga, T., Lavigne, V., Boidron, J.N., & Dubourdieu, D. (1995). Identification of a powerful aromatic component of *Vitis vinifera* L. var. Sauvignon wines: 4-Mercapto-4-methylpentan-2-one. *Flav. Frag. J., 10*, 385–392.

Demole, E., Enggist, P., & Ohloff, G. (1982). 1-p-Menthene-thiol, a powerful flavor impact constituent of grapefruit juice. *Helv. Chim. Acta, 65*, 1785–1794.

Engel, K. H., & Tressl, R. (1991). Identification of new sulfur-containing volatiles in yellow passion fruits (*Passiflora edulis f. flavicarpa*). *J. Agric. Food Chem.*, 39, 2249–2252.

Escher, S., Niclass, Y., van de Waal, M., & Starkenmann, C. (2006). Combinatorial synthesis by nature: volatile organic sulfur-containing constituents in *Ruta Chalepensis* L. *Chemistry & Biodiversity, 3*, 943–957.

Fretz C., Luisier J.L., Tominaga T., Dubourdieu D., & Amado R. (2005). 3-Mercaptohexanol: An aroma impact compound of Petite Arvine wine. *Am. J. Enol. Vitic., 56*, 407–410.

Galet, P. (1968). *Précis d'Ampélographie Pratique*, Paul Déhan Ed., Montpellier, France.

Gasser, U., & Grosch, W. (1988). Identification of volatile flavour compounds with hight aroma values from cooked beef. *Z. Lebensm. Unters. Forsch., 186*, 489–494.

Guth, H. (1997). Quantitative and sensory studies of character impact odorants of different white wine varieties. *J. Agric. Food Chem., 45*, 3027–3032.

Guth, H., & Grosch, W. (1994). Identification of the character impact odorants of stewed beef juice by instrumental analysis and sensory studies. *J. Agric. Food Chem., 42*, 2862–2866.

Howell, K. S., Klein, M., Swiegers, J. H., Hayasaka, Y., Elsey, G. M., Fleet, G. H., Hoj, P. B., Pretorius, I. S., & de Barros Lopes, M. A. (2005). Genetic determinants of volatile-thiol release by *Saccharomyces cerevisiae* during wine fermentation. *Appl. Environ. Microbiol., 71*, 5420–5426.

Idstein, H., & Schreier, P. (1985). Volatile constituents from guava (Psidium guajava, L.) fruit. *J. Agric. Food Chem., 33*, 138–143.

Kolor, M.G. (1983). Identification of an important new flavor compound in concord grape: Ethyl 3-mercaptopropionate. *J. Agric. Food Chem., 31*, 1125–1127.

Kotseridis, Y., & Baumes, R. (2000). Identification of impact odorants in Bordeaux red grape juice, in the commercial yeast used for its fermentation, and in the produced wine. *J. Agric. Food Chem.*, 48, 400–406

Lavigne, V., Henry, R., & Dubourdieu, D. (1998). Identification et dosage de composés soufrés intervenant dans l'arôme ≪ grillé ≫ du vin. *Sci. Aliments, 18*, 175–191.

Lopez, R., Ortin, N., Perez-Trujillo, J.P., Cacho, J., & Ferreira, V. (2003). Impact odorants of different young white wines from the Canary Islands. *J. Agric. Food Chem., 51*, 3419–3425.

Masneuf, I., Murat, M.L., Naumov, G.I., Tominaga, T., & Dubourdieu, D. (2002). Hybrids Saccharomyces cerevisiae × Saccharomyces bayanus var-uvarum having a higt liberating ability of some sulfur varietal aromas of *Vitis vinifera* Sauvignon blanc wines. *J. Int. Sci. Vigne Vin., 36*, 205–212.

Murat, M.L., Tominaga, T., & Dubourdieu, D. (2001a). Impact of some components on Bordeaux rosés and clairets. *J. Int. Sci. Vigne Vin., 35*, 99–105.

Murat, M.L., Tominaga, T., & Dubourdieu, D. (2001b). Assessing the aromatic potential of Cabernet Sauvignon and Merlot musts used to produce rose wines by assaying the cysteinylated precursor of 3-mercaptohexan-1-ol. *J. Agric. Food Chem., 49*, 5412–5417.
Murat, M.L., Masneuf, I., Darriet, Ph., Lavigne, V., Tominaga, T., & Dubourdieu, D. (2001c). Effect of Saccharomyces cerevisiae yeast strains on the liberation of volatile thiols in Sauvignon blanc wine. *Am. J. Enol. Vitic., 52*, 136–139.
Murat, M. L., Tominaga, T., Saucier, C., Glories, Y., & Dubordieu, D. (2003). Effect of anthocyanins on stability of the key odorous compound 3-mercaptohexan-1-ol in Bordeaux rosé wines. *Am. J. Enol. Vitic., 54*, 135–138.
Peynaud, E. (1980). Le Goût du Vin. Ed., Dunod, Paris.
Peyrot des Gachons, C., Tominaga, T., & Dubourdeiu, D. (2000). Measuring the aromatic potential of *Vitis vinifera* L. cv. Sauvignon blanc grapes by assaying *S*-cysteine conjugates, Precursors of the volatile thiols responsible for their varietal aroma. *J. Agric. Food Chem., 48*, 3387–3391.
Peyrot des Gachons, C., Tominaga, T., & Dubourdeiu, D. (2002a). Localization in the berry of S-cysteine conjugates, aroma precursors of *Vitis vinifera* L. cv. Sauvignon blanc grapes. The effect of skin contact on the aromatic potential of the musts. *Am. J. Enol. Vitic., 53*, 144–146.
Peyrot des Gachons, C., Tominaga, T., & Dubourdeiu, D. (2002b). Sulfur aroma precursor present in S-glutathione conjugate form: Identification of S-3-(hexan-1-ol)-glutathione in must from *Vitis vinifera* L. Sauvignon blanc. *J. Agric. Food Chem., 50*, 4076–4079.
Reichstein, T., & Staudinger, H. (1926). British Patent 260 960.
Rigaud, J., Etievant, P., Henry, R., & Latrasse, A. (1986). Le méthoxy-4-méthyl-2-butanethiol-2, un constituant majeur de l'arôme du bourgeon de cassis. *Sci. Aliment, 6*, 213–220.
Sarrazin, E., Shinkaruk, S., Tominaga, T., Bennetau, B., Frerot, E., & Dubourdieu, D. (2007). Odorous impact of Volatile thiols on the aroma of young botrytized sweet wines: Identification and quantification of new sulfanyl alcohols. *J. Agric. Food Chem., 55*, 1437–1444.
Schieberle, P. (1991). Primary odorants in popcorn. *J. Agric. Food Chem., 39*, 1141–1144.
Schneider, P. (2001). Contribution à la connaissance de l'arôme et du potentiel aromatique des vins du Melon B. (Vitis vinifera L.) et des vins de Muscadet, Thèse de Doctorrat, Université de Montpellier II.
Schreier, P., & Drawert, F. (1974). Gaschromatographisch-massenspectrometrische. Untersuchung flüchtiger Inhaltsstoffe des Weines. V. mitteilung. Alkohole, hydroxy-ester, lactones und andere polare komponenten des weinaromas. *Chem. Mikrobiol. Technol. Lebensm., 3*, 154–160.
Shankaranarayana, M.L., Raghavan, B., Abraham, K.O., & Natarajan, C.P. (1982). Sulphur compounds in flavours. In Food Flavors part A. ed. by Morton, I.D., Macleod, A.J. Elsevier Science publishers, p. 169.
Takoi, K., Tominaga, T., Degueil, M., Sakata, D., Kurihara, T., Shinkaruk, S., Nakmura, T., Maeda, K., Akiyama, H., Watari, J., Bennetau, B., Duburdieu, D. (2007). Proceedings of the 31st International Congress of the European Brewery Convention. p. 241–251. Venice, Italy
Thibon, C., Tominaga, T., & Dubourdieu, D. (2007). Distribution stéréoisomèrique du précurseur du 3-sulfanylhexanol dans les moûts sains ou issue de raisins botrytisés de Vitis vinifera L. cv. Sauvignon blanc et Sémillon. 8 th International Syposium of Oenology. Bordeaux Lac. July.
Thibon, C., Shinkaruk, S., Tominaga, T., Bennetau, B., & Dubourdieu, D. (2008a). Analysis of the diastereoisomers of the cysteinylated aroma precursor of 3-sulfanylhexanol in Vitis vinifera grape must by gas chromatography coupled with ion trap tandem mass spectrometry. *J. Chromatogr. A., 1183*, 150–157.
Thibon, C., Marullo, P., Claisse, O., Cullin, C., Dubourdieu, D., & Tominaga, T. (2008b). Nitrogen catabolic repression controls the release of volatile thiols by *Saccharomyces cerevisiae* during wine fermentation. *FEMS Yeast Research.* (DOI 10 1111/ j. 1567-1364.2008.00381).
Tominaga, T. (1998). Recherche sur l'arôme des vins de Vitis vinifera L. cv. Sauvignon blanc et sa genèse à partir du précurseur inodore dans le moût. Thesis. Université Victor Segalen Bordeaux 2.
Tominaga, T., & Dubourdieu, D. (1997). Identification of 4-mercapto-4-methylpentan-2-one from the Box Tree (*Buxus sempervirens* L.) and Broom (*Sarothamnus scoparius* (L.) Koch.). *Flav. Frag. J., 12*, 373–376.

Tominaga, T., & Dubourdieu, D. (2000). Identification of cysteinylated aroma precursors of some volatil thiols in Passion fruit juice. *J. Agric. Food Chem., 48*, 2874–2876.

Tominaga, T., & Dubourdieu, D. (2006). A novel method for quantification of 2-methyl-3-furanthiol and 2-furanmethanethiol in wines made from *Vitis Vinifera* grape varieties. *J. Agric. Food Chem., 54*, 29–33.

Tominaga, T., Masneuf, I., & Dubourdieu, D. (1995). Mise en évidence d'un S-conjugué de la cystéine, précurseurs d'arôme du Sauvignon. *J. Int. Sci. Vigne et Vin, 29*, 227–232.

Tominaga, T., Darriet, P., & Dubourdieu, D. (1996). Identification de l'acétate de 3-mercaptohexanol, composé à forte odeur de buis, intervenant dans l'arôme des vins de Sauvignon. *Vitis, 35*, 4, 207–210.

Tominaga, T., Murat, M.L., & Dubourdieu, D. (1998a). Development of a method for analyzing the volatile thiols involved in the characteristic aroma of wines made from *Vitis vinifera* L. Cv. Sauvignon blanc. *J. Agric. Food Chem., 46*, 1044–1048.

Tominaga, T., Furrer, A., Henry, R., & Dubourdieu, D. (1998b). Identification of new volatile thiols in the aroma of *Vitis vinifera* L. var. Sauvignon blanc wines. *Flav. Frag. J., 13*, 159–162.

Tominaga, T., Peyrot des Gachons, C., & Dubourdieu, D. (1998c). A new type of flavor precursors in *Vitis vinifera* L. cv. Sauvignon blanc: S-cysteine conjugates. *J. Agric. Food Chem., 46*, 5215–5219.

Tominaga T., Baltenweck-Guyot R., Peyrot des Gachons C., & Dubourdieu D. (2000a). Contribution of volatile thiols to the aromas of white wines made from several *Vitis vinifera* grape varieties. *Am. J. Enol. Vitic., 51*, 178–181.

Tominaga, T., Blanchard, L., Darriet, Ph., & Dubourdieu, D. (2000b). A powerful aromatic volatile thiol, 2-furanmethanethiol, exhibiting roast coffee aroma in wines made from several vitis vinifera grape varieties. *J. Agric. Food Chem., 46*, 1799–1802.

Tominaga, T., Masneuf, I., & Dubourdieu, D. (2001). Powerful aromatic volatile thiols in wines made from several *Vitis vinifera* grape varieties and their releasing mechanism. Symposium on Nutraceutical Beverages: Chemistry and Flavor Characteristics, ACS National Meeting in Chicago, Il, August.

Tominaga, T, Guimbertau, G., & Dubourdieu, D. (2003a). Role of certain volatile thiols in the bouquet of aged champagne wines. *J. Agric. Food Chem., 51*, 1016–1020.

Tominaga, T, Guimbertau, G., & Dubourdieu, D. (2003b). Contribution of benzenemethanethiol to smoky aroma of certain *Vitis vinifera* L. wines. *J. Agric. Food Chem., 51*, 1373–1376.

Tominaga, T., Niclass, Y., Frerot, E., & Dubourdieu, D. (2006). Stereoisomeric distribution of 3-merccaptohexan-1-ol and 3-mercaptohexyl acetate in dry and sweet white wines made from Vitis vinifera (Var. Sauvignon blanc and Semillon). *J. Agric. Food Chem., 54*, 7251–7255.

Tressl, R., & Silwar, R. (1981). Investigation of sulfur-containing components in roasted coffee. *J. Agric. Food Chem., 29*, 1078–1082.

Vermeulen, C., Lejeune, I., Tran, T.T.H., & Collin, S. (2006). Ocurrence of polyfunctional thiols in lager beers. *J. Agric. Food Chem., 54*, 5061–5068.

Weber, B., Haag, H.-P., & Mosandl, A. (1992). Stereoisomere aromastoffe. LIX. 3-mercaptohexyl- und 3-methylthiohexylalkanoate-Strukture und eigenschaften des enantiomeren. *Z. Lebensm. Unters. Forsch., 195*, 426–428.

Widder, S., Sabater Lüntzel, C., Dittner, T., & Pichenhagen, W. (2000). 3-sulfanyl-2-methylpentan-1-ol, a new powerful aroma compound. *J. Agric. Food Chem., 48*, 418–423.

Chapter 8C
Volatile Compounds and Wine Aging

M. Soledad Pérez-Coello and M. Consuelo Díaz-Maroto

Contents

8C.1 Introduction

Oak wood barrel aging of wine from the end of fermentation to bottling is a common practice at wineries, chiefly for red wines but sometimes also for white wines.

Bottle aging usually takes place after the barrel aging period, especially in the case of red wines. Although some white wines like Riesling, Chenin Blanc or Colombar wines are consumed one or two years after bottling, achieving during this time a characteristic "bouquet of aging". The same is not true for wines from other varieties in which oxidation of the terpenes and hydrolysis of acetates and ethyl esters of fatty acids contribute to the loss of the flowery and fruity character of young white wines (Rapp and Marais 1993; González-Viñas et al. 1996).

Generally speaking, aged wines are associated with wines that have spent a period of time in an oak barrel acquiring a special aroma that is well appreciated by the consumer, reaching a high price in the marketplace. For these reasons, this chapter is focussed on all those aspects that may affect the volatile composition of oak wood, and consequently of oak aged wine.

Oak barrels are regarded as being conducive to the sensory development of wines, though the events taking place in the barrel are extremely complex. A series of

M.S. Pérez-Coello (✉)
Área de Tecnología de los Alimentos, Facultad de Ciencias Químicas, Campus Universitario 10, Universidad de Castilla-La Mancha, 13071 Ciudad Real, Spain
e-mail: soledad.perez@uclm.es

M.V. Moreno-Arribas, M.C. Polo (eds.), *Wine Chemistry and Biochemistry*,
DOI 10.1007/978-0-387-74118-5_13, © Springer Science+Business Media, LLC 2009

stabilizing reactions that affect wine colour and clarity, structural changes in the phenolics and new aroma formation all take place during oak wood aging. None of these reactions occurs in tanks or bottles, which are inert, do not interact with the wine, and do not allow the controlled oxidation that takes place in oak barrels.

The release of oak wood aromas is one of the primary reasons why wines are made to undergo aging in the wood. This makes it necessary to take into account the different factors regulating the release of volatile components from the oak to the wine. These factors include the type of oak employed and the geographical origin of the oak (Mosedale and Ford 1996; Chatonnet and Dubourdieu 1998; Pérez-Coello et al. 1999; Fernández de Simón et al. 2003), the drying treatment to which the oak has been subjected (Sefton et al. 1993; Masson et al. 2000b; Doussot et al. 2002), the degree of toasting (Chatonnet 1999; Hale et al. 1999; Cadahía et al. 2003), and the length of time a wine spends in the barrel, along with barrel age and usage, i.e., the number of times a barrel has been used (Singleton 1995; Escalona et al. 2002; Pérez-Prieto et al. 2002).

Several chemical reactions that happen in the barrel during aging can also affect the final volatile composition of wines. Among these reactions is found the oxidation due to the entry of oxygen across the pores of the wood, which modify both the wine volatile compounds and those extracted from wood. On the other hand, microorganisms, proceeding from the wine or from the barrel, especially lactic bacteria and yeasts, can produce important volatile compounds from a sensorial point of view, like vinylphenol and vinylguaiacol (Jarauta et al. 2005).

Finally, the adsorption of certain wine volatiles by the lees and the barrel itself (Chatonnet et al. 1992a; Ramírez Ramírez et al. 2001; Jiménez Moreno and Ancín Azpilicueta 2007), together with the possible changes that can occur during the additional bottle aging period (Pérez-Prieto et al. 2003), will also cause modifications in the final aromatic composition of the wine.

The long aging process normally increases the price of the final product, since the wine has to be left in the barrels before it can be brought to market. For this reason, the use of oak chips or staves as an alternative for the aging process has been widely considered in the last few years (Ducournau et al. 1999; Gutiérrez Afonso 2002; Frangipane et al. 2007). However, since oak barrel aging involves something more than the simple transfer of compounds from the wood to the wine, this new type of skill will have to be optimised to obtain a quality product.

8C.2 Volatile Composition of Oak Wood and Sources of Variation

Only oak heartwood is used to manufacture barrels. The basic composition varies little from one species of oak to another. Oak wood consists mainly of cellulose (50%) and hemicellulose (20%), which provide the framework and matrix for the wood. Lignin is the other large polymer that makes up 30% of dry wood. It is a three-dimensional, reticulated polymer comprising phenylpropane with units of guaiacyl (2-methoxyphenol) and syringyl (2,6-dimethoxyphenol) that are cross-linking by oxidation. Lignin is mainly present in the cell walls.

Cellulose, hemicellulose, and lignin are insoluble substances, but some of their monomers can be partially extracted in hydroalcoholic solutions like wine. Some of the compounds thus extracted have interesting organoleptic qualities (Boidron et al. 1988).

The remaining wood components make up the fraction that can be extracted with different solvents, mainly comprising volatile and non-volatile acids, sugars, steroids, hydrolysable tannins (ellagitannins), volatile phenols, terpenes, lactones, etc. This review will deal with the volatile components that can leach out into the wine during barrel aging and as a result may affect the sensory attributes of wine.

The two oak lactone isomers (*cis*- and *trans*-β-methyl-γ-octalactone) were first identified in oak wood by Masuda and Nishimura (1971). They are one of the main volatile components present in the extractible fraction of oak wood. Studies have revealed the detection thresholds for these two isomers to be 0.092mg/L for the *cis* isomer and 0.49mg/L for the *trans* isomer in white wines (Chatonnet 1991). Their aroma has been identified as the source of the typical odour of oak barrel aging and has been described as "oak" or "coconut".

The major phenolic aldehydes, vanillin and syringaldehyde and their derivatives (coniferaldehyde, sinapaldehyde, acetovanillone, acetosyringone, and others) come from fragments of lignin arising through hydrolysis, pyrolysis, and oxidation reactions. Vanillin is the only one of these that influences wine aroma. It has a characteristic vanilla flavour with an olfactory detection threshold of 0.32mg/L in red wine (Boidron et al. 1988). Guaiacol and its 4-ethyl and 4-vinyl derivatives along with eugenol and isoeugenol are the main volatile phenols that have a sensory impact on wines aged in the wood. The former group contributes spicy, toasted, and smoky aromas, while the latter impart characteristic spicy, clove, and oak aromas (Boidron et al. 1988).

Terpenes and norisoprenoids, including linalool and its derivatives, β-terpineol, geranyl acetone, β-ionone and its derivatives, 3-OH-β-damascone, blumenol, vomifoliol, spathulenol, and vulgarol have been identified in oak wood samples having different origins. These substances were more abundant in samples of American oak than in samples of French oak (Sefton et al. 1990; Pérez-Coello et al. 1998). Certain of these components have interesting fruity and floral aromas in the case of terpenes and tobacco, or woody aromas for some norisoprenoids. Many of them are present in the grapes and in wine, but there has been little research on their presence in wines during aging, which gives rise to a series of uncertainties as to their effects on the aroma of oak wood barrel aged wines.

Other volatile compounds present in oak wood can transmit unpleasant aromas to the wine, such as the "sawdust" aroma of dry wood that is perceptible in some wines aged in new barrels. The substances that cause these aromas have been identified in both American and European oak wood, for instance, (*E*)-2-nonenal, 3-octen-1-one, (*E*)-2-octenal, and 1-decanal. Their connection with the "sawdust" aroma has been established by olfactometry, and they have been identified in wines suffering from this flaw, though toasting the wood (Chatonnet and Dubourdieu 1998).

The volatile composition of each barrel depends on its origin (species, geographic region, tree age, etc.), as well as on the different barrel manufacturing and processing methods employed.

Quercus alba, also known as American white oak, and two European species, *Quercus robur* L. and *Quercus petraea* Liebl, the former also known as pedunculate oak and the latter as sessile oak, are the species most frequently used in coopering.

On the whole, American oak is richer in oak lactones, especially in the *cis* isomer, while European oak contains higher quantities of ellagitannins (Singleton 1995).

In the case of French oak, there are traditional differences between the various regions, which ordinarily contain a single species, either *Quercus petraea* or *Quercus robur*. These regions are Allier, Limousin, Nevers, Vosges, Troncais, and others. The oak woods from these regions can be differentiated on the basis of their differing oak lactone, vanillin, and eugenol contents (Chatonnet 1998; Pérez-Coello et al. 1999).

However, large-scale studies have confirmed that there is high variability with geographic origin, environmental factors, and growth factors. Using a sample set of 400 American, French, and Eastern European trees, Prida and Puech (2006) concluded that oak lactones and ellagitannins were the best substances to use as variables for distinguishing between species, whereas eugenol, 2-phenylethanol, and the phenolic aldehydes vanillin and syringaldehyde, which appeared at higher concentrations in oak wood samples from Eastern Europe, were the best components to use as variables for distinguishing geographic origins.

Factors such as tree age decrease the amount of extractible substances in the heartwood. At the same time, growth rate affects wood density. Growth rate depends on the climate, type of soil, distance between trees, etc. (Singleton 1995). Coopers prefer slow-growth ("fine-grain") trees, because they are easier to cut and bend and contain larger amounts of extractible substances and fewer ellagitannins. In any event, recent studies didn't find any correlation between the volatiles content and ring width and have shown the botanical species to be the principal factor affecting the amount of volatile substances present in oak wood (Doussot et al. 2000; Prida et al. 2007).

Ordinarily, *Q. petraea* is the European species that contains the highest quantities of oak lactones, furfurals, and phenolic aldehydes, though there is high diversity according to geographical origin (Doussot et al. 2000, 2002; Guchu et al. 2006a; Prida et al. 2007).

New types of oak wood from Eastern Europe (Ukraine, Russia, Romania, Hungary, and others), also from the species *Q. petraea* and *Q. robur*, have lately entered the marketplace, alongside the French and American oak woods traditionally employed for coopering. The volatile composition of this wood is similar to that of French oak, and these sources may soon be competing with wood from traditional sources. Little information on the composition of Eastern European oak wood is available, though some recent work suggests that it contains higher amounts of phenolic aldehydes and volatile phenols than French oak wood from the same species and that its oak lactone content is intermediate between French and American oak (Prida and Puech 2006).

In terms of its volatile composition, Spanish oak is comparable to French oak of the same species. There are also two native species, *Q. pyrenaica* and *Q. faginea*, which have amounts of the oak lactones, vanillin, and eugenol that are statistically comparable to those in French oak, though wines aged in *Q. faginea* barrels have received particularly poor ratings by taste panels, earning low scores for the characteristic oak wood sensory attributes (Fernández de Simón et al. 2003, 2006; Cadahía et al. 2003).

The ratio between the *cis* and *trans* isomers of oak lactones has been used as a means of differentiating oaks of different origins. The oak lactone isomer ratio is higher (between 5 and 9) in American oak. The ratio differs somewhat for French oak, between 1 and 5, depending on the geographic region of origin (Vosges, Centre, and Allier) (Waterhouse and Towey 1994; Pérez-Coello et al. 1999, 2000a), with *Q. robur* having particularly low levels of both isomers (Feuillat et al. 1997).

For barrel manufacturing, in its turn, the wood has to undergo a series of preliminary processing steps. First of all, the wood for the staves used to make the barrels has to be dried to decrease the moisture content of the newly cut wood. Traditionally, drying took place outdoors in the open air for at least three years, but now to save time artificial kiln drying is commonplace, or a combination of the two may be employed.

The quantity of volatile compounds in newly harvested wood is quite low. These components form as a result of oxidation taking place during drying. Traditional drying through outdoor seasoning increases levels of aromatic aldehydes in the wood, especially when seasoning is carried out in countries with a hot climate, like Australia (Chatonnet et al. 1994; Spillman et al. 2004a). Some researchers have reported lower levels of eugenol and oak lactones and a variable effect on vanillin concentrations depending on the temperature used in kiln drying (Sefton et al. 1993; Masson et al. 2000).

Toasting the inner surface of the staves is a common practice employed in barrel making. Toasting helps the wood to bend when the barrels are being made and also offers other advantages, e.g., pyrolysis of the lignin helps increase levels of phenolic aldehydes and other components with desirable sensory effects (Maga 1984). French coopers perhaps use toasting most of all to lower the tannin levels that could increase wine astringency.

The level of toasting is decided on empirically by coopers based on the degree of darkening of the wood and toasting time. Chatonnet and Boidron (1989) measured the temperature on the toasted surfaces of wood undergoing different levels of toasting: light toasting for 5min reached a temperature of 180°C, whereas 230°C were reached for heavy toasting (17.5min).

Heating the wood during toasting alters the volatile composition of the wood, and for this reason solid-phase microextraction (SPME) followed by gas chromatography has been put forward as a quick and simple method of analysing volatile compounds and providing information about the volatile composition of the wood that can be correlated with the toasting level (Chatonnet et al. 1994; Díaz-Maroto et al. 2004). The heat breakdown of the lignin that takes place during medium toasting brings about an increase in the volatile phenols (vanillin, syringaldehyde, guaiacol, eugenol). On the other hand, temperatures higher than 250–300°C reduce the content of these compounds as a result of losses due to carbonization (Pérez-Coello et al. 1997; Chatonnet 1998,1999).

Toasting can raises the levels of the oak lactones by releasing them from their precursor, first isolated by Otsuka et al. (1980), recently joined by one additional precursor, 6′-*O*-gallate derivative of β-D-glucopyranosyloxy-3-methyloctanoic acid (Masson et al. 2000a).

Heating the wood also results in the formation of carbohydrate (cellulose and hemicellulose) breakdown products, e.g., the furanic derivatives furfural, 5-methyl-furfural, furfuryl alcohol, etc. (Chatonnet 1998).

Heating also produces a series of other substances as a result of Maillard reactions. These include cyclotene, maltol and its derivatives (hydroxymaltol, dihydro-maltol), 2,3-dihydro-5-hydroxy-2-methyl 4(*H*) pyranone (DHM) and its 5-hydroxy derivative (DDMP) and furaneol. The sensory impact of these components can play an important role in developing the "toasted", "burnt", or "caramel" aromas of some barrel-aged wines (Cutzach et al. 1997,1999).

Barrel toasting is of fundamental importance in the leaching of volatile components into the wine and in many cases may mask the attributes intrinsic to the species or geographical region of origin of the oak (Francis et al. 1992).

As a consequence, the variability inherent to oak wood is augmented by further variability at the barrel cooperage, resulting in very high levels of variation even within the same batch of manufactured barrels. On examining the furfural, oak lactone, syringol, and guaiacol contents of wines aged in American and French oak wood barrels manufactured by four different cooperages, Towey and Waterhouse (1996) recorded levels of variation ranging between 15% and 40%, depending on the substance considered.

8C.3 Sensory Relevance of the Volatile Components of Oak Wood

Not all of the volatile components present in oak wood play a decisive role in determining the aroma of aged wines. While concentrations in the oak wood will be a conditioning factor, final aroma is the result of each component's olfactory detection threshold and the synergistic effects between it and the other wine components.

In a study of Pinot Noir wines aged in different types of oak wood, the wines that earned the highest scores for the attributes "woody", "toasted", "vanilla" and "coconut" contained the highest quantities of the *cis* oak lactone isomer (Sauvageot and Feuillat 1999). This component exerts the greatest sensory effect in wines aged in oak wood, and the *cis* isomer is present in quantities greater than the detection threshold even in species low in lactones, like *Q. robur* (Díaz-Maroto et al. 2004).

Chardonnay wines in contact with Hungarian and Russian oak wood with low oak lactone contents scored well for oak wood sensory attributes. GC-sniffing analysis of these woods concluded that such other components as eugenol and the *cis* and *trans* isomers of isoeugenol, together with guaiacol and its derivatives, also contributed to the "spicy", "clove", "cinnamon", and "woody/oaky" aromas. Oak lactones, on the other hand, elicited the odour descriptors "sweet", "vanilla", and "oaky" (Díaz-Maroto et al. 2008).

Table 8C.1 summarizes the oak wood components that are significant from a sensory standpoint together with the corresponding sensory descriptors as determined by GC-sniffing of different oak extracts, with an indication of the effect of toasting on the relative odour intensity of these components. The table shows furfural and its derivatives, guaiacol, cyclotene, and phenylacetaldehyde to be the components having toasting-related odour descriptors. These components together with vanillin, eugenol and derivatives increased their odour intensity during toasting.

In contrast, odour intensity of other components related to "off flavours" ("herbal", "sawdust", "greasy", and "cucumber") can be seen to decrease with toasting. The same happened with the intensity of fruity and floral aromas of linalool oxide, β-damascenone, or 2-phenylethanol, and the oak lactones odour descriptors.

Figure 8C.1 uses a spider plot to compare the sensory scores awarded to the Chardonnay wines in contact with American and Hungarian toasted wood chips. It shows that, despite differences in oak lactone content, the wood attributes were similar in these wines. Toasting increase the similarities between wines treated with oak wood of different origins, masking the natural fruity and floral attributes of Chardonnay wine. None of the "off flavours" in the woods used for maceration that were detected by sniffing (Table 8C.1) were later detected in the wines, which suggests either that concentrations of the components responsible were below the detection thresholds or that these aromas were masked by other stronger attributes.

Spillman et al. (2004b) studied the sensory effects of the oak wood components in wines made from Chardonnay and Cabernet Sauvignon grapes and recorded different effects in each type of wine. The "smoky" attribute was most highly correlated with the wood components in the Chardonnay wines, but this descriptor was not correlated with any of the components in the Cabernet Sauvignon wines. On the other hand, the Cabernet Sauvignon wines tended to have highly complex aroma profiles, and a number of the components present in these wines (*cis* oak lactones, 4-methylguaiacol, vanillin, eugenol, furfural derivatives) were positively correlated with the "vanilla", "coffee", and "dark chocolate" descriptors.

Similarly, such flaws as "green apple" and "earthy" aromas were negatively correlated with oak wood components, and this attests to the ability of these oak wood components to mask vegetative aromas and "off flavours" in wine. However, other authors found that the descriptor "wood-vanilla-cinnamon", due to oak extracted compounds in Spanish red wines, was negatively influenced by the presence of 4-ethylphenol, 4-ethylguaiacol and phenylacetaldehyde (Aznar et al. 2003).

Table 8C.1 Odour descriptors obtained by gas chromatography-sniffing of different toasted and no toasted oak wood extracts

Compounds that decreased their odour intensity with oak wood toasting	Odour descriptors obtained by GC-sniffing	Compounds that increased their odour intensity with oak wood toasting	Odour descriptors obtained by GC-Sniffing
Hexanal	Grassy, green	Furfural	Slightly toasty, caramel
Hexanol	Grassy, green	1-(2-Furanyl)ethanone	Toasty, toasted grain
Heptanal	Rancid, strong smell	5-Methylfurfural	Spicy, toasty, sweet
Nonanal	Herbal, floral	Guaiacol	Spicy, toasty, smoky/burnt
3-Octen-1-one	Mushroom	4-Methylguaiacol	Spicy, lightly green, phenolic
2-Octenal	Penetrating, green leaf	Eugenol	Spicy, clove, cinnamon
2,4-Nonadienal	Fatty, cucumber	*cis*-Isoeugenol	Spicy, clove, woody/oak
2,6-Nonadienal	Cucumber	*trans*-Isoeugenol	Spicy, clove, woody/oak
trans-2-Nonenal	Greasy, saw dust, cucumber	Vanillin	Sweet, vanilla
Decanal	Citrus, fruity	Phenylacetaldehyde	Toasty
Hexanoic acid	Faintly cheesy, sweaty	Benzaldehyde	Bitter almonds, nutty
Octanoic acid	Sweaty, penetrating	Cyclotene	Sweet, toasty, caramel
β-Damascenone	Sweet, fruity, peach		
Linalool oxide	Floral, rosy		
Phenyl ethanol	Floral, rosy		
trans-Oak lactone	Vanilla, oaky, clove, coconut		
cis-Oak lactone	Vanilla , oaky , clove, coconut		

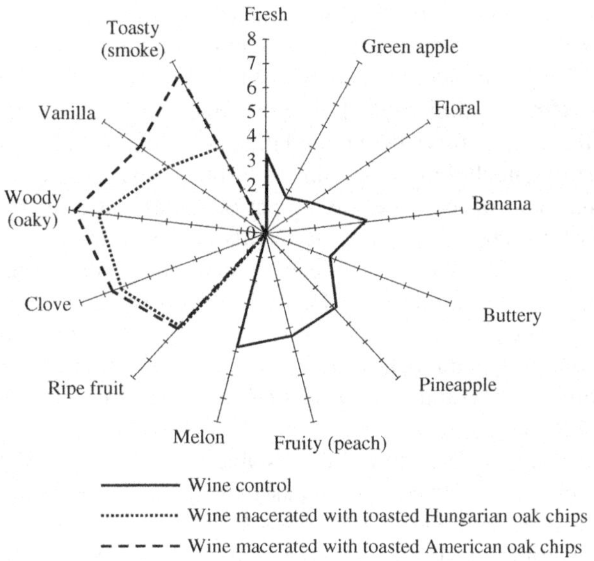

Fig. 8C.1 Sensory characteristics of a Chardonnay wine and the same wine macerated for 24 days with toasted Hungarian and American oak chips

8C.4 Changes in the Wine Volatile Composition During Oak-Barrel Aging

The release of volatiles from the wood into the wine during the barrel aging of wine is regulated by diffusion kinetics (Rous and Alderson 1983). As a general rule, extraction of volatiles is highest at the beginning of aging and then gradually tapers off with aging time and barrel age (Puech 1987; Towey and Waterhouse 1996; Gómez-Plaza et al. 2004). Small amounts of such compounds as guaiacol, furfural, and 5-methylfurfural are present in wines aged in previously used barrels. Other compounds like eugenol and 4-methylguaiacol are not detectable in these wines, but on the other hand concentrations of oak lactones increased in the second year of barrel use and then begin to fall back in the third year (Towey and Waterhouse 1996).

One recent study showed that extraction of volatile substances from the wood dropped off sharply after 12 months of aging, though there was high variability between wines, highlighting the influence of such major oenological factors as the tartaric acid content and the potassium content and also the SO_2 concentration in the wines, since SO_2 can combine with certain furanic derivatives and phenolic aldehydes (Ancín et al. 2004; Ortega-Heras et al. 2007).

Nevertheless, aging time and barrel age should not be looked at separately, because differences in the release of volatiles have been shown to be closely linked to barrel age only for short aging times of six to nine months, with the effect of barrel age tailing off considerably for aging times longer than 12 months (Garde-Cerdán and Ancín-Azpilicueta 2006).

Once the oak components have been extracted, they undergo a series of microbiological alterations. These include the reduction of furanic and phenolic aldehydes to their respective alcohols and subsequent formation of their ethyl esters (Boidron et al. 1988; Spillman et al. 1998b). Furfural is especially sensitive to such changes, forming furfuryl alcohol and the corresponding ethyl ester. Thus, the presence of furfuryl alcohol indicates that microbial activity has taken place in the barrels for at least part of the wine's aging period (Spillman et al. 1998a). Vanillin and 5-methylfurfural undergo alterations of this type to a lesser extent (Spillman et al. 1998), which limits the sensory impact of these compounds, since the alcohols have higher olfactory detection thresholds than the corresponding aldehydes.

This effect is particularly important for wines fermented in the barrel, a practice that is widely used to add a touch of oak to white wines while avoiding the oxidation effects commonly encountered during prolonged aging (Aleixandre et al. 2003; Herjavec et al. 2007).

Eugenol and guaiacol seem to be more stable than phenolic aldehydes, and to date no alterations during wine aging in oak barrels have been reported; hence the concentration of these compounds depends solely on the extent to which they are extracted (Pérez-Prieto et al. 2003; Gómez-Plaza et al. 2004; Garde-Cerdán and Ancín-Azpilicueta 2006). Their concentration in wine increases over the initial months of aging and the levels off and remains virtually constant after 12 months in the barrel (Garde-Cerdán et al. 2002).

Similarly, *cis* and *trans* oak lactones undergo little alteration during aging, and their concentration profile in wine is similar to that of eugenol (Spillman et al. 1998), though it has been reported the possible release of oak lactones by their precursor during the aging period (Wilkinson et al. 2004).

Vinylphenol and vinylguaiacol are present at low concentrations in oak wood, but the amounts found in wine are chiefly a product of yeast metabolism, including the yeasts taking part in alcoholic fermentation, which are able to produce these compounds by decarboxylizing the coumaric acid and ferulic acid in the grapes. Vinylguaiacol concentrations in young wines can reach the olfactory detection threshold of 380 μg/L in red wines and 440 μg/L in white wines (Boidron et al. 1988), affording a spicy aroma. Both these components are highly reactive at wine pH, giving rise to ethanol adducts and stable wine pigments in red wines (Hakansson et al. 2003), and hence concentrations tend to be lower in red wines aged for longer periods of time (Spillman et al. 2004a).

The formation of 4-ethylphenol and 4-ethylguaiacol contaminating microorganisms represents a more serious problem. These two compounds are present at trace quantities in oak wood but can reach values approaching their olfactory detection thresholds (605 μg/L for the former and 110 μg/L for the latter in red wines) in wines aged for lengthy periods, giving rise to unpleasant "horse stable" and "medicinal" aromas (Chatonnet et al. 1992b).

Wineries commonly reuse their barrels. The useful lifetime of a barrel, however, is between five and six years, not just because of the depletion of volatile components that can leach into the wine but also because of the growth of unwanted microorganisms such as *Brettanomyces/Dekkera*. This yeast can synthesize

ethylphenols, mainly 4-ethylphenol and 4-ethylguaiacol, from the hydroxycinnamic acid derivatives in the grapes (*p*-coumaric acid and ferulic acid) (Chatonnet et al. 1992b). The ethylphenol content in wine increases with aging time, especially when aging takes place in used barrels. Ethylphenols have also been recorded at lower concentrations in wines aged in new barrels (Chatonnet 1999; Pollnitz et al. 2000; Díaz-Plaza et al. 2002; Pérez-Prieto et al. 2003; Garde-Cerdán and Ancín-Azpilicueta
2006). The highly porous nature of oak wood is favourable to *Brettanomyces* growth, even in new barrels, and these yeasts are extremely hard to get rid of. Different methods of sanitizing barrels have been developed, from ozone sterilization and the use of SO_2 ($\geq$7g of SO_2 gas per barrel), to treating barrels with alternating cycles of cold and hot (70°C) water and low-pressure steam (10min) (Henick-Kling et al. 2000; Malfeito-Ferreira 2005). Ozone sterilization results in smaller losses of desirable volatile substances such as vanillin, oak lactones, and eugenol than the hot water treatment commonly employed (Marko et al. 2005).

A series of chemical reactions, in particular oxidation, take place during barrel aging, affecting both intrinsic wine components and those extracted from the oak. Towey and Waterhouse (1996) did not observe any alterations in the fermentation components during barrel aging, but other researchers have reported higher acid and ester contents in wines aged in used barrels (Pérez-Prieto et al. 2002). The evaporation rate in these barrels is lower, because used barrels are less porous due to deposits of mineral salts and colour compounds. Oxidation of certain wine components inside the barrel gives rise to such important substances as acetaldehyde and phenylacetaldehyde while resulting in decreases in such alcohols (Jarauta et al. 2005). The phenylacethaldehyde has been related to oxidation aromas that present some aged wines (Aznar et al. 2003). The formation of acetaldehyde during wine aging is compensated by condensation reactions with polyphenols.

Wine lees and the barrel itself can adsorb certain wine volatiles and for that reason may play an extremely important role in determining the aromatic composition of the wine (Chatonnet et al. 1992a; Ramírez Ramírez et al. 2001; Jiménez Moreno and Ancín Azpilicueta 2007). Ramírez Ramírez et al. (2001) demonstrated that oak wood could adsorb such volatile components as linalool and ethyl octanoate in a model wine solution.

Lees aging is widely used, particularly for white wines or wines aged in new barrels. Adsorption by the lees of substances released by the wood holds the contribution of wood aromas down and prevents them from masking the other wine aromas present. Eugenol, 4-propylguaiacol, 4-methylguaiacol, furfural, and 5-methylfurfural have high affinities for the lees, whereas high quantities of lees are needed if there is to be any decrease in the concentrations of oak lactones and phenolic aldehydes, namely, vanillin and syringaldehyde (Jiménez Moreno and Ancín Azpilicueta 2007).

At the end of barrel aging, wines ordinarily undergo further aging in the bottle. During this period further chemical alterations that result in their final concentrations in the wine. For instance, concentrations of 5-methylfurfural and vanillin have been observed to decrease while wine is in the bottle, probably due to the formation of their corresponding alcohols and ethyl esters. On the other hand, oak lactone, furfural, and ethylphenol contents increase, and the concentrations of still other components like furfuryl alcohol and guaiacol do not change very appreciably (Pérez-Prieto et al. 2003). The increase of furfural during bottle aging has been previously observed, since it does not originate exclusively from oak but can also be formed from other hexoses and pentoses in the wine (Schreier 1979; Pérez-Coello et al. 2003).

8C.5 New Trends in Wine Aging

Oak barrel aging present some inconveniences, such as the high price of barrels, the amount of time the wine has to be left in the barrels before it can be brought to market, monitoring of the wine during aging, and others. This is especially true for white wines, because oxygenation through the pores in the wood is undesirable for these wines. Depending on the extraction conditions, results similar to those achieved by conventional barrel aging can be achieved using oak chips and relatively short contact times (Wilker and Gallander 1988).

Just a few years ago the use of oak chips was legal only in certain non-European countries. However, for the reasons just stated, the European Commission has provided for the use of oak chips in the production of wines, obligating producers who want to use this technique to write it in the labels (Regulation N. 1507/2006 of October 11th, 2006). The said resolution has, nonetheless, given rise to additional difficulties connected with wood chip use, namely, different oak origins, different chip sizes, and different processing methods, all of which only increases the variability in the volatiles composition already observed in barrel aging, especially since oak chip production is not standardized (Ducournau et al. 1999).

Oak chips were first used for white wines, to which they were added during fermentation or afterwards, during a short period of aging. This method appears to hold out special promise for neutral grape varieties with weak floral or fruity aromas. Using wood chips imparts "oak", "spicy", and "vanilla" aromas to the wines, improving their sensory impressions. Additionally, compared with barrel aging, the fruity attributes intrinsic to the grape variety used are still expressed by the wines, because the volatile components do not undergo oxidation while the wines are in contact with the wood chips (Wilker and Gallander 1988; Gutiérrez Afonso 2002).

Fermentation rates have been observed to increase when fermentation is carried out in the presence of wood chips, presumably because the oak chips act as carriers for the yeasts, resulting in the formation of higher quantities of acetates and fatty acid esters, which may help impart a more fruity aroma to wood chip-treated wines (Pérez-Coello et al. 2000b).

Wood chips are ordinarily used in amounts ranging between 2g/L and 8g/L for white wines. Higher quantities cause the wines to lose consumer appeal because of the intensity of the wood attributes (Pérez-Coello et al. 2000a).

The type of oak wood used is also important, since the volatiles released by different types of oak usually differ hence so do the resulting wines. Tasters tend to prefer a balance between the "oak" aromas ordinarily produced by the oak lactones and the grape variety's own fruity aromas (Pérez-Coello et al. 2000b).

On the other hand, using toasted oak chips has been reported to cancel out the varietal attributes by increasing the wood attributes and adding toasted and smoky aromas that can mask the white wines fruity aromas (Guchu et al. 2006b).

Chip size, wood pieces vs wood chips, and the time the wood is in contact with the wine are all major factors affecting the volatiles released by the wood and hence the final volatiles content and the sensory characteristics of the wine. Barrel-aged wines were discriminated from oak chips treated wines from the same variety on the basis of their syringaldehyde and eugenol contents, while wine treated with small and big chips were discriminated on the basis of guaiacol (Arapitsas et al. 2004).

Wine treated with oak chips matures quickly, and more phenolic compounds can be extracted from wood chips than from oak staves and oak barrel aging (Alamo-Sanza et al. 2004).

Comparing the aging of a Merlot wine in contact with different types of French oak chips and the barrel aging of the same wine demonstrated that the wood chips released more oak lactones, furfural, and eugenol than new barrels. On the other hand, oak chips aging yielded more uniform wines with fewer differences according to oak origin (Frangipane et al. 2007).

Adding small doses of oxygen (microoxygenation) while wines are in contact with oak chips can yield wines that are much like red wines that have been aged in the barrel by producing changes similar to those brought about by the oxidation that takes place during barrel aging. More research on these treatments and their effect on wine aroma is needed.

The complexity of the oak aged wine aroma justifies the numerous studies that have been realised on the different aspects considered in this review. Factors that affect volatile composition of the oak woods and their treatments, the shelf-life of the barrels, and the chemical reactions that happens into the barrel or those induced by microorganisms. The sensory impact of the compounds ceded by the wood and their synergist effects with those of the wine make difficult the standardisation of an oak aged wine. On the other hand, the possibility of using oak wood chips or staves as an alternative to oak barrel aging, together with their effectiveness for obtaining a quality product, are aspects that will have to be developed in the near future.

References

Alamo-Sanza, M., Nevares-Domínguez, I., & García-Merino, S. (2004). Influence of different aging systems and oak woods on aged wine color and anthocyanin composition. *Eur. Food Res. Technol., 219*, 124–132.

Aleixandre, J. L., Padilla, A. I., Navarro, L. L., Suira, A., García, M. J., & Álvarez, I. (2003). Optimization of making barrel-fermented dry muscatel wines. *J. Agric. Food Chem., 51*, 1889–1893.

Ancín, C., Garde, T., Torrea, D., & Jiménez, N. (2004). Extraction of volatile compounds in model wine from different oak woods. Effect of SO_2. *Food Res. Int., 37*, 375–383.

Arapitsas, P., Antonopoulos, A., Stefanou, E., & Dourtoglou, V. G. (2004). Artificial aging of wines using oak chips. *Food Chem., 86*, 563–570.

Aznar, M., López, R., Cacho, J., & Ferreira, V. (2003). Prediction of aged red wine aroma properties from aroma chemical composition. Partial least squares regression models. *J. Agric. Food Chem., 51*, 2700–2707.

Boidron, J. N., Chatonnet, P., & Pons, M. (1988). Effects of wood on aroma compounds of wine. *Conn. Vigne Vin, 22*, 275–294.

Cadahía, E., Fernandez de Simon, B., & Jalocha, J. (2003). Volatile compounds in Spanish, French and American oak woods after natural seasoning and toasting. *J. Agric. Food Chem., 51*, 5923–5932.

Chatonnet, P. (1991). Incidence du bois de chêne sur la composition chimique et les qualités organoleptiques des vins. Applications technologiques. Thesis. Université de Bordeaux II, France.

Chatonnet, P. (1998). Volatile and odoriferous compounds in barrel-aged wines: Impact of cooperage techniques and aging conditions. In A. L. Waterhouse & S. E. Ebeler (Eds.), *Chemistry of wine flavour*, (pp. 180–207). American Chemical Society: Washington.

Chatonnet, P. (1999). Discrimination and control of toasting intensity and quality of oak wood barrels. *Am. J. Enol. Vitic., 50*, 479–494.

Chatonnet, P., & Boidron, J. N. (1989). Incidence du traitement thermique du bois de chêne sur sa composition chimique. 1*ere* partie: Définition des paramètres thermiques de la chauffe des fûts en tonnellerie. *Conn. Vigne Vin, 23*, 77–87.

Chatonnet, P., & Dubourdieu, D. (1998). Comparative study of the characteristics of American white oak (*Quercus alba*) and European oak (*Quercus petraea* and *Q. robur*) for production of barrels used in barrel aging of wines. *Am. J. Enol. Vitic., 49*, 79–85.

Chatonnet, P., Dubourdieu, D., & Boidron, J. N. (1992a). Incidence des conditions de fermentation et d`élevage des vins blancs secs en barriques sur leur composition en substances cédées par le bois de chêne. *Sci. Aliment., 120*, 19–24.

Chatonnet, P., Dubourdieu, D., Boidron, J. N., & Pons, M. (1992b). The origin of ethylphenols in wines. *J. Scic. Food Agric., 60*, 165–178.

Chatonnet, P., Boidron, J. N., Dubourdieu, D., & Pons, M. (1994). Évolution de certains composés volatils du bois de chêne au cours de son séchage-premiers résultat. *J. Inter. Sci. Vigne Vin, 28*, 359–380.

Chatonnet, P.; Masneuf, I., Gubbiotti, M. C., & Dubourdieu, D. (1999). Prévention et détection des contaminants par *Brettanomyces* au cours de la vinification et de l`élevage des vins. *Rev. Fran. Oenol., 179*, 20–24.

Cutzach, I., Chatonnet, P., Henry, R., & Dubourdieu, D. (1997). Identification of volatile compounds with a "toasty" aroma in heated oak used in barrelmaking. *J. Agric. Food Chem., 45*, 2217–2224.

Cutzach, I., Chatonnet, P., Henry, R., & Dubourdieu, D. (1999). Identifying new volatile compounds in toasted oak. *J. Agric. Food Chem., 47*, 1663–1667.

Díaz-Maroto, M. C., Sánchez-Palomo, E., & Pérez-Coello, M. S. (2004). Fast screening method for volatile compounds of oak wood used for aging wines by headspace SPME-GC-MS (SIM). *J. Agric. Food Chem., 52*, 6857–6861.

Díaz-Maroto, M. C., Guchu, E., Castro-Vázquez, L., de Torres, C., & Pérez-Coello, M. S. (2008). Aroma active compounds of American oak (*Quercus alba)* and French, Hungarian and Russian oak woods (*Quercus petraea)* studied by Gas Chromatography-Mass Spectrometry and Gas Chromatography-Olfactometry. *Flavour Fragr. J., 23*, DOI 10.1002/ffj.1858.

Díaz-Plaza, E. M., Reyero, J. R., Pardo, F., & Salinas, M. R. (2002). Comparison of wine aromas with different tannic content aged in French oak barrels. *Anal. Chim. Acta, 458*, 139–145.

Doussot, F., Pardon, P., Dedier, J., & de Jeso, B. (2000). Individual, species and geographic origin influence on cooperage oak extractible content (*Quercus robur* L. and *Quercus petraea* Liebl.). *Analusis, 28*, 960–965.

Doussot, F., de Jeso, B., Quideau, S., & Pardon, P. (2002). Extractives content in cooperage oak wood during natural seasoning and toasting; Influence of tree species, geographic location and single-tree effects.*J. Agric. Food Chem., 50*, 5955–5961.

Ducournau, P., Chassin, M., & Lemaire, T. (1999). Chips as a factor of quality. *Am. J. Enol. Vitic., 50*, 545.

Escalona, E., Birkmyre, L., Piggot, J. R., & Paterson, A. (2002). Effect of maturation in small oak casks on the volatility of red wine aroma compounds. *Anal. Chim. Acta, 458*, 45–54.

Fernández de Simón, B., Cadahía, E., & Jalocha, J. (2003). Volatile compounds in a Spanish red wine aged in barrels made of Spanish, French, an American oak wood. *J. Agric. Food Chem., 51*, 7671–7678.

Fernández de Simón, B., Sanz, M., Cadahía, E., Poveda, P., & Broto, B. (2006). Chemical characterization of oak heartwood from Spanish forests of *Quercus pyrenaica* (Wild.). Ellagitannins, low molecular weight phenolic, and volatile compounds. *J. Agric. Food Chem., 54*, 8314–8321.

Feuillat, F., Moio, L., Guichard, E., Marinov, M., Fournier, N., Puech, J. L. (1997). Variation in the concentration of ellagitannins and *cis*- and *trans*-β-methyl-γ-octalactone extracted from oak wood (*Quercus robur* L., *Quercus petraea* Liebl.) under model wine cask conditions. *Am. J. Enol. Vitic., 48*, 509–515.

Francis, I. L., Sefton, M. A., & Williams, P. J. (1992). A study by sensory descriptive analysis of the effects of oak origin, seasoning, and heating on the aromas of oak model wine extracts. *Am. J. Enol. Vitic., 43*, 23–30.

Frangipane, M. T., De Santis, D., & Ceccarelli, A. (2007). Influence of oak woods of different geographical origins on quality of wines aged in barriques and using oak chips. *Food Chem., 103*, 46–54.

Garde Cerdán, T., & Ancín-Azpilicueta, C. (2006). Effect of oak barrel type on the volatile composition of wine: Storage time optimization. *Lebens. Wissen. Technol., 39*, 199–205.

Garde-Cerdán, T., Torrea-Goñi, D., & Ancín-Azpilicueta, C. (2002). Changes in the concentration of volatile oak compounds and esters in red wine stored for 18 months in re-used French oak barrels. *Aust. J. Grape Wine Res., 8*, 140–145.

Gómez-Plaza, E., Pérez-Prieto, L. J., Fernández-Fernández, J. I., & López-Roca, J. M. (2004). The effect of successive uses of oak barrels on the extraction of oak-related volatile compounds from wine. *Int. J. Food Sci. Technol., 39*, 1069–1078.

González-Viñas, M. A., Pérez-Coello, M. S., Salvador, M. D., Cabezudo, M. D., & Martín-Álvarez, P. J. (1996). Changes in GC volatiles of young Airén wines during the storage in bottle. *Food Chem., 56*, 399–403.

Guchu, E., Díaz-Maroto, M. C., Díaz-Maroto, I. J., Vila-Lamerio, P., & Pérez-Coello, M. S. (2006a). Influence of the species and geographical location on volatiles composition of Spanish oak wood (*Quercus petraea* Liebl. and *Quercus robur* L.). *J. Agric. Food Chem., 54*, 3062–3066.

Guchu, E., Díaz-Maroto, M. C., Pérez-Coello, M. S., González-Viñas, M. A., & Cabezudo Ibáñez, M. D. (2006b). Volatile composition and sensory characteristics of Chardonnay wines treated with American and Hungarian oak chips. *Food Chem., 99*, 350–359.

Gutiérrez Afonso, V. L. (2002). Sensory descriptive analysis between white wines fermented with oak chips and in barrels. *J. Food Sci., 67*, 2415–2419.

Hakansson, A. E., Pardon, K., Hayasaka, Y., de Sa, M., & Herderich, M. (2003). Structure and colour properties of new red wine pigments. *Tetrahedrom Letters, 44*, 4887–4891.

Hale, M. D., McCafferty, K., Larmie, E., Newton, J., & Swan, J. S. (1999). The influence of oak seasoning and toasting parameters on the composition and quality of wine. *Am. J. Enol. Vitic., 50*, 495–502.

Henick-Kling, T., Egli, C., Licker, J., Mitrakul, C., & Acree, T. E. (2000). In *Proceedings of the 5th international symposium on cool climate viticulture and oenology, Melbourne, Australia, 16–20 January.*

Herjavec, S., Jeromel, A., Da Silva, A., Orlic, S., & Redzepovic, S. (2007). The quality of white wines fermented in Croatian oak barrels. *Food Chem., 100*, 124–128.

Jarauta, I., Cacho, J., & Ferreira, V. (2005). Concurrent phenomena contributing to the formation of the aroma of wine during aging in oak wood: An analytical study. *J. Agric. Food Chem., 53*, 4166–4177.

Jiménez Moreno, N., & Ancín Azpilicueta, C. (2007). Binding of oak volatile compounds by wine lees during simulation of wine ageing. *Lebens. Wissen. Technol., 40*, 619–624.

Maga, J. A. (1984). Flavor contribution of wood in alcoholic beverages. In J. Adda, (Ed.), *Progress in Flavour Research: Proceedings of the 4th Weurman Flavour Research Symposium*, Dourdan, France, (pp. 409–416). Elsevier Science, Amsterdam.

Malfeito-Ferreira, M. (2005). Avances recientes en el control de *Brettanomyces/Dekkera bruxellensis* en vinos. In *Enotour 2005 Agrovin, Zaragoza, Spain, 28–30 June.*

Marko, S. D., Dormedy, E. S., Fugelsang, K. C., Dormedy, D. F., Gump, B., & Wample, R. L. (2005). Analysis of oak volatiles by gas chromatography-mass spectrometry after ozone sanitization. *Am. J. Enol. Vitic., 56*, 46–51.

Masson, E., Baumes, R., Le Guernevé, C., & Puech, J. L. (2000a). Identification of a precursor of β-methyl-γ-octalactone in the wood of sessile oak (*Quercus petraea* (Matt.) Liebl.). *J. Agric. Food Chem., 48*, 4306–4309.

Masson, E., Baumes, R., Moutounet, M., & Puech, J. L. (2000b). The effect of kiln drying on the levels of ellagitannins and volatile compounds of European oak (*Quercus petraea Liebl.)* stave wood. *Am. J. Enol. Vitic., 51*, 201–214.

Masuda, M., & Nishimura, K. (1971). Branched nonalactones from some *Quercus* species. *Phytochemistry, 10*, 1401–1402.

Mosedale, J. R., & Ford, A. (1996). Variation of the flavour and extractives of European oak wood from two French forests. *J. Scic. Food Agric., 70*, 273–287.

Ortega-Heras, M., González-Sanjosé, M. L., & González-Huerta, C. (2007). Consideration of the influence of aging process, type of wine and oenological classic parameters on the levels of wood volatile compounds present in red wines. *Food Chem., 103*, 1434–1448.

Otsuka, K. I., Sato, K., Yamashita, T. (1980). Structure of a precursor of β-methyl-γ-octalactone, an aging flavour compound of distilled liquors. *J. Fermen. Technol., 58*, 395–398.

Pérez Prieto, L. J., López Roca, J. M., Martínez Cutillas, A., Pardo Mínguez, F., & Gómez Plaza, E. (2002). Maturing wines in oak barrels. Effects of origin, volume, and age of the barrel on the wine volatile composition. *J. Agric. Food Chem., 50*, 3272–3276.

Pérez Prieto, L. J., López Roca, J. M., Martínez Cutillas, A., Pardo Mínguez, F., & Gómez Plaza, E. (2003). Extraction and formation dynamic of oak-related volatile compounds from different volume barrels to wine and their behavior during bottle storage. *J. Agric. Food Chem., 51*, 5444–5449.

Pérez-Coello, M. S., Sanz, J., & Cabezudo, M. D. (1997). Analysis of volatile components of oak wood by solvent extraction and direct thermal desorption-gas chromatography-mass spectrometry. *J. Chromatogr. A, 778*, 427–434.

Pérez-Coello, M. S., Sanz, J., & Cabezudo, M. D. (1998). Gas chromatographic-mass spectrometric analysis of volatile compounds in oak wood used for ageing of wines and spirits. *Chromatographia, 47*, 427–432.

Pérez-Coello, M. S., Sanz, J., & Cabezudo, M. D. (1999). Determination of volatile compounds in hydroalcoholic extracts of French and American oak woods. *Am. J. Enol. Vitic., 50*, 162–165.

Pérez-Coello, M. S., González-Viñas, M. A., García-Romero, E., Cabezudo, M. D., & Sanz, J. (2000a). Chemical and sensory changes in white wines fermented in the presence of oak chips. *Int. J. Food Sci. Technol., 35*, 23–32.

Pérez-Coello, M. S., Sánchez, M. A., García, E., González-Viñas, M. A., Sanz, J., & Cabezudo, M. D. (2000b). Fermentation of white wines in the presence of wood chips of American and French oak. *J. Agric. Food Chem., 48*, 885–889.

Pérez-Coello, M. S., González-Viñas, M. A., García-Romero, E., Díaz-Maroto, M. C., & Cabezudo, M. D. (2003). Influence of storage temperature on the volatile compounds of young white wines. *Food Control, 14*, 301–306.

Pollnitz, A. P.; Pardon, K. H., & Sefton, M. A. (2000). Quantitative analysis of 4-ethylphenol and 4-ethylguaiacol in red wines. *J. Chromatog. A, 874*, 101–109.

Prida, A., & Puech, J. L. (2006). Influence of geographical origin and botanical species on the content of extractives in American, French, and East European oak woods. *J. Agric. Food Chem., 54*, 8115–8126.

Prida, A., Ducousso, A., Petit, R. J., Nepveu, G. & Puech, J. L. (2007). Variation in wood volatile compounds in a mixed oak stand: strong species and spatial differentiation in whisky-lactone content. *Ann. Forest Sci., 64*, 313–320.

Puech, J. L. (1987). Extraction of phenolic compounds from oak wood in model solutions and evolution of aromatic aldehydes in wines aged in oak barrels. *Am. J. Enol. Vitic., 38*, 236–238.

Ramírez Ramírez, G., Lubbers, S., Charpentier, C., Feuillat, M., Voilley, A., & Chassagne, D. (2001). Aroma compound sorption by oak wood in a model wine. *J. Agric. Food Chem., 49*, 3893–3897.

Rapp, A., & Marais, J. (1993). The shelf life of wine: Changes in aroma substances during storage and ageing of white wines. In: Shelf Life Studies of Foods and Beverages. Chemical, Biological, Physical and Nutritional Aspects. Ed. (pp. 891–923). Charalambous, Elsevier, Amsterdam,.

Rous, C., & Alderson, B. (1983). Phenolic extraction curves for white wine aged in French and American oak barrels. *Am. J. Enol. Vitic., 34*, 211–215.

Sauvageot, F., & Feuillat, F. (1999). The influence of oak wood (*Quercus robur* L., *Q. petraea* Liebl.) on the flavor of Burgundy Pinot noir. An examination of variation among individual trees. *Am. J. Enol. Technol., 50*, 447–455.

Schreier, P. (1979). Flavor composition of wines: a review. *Crit. Rev. Food Sci. Nut., 12,* 59–111.

Sefton, M. A., Francis, I. L., & Williams, P. J. (1990). Volatile norisoprenoid compounds as constituents of oak woods used in wine and spirit maturation. *J. Agric. Food Chem., 38*, 2045–2049.

Sefton, M. A., Francis, I. L., Pocock, K. F., & Williams, P. J. (1993). The influence of natural seasoning on the concentrations of eugenol, vanillin and *cis-trans-*β-methyl-γ-octalactone extracted from French and American oak wood. *Sci. Aliments, 13*, 629–643.

Singleton, V. L. (1995). Maturation of wines and spirits: comparisons, facts, and hypotheses. *Am. J. Enol. Vitic., 46*, 98–115.

Spillman, P. J., Iland, P. G., & Sefton, M. A. (1998a). Accumulation of volatile oak compounds in a model wine stored in American and Limousin oak barrels. *Aust. J. Grape Wine Res., 4*, 67–73.

Spillman, P. J., Pollnitz, A. P., Liacopoulos, D., Pardon, K. H., & Sefton, M. A. (1998b). Formation and degradation of furfuryl alcohol, 5-methylfurfuryl alcohol, vanillyl alcohol, and their ethyl ethers in barrel-aged wines. *J. Agric. Food Chem., 46*, 657–663.

Spillman, P. J., Sefton, M. A., & Gawel, R. (2004a). The effect of oak wood source, location of seasoning and coopering on the composition of volatile compounds in oak-matured wines. *Aust. J. Grape Wine Res., 10*, 216–226.

Spillman, P. J., Sefton, M. A., & Gawel, R. (2004b). The contribution of volatile compounds derived during oak barrel maturation to the aroma of a Chardonnay and Cabernet Sauvignon wine. *Aust. J. Grape Wine Res., 10*, 227–235.

Towey, J. P., & Waterhouse, A. L. (1996). The extraction of volatile compounds from French and American oak barrels in Chardonnay during three successive vintages. *Am. J. Enol. Vitic., 47*, 163–172.

Waterhouse, A. L., & Towey, J. P. (1994). Oak lactone isomer ratio distinguishes between wines fermented in American and French oak barrels. *J. Agric. Food Chem., 42*, 1971–1974.

Wilker, K. L., & Gallander, J. F. (1988). Comparison of Seyval blanc wine aged in barrels and stainless steel tanks with oak chips. *Am. J. Enol. Vitic., 39*, 38–43.

Wilkinson, K. L., Elsey, G. M., Prager, R. H., Pollnitz, A. P., & Sefton, M. A. (2004). Rates of formation of *cis-* and *trans-*oak lactone from 3-methyl-4-hydroxyoctanoic acid. *J. Agric. Food Chem., 52*, 4213–4218.

Chapter 8D
Yeasts and Wine Flavour

Maurizio Ugliano and Paul A. Henschke

Contents

P.A. Henschke (✉)
The Australian Wine Research Institute, Urrbrae, Adelaide, South Australia
e-mail: paul.henschke@awri.com.au

M.V. Moreno-Arribas, M.C. Polo (eds.), *Wine Chemistry and Biochemistry*,
DOI 10.1007/978-0-387-74118-5_14, © Springer Science+Business Media, LLC 2009

8D.1 Introduction

Vinification is a complex process involving interactions between microbes and grape must, resulting in wine. The key microorganisms are fermentative yeasts, which transform an acidic, sweet, low flavoured grape must into a distinctive, highly flavoured alcoholic beverage. In essence, this process is an alcoholic fermentation involving biochemical conversion of hexose sugars to ethanol and carbon dioxide. The central glycolytic pathway, which is coupled to the reductive formation of ethanol, provides energy, precursor molecules and reducing power for cell growth, maintenance and reproduction. In addition, the glycolytic and associated pathways form volatile and non-volatile metabolites that contribute to wine flavour. The species of yeast(s), fermentation conditions and nutrient content of the must modulate the production of these compounds, and consequently, a wine's flavour profile (Lambrechts and Pretorius 2000; Fleet 2003; Romano et al. 2003b; Swiegers et al. 2000).

Over recent decades, research into the role of yeast in the development of wine flavour has revealed complex interactions between this microbe and grape compounds; many of these interactions contribute to the appearance, aroma, flavour and texture of wine. When wood is used in fermentation, some wood-derived flavour compounds can also be modified by yeasts. Together, all these compounds that are present in must, and produced and modified during fermentation, and by other processes, contribute to the distinctive 'varietal' character of wine.

Although many wines are still made with little intervention and rely on the indigenous yeasts present in must, modern winemaking is founded on the use of selected strains of *Saccharomyces cerevisiae,* that have reliable fermentation properties coupled to good flavour characteristics. As our understanding of the impacts that yeasts have on wine flavour has improved, strains have been selected that are less susceptible to off-flavour formation and better augment different components of wine flavour. Chemical identification of key flavour compounds and definition of the genetic and biochemical basis of the associated metabolic pathways, and their regulation, has often lagged behind the use of selected strains, but remains an essential step in the development of yeasts with improved flavour properties that are better adapted to grape must (Henschke 1997; Pretorius 2000; Swiegers et al. 2005). Characterisation of *Saccharomyces* species has further expanded winemakers' ability to modulate wine flavour due to greater genetic and metabolic variability. Driven by the search for even greater diversity of wine flavour, non-*Saccharomyces* species, which originate from grapes, are also being evaluated, and novel inoculation strategies developed, to exploit their flavour complexing properties (Jolly et al. 2006). Wild yeasts, although representing an undefined mixture of species and strains, produce flavour profiles, which cannot be easily achieved by other techniques.

This chapter will summarise the microbiological, physiological and biochemical interactions between grape compounds and yeast metabolism that lead to the development of wine flavour. The role of different yeasts and inoculation strategies in the modulation of wine flavour will be highlighted, and the potential for improving wine flavour by advanced screening and genetic techniques will be briefly discussed.

The term 'flavour' will be used in its broadest sense to incorporate aroma, taste, and mouth-feel. Yeast taxonomy is according to Kurtzman and Fell (1998), with the exception of the former species *Saccharomyces uvarum*, for which species status is still unresolved (Nguyen and Gaillardin 2005), and will therefore be recorded as *Saccharomyces* species *bayanus/uvarum* Yeast gene and enzyme notations are according to the *Saccharomyces Genome* Database (http://yeastgenome.org).

8D.2 Overview of Yeast Interactions with Grape Compounds

In standard winemaking conditions, fermentation can only commence when sugars and other nutrients necessary for yeast growth are released from the berry by crushing. The yeasts associated with the grapes and processing equipment, often referred to as indigenous, 'native' or 'wild' yeasts, can initiate fermentation or, more commonly, a starter culture of a selected yeast is added. Sulfur dioxide is commonly added as an antioxidant and antimicrobial compound, which reduces the populations of indigenous yeasts and bacteria. Crushing also releases volatile and non-volatile compounds associated with the berry pulp into the must, whereas maceration, often with added exogenous pectinolytic enzymes, is required to facilitate release of flavour and precursor flavour compounds associated with berry skin, and to a lesser extent seeds. Must compounds can be considered to fall into three broad functional groups, 'nutrients', 'flavour precursors' and 'non-precursor flavour-active compounds' (Fig 8D.1). The action of yeasts on many of these compounds results in the complex chemistry of wine, which forms the basis of 'appearance', 'fermentation bouquet', 'varietal character' and 'mouth-feel' of wine. Known key compounds are summarised in Table 8D.1.

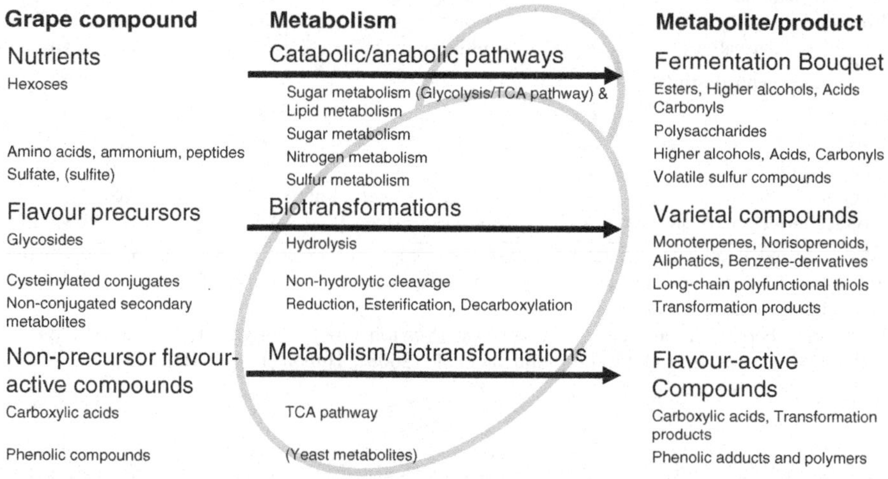

Fig. 8D.1 Interactions between yeasts and grape compounds

Table 8D.1 Role of yeast in the formation of the main sensorially-active compounds in wine

Compound(s)	Origin [a]	Sensory attribute	Effect of yeast metabolism	Effect of yeast nutrients
Volatile compounds				
Acetate esters Fatty acid ethyl esters Branched chain ethyl esters	A	Flowery, fruity, estery	Yes	Yes
Higher alcohols	A, B	Alcohol, herbaceous	Yes	Yes
Volatile fatty acids	A	Sour, sweat, cheese	Yes	Yes
Monoterpene alcohols	A, B, C	Flowery, citrus	Yes	Yes
cis-Rose oxide	B, C	Flowery	Yes	Unknown
Rotundone	C	Spicy	Unknown	Unknown
β-Damascenone	B, C	Quince paste, stewed apple	Yes	Unknown
TDN and vitispiranes	B	Kerosene	Unknown	Unknown
β-Ionone	B	Violet	Unknown	Unknown
Methoxypyrazines	C	Capsicum	Unknown	Unknown
H_2S and mercaptans	A	Rotten egg, onion, garlic, cabbage	Yes	Yes
Sulfides	A, B	Asparagus, truffle, blackcurrant	Yes	Yes
4-Mercapto-4-methylpentan-2-one	B	Box tree	Yes	Unknown
3-Mercaptohexan-1-ol	B	Green mango, box tree	Yes	Yes
3-Mercaptohexyl acetate	B	Tropical fruit	Yes	Yes
Furfurylthiol	A	Coffee-like	Yes	Yes
Methional	A	Boiled potatoes	Yes	Yes
Methionol	A	Boiled potatoes	Yes	Yes
Diacetyl	A	Buttery	Yes	Yes
Sotolon	B	Spicy	Yes	Unknown
Volatile phenols	B	Spicy, clove-like	Yes	Unknown
Aliphatic aldehydes	A, B	Green, grassy	Yes	Unknown
Aliphatic lactones	B	Sweet, apricot	Yes	Unknown
Whiskey lactone	D	Coconut	None	None
Vanillin and derivatives	B, D	Vanilla, spicy, sweet	Yes	Unknown
Non-volatile compounds				
Sugars	C	Sweet	Yes	Yes
Polyols	A	Sweet	Yes	Yes
Organic acids	C	Acid	Yes	Yes
Phenolics	C	Colour, astringent, bitter	Yes	Yes
Polysaccharides/mannoproteins	A,C	Modifies astrigency	Yes	Unknown

[a]A: compound largely synthesized by yeast; B: compound present in the form of non-volatile precursors; C: compound present in significant concentrations in grapes; D: compound derived from other sources, e.g. oak wood

References: Ferreira et al. (2008); Francis and Newton (2005); Smyth et al. (2005); Lavigne et al. (2008); Swiegers et al. (2005); Subileau et al. (2008); Thibon et al. (2008).

8D.2.1 Nutrients

Saccharomyces species can grow on a minimal range of organic and inorganic nutrients, with hexose sugars the preferred source of carbon and energy under anaerobic conditions. Sources of nitrogen, phosphate, sulfate, various minerals K^+, Mg^{2+} and Zn^{2+}, and trace elements, provide the necessary nutrients for growth and reproduction. Some *Saccharomyces cerevisiae* strains require several vitamins, typically biotin. In highly clarified (lipid-free) fermentations, *Saccharomyces* species produce limited growth in the complete absence of oxygen, whereas most non-*Saccharomyces* require oxygen. Exposure of fermentation to oxygen is normally limited, to prevent unwanted oxidation reactions. However, small additions stimulate growth and fermentation by balancing metabolic redox and increasing tolerance to the inhibitory effects of ethanol by allowing the synthesis of sterols and unsaturated fatty acids that improve membrane function (Rosenfeld et al. 2003). Oxygen and exogenous lipids (grape solids) affect metabolism and, consequently, modulate wine flavour.

Grape must typically contains sufficient essential nutrients to enable the adequate growth of yeast, which is necessary to complete the fermentation of sugars in an essentially oxygen-free environment. Supplementation experiments reveal that most nutrients are not present in optimal amounts; sugars are a major exception. Glucose and fructose, the predominant sugars, are fermented to ethanol and CO_2 to provide energy and carbon compounds for yeast growth. Sugar concentration largely determines wine ethanol content, which can range from 8 vol.% to >15 vol.%. The catabolism of sugars leads to formation of the non-volatile compounds, polyols and carboxylic acids, and the volatile compounds, higher alcohols, ethyl and acetate esters, carbonyls, volatile fatty acids, and volatile sulfur compounds. Polymers, mainly mannoproteins, are also produced in small amounts (Table 8D.1) (Swiegers et al. 2005).

Yeasts assimilate a variety of nitrogenous compounds, predominantly primary amino acids, ammonium ions and small peptides. The types and concentrations of extracellular nitrogenous compounds regulate yeast growth and metabolism, including sugar, nitrogen and sulfur pathways, and therefore affect the production of non-volatile and volatile metabolites, many of which have sensory implications. Nitrogen is a limiting nutrient in many wine regions, and consequently restricts growth, and fermentation rate and completion. Inadequate nitrogen is often associated with high concentrations of higher (fusel) alcohols and increased risk of sulfidic off-flavours (principally H_2S and mercaptans), whereas excessive nitrogen can lead to estery off-flavour (Bell & Henschke 2005).

Because grape must contain little organic sulfur, in the form of S-amino acids and glutathione, yeast assimilates sulfate, and sulfite when added, for the synthesis of organic sulfur compounds. Sulfur metabolism is important in winemaking since, depending on the yeast strain and nutrient composition, it is the source of mostly unpleasant compounds, including H_2S and mercaptans (Rauhut 1993).

Phosphate is an essential nutrient for energy metabolism and consequently regulates many metabolic pathways; however, no information is available on how phosphate affects flavour metabolism. Similarly, metal ions primarily function as cofactors in enzymatic reactions but again the link between availability and flavour metabolism is essentially unknown. Conversely, certain vitamins, particularly biotin and pantothenic acid, are known to affect the formation of aroma compounds due to their variable content in must.

Yeast characteristics are largely determined by its genetic constitution, nevertheless the balance of nutrients in each grape must can strongly affect growth and metabolite production, which is a determinant of wine composition, flavour and style. Few wine strains are well characterised and little information exists to guide winemakers on using nutrients to modulate flavour. Apart from the routine measurement of sugar and, to a growing extent, α-amino nitrogen concentration, few rapid methods for measuring other important nutrients are available to winemakers. For this reason proprietary nutrient mixtures are widely used, according to wine regulations. Because the balance of nutrients, such as sugar, amino acids and vitamins, appear to depend on the viticultural conditions, regions which favour a particular balance of nutrients might owe some of the regional characteristics of wine, to nutrient composition, as well as the winery microflora, at least when winemaker intervention is minimal.

8D.2.2 Biotransformation of Flavour Precursors

Apart from sweetness, acidity and astringency, grapes and musts exhibit relatively low flavour when compared to wines. Furthermore, wines made from most grape varieties posses a characteristic aroma and flavour, which is attributed to, at least in part, grape-derived flavour-active precursor compounds (Dubourdieu et al. 2006; Francis et al. 1999). During fermentation and subsequent wine aging these non-volatile, odourless compounds undergo transformation into volatile, aromatic compounds. A variety of yeast biochemical mechanisms are involved, including hydrolysis, transformation reactions such as reduction, esterification, and decarboxylation, oxidation and metabolite-induced condensation reactions. Examples of these reactions are given in Table 8D.2, and will be discussed in more detail later in this chapter.

The best studied enzyme-catalysed reactions relate to two important classes of odourless flavour-precursor compounds, namely glycosides and cysteinylated conjugates. Grape varieties related to *Vitis vinifera* cv. Muscat contain terpenoid glycosides, which give rise to the intense floral Muscat aroma after fermentation (Cordonnier and Bayonove 1974; Strauss et al. 1986). Yeasts possess hydrolytic enzymes with β-D-glucopyranosidase (β-glucosidase) and β-glucanase activities, which are capable of enhancing the hydrolysis of glycosides during fermentation (Darriet et al. 1988; Gil et al. 2005; Ugliano et al. 2006). In addition to containing glycosides, *Vitis vinifera* L. cv. Sauvignon Blanc contains non-volatile *S*-cysteinyl-conjugates, which are responsible for the characteristic passion fruit/grape fruit and box-hedge aromas that develop during fermentation (Dubourdieu et al. 2006).

Table 8D.2 Summary of yeast metabolic interactions with grape compounds

Reaction	Enzyme/metabolite	Precursor(s)	Product	Ref.
Hydrolysis	β-Glucosidase	Linalyl-glucoside	Linalool	1
	β-Glucanase[a]	Malvidin-3-*O*-glucoside	Malvidin	2
Non-hydrolytic cleavage	Carbon-sulfur lyase	3-(Hexan-1-ol)-L-cysteine	3-mercapto-hexanol	3a,b
Reduction	Alcohol dehydrogenase	Vanillin	Vanillic alcohol	4
		Geraniol	Citronellol	5
Decarboxylation	Phenyl acrylic acid decarboxylase	*p*-Coumaric acid	4-Vinyl phenol	6
Esterification	Alcohol acyltransferase	3-Mercapto-hexanol	3-Mercapto-hexylacetate	7
Metabolite-induced condensation or adduct reactions	Acetaldehyde[a]	Catechin and malvidin-3-*O*-glucoside	Catechin-ethyl-malvidin-3-glucoside	8
	Pyruvic acid[a]	Malvidin-3-*O*-glucoside	Vitisin A	9

[a] Extracellular reaction; References: (1) Ugliano et al. (2006); (2) Gil et al. (2005); (3a) Tominaga et al. (1995); (3b) Howell et al. (2005); (4) Chatonnet et al. (1992a); (5) Gramatica et al. (1982); (6) Clausen et al. (1994); (7) Swiegers et al. (2006); (8) Timberlake and Bridle (1976); (9) Fulcrand et al. (1998)

Carbon-sulfur lyase, present in *Eubacterium limosum* cell-free extracts, can liberate volatile long-chain polyfunctional thiols, such as 4-MMP and 3-MH, from *S*-cysteine conjugates. Gene knock-out and expression studies in yeast support the role of carbon-sulfur lyases (Tominaga et al. 1995; Howell et al. 2005; Swiegers et al. 2007).

Grape compounds which can enter the yeast cell either by diffusion of the undissociated lipophilic molecule or by carrier-mediated transport of the charged molecule across the cell membrane are potentially subject to biochemical transformations by enzymatic functions. A variety of biotransformation reactions of grape compounds that have flavour significance are known. One of the earlier studied biotransformations in yeast relates to the formation of volatile phenols from phenolic acids (Thurston and Tubb 1981). Grapes contain hydroxycinnamic acids, which are non-oxidatively decarboxylated by phenyl acryl decarboxylase to the vinyl phenols (Chatonnet et al. 1993; Clausen et al. 1994).

Reductive enzyme catalysed reactions of grape-derived compounds can result, as they do for yeast metabolites, by interaction with dehydrogenases and reductases. Dehydrogenases, such as alcohol dehydrogenase appear to have wide specificity. Using the redox cofactor NADH to reduce acetaldehyde to ethanol, it is likely that other compounds such as geraniol can be reduced to citronellol (Gramatica et al. 1982). Vanillin, a highly aromatic compound derived from oak wood, is reduced to vanillic alcohol by yeast, accounting for the lower intensity of oak aroma of wine fermented in oak wood compared to wine stored in oak wood after fermentation (Chatonnet et al. 1992a). Reductive metabolism is favoured by yeast under anaerobic conditions in order to enhance regeneration of NAD^+, most likely catalysed by alcohol dehydrogenase, to facilitate the fermentation of sugars.

Esterification reactions involving alcohols are also favoured by yeast but the reason why this reaction occurs is not clear. It could involve detoxification or to produce attractant molecules to assist yeast dispersal by insects. Acetate esters, such as ethyl acetate are formed from ethanol and acetyl-CoA by alcohol acyltransferase (Fujii et al. 1994). An example of a grape derived compound is the esterification of 3-MH to the more aroma active 3-MHA (Swiegers et al. 2006).

The occurrence of many of these reactions depends on the presence of appropriate precursor in the grape must and together with the concentration of precursor and the ability of the yeast to (i) produce an appropriate extracellular enzyme for non-transportable precursors, (ii) take up the precursor, and (iii) transform the precursor; different combinations of yeast and grape variety can lead to small or large differences in the aroma and flavour profile of the wine.

8D.2.3 Non-precursor Flavour Compounds

Some grape-derived compounds, which are released from the berry pulp during crushing and from the skin, are not metabolised by microorganisms during fermentation, and therefore they accumulate in wine in a chemically unmodified form. Examples are certain carboxylic acids, phenolics and aroma volatiles.

L-Tartaric acid, the major acid of *Vitis vinifera* grapes, together with L-malic acid and smaller amounts of citric acid form the acidic structure of wine and are responsible for balance and mouth-feel (Jackson 2000); tartaric acid is stable to degradation by fermentation microorganisms whereas malic acid can be partially metabolised by yeast and malic and citric acids are metabolised by lactic acid bacteria during the malolactic fermentation, thereby reducing wine titratable acidity and increasing pH (Swiegers et al. 2005).

Grapes are rich in phenolic compounds that are responsible for colour, astringency and bitterness in wine. The latter two flavour attributes are contributed by a large group of compounds derived from flavan-3-ols, which are represented by monomers, oligomers and polymers (condensed tannins or proanthocyanidins). A smaller group of phenolics are the flavonols, of which quercetin is best known for its antioxidant and beneficial pharmacological properties of wine. These groups of compounds are largely unaffected by yeast metabolism; however, a proportion become bound to the cell wall of yeast. The pigmented polyphenols, anthocyanins, which give the red/purple colour of wine, and the flavan-3-ols, monomeric precursors of tannins, react with yeast metabolites to form stable pigments. A small proportion of pigmented phenolics bind to the yeast cell wall. Phenolic acids are, on the other hand, metabolised intracellularly and their products secreted (Chatonnet et al. 1993; Kennedy et al. 2006). Metabolism of phenolics is discussed in more detail in Sect. 8D.5.3

Methoxypyrazines represent an important group of flavour compounds, which contribute the vegetative, herbaceous, capsicum-like aromas of certain grape varieties, especially Cabernet Sauvignon and Sauvignon Blanc (Lacey et al. 1991

Hashizume and Samuta 1999). 3-Isobutyl-2-methoxypyrazine (IBMP) is the dominant compound and is located in the grape skin, and being water soluble is rapidly extracted into the must. Although there is unproven evidence that yeast can metabolise this compound, sensory studies suggest that its aroma impact is dependent on the presence of other compounds that partially mask its aroma, such as phenyl ethanol and its ester, which are produced in higher amounts by *Saccharomyces bayanus* (Treloar and Howell 2006; Eglinton and Henschke, unpublished).

8D.3 Yeast Formation of Major Wine Compounds

8D.3.1 Ethanol

8D.3.1.1 Significance

Ethanol concentration of dry white and red wines can vary from 8 vol.% to 16 vol.% and signifies the style of wine and the degree of maturity of the grapes from which the wine was prepared. The latter aspect indicates the extent to which a wine might exhibit greener, ripe or over-ripe flavours. Ethanol content does in itself affect the chemical, physical and sensory properties of wine. Ethanol content can impact on perceived (alcohol) hotness, body and perceived viscosity, and lesser effects on sweetness, acidity, aroma, flavour intensity and textural properties have been noted (Gawel et al. 2007a, b). Ethanol concentration also affects the perception of wine aroma compounds but not their partition coefficients (Guth 1997). Despite lower perceived aroma intensities due to higher odour thresholds, the alcohol content of wines has progressively increased by 1–1.5 vol.% over recent decades, especially in New World wines (Godden and Gishen 2005). This trend has exploited the higher flavour intensity of rich, ripe flavours that grapes harvested at higher sugar maturities can produce.

8D.3.1.2 Metabolism

Under winemaking conditions, ethanol and CO_2 are the major products of alcoholic fermentation of grapes sugars by *Saccharomyces cerevisiae*. Even when oxygen is present, the high sugar concentration of grape must locks Crab-tree positive yeast into the fermentative mode of metabolism. Some sugar-derived carbon is however consumed in metabolic reactions required for generating ATP energy and biosynthesis of cell mass (Verduyn et al. 1990a). Other quantitatively important metabolites produced during fermentation include polyols, especially glycerol and 2,3-butanediol, and organic acids, such as succinic, keto and acetic acids. The numerous minor metabolites (higher alcohols, esters, volatile fatty acids, carbonyls), many of which contribute flavour-active properties, represent <1% of sugar carbon. Due to these metabolic losses, the complete fermentation of hexose sugars only yields 94–96% of the theoretical maximum ethanol yield. A widely adopted conversion figure for sugar to alcohol is: 16.83 g/L sugar produces 1 vol. %alcohol (Ribéreau-Gayon et al. 2000).

The question of strain variability in ethanol yield has assumed high interest in recent years as a potential tool for controlling the alcohol content of wine. However, published data show that the ethanol production by 56 commercial wine yeast strains in a simulated white wine fermentation of 200 g/L sugar only ranged from 11.75 vol.% to 12.09 vol.%, that is they only varied by 0.34% v/v alcohol (Palacios et al. 2007). This variation between strains is of limited commercial significance. Larger reductions in ethanol yield are likely to depend on genetic strategies that divert sugar-carbon away from ethanol production (de Barros Lopes et al. 2003).

8D.3.1.3 Modulating Factors

Few studies on the impact of fermentation conditions on ethanol yield have been reported. Regarding nutrients, the choice of nitrogen source can significantly modulate ethanol yield by altering metabolite formation, especially glycerol and organic acids. Growth of yeast on amino acids, as the main nitrogen source, limits the need for amino acid biosynthesis, and hence little associated NADH is generated. This limits the need for NADH reoxidation, which is coupled to glycerol production. On the other hand, growth on ammonium salts requires de novo amino acid synthesis, which results in significant glycerol production in order to maintain redox balance. The extra glycerol produced results in a 14% decrease in ethanol yield in chemically defined media (Albers et al. 1996). Conflicting results have been reported in grape juice-like media (Hernández-Orte et al. 2006; Vilanova et al. 2007) indicating further work should be carried out with authentic grape musts.

Fermentation temperature and fermentor design can induce evaporative alcohol losses but these are relatively small. Based on current state of knowledge, choice of yeast and fermentation conditions constrain wine style and do not provide a reliable approach for achieving at least a one percent or greater loss of potential alcohol content of wine (Boulton et al. 1998; de Barros Lopes et al. 2003; Palacios et al. 2007). Several physical techniques, such as spinning-cone and reverse osmosis, and nanofiltration can be used to remove sufficient sugar or alcohol for the production of reduced or low alcohol products (Ferrarini et al. 2008; Grosser 2008).

8D.3.2 Glycerol and Other Polyols

8D.3.2.1 Significance

Glycerol is the major polyol produced by fermentation yeast, with a smaller contribution by 2,3-butanediol. Glycerol ranges from 5 g/L to 14 g/L in dry and semi-dry wines, with less in flor Sherry wines and up to 25 g/L in botrytised wines (Rankine and Bridson 1971; Nieuwoudt et al. 2002). Red wines contain higher concentrations than white wines, with mean values of 10.49 g/L and 6.82 g/L, respectively

(Nieuwoudt et al. 2002). The sensory impact of glycerol is not entirely clear. Effects on perceivable viscosity and body depend on the wine and glycerol concentrations (Gawel et al. 2007b; Noble and Bursick 1984). Having a sensory threshold of about 5.2 g/L for sweetness, glycerol can be expected to contribute this attribute in dry wines, although this effect is probably not consistent, as is the impact on acidity, and aroma and flavour intensity (Nieuwoudt 2004; Gawel et al. 2007b). Although 2,3-butanediol has a mildly bitter sweet taste, it probably has limited impact on wine flavour due to its relatively high odour threshold (Jackson 2000).

8D.3.2.2 Metabolism

Glycerol production provides a major route for the reoxidation of NADH generated from cellular anabolic reactions under anaerobic growth conditions. It is also a compatible intracellular solute generated to offset osmotic stress. Glycerol is derived from the glycolytic intermediate, dihydroxyacetonephosphate, by reduction to glycerol-3-phosphate followed by dephosphorylation to glycerol. The first step is carried out by two NADH-dependent glycerol-phosphate dehydrogenase isoenzymes (Gpd1,2p), encoded by *GPD1* and *GPD2* genes (Albertyn et al. 1994; Eriksson et al. 1995) (Fig 8D.2). This step is rate limiting for glycerol production since over expression of either gene increases glycerol production. Intracellular glycerol concentration is also regulated by the glycerol permease Fps1p, which controls glycerol release from the cell. *GPD1* is expressed in response to hyperosmotic conditions in order to protect the cell from dehydration whereas *GPD2* is expressed under anaerobic growth conditions as a key step in the reoxidation of NADH generated by anabolic metabolism (amino acid biosynthesis) and as the first step in the biosynthesis of triacylglycerols and glycerophospholipids for membrane growth. Several studies suggest that many redox reactions are involved in restoring the high NAD^+:NADH balance generated from non-growth related glycerol production, with the main NAD^+ sink being the oxidation of acetaldehyde to acetic acid by $NAD(P)^+$-dependent aldehyde dehydrogenases (Eglinton et al. 2002; Remize et al. 1999). Other significant NADH reoxidation steps with flavour implications include the production of 2,3-butanediol and L-malic and succinic acids, as summarised in (Fig 8D.2).

8D.3.2.3 Modulating Factors

Wine yeast vary widely in glycerol production in dry wines, ranging from 4.2 g/L to 10.4 g/L (Radler and Schütz 1982; Rankine and Bridson 1971). Compared to mesophilic *Saccharomyces cerevisiae* strains, thermotolerant strains and those of the cryotolerant yeast *Saccharomyces bayanus/uvarum* produce substantially more glycerol under similar conditions (Antonelli et al. 1999; Giudici et al. 1995; Rainieri et al. 1998). Some non-*Saccharomyces* yeasts, particularly *Candida stellata* and *Saccharomycodes ludwigii*, produce comparatively high amounts of glycerol (Ciani

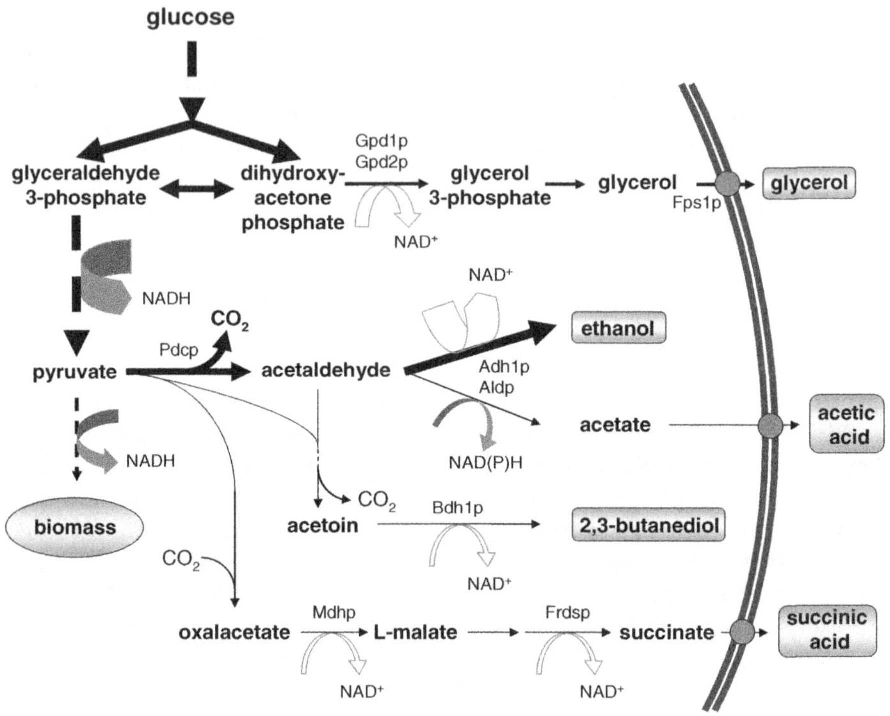

Fig. 8D.2 Production of glycerol and associated redox reactions with flavour implications

Ethanol production is essentially redox neutral; however metabolism associated with biomass production generates nett NADH, which is oxidised largely by glycerol production. Other important NADH oxidising reactions with flavour implications are the production of 2,3-butanediol, L-malic acid and succinic acid. When glycerol production is stimulated by non-growth associated reactions (i.e. osmotic stress) NAD^+ reduction occurs by other reactions including the oxidation of acetaldehyde to acetic acid

and Maccarelli 1998). Strain variability is largely due to differences in the regulation of glyceraldehyde-3-phosphate dehydrogenase activity.

Glycerol production is modulated by fermentation conditions, especially those that affect growth or physiological stress. Higher maturity grapes, having higher sugar concentration, increases glycerol production by hyperosmotic stress (Pigeau and Inglis 2005; Rankine and Bridson 1971). Nutrient availability appears to affect glycerol mainly through redox maintenance mechanisms. Unlike aerobic growth which use O_2as the terminal electron acceptor, anaerobic growth depends on glycerol production to restore NAD^+:NADH balance (Verduyn et al. 1990b). Nitrogen availability generally stimulates biomass formation and hence NADH production, although the effects are strain dependent (Albers et al. 1998; Radler and Schütz 1982; Vilanova et al. 2007). Growth on inorganic nitrogen (ammonium salts) compared to mixtures of amino acid also generates NADH due to amino acid biosynthetic reactions, which stimulates glycerol production (Albers et al. 1996). Thiamin deficiency, which can be induced by significant growth of wild yeast or over use of SO_2, stimulates glycerol production (Bataillon et al. 1996). Due to its essential role as cofactor in the conversion of acetaldehyde to ethanol by alcohol dehy drogenase (Adhp), NADH oxidation depends on Gpdp activity, thereby increasing glycerol formation (Radler and Schütz 1982). SO_2 addition to must can also stimulate glycerol production by forming a hydroxysulfonate adduct with acetaldehyde, again limiting NADH oxidation by Adhp, although the increase in glycerol is relatively small at the rates of SO_2 addition in winemaking (Rankine and Bridson 1971). Higher fermentation temperature, 25 °C compared to 15 °C, and higher pH, 3.8 compared to 3.3, can increase glycerol production to a small extent (Rankine and Bridson 1971).

8D.3.3 Non-volatile Organic Acids

8D.3.3.1 Significance

Acidity and pH constitute fundamental importance to the sensory perception of wine, essentially defining its structure and balance. Excessive acidity increases the perception of sourness whereas low acidity decreases flavour harmony. The perception of sourness and astringency of organic acids depends on concentration, pH and anion species (Sowalsky and Noble 1998). Wine acidity is largely determined by the major grape acids L(+)-tartaric and L(–)-malic, with smaller contributions from citric and D(–)-lactic acids. Grape acids can range from <5 g/L for over-ripened grapes and exceed 10 g/L in early season grapes from cool regions intended for sparkling wine, with 6–8 g/L usually being preferred for dry wines. Wine also contains various non-volatile acids of microbial origin succinic, keto acids pyruvic and α-ketoglutaric, L(+)-lactic and lesser amounts of other non-volatile acids, which contribute to acidity (Boulton et al. 1998; Fowles 1992). Volatile fatty acids of either grape or microbial origin, which can further contribute to acidity and aroma, are discussed in Sect. 8D.4.3

8D.3.3.2 Metabolism and Modulating Factors

Organic acid metabolism serves several important functions, as precursors for biosynthetic pathways and in maintenance of redox balance. Of the major grape acids, tartaric acid is not metabolised by *Saccharomyces cerevisiae* whereas L-malic acid can be partially degraded (3–45%) by most strains, but this pathway is restricted by diffusion of the non-dissociated acid into the cell and active transport of malate out of the cell (Salmon 1987). Several wine strains of *Saccharomyces cerevisiae* have a variable ability to produce L-malic acid, which can cause difficulties in predicting and controlling titratable acidity in new wines (Holgate 1997). Some strains of *Saccharomyces bayanus/uvarum* produce malic acid, though this properties tends to be strain dependent (Antonelli et al. 1999; Giudici et al. 1995). L-Malate is formed

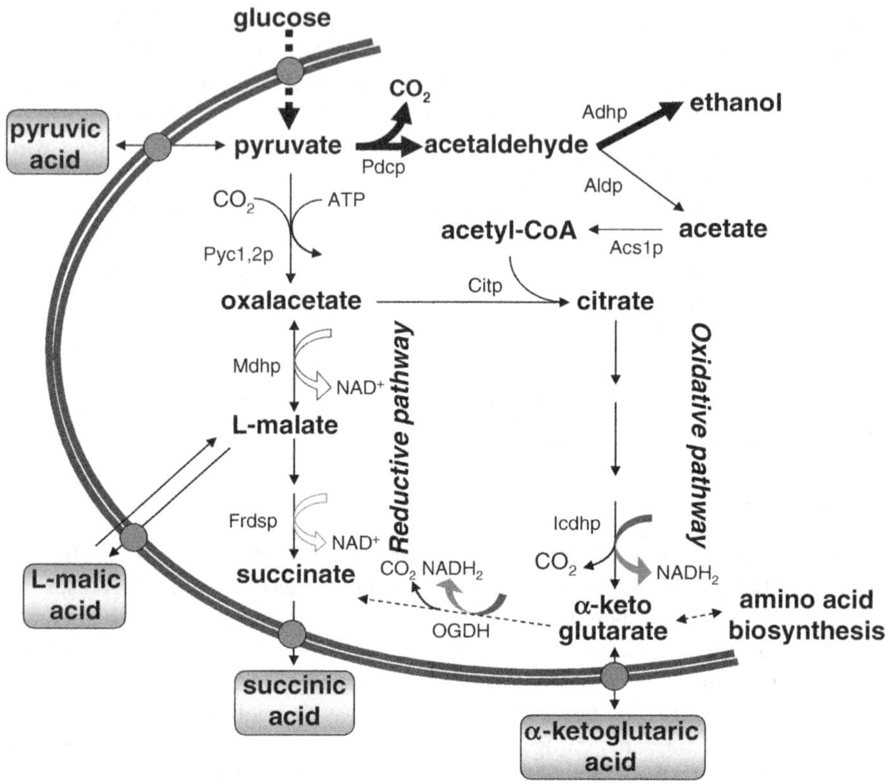

Fig. 8D.3 Metabolism of organic acids

With the exception of succinic acid, most organic acids are metabolic intermediates which provide precursors for anabolic pathways and/or provide redox balance. Pyruvic acid, the product of glycolysis and major source of redox during ethanol production (Fig 8D.2), is a key precursor for anabolic and associated redox reactions, hence little is secreted. L-Malate and succinate, catalysed by malate dehydrogenase (Mdhp) and fumarate reductase (Frdsp), respectively, are produced by the reductive arm of the tricarboxylic acid pathway, thereby providing additional NAD^+. Succinate is excreted as succinic acid. α-Ketoglutarate, a key compound for sequestering ammonia in amino acid biosynthesis, is a product of the oxidative arm of the tricarboxylic acid pathway, being derived from citrate, which is formed by the condensation of pyruvate derivatives oxalacetate and activated acetate. A small amount of α-ketoglutarate is secreted and surplus accumulated acid appears to be oxidatively decarboxylated to succinate by the glutarate dehydrogenase complex (OGDH)

by the reduction of oxalacetate catalysed by malate dehydrogenase (Mdhp), which is derived from pyruvic acid by pyruvate carboxylase (Pyc1,2p); this step provides oxidation of NADH (Radler 1993) (Fig 8D.3). The first step of L-malate degradation is catalysed by malate dehydrogenase followed by the oxidative reactions of the tri-carboxylic acid pathway; this pathway provides carbon skeletons for biosynthetic reactions beyond pyruvic acid, when pyruvate carboxylase (Pyc1,2p) activity is low (Salmon et al. 1987). Regulation of formation/degradation is genetically determined but environmental conditions modulate L-malate metabolism; production is favoured by low nitrogen availability and low pH whereas low sugar concentration can promote malate consumption (Radler 1993; Salmon et al. 1987).

Succinic acid, which has an 'unusual salty, bitter taste', is the major organic acid produced by yeast metabolism, (Coulter et al. 2004; Radler 1993). *Saccharomyces cerevisiae* strains produce succinic acid in highly variable amounts of up to 2 g/L whereas *Saccharomyces bayanus/uvarum* produce slightly greater amounts (Antonelli et al. 1999; Eglinton et al. 2000; Giudici et al. 1995; Radler 1993). Succinate is primarily formed via the reductive arm of the tricarboxylic pathway (Fig 8D.3). Oxalacetate is first formed from pyruvate by a carboxylation reaction (Pyc1,2p), oxalacetate is reduced to L-malate, and hydrated to form fumarate, from which succinate is formed in a reductive step catalysed by fumarate reductase (Frdsp) (Enomoto et al. 2002). Succinate formation from pyruvate provides an alternative pathway for the oxidation of reduced dinucleotides formed by anabolic reactions required for cell growth under anaerobic conditions (Salmon et al. 1987; Camarasa et al. 2003). A variety of fermentation conditions affect succinic acid accumulation during fermentation, including fermentation temperature and must clarity and composition, including sugar concentration, assimilable nitrogen, biotin, pH, acidity and SO_2 (Coulter et al. 2004). Abnormal amounts of succinic acid result from high levels of γ-amino butyric acid, which can form in grape must under certain conditions (Bach et al. 2004). Succinate can also form by the oxidative decarboxylation of α-ketoglutarate/glutamate catalysed by the oxo-glutarate dehydrogenase (OGDH) complex (Camarasa et al. 2003).

Several important keto acids are formed in relatively small amounts, notably pyruvic acid and α-ketoglutaric acid, which have implications for wine chemical and microbiological stability but also play a role in formation of stable wine pigments. Keto acids can bind SO_2, thereby lowering its efficacy as an antioxidant and antimicrobial (Rankine 1968). Keto acids can react with anthocyanins to form pyranoanthocyanins, which are more stable to oxidative degradation than anthocyanin pigments (Asenstorfer et al. 2003; Bakker and Timberlake 1997; Benabdeljalil et al. 2000). Keto acids can form from sugar catabolism and from their respective amino acids, alanine and glutamic acid, respectively, by the Ehrlich pathway. Strains vary in keto acid formation which is further modulated by assimilable nitrogen composition of must (Radler 1993).

8D.4 Yeast Volatile Aroma Compounds

The anaerobic fermentation of sugars by *Saccharomyces* wine yeasts generates a variety of volatile metabolites that contribute to the sensory profile of wine. The important compounds include esters, higher alcohols, volatile fatty acids, carbonyls, and volatile sulfur compounds. The accumulation of these compounds in wine depends on the strain of yeast, must composition (chemical, physical and nutrient composition) and fermentation conditions. In addition, a variety of volatile compounds are also released from non-volatile grape flavour precursor compounds by yeast enzymes. Examples are monoterpenes and C_{13}-norisoprenoids, which are released from glycosidic precursors, and long-chain polyfunctional thiols, which are derived from *S*-cysteinyl conjugates. Some phenolic compounds, which undergo transformation reactions or react with yeast metabolites, affect the colour, aroma and flavour of wine. The significance of these compounds to wine flavour, their metabolism, and factors that affect their formation are discussed in Sect. 8D.5

8D.4.1 Esters

8D.4.1.1 Significance

Fermentation-derived esters are largely responsible for wine fruitiness, and therefore they play an important role in the sensory composition of young red and white wines. Although at wine pH a significant portion of esters hydrolyses during aging, the major fermentation-derived esters are still present in concentration higher than their odour threshold in 1–2 years old wines (Escudero et al. 2007; Moio et al. 2004; Rapp and Marais 1993), which suggests their involvement in the sensory composition of aged wines as well. The two main groups of fermentation-derived esters that have been long associated with wine fruitiness are acetate esters [ethyl acetate, 2-methylpropyl acetate (isobutyl acetate), 2- and 3-methylbutyl acetate (active amyl and isoamyl acetate, respectively), hexyl acetate, and 2-phenylethyl acetate], and ethyl fatty acid esters (ethyl C_3- ethyl C_{12}). The organoleptic synergy existing between different esters determines the overall sensory characteristics of esters mixtures. Van der Merwe and van Wyk (1981) demonstrated that acetates are more important than ethyl fatty acid esters for the perceived aroma of wine, and that the odour intensity of a mixture of esters is higher that that of the individual components. More recently, the role played by branched-chain esters, namely ethyl 2-methylpropanoate, ethyl 2-methylbutanoate, and ethyl 3-methylbutanoate, in the aroma of wine has been investigated. These esters are characterized by strawberry-like aromas, and probably contribute to the red-fruits notes of some red wines (Piombino et al. 2004). Odor threshold values fall between 3 μg/L for ethyl 3-methylbutanoate (Guth 1997) and 18 μg/L for ethyl 2-methylbutanoate (Ferreira et al. 2000), indicating that these compounds are among the most powerful odorants in the class of esters. However, as they occur in much lower concentrations compared to ethyl fatty acid esters and acetates, their impact on wine aroma is still to be established (Escudero et al. 2007).

In a recent study, using multivariate statistical analysis of quantitative sensory descriptive analysis and precise chemical compositional data, Smyth et al. (2005) found that the importance of individual yeast esters to the aroma profile of wine can vary with the type of wine. In the case of unwooded Chardonnay wines, for example, the ethyl esters ethyl hexanoate, ethyl octanoate, ethyl decanoate, the branched-chain esters ethyl 2-methyl propanoate, ethyl 2-methyl butanoate, ethyl 3-methyl butanoate, and the acetate esters hexyl acetate, 2-methylbutyl acetate and 3-methylbutyl acetate were likely to be important. Conversely, in the case of Riesling, ethyl-2-methyl propanoate, ethyl-2-methyl butanoate, ethyl-3-methyl butanoate, ethyl octanoate, ethyl decanote, hexyl acetate, 2-methylbutyl acetate and 3-methylbutyl acetate were found to be important.

8D.4.1.2 Metabolism

Acetate esters are formed through the condensation of higher alcohols with acetyl-CoA, catalysed in the cell by alcohol acetyltransferase enzymes (Malcorps et al. 1991; Mason and Dufour 2000). The final concentration of these compounds is the result of the balance between alcohol acetyltransferase enzymes promoting their synthesis, and esterase enzymes promoting their hydrolysis (Fig. 8D.4) (Fukuda et al. 1998; Plata et al. 2005). The *ATF1* and *ATF2* genes encode for alcohol acetyltransferase enzymes (Atf1p and Atf2p) that are responsible for the synthesis of acetate esters by *Saccharomyces* yeast (Yoshimoto et al. 1998; Lilly et al. 2006a). In gene over-expression and deletion studies *ATF1,* with a minor role by*ATF2,* contributed to the formation of ethyl acetate, 3-methylbutyl acetate, and 2-phenylethyl acetate, which are associated with fruity and floral characters in wine (Lilly et al. 2000, 2006a; Verstrepen et al. 2003b). Deletion of both *ATF* genes eliminated 3-methylbutyl acetate but not ethyl acetate formation indicating that another biosynthetic gene exists for an ethanol acetyltransferase. On the other hand, overexpression of the esterase gene *IAH1* lowered formation of ethyl acetate, 3-methylbutyl acetate, 2-phenylethyl acetate and hexyl acetate (Fukuda et al. 1998). Variations in the expression profile of these ester synthetic and hydrolytic genes are believed to provide the basis for modulated production of acetate esters by various yeasts strains under different fermentation conditions.

In contrast with the considerable knowledge available on the formation of acetate esters during fermentation, the enzymology and genetics of ethyl fatty acid esters formation is still poorly understood. The low molecular weight ester ethyl propanoate is formed from propanoic acid derived from α-ketobutyrate (Eden et al. 2001). Ethyl butanoate is also formed, in part, from butanoic acid derived from the same pathway. However, the largest portion of ethyl fatty acid esters is thought to be formed enzymatically through esterification of the activated fatty acids (acyl-CoA) formed during the early stages of lipid biosynthesis (Suomalainen 1981). Recently, two enzymes (hexanoyl transferase or acyl-coenzymeA:ethanol *O*-acyltransferase) responsible for the formation of ethyl esters of medium chain fatty acids, namely Eht1p and Eeb1p, have been identified, although others are predicted to exist (Mason and Dufour 2000; Lilly et al. 2006a; Saerens 2006). In vitro studies show that Eht1p has preference for short-chain substrates whereas Eeb1p accepts longer-chain substrates, although no preference was observed by in vivo studies. In wine strains Eht1p has been shown to catalyse formation of ethyl hexanoate, ethyl octanoate and ethyl decanoate, which have an apple, fresh fruit aroma. Interestingly, while deletion of one or both of the genes encoding for these enzymes resulted in reduced formation of ethyl esters, their overexpression had limited or no effect on the final concentration of these metabolites, however, depending on the yeast strain (Lilly et al. 2006a; Saerens 2006). Several possible explanations are that these enzymes might possess bifunctional synthetic and hydrolytic activities or that the levels of fatty acid precursor are limiting (Saerens 2006, 2008).

8D.4.1.3 Modulating Factors

Yeasts vary widely, both between and within species, in their ability to produce esters (Antonelli et al. 1999; Heard 1999; Houtman and Du Plessis 1986; Miller et al. 2007; Plata et al. 2003; Rojas et al. 2001; Soles et al. 1982; Swiegers et al. 2008c). Of the principal esters, ethyl acetate is produced in higher amounts by non-*Saccharomyces* species, including *Hanseniaspora uvarum (Kloeckera apiculata), Hanseniaspora guilliermondii, Issatchenkia orientalis* (*Candida krusei*), *Issatchenkia terricola, Metschnikowia pulcherrima,* and *Pichia anomala,* which produce higher concentrations than *Saccharomyces cerevisiae, Torulaspora delbrueckii, Kluyveromyces thermotolerans* and *Candida stellata,* whereas more 3-methylbutyl acetate is produced by *Saccharomyces cerevisiae, Hanseniaspora uvarum, Kluyveromyces thermotolerans and Pichia anomala* than the remaining non-*Saccharomyces* species mentioned. This pattern of ester production tends to preclude the use of many non-*Saccharomyces* strains as key yeasts in wine production; however cofermentation with *Saccharomyces cerevisiae* moderates the production of esters and other compounds, rendering such wines with a greater diversity of acceptable flavours (Sect. 8D.6.4). Strains vary in the pattern of acetates and ethyl esters produced as well as in the overall concentration of esters (Soles et al. 1982), which contribute to sensorially significant differences in young wines, with some differences still apparent up to two years of storage (Dumont and Dulau 1997; Jane et al. 1996).

Ester production is susceptible to variations in nutrient and fermentation conditions; however, acetates and ethyl esters do not always vary in the same way, reflecting their different metabolic pathways. Variations in juice contents of sugar, oxygen, lipids, and assimilable nitrogen, juice turbidity and fermentation temperature are important factors, as observed under brewing conditions (Verstrepen et al. 2003a).

Grape maturity at harvest modulates the relative abundance of different ester types produced during fermentation, such that acetate esters decrease whereas ethyl esters increase with increasing maturity (Houtman et al. 1980). The addition of sugar to de-aerated, clarified grape juice, however, stimulates both types of esters in a similar manner, whereas added sugar to wort fermented with ale yeast stimulates acetates production with little influence on ethyl esters (Plata et al. 2005; Saerens et al. 2008). The level of grape solids, a source of lipids, in grape juice affects yeast growth, fermentation rate and accumulation of esters; a small addition of freshly settled juice solids (1–2 vol.%) to membrane-filtered juice stimulates ester production which however decreases at higher addition levels (>5 vol.%) (Houtman and Du Plessis 1981).

Lipid composition of fermentation media markedly affects yeast cellular composition due to the anaerobically-induced requirement of *Saccharomyces cerevisiae* for sterol and unsaturated fatty acid necessary for growth (Henschke and Jiranek 1991). Lipid availability also influences the production of esters and other metabolites of sensory importance (Anderson and Kirsop 1974; Rossi and Bertuccioli 1992). Addition of fatty acids and triglyceride mixtures to a fermentation medium promotes increase in the yeast intracellular concentration of C_8-C_{16} fatty acids, which results in increased concentrations of acetates and ethyl fatty acid esters. Grape-derived short to long chain fatty acids vary according to grape variety and cultural conditions, and influence the formation of their respective esters during fermentation, although higher levels of linoleic acid ($C_{18:2}$) in grape must are associated with lower formation of total fatty acid esters by yeast (Yunoki et al. 2007). One interesting observation can be made regarding the regulation of ethyl fatty acid esters compared to acetates. Whereas the final concentration of acetates does not always depend on the concentration of their corresponding higher alcohol, formation of ethyl fatty acid esters is closely linked to the availability of their precursor fatty acids, so that the conditions promoting formation of esters are usually also favorable for the respective fatty acid (Saerens et al. 2008).

Addition of ergosterol, the principal yeast sterol, to clarified grape juice markedly stimulates synthesis of acetate esters, especially 3-methylbutyl, 2-phenylethyl and *n*-hexyl acetates but ethyl acetate to a lesser extent. Ergosterol is believed to stimulate acetate production by stimulating acetyl-CoA synthesis (Houtman et al. 1980). In winemaking, fatty acids are derived from grape pulp and waxes present on the grape berry cuticle and grapes contain a variety of phytosterol that yeast can accumulate (Jackson 2000; Luparia et al. 2004). Therefore, clarification methods can be used to lower or modulate concentration of lipids in grape juices (Delfini et al. 1992; Delfini and Costa 1993). Exogenous sources of lipids, such as inactive dry yeast additives, which can be added during yeast rehydration, could provide another level of control of flavour-active compounds (Belviso et al. 2004; Swiegers et al. 2008b; Trioli 1996).

Oxygen availability is an important factor affecting the formation of esters during fermentation (Fujii 1997; Houtman et al. 1980; Mauricio et al. 1995). Aeration of clarified grape juice stimulates yeast growth and ester production whereas in slightly turbid, settled juice esters are decreased, as is observed in wort fermentation. In the case of acetate esters, aeration of the fermentation medium represses *ATF1* gene expression and consequently decreases the formation of acetates (Fujii 1997; Plata et al. 2005). According to Bardi et al. (1998), the inverse relationship between production of ethyl fatty acid esters and oxygen availability can be considered to be a consequence of the arrest of lipid metabolism due to lack of oxygen. Under these conditions, an excess of acyl-CoA accumulates in the cell, and esters are then formed in order to recover free coenzyme A. On the other hand, aerated fermentation tends to stimulate formation of acetates by several non-*Saccharomyces* yeasts (Rojas et al. 2001).

Empirically, it has long been known that lower fermentation temperature preserves the fruity characters typical of young white wines. Temperature affects the accumulation of acetate and ethyl fatty acid esters during fermentation (Killian and Ough 1979), although the response of individual esters appears to be linked to both evaporative losses and changes in yeast metabolism. Altered fatty acid biosynthesis, as reflected by differences in membrane fatty acyl composition, is involved (Boulton et al. 1998; Torija et al. 2003; Watson 1987). In a model study, lower fermentation temperature (15 °C) favoured ethyl esters whereas higher temperature (28 °C), typical of red wine fermentation, favoured 2-methyl acetate, ethyl 2-methylbutanoate, 2-phenyl ethanol and 2-phenylethyl acetate (Molina et al. 2007). Similar results were obtained in grape juice (Beltran et al. 2008). These ester profiles could be linked to different expression profiles of ester synthase genes *ATF1* and *EHT1*, and the ester degradation gene *IAH1* under different temperature conditions (Molina et al. 2007).

The nitrogen content and composition of must can strongly affect the accumulation of volatile esters during fermentation (Bell and Henschke 2005; Carrau et al. 2008; Garde-Cerdán and Ancín-Azpilicueta 2008; Guitart et al. 1999; Hernández-Orte et al. 2002; Vilanova et al. 2007). Total and individual esters tend to increase with increasing must amino nitrogen, although the responses of the various esters to individual amino acids is not yet clear (Hernández-Orte et al. 2002). Addition of ammonium salts, a common practice in the wine industry, strongly stimulates the production of esters. The stimulation of ester biosynthesis with nitrogen supplements, which is observed under various winemaking conditions, appears to confirm the link between increased nitrogen availability and expression of *ATF1* gene reported by some authors in model fermentations (Yoshimoto et al. 1998). Stimulated production of medium chain fatty acids by assimilable nitrogen might account for large increases in corresponding esters.

The metabolism of branched-chain esters is largely unknown. Based on their structure, it is reasonable to assume that they are formed through esterification of the branched-chain acids formed during amino acid metabolism. It has been reported that nitrogen supplementation of low nitrogen must with ammonium salts has a negative impact on the formation of these compounds during fermentation (Hernández-Orte et al. 2006; Vilanova et al. 2007).

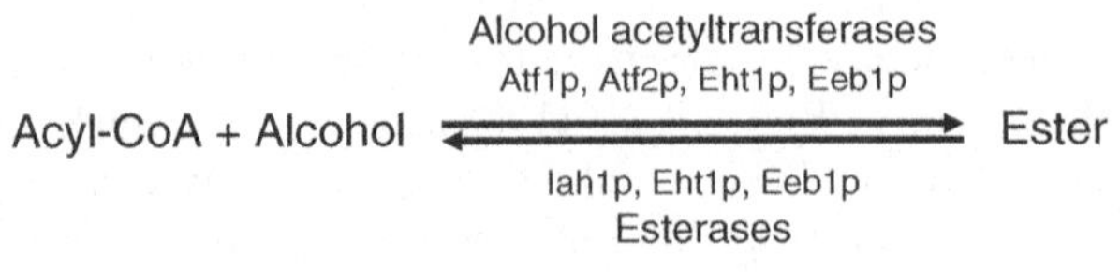

Fig. 8D.4 The biosynthetic and degradation pathway for esters

Fatty acids, following activation with coenzyme A, condense with alcohols catalysed by alcohol acetyltransferases. Esterases hydrolyse esters to their constituent acid and alcohol. Several enzymes have dual synthetic and esterase activity

8D.4.2 Higher Alcohols

8D.4.2.1 Significance

Higher alcohols, sometimes referred to as fusel alcohols, are, from a quantitative point of view, the most important group of volatile compounds produced by yeast during alcoholic fermentation of sugars. They are characterised by containing more than two carbon atoms and include the branched-chain alcohols 2-methylpropanol (isobutanol), 2-methylbutanol (amyl alcohol), and 3-methylbutanol (isoamyl alcohol), and the aromatic alcohols 2-phenylethanol and tyrosol. According to Rapp and Versini (1996) concentrations of higher alcohols below 300 mg/L add desirable complexity to wine, whereas higher concentrations can be detrimental to wine quality. Conversely, 2-phenylethanol is generally a positive contributor to wine aroma, being characterised by a pleasant rose-like aroma (Swiegers et al. 2005). The concentrations of higher alcohols that act positively or negatively on wine aroma are likely to depend on aroma intensity and style of wine. Correlation of the composition of wine volatile compounds with data obtained by formal descriptive sensory analysis of white wines suggests that for Chardonnay wines 2-methyl propanol and 2- and 3-methylbutanol were important whereas for Riesling wines none of the higher alcohols measured were important to the aroma profile (Smyth et al. 2005).

8D.4.2.2 Metabolism

Higher alcohols are formed by decarboxylation and subsequent reduction of α-keto-acids produced as intermediates of amino acids biosynthesis and catabolism (Fig 8D.5). The latter pathway has been named after Ehrlich, its discoverer. Studies with labelled amino acids have shown that the greater proportion of higher alcohols is synthesized from the α-keto acids derived from glycolysis and destined for amino acids biosynthesis (Äyräpää 1971; Chen 1978). Accordingly, amino acid biosynthesis accounts for most of the higher alcohols formed during fermentation. This is consistent with the observations that consumption of the structurally related amino acids is not correlated with the final concentration of the corresponding higher alcohols.

The structurally related amino acids for the branched chain aliphatic alcohols 2-methylpropanol, 2-methylbutanol and 3-methylbutanol are valine, leucine and isoleucine, respectively. For the aromatic alcohols 2-phenylethanol, tyrosol and tryptophol, they are phenylalanine, tyrosine and tryptophan, respectively, and for the thioalcohol methionol, the corresponding amino acid is methionine. Uptake of the branched-chain amino acids is mediated by several transport proteins: the branched-chain amino acid permease Bap2p and Bap3p, and the general amino acid permease Gap1p. The aromatic amino acids are transported by Tat1p and Tat2p, and also Gap1p and Bap2p, whereas methionine is transported by Mup1p, Mup3p and Gap1 (Regenberg et al. 1999). The first step in the Ehrlich degradation pathway involves transamination to form the α-keto acids, catalysed by branched-chain (Bat1p and Bat2p) and aromatic (Aro8p and Aro9p) amino acid transferases, (Hazelwood

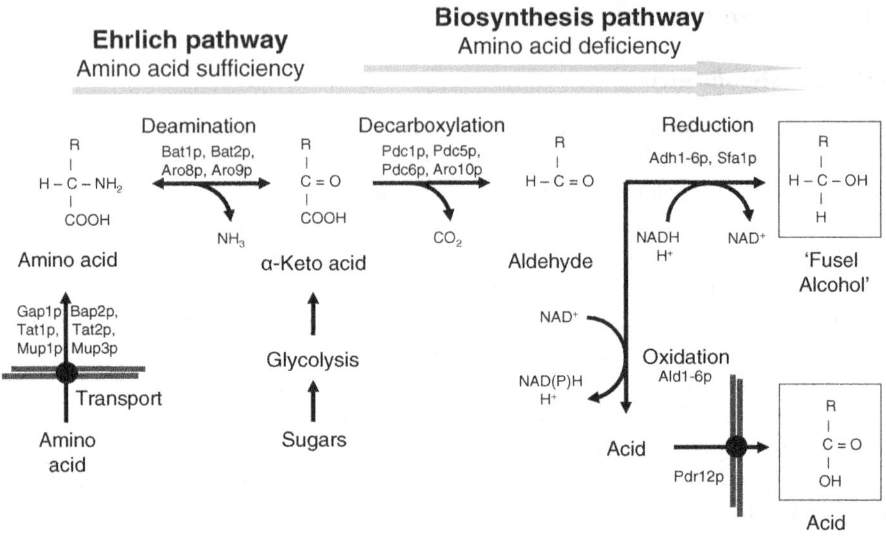

Fig. 8D.5 Formation of higher alcohols from sugar and amino acids by the Ehrlich pathway
A deficiency of amino acids during growth activates their synthesis from α-ketoacids, derived from sugars via glycolysis. If insufficient nitrogen is available for transamination reactions, surplus α-ketoacids are excreted as higher alcohols (Biosynthetic pathway). During amino acid sufficiency, transamination of amino acids can produce a surplus of α-ketoacids, some of which are decarboxylated and reduced to alcohols (Ehrlich pathway)

et al. 2008) (Fig 8D.5). Pyruvate decarboxylases (Pdc1p, Pdc5p, Pdc6p, Aro10p) convert α-keto acids to their corresponding aldehydes, which are then reduced to alcohols by alcohol dehydrogenases (Adh1p-Adh6p, Sfa1p). Catalytic oxidation of aldehydes by aldehyde dehydrogenases (Ald1p-Ald6p) forms the corresponding acids, which are then removed from the cell by the weak organic acid permease Pdr12p (Hazelwood et al. 2008). Over-expression of the *BAT1* gene in a commercial wine yeast (VIN13) increased 3-methybutanol, its acetate ester, 2-methylpropanol and 2-methylpropanoic acid, whereas over-expression of the *BAT2* gene increased 2-methylpropanol, 2-methylpropanoic acid and propanoic acid. Deletion of the *BAT2* gene decreased these compounds. Altered expression of these transaminase genes might provide a means for modulating the aroma of wine (Lilly et al. 2006ab).

The production of *n*-propanol is directly related to initial nitrogen and yeast growth, and appears not to be influenced by the structurally-related amino acids, threonine and α-aminobutyric acid (Rapp and Versini 1996). A negative relationship has been observed between *n*-propanol production and H_2S formation (Giudici et al. 1993) but impact on wine quality was not reported. Methionol production is related to methionine concentration, and is therefore limited by the generally low methionine in grape must. Hexanol is believed to be reductively formed by yeast from hexanal, which in turn is formed from linoleic ($C_{18:2}$) acid during must processing (Ribéreau-Gayon et al. 2000a).

8D.4.2.3 Modulating Factors

Many factors affect formation of higher alcohols during fermentation, including yeast species and strain, initial sugar, fermentation temperature, the pH and composition of grape juice, assimilable nitrogen, aeration, level of solids, grape variety and skin contact time (Fleet and Heard 1993; Houtman and Du Plessis 1981; Houtman et al. 1980). Choice of yeast has a considerable impact on higher alcohol content of wine due to wide variability amongst strains of *Saccharomyces* species with *Saccharomyces cerevisiae* generally producing lesser concentrations of higher alcohols than cryotolerant *Saccharomyces bayanus/uvarum* strains (Antonelli et al. 1999; Massoutier et al. 1998). Most non-*Saccharomyces* species are reported to produce lesser concentrations of higher alcohols than *Saccharomyces cerevisiae* although in fermentation with mixed yeast species higher concentrations of higher alcohols can be produced (Heard 1999). *Metschnikowia pulcherrima* has a notably high production of 2-phenyl ethanol (Clemente-Jimenez et al. 2004).

The assimilable nitrogen composition of grape juice has a strong influence on the production of higher alcohols during fermentation. When initial nitrogen concentrations are very low, increasing nitrogen results in greater concentrations of higher alcohols (Äyräpää 1971; Carrau et al. 2008; Garde-Cerdán and Ancín-Azpilicueta 2008; Vilanova et al. 2007). According to Oshita et al. (1995), under these conditions, surplus keto acids cannot be converted into amino acids, due to the lack of available nitrogen, and are therefore excreted as higher alcohols. Conversely, at high initial juice nitrogen concentrations, increased availability of nitrogen causes a reduction in higher alcohol production because most of the keto acids produced are directly converted into the corresponding amino acids.

The degree of juice turbidity increases the production of higher alcohols, which is linked to biomass production (Houtman and Du Plessis 1981; Klingshirn et al. 1987); red wines, produced by fermentation on grape solids typically contain more higher alcohols. Similarly, chemically inert solids stimulate higher alcohols production, although the mechanism of action is not clear. Aeration and increased fermentation temperature are believed to stimulate the production of higher alcohols by stimulating nutrient (nitrogen) uptake and growth.

8D.4.3 Volatile Fatty Acids

8D.4.3.1 Significance

Wine contains a mixture of straight chain fatty acids, usually referred to as short chain (C_2–C_4), medium chain (C_6–C_{10}), long chain (C_{12}–C_{18}), and a group of branched-chain fatty acids that include 2-methyl propanoic, 2-methyl butanoic, and 3-methyl butanoic acids. Acetic acid is quantitatively and sensorially the most important volatile fatty acid produced during alcoholic fermentation. Accounting for more that 90% of the total wine volatile acidity, acetic acid plays the most important role in wine quality (Eglinton and Henschke 1999). Concentrations vary widely (<0.2 to >2.0 g/L, expressed as acetic acid) and depend on wine type. Dry white wines typically have lowest concentrations whereas sweet whites, especially when prepared from *Botrytis*-affected grapes, tend to have the highest concentrations. Wine law controls the maximum concentration of volatile acids allowed in wine, which in the European Community cannot exceed 1.5 g/L, expressed as acetic acid.

The flavour threshold for acetic acid depends on wine type and style, and ranges from 0.4 to 1.1 g/L (Dubois 1994). At threshold concentration it provides warmth to the palate and, as the concentration increases, it imparts a sourness/sharpness to the palate and a vinegary odour at higher concentration. As the fatty chain length increases, volatility decreases and the odour changes from sour to rancid and cheese (Francis and Newton 2005). Sensory studies show that hexanoic, octanoic, and decanoic acids can contribute to the aroma of some white wines (Smyth et al. 2005). The branched-chain fatty acids can also contribute to the fermentation bouquet of wine, with the concentration of 2-methylpropanoic acid typically exceeding its odour threshold (Francis and Newton 2005).

8D.4.3.2 Metabolism

Acetic acid formation fulfils several metabolic roles including providing a precursor for the synthesis of acetyl-CoA and a redox sink for anabolic and physiological stress reactions. Acetate is formed by the action of aldehyde dehydrogenases from acetaldehyde, which is derived by the decarboxylation of pyruvate (Fig 8D.6). Five genes (*ALD2-6*), encoding different isozymes, have been described and partially characterised in *Saccharomyces cerevisiae*, but their physiological functions under wine fermentation conditions are still somewhat unclear (Eglinton et al. 2002; Meaden et al. 1997; Navarro-Avino et al. 1999; Remize et al. 2000; Saint-Prix et al. 2004; Pigeau and Inglis 2007). Under dry wine fermentation conditions (20% sugar; anaerobic conditions) Ald6p, a cytosolic $NADP^+$-dependent aldehyde dehydrogenase, is the major isozyme that produces acetic acid and provides redox balance. Lesser contributions are made by the mitochondrial $NADP^+$-dependent Ald5p isozyme, and depending on the strain and conditions, also the mitochondrial $NAD(P)^+$-dependent Ald4p isozyme. Under high osmotic stress conditions, as imposed by Eiswein and Icewine fermentation, *ALD3,* encoding the cytosolic NAD^+-dependent aldehyde dehydrogenase, is differentially expressed, suggesting a role in the higher production of acetic acid in these wines over and above that produced by conventional wine fermentation. In this case, *ALD3* might provide a redox role to balance the extra demand for NADH generated by the concomitantly higher production of glycerol, which is induced in response to the high osmotic stress imposed by very high sugar concentrations.

Straight-chain fatty acids (C_4–C_{12}) are by-products of saturated fatty acid metabolism. Malonyl-CoA is first synthesised from acetyl-CoA by acetyl-CoA carboxylase. Subsequent reactions are catalysed by the synthase enzyme complex, which increases chain length sequentially by two C units. C_{16} and C_{18} fatty acids

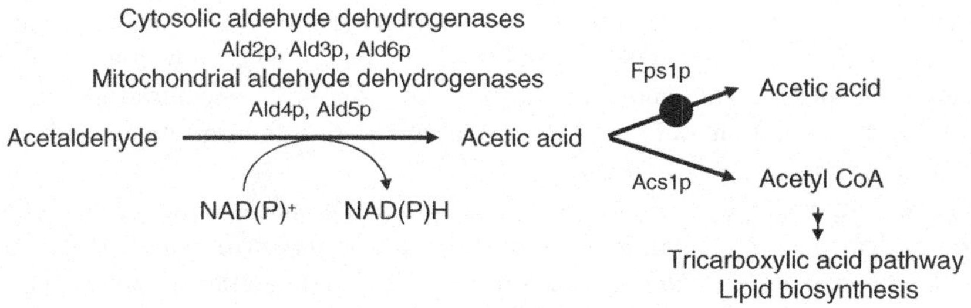

Fig. 8D.6 Acetic acid metabolism in *Saccharomyces cerevisiae*

Acetaldehyde is oxidised to acetic acid by co-reduction of $NAD(P)^+$ catalysed by cytosolic and mitochondrial aldehyde dehydrogenases (Ald2-Ald6p). Acetic acid is activated with coenzyme A for lipid biosynthetic reactions, including fatty acids. Surplus acetic acid generated from redox balancing reactions (growth and stress induced) is secreted into the fermentation medium. The different aldehyde dehydrogenase genes are expressed according to their physiological role, with *ALD6* being the major isogene under dry wine fermentation conditions

are the predominant final products, which are incorporated into phospholipids, the backbone of cell membranes. Growth limiting factors, which inhibit acetyl-CoA carboxylase, a key enzyme in regulating fatty acid synthesis, causes early release of the fatty acid from the fatty acid synthase complex. This results in the generation of short and medium chain fatty acids (Wakil et al. 1983; Bardi et al. 1999).

Branched chain fatty acids, such as 2-methylpropanoic, and 2-methylbutanoic and 3-methylbutanoic acids, are not products of the fatty acid synthetic pathway. They are instead derived from oxidation of the aldehydes formed from α-keto acids during amino acid metabolism (Fig 8D.5). The mechanism of regulation is not known.

8D.4.3.3 Modulating Factors

Yeast strain, and nutrient status of the must and fermentation conditions, many of which affect growth or induce physiological stress, modulate the accumulation of acetic and other fatty acids in wine. Reported factors include must sugar concentration, nutrient balance, inoculum level, fermentation temperature, pH and aeration (Delfini and Costa 1993; Henschke and Jiranek 1993; Shimazu and Watanabe 1981). The effects of osmotic stress, as induced by sugar concentration, on acetic acid production are discussed in Sect. 8D.3.2.

Strains of *Saccharomyces cerevisiae* vary widely in acetic acid production (ranging from 37 mg/L to 999 mg/L; mean 300 mg/L) such that strain selection has been the most important factor for controlling the acetic acid content of wines during fermentation (Giudici and Zambonelli 1992; Heard 1999; Paraggio and Fiore 2004; Radler 1993; Reynolds et al. 2001; Shimazu and Watanabe 1981). Strains selected for winemaking show minimal variation (Reynolds et al. 2001) although strain interaction with musts can be considerable (Paraggio and Fiore 2004). Cryotolerant *Saccharomyces bayanus/uvarum* strains are characterised by producing very low amounts of acetic acid, typically <150 mg/L (Antonelli et al. 1999; Eglinton et al. 2000; Giudici et al. 1995) and might be useful when low concentrations are required, such as when musts already contain excessive concentrations, and for malolactic fermentation and long term barrel maturation during which acetic acid content often increases.

Acetic acid production amongst non-*Saccharomyces* yeasts varies widely. *Torulaspora delbrueckii* (*Candida colliculosa*), *Metschnikowia pulcherrima* and *Issatchenkia orientalis (Candida krusei)* produce low to moderate amounts whereas apiculate yeasts, *Candida stellata* and *Pichia anomala* produce moderate to high amounts and *Brettanomyces* species and *Zygosaccharomyces bailii* very high amounts (Giudici and Zambonelli 1992; Heard 1999; Shimazu and Watanabe 1981). As is the case for *Saccharomyces* species, intraspecific strain variability is high, meaning that strain selection is essential for winemaking purposes. Laboratory cofermentation studies with mixed cultures of yeasts, which produce different concentrations of acetic acid in monoculture, typically show lower levels in the final wine than expected and are usually similar to that of the lowest producer yeast in the cofermentation (for examples see Ciani and Comitini 2006; Grossmann et al. 1996; Soden et al. 2000). As has been observed with acetaldehyde and redox balance in cofermentation, diffusion of various metabolites between yeasts with different 'metabolic tuning' can result in metabolite concentrations different from those that would be achieved by blending wines (Cheraiti et al. 2005; Howell et al. 2006).

The assimilable nitrogen concentration of must, across the range 50–500 mg/L yeast assimilable nitrogen, has a relatively strong effect on acetic acid accumulation during fermentation. Lowest acetic acid concentrations occur around 200–250 mg/L yeast assimilable nitrogen with increases of up to twofold at nitrogen concentrations well outside this range (Bely et al. 2003; Vilanova et al. 2007), apparently irrespective of initial sugar concentration or osmotic stress. Nitrogen additions made at inoculation rather than later during fermentation are more effective in preventing acetic acid accumulation.

Some strains of yeast have partial requirements for the vitamins, nicotinic acid, inositol and panthothenic acid, which can affect acetate metabolism (Nordström 1964). Nicotinic acid addition to musts can markedly increase acetic acid production by yeast with the extent of production depending on yeast strain, fermentation temperature, sugar concentration and must composition (Eglinton and Henschke 1993; Monk and Cowley 1984). Many strains of *Saccharomyces cerevisiae* cannot synthesise pyridine nucleotides under anaerobic conditions and are therefore dependent on the nicotinic acid content of must. Increasing nicotinic acid content of must is presumed to act by optimising the pyridine nucleotide content of yeast, stimulating NADH production for growth, and consequently increasing acetic acid production.

Fermentation temperature affects acetic acid production, which is maximal around the optimum temperature for biomass production ($\approx$25 °C; Monk and Cowley 1984). As yeast growth becomes progressively inhibited by growth at suboptimal temperatures biomass production decreases as does acetate accumulation; nevertheless acetate yield remains constant for each gram of yeast protein formed. At very low temperatures, however, yeast growth becomes uncoupled from sugar metabolism resulting in low biomass yield but high acetate production.

Volatile fatty acids are by-products in the formation of long-chain fatty acids, which are required for cell membrane phospholipid biosynthesis. The biosynthesis of volatile fatty acids is generally controlled by the same factors that control the formation of ethyl fatty acid esters, that is, oxygen, ergosterol and various insoluble solids (grape solids, clarification solids, yeast hulls) tends to suppress production whereas sugar concentration and clarification are stimulatory (Bardi et al. 1999; Delfini et al. 1992, 1993; Edwards et al. 1990; Houtman et al. 1980).

8D.4.4 Carbonyl Compounds

8D.4.4.1 Significance

Yeasts produce various carbonyl compounds from sugar metabolism, including aldehydes, ketones and keto acids; the latter are discussed in Sect. 8D.3.3. Acetaldehyde is quantitatively the most important saturated aldehyde. It has a sensory threshold of 100 mg/L wine, and ranges in concentration from 10 mg/L to 75 mg/L in dry wines to which it contributes a 'bruised apple' and 'nutty' character when present at sensorially detectable concentrations (Schreier 1979). In addition, higher saturated aldehydes (C_3–C_9) contribute herbaceous, grassy, green, fatty, fruity and pungent flavours (Ebeler and Spaulding 1998). Hexanal, *cis*-3-hexenal and *trans*-2-hexenal contribute a green character evident in grape must. From a sensory point of view, high concentrations of acetaldehyde are generally associated with oxidation off-flavors (aldehydic) in dry wines, although this compound plays a key role in the aroma of particular wines made by oxidative processes, such as Spanish sherries, French *vin jaune*, Sicilian Marsala and Sardinian Vernaccia (Jackson 2000; Schreier 1979). Evident aldehydic character in young wines can be suppressed by treatment with sulfite, which binds to acetaldehyde to form the non-volatile adduct acetaldehyde hydroxysulfonate. Unintended exposure of wines to oxygen (air), such as wine stored under ullage or aeration during wine transfer operations, increases acetaldehyde content, which leads to higher concentrations of bound sulfite. This occurrence is both undesirable from a consumer perspective and total SO_2 content of wine is subject to wine regulatory requirements. Furthermore, bound SO_2 provides limited protective antioxidant and antimicrobial activity.

The most significant ketone produced by yeast is diacetyl (2,3-butanedione), a vicinal diketone, although malolactic fermentation is a more important source, when it is used in wine production. Having a sensory threshold of 0.2–2.9 mg/L, according to the type of wine, it is characterised by a 'nutty', 'toasty' or 'buttery' aroma depending on concentration (Martineau et al. 1995). Dry white wines tend to contain lower concentrations (0.1–2.3 mg/L) than red wines (0–7.5 mg/L) (Bartowsky et al. 2002; Martineau et al. 1995). Acetoin, which produces a buttery flavour, is formed by partial reduction of diacetyl, and is itself reduced to 2,3-butanediol. Acetoin is usually present at concentrations (<80 mg/L) much lower than its sensory threshold of 150 mg/L (Romano and Suzzi 1996).

Lactones are another group of carbonyl compounds formed by yeast during alcoholic fermentation. Concentrations are expecially high during oxidative maturation of wine with flor yeast (Wurz et al. 1988). Gamma lactones, including γ-butyrolactone, are quantitatively the most important. The sensory role of lactones is not clear, although it has been suggested that these compounds contribute to the aroma characteristics of Sherry wines. The role played by yeast in the formation of other aliphatic γ-lactones (C_8; C_{12}) in wine (Ferreira et al. 2004) is not clear.

8D.4.4.2 Metabolism

Yeast-derived saturated short-medium chain and branched-chain aldehydes are formed from sugar metabolism, fatty acid metabolism and branched-chain amino acid metabolism (Fig 8D.7). In addition, hexanal, as well as hexenal isomers, are formed during the pre-fermentative stages of winemaking by the sequential action of grape lipoxygenase and hydroperoxide cleavage enzyme on linoleic and linolenic acid, respectively (Crouzet 1986).

The major carbonyl, acetaldehyde, is formed as an intermediate in ethanol production. In fermenting yeast, pyruvic acid is principally decarboxylated to acetaldehyde by pyruvate decarboxylase (Pdcp), which is encoded by the genes *PDC1,5,6* (Flikweert et al. 1996). Acetaldehyde accumulation during fermentation is affected by its rate of formation and the various reactions that consume it, with lowest levels being found at end of fermentation. These reactions are summarised in Fig 8D.7. The rate of acetaldehyde formation can be adversely affected by a deficiency of thiamine, a cofactor of Pdcp; thiamine depletion mainly results from wild yeast growth during must processing (Bataillon et al. 1996). The main pathway for acetaldehyde consumption is reduction to ethanol by alcohol dehydrogenase, which is encoded by *ADH1*. This dehydrogenase reaction, which maintains redox balance in the cell by oxidising NADH to NAD^+, is essential for sugar metabolism (Flikweert et al. 1996). Acetaldehyde is also oxidised to acetic acid by aldehyde dehydrogenase, encoded by *ALD4-6*, as precursor for lipid biosynthesis and for regenerating NAD(P)H required for biosynthetic and redox balancing reactions during cell growth.

Diacetyl, and its reduction products, acetoin and 2,3-butanediol, are also derived from acetaldehyde (Fig 8D.7), providing additional NADH oxidation steps. Diacetyl, which is formed by the decarboxylation of α-acetolactate, is regulated by valine and threonine availability (Dufour 1989). When assimilable nitrogen is low, valine synthesis is activated. This leads to the formation of α-acetolactate, which can be then transformed into diacetyl via spontaneous oxidative decarboxylation. Because valine uptake is suppressed by threonine, sufficient nitrogen availability represses the formation of diacetyl. Moreover, the final concentration of diacetyl is determined by its possible stepwise reduction to acetoin and 2,3-butanediol, both steps being dependent on NADH availability. Branched-chain aldehydes are formed via the Ehrlich pathway (Fig 8D.7) from precursors formed by combination of acetaldehyde with pyruvic acid and α-ketobutyrate (Fig 8D.7).

Finally, acetaldehyde can become bound to SO_2, derived from the sulfate reduction pathway or added by winemakers as an antioxidant and antimicrobial compound prior to fermentation (refer to Sect. 8D.4.5). Prefermentation additions of SO_2 increase the concentration of the acetaldehyde-hydroxysulfonate adduct and

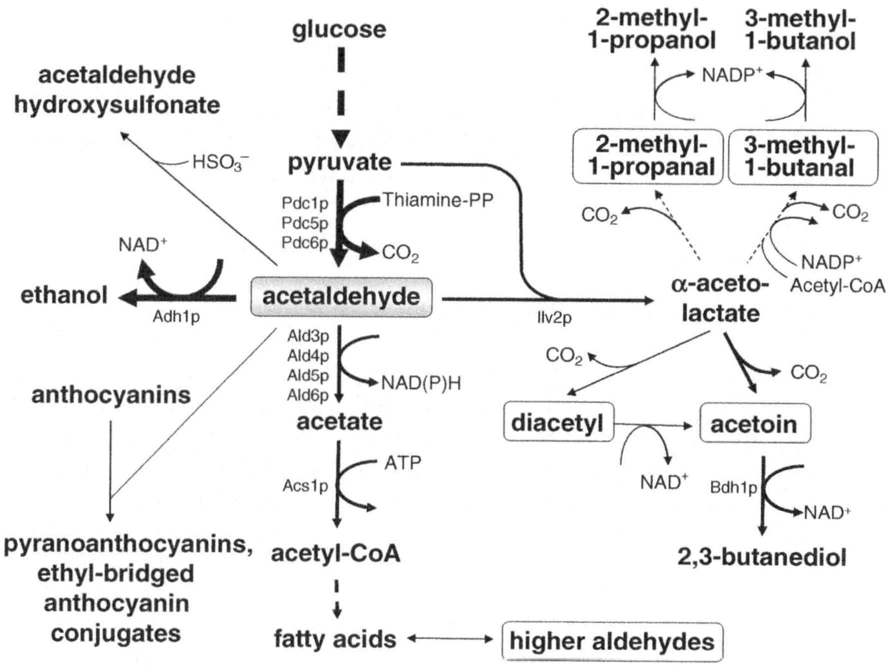

Fig. 8D.7 Aldehyde metabolism, showing key role of acetaldehyde in formation of branched-chain and higher aldehydes, and diacetyl and acetoin

Acetaldehyde is formed from pyruvate catalysed by pyruvate decarboxylases (Pdc1p,5p,6p), with most being reduced to ethanol by alcohol dehydrogenase (Adh1p). Higher aldehydes, excluding hexanal, are formed during fatty acid biosynthesis from acetaldehyde, which is first oxidised to acetate by aldehyde dehydrogenases (Ald3–Ald6p) and activated with ATP by acetyl-CoA synthase (Acs1p). Branched-chain aldehydes, and diacetyl and acetoin are formed by decarboxylation reactions from α-acetolactate, which is derived by condensation of pyruvate and acetaldehyde thiamine-PP complex, catalysed by acetolactate synthase (Ilv2p). As fermentation completes, free aldehydes are largely reduced by dehydrogenase reactions to corresponding alcohols. Extracellular acetaldehyde can react with bisulfite ion to form acetaldehyde-hydroxysulfonate, and can react with anthocyanins to form stable pigments, pyranoanthocyanins and ethyl-bridged anthocyanins conjugates with proanthocyanidins (tannins)

the higher aldehydes [branched-chain aldehydes (2-methyl 1-propanal, 3-methyl 1-butanal), and the C_1, C_3, C_5, C_6 and C_7 saturated aldehydes] (Frivik and Ebeler 2003). The saturated higher aldehydes, usually produced in trace amounts, are derived from the biosynthesis of fatty acids from acetyl-CoA, which is derived from acetic acid (Fig 8D.7). The C_6 aldehydes, hexanal, *cis*-3-hexenal and *trans*-2-hexenal, are derived by enzymatic oxidation reactions of grape C_{18} unsaturated fatty acids during must processing (Ribéreau-Gayon et al. 2000a).

Acetaldehyde plays an important role during red wine fermentation in which it can react directly with anthocyanins pigments to form vinyl adducts (pyroanthocyanins) (Fulcrand et al. 1998), and can participate in the condensation of proanthocyanidins (tannins) and anthocyanins to form ethyl-bridge linked dimers

(Timberlake and Bridle 1976). These pigments are chemically more stable than anthocyanins and contribute to the colour of aged red wines.

Formation of lactones can occur though chemical cyclisation, although enzymatic formation is generally more frequent in nature. Glutamic acid is the presumed precursor for γ-lactones formation during fermentation (Wurz et al. 1988). The mechanisms regulating the formation of lactones in wine are not known.

8D.4.4.3 Modulating Factors

Yeast species and strain, juice clarification, fermentation temperature, medium composition, extreme aeration of fermentation and SO_2 addition can affect the formation of aldehydes during fermentation (Romano et al. 1994). Aldehydes are unstable compounds, which are readily reduced to corresponding alcohols by yeast dehydrogenases and reductases. Therefore, wine aldehyde content is relatively low immediately after fermentation unless fermentation arrested either due to yeast physiological factors (stuck fermentation) or intervention by the winemaker. SO_2 addition to must, before fermentation, has a strong influence on total acetaldehyde content of wine by the formation of the stable adduct acetaldehyde-hydroxysulfonate. Acetaldehyde production is highly variable amongst strains of *Saccharomyces cerevisiae*, which can range from 6 mg/L to 190 mg/L, with sulfite-resistant and high sulfite producing strains typically accumulating more acetaldehyde as a result of adduct formation.

8D.4.5 Volatile Sulfur Compounds

8D.4.5.1 Significance

Volatile sulfur compounds are potent aroma compounds that have very low sensory threshold values, in the low ng/L to μg/L range (Mestres et al. 2000; Rauhut 1993). A large number of sulfur compounds appear in association with fermentation, those that have received more attention are summarised in Table 8D.3. The types of compounds and their concentrations in wine are highly variable. The majority of sulfur compounds are associated with off-flavours and considered by many as undesirable or a defect in wine by contributing aromas of rotten egg, cooked cabbage, onion and rubber (Bell and Henschke 2005; Fedrizzi et al. 2007; Mestres et al. 2000; Park et al. 1994; Rauhut 1993; Rauhut et al. 1996, 1998; Ribéreau-Gayon et al. 2000a, 2000b; Swiegers and Pretorius 2007). One exception is represented by the long-chain polyfunctional sulfur compounds, which at low concentrations are characterised by pleasant, fruity aromas, although high concentrations of these compounds can also be objectionable (Dubourdieu et al. 2006; Mestres et al. 2000; Swiegers et al. 2005; Swiegers and Pretorius 2007). Dimethyl sulfide (DMS) also appears to be an exception in that the asparagus, corn, molasses aromas can be considered pleasant in some wines when present at low suprathreshold concentrations. Odours of quince, truffles and metallic have also been reported to depend on DMS concentration, wine type and age. DMS appears to be especially

Table 8D.3 Volatile sulfur compounds produced in wine by yeast metabolism[a]

Sulfur compound	Concentration (μg/L)	Odour
Sulfides		
Hydrogen sulfide	nd–370	Rotten egg
Dimethyl sulfide	nd–480	Asparagus, cabbage, cooked corn
Carbon disulfide	nd–18	Rubber, cabbage, chokingly repulsive
Diethyl sulfide	nd–10	Garlic
Dimethyl disulfide	nd–22	Vegetable, cabbage, onion-like (high conc.)
Diethyl disulfide	nd–80	Bad smelling, onion
Mercaptans		
Methanethiol	nd–16	Cooked cabbage
Ethanethiol	nd–12	Onion, rubber, natural gas, fecal
Thioacetates		
Methyl thioacetate	nd–20	Sulfurous
Ethyl thioacetate	nd–56	Sulfurous
Thioalcohols		
2-Mercaptoethanol	nd–180	Poultry
2-(Methylthio)-1-ethanol	nd–70	French bean
3-(Methylthio)-1-propanol (methionol)	nd–4500	Potato, cauliflower, cooked cabbage
4-(Methylthio)-1-butanol	nd–180	Onion, garlic, earthy
Thiazoles		
Benzothiazole	nd–14	Rubber
2-Methyltetrahydro-thiophan-3-one		Metallic, natural gas
Long-chain polyfunction volatile thiols		
4-Mercapto-4-methyl-pentan-2-one	nd–0.03	Cat urine, box-tree, broom, blackcurrant
3-Mercaptohexanol	nd–5	Box tree, broom, passionfruit, grapefruit
3-Mercaptohexyl acetate	nd–0.2	Box tree, broom, passion fruit

[a] Compiled from Mestres et al. (2000); Fedrizzi et al. (2007); Swiegers et al. (2005)

important in red wines by acting as an enhancer of the berry fruit aroma (Escudero et al. 2007; Mestres et al. 2000; Segurel et al. 2004; Silva Ferreira et al. 2003).

Nevertheless, the impact of volatile sulfur compounds on wine aroma has been reassessed in the recent literature, and some have argued that several volatile sulfur compounds can contribute varietal character, at least in some varieties when present at near-threshold concentrations (De Mora et al. 1987; Fedrizzi et al. 2007; Rauhut et al. 1998). For example, sulfides, disulfides, benzothiazole and thioalcohols were found to be in higher concentrations in some Merlot wines whereas thiols and thioacetates were more abundant in Marzemino and Teroldego wines, indicating that varietal character might depend on several 'objectionable' volatile sulfur compounds to a greater extent than was previously thought (Fedrizzi et al. 2007). H_2S, though very evident when being produced during fermentation, is highly volatile and quickly lost from young wines, especially those subjected to aerative or sparging treatment during racking procedures (Moreira et al. 2002). The mercaptans, methanethiol and ethanethiol and 2-mercaptoethanol are more persistent in wine to which they elicit objectionable odours of cooked cabbage, onion, rubber, and poultry. Thioesters can also form during fermentation, and slowly hydrolyse under the acidic conditions of wine to release the parent objectionable mercaptan (Rauhut et al. 1998). Quantitatively, the most important low volatility sulfur compound is methionol, which possesses a potato, cauliflower odour. Unlike the lower boiling point volatile sulfur compounds, methionol can be present in wine in the low mg/L range, up to 5 mg/L.

8D.4.5.2 Metabolism and Modulating Factors

The appearance of volatile sulfur compounds in freshly fermented must is a complex, multifaceted chemical and biochemical process. Most research has focused on the induction of H_2S metabolism in yeast, especially in response to changes in nutrient composition (Eschenbruch 1974; Henschke and Jiranek 1991; Rauhut 1993; Jiranek et al. 1995a, 1996; Linderholm et al. 2006; Rankine 1963; Swiegers and Pretorius 2007; Spiropoulos et al. 2000; Vos and Gray 1979). Yeasts produce H_2S and other volatile sulfur compounds by a number of pathways, which include chemical reduction of elemental S, degradation of sulfur amino acids or reduction of sulfite or sulfate. Apart from the reduction of inorganic sulfur to H_2S, few of the chemical and/or metabolic pathways for the formation of other volatile sulfur compounds have been reported or verified in wine fermentation.

Hydrogen Sulfide

Elemental S (crystalline or colloidal form) can accumulate as a residue in must as the result of its use as a vineyard agrochemical, where it is used to control grape vine powdery mildew *Erysiphe necator* and various pests. Direct reduction of S to H_2S is induced by the highly reductive conditions that exist at the yeast cell surface during fermentation. This mechanism is of little practical importance except when the application of elemental S is not used according to manufacturers' recommendations, such as application within the recommended with-holding period before grape harvest (Rankine 1963; Rauhut and Kürbel 1994; Thomas et al. 1993).

Along with elemental S, grape musts also contain other forms of inorganic sulfur. Sulfate is indeed usually present in excess amounts (up to 700 mg/L) and sulfite (up to 100 mg/L) is often added as an antioxidant and antimicrobial compound (Henschke and Jiranek 1991). These two forms of inorganic S have been shown to be the main sources of H_2S formed during fermentation. Sulfate is accumulated by specific transporters (Sul1p and Sul2p), activated with 2 moles of ATP and reduced to H_2S by the sulfate reductive assimilation pathway (Fig 8D.8). The last step of this pathway reduces sulfite to H_2S, catalysed by sulfite reductase, which is encoded by the *MET5* and *MET10* genes. Sulfite, when present in the must, enters the cell by diffusion across the plasma membrane (Stratford and Rose 1985a) and can be directly reduced to sulfide (Hallinan et al. 1999; Jiranek et al. 1996; Stratford and Rose 1985b). Sulfite is a favoured source of sulfur in many yeast and

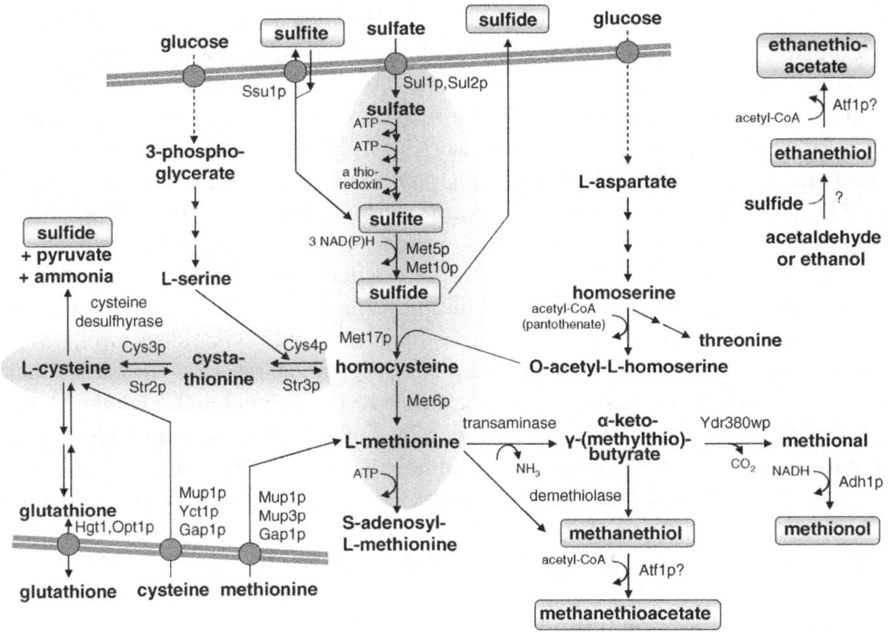

Fig. 8D.8 Sulfur metabolism in *Saccharomyces cerevisiae* yeast

Cysteine and methionine are accumulated by general and specific permeases (Mup1p, Mup3p, Gap1p, Yct1p) according to nutrient composition of must, but typically low concentrations cause early induction of the sulfate assimilation pathway. Sulfate is actively accumulated (Sul1p, Sul2p) and reduced to sulfite, which is then reduced to sulfide by sulfite reductase (Met5p and Met10p). Extracellular sulfite can also be reduced to sulfide. Sulfide is sequestered by *O*-acetyl-L-homoserine to form methionine and also by serine to form cysteine. Relative to the rate of sulfide formation, a shortage of precursor compounds (*O*-acetyl-L-homoserine and serine) or cofactors (pantothenate or acetyl-CoA) can lead to surplus H_2S. Under stress conditions caused by nutrient starvation, glutathione can be used as a source of nitrogen with information of H_2S. Methionine can be deaminated and reductively decarboxylated to methionol. Methanethiol can be formed from methionine by demethiolase and esterified to methanethioacetate. Sulfide can also react with ethanol/acetaldehyde to form ethanethiol, and esterified to ethanethioacetate

can lead to abundant H_2S production because its uptake is essentially unregulated. H_2S is sequestered by the organic nitrogen precursor, *O*-acetyl-L-homoserine, to form homocysteine. This latter compound is condensed with serine to produce cysteine, which can be incorporated into the sulfur reserve and antioxidant compound glutathione. Homocysteine is also converted to methionine from which Met-tRNA and *S*-adenosylmethionine are produced. Together with cysteine these compounds regulate the sulfate assimilation pathway.

Degradation of the sulfur amino acids cysteine and methionine to H_2S and other volatile sulfur compounds has been observed under laboratory conditions but their roles under wine fermentation conditions are less clear (Eschenbruch 1974; Jiranek et al. 1995a; Moreira et al. 2002; Perpète et al. 2006; Rankine 1963; Vos and

Gray 1979). Cysteine is a powerful source of H_2S when added to a yeast culture, irrespective of nitrogen limitation, although the response is yeast strain dependent (Duan et al. 2004; Eschenbruch and Bonish 1976; Jiranek et al. 1995; Moreira et al. 2002; Rankine 1963). Under nitrogen limitation, cysteine is accumulated by various specific and general permeases (Yct1p, Gap1p, Mup1p), depending on amino acid composition of the medium (Kaur and Bachhawat 2007), and degraded to H_2S, pyruvate and ammonia by cysteine desulfhydrase (Tokuyama et al. 1973) (Fig 8D.8). The concentration of free cysteine in grape juice is typically insufficient (<20 mg/L; Amerine et al. 1980) to explain the prolonged formation of large amounts of H_2S often observed during fermentation. However, cysteine can be derived from other sulfur containing compounds. Hydrolytic release from grape proteins by yeast proteases has been postulated (Vos and Gray 1979) or intracellular cysteine can be derived from the major sulfur reserve compound, glutathione, by enzymatic hydrolysis induced by nitrogen or sulfur limitation, which also yields glutamate and glycine (Mehdi and Penninckx 1997). Under nitrogen limitation, cysteine can be further degraded as a source of nitrogen by cysteine desulfhydrase (Hallinan et al. 1999). Cysteine might also serve as precursor for dimethyl sulfide and 2-mercaptoethanol (Moreira et al. 2002; Ribéreau-Gayon et al. 2000b). The roles of these various mechanisms as sources of H_2S in wine fermentation are yet to be demonstrated conclusively.

Methionine is an important though limited source of organic S and N in grape must and serves several roles in yeast metabolism (Henschke and Jiranek 1991; Thomas and Surdin-Kerjan 1997) (Fig 8D.8). Its depletion during the early stages of growth activates the sulfate reductive assimilation pathway to allow biosynthesis of the S-amino acids, cysteine and methionine. Upon limitation of assimilable nitrogen, insufficient *O*-acetyl-L-homoserine is apparently formed to efficiently sequester the H_2S formed from sulfate reduction, leading to its diffusion out of the cell. Supplementation of nitrogen-limited or starved cells with methionine suppresses H_2S liberation from the culture of many but not all wine yeast studied (Duan et al. 2004; Jiranek et al. 1995a,b; Moreira et al. 2002; Spiropoulos et al. 2000) and hence methionine supplementation does not appear to offer a satisfactory option for controlling H_2S production. Furthermore, in some wine strains ammonium supplementation is ineffective when methionine is low.

Nitrogen availability is widely accepted to represent a critical factor in regulating the formation of H_2S during fermentation. Because sulfide is a metabolic intermediate in the synthesis of organic sulfur compounds, shortage of nitrogen results in H_2S accumulation, due to restricted synthesis of *O*-acetyl-L-homoserine, the sulfide acceptor in the sulfate reduction pathway. Furthermore, sulfite reductase activity is only slowly down-regulated allowing H_2S production to continue (Jiranek et al. 1996; Stratford and Rose 1985b). Supplementation of grape juice with preferential nitrogen sources, such as ammonium salts, is therefore frequently carried out in the winery to control H_2S formation due to nitrogen starvation. Other nutrients, particularly vitamins such as biotin, pantothenic acid, and pyridoxine, affect H_2S formation. Supplementation with pantothenic acid has been shown to reduce the ability of some strains to form H_2S (Edwards and Bohlscheid 2007; Tokuyama et al. 1973). Deficiencies in pyridoxine, a cofactor in the methionine biosynthetic pathway, can also result in large production of H_2S, particularly in the case of yeast strains that are unable to synthesize this vitamin (Monk 1986). Lack of biotin can also cause an increase in the formation of H_2S (Thomas and Surdin-Kerjan 1997), possibly due to the role of this vitamin as cofactor in the biosynthesis of *O*-acetyl-L-homoserine (Bohlscheid et al. 2007). Addition of commercially available fermentation nutrients containing the above-mentioned vitamins is a common measure to reduce the risk of H_2S production in the wine industry.

Nevertheless, under winemaking conditions, nutrient supplementation has not eliminated the risk of H_2S production (Henschke and Jiranek 1991; Jiranek et al. 1995a; Park et al. 2000; Spiropoulos et al. 2000). Recent genetic studies highlight the complexity of regulation of the sulfate reductive assimilation pathway (Linderholm et al. 2006, 2008; Spiropoulos et al. 2000). Controlling sulfite reductase activity is an obvious target for lowering excessive H_2S production to acceptable levels but so far no commercial strains have been developed (Sutherland et al. 2003; Zambonelli et al. 1975). An inability to control H_2S liberation by over-expression of genes associate with H_2S sequestration (*MET17*) and *S*-amino acid biosynthesis (*CYS4* and *MET6*) suggests that metabolite flux might be a limiting factor rather than inadequate enzyme activities. Analysis of the S-amino acids biosynthetic pathway, by studying various mutants carrying gene defects, suggests that some of these genes, or substrates or products of their corresponding protein products, might play key roles in regulating sulfate reduction (Linderholm et al. 2008). Interactions between the sulfate assimilation pathway and amino acid pathways and various metabolites (acetaldehyde) provide an insight into the complexity of pathway regulation (Aranda and del Olmo 2004; Backhus et al. 2001; Marks et al. 2003).

Organic Volatile Sulfur Compounds

Conditions that favour H_2S production also favour production of other volatile sulfur compounds, methanethiol and methanethioacetate (Rauhut et al. 1996), suggesting a metabolic link to methionine catabolism. Catabolism of this amino acid essentially follows the Ehrlich pathway, as shown in Fig 8D.5. The first step involves transamination to yield the keto acid α-keto-gamma-(methylthio)butyric acid, which is decarboxylated to 3-(methylthio)-1-proprionaldehyde specifically by Ydr380wp, and reduced to 3-(methylthio)-1-propanol (methionol) by alcohol dehydrogenases (Fig 8D.8). The production of methionol is likely to be regulated in a similar manner to that of other higher alcohols, that is, higher concentrations are formed in low to moderate nitrogen musts (Hernández-Orte et al. 2005). Both methionine and α-keto-γ-(methylthio)butyric acid can act as a source of methanethiol through a demethiolase step, which also produces alpha-ketobutyric acid (Perpète et al. 2006). Yeast strains with intense H_2S production also produce higher amounts of thioacetic acid esters of methanethiol and ethanethiol (Rauhut et al. 1996); it is likely that these thiols are esterified by alcohol acetyltransferase in the same way as ethyl acetate is formed from and acetic acid. The formation of acetic acid esters is a major problem in winemaking since these compounds are not removed from wine during fining with Cu^{2+} salts that can be used in some countries to remove H_2S and mercaptans. As is the case with other acetate esters, they undergo slow acid-catalysed hydrolysis to release the constituent mercaptan, which having a lower sensory threshold, can lead to off-flavour formation during storage in bottled wine.

8D.5 Biotransformation of Grape and Oak Flavour Compounds

8D.5.1 Monoterpenes, Norisoprenoids, Aliphatics, and Volatile Phenols

8D.5.1.1 Significance

It is well known that the secondary metabolites of grapes provide the basis of varietal character in wine. The important secondary metabolites are represented by several groups of compounds that contribute to the distinctive aroma profile of wines made from particular varieties of *Vitis vinifera*. They include terpenes, C_{13}-norisoprenoids, aliphatics, benzene-derivatives, volatile phenols and long-chain polyfuntional thiols. While constituent aroma compounds within these groups occur in most grape varieties, it is only when one or more of these compounds occur at concentrations well above their odour threshold that a distinctive varietal aroma emerges.

Monoterpenes are regarded as key odorants in aromatic grape varieties of *Vitis vinifera*, such as Muscat of Alexandria, Riesling and Gewürztraminer, to which they impart their characteristic floral aromas. The monoterpenes linalool, geraniol, nerol, citronellol, and α-terpineol are the more important aroma active compounds of this group, and contribute floral, fruity and citrus attributes (Strauss et al. 1986). In the case of non-aromatic varieties, the contribution of monoterpenes to the aroma character of wines is more subtle, although it is generally accepted that linalool, the most powerful odorant in this group of compounds, is an important component in the aroma of many white wines.

Several C_{13}-norisoprenoids, including β-damascenone, β-ionone, and 1,1,6-trimethyl-1,2-dihydronaphtalene (TDN), are powerful odorant in wine. β-Damascenone, which can be present in young wine at up to 180 times its odour threshold of 0.05 μg/L, has a stewed apple, rose, honey aroma (Francis and Newton 2005). β-Ionone, which has violet, flower, and raspberry aroma, is typically present at higher concentration in red wines. TDN develops a kerosene-like odour in aged whites, particularly Riesling (Francis and Newton 2005).

A variety of benzene-derivatives are found in many grape varieties, including vinyl phenols, benzyl alcohol, 2-phenyl ethanol and raspberry ketone. Vinyl phenols are characterised by spice and clove-like, 2-phenyl ethanol by rose and lilac, and raspberry ketone by a raspberry attribute (Francis and Newton 2005). It should be noted that, although a portion of 2-phenyl ethanol can derive from glycoside hydrolysis, a greater proportion of this compound is formed in the metabolism of the amino acid phenylalanine (Ugliano et al. 2006).

The most important aliphatic compounds comprise the C_6 compounds, including hexanol, which give a green (cut grass) character in wine. These C_6 compounds are found in wines of many varieties.

8D.5.1.2 Metabolism and Modulating Factors

With the exception of monoterpenes in aromatic grape varieties, few of the aroma-active secondary metabolites in mature grapes are present in sufficient concentration to provide a distinctive aroma or taste. Most of these compounds exist as non-volatile and odourless glycoconjugates. These glycoconjugates are either glucosides, disaccharides or trisaccharides, with the disaccharide glycosides representing the major source of aroma compounds. They all contain a glucosyl moiety, but for the disaccharide glycosides, the glucose moiety is further substituted with α-L-arabinofuranosyl, α-L-rhamnopyranosyl, β-D-xylopyranosyl or β-apiofuranosyl sugars (Williams et al. 1982).

Hydrolytic release of the sugar moiety(ies) is necessary to liberate the aglycon, which then becomes volatile and aroma-active. Several mechanism have been proposed, including acid-induced hydrolysis and grape endogenous and microbial hydrolytic enzymes (Fig 8D.9). Acid-catalysed hydrolysis, as the result of the mild acidic conditions of must, has generally been observed to be too slow to explain major release during fermentation (Williams et al. 1982; Ugliano et al. 2006). Nevertheless, acidic hydrolysis is considered the important mechanism operating during wine aging. Yeast associated hydrolase activities, on the other hand, induce a significant release of aglycons, which vary according to the structures of the sugars and aglycon moieties (Ugliano et al. 2006). Various extracellular enzymes with hydrolytic activities, such as β-glucosidase, α-arabinosidase, α-rhamnosidase, α-xylosidase or α-apiosidase, have been described in *Saccharomyces cerevisiae* and non-*Saccharomyces* species (Charoenchai et al. 1997; Darriet et al. 1988; Ugliano et al. 2006). Recent evidence suggests that exo-β-glucanase could be responsible for yeast-associated hydrolysis of glycoconjugates (Gil et al. 2005).

In addition to glycoside hydrolysis, some *S. cerevisiae* and non-*Saccharomyces* species can, under certain conditions, synthesize detectable amounts of monoterpenes in the absence of grape precursors (Chambon et al. 1990; Carrau et al. 2005). They are believed to derive from the sterol biosynthetic pathway in which geranyl-PP is synthesised from the intermediate isopentenyl-PP by geranyl-PP synthase. The various monoterpenes (C_{10}) geraniol, nerol, citronellol, linalool and α-terpineol, are then produced by various chemical, or possibly enzymatic transformation reactions, involving isomerisations, reductions and cyclisations. The sesquiterpene (C_{15}) farnesol is derived from the sterol pathway intermediate farnesyl-PP and is partially isomerised to nerolidol. Carrau et al. (2005) recently hypothesized that biosynthesis of the C_{10} terpenes is derived from the leucine-mevalonic acid pathway and, thus, is independent of sterol metabolism. The same authors also showed that factors such as assimilable nitrogen and oxygen differentially regulated production of monoterpenes and sesquiterpenes by *Saccharomyces cerevisiae*. Combined high nitrogen and microaerobiotic conditions stimulated monoterpene production whereas low nitrogen and anaerobic conditions favoured sesquiterpenes.

The monoterpene, citronellol, has been shown to form by the yeast-catalysed reduction of free geraniol present in the fermenting medium (Gramatica et al. 1982). Ugliano et al. (2006) suggested that, in low monoterpene grape varieties, in which low concentrations of free geraniol are present, citronellol can be formed by the yeast-driven reduction of free geraniol derived from glycoside hydrolysis (Fig 8D.9). The presence of ergosterol under aerobic conditions may favour the retention of geraniol and produce a higher ratio of citronellol to geraniol during fermentation, enhancing the aromatic quality of wine (Vaudano et al. 2004).

It is generally accepted that C_{13}-norisoprenoids, including the powerful odorants β-damascenone, β-ionone, and 1,1,6- trimethyl-1,2-dihydronaphtalene (TDN) are formed through acid-catalyzed transformation of grape-derived precursors (Sefton 1998). However, recent studies have shown that yeast strain can affect the concentration of β-damascenone and β-ionone at the end of fermentation (Loscos et al. 2007; Hernández-Orte et al. 2008). Enzymatic hydrolysis of different precursors followed by acid-catalyzed transformations of the aglycon has been indicated as a possible pathway for the formation of these compounds during fermentation (Hernández-Orte et al. 2008), although the actual mechanisms involved in the yeast-mediated formation of norisoprenoids have still to be clarified. Though

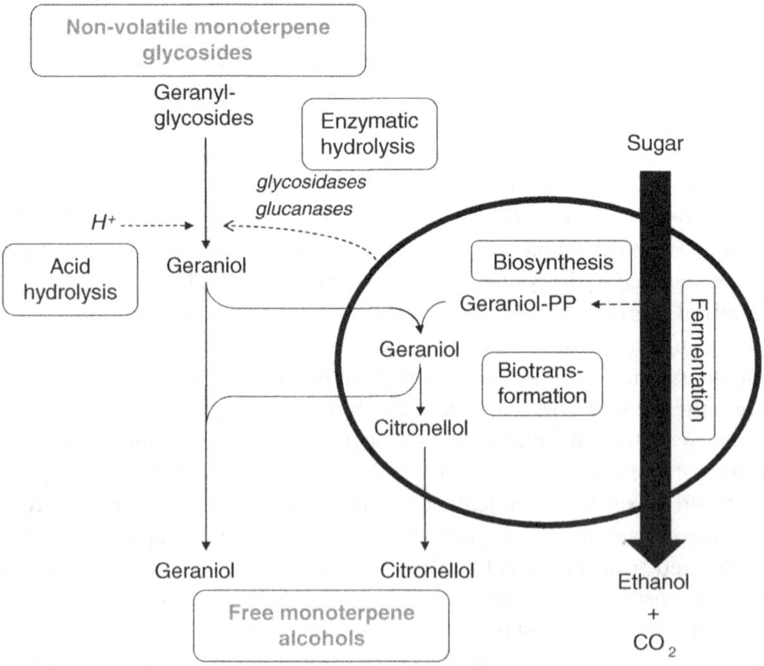

Fig. 8D.9 Acid-catalysed (H^+) and enzyme-catalysed hydrolytic release of grape glycosidically-bound geraniol and its biosynthesis and biotransformation by *Saccharomyces cerevisiae*

not yet demonstrated, it is likely that aliphatic-, shikimate-, benzene-derivatives- and phenol-glycoconjugates are similarly hydrolysed predominantly by yeast hydrolases during fermentation. In addition to this mechanism, it is well documented that the volatile phenols, 4-vinyl phenol and 4-vinyl guaiacol can be formed by *Saccharomyces cerevisiae* through non-oxidative decarboxylation of hydroxycinnamic acid precursors, *p*-coumaric and ferulic acids, respectively (Chatonnet et al. 1993). This enzymatic activity is naturally inhibited by catechins, which explains why, in comparison to white wines, red and rose wines typically exhibit much lower concentrations of vinyl phenols, despite having more precursors (Chatonnet et al. 1993). Reaction with anthocyanins to form pyranoanthocyanins might also contribute to lower vinyl phenols in red wines.

8D.5.2 Long-Chain Polyfunctional Thiols

8D.5.2.1 Significance

The long-chain polyfunctional volatile sulfur compounds, represented by 4-mercapto-4-methylpentan-2-one (4-MMP), 4-mercapto-4-methylpentan-2-ol (4-MMPOH), 3-mercaptohexan-1-ol (3-MH) and 3-mercaptohexyl acetate (3-MHA), are amongst the most potent aroma compounds identified in wine with sensory thresholds of 0.8–60 ng/L (Dubourdieu et al. 2006). These potent volatile thiols are major contributors to the varietal aroma of Sauvignon Blanc wines, and are associated with aroma characteristics such as box tree, blackcurrant, grapefruit, passionfruit and citrus zest (Darriet et al. 1995; Tominaga et al. 1995, 1998; Dubourdieu et al. 2006). The compound 4-MMP is also present in wines made from Scheurebe, Gewürztraminer, Riesling, Colombard, Petit Manseng, Semillon, Cabernet Sauvignon and Merlot.

8D.5.2.2 Metabolism and Modulating Factors

Free forms of the long-chain polyfunctional thiols 4-MMP, 3-MH and 3-MHA are virtually absent in grapes. However, a large number of *Vitis vinifera* varieties contain variable amounts of 4-MMP and 3-MH in the form of non-volatile *S*-cysteinyl- and related conjugates. *Saccharomyces cerevisiae* can hydrolyse these cysteine conjugates, liberating free 4-MMP and 3-MH (Fig 8D.10) (Dubourdieu et al. 2006; Howell et al. 2004b; Murat et al. 2001; Swiegers et al. 2008c). The mechanism of uptake of the non-volatile thiol precursors and the factors that control this are still to be determined. However, ammonium appears to suppress thiol production suggesting that precursors are excluded from the cell by nitrogen catabolite repression control, which regulates the uptake of nitrogen sources (Subileau et al. 2008; Thibon et al. 2008).

Release of the free thiol from the cysteinyl-conjugate appears to involve carbon-sulfur lyase activity (Howell et al. 2005; Tominaga et al. 1995). Overexpression and deletion experiments in *Saccharomyces cerevisiae* have shown that tryptophane

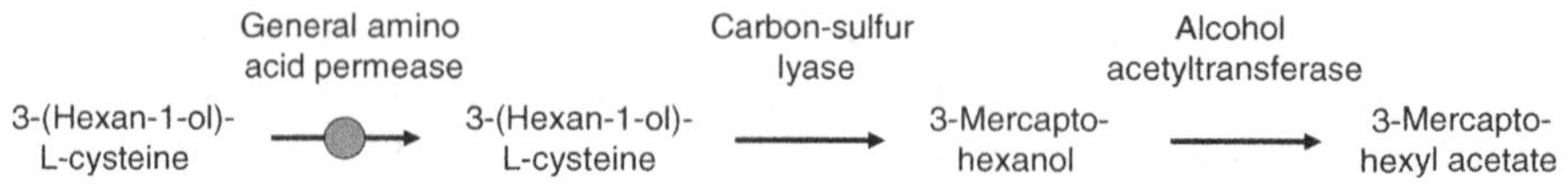

Fig. 8D.10 Enzymatic cleavage of S-cysteinyl conjugate and subsequent esterification

lyase is one of the enzymes potentially involved in this clevage process, as overexpression of this gene resulted in enhanced release of 4-MMP and 3-MH (Swiegers et al. 2007).

Formation of 3-MHA occurs thorough a more complex mechanism, that involves first liberation of 3-MH from the cysteinyl-conjugate precursor, followed by yeast-driven esterification with acetic acid (Fig 8D.10). The formation of 3-MHA from 3-MH occurs through the same pathway leading to the formation of acetate esters, since over expression of the alcohol acetyltransferase gene *ATF1* increased formation and overexpression of esterase gene *IAH1* decreased formation (Swiegers et al. 2006). 3-MH can also form chemically by reaction between H_2S produced by the yeast and carbonyl compounds present in the must, such as 2-hexenal. This pathway only accounts for 10% of the 3-MH typically formed in fermentation (Schneider et al. 2006).

Because the ability of different strains of *S. cerevisiae* to liberate long-chain polyfunctional thiols from their precursors is genetically determined, selection of yeast strain is a powerful tool for controlling the release of 4-MMP and 3-MH during fermentation (Dubourdieu et al. 2006; Murat et al. 2001; Swiegers et al. 2008c). Similarly, the ability to form 3-MHA from 3-MH depends on genetic characteristics of individual strains. Strain characterisation studies have indicated that some yeast strain exhibit higher ability to hydrolyse *S*-cysteinyl-conjugates, while other strains are characterised by increased ester synthetic activity (Swiegers et al. 2008a). Based on these findings, the use of mixed cultures containing two or more yeasts, one with high cysteine-lyase activity and the other with high acetate production has been proposed as a tool to modulate the composition of the pool of long-chain polyfunctional thiols formed during fermentation.

8D.5.3 Anthocyanins and Tannins

8D.5.3.1 Significance

Grape phenolics compounds are important to wine colour, flavour, astringency and bitterness, with red wines generally containing 1200–1800 mg gallic acid equivalents/L of total phenolics, six- to ninefold more than present in white wines (Kennedy et al. 2006). Hydroxycinnamic acids (non-flavonoid phenolics) are major phenolic compounds of white wines and are responsible for their colour. Other non-flavonoid phenolics contribute flavour, such as vanillin, vinyl phenols and gallic acid. Vinyl and ethyl phenols, which can be present to variable extents, elicit phenolics, medical, 'Bandaid', barnyard and spicy characters in wine, which are generally considered to lower wine quality. Ethyl phenols indicate growth of *Dekkera bruxellensis*, a wine spoilage yeast (Chatonnet et al. 1992b; Suárez et al. 2007). In red wines, colour is based on anthocyanin pigments and derivatives. Monomeric pigments are the major determinant of colour in young red wines but decline as the wine ages through chemical modifications and condensation reactions with flavan-3-ols and proanthocyanidins. These reactions form more stable pigments, pyranoanthocyanins and polymeric pigments. Red wines also contain up to 10-fold higher concentrations of flavan-3-ol monomers (catechins) than white wines that provide bitterness and probably contribute astringency. Red wine astringency is, however, primarily determined by the types and content of proanthocyanidins, generally referred to as tannins in the wine literature (Kennedy et al. 2006).

8D.5.3.2 Metabolism

Grapes contain several hydroxycinnamic acids, *p*-coumaric, caffeic, ferulic and sinapic acids, which exist as free acids and esterified with tartaric acid. *Saccharomyces* species can take up free acids to produce the corresponding vinyl phenol catalysed by hydroxycinnamate decarboxylase (phenylacrylic acid decarboxylase; Pad1p) (Fig 8D.11) (Chatonnet et al. 1992b; Chatonnet et al. 1993; Edlin et al. 1995). Vinyl phenols are unstable and highly reactive. *Dekkera bruxellensis* is one of few wine microorganisms that can further reduce vinyl phenols to highly stable ethyl phenols in wine. Vinyl phenols can also react with anthocyanins to form vinyl derivatives, a reaction that is favoured by fermentation yeast having hydroxycinnamate decarboxylase activity (Morata et al. 2006).

Various yeast metabolic interactions with grape phenolics, involving several vinyl and carbonyl metabolites, can affect red wine colour and astringency (Fig 8D.11). Acetaldehyde, pyruvic acid, α-ketoglutaric acid and vinyl phenols can react with anthocyanins to form pyranoanthocyanins, such as Vitisins (Morata et al. 2006). These pigments, which are more stable than anthocyanins to pH and SO_2 bleaching effects, have different spectral properties to anthocyanins and contribute to the less purple and more red orange colour of aging wines (Hayasaka et al. 2007). Acetaldehyde can also form ethyl-linked conjugates of anthocyanins with flavanol monomers, oligomers and polymers (tannins), which also stabilise wine colour. Yeast polysaccharides, in addition to grape polysaccharides, interact with anthocyanins to improve colour stability (Escot et al. 2001). Several yeast mediated reactions can also negatively impact on wine colour development, such as by binding anthocyanins to the cell wall (2–6% of total anthocyanins) (Morata et al. 2003) or by producing anthocyanin-β-D-glucosidase or possibly exoglucanase that releases the corresponding, but unstable, anthocyanidin (Manzanares et al. 2000; Gil et al.2005).

These various reactions involving yeast metabolites and non-pigmented phenolics are also believed to lead to changes that affect the astringency of tannins (Eglinton et al. 2005). Sensory evaluation of wines made with two *Saccharomyces* yeast which differed in acetaldehyde production led to differences in mouth-feel attributes associated with tannins, namely grainy, silky, velvet, drying and pucker.

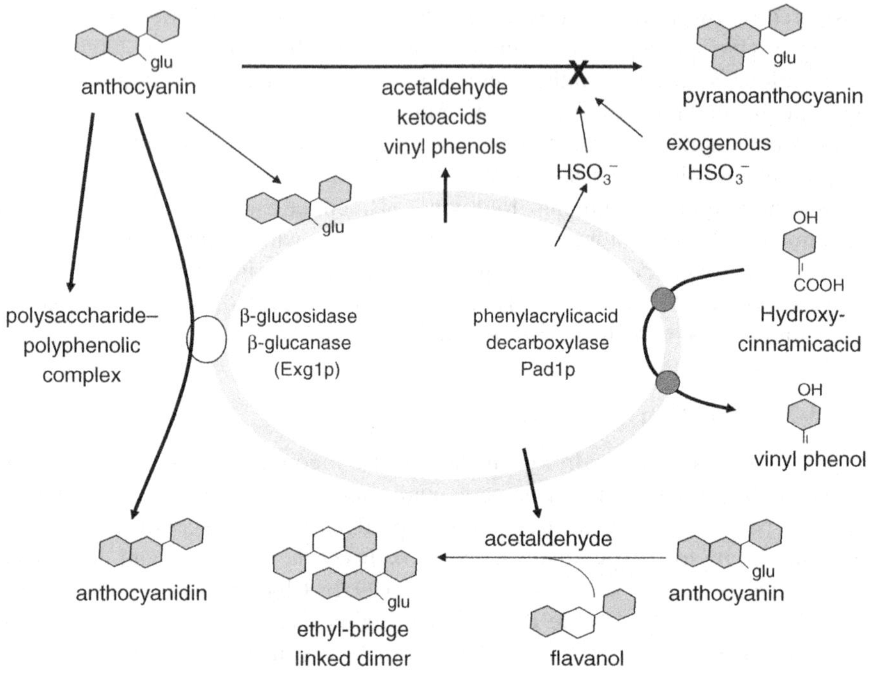

Fig. 8D.11 Modifications of grape phenolic compounds by *Saccharomyces*

Anthocyanins react with several yeast metabolites (e.g. acetaldehyde, ketoacids, vinyl phenols) to form stable pyranoanthocyanin pigments; bisulfite can inhibit these reactions by forming adducts with anthocyanins and carbonyls. Acetaldehyde can condense anthocyanins and flavanols to form stable ethyl-bridge linked polymeric pigments. Anthocyanins can bind to yeast cell walls, complex with yeast polysaccharides (mannoproteins), and form unstable anthocyanidin aglycones after hydrolysis by β-glucosidase or β-glucanase (Exg1p). Hydroxycinnamic acids can be decarboxylated to volatile vinyl phenols by phenylacrylic acid decarboxylase (Pad1p)

The chemical basis for these apparent changes is unknown but is likely to involve yeast mediated changes to tannin structure.

8D.5.3.3 Modulating Factors

The depth of colour and hue of red wine is largely dependent on grape variety, viticultural factors and choice of winemaking conditions. Nevertheless, the choice of fermentation yeast has a larger impact on red wine colour and mouth-feel properties than previously believed (Bartowsky et al. 2004; Caridi et al. 2004; Castino 1982; Eglinton et al. 2005; Medina et al. 2005). A survey of 17 *Saccharomyces cerevisiae* wine fermentation yeast showed that the colour density of 4–6 week old red wines varied by 38% (6.8–11.0 AU); however, the colour stability of aged wines was not necessarily maintained according to yeast strain, suggesting that colour stability is a complex phenomenon. Relative colour properties of yeast strains were, however, maintained for grapes of different phenolics composition, such as from different viticultural regions. *Saccharomyces bayanus* strains compared to species *cerevisiae* appear to have an especially notable impact on red wine properties, which is apparent immediately after fermentation and remains evident after maturation for at least one year (Eglinton et al. 2005; Hayasaka et al. 2007). Variations in the production of the various vinyl and carbonyl metabolites are likely to explain, at least in part, differences between strains (Caridi et al. 2004; Hayasaka et al. 2007; Medina et al. 2005; Morata et al. 2006). Fermentation conditions, which alter the maceration time on grape skins, act by not only affecting extraction of anthocyanins and tannins, but also the production of important yeast metabolites. Fermentation temperature, aeration, pH, nutrients, use of SO_2 in the case of carbonyls, all affect yeast metabolite formation and modulate wine colour and stability. These factors are also likely to explain changes in the intensity and mouth-feel properties of astringency.

8D.5.4 Oak Derived Volatile Compounds

8D.5.4.1 Significance

Oak barrels have long been used in winemaking for their convenience as containers to ferment, store, and transport wine. Nevertheless, during fermentation and maturation in contact with oak, wine undergoes a series of complex transformations, which can result in significant changes to its aroma composition and quality. In particular, toasted oak staves, commonly employed in cask manufacturing, contain significant amounts of potent aroma compounds, such as vanillin and vanillin derivatives, volatile phenols, and lactones. The accumulation of these compounds in wine during barrel storage results in increased intensity of sensory descriptors, such as vanilla, spice, and coconut, which are characteristic of these compounds. Optimal balance between oak-derived sensory attributes and grape- and fermentation-derived aromas is particularly important for the achievement of the desired level of wine aroma complexity.

8D.5.4.2 Metabolism and Modulating Factors

When fermentation is carried out in oak barrels, some of the compounds migrating from the wood to fermenting grape juice can be transformed by yeast through enzymatic reactions. The main reactions are reductive, such as conversion of carbonyl compounds to the corresponding alcohols. Amongst these, the powerful odorant vanillin can be reduced to vanillic alcohol during fermentation (Chatonnet et al. 1992a). As the odour threshold for vanillic is much lower then vanillin, it has been suggested that this transformation contributes to reducing the 'woody' character of wines that have been fermented in barrel (Chatonnet et al. 1992a). A similar reaction results in the formation of furfuryl alcohol from furfural during barrel fermentation (Marsal et al. 1988). 2-Furanmethanethiol (furfuryl mercaptan) can also be formed from furfural, possibly catalysed by yeast cysteine desulfhydrase, which releases H_2S from cysteine under conditions of low assimilable nitrogen (Tominaga et al. 2000). This compound has a roasted coffee-like aroma.

8D.6 Yeast Modulation of Wine Flavour

From the proceeding discussion on yeast metabolism in this chapter, it is clear that the scope for yeast to modulate the non-volatile and volatile composition of wine is considerable. And while the impact of many compounds on their own may not be detectable, some chemically-alike compounds act together, such as esters, to make a significant sensory impact (Francis and Newton 2005; Ferreira et al. 2008; Guth 1997; Van der Merwe and van Wyk 1981). Many of the metabolic compounds produced during sugar metabolism contribute to the appearance, aroma, flavour and mouth-feel properties of wine. Furthermore, the profile of sensorially important metabolites can vary considerably with the fermentation conditions under which the yeast is subjected, that is, the physicochemical and nutrition parameters. Finally, the release of a wide range of grape secondary metabolites present in grape must as non-volatile flavour precursors, principally glycosides and *S*-cyteinyl and related conjugates, depends on the strain of yeast (Dubourdieu et al. 2006; Ferreira et al. 2008; Jolly et al. 2006; Lambrechts and Pretorius 2000; Swiegers et al. 2005).

In addition to the choice of yeast strain, the method of inoculation can be used to modulate wine flavour in ways not readily achievable with conventional yeasts, which are typically used in monoculture. A combination of alternative yeast species and inoculation strategies can lead to wines with very different chemical and flavour profiles, such as greater complexity and diversity of flavours, and enhanced mouth-feel and persistence of flavour (Table 8D.4).

Allowing musts to ferment with indigenous yeasts can potentially produce a high diversity of flavours but the flavour profile is dependent on the yeast species and numbers present in the must, as well as the physico-chemical and nutrient composition. Although, from a microbiological perspective, depending on indigenous

Table 8D.4 Types of yeasts and starter cultures used in winemaking

Starter culture	Yeast species	Wine characteristics
Indigenous yeast	Multiple species and strains of non-*Saccharomyces* and *Saccharomyces*	Complexity and diversity of flavours, enhanced mouth-feel
Selected strains of *Saccharomyces cerevisiae*	Predominantly *Saccharomyces cerevisiae* selected strain	Intensity and diversity of fruity flavours or enhancement of varietal attributes
Saccharomyces sensu stricto species	Predominantly strain of selected non-*Saccharomyces cerevisiae* species	Increased diversity of flavours, especially mouth-feel properties
Non-*Saccharomyces* species	Predominantly non-*Saccharomyces* species and indigenous or added *Saccharomyces cerevisiae*	More controlled complexity and diversity of flavours, and enhanced mouth-feel

yeasts is a relatively unreliable fermentation strategy, a growing proportion of wines in the New World are being produced in this way. The basis of modern winemaking has been the so-called 'pure culture inoculation technology' that commenced with isolation of yeasts by Müller-Thurgau (1896), but in winemaking this practice only became widely established in the second half of the twentieth century (Rankine 1953). This biotechnological process involves minimizing the presence of indigenous yeasts by chemical and/or physical processes and adding a starter culture of a selected strain of *Saccharomyces cerevisiae*. The advent of active dried yeast has revolutionised the availability of yeast selections with some 200 strains used globally, mostly *Saccharomyces cerevisiae* (Fernandez-Espinar et al. 2001; Henschke 2007). Several *Saccharomyces* sensu stricto species, including *uvarum, bayanus* and their *cerevisiae* hybrids, *paradoxus*, and hybrids of *cariocanus* and *kudriavzevii* with *cerevisiae,* have also been recently produced. A few selections of non-*Saccharomyces* species, *Candida stellata, Torulaspora delbrueckii* and *Kluyveromyces thermotolerans*, mainly in mixtures with *Saccharomyces cerevisiae*, have also recently become available for commercial wine production (Antonelli et al. 1999; Bellon et al. 2008; Henschke 2007; see also the websites of yeast producers, such as www.chr-hansen.com; www.lallemandwine.com; www.maurivinyeast.com).

8D.6.1 Indigenous Yeasts

The simplest form of fermentation, in which no intervention is needed except for preparation of the must, is to rely on the indigenous strains of yeasts that are associated with the grapes, and harvest, must processing and fermentation equipment. The diversity of populations of yeast species and strains in the must or juice at the beginning of fermentation is remarkably variable. The dynamics of growth and decline of each yeast species and strains is likewise highly variable (Fleet 2003; Fleet and Heard, 1993; Fleet et al. 2002), Therefore, in general terms, the ecology of fermentation can be summarised as: the highly oxidative (non-fermentative) species undergo little or no growth and decline early in fermentation, the weakly fermentative species, which are usually present in greater initial numbers, grow to a greater extent but eventually become superseded by the physiologically better adapted and most strongly fermentative yeasts *Saccharomyces* species. *Saccharomyces cerevisiae* is the most commonly encountered species whereas in cool climate winemaking regions *Saccharomyces bayanus* (var. *uvarum)* can occur more frequently, due to its cryotolerant properties (Le Jeune et al. 2007; Naumov et al. 2000). *Saccharomyces paradoxus* is commonly found in Croatian musts (Redzepovic et al. 2002).

Because of the high variability of yeast species and their population sizes, and the difficulty in microbial profiling of fermentation, few studies have reported on wine chemical composition (Beh et al. 2006). General characteristics are recorded in Table 8D.5. These data suggest that the diminished role of *Saccharomyces* allows some of the non-*Saccharomyces* characteristics to be more evident and to

Table 8D.5 A generalized chemical composition of wine made with indigenous yeasts

Wine composition	Concentration from indigenous yeast fermentation wines relative to conventional wines
Residual sugar	1–5 g/L higher
Alcohol	0–1 vol.% lower
Glycerol	0–5 g/L higher
Titratable acidity	0–2 g/L higher
Acetic acid	Up to twofold higher
Sulfur dioxide (total)	Up to twofold lower
Acetaldehyde	Up to twofold lower
Tannins	Lower

modulate wine composition, thereby affecting most of the major constituents of wine. The diversity of aroma compounds in wines also reflects the diversity of yeasts in indigenous yeast fermentations. This fact is readily apparent in Fig 8D.12, which compares Chardonnay wines made with indigenous yeasts with those made with monocultures of *Saccharomyces cerevisiae*, in the same juices (Varela et al. 2008). This figure highlights the anecdotal observation that wines made with *Saccharomyces cerevisiae* are more alike than those made with indigenous yeasts. That is, in this experiment, wines made by inoculation were associated with the esters

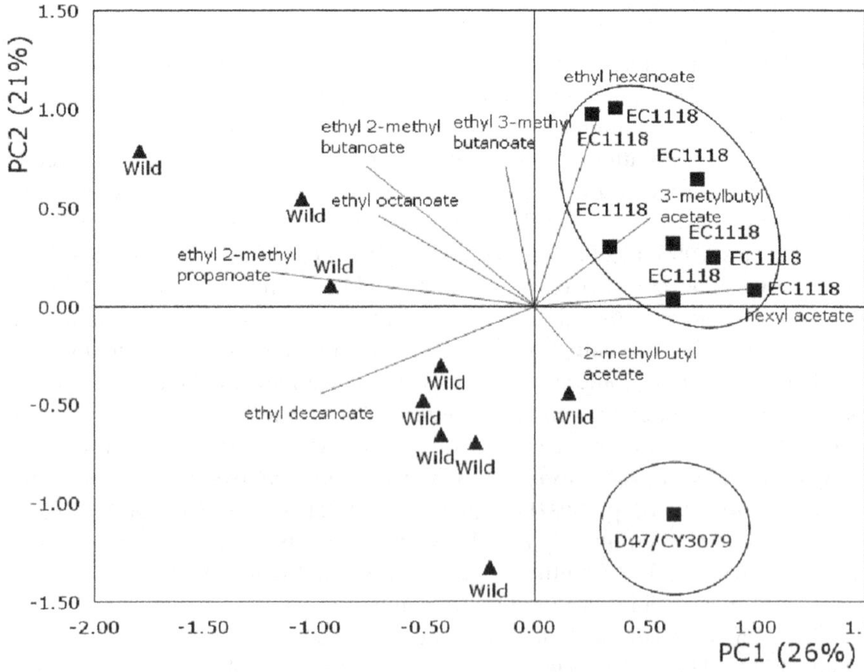

Fig. 8D.12 Association between fermentation esters and indigenous yeasts (Wild) and *Saccharomyces cerevisiae* (EC1118, D47/CY3079) (Varela et al., 2008)

ethyl hexanoate, 3-methylbutyl acetate and hexyl acetate, to produce a clear cluster by principal component analysis, whereas the indigenous yeast wines showed no clear associations with volatile metabolites of yeast origin. Similar observations have been reported by others (Garde-Cerdán and Ancín-Azpilicueta 2006).

A consensus of the principal sensory attributes of indigenous yeast wines is not possible since almost no controlled studies have been published. Based on comments and articles by winemakers, wines made with 'wild yeasts' can be summarised by terms such as complex, diverse, funky, mineral, creamy texture, greater palate weight and flavour persistence (for example Bellon et al. 2008; Goldfarb 1994; Ramey 1996; Ross 1997). Unlike inoculated yeast wines, which tend to have well defined fruity aromas, indigenous yeast wines tend to show a broader, less well defined fruity to savoury aroma profile. In one of several reports, Riesling wines made by uninoculated fermentation, for example, had higher sensory scores for 'spice', 'apple', 'melon', 'pear', 'diacetyl' and 'H_2S' compared to 'paper', 'oxidised', 'sweat', 'acetic' and 'overall fruit' for wine made by inoculation (Egli et al. 1998; Henick-Kling et al. 1998).

8D.6.2 Monocultures of Saccharomyces *Species*

The species *Saccharomyces cerevisiae*, being highly adapted to fermenting grape must in monoculture, has become the preferred species for global wine production (Henschke 1997; Pretorius 2000). The number of strains developed commercially has grown dramatically over the past five decades. This growth is, in part, due to the progressive understanding of the numerous roles that yeast play in fermentation, and in the evolution and modulation of many key wine sensory characteristics. Furthermore, growth in the selection of autochthonous strains, which are intended to enhance regional attributes in wines, has greatly expanded the choice of strains (Clemente-Jimenez et al. 2004; Dumont and Dulau 1997; Henschke 1997; Lambrechts and Pretorius 2000; Pretorius 2000; Pretorius et al. 2006; Swiegers et al. 2005; Swiegers and Pretorius 2007; Romano et al. 2003a,b). The extent to which the commercialized strains can engender chemical differences in wine is generally well shown (Table 8D.6) (Dubourdieu et al. 2006; Heard 1999; Lambrechts and Pretorius 2000; Kunkee and Vilos, 1994; Reynolds et al. 2001; Swiegers et al. 2008c); however strain impact on sensory attributes is largely unknown since there is almost a complete absence of well-designed, systematic studies published (reviewed by Thorngate 1998; Dumont and Dulau 1997; Egli et al. 1998; Estévez et al. 2004; Henick-Kling et al. 1998; Jane et al. 1996; Reynolds et al. 2001; Reynolds et al. 2007; Swiegers et al. 2008a,c). Many of the grape and yeast compounds with well documented sensory properties that vary according to yeast species and strain have been discussed in Sects. 8D.3, 8D.4 and 8D.5.

In addition to *Saccharomyces cerevisiae*, other species of the *Saccharomyces* sensu stricto group have recently become of interest to research and winemaking, in part because of their different capacities to modulate wine composition, and flavour

Table 8D.6 The diversity of volatile metabolites produced by different species of wine yeasts[a]

Genus, species	Ethanol	Ethyl acetate	3-Methyl-butyl acetate	Ethyl octanoate	2-Methyl propanol	3-Methyl-butanol	2-Phenyl ethanol	Acetic acid	Acetaldehyde
Saccharomyces cerevisiae	6–23	10—100	0.1–16	Tr-2.1, 34	5–78	17–490	5–190	0.1–0.9	50–120
Torulaspora delbrueckii (Candida colliculosa)	10.5–12.5	20—74	0–0.1	–	52–212	106–249	21–30, 111	0.1–0.3, 0.86	5–23,122
Kluyveromyces thermotolerans	13.5	59—87	3–8	0.4	–	–	6–7	0.18–0.45	15–24
Candida stellata	4.5–8.5, 13.5	7—75	0.1–0.4	0–0.1	10–34	18–91	6–35	0.6–1.3	30–188
Issatchenkia orientalis (Candida krusei)	1–6.5	220—730	<0.1	–	38–106	22–119	54	0.3–1	41–73
Metschnikowia pulcherrima	2–5.5	62—676	0.1–0.8	0—129	9–123	0–243	22, 249	0.1–0.4	23–73
Hanseniaspora uvarum (Kloeckera apiculata)	3–6.5	25—650	0.3–5	0	4–38	4–115	7–48	0.2–2.5	6–260
Hanseniaspora guilliermondii	9.6	400	1	–	56	100	15	0.53	19
Pichia anomala	0.2–7.5	137—2150	0.7–11	–	4–29	11–84	27–47	0.9–2	18
Pichia fermentans	6	52	–	227	4	0	152		101
Issatchenkia terricola	8	222	–	0	18	140	121		183

[a]Adapted from Heard (1999); Ciani et al. (2006); Ciani and Maccarelli (1998); Clemente-Jimenez et al. (2004); Herraiz et al. (1990); Jolly et al. (2003); Moreno et al. (1991); Rojas et al. (2003); Soden et al. (2000); Zeeman et al. (1982); Zohre and Erten (2002)

Table 8D.7 General properties of *Saccharomyces* species *cerevisiae, bayanus/uvarum* and *paradoxus*[a]

Property	*Saccharomyces cerevisiae*	*Saccharomyces bayanus/uvarum*	*Saccharomyces paradoxus*
Fermentation temperature (°C)	10–35	6–30	6–30
Optimum growth temperature (°C)	> 30 'Mesophilic'	25–30 Often 'cryotolerant'	25–30 Undefined
Formation of:			
Acetic acid	Low – high	Low	Low – moderate
Ethanol	Wide range	<*S. cerevisiae*	≤*S. cerevisiae*
Glycerol	Wide range	>*S. cerevisiae*	>*S. cerevisiae*
Malic acid	Neutral/degrade	Neutral/produce	Neutral/degrade
Succinic acid	Low – medium	Medium – high	Medium

[a] References: Feuillat et al. (1997); Orlic et al. (2007)

when determined (Table 8D.7). Those *Saccharomyces* species of active interest are *bayanus/uvarum, paradoxus, cariocanus,* and *kudriavzevii* and various interspecific hybrids with *cerevisiae* (see for example Antonelli et al. 1999; Bellon et al. 2008; Dubourdieu et al. 2006; Eglinton et al. 2005; González et al. 2007; Henschke 2007; Lopandic et al. 2007; Masneuf et al. 2002; Orlic et al. 2007).

8D.6.2.1 Major Compounds

Saccharomyces species can modulate all the major constituents of wine, including residual sugar, ethanol, polyols, acids and phenolics (Table 8D.8). These compounds are essential to the flavour balance of wine by providing sweetness, sourness, astringency, and sometimes bitterness, an undesirable flavour in wine. The major emphasis on strain selection is to ensure that strains have a high capacity to produce low residual sugar, less than perceptible (<4 g/L) and often less than 0.5 g/L, for wines made from high maturity grapes, and for wines intended for maturation under conditions of increased risk to microbial spoilage, such as prolonged storage in wooden barrels. A strong demand for selecting *Saccharomyces* strains with lowered efficiency in the conversion of sugar to ethanol has arisen in

Table 8D.8 Major compounds in wine affected by yeast

Compound class	Wine sensory characteristics	Concentration range in dry wines
Residual sugar	Sweetness	Variable
Ethanol	Sweetness, 'hotness'	8–>16 vol.%
Polyols	Sweetness	4–14 g/L
Acids	Sourness	5–9 g/L
Phenolics	Colour, astringency, bitterness	0.2–2 g/L

the warmer viticultural regions but recent data suggest that the high conversion efficiency trait is highly conserved in *Saccharomyces cerevisiae* (Palacios et al. 2007). *Saccharomyces bayanus/uvarum* strains, however, show a small reduction in ethanol production efficiency but this is probably of limited commercial significance. Polyol production varies widely amongst *Saccharomyces* strains, and is notably higher in *bayanus, uvarum, paradoxus* and *kudriavzevii* species, and partially accounts for the lower ethanol yield (Antonelli et al. 1999; Castellari et al. 1994; Eglinton et al. 2000; Feuillat et al. 1997; Giudici et al. 1995; González et al. 2007; Nieuwoudt et al. 2002; Orlic et al. 2007; Radler and Schütz 1982; Rankine and Bridson 1971).

Although non-volatile acids are important to wine flavour balance, strain variability in acid metabolism by *Saccharomyces cerevisiae* is generally limited, though with several important exceptions. Several *Saccharomyces cerevisiae* and *paradoxus* strains can degrade a higher proportion of L-malic acid, thereby lowering wine acidity, whereas other strains of these species, including cryotolerant *Saccharomyces bayanus/uvarum* strains, produce elevated amounts of succinic acid, and sometimes malic acid. While increased acid production also partially contributes towards lower ethanol yield, it is beneficial in improving acidity in low acid musts. This property, however, appears to be erratic in *Saccharomyces cerevisiae*, that is, acid production depends on poorly understood must factors, and therefore causes difficulties in managing wine acidity in commercial scale volumes. Acetic acid production is lower in non-*Saccharomyces cerevisiae* species, especially *bayanus/uvarum*. This characteristic is of practical importance in wines subjected to malolactic fermentation and wood aging, and in the production of sweet wines, during which volatile acidity tends to increase (Antonelli et al. 1999; Castellari et al. 1994; Eglinton et al. 2000; Feuillat et al. 1997; Giudici et al. 1995; Holgate 1997; Muratore et al. 2007; Redzepovic et al. 2003).

The impact of yeasts on the phenolics compounds of wine has long been recognized (Castino 1982; Wenzel 1989) but with little specific information until suitable analytical methods had been developed. *Saccharomyces cerevisiae* strains affect anthocyanin, polymeric pigments and tannin profiles of red wines (see for example Bartowsky et al. 2004; Caridi et al. 2004; Medina et al. 2005; Morata et al. 2006). The complex interactions between yeast and wine phenolics has been discussed in Sect. 8D.5.3. Colour density of Shiraz red wines was reported to vary by 38% in a study of 17 strains, showing considerable scope for strain selection. A survey of non-*Saccharomyces* yeasts generally revealed greater colour loss than by *Saccharomyces cerevisiae* strains (Gockowiak and Henschke, unpublished data). However, *Saccharomyces bayanu/uvarums* and hybrid strains can increase the polymeric pigments fraction, which is associated with chemically more stable pigments (Bellon et al. 2008; Caridi et al. 2002; Eglinton et al. 2005; Hayasaka et al. 2007). Increased production of carbonyls, such as acetaldehyde, is associated with stable pigment formation. Changes to the tannins fraction and the mouthfeel properties of red wine, texture and astringency, are also affected by yeast species and strains; the mechanisms are likely to involve similar reactions shown for anthocyanins.

8D.6.2.2 Volatile Compounds

The aroma profile of wines made with strains of *Saccharomyces cerevisiae*, selected for (white) wine production, can be broadly classified into two groups, those that are described as fruity/estery and others for which varietal character is enhanced. The fruity/estery group can be further sub-divided into high and low ester producers, with the latter group sometimes referred to as 'neutral' strains. This classification is based on sensory perception by winemakers world-wide and made over several decades, and this information has been collated by various research institutes and wine yeast producing companies. Nevertheless, this empirical based information has some chemical basis, bearing in mind the variability introduced by fermentation conditions and must composition, especially nutrients that stimulate ester production (see for example Carrau et al. 2008; Delteil and Jarry 1992; Dubourdieu et al. 2006; Dumont and Dulau 1997; Estévez et al. 2004; Houtman and Du Plessis 1986; Lurton et al. 1995; Mateos et al. 2006; Loscos et al. 2007; Miller et al. 2007; Rankine 1977; Reynolds et al. 2001; Soles et al. 1982; Swiegers et al. 2008; Torrea et al. 2003; Ubeda-Iranzo et al. 2000; Vila et al. 2000; Vilanova et al. 2007; Zeeman et al. 1982).

Examples of neutral yeasts are the various clones of the group that comprise IOC Prise de Mousse, Lalvin EC1118, Mauurivin PDM, and many others. These clones are highly versatile and can be used in the production of every wine type, from sparkling, dry white, red and rosé, sweet white and red, to botrytis-affected, Eiswein and Ice-wine, excepting flor yeast wines. Estery yeasts typically produce higher concentrations of some or most of the principal sensorially important esters (ethyl acetate, 3-methylbutyl acetate, phenylethyl acetate, hexyl acetate and the ethyl esters of hexanoic, octanoic, decanoic and dodecanoic acids) and average concentrations of higher alcohols, which otherwise tend to mask aroma intensity (van der Merwe and van Wyk 1981). Examples of estery yeast are Enoferm M1 and M2 (Lallemand), and Maurivin AWRI 350, AWRI 796 and AWRI R2 (Mauri Yeast). Varietal enhancing yeast are those that, relative to ester production, enhance the release and/or modification of grape-derived flavour compounds, to yield varietal related aroma compounds, such as the fruity, long-chain, polyfunctional thiols 4-mercapto-4-methylpentan-2-one (4-MMP), 3-mercaptohexan-1-ol (3-MH) and 3-mercaptohexyl acetate (3-MHA) (Dubourdieu et al. 2006; Swiegers et al. 2008).

The aroma profiles of non-*Saccharomyces cerevisiae* wines are distinctly different when compared to *Saccharomyces cerevisiae* wines. For example, Chardonnay wines made with *Saccharomyces bayanus* are more associated with savory- and cooked-like aroma attributes including 'cooked orange peel', 'honey', 'yeasty', 'nutty' and 'aldehyde' whereas *Saccharomyces cerevisiae* are more estery/fruity with 'estery', 'ethyl acetate', 'floral', 'pineapple', 'peach' and 'citrus' attributes (Eglinton et al. 2000, 2005). These differences might be associated with higher concentrations of phenyl ethanol, phenyl ethyl acetate, ethyl lactate and acetaldehyde, and lower ethyl acetate, acetic acid and the remaining higher alcohols produced by *Saccharomyces bayanus* (Antonelli et al. 1999; Eglinton et al. 2000; Feuillat et al. 1997). In addition, strains of *Saccharomyces bayanus/uvarum,* and their hybrids with *Saccharomyces cerevisiae*, have improved ability to hydrolyse *S*-cysteinylated conjugates to increase the concentration of the tropical fruity long-chain polyfunctional thiols (Dubourdieu et al. 2006), and mask the green/vegetative aromas, possibly due to methoxypyrazines (Bellon et al. 2008; Eglinton and Henschke, unpublished). Consequently, selected strains of *Saccharomyces bayanus/uvarum* and hybrids are becoming useful for improving the flavour of Sauvignon Blanc, Semillon, Chardonnay, Cabernet Sauvignon and Merlot wines.

Little information is yet available on other *Saccharomyces* sensu stricto species. *Saccharomyces paradoxus* wines are augmented in 'floral' and 'citrus' attributes, and contain lower concentrations of ethyl acetate, 3-methylbutanol, total volatile esters, higher concentrations of total higher alcohols and variable formation of volatile fatty acids, when compared to the reference strain of *Saccharomyces cerevisiae* (Orlic et al. 2007). No aroma descriptions have been published for *Saccharomyces kudriavzevii* wines but the one strain tested in a red and white juice produced less total esters and more higher alcohols than the reference strain of *Saccharomyces cerevisiae* (González et al. 2007). Two *Saccharomyces kudriavzevii* x *Saccharomyces cerevisiae* hybrids (Lallemand W27 and W46) tended to show a more moderate behaviour although with slightly higher amounts of higher alcohols than the reference yeast. Another hybrid (Maurivin AWRI 1503) has been reported to produce 'estery', 'floral', 'tropical', 'citrus', 'nutty' and 'waxy' aromas in Chardonnay wines, whereas a *Saccharomyces cariocanus* x *Saccharomyces cerevisiae* hybrid (Maurivin AWRI 1502) gave a different profile described as 'peach', 'pear', 'nectarine', 'violets', 'lemon fruit/zest', 'creamy', 'mineral' and 'matchstick'; the non-*Saccharomyces* parents of these hybrids failed to ferment grape must (Bellon et al. 2008; Henschke 2007).

8D.6.3 Mixed Cultures of Saccharomyces

Winemakers have been combining strains of *Saccharomyces cerevisiae* for several decades on the basis of observations that mixed-cultures produce more flavour diversity and balanced wines, by introducing a greater range of flavour notes and moderating the intensity of distinctive estery/fruity notes, such as pineapple and banana in Chardonnay wines. Little research has, however, been conducted into the effects on yeast growth, wine composition and flavour profiles resulting from fermentations made with mixed cultures of *Saccharomyces* strains (Cheraiti et al. 2005; Favale et al. 2007; Fleet 2003; Grossmann et al. 1996; Howell et al. 2004a, 2006). The differences in yeast growth patterns of genetically distinct strains, that are observed when they are cultivated individually and in mixtures, suggest that metabolic interactions can occur between different yeasts, and that they can be antagonistic, neutral or stimulatory.

Diffusable metabolites, such as acetaldehyde, are likely candidate effector molecules. Acetaldehyde can, for example, be produced/utilised at different rates by different strains, and can affect the redox status of each yeast, as has been observed when an exogenous source of acetaldehyde is introduced into a culture (Stanley et al. 1993). Exogenous acetaldehyde stimulates fermentation by coupled oxidation of NADH, which is catalysed by alcohol dehydrogenase. Cofermentation of two metabolically distinct strains not only affects the growth rate of each strain, but also metabolite formation, since the production of many metabolites are potentially modulated by cellular redox balance (Cheraiti et al. 2005; Eglinton et al. 2002) (Fig 8D.2). In a cofermentation study of three *Saccharomyces cerevisiae* strains, the final concentration of acetaldehyde was low compared to that produced by the three strains when cultivated individually, pyruvate was intermediate and sulfite and acetic acid were formed in much lower concentrations. Acetate esters were found at intermediate concentration in mixed cultures compared to monocultures (Grossmann et al. 1996). This yeast preparation of three strains is produced commercially by Siha as Varioferm.

Favale et al. (2007), who investigated cofermentation and sequential fermentation of a *Saccharomyces cerevisiae* and a *Saccharomyces bayanus* hybrid by comprehensive analysis of the volatile compounds formed, found somewhat different effects. Ethyl esters and higher alcohols tended to be produced in concentrations similar to the highest yeast producer in monoculture, whereas ethanol, glycerol and fatty acids tended to be produced in intermediate concentrations. These differences suggest that wine composition changes will depend on, at least in part, the choice of yeasts. Furthermore, sequential culture fermentation introduces additional differences that further modulate wine composition.

Howell et al. (2006) also carried out extensive metabolite profiling of the volatile metabolites produced in wines made by cofermentation with *Saccharomyces* strains. They found that cofermentation not only produced a metabolite profile different from wines made by monoculture fermentation, but that blends of the monoculture fermentation wine were also notably different. These various finding therefore strengthen the anecdotal observations of winemakers that cofermentation with two or more strains can potentially increase the flavour complexity of wines. These studies also suggest that cofermentation of yeasts with greater metabolic dissimilarity are likely to produce greater differences in the profile of volatile and non-volatile metabolites, as has been observed when non-*Saccharomyces* yeasts are cofermented with *Saccharomyces* (Sect. 8D.6.4). Additional studies are needed to understand better the complex metabolic interactions and to describe the major impacts on wine composition and flavour.

Cofermentation with two or more strains has recently been exploited to enhance the aromatic profile of Sauvignon Blanc wines, in which a non-volatile *S*-cysteinylated precursor (3-(hexan-1-ol)-L-cysteine) is first hydrolysed by a carbon-sulfur lyase and subsequently esterified by alcohol acetyltransferase (Sect. 8D.5.2). Strains of *Saccharomyces cerevisiae* vary in ability to carry out the two reactions (Dubourdieu et al. 2006; Swiegers and Pretorius 2007). Cofermentation with two strains, one having higher hydrolytic function (release of 3-MH) and the other higher esterification activity, substantially enhanced formation of 3-MHA, thereby increasing the passion-fruit aroma when compared to monoculture wines. The authors showed that interaction between the two strains produced more 3-MH and 3-MHA than could be achieved by blending wines made by monoculture fermentation (Swiegers et al. 2008a). Alchemy I and II mixed yeast cultures for enhancing white wine aroma have been commercialized by Anchor Yeast, South Africa.

*8D.6.4 Non-*Saccharomyces *Species and Cofermentation*

Grapes and associated processing equipment contain a variety of yeasts that accumulate in the must or juice. Whether or not the must or juice is inoculated with a starter culture, some species and strains will grow according to their adaptability to the must/juice composition and fermentation conditions, and metabolise grape-derived compounds, to produce a variety of volatile and non-volatile metabolites and carry out transformation reactions. Depending on the vigour of the *Saccharomyces* starter culture and the growth and metabolic activity of the non-*Saccharomyces* yeasts during fermentation, wine composition will to various degrees reflect the metabolic activities of all yeasts present. Fermentations in which non-*Saccharomyces* species produce significant populations will have greater metabolic impact on wine composition and flavour.

Many of the persistent and numerically dominant non-*Saccharomyces* species, observed in fermentations made with indigenous yeasts, have been isolated and characterised, and in some cases experimental wines have been made to evaluate their potential sensory contribution (Bisson and Kunkee 1993; Ciani and Maccarelli 1998; Clemente-Jimenez et al. 2004; Fleet 2003; Fleet and Heard 1993; Heard 1999; Henschke et al. 2002; Hernández-Orte et al. 2008; Jolly et al. 2003, 2006; Moreno et al. 1991; Moreira et al. 2002; Mateo et al. 1991; Plata et al. 2003; Romano et al. 2003b; Rojas et al. 2001; Soden et al. 1999; Zeeman et al. 1982). Table 8D.9 records some of the aroma descriptors associated with wines made with several species of non-*Saccharomyces* yeasts. Despite considerable genetic distance between non-*Saccharomyces* and *Saccharomyces* yeasts, they share some aroma attributes but in addition non-*Saccharomyces* yeasts produce a diversity of novel aromas, some of which can be perceived as positive whereas other are distinctly negative. Clearly, strain selection, as it is for *Saccharomyces* species, is important. The nature of these aroma attributes suggests that these yeasts could be useful in wine production by providing extensive wine blending options in order to increase aroma diversity, such that the aroma notes do not become dominant and impart a monodimensional character. For example, the production of phenyl ethanol and its acetate by some *Saccharomyces bayanus* strains can mask more delicate aroma attributes (Dubourdieu et al. 2006).

A practical problem with non-*Saccharomyces* yeasts is that few strains are capable of completing fermentation, and in many cases only a small proportion of grape sugar will be fermented. Two strategies have evolved to enable complete fermentation; these are cofermentation with a robust *Saccharomyces* strain and sequential fermentation, in which the non-*Saccharomyces* yeast and *Saccharomyces* strain are inoculated successively, in order to complete fermentation. Several studies have

Table 8D.9 Comparison of aroma descriptors[a] for Chardonnay wines made with *Saccharomyces cerevisiae* and non-*Saccharomyces* species

	Aroma descriptors	
Aroma group	*Saccharomyces cerevisiae*	*Non-Saccharomyces species*[b]
Fruity	Tropical fruit, ethyl acetate, pineapple, banana, pear	Stone fruit, apple cider, toffee apple, quince, dried pear, floral, fruity ester, banana ester,ethyl acetate, cooked apricot,pineapple
Citrus	Citrus, lime	Citrus peel, cooked orange, lime
Aged	Honey	Brandy, caramel, honey
Microbiological		Sweat, dirt, malt, bread, dough, meat, salami, sauerkraut

[a]Compiled from Jane et al. (1996); Henschke et al. (2002); Soden et al. (1999)
[b]*Candida krusie, Candida stellata and Torulaspora delbrueckii*

reported the production of wines with different characteristics according to the inoculation procedure (see, for example, Ciani et al. 2006; Gill et al. 1996; Heard 1999; Herraiz et al. 1990; Jolly et al. 2003, 2006; Soden et al. 2000; Zironi et al. 1993; Zohre and Erten 2002). Successful co-fermentation depends on knowledge of the physiological properties of each yeast, especially compatibility, and consequential effects on growth rate and biomass development. Suppression of one yeast by the other can result in reduced metabolic activity and hence lessened impact on the wine characteristics. Indeed, sequential fermentation can be used to favour weak fermentative strains by delaying inoculation of *Saccharomyces cerevisiae. Candida stellata* only impacted on the chemical and sensory composition of wine when inoculated prior to *Saccharomyces cerevisiae* compared to coinoculation (Soden et al. 2000). On the other hand, coinoculation of a weakly fermentative yeast at high ratio to a strongly fermentative yeast can also achieve greater impact of the former yeast. In practice, sequential fermentations are easier to manage in the winery and although the wine can be out of balance, it nevertheless provides a useful aroma component for improving the complexity of the final blend. In our experience, few non-*Saccharomyces* ferments require inoculation with a *Saccharomyces* yeast due to the presence of a sufficient *Saccharomyces* indigenous population in the juice or must.

Several examples of how the novel properties of non-*Saccharomyces* yeasts can be used to improve the composition and aroma properties of wine will be discussed according to yeast species. The salient characteristics of several non-*Saccharomyces* species are summarised in Table 8D.10.

Torulaspora delbrueckii (anamorp *Candida colliculosa;* formerly *Saccharomyces rosei)* has a moderate tolerance to ethanol (<12.5 vol.%; Table 8D.6) and produces wines which resemble those made with *Saccharomyces cerevisiae*. Production of higher alcohols is however highly variable and strain dependent. Because this yeast produces comparatively low concentrations of acetic acid, ethyl acetate, acetaldehyde and acetoin, its potential suitability for wine production has been suggested by several studies (Cabrera et al. 1988; Ciani and Ferraro 1998; Herraiz et al. 1990;

Table 8D.10 Principal characteristics of selected non-*Saccharomyces* yeasts

Genus species	Characteristics
Torulaspora delbrueckii (Candida colliculosa)	Osmotolerant, low volatile acidity
Kluyveromyces thermotolerans	Bioacidification, aroma enhancement
Candida stellata	High glycerol, preferential fermentation of fructose
Issatchenkia orientalis (Candida krusei)	Malic acid degradation, aroma enhancement
Metschnikowia pulcherrima (Candida pulcherrima)	Aroma enhancement
Hanseniaspora uvarum (Kloeckera apiculata)	Flavour complexity (potential to increase volatile acidity)
Schizosaccharomyces malidevorans (Schizosaccharomyces pombe)	Deacidification (malic acid)
Dekkera bruxellensis (Brettanomyces bruxellensis)	Flavour complexity
Pichia anomala, Pichia fermentans, Pichia membranifaciens, Issatchenkia terricola, Pichia subpelliculosa, Williopsis staurnus	Aroma enhancement

Jolly et al. 2003b; Moreno et al. 1991). *Torulaspora delbrueckii* has been isolated from high sugar musts and this osmotolerant property is exploited in the fermentation of botrotized wines (Henschke and Dixon 1990; Lafon-Lafourcade et al. 1981; Bely et al. 2008). Under hyperosmotic stress conditions *Torulaspora delbrueckii* typically retains the ability to produce high glycerol and low acetic acid when compared to *Saccharomyces*, although cofermentation with *Saccharomyces* is necessary to yield high ethanol concentrations. *Torulaspora delbrueckii* is a component of a mixed yeast culture with *Saccharomyces cerevisiae* and *Kluyveromces thermotolerans* that has been commercialized by Chr. Hansen (Viniflora® HARMONY.nsac and MELODY.nsac).

Kluyveromces thermotolerans is an acid-producing yeast, which has a moderate to high tolerance to ethanol (<13.5 vol.%) when compared to other non-*Saccharomyces* species. Strains of this species can produce up to 7 g/L of lactic acid, and has the potential to balance low acidity musts. Strains of *Kluyveromces thermotolerans* are generally characterised by moderate ester, higher alcohol and acetic acid production and low acetaldehyde and off-flavour production, suggesting suitability to wine production (Ciani et al. 2006; Kapsopoulou et al. 2007; Mora et al. 1990; Zeeman et al. 1982).*Kluyveromyces thermotolerans* is a component of a mixed yeast culture with *Saccharomyces cerevisiae* commercialized by Chr. Hansen (Viniflora® SYMPHONY.nsac).

Candida stellata is widely distributed in musts world-wide, and while it can consume 2-ketogluconate, a sulfite binding substance produce in grapes and musts by *Botrytis cinerea* and acetic acid bacteria, some strains are tolerant to sulfite and can produce large concentrations of 2-methyl-1-propanol, rendering such strains of this yeast undesirable in fermentation (Holloway et al. 1992;). However, more recent studies on other isolates have revealed several potentially useful winemaking characteristics, including high glycerol and succinic acid production, a small reduction in final ethanol, preference for fructose consumption fermentation, moderate ester and higher alcohol production, and production of novel wine aromas; acetic acid and acetaldehyde are highly variable amongst strains (Table 8D.6) (Ciani and Maccarelli 1998; Henschke et al. 2002; Jolly et al. 2003b; Soden et al. 2000). Strain selection and sequential fermentation with *Saccharomyces cerevisiae* is especially important to avoid acetoin and aldehyde off-flavours and to enhance floral and estery aromas in low aromatic grape varieties.

Issatchenkia orientalis (anamorph *Candida krusei)* is occasionally isolated from grape musts and the early stages of fermentation, in which it can tolerate up to 6.5 vol.% ethanol (Table 8D.6). It produces relatively high concentrations of esters, higher alcohols and succinic acid together with moderate concentrations of acetic acid and acetaldehyde. In sequential fermentation with *Saccharomyces cerevisiae* it produced good acidity, very low volatile acidity, intense fruity ester production, and in some wines, a desirable ‘wild yeast’ fermentation character. A recently studied isolate of this thermotolerant, acidphilic yeast was found to degrade L-malic acid, and in cofermentation with *Saccharomyces cerevisiae* could produce wine with acceptable quality (Bellon et al. 2008; Henschke et al. 2002; Kim et al. 2008). Appropriate selections of this yeast appear to offer white and red wine making potential.

*Metschnikowia pulcherrima (*anamorph *Candida pulcherrima)* is commonly isolated from grapes and musts, and can dominate in the early stages of fermentation. In monoculture some strains produce very high concentrations of esters, especially ethyl acetate, which has a fruity, solvent character, and ethyl octanoate (ethyl caprylate) a fruity, pear-like aroma (Table 8D.6). Considerable 2-phenylethyl alcohol, which has a rose aroma, and diacetyl a buttery aroma, can also be produced in high concentrations. Wines made with this species are overly estery; however these characteristics can be used to intensify wine aroma. However, having a low tolerance to ethanol, this yeast has been used in coculture with *Saccharomyces cerevisiae.* Surprisingly, the composition of volatile compounds varies little from the corresponding *Saccharomyces cerevisiae* strain wines, but in sensory tests these wines were preferred (Clemente-Jimenez et al. 2004; Jemec and Raspor 2005; Jolly et al. 2003b; Zohre and Erten 2002). This yeast is a good example of how their characteristics can be beneficially moderated by cofermentation with *Saccharomyces cerevisiae.*

Hanseniaspora uvarum (anamorph *Kloeckera apiculata*) is commonly the major yeast present on the grape berry and in musts and juices, but due to low tolerance to ethanol, populations decline quickly in the presence of *Saccharomyces cerevisiae.* Strains are typically characterised by low fermentative ability and high production of acetic acid, ethyl acetate and acetaldehyde, which render such strains more suitable to vinegar production. Nevertheless, Ciani and Maccarelli (1998) surveyed 37 isolates and found considerable variability, with some strains producing concentrations of these compounds approaching concentrations present in wines made with *Saccharomyces cerevisiae* (Table 8D.6). Cofermentation fermentation with *Saccharomyces cerevisiae* can produce wines with an acceptable balance of volatile and non-volatile compounds and sensory scores (Ciani et al. 2006; Jemec and Raspor 2005; Jolly et al. 2003b; Zohre and Erten 2002).

Hanseniaspora guilliermondii can produce more ethanol than *Hanseniaspora uvarum* and has a better balance of volatile and non-volatile compounds, including production of 2-phenylethyl acetate, which elicits a fruity, honey, rose-like aroma (Table 8D.6) (Rojas et al. 2003; Swiegers et al. 2005). Cofermentation with *Saccharomyces cerevisiae* has potential for aroma enhancement of wine.

Dekkera species participate in the production of Belgium, acidic, lambic beers; however, this species, especially *Dekkera bruxellensis*, which is highly adapted to growth in wine, is generally considered to be a spoilage yeast although some consider that it adds favour complexity. These yeast produce volatile phenols, especially ethyl phenol, ethyl guaiacol and ethyl catechol, by the reductive decarboxylation of the grape hydroxycinnamic acids *p*-coumaric, ferulic and caffeic acids, respectively. In combination with volatile fatty acids, such as isovaleric acid, a variety of aromas are produced in wine, including medicinal, Bandaid®, barnyard, leathery, horsey, smoky and spicy. The nitrogen-heterocycles, 2-acetyltetrathydropyridine and 2-acetyl-1-pyrroline, which give the so-called 'mousy' off-flavour, a persistent and very disagreeable taste, can also be produced. Strain variability in volatile phenol production is observed such that those producing smokey or spicy aromas could be used to add complexity but the control of this species in wineries remains a major problem globally (Chatonnet et al. 1992b; Fugelsang and Zoecklein 2003; Grbin and Henschke 2000; Suárez et al. 2007).

Schizosaccharomyces malidevorans has been proposed for the biodeacidification of grape must since it strongly degrades L-malic acid to a greater extent than by closely related *Schizosaccharomyces pombe* strains. Anaerobically, malic acid is degraded to ethanol and CO_2, thereby giving a strong reduction in acidity. This yeast produces H_2S, limiting its use in winemaking. However, a mutant strain, which utilises malic acid at a greater rate and sugar to a limited extent, has been successfully used in red and white wine cofermentation trials with *Saccharomyces cerevisiae* (Rankine 1966; Thornton and Rodriguez 1996).

Several other non-*Saccharomyces* yeasts, including *Pichia anomala (*synonyms *Candida pelliculosa, Hansenula anomala, Candida beverwijkiae), Pichia fermentans (*anamorph *Candida lambica), Pichia membranifaciens* (anamorph *Candida valida*), *Issatchenkia terricola, Pichia subpelliculosa (*synonym *Hansenula subpelliculosa)* and *Williopsis staurnus (*synonym *Hansenula saturnus*) have been partially evaluated for aroma enhancement of wines (Table 8D.6) (Clemente-Jimenez et al. 2004; Jemec and Raspor 2005; Jolly et al. 2006; Rojas et al. 2003). Yeasts *Pichia anomala* and *Issatchenkia terricola* have high acetate ester production but in cofermentation with *Saccharomyces cerevisiae* more balanced wines can be produced. The high production of fruity-like ethyl esters of medium chain fatty acids, such as the ethyl octanoate, and the rose-like aromatic alcohol 2-phenyl ethanol by *Pichia fermentans* are potentially useful for aroma enhancement in cofermentations. Several weakly fermentative yeasts *Pichia anomala, Pichia membranifaciens, Pichia subpelliculosa* and *Williopsis staurnus* have also been evaluated for production of low alcohol wines, which under aerobic conditions increases biomass and increases ester production (Erten and Campbell 2001).

The potential for non-*Saccharomyces* yeasts to enhance wine aroma intensity and flavour complexity is considerable. Few strains are capable of completing fermentation in monoculture and therefore are used in cofermentation, usually with *Saccharomyces cerevisiae*. The extent of flavour enhancement can be modulated by using different inoculation strategies. In practice, it is difficult to obtain balanced wines by cofermentation with *Saccharomyces cerevisiae*; however, wines made with high non-*Saccharomyces* yeast character, although normally very unbalanced, make useful blending options. The pace of research on this topic is now considerable and new yeasts with better properties are to be expected. Sensory evaluations, which are strongly lacking in current research, will be essential for successful industry adoption. Although the isolates studied are normally obtained from grape must, health and safety aspects should be observed when producing high cell density cultures. A significant technical problem has been the successful development of active dried yeast preparations.

8D.7 Future Directions

The past decade has witnessed a remarkable expansion in knowledge of the roles that yeasts play in the development of wine sensory properties. Until recently, yeasts were assumed to provide a generic vinous character upon which the highly distinctive secondary compounds, derived from the grape or formed through purely chemical processes, defined the varietal and regional characteristics of the wine. With the recent discovery that yeasts are responsible for unlocking from non-volatile precursors several aroma compounds involved in wine varietal attributes, such as monoterpenes, norisoprenoids, and long-chain polyfuntional thiols, we are learning that yeasts play a much more complex role in determining wine varietal character. These important finding have enabled a more detailed understanding of yeast transformation reactions and mechanisms involved in the development of several key varietal odorants (Dubourdieu et al. 2006; Loscos et al. 2007; Swiegers and Pretorius 2007; Ugliano and Moio 2008).

Progress in this field has been driven, at least in part, by advances in analytical techniques. The use of simple analytical strategies, based on HPLC and GC, has provided an important but limited understanding of wine volatile composition and of the importance of individual components to the overall flavour composition of wine. Advances in GC-Olfactometry and GC-MS are now allowing identification and characterization of many new wine components. Compounds that are considered of primary importance in wine flavour are only now being discovered, as in the case of rotundone, responsible for the peppery character in wine, and the esters cyclohexanoate and 2-, 3-, and 4-methylpentanoate, considered to contribute to the berry fruit character of wine (Campo et al. 2007; Escudero et al. 2007; Wood et al. 2008). Nevertheless, the effect of yeast and fermentation on these compounds is still to be established. In addition, the role of some volatile compounds, originally considered spoilage off-flavours, such as volatile sulfur compounds, is being reassessed (Fedrizzi et al. 2007).

Another poorly explored area concerns grape and wine phenolics. Although yeast interaction with phenolics was recognized several decades ago, again progress has depended on development of suitable analytical techniques. Some understanding of the interactions that yeast play in colour development of wines has emerged (Hayasaka et al. 2007; Morata et al. 2006) but almost nothing is known about interactions with tannins, which are responsible for flavour, astringency and bitterness (Eglinton et al. 2005). The complex polymeric nature of phenolic compounds, and the need for sensory evaluation techniques, has slowed progress on these fundamental charactersistics of wine.

The published literature on the effects of microbial activities on wine chemical composition is now considerable. Understanding the significance of wine chemistry is, however, heavily dependent on complex analytical strategies which combine extensive chemical characterization and sensory descriptive analysis. However, sensory analysis is extremely resource-intense, requiring many hours of panelists' time. This prevents widespread application of these powerful analytical tools. Advanced statistical techniques have been developed that are closing the gap between chemical and sensory techniques. Such techniques allow the development of models, which should ultimately provide a sensory description based on chemical data. For example, Smyth et al. (2005) have developed reasonable models which can reveal the most likely compounds that relate to particular attributes that characterise the overall sensory profile of a wine. For wines such as Riesling and Chardonnay, the importance of several yeast volatile compounds has been indicated. Such information will allow yeast studies to target key compounds better rather than just those that are convenient to measure.

Discovery of several yeast-catalysed transformation reactions of grape compounds, often leading to enhanced flavour character and intensity, has opened another area for research. At this time most effort has been focused on glycosides and cysteinylated precursors. Glycosides were previously thought to be largely hydrolysed by the mild acidic conditions of grape musts and wine, with yeasts playing no significant role in the release of their volatile moiety. Although the mechanism of yeast catalysed hydrolysis is still not yet resolved, clear evidence now exists that yeasts are the driving force behind the release of glycosidically-bound aroma compounds during winemaking. Furthermore, some released aglycons can be transformed by yeast enzymes, resulting in a change to their odour intensity and/or quality. For example, geraniol can be reduced by yeast to citronellol, changing the aroma character of wine (Ugliano et al. 2006; Loscos et al. 2007). The liberation of 3-mercaptohexanol and its subsequent esterification to the more powerful odorant 3-mercaptohexylacetate, which gives a tropical, passion-fruit aroma, is another example of the key role of yeast transformation reactions in the appearance of highly distinctive aroma characters during fermentation (Swiegers et al. 2006).

Selection of new flavour strains is dependent on the chemical nature of the odorant of interest, when it is known. Flavour compounds that have convenient methods for screening, such as H_2S, which can be assessed by plating onto agar containing bismuth salts, allow rapid and efficient selection. On the other hand, screening yeasts for odorants for which a rapid screen is not available requires individual fermentations evaluated by chemical and/or sensory techniques. Nevertheless, the process of screening and evaluation of new yeasts remains a relatively inefficient process, since various fermentation characteristics are also assessed in addition to the target flavour property. Basic fermentation properties include vigour, attenuation, nutrient requirements, flocculation, and foaming, and freedom from off-flavour formation, such as H_2S, acetic acids, high ethyl acetate and high SO_2. Use of robotic technologies will greatly facilitate high throughput screening; however key developments are still required for multiple characterisations of large numbers of strains. High throughput metabolomic approaches, with which the volatile fraction of fermentation metabolites produced can be determined by GC techniques (Howell et al. 2006) or the non-volatiles by NIR techniques (Cozzolino et al. 2006) coupled to powerful chemometric tools, are still not suitable for routine work or are inaccessible to many researchers.

Advanced yeast genetic techniques will bring about the greatest developments in yeast and wine flavour evolution (Pretorius 2000; Pretorius and Bauer 2002). Screening yeast using gene arrays techniques, for appropriate gene markers which are associated with flavour metabolism could reduce the need for extensive metabolite screening (see for example Marullo et al. 2007). Several key genes relating to ester and long-chain polyfunctional thiol formation, degradation and transformation have recently been discovered and manipulated to improve the flavour modulation properties of yeasts (Howell et al. 2005; Subileau et al. 2008; Swiegers et al. 2006, 2007; Thibon et al. 2008). More appropriate regulation of gene expression, or modulation of a metabolic pathway, will further facilitate creation of strains with optimised flavour profiles. Controlling production of off-flavours such as H_2S and undesirable mercaptans are important targets (Linderholm et al. 2008; Sutherland et al. 2003), as is the control of acetic acid production (Pigeau and Inglis 2005; Saint-Prix et al. 2004). Disconnecting environmental factors from flavour metabolic pathways will produce strains with more stable flavour properties in musts of variable nutrient content, and lower the risk of nutrient linked off-flavour formation. Such targets are likely to require application of systems biology approaches that can integrate genome expression profiles and enzyme catalytic reactions (Borneman et al. 2007). Further ahead, the introduction of new metabolic pathways into yeast could be envisaged, leading to the production of compounds which are otherwise grape-derived. Such an example is the cloning of key genes into yeast to produce resveratrol (Becker et al. 2003), although this particular compound has no flavour significance. Strains developed by genetic engineering approaches will have limited impact in those countries in which consumers and governments regulate their use in grape and wine production; those countries continue to develop strains for wine production by conventional genetic breeding techniques (Chambers et al. 2008).

Another recent growth area has been the selection of autochthonous yeasts, which can promote the regional characteristics of wines in order to overcome the apparent problem of homogenous or industrial wines. A recent focus has been to select yeasts on the basis of several yeast volatile aroma compounds but with improved understanding of the key odorants of wine the focus will move to those grape-derived compounds that drive varietal character, as summarised in Table 8D.1. These strains will therefore optimise the largely silent code locked away in the non-volatile grape precursor compounds, providing the consumer with a much wider choice of wine flavours and styles.

Acknowledgments We thank our former and present colleagues for many helpful discussions relating to this work, including Eveline Bartowsky, Jenny Bellon, Paul Chambers, Peter Costello, Adrian Coulter, Jeff Eglinton, Leigh Francis, Peter Godden, Mark Sefton, Jan Swiegers, Diego Torrea, Cristian Varela, Mar Vilanova, Markus Herderich, Isak Pretorius and Luigi Moio. Research at The Australian Wine Research Institute is supported by Australia's grapegrowers and winemakers through their investment agency the Grape and Wine Research and Development Corporation, with matching funds from the Australian Government.

References

Albers, E., Larsson, C., Lidén, G., Niklasson, C., & Gustafsson, L. (1996) Influence of the nitrogen source on *Saccharomyces cerevisiae* anaerobic growth and product formation. *Applied and Environmental Microbiology, 62*, 3187–3195.

Albers, E., Lidén, G., Larsson, C., & Gustafsson, L. (1998) Anaerobic redox balance and nitrogen metabolism in *Saccharomyces cerevisiae. Recent Research Developments in Microbiology, 2*, 253–279.

Albertyn, J., Hohmann, S., Thevelein, J. M., & Prior, B. A. (1994) *GPD1*, which encodes glycerol-3-phosphate dehydrogenase, is essential for growth under osmotic stress in *Saccharomyces cerevisiae*, and its expression is regulated by the high-osmolarity glycerol response pathway. *Molecular Cell Biology, 14*, 4135–4144.

Amerine, M. A., Berg, H. W., Kunkee, R. E., Ough, C. S., Singleton, V. L., & Webb, A. D. (1980) The technology of winemaking, 4th edition. Westport, Connecticut: AVI Publishing Company, Inc.

Anderson, R. G., & Kirsop, B. H. (1974) Control of volatile ester synthesis during fermentation of wort of high specific gravity. *Journal of the Institute of Brewing, 80*, 48–55.

Antonelli, A., Castellari, L., Zambonelli, C., & Carnacini, A. (1999) Yeast influence on volatile composition of wines. *Journal of Agricultural and Food Chemistry, 47*, 1139–1144.

Aranda, A., & del Olmo, M. (2004) Exposure to acetaldehyde in yeast determines an induction of sulfur amino acid metabolism and polyamine transporter genes, which depends on Met4p and Haa1p transcription factors, respectively. *Applied & Environmental Microbiology, 70*, 1913–1922.

Äyräpää, T. (1971) Biosynthetic formation of higher alcohols by yeast. Dependence on the nitrogen nutrient level of the medium. *Journal of the Institute of Brewing, 77*, 266–275.

Asenstorfer, R. E., Markides, A. J., Iland, P. G., & Jones, G. P. (2003) Formation of vitisin A during red wine fermentation and maturation. *Australian Journal of Grape and Wine Research, 9*, 40–46.

Bach, B., Camarasa, C., Rossignol, T., Blondin, B., & Dequin, S. (2004) γ-Aminobutyrate assimilation in *Saccharomyces cerevisiae* during wine fermentation. Proceedings of Physiology of Yeasts and Filamentous Fungi (PYFF2), Anglet, France. p. 185 (abstract).

Backhus, L. E., DeRisi, J., Brown, P. O., & Bisson, L. F. (2001) Functional genomic analysis of a commercial wine strain of *Saccharomyces cerevisiae* under differing nitrogen conditions. *FEMS Yeast Research, 1*, 111–125

Bakker, J., & Timberlake, C. F. (1997) Isolation, identification and characterization of new color-stable anthocyanins occurring in some red wines. *Journal of Agricultural and Food Chemistry, 45*, 35–43.

Bataillon, M., Rico, A., Sablayrolles, J.-M., Salmon, J.-M., & Barre, P. (1996) Early thiamin assimilation by yeasts under enological conditions: impact of alcoholic fermentation kinetics. *Journal of Fermentation and Bioengineering, 82*, 145–150.

Bardi, L., Crivelli, C., & Marzona, M. (1998) Esterase activity and release of ethyl esters of medium-chain fatty acids by *Saccharomyces cerevisiae* during anaerobic growth. *Canadian Journal of Microbiology, 44*, 1171–1176.

Bardi, L., Cocito, C., & Marzona, M. (1999) *Saccharomyces cerevisiae* cell fatty acid composition and release during fermentation without aeration and in absence of exogenous lipids. *International Journal of Food Microbiology, 47*, 133–140.

Bartowsky, E. J., Francis, I. L., Bellon, J. R., & Henschke, P. A. (2002) Is buttery aroma perception in wines predictable from diacetyl concentration? *Australian Journal of Grape and Wine Research, 8*, 180–185.

Bartowsky, E. J., Dillon, S. J., Henschke, P. A., Markides, A. J., Dumont, A., Pretorius, I. S., Ortiz-Julien, A., & Herderich, M. (2004) The potential of *Saccharomyces cerevisiae* wine yeast to improve red wine colour. *Australian & New Zealand Grapegrower & Winemaker, 490*, 83–85.

Becker, J. V., Armstrong, G. O., van der Merwe, M. J., Lambrechts, M. G., Vivier, M. A., & Pretorius, I. S. (2003) Metabolic engineering of Saccharomyces cerevisiae for the synthesis of the wine-related antioxidant resveratrol. *FEMS Yeast Research, 4*, 79–85.

Beh, A. L., Fleet, G. H., Prakitchaiwattana, C., & Heard, G. M. (2006) Evaluation of molecular methods for the analysis of yeasts in foods and beverages. *Advances In Experimental Medicine And Biology: Advances In Food Mycology, 571*, 69–106.

Bell, S.-J., & Henschke, P. A. (2005) Implications of nitrogen nutrition for grapes, fermentation and wine. *Australian Journal of Grape and Wine Research, 11*, 242–295.

Bellon, J., Rose, L., Currie, B., Ottawa, J., Bell, S., Mclean, H., Rayment, C., Treacher, C., & Henschke, P. (2008) Summary from the winemaking with non conventional yeasts workshops, 13th AWITC. *Australian & New Zealand Grapegrower & Winemaker, 528*, 72–77.

Beltran, G., Novo, M., Guillamon, J. M., Mas, A., & Rozes, N. (2008) Effect of fermentation temperature and culture media on the yeast lipid composition and wine volatile compounds. *International Journal of Food Microbiology, 121*, 169–177.

Belviso, S., Bardi, L., Bartolini, A. B., & Marzona, M. (2004) Lipid nutrition of *Saccharomyces cerevisiae* in winemaking. *Canadian Journal of Microbiology, 50*, 669–674.

Bely, M., Rinaldi, A., & Dubourdieu, D. (2003) Influence of assimilable nitrogen on volatile acidity production by *Saccharomyces cerevisiae* during high sugar fermentation. *Journal of Bioscience and Bioengineering, 96*, 507–512.

Bely, M., Stoeckle, P., Masneuf-Pomarede, I., & Dubourdieu, D. (2008) Impact of mixed Torulaspora delbrueckii-*Saccharomyces cerevisiae* culture on high-sugar fermentation. *International Journal of Food Microbiology, 122*, 312–320.

Benabdeljalil, C., Cheynier, V., Fulcrand, H., Hakiki, A., Mosdaddak, M., & Moutounet, M. (2000) Evidence of new pigments resulting from reactions between anthocyanins and yeast metabolites. *Science des Aliments, 20*, 203–220.

Bisson L. F., & Kunkee, R. E. (1993) Microbial interactions during wine production. In: J. G. Zeikus, & E. A. Johnson (Eds.), *Mixed Cultures in Biotechnology* (pp. 37–68). New York: McGraw-Hill.

Bohlscheid, J. C., Fellman, J. K., Wand, X. D., Ansen, D., & Edwards, C. G. (2007) The influence of nitrogen and biotin interactions on the performance of *Saccharomyces* in alcoholic fermentation. *Journal of Applied Microbiology, 102*, 390–400.

Borneman, A. R., Chambers, P. J., & Pretorius, I. S. (2007) Yeast systems biology: Modelling the winemaker's art. *TRENDS in Biotechnology, 25*, 349–355.

Boulton, R. B., Singleton, V. L., Bisson, L. F., & Kunkee, R. E. (1998) Principles and Practices of Winemaking. USA: Aspen Publishers, Inc.

Cabrera, M. J., Moreno, J., Ortega, J. M., Medina, M. (1988) Formation of ethanol, higher alcohols, esters, and terpenes by 5 yeast strains in musts from Pedro Ximenez grapes in various degrees of ripeness. *American Journal of Enology and Viticulture, 39*, 283–287.

Camarasa, C., Grivet, J., & Dequin, S. (2003) Investigation by ^{13}CNMR and tricarboxylic acid (TCA) deletion mutant analysis of pathways for succinate formation in *Saccharomyces cerevisiae* during anaerobic fermentation. *Microbiology, 149*, 2669–2678.

Campo, E., Cacho, J., & Ferreira, V. (2007) Solid phase extraction, multidimensional gas chromatography mass spectrometry determination of four novel aroma powerful ethyl esters. Assessment of their occurrence and importance in wine and other alcoholic beverages. *Journal of Chromatography A, 1140*, 180–188.

Caridi, A., Cufari, A., & Ramondino, D. (2002) Winemaking from Gaglioppo grapes with hybrid strains of *Saccharomyces*. *Folia Microbiologica, 47*, 407–408.

Caridi, A., Cufari, A., Lovino, R., Palumbo, R., & Tedesco, I. (2004) Influence of yeast on polyphenol composition of wine. *Food Technology and Biotechnology, 42*, 37–40.

Carrau, F. M., Medina, K., Boido, E., Farina, L., Gaggero, C., Dellacassa, E., Versini, G., & Henschke, P. A. (2005) De novo synthesis of monoterpenes by *Saccharomyces cerevisiae* wine yeasts. *FEMS Microbiology Letters, 243*, 107–115.

Carrau, F. M., Medina, K., Farina, L., Boido, E., Henschke, P. A., & Dellacassa, E. (2008) Production of fermentation aroma compounds by *Saccharomyces cerevisiae* wine yeasts: effects of yeast assimilable nitrogen on two model strains. *FEMS Yeast Research* DOI: 10.1111/j.1567-1364.2008.00412.x.

Castellari, L., Ferruzzi, M., Magrini, A., Giudici, P., Passarelli, P., & Zambonelli, C. (1994) Unbalanced wine fermentations by cryotolerant vs. non-cryotolerant *Saccharomyces cerevisiae* strains. *Vitis, 33*, 49–52.

Castino, M. (1982) Lieviti e polifenoli. *Revista Vitic. Enol. 34*, 333–348.

Chambers, P. J., Bellon, J. R., Schmidt, S. A., Varela, C., & Pretorius, I. S. (2008) Non-genetic engineering approaches for isolating and generating novel yeast for industrial applications. In T. Satyanarayana & G. Kunze (Eds.), *Yeasts Biotechnology: Diversity and Applications, Part II Genetic and Molecular Insights* (pp. 431–435). Springer.

Chambon, C., Ladeveze, V., Oulmouden, A., Servouse, M., & Karst, F. (1990) Isolation and properties of yeast mutants affected in farnesyl diphosphate synthetase. *Current Genetics, 18*, 41–46.

Charoenchai, C., Fleet, G. H., Henschke, P. A., & Todd, B. E. N. (1997) Screen of non-*Saccharomyces* wine yeasts for the presence of extracellular hydrolytic enzymes. *Australian Journal of Grape and Wine Research, 3*, 2–8.

Chatonnet, P., Dubourdieu, D., & Boidron, J. N. (1992a) Incidence des conditions de fermentation et d'elevage des vins blanc secs en barriques sur leur composition en substances cedees par le bois de chene. *Science des Aliments, 12*, 665–685.

Chatonnet, P., Dubourdieu, D., Boidron, J. N., & Pons, M. (1992b) The origin of ethylphenols in wines. *Journal of the Science of Food and Agriculture, 60*, 165–178.

Chatonnet, P., Dubourdieu, D., Boidron, J., & Lavigne, V. (1993) Synthesis of volatile phenols by *Saccharomyces cerevisiae* in wines. *Journal of the Science of Food and Agriculture, 62*, 191–202.

Chen, E. C.-H. (1978) The relative contribution of Ehrlich and biosynthetic pathways to the formation of fusel alcohols. *Journal of the American Society of Brewing Chemists, 36*, 39–43.

Cheraiti, N., Guezenec, S., & Salmon, J. M. (2005) Redox interactions between *Saccharomyces cerevisiae* and *Saccharomyces uvarum* in mixed culture under enological conditions. *Applied & Environmental Microbiology, 71*, 255–260.

Ciani, M., & Comitini, F. (2006) Influence of temperature and oxygen concentration on the fermentation behaviour of *Candida stellata* in mixed fermentation with *Saccharomyces cerevisiae*. *World Journal of Microbiology & Biotechnology, 22*, 619–623.

Ciani, M., & Maccarelli, F. (1998) Oenological properties of non-*Saccharomyces* yeasts associated with wine-making. *World Journal of Microbiology & Biotechnology, 14*, 199–203.

Ciani, M., Beco, L., & Comitini, F. (2006) Fermentation behaviour and metabolic interactions of multistarter wine yeast fermentations. *International Journal of Food Microbiology, 108*, 239–245.

Clausen, M., Lamb, C. J., Megnet, R., & Doerner, P. W. (1994) PAD1 encodes phenylacrylic acid decarboxylase which confers resistance to cinnamic acid in *Saccharomyces cerevisiae*. *Gene, 142*, 107–112.

Clemente-Jimenez, J. M., Mingorance-Cazorla, L., Martínez-Rodríguez, S., Las Heras-Vázquez, F. J., & Rodríguez-Vico, F. (2004) Molecular characterization and oenological properties of wine yeasts isolated during spontaneous fermentation of six varieties of grape must. *Food Microbiology, 21*, 149–155.

Cordonnier, R., & Bayonove, C. (1974) Mise en évidence dans la baie de raisin, variété Muscat d'Alexandrie, de monoterpènes liés révélables par une ou plusieurs enzymes du fruit. *Compte rendus Académie Science Paris, 278*, 3387–3390 Série D.

Coulter, A. D., Godden, P. W., & Pretorius, I. S. (2004) Succinic acid – How it is formed, what is its effect on titratable acidity, and what factors influence its concentration in wine? *Australian and New Zealand Wine Industry Journal, 19*, 16–20, 21–25.

Cozzolino, D., Flood, L., Bellon, J., Gishen, M., & De Barros Lopes, M. (2006) Combining near infrared spectroscopy and multivariate analysis as a tool to differentiate different strains of *Saccharomyces cerevisiae*: a metabolomic study. *Yeast, 23*, 1089–1096.

Crouzet, J. (1986) Les enzymes et l'arôme des vins. *Rev. Fran. Œnol., 102*, 42–49.

Darriet, P., Boidron, J.-N., & Dubourdieu, D. (1988) Hydrolysis of the terpenic heterosides of small muscat grapes using the periplasmic enzymes of *Saccharomyces cerevisiae. Connaissance de la Vigne et du Vin, 22*, 189–196.

Darriet, P., Tominaga, T., Lavigne, V., Boidron, J., & Dubourdieu, D. (1995) Identification of a powerful aromatic compound of *Vitis vinifera* L. var. Sauvignon wines: 4-Mercapto-4-methylpentan-2-one. *Flavour and Fragrance Journal, 10*, 385–392.

de Barros Lopes, M., Eglinton, J., Henschke, P., Høj, P., & Pretorius, I. (2003) The connection between yeast and alcohol reduction in wine: managing the double-edged sword of bottled sunshine. *Australian and New Zealand Wine Industry Journal, 18*, 17–18, 20, 22.

De Mora, S. J., Knowles, S. J., Eschenbruch, R., & Torrey, W. J. (1987) Dimethyl sulphide in some Australian red wines. *Vitis, 26*, 79–84.

Delfini, C., & Costa, A. (1993) Effects of the grape must lees and insoluble materials on the alcoholic fermentation rate and the production of acetic acid, pyruvic acid, and acetaldehyde. *American Journal of Enology and Viticulture, 44*, 86–92.

Delfini, C., Conterno, L., Giacosa, D., Cocito, C., Ravaglia, S., & Bardi, L. (1992) Influence of clarification and suspended solid contact on the oxygen demand and long-chain fatty acid contents of free run, macerated and pressed grape musts, in relation to acetic acid production. *Viticultural and Enological Sciences, 47*, 69–75.

Delteil, D., & Jarry, J. M. (1992) Characteristic effects of two commercial yeast strains on Chardonnay wine volatiles and polysaccharide composition. *Australian & New Zealand Wine Indsurty Journal, 7*, 29–33.

Duan, W., Roddick, F. A., Higgins, V. J., & Rogers, P. J. (2004) A parallel analysis of H_2S and SO_2 formation by brewing yeast in response to sulfur-containing amino acids and ammonium ions. *Journal of the American Society of Brewing Chemists, 62*, 35–41.

Dubois, P. (1994) Les arômes des vins et leurs défauts. *Revue Française d'Oenologie, 145*, 27–40.

Dubourdieu, D., Tominaga, T., Masneuf, I., Peyrot des Gachons, C., & Murat, M. L. (2006) The Role of Yeasts in Grape Flavor Development during Fermentation: The Example of Sauvignon blanc. *American Journal of Enology and Viticulture, 57*, 81–88

Dufour, J.-P. (1989) Effect of wort amino acids on beer quality. *Louvain Brewing Letters, 2*, 11–19.

Dumont, A., & Dulau, L. (1997) The role of yeasts in the formation of wine flavours. In M. Allen, P. Leske, & G. Baldwin (Eds.), *Advances in juice clarification and yeast inoculation*: proceedings of a seminar; 15 August 1996; Melbourne, Victoria (pp. 14–16). Adelaide S.A: Australian Society of Viticulture and Oenology.

Ebeler, S. E., & Spaulding, R. S. (1998) Characterization and measurement of aldehydes in wine. In A. L. Waterhouse & S. E. Ebeler (Eds.), *Chemistry of wine flavor* (pp. 166–179). Washington, DC: American Chemical Society.

Eden, A., Van Nedervelde, L., Drukker, M., Benvenisty, N., & Debourg, A. (2001) Involvement of branched-chain amino acid aminotransferases in the production of fusel alcohols during fermentation in yeast. *Applied Microbiology and Biotechnology, 55*, 296–300.

Edlin, D. A. N., Narbad, A., Dickinson, J. R., & Lloyd, D. (1995) The biotransformation of simple phenolic-compounds by *Brettanomyces anomalus*. *FEMS Microbiology Letters, 125*, 311–315.

Edwards, C., & Bohlscheid, J. C. (2007) Impact of pantothenic acid addition on H_2S production by *Saccharomyces cerevisiae* under fermentative conditions. *Enzyme & Microbial Technology, 41*, 1–4.

Edwards, C. G., Beelman, R. B., Bartley, C. E., & Mcconnell, A. L. (1990) Production of decanoic acid and other volatile compounds and the growth of yeast and malolactic bacteria during vinification. *American Journal of Enology and Viticulture, 41*, 48–56.

Egli, C. M., Edinger, W. D., Mitrakul, C. M., & Henick-Kling, T. (1998) Dynamics of indigenous and inoculated yeast populations and their effect on the sensory character of Riesling and Chardonnay wines. *Journal of Applied Microbiology, 85*, 779–789.

Eglinton, J. M., & Henschke, P. A. (1993) Can the addition of vitamins during fermentation be justified? *Australian Grapegrower & Winemaker, 352*, 47–49, 51–52.

Eglinton, J. M., & Henschke, P. A. (1999) The occurrence of volatile acidity in Australian wines. *Australian Grapegrower & Winemaker, 426a*, 7–12.

Eglinton, J. M., McWilliam, S. J., Fogarty, M. W., Francis, I. L., Kwiatkowski, M. J., Høj, P. B., & Henschke, P. A. (2000) The effect of *Saccharomyces bayanus*-mediated fermentation on the chemical composition and aroma profile of Chardonnay wine. *Australian Journal of Grape and Wine Research, 6*, 190–196.

Eglinton, J. M., Heinrich, A. J., Pollnitz, A. P., Langridge, P., Henschke, P. A., & de Barros Lopes, M. (2002) Decreasing acetic acid accumulation by a glycerol overproducing strain of *Saccharomyces cerevisiae* by deleting the *ALD6* aldehyde dehydrogenase gene. *Yeast, 19*, 295–301.

Eglinton, J. M., Francis, I. L., & Henschke, P. A. (2005) Selection and potential of Australian *Saccharomyces bayanus* yeast for increasing the diversity of red and white wine sensory properties. Yeast's contribution to the sensory profile of wine: maintaining typicity and biodiversity in the context of globalization: proceedings of Les XVIIes Entretiens Scientifiques Lallemand; 27–28 April 2005, La Rioja. (pp. 5–12) Blagnac Cedex, France: Lallemand.

Enomoto, K., Arikawa, Y., & Muratsubaki, H. (2002) Physiological role of soluble fumarate reductase in redox balancing during anaerobiosis in *Saccharomyces cerevisiae*. *FEMS Microbiology Letters, 215*, 103–108.

Eriksson, P., Andre, L., Ansell, R., Blomberg, A., & Adler, L. (1995) Cloning and characterization of GPD2, a 2nd gene encoding sn-glycerol 3-phosphate dehydrogenase (NAD(+)) in *Saccharomyces cerevisiae*, and its comparison with *GPD1*. *Molecular Microbiology, 17*, 95–107.

Erten, H., & Campbell, L. (2001) The production of low-alcohol wines by aerobic yeasts. *Journal of the Institute of Brewing, 107*, 207–215.

Eschenbruch, R. (1974) sulfite and sulfide formation during winemaking – A review *American Journal of Enology and Viticulture, 25*, 157–161.

Eschenbruch, R., & Bonish, P. (1976) Production of sulfite and sulfide by low- and high-sulfite forming wine yeasts. *Archives in Microbiology, 107*, 299–362.

Escot, S., Feuillat, M., Dulau, L., & Charpentier, C. (2001) Release of polysaccharides by yeasts and the influence of released polysaccharides on colour stability and wine astringency. *Australian Journal of Grape and Wine Research, 7*, 153–159.

Escudero, A., Campo, E., Fariña, L., Cacho, J., & Ferreira, V. (2007) Analytical characterization of the aroma of five premium red wines. Insights into the role of odor families and the concept of fruitiness of wines. *Journal of Agricultural and Food Chemistry, 55*, 4501–4510.

Estévez, P., Gil, M. L., & Falque, E. (2004) Effects of seven yeast strains on the volatile composition of Palomino wines. *International Journal of Food Science and Technology, 39*, 61–69.

Favale, S., Pietromarchi, P., & Ciolfi, G. (2007) Metabolic activity and interactions between two strains, *Saccharomyces cerevisiae* r.f. *bayanus* (SBC2) and *Saccharomyces cerevisiae* r.f.*uvarum* (S6u), in pure and mixed culture fermentations. *Vitis, 46*, 39–43.

Fedrizzi, B., Magno, F., Badocco, D., Nicolini, G., & Versini, G. (2007) Aging effects and grape variety dependence on the content of sulfur volatiles in wine. *Journal of Agricultural and Food Chemistry, 55*, 10880–10887.

Fernandez-Espinar, M. T., Lopez, V., Ramon, D., Bartra, E., & Querol, A. (2001) Study of the authenticity of commercial wine yeast strains by molecular techniques. *International Journal of Food Microbiology, 70*, 1–10.

Ferrarini, R., Buiani, M., & Bocca, E. (2008) Membrane processing across the vinification chain. In M. Allen, W. Cameron, R. Johnstone, & P. Pattison (Eds.), Proceedings of ASVO seminar – *Towards best practice through innovation in winery processing* (pp. 21–26). Adelaide, South Australia: Australian Society of Viticulture and Oenology.

Ferreira, V., Lopez, R., & Cacho, J. F. (2000) Quantitative determination of the odorants of young red wines from different grape varieties*Journal of the Science of Food and Agriculture, 80*, 1659–1667.

Ferreira, V., Jarauta, I., Ortega, L., & Cacho, J. (2004) Simple strategy for the optimization of solid-phase extraction procedures through the use of solid–liquid distribution coefficients: Application to the determination of aliphatic lactones in wine. *Journal of Chromatography A, 1025*, 147–156.

Ferreira, V., Escudero, A., Campo, E., & Cacho, J. (2008) The chemical foundations of wine aroma – a role game aiming at wine quality, personality and varietal expression. In R. J. Blair, P. J. Williams, & I. S. Pretorius (Eds.), Proceedings of the thirteenth Australian Wine Industry Technical Conference, Adelaide, South Australia, 28 July – 2 August 2007 (pp. 142–150) Adelaide, South Australia: Australian Wine Industry Technical Conference Inc.

Feuillat, M., Charpentier, C., & Massoutier, C. (1997) Intérêt œnological des souches de levures *Saccharomyces* cryotolérantes. *Revue des Œnologues, 85*, 18–21.

Fleet, G. H. (2003) Yeast interactions and wine flavour. *International Journal of Food Microbiology, 86*, 11–22.

Fleet, G. H., & Heard, G. M. (1993) Yeasts – Growth during fermentation. In G. H. Fleet (Ed.), *Wine microbiology and biotechnology* (pp. 27–54). Chur, Switzerland: Harwood Academic Publishers.

Fleet, G. H., Prakitchaiwattana, C., Beh, A. L., & Heard, G. (2002) The yeast ecology of wine grapes. In M. Ciani (Ed.), *Biodiversity and biotechnology of wine yeast* (pp. 1–17). Kerala, India: Research Signpost.

Flikweert, M. T., Vander Zanden, L., Janssen, W. M. T. M., Steensma, H. Y., van Dijken, J. P., & Pronk, J. T. (1996) Pyruvate decarboxylase: An indispensable enzyme for growth of *Saccharomyces cerevisiae* on glucose. *Yeast, 12*, 247–257.

Fowles, G. W. A. (1992) Acids in grapes and wine: A review. *Journal of Wine Research, 3*, 25–41.

Francis, I. L., & Newton, J. L. (2005) Determining wine aroma from compositional data. *Australian Journal of Grape and Wine Research, 11*, 114–126.

Francis, I. L., Kassara, S., Noble, A. C., & Williams, P. J. (1999) The contribution of glycoside precursors to Cabernet Sauvignon and Merlot aroma: sensory and compositional studies. In A. L. Waterhouse & S. E Ebeler (Eds.), *Chemistry of wine flavour* (pp. 13–30). Washington, DC: American Chemical Society.

Frivik, S. K., & Ebeler, S. E. (2003) Influence of sulfur dioxide on the formation of aldehydes in white wine. *American Journal of Enology and Viticulture, 54*, 31–38.

Fugelsang, K. C., & Zoecklein, B. W. (2003) Population dynamics and effects of *Brettanomyces bruxellensis* strains on Pinot noir (Vitis vinifera L.) wines. *American Journal of Enology and Viticulture, 54*, 294–300.

Fujii, T., Nagasawa, N., Iwamatsu, A., Bogaki, T., Tamai, W., & Hamachi, M. (1994) Molecular-cloning, sequence-analysis, and expression of the yeast alcohol acetyltransferase gene. *Applied and Environmental Microbiology, 60*, 2786–2792.

Fujii, T., Kobayashi, O., Yoshimoto, H., Furukawa, S., & Tamai, Y. (1997) Effect of aeration and unsaturated fatty acids on expression of the *Saccharomyces cerevisiae* alcohol acetyltransferase gene.*Applied and Environmental Microbiology, 63*, 910–915.

Fukuda, K., Yamamoto, N., Kiyokawa, Y., Yanagiuchi, T., Wakai, Y., Kitamoto, K., Inoue, Y., & Kimura, A. (1998) Balance of activities of alcohol acetyltransferase and esterase in *Saccharomyces cerevisiae* is important for production of isoamyl acetate. *Applied and Environmental Microbiology, 64*, 4076–4078.

Fulcrand, H., Benabdeljalil, C., Rigaud, J., Cheynier, V., & Moutounet, M. (1998) A new class of wine pigments generated by reaction between pyruvic acid and grape anthocyanins. *Phytochemistry, 47*, 1401–1407.

Garde-Cerdán, T., & Ancín-Azpilicueta, C. (2006) Contribution of wild yeasts to the formation of volatile compounds in inoculated wine fermentations. *European Food Research and Technology, 222*, 15–25.

Garde-Cerdán, T., & Ancín-Azpilicueta, C. (2008) Effect of the addition of different quantities of amino acids to nitrogen-deficient must on the formation of esters, alcohols, and acids during wine alcoholic fermentation. *LWT-Food Science and Technology, 41*, 501–510.

Gawel, R., Francis, L., & Waters, E. J. (2007a) Statistical correlations between the in-mouth textural characteristics and the chemical composition of Shiraz wines. *Journal of Agricultural and Food Chemistry, 55*, 2683–2687.

Gawel, R., Van Sluyter, S., & Waters, E. J. (2007b) The effects of ethanol and glycerol on the body and other sensory characteristics of Riesling wines. *Australian Journal of Grape and Wine Research, 13*, 38–45.

Gil, J. V., Mateo, J. J., Jimenez, M., Pastor, A., & Huerta, T. (1996) Aroma compounds in wine as influenced by apiculate yeasts. *Journal of Food Science, 61*, 1247–1249.

Gil, J. V., Manzanares, P., Genoves, S., Valles, S., & Gonzalez-Candelas, L. (2005) Overproduction of the major exoglucanase of *Saccharomyces cerevisiae* leads to an increase in the aroma of wine. *International Journal of Food Microbiology, 103*, 57–68.

Giudici, P., & Zambonelli, C. (1992) Biometric and genetic-study on acetic acid production for breeding of wine yeast. *American Journal of Enology and Viticulture, 43*, 370–374.

Giudici, P., Zambonelli, C., & Kunkee, R. E. (1993) Increased production of n-propanol in wine by yeast strains having an impaired ability to form hydrogen-sulfide.*American Journal of Enology and Viticulture, 44*, 17–21.

Giudici, P., Zambonelli, C., Passarelli, P., & Castellari, L. (1995) Improvement of wine composition with cryotolerant *Saccharomyces* strains. *American Journal of Enology and Viticulture, 46*, 143–147.

Godden, P., & Gishen, M. (2005) Trends in the composition of Australian wine. *Australian & New Zealand Wine Industry Journal, 20*(5), 21–46.

Goldfarb, A. (1994) Wild or natural, native yeasts have a role in modern winemaking. *Wine and Vines, 75*, 27–30.

González, S. S., Gallo, L., Climent, M. D., Barrio, E., & Querol, A. (2007) Enological characterization of natural hybrids from *Saccharomyces cerevisiae* and *S. kudriavzevii. International Journal of Food Microbiology, 116*, 11–18.

Gramatica, P., Manitto, P., Ranzi, B. M., Delbianco, A., & Francavilla, M. (1982) Stereospecific reduction of geraniol to (*R*)-(+)-citronellol by *Saccharomyces cerevisiae. Experientia, 38*, 775–776.

Grando, M. S., Versini, G., Nicolini, G., & Mattivi, F. (1993) Selective use of wine yeast strains having different volatile phenols production. *Vitis, 32*, 43–50.

Grbin, P. R., & Henschke, P. A. (2000) Mousy off-flavour production in grape juice and wine by *Dekkera* and *Brettanomyces* yeasts. *Australian Journal of Grape and Wine Research, 6*, 255–262.

Grosser, T. (2008) Membrane technology for moderation of alcohol content. In M. Allen, W. Cameron, R. Johnstone, & P. Pattison (Eds.), Proceedings of ASVO seminar – *Towards best practice through innovation in winery processing* (pp. 33–34). Adelaide, South Australia: Australian Society of Viticulture and Oenology.

Grossmann, M., Linsenmeyer, H., Muno, H., & Rapp, A. (1996) Use of oligo-strain yeast cultures to increase complexity of wine aroma. *Viticultural and Enological Science, 51*, 175–179.

Guitart, A., Hernández-Orte, P., Ferreira, V., Pena, C., & Cacho, J. (1999) Some observations about the correlation between the amino acid content of musts and wines of the Chardonnay variety and their fermentation aromas. *American Journal of Enology and Viticulture, 50*, 253–258.

Guth, H. (1997) Quantitation and sensory studies of character impact odorants of different white wine varieties. *Journal of Agricultural and Food Chemistry, 45*, 3027–3032.

Hallinan, C. P., Saul, D. J., & Jiranek, V. (1999) Differential utilisation of sulfur compounds for H_2S liberation by nitrogen-starved wine yeasts. *Australian Journal of Grape and Wine Research, 5*, 82–90.

Hayasaka, Y., Birse, M., Eglinton, J., & Herderich, M. (2007) The effect of *Saccharomyces cerevisiae* and *Saccharomyces bayanus* yeast on colour properties and pigment profiles of a Cabernet Sauvignon red wine. *Australian Journal of Grape and Wine Research, 13*, 176–185.

Hazelwood, L. A., Daran, J.-M., van Maris, A. J. A., Pronk, J. T., & Dickinson, J. R. (2008) The Ehrlich pathway 1 for fusel alcohol production: a century of research on yeast metabolism *Applied and Environmental Microbiology* doi:10.1128/AEM.02625-07.

Heard, G. (1999) Novel yeasts in winemaking – looking to the future. *Food Australia, 51*, 347–352.

Henschke, P. A. (1997) Wine yeast. In F. K. Zimmermann & K.-D. Entian (Eds.), *Yeast sugar metabolism: biochemistry, genetics, biotechnology, and applications* (pp. 527–560). Lancaster, PA: Technomic.

Henick-Kling, T., Edinger, W., Daniel, P., & Monk, P. (1998) Selective effects of sulfur dioxide and yeast starter culture addition on indigenous yeast populations and sensory characteristics of wine. *Journal of Applied Microbiology, 84*, 865–876.

Henschke, P. A. (2007) Yeast strains available for winemaking 2007/2008. *Technical Review (The Australian Wine Research Institute), 171*, 9–29.

Henschke, P. A., & Dixon, G. (1990) Effect of yeast strain on acetic acid accumulation during fermentation of Botrytis affected grape juice. In P. J. Williams, D. M. Davidson, T. H. Lee, (Eds.), *Proceedings of the seventh Australian wine industry technical conference*; 13-17 August 1989; Adelaide, SA. (pp. 242-244) Adelaide, SA: Australian Industrial Publishers.

Henschke, P. A., & Jiranek, V. (1991) Hydrogen sulfide formation during fermentation: Effect of nitrogen composition in model grape must. In J. Rantz (Ed.), *Proceedings of the International Symposium on Nitrogen in Grapes and Wine*, Seattle, USA (pp. 172–184). Davis, CA: American Society for Enology and Viticulture.

Henschke, P. A., & Jiranek, V. (1993) Yeasts – metabolism of nitrogen compounds. In G. H. Fleet (Ed.) *Wine microbiology and biotechnology* (pp. 77–164). Chur, Switzerland: Harwood Academic Publishers.

Henschke, P. A., & Rose, A. H. (1991) Plasma membranes. In A. H. Rose & J. S. Harrison (Eds.), *The Yeasts. Volume 4. Yeast organelles. 2nd ed.* (pp. 297–345). London: Academic Press.

Henschke, P. A., Eglinton, J. M., Costello, P. J., Francis, I. L., Gockowiak, H., Soden, A., & Høj, P. B. (2002) Winemaking with selected strains of non-*Saccharomyces cerevisiae* yeasts. Influence of *Candida stellata* and *Saccharomyces bayanus* on Chardonnay wine composition and flavour. In H. Trogus, J. Gafner, & A. Sütterlin (Eds.), Proceedings of the 13th International Oenology Symposium; 9–12 June 2002; Montpellier, France: (pp. 459–481). Kaiserstuhl, Breisach, Germany: International Association of Oenology, Management and Wine Marketing.

Hernández-Orte, P., Cacho, J. F., & Ferreira, V. (2002) Relationship between varietal amino acid profile of grapes and wine aromatic composition. Experiments with model solutions and chemometric study. *Journal of Agricultural and Food Chemistry, 50*, 2891–2899.

Hernández-Orte, P., Ibarz, M. J., Cacho, J., & Ferreira, V. (2005) Effect of the addition of ammonium and amino acids to musts of Airen variety on aromatic composition and sensory properties of the obtained wine. *Food Chemistry, 89*, 163–174.

Hernández-Orte, P., Bely, M., Cacho, J., & Ferreira, V. (2006) Impact of ammonium additions on volatile acidity, ethanol, and aromatic compound production by different *Saccharomyces cerevisiae* strains during fermentation in controlled synthetic media. *Australian Journal of Grape and Wine Research, 12*, 150–160.

Hernández-Orte, P., Cersosimo, M., Loscos, N., Cacho, J., Garcia-Moruno, E., & Ferreira, V. (2008) The development of varietal aroma from non-floral grapes by yeasts of different genera. *Food Chemistry, 107*, 1064–1077.

Herraiz, T., Reglero, G., Herraiz, M., Martinalvarez, P. J., & Cabezudo, M. D. (1990) The influence of the yeast and type of culture on the volatile composition of wines fermented without sulfur-dioxide. *American Journal of Enology and Viticulture, 41*, 313–318.

Holgate, A. (1997) The influence of yeast strain on the acid balance of red wine. In: M. Allen, P. Leske, & G. Bladwin (Eds.), *Proceedings Advances in juice clarification and yeast inoculation*, Melbourne, Australia. (pp. 39–41). Adelaide, SA: Australian Society of Viticulture and Oenology.

Holloway, P., Vantwest, R. A., Subden, R. E., & Lachance, M. A. (1992) A strain of *Candida stellata* of special interest to oenologists. *Food Research International, 25*, 147–149.

Houtman, A. C., & Du Plessis, C. S. (1981) The effect of juice clarity and several conditions promoting yeast growth on fermentation rate, the production of aroma components and wine quality. *South African Journal of Enology and Viticulture, 2*, 71–81.

Houtman, A. C., & Du Plessis, C. S. (1986) The effect of grape cultivar and yeast strain on fermentation rate and concentration of volatile components in wine. *South African Journal of Enology and Viticulture, 7*, 14–20.

Houtman, A. C., Morais, J., & Du Plessis, C. S. (1980) Factors affecting the reproducibility of fermentation of grape juice and of the aroma composition of wines. I Grape maturity, sugar, inoculation concentration, aeration, juice turbidity and ergosterol. *Vitis, 19*, 37–54.

Howell, K. S., Bartowsky, E. J., Fleet, G. H., & Henschke, P. A. (2004a) Microsatellite PCR profiling of *Saccharomyces cerevisiae* strains during wine fermentation. *Letters in Applied Microbiology, 38*, 315–320.

Howell, K. S., Swiegers, J. H., Elsey, G. M., Siebert, T. E., Bartowsky, E. J., Fleet, G. H., Pretorius, I. S., & de Barros Lopes, M. A. (2004b) Variation in 4-mercapto-4-methyl-pentan-2-one release by *Saccharomyces cerevisiae* commercial wine strains. *FEMS Microbiology Letters, 240*, 125–129.

Howell, K. S., Klein, M., Swiegers, J. H., Hayasaka, Y., Elsey, G. M., Fleet, G. H., Høj, P. B., Pretorius, I. S., & de Barros Lopes, M. A. (2005) Genetic determinants of volatile-thiol release by *Saccharomyces cerevisiae* during wine fermentation. *Applied and Environmental Microbiology, 71*, 5420–5426.

Howell, K. S., Cozzolino, D., Bartowsky, E. J., Fleet, G. H., & Henschke, P. A. (2006) Metabolic profiling as a tool for revealing *Saccharomyces* interactions during wine fermentation. *FEMS Yeast Research, 6*, 91–101.

Jackson, R.S. (2000) Wine Science. Principles and Applications (London, UK: Academic Press).

Jane, T. M., Gawel, R., & Henschke, P. A., (1996) Yeast strain can affect the flavour of Chardonnay wine. In C. S. Stockley, A. N. Sas, R. S. Johnstone, & T. H. Lee (Eds.), *Proceedings of the ninth Australian wine industry technical conference* (p. 223). Adelaide SA: Winetitles.

Jemec, K. P., & Raspor, P. (2005) Initial *Saccharomyces cerevisiae* concentration in single or composite cultures dictates bioprocess kinetics. *Food Microbiology, 22*, 293–300.

Jiranek, V., Langridge, P., & Henschke, P. A. (1995a) Regulation of hydrogen sulfide liberation in wine-producing *Saccharomyces cerevisiae* strains by assimilable nitrogen. *Applied and Environmental Microbiology, 61*, 461–467.

Jiranek, V., Langridge, P., & Henschke, P. A. (1995b) Validation of bismuth-containing indicator media for predicting H_2S producing potential of *Saccharomyces cerevisiae* wine yeasts under enological conditions. *American Journal of Enology and Viticulture, 46*, 269–273.

Jiranek, V., Langridge, P., & Henschke, P. A. (1996) Determination of sulphite reductase activity and its response to assimilable nitrogen status in a commercial *Saccharomyces cerevisiae* wine yeast. *Journal of Applied Bacteriology, 81*, 329–336.

Jolly, N. P., Augustyn, O. P. H., & Pretorius, I. S. (2003) The effect of non-*Saccharomyces* yeasts on fermentation and wine quality. *South African Journal of Enology and Viticulture, 24*, 55–62.

Jolly, N., Augustyn, O., & Pretorius I. S. (2006) The role and use of non-*Saccharomyces* yeasts in wine production. *South African Journal of Enology and Viticulture, 27*, 15–39.

Kapsopoulou, K., Mourtzini, A., Anthoulas, M., & Nerantzis, E. (2007) Biological acidification during grape must fermentation using mixed cultures of *Kluyveromyces thermotolerans* and *Saccharomyces cerevisiae*. *World Journal of Microbiology & Biotechnology, 23*, 735–739.

Kaur, J., & Bachhawat, A. K. (2007) Yct1p, a novel, high-affinity, cysteine-specific transporter from the yeast *Saccharomyces cerevisiae*. *Genetics, 176*, 877–890.

Kennedy, J. A., Saucier, C., & Glories, Y. (2006) Grape and wine phenolics: History and perspective. *American Journal of Enology and Viticulture, 57*, 239–248.

Killian, E., & Ough, C. S. (1979) Fermentation esters – formation and retention as affected by fermentation temperature. *American Journal of Enology and Viticulture, 30*, 301–305.

Kim, D.-H., Hong, Y.-A., & Park, H.-D. (2008) Co-fermentation of grape must by *Issatchenkia orientalis* and *Saccharomyces cerevisiae* reduces the malic acid content in wine. Biotechnology Letters, DOI 10.1007/s10529-008-9726-1.

Klingshirn, L. M., Liu, J. R., & Gallander, J. F. (1987) Higher alcohol formation in wines as related to the particle size profiles of juice insoluble solids. *American Journal of Enology and Viticulture, 38*, 207–210.

Kunkee, R. E., & Vilas, M. R. (1994) Toward an understanding of the relationship between yeast strain and flavour production during vinifications: flavor effects in vinifications of a nondistinct variety of grapes by several strains of wine yeast. *Wein-Wissenschaft, 49*, 46–50.

Kurtzman, C. P., & Fell, J. W. (1998) The yeasts: A taxonomic study. Amsterdam: Elservier.

Lafon-Lafourcade, S., Lucmaret, V., Joyeux, A., & Ribéreau-Gayon, P. (1981) Utilisation de levains mixtes dans l'élaboration des vins de pourriture noble, en vue de rduire l'acidité volatile. *Compte Rendu de l'Acadmie d'Agriculture de France, 67*, 616–622.

Lambrechts, M. G., & Pretorius, I. S. (2000) Yeast and its importance to wine aroma – A review. *South African Journal of Enology and Viticulture, 21*, 97–129.

Lavigne, V., Pons, A., Darriet, P., & Dubourdieu, D. (2008) Changes in the sotolon content of dry white wines during barrel and bottle aging. *Journal of Agricultural and Food Chemistry, 56*, 2688–2693.

Le Jeune, C., Lollier, M., Demuyter, C., Erny, C., Legras, J. L., Aigle, M., & Masneuf-Pomarede, I. (2007) Characterization of natural hybrids of *Saccharomyces cerevisiae* and *Saccharomyces bayanus* var. *uvarum*. *FEMS Yeast Research, 7*, 540–549.

Lilly, M., Lambrechts, M. G., & Pretorius, I. S. (2000) Effect of increased yeast alcohol acetyltransferase activity on flavor profiles of wine and distillates. *Applied and Environmental Microbiology, 66*, 744–753.

Lilly, M., Bauer, F. F., Lambrechts, M. G., Swiegers, J. H., Cozzolino, D., & Pretorius, I. S. (2006a) The effect of increased yeast alcohol acetyltransferase and esterase activity on the flavour profiles of wine and distillates. *Yeast, 23*, 641–659.

Lilly, M., Bauer, F. F., Styger, G., Lambrechts, M. G., & Pretorius, I. S. (2006b) The effect of increased branched-chain amino acid transaminase activity in yeast on the production of higher alcohols and on the flavour profiles of wine and distillates. *FEMS Yeast Research, 6*, 726–743.

Linderholm, A. L., Olineka, T. L., Hong, Y., & Bisson, L. F. (2006) Allele diversity among genes of the sulfate reduction pathway in wne srains of *Saccharomyces cerevisiae*. *American Journal of Enology and Viticulture, 57*, 431–440

Linderholm, A. L., Findleton, C. L., Kumar, G., Hong, Y., & Bisson, L. F. (2008) Identification of genes affecting hydrogen sulfide formation in *Saccharomyces cerevisiae*. *Applied Environmental Microbiology, 74*, 1418–1427.

Lopandic, K., Gangl, H., Wallner, E., Tscheik, G., Leitner, G., Querol, A., Borth, N., Breitenbach, M., Prillenger, H., Tiefenbrunner, W. (2007) Genetically different wine yeasts isolated from Austrian vine-growing regions influence wine aroma differently and contain putative hybrids between *Saccharomyces cerevisiae* and *Saccharomyces kudriavzevii*. *FEMS Yeast Research, 7*, 953–965.

Loscos, N., Hernández-Orte, P., Cacho, J., & Ferreira, V. (2007) Release and formation of varietal aroma compounds during alcoholic fermentation from nonfloral grape odorless flavor precursors fractions. *Journal of Agricultural and Food Chemistry, 55*, 6674–6684.

Lurton, L., Snakkers, G., Roulland, C., Galy, B., & Versavaud, A. (1995) Influence of the fermentation yeast-strain on the composition of wine spirits. *Journal of the Science of Food and Agriculture, 67*, 485–491.

Luparia, V., Soubeyrand, V., Berges, T., Julien, A., & Salmon, J. M. (2004) Assimilation of grape phytosterols by *Saccharomyces cerevisiae* and their impact on enological fermentations. *Applied Microbiology and Biotechnology, 65*, 25–32.

Malcorps, P., Cheval, J. M., Jamil, S., & Dufour, J.-P. (1991) A new model for the regulation of ester synthesis by alcohol acetyltransferases in *Saccharomyces cerevisiae. Journal of the American Society of Brewing Chemists, 49*, 47–53.

Marullo, P., Yvert, G., Bely, M., Aigle, M., & Dubourdieu, D. (2007) Efficient use of DNA molecular markers to construct industrial yeast strains. *FEMS Yeast Research, 7*, 1295–1306.

Manzanares, P., Rojas, V., Genoves, S., & Valles, S. (2000) A preliminary search for anthocyanin-beta-D-glucosidase activity in non-*Saccharomyces* wine yeasts. *International Journal of Food Science and Technology, 35*, 95–103.

Marks, V. D., van der Merwe, G. K., & van Vuuren, H. J. J. (2003) Transcriptional profiling of wine yeast in fermenting grape juice: Regulatory effect of diammonium phosphate. *FEMS Yeast Research, 3*, 269–287.

Marsal, F., Sarre, Ch., Dubourdieu, D., & Boidron, J.-N. (1988) Role de la levure dans la transformation de certains constituants volatils du bois de chene au cours de l'elaboration en barrique des vins blancs secs. *Connaissance de la Vigne et du Vin, 22*, 33–38.

Martineau, B., Acree, T. E., & Henick-Kling, T. (1995) Effect of wine type on the detection threshold for diacetyl. *Food Research International, 28*, 139–143.

Masneuf, I., Murat, M. L., Naumov, G. I., Tominaga, T., & Dubourdieu, D. (2002) Hybrids *Saccharomyces cerevisiae* x*Saccharomyces bayanus* var. *uvarum* having a high liberating ability of some sulfur varietal aromas of *Vitis vinifera* Sauvignon blanc wines. *Journal International Des Sciences de la Vigne et du Vin, 36*, 205–212.

Mason, A. B., & Dufour, J. P. (2000) Alcohol acetyltransferases and the significance of ester synthesis in yeast. *Yeast, 16*, 1287–1298.

Massoutier, C., Alexandre, H., Feuillat, M., & Charpentier, C. (1998) Isolation and characterization of cryotolerant *Saccharomyces* strains. *Vitis, 37*, 55–59.

Mateo, J. J., Jimenez, M., Huerta, T., & Pastor, A. (1991) Contribution of different yeasts isolated from musts of monastrell grapes to the aroma of wine. *International Journal of Food Microbiology, 14*, 153–160.

Mateos, J. A. R., Perez-Nevado, F., & Fernandez, M. R. (2006) Influence of *Saccharomyces cerevisiae* yeast strain on the major volatile compounds of wine *Enzyme and Microbial Technology, 40*, 151–157.

Mauricio, J. C., Bravo, M., Moreno, J., Medina, M., & Ortega, J. M. (1995) Influence of different fermentation conditions on the production of medium chain fatty acids (C6, C8, C10) and their respectiveethyl esters by *Saccharomyces cerevisiae. Acta Horticulturae, 388*, 209–214.

Meaden, P. G., Dickinson, F. M., Mifsud, A., Tessier, W., Westwater, J., Bussey, H., & Midgley, M. (1997) The *ALD6* gene of *Saccharomyces cerevisiae* encodes a cytosolic, Mg^{2+}-activated acetaldehyde dehydrogenase. *Yeast, 13*, 1319–1327.

Medina, K., Boido, E., Dellacassa, E., & Carrau, F. (2005) Yeast interactions with anthocyanins during red wine fermentation. *American Journal of Enology and Viticulture, 56*, 104–109.

Mehdi, K., & Penninckx, M. J. (1997) An important role for glutathione and gamma-glutamyltranspeptidase in the supply of growth requirements during nitrogen starvation of the yeast *Saccharomyces cerevisiae. Microbiology-UK, 143*, 1885–1889.

Mestres, M., Busto, O., & Guasch, J. (2000) Analysis of organic sulfur compounds in wine aroma. *Journal of Chromatography A, 881*, 569–581.

Miller, A. C., Wolff, S. R., Bisson, L. F., & Ebeler, S. E. (2007) Yeast strain and nitrogen supplementation: Dynamics of volatile ester production in Chardonnay juice fermentations. *American Journal of Enology and Viticulture, 58*, 470–483.

Moio, L., Ugliano, M., Genovese, A., Gambuti, A., Pessina, R., & Piombino, P. (2004) Effect of antioxidant protection of must on volatile compounds and aroma shelf life of Falanghina (*Vitis vinifera* L.) wine. *Journal of Agricultural and Food Chemistry, 52*, 891–897.

Molina, A. M., Swiegers, J. H., Varela, C., Pretorius, I. S., & Agosin, E. (2007) Influence of wine fermentation temperature on the synthesis of yeast-derived volatile aroma compounds. *Applied Microbiology and Biotechnology, 77*, 675–687.

Monk, P.R. (1986) Formation, utilization, and excretion of hydrogen sulphide by wine yeast. *Australian & New Zealand Wine Industry Journal, 1*, 11–16.

Monk, P. R., & Cowley, P. J. (1984) Effect of nicotinic acid and sugar concentration of grape juice and temperature on accumulation of acetic acid during yeast fermentation. *Journal of Fermentation Technology, 62*, 515–521.

Morata. A., Gomez-Cordoves, M. C., Suberviola, J., Bartolome, B., Colomo, B., & Suarez, J. A. (2003) Adsorption of anthocyanins by yeast cell walls during the fermentation of red wines. *Journal of Agricultural and Food Chemistry, 51*, 4084–4088.

Morata, A., Gomez-Cordoves, M. C., Calderon, F., & Suarez, J. A. (2006) Effects of pH, temperature and SO_2 on the formation of pyranoanthocyanins during red wine fermentation with two species of *Saccharomyces*. *International Journal of Food Microbiology, 106*, 123–129.

Moreira, N., Mendes, F., Pereira, O., Guedes de Pinho, P., Hogg, T., & Vasconcelos, I. (2002) Volatile sulphur compounds in wines related to yeast metabolism and nitrogen composition of grape musts. *Analytica Chimica Acta, 458*, 157–167.

Moreno, J. J., Millan, C., Ortega, J. M., & Medina, M. (1991) Analytical differentiation of wine fermentations using pure and mixed yeast cultures. *Journal of Industrial Microbiology, 7*, 181–189.

Müller-Thurgau, L. (1896) Uber den Ursprung der Weinhefe und Hieran sich knuepfende praktische Folgerungen. *Weinbau Weinhandel, 7*, 40–41.

Murat, M. L., Masneuf, I., Darriet, P., Lavigne, V., Tominaga, T., & Dubourdieu, D. (2001) Effect of *Saccharomyces cerevisiae* yeast strains on the liberation of volatile thiols in Sauvignon blanc wine. *American Journal of Enology and Viticulture, 52*, 136–139.

Muratore, G., Asmundo, C. N., Lanza, C. M., Caggia, C., Licciardello, F., & Restuccia, C. (2007) Influence of *Saccharomyces uvarum* on volatile acidity, aromatic and sensory profile of Malvasia delle Lipari wine. *Food Technology and Biotechnology, 45*, 101–106.

Naumov, G. I., Masneuf, I., Naumova, E. S., Aigle, M., & Dubourdieu, D. (2000) Association of *Saccharomyces bayanus* var. *uvarum* with some French wines: genetic analysis of yeast populations. *Research in Microbiology, 151*, 683–691.

Navarro-Avino, J. P., Prasad, R., Miralles, V. J., Benito, R. M., & Serrano, R. (1999) A proposal for nomenclature of aldehyde dehydrogenases in *Saccharomyces cerevisiae* and characterization of the stress-inducible *ALD2* and *ALD3* genes. *Yeast, 15*, 829–842.

Nguyen, H. V., & Gaillardin, C. (2005) Evolutionary relationships between the former species *Saccharomyces uvarum* and the hybrids *Saccharomyces bayanus* and *Saccharomyces pastorianus*; reinstatement of *Saccharomyces uvarum* (Beijerinck) as a distinct species. *FEMS Yeast Research, 5*, 471–483.

Nieuwoudt, H. H. (2004) Glycerol and wine. (Doctoral dissertation, University of Stellenbosch, 2004) p. 115.

Nieuwoudt, H. H., Prior, B. A., Pretorius, I. S., & Bauer, F. F. (2002) Glycerol in South African table wines: An assessment of its relationship to wine quality. *South African Journal of Enology and Viticulture, 23*, 22–30.

Noble, A. C., & Bursick, G. F. (1984) The contribution of glycerol to perceived viscosity and sweetness in white wine. *American Journal of Enology and Viticulture, 35*, 110–112.

Nordström, K. (1964) Formation of ethyl acetate in fermentation with brewer's yeast. V. Effect of some vitamins and mineral nutrients. *Journal of the Institute of Brewing, 70*, 209–221.

Orlic, S., Redzepovic, S., Jeromel, A., Herjavec, S., & Iacumin, L. (2007) Influence of indigenous *Saccharomyces paradoxus* strains on Chardonnay wine fermentation aroma.*International Journal of Food Science and Technology, 42*, 95–101.

Oshita, K., Kubota, M., Uchida, M., & Ono, M. (1995) Clarification of the relationship between fusel alcohol formation and amino acid assimilation by brewing yeast using ^{13}C-labelled amino acid. *Proceedings european brewing convention. Brussels* (pp. 387–394). Oxford, UK: Oxford University Press.

Palacios, A., Raginel, F., & Ortiz-Julien, A. (2007) Can the selection of *Saccharomyces cerevisiae* yeast lead to variations in the final alcohol degree of wines? *The Australian & New Zealand Grapegrower & Winemaker, (Dec.), 71*, 73–75.

Paraggio, M., & Fiore, C. (2004) Screening of *Saccharomyces cerevisiae* wine strains for the production of acetic acid. *World Journal of Microbiology & Biotechnology, 20*, 743–747.

Park, S. K., Boulton, R. B., Bartra, E., & Noble, A. C. (1994) Incidence of volatile sulfur compounds in California wines. A preliminary survey. *American Journal of Enology and Viticulture, 45*, 341–344.

Park, S. K., Boulton, R. B., & Noble, A. C. (2000) Formation of hydrogen sulfide and glutathione during fermentation of white grape musts. *American Journal of Enology and Viticulture, 51*, 91–97.

Perpète, P., Duthoit, O., De Maeyer, S., Imray, L., Lawton, A. I., Stavropoulos, K. E., Gitonga, V. W., Hewlins, M. J. E., & Dickinson, J. R. (2006) Methionine catabolism in *Saccharomyces cerevisiae*. *FEMS Yeast Research, 6*, 48–56.

Pigeau, G. M., & Inglis, D. L. (2005) Upregulation of *ALD3* and *GPD1* in *Saccharomyces cerevisiae* during Icewine fermentation. *Journal of Applied Microbiology, 99*, 112–125.

Pigeau, G. M., & Inglis, D. L. (2007) Response of wine yeast (*Saccharomyces cerevisiae*) aldehyde dehydrogenases to acetaldehyde stress during Icewine fermentation. *Journal of Applied Microbiology, 103*, 1576–1586.

Piombino, P., Pessina, R., Genovese, A., Le Quere, J.-L. & Moio, L. (2004) Gli odori di frutti di bosco dell'aroma del vino. Parte II: analisi strumentale. *L'Enologo, 50*, 97–101.

Plata, C., Millán, C., Mauricio, J. C., & Ortega, J. M. (2003) Formation of ethyl acetate and isoamyl acetate by various species of wine yeasts *Food Microbiology, 20*, 217–224.

Plata, C., Mauricio, J.C., Millan, C., & Ortega, J. M. (2005) Influence of glucose and oxygen on the production of ethyl acetate and isoamyl acetate by a *Saccharomyces cerevisiae* strain during alcoholic fermentation. *World Journal of Microbiology & Biotechnology, 21*, 115–121.

Pretorius, I. S. (2000) Tailoring wine yeast for the new millennium: novel approaches to the ancient art of winemaking. *Yeast, 16*, 675–729.

Pretorius, I. S., Bartowsky, E. J., de Barros Lopes, M., Bauer, F. F., du Toit, M., van Rensburg, P., & Vivier, M. A. (2006) The tailoring of designer grapevines and microbial starter strains for a market-directed and quality-focused wine industry. In H. Y. Hui, E. Castell-Perez, L. M. Cunha, C. I. Guerrero-Legarreta, H. H. Liang, Y. M. Lo, D. L. Marshall, W. K. Nip, F. Shahidi, F. Sherkat, R. J. Winger, K. L.Yam, (Eds.), *Handbook of Food Science, Technology and Engineering; Volume 4. Food Technology and Food Processing*. (pp. 174-1–174-24) New York, NY: CRC Press.

Pretorius, I. S., & Bauer, F. F. (2002) Meeting the consumer challenge through genetically customized wine-yeast strains. *Trends in Biotechnology, 20*, 426–432.

Radler, F. (1993) Yeasts – metabolism of organic acids. In G. H. Fleet (Ed.), *Wine Microbiology and Biotechnology* (pp. 165–182). Singapore: Harwood Academic Publishers.

Radler, F., & Schütz, H. (1982) Glycerol production of various strains of *Saccharomyces*. *American Journal of Enology and Viticulture, 33*, 36–40.

Rainieri, S., Zambonelli, C., Tini, V., Castellari, L., & Giudici, P. (1998) The enological traits of thermotolerant *Saccharomyces* strains. *American Journal of Enology and Viticulture, 49*, 319–324.

Ramey, D. (1996) Low input winemaking—let nature do the work. In C. S. Stockley, A. N., Sas, R. S. Johnstone, & T. H. Lee, (Eds.) *Proceedings of the Ninth Australian Wine Industry Technical Conference*; 16–19 July 1995; Adelaide, South Australia (pp. 26–29). Adelaide, South Australia: Winetitles.

Rankine, B. C. (1953) Quantitative differences in products of fermentation by different strains of wine yeasts. *Australian Journal of Applied Science, 4*, 590–602.

Rankine, B. C. (1963) Nature, origin and prevention of hydrogen sulphide aroma in wines. *Journal of the Science of Food and Agriculture, 14*, 79–91.

Rankine, B. C. (1966) Decomposition of L-malic acid by wine yeasts. *Journal of the Science of Food and Agriculture, 17*, 312–316.

Rankine, B. C. (1968) The importance of yeasts in determining the composition and quality of wines. *Vitis, 7*, 22–49.

Rankine, B., & Bridson, D. (1971) Glycerol in Australian wines and factors influencing its formation. *American Journal of Enology and Viticulture, 23*, 6–12.

Rankine, B. C. (1977) Modern developments in selection and use of pure yeast cultures for winemaking. *Australian Wine, Brewing and Spirit Review 96(8)*, 31–33 and *96(9)*, 32–34.

Rapp, A., & Marais, J. (1993) The shelf life of wine: changes in aroma substances during storage and aging of white wines. In G. Charalambous (Ed.), *The Shelf Life of Foods and Beverages. Chemical, Biological, Physical and Nutritional Aspects* (pp. 891–921). Amsterdam, The Netherlands: Elsevier Science Publisher.

Rapp, A., & Versini, G. (1996) Influence of nitrogen on compounds in grapes on aroma compounds in wines.*Journal International des Sciences de la Vigne et du Vin, 51*, 193–203.

Rauhut, D. (1993) Yeasts – Production of sulfur compounds. In G. H. Fleet (Ed.), *Wine microbiology and biotechnology* (pp. 183–223). Chur, Switzerland: Harwood Academic Publishers.

Rauhut, D., & Kürbel, H. (1994) The production of H_2S from elemental sulfur residues during fermentation and its influence on the formation of sulfur metabolites causing off-flavours in wine. *Viticultural and Enological Science, 49*, 27–36.

Rauhut, D., Kürbel, H., Dittrich, H. H, & Grossmann, M. (1996) Properties and differences of commercial yeast strains with respect to their formation of sulfur compounds. *Viticultural and Enological Science, 51*, 187–192.

Rauhut, D., Kürbel, H., MacNamara, K., & Grossmann, M. (1998) Headspace GC-SCD monitoring of low volatile sulfur compounds during fermentation and in wine. *Analusis, 26*, 142–145.

Redzepovic, S., Orlic, S., Sikora, S., Majdak, A., & Pretorius, I. S. (2002) Identification and characterization of *Saccharomyces cerevisiae* and *Saccharomyces paradoxus* strains isolated from Croatian vineyards. *Letters in Applied Microbiology, 35*, 305–310.

Redzepovic, S., Orlic, S., Majdak, A., Kozina, B., Volschenk, H., Viljoen-Bloom, M. (2003) Differential malic acid degradation by selected strains of *Saccharomyces* during alcoholic fermentation. *International Journal of Food Microbiology, 83*, 49–61.

Regenberg, B., During-Olsen, L., Kielland-Brandt, M. C., & Holmberg, S. (1999) Substrate specificity and gene expression of the amino-acid permeases in *Saccharomyces cerevisiae. Current Genetics, 36*, 317–328.

Remize, F., Roustan, J. L., Sablayrolles, J. M., Barre, P., & Dequin, S. (1999) Glycerol overproduction by engineered *Saccharomyces cerevisiae* wine yeast strains leads to substantial changes in byproduct formation and to a stimulation of fermentation rate in stationary phase. *Journal of Applied and Environmental Microbiology, 65*, 143–149.

Remize, F., Andrieu, E., & Dequin, S. (2000) Engineering of the pyruvate dehydrogenase bypass in *Saccharomyces cerevisiae*: Role of the cytosolic Mg^{2+} and mitochondrial K^{+} acetaldehyde dehydrogenases Ald6p and Ald4p in acetate formation during alcoholic fermentation. *Applied and Environmental Microbiology, 66*, 3151–3159.

Reynolds, A. G., Edwards, C. G., Cliff, M. A., Thorngate, J. H., & Marr, J. C. (2001) Evaluation of yeast strains during fermentation of Riesling and Chenin blanc musts. *American Journal of Enology and Viticulture, 52*, 336–344.

Reynolds, A. G., Schlosser, J., Power, R., Roberts, R., Willwerth, J., & de Savigny, C. (2007) Magnitude and interaction of viticultural and enological effects. I. Impact of canopy management and yeast strain on sensory and chemical composition of Chardonnay Musque. *American Journal of Enology and Viticulture, 58*, 12–24.

Ribéreau-Gayon, P., Dubourdieu, D., Donéche, B., & Lonvaud, A. (2000a) Handbook of Enology, Volume 1. *The microbiology of wines and vinification.* (p. 497) Chichester, UK: John Wiley & Sons Ltd.

Ribéreau-Gayon, P., Glories, Y., Maujean, A., & Dubourdieu, D. (2000b) Handbook of Enology, Volume 2. *The chemistry of wine stabilisation and treatments.* (p. 441) Chichester, UK: John Wiley & Sons Ltd.

Rojas, V., Gil, J. V., Pinaga, F., & Manzanares, P. (2001) Studies on acetate ester production by non-*Saccharomyces* wine yeasts. *International Journal of Food Microbiology, 70*, 283–289.

Rojas, V., Gil, J. V., Pinaga, F., & Manzanares, P. (2003) Acetate ester formation in wine by mixed cultures in laboratory fermentations. *International Journal of Food Microbiology, 86*, 181–188.

Romano, P., & Suzzi, G. (1996) origin and production of acetoin during wine yeast fermentation. *Applied and Environmental Microbiology, 62*, 309–315.

Romano, P., Suzzi, G., Turbanti, L., & Polsinelli, M. (1994) Acetaldehyde production in *Saccharomyces cerevisiae* wine yeasts. *FEMS Microbiology Letters, 118*, 213–218.

Romano, P., Caruso, M., Capece, A., Lipani, G., Paraggio, M., & Fiore, C. (2003a) Metabolic diversity of *Saccharomyces cerevisiae* strains from spontaneously fermented grape musts. *World Journal of Microbiology & Biotechnology, 19*, 311–315.

Romano P, Fiore C, Paraggio M, Caruso M, & Capece A (2003b) Function of yeast species and strains in wine flavour. *International Journal of Food Microbiology, 86*, 169–180.

Rosenfeld, E., Beauvoit, B., Blondin, B., & Salmon, J. M. (2003) Oxygen consumption by anaerobic *Saccharomyces cerevisiae* under enological conditions: Effect on fermentation kinetics. *Applied and Environmental Microbiology, 69*, 113–121.

Ross, J. P. (1997) Going wild: wild yeast in winemaking. *Wine and Vines, Sept.*, 16–21.

Rossi, I., & Bertuccioli, M. (1992) Influences of lipid additions on fatty acid composition of *Saccharomyces cerevisiae* and aroma characteristics of experimental wines. *Journal of the Institute of Brewing, 98*, 305–314.

Saerens, S. M. G., Verstrepen, K. J., Van Laere, S. D. M., Voet, A. R. D., Van Dijck, P., Delvaux, F. R., Thevelein, J. M. (2006) The *Saccharomyces cerevisiae EHT1* and *EEB1* genes encode novel enzymes with medium-chain fatty acid ethyl ester synthesis and hydrolysis capacity. *Journal of Biological Chemistry, 281*, 4446–4456.

Saerens, S. M. G., Delvaux, F., Verstrepen, K. J., Van Dijck, P., Thevelein, J. M., Delvaux, F. R. (2008) Parameters affecting ethyl ester production by *Saccharomyces cerevisiae* during fermentation. *Applied and Environmental Microbiology, 74*, 454–461.

Saint-Prix, F., Bonquist, L., & Dequin, S. (2004) Functional analysis of the *ALD* gene family of *Saccharomyces cerevisiae* during anaerobic growth on glucose: The $NADP^{(+)}$-dependent Ald6p and Ald5p isoforms play a major role in acetate formation. *Microbiology, 150*, 2209–2220.

Salmon, J. M. (1987) L-malic acid permeation in resting cells of anaerobically grown *Saccharomyces cerevisiae*. *Biochimica et Biophysica ACTA, 901*, 30–34.

Salmon, J. M., Vezinhet, F., & Barre, P. (1987) Anabolic role of L-malic acid in *Saccharomyces cerevisiae* in anaerobiosis during alcoholic fermentation. *FEMS Microbiology Letters, 42*, 213–220.

Schneider, R., Charrier, F., Razungles, A., & Baumes, R.(2006) Evidence for an alternative biogenetic pathway leading to 3-mercaptohexanol and 4-mercapto-4-methylpentan-2-onein wines. *Analytica Chimica Acta, 563*, 58–64.

Schreier, P. (1979) Flavour composition of wines. *CRC Critical Reviews in Food Science and Nutrition, 12*, 59–111.

Sefton, M. A. (1998) Hydrolytically-released volatile secondary metabolites from a juice sample of *Vitis vinifera* grape cvs Merlot and Cabernet Sauvignon. *Australian Journal of Grape and Wine Research, 4*, 30–38.

Segurel, M. A., Razungles, A. J., Riou, C., Salles, M., & Baumes, R. L. (2004) Contribution of dimethyl sulfide to the aroma of Syrah and Grenache Noir wines and estimation of its potential in grapes of these varieties. *Journal of Agricultural and Food Chemistry, 52*, 7084–7093.

Shimazu, Y., & Watanabe, M. (1981) Effects of yeast strains and environmental conditions on formation of organic acids in must fermentations. *Journal of Fermentation Technology, 59*, 27–32.

Silva Ferreira, A. C., Rodrigues, P., Hogg, T., & De Pinho, P. G. (2003) Influence of some technological parameters on the formation of dimethyl sulfide, 2-mercaptoethanol, methionol, and dimethyl sulfone in port wines. *Journal of Agricultural and Food Chemistry, 51*, 727–732.

Smyth, H., Cozzolino, D., Herderich, M. J., Sefton, M. A., & Francis, I. L. (2005) Relating volatile composition to wine aroma: Identification of key aroma compounds in Australian white wines. In R. J Blair, P. J. Williams, & I. S. Pretorius (Eds.), *Proceedings of the Twelfth Australian Wine Industry Technical Conference, Melbourne, Australia* (pp. 31–33). Australian Wine Industry Technical Conference Inc.: Adelaide, SA.

Soden, A., Francis, I. L., Gockowiak, H., Lee, T. H., & Henschke, P. A. (1999) The use of non-*Saccharomyces* yeasts in winemaking. In R. J. Blair, A. N. Sas, P. F. Hayes, P. F. & P. B. Høj (Eds.), *Proceedings of the tenth Australian wine industry technical conference*; 2–5 August 1998 (pp. 166–171). Adelaide, SA: Australian Wine Industry Technical Conference Inc.

Soden, A., Francis, I. L., Oakey, H., & Henschke, P. A. (2000) Effects of co-fermentation with *Candida stellata* and *Saccharomyces cerevisiae* on the aroma and composition of Chardonnay wine. *Australian Journal of Grape and Wine Research, 6*, 21–30.

Soles, R. M., Ough, C. S., Kunkee, R. E. (1982) Ester concentration differences in wine fermented by various species and strains of yeasts. *American Journal of Enology and Viticulture, 33*, 94–98.

Sowalsky, R. A., & Noble, A. C. (1998) Comparison of the effects of concentration, pH and anion species on astringency and sourness of organic acids. *Chemical Senses, 23*, 343–349.

Spiropoulos, A., Tanaka, J., Flerianos, I., & Bisson, L. F. (2000) Characterization of hydrogen sulfide formation in commercial and natural wine isolates of *Saccharomyces*. *American Journal of Enology and Viticulture, 51*, 233–248.

Stanley, G. A., Douglas, N. G., Every, E. J., Tzanatos, T., & Pamment, N. B. (1993) inhibition and stimulation of yeast growth by acetaldehyde. *Biotechnology Letters, 15*, 1199–1204.

Stratford, M., & Rose, A. H. (1985a) Transport of sulfur dioxide by *Saccharomyces cerevisiae*. *Journal of General Microbiology, 132*, 1–6.

Stratford, M., & Rose, A. H. (1985b) Hydrogen sulfide production from sulfite by *Saccharomyces cerevisiae*. *Journal of General Microbiology, 131*, 1417–1424.

Strauss, C. R., Wilson, B., Gooley, P.R., & Williams, P. J. (1986) Role of monoterpenes in grape and wine flavor. ACS Symposium Series, *317*, 222–242.

Suárez, R., Suárez-Lepe, J. A., Morata, A., & Calderon, F. (2007) The production of ethylphenols in wine by yeasts of the genera *Brettanomyces* and *Dekkera*: A review. *Food Chemistry, 102*, 10–21.

Subileau, M., Schneider, R., Salmon, J.-M., & Degryse, E. (2008) Nitrogen catabolite repression modulates the production of aromatic thiols characteristic of Sauvignon Blanc at the level of precursor transport. *FEMS Yeast Research, 8*, 771–780.

Suomalainen, H. (1981) Yeast esterases and aroma esters in alcoholic beverages. *Journal of the Institute of Brewing, 87*, 296–300.

Sutherland, C. M., Henschke, P. A., Langridge, P., & de Barros Lopes, M. (2003) Subunit and cofactor binding of *Saccharomyces cerevisiae* sulfite reductase—towards developing wine yeast with lowered ability to produce hydrogen sulfide. *Australian Journal of Grape & Wine Research, 9*, 186–193.

Swiegers, J. H., & Pretorius, I. S. (2007) Modulation of volatile sulfur compounds by wine yeast. *Applied Microbiology & Biotechnology, 74*, 954–960.

Swiegers, J. H., Bartowsky, E. J., Henschke, P. A., & Pretorius, I. S. (2005) Yeast and bacterial modulation of wine aroma and flavour. *Australian Journal of Grape and Wine Research, 11*, 139–173.

Swiegers, J. H., Willmott, R., Hill-Ling, A., Capone, D. L., Pardon, K. H., Elsey, G. M., Howell, K. S., de Barros Lopes, M. A., Sefton, M. A., Lilly, M., & Pretorius, I. S. (2006) Modulation of volatile thiol and ester aromas by modified wine yeast. Flavour Science: recent advances and trends. In W. L. P. Bredie & M. A. Petersen (Eds.), *Proceedings of the 11th Weurman Flavour*

Research Symposium; 21–24 June, 2005, Roskilde, Denmark. Amsterdam, The Netherlands: Elsevier, pp. 113–116.

Swiegers, J. H., Capone, D. L., Pardon, K. H., Elsey, G. M., Sefton, M. A., Francis, I. L., & Pretorius, I. S. (2007) Engineering volatile thiol release in *Saccharomyces cerevisiae* for improved wine aroma. *Yeast, 24*, 561–574.

Swiegers J. H., Cordente, A. G., Willmott, R. L., King, E. S., Capone, D. L., Francis, I. L., &. Pretorius, I. S. (2008a) Development of flavour-enhancing wine yeast. In R. J. Blai, P. J. Williams, I. S. Pretoris (Eds.), Proceedings of the 13th Australian Wine Industry Technical Conference, 2007, (pp. 184–188). Adelaide, SA: Australian Wine Industry Technical Conference Inc.

Swiegers J. H., Ugliano, M., van der Westhuizen, T., & Bowyer, P. (2008b) Impact of yeast rehydration on the aroma of Sauvignon Blanc wine. *Australian & New Zealand Grapegrower & Winemaker, 528*, 68–71.

Swiegers, J. H., Kievit, R. L., Siebert, T., Lattey, K. A., Bramley, B. R., Francis, I. L., King, E. S., & Pretorius, I. S. (2008c) The influence of yeast strain on the aromatic character of Sauvignon Blanc wine. *Food Microbiology* (in press).

Thibon, C., Marullo, P., Claisse, O., Cullin, C., Dubourdieu, D., & Tominaga, T. (2008) Nitrogen catabolic repression controls the release of volatile thiols by *Saccharomyces cerevisiae* during wine fermentation. *FEMS Yeast Research* DOI: 10.1111/j.1567-1364.2008.00381.x

Thomas, C. S., Boulton, R. B., Silacci, M. W., & Gubler, W. D. (1993) The effect of elemental sulfur, yeast strain, and fermentation medium on hydrogen sulfide production during fermentation. *American Journal of Enology and Viticulture, 44*, 211–216.

Thomas, D., & Surdin-Kerjan, Y. (1997) Metabolism of sulfur amino acids in *Saccharomyces cerevisiae*. *Microbiology and Molecular Biology Reviews, 61*, 503–532.

Thornton, R. J., & Rodriguez, S. B. (1996) Deacidification of red and white wines by a mutant of *Schizosaccharomyces malidevorans* under commercial winemaking conditions. *Food Microbiology, 13*, 475–482.

Treloar, J.D. and Howell, G.S. (2006) Influence of yeast and malolactic bacteria strain choice on 3-isobutyl-2-methoxypyrazine concentration in Cabernet Sauvignon and Franc wines. Abstracts of Wine Making Workshop, Vegetable & Farm Market EXPO, Great Lakes Fruit, Grand Rapids, MI, USA: <http://www.glexpo.com/abstracts/2006abstracts/WineMaking2006.pdf>.

Timberlake, C. F., & Bridle, P. (1976) Interactions between anthocyanins, phenolic compounds, and acetaldehyde and their significance in red wines. *American Journal of Enology and Viticulture, 27*, 97–105.

Tokuyama, T., Kuraishi, H., Aida, K., & Uemura, T. (1973) Hydrogen sulfide evolution due to a pantothenic acid deficiency in the yeast requiring this vitamin, with special reference to the effect of adenosine triphosphate on yeast cysteine desulfhyrase. *Journal of General and Applied Microbiology, 19*, 439–466.

Tominaga, T., Masneuf, I., & Dubourdieu, D. (1995) A S-cysteine conjugate precursor of aroma of white sauvignon. *Journal International des Sciences de la Vigne et du Vin, 29*, 227–232.

Tominaga, T., Furrer, A., Henry, R., & Dubourdieu, D. (1998) Identification of new volatile thiols in the aroma of *Vitis vinifera* L. var. Sauvignon blanc wines. *Flavour & Fragrance Journal, 29*, 227–232.

Tominaga, T., Blanchard, L., Darriet, P., & Dubourdieu, D. (2000) A powerful aromatic volatile thiol, 2-furanmethanethiol, exhibiting roast coffee aroma in wines made from several Vitis vinifera grape varieties. *Journal of Agricultural and Food Chemistry, 48*, 1799–1802.

Torija, M. J., Beltran, G., Novo, M., Poblet, M., Guillamon, J. M., Mas, A., & Rozes, N. (2003) Effects of fermentation temperature and *Saccharomyces* species on the cell fatty acid composition and presence of volatile compounds in wine. *International Journal of Food Microbiology, 85*, 127–136.

Torrea, D., Fraile, P., Garde, T., & Ancin, C. (2003) Production of volatile compounds in the fermentation of chardonnay musts inoculated with two strains of *Saccharomyces cerevisiae* with different nitrogen demands. *Food Control, 14*, 565–571.

Trioli, G. (1996) Effect of Fermaid addition to white grape juice on the behaviour of several commercial yeast strains. *Wein–Wissenschaft, 51*, 204–209.

Ubeda-Iranzo, J. F., González-Magaña, F., & González-Viñas, M. A. (2000) Evaluation of the formation of volatiles and sensory characteristics in the industrial production of white wines using different commercial strains of the genus *Saccharomyces*. *Food Control, 1*, 143–147.

Ugliano, M., Bartowsky, E. J., McCarthy, J., Moio, L., & Henschke, P. A. (2006) Hydrolysis and transformation of grape glycosidically bound volatile compounds during fermentation with three *Saccharomyces* yeast strains*Journal of Agricultural and Food Chemistry, 54*, 6322–6331.

Ugliano, M. & Moio, L. (2008). Free and hydrolytically released volatile compounds of *Vitis vinifera* L. cv. Fiano grapes as odour-active constituents of Fiano wine. *Analytica Chimica ACTA, 621*, 79–85.

van der Merwe, C. A., & van Wyk, C. J. (1981) The contribution of some fermentation products to the odor of dry white wines. *American Journal of Enology and Viticulture, 32*, 41–46.

Varela, C., Siebert, T., Cozzolino, D., Rose, L., Mclean, H., & Henschke, P. A. (2008) Discovering a chemical basis for differentiating wines made by fermentation with 'wild' and inoculated yeasts: role of yeast volatile compounds; submitted.

Vaudano, E., Moruno, E. G., & Di Stefano, R. (2004) Modulation of geraniol metabolism during alcohol fermentation. *Journal of the Institute of Brewing, 110*, 213–219.

Verduyn, C., Postma, E., Scheffers, W. A., & van Dijken, J. P. (1990a) Physiology of *Saccharomyces cerevisiae* in anaerobic glucose-limited chemostat cultures. *Journal of General Microbiology, 136*, 395–403.

Verduyn, C., Postma, E., Scheffers, W. A., & van Dijken, J. P. (1990b) Energetics of *Saccharomyces cerevisiae* in anaerobic glucose-limited chemostat cultures. *Journal of General Microbiology, 136*, 405–412.

Verstrepen, K., Derdelinckx, G., Dufour, J.-P., Winderickx, J., Thevelein, J. M., Pretorius, I. S., & Delvaux, F. R. (2003a) Flavour active esters: Adding fruitiness to beer – A practical review. *Journal of Bioscience and Bioengineering, 96*, 110–118.

Verstrepen, K. J., Van Laere, S. D. M., Vanderhaegen, B. M. P., Derdelinckx, G., Dufour, J. P., Pretorius, I. S., Winderickx, J., Thevelein, J. M., & Delvaux, F. R. (2003b) Expression levels of the yeast alcohol acetyltransferase genes ATF1, Lg-ATF1, and ATF2 control the formation of a broad range of volatile esters. *Applied and Environmental Microbiology, 69*, 5228–5237.

Vila, I., Sablayrolles, J.-M., Gerland, C., Baumes, R., Bayonove, C., & Barre, P. (2000) Comparison of 'aromatic' and 'neutral' yeast strains: influence of vinification conditions. *Viticultural and Enological Science, 55*, 59–66.

Vilanova, M., Ugliano, M., Varela, C., Siebert, T., Pretorius, I. S., & Henschke, P. A. (2007) Assimilable nitrogen utilisation and production of volatile and non-volatile compounds in chemically defined medium by *Saccharomyces cerevisiae* wine yeasts. *Applied Microbiology and Biotechnology, 77*, 145–157.

Vos, P. J. A., & Gray, R. S. (1979) The origin and control of hydrogen sulphide during fermentation of grape must. *American Journal of Enology and Viticulture, 30*, 187–197.

Wakil, S. J., Stoops, J. K., & Joshi, V. (1983) Fatty acid synthesis and its regulation. *Annual Review of Biochemistry, 52*, 537–579.

Watson, K. G. (1987) Temperature relations. In A. H. Rose & J. S. Harrison (Eds.), *The Yeasts. Volume 2. Yeast and environment. 2nd ed.* (pp. 41–72) London: Academic Press.

Wenzel, K. (1989) The selection of a yeast mutant to reduce color losses during red wine fermentation. *Vitis, 28*, 111–120.

Williams, P. J., Strauss, C. R., Wilson, B., & Massy Westropp, R. A. (1982) Studies on the hydrolysis of *Vitis vinifera* monoterpene precursor compounds and model monoterpene β-D-glucosides rationalizing the monoterpene composition of grapes. *Journal of Agricultural and Food Chemistry, 30*, 1219–1223.

Wood, C., Siebert, T. E., Parker, M., Capone, D. L., Elsey, G. M., Pollnitz, A. P., Eggers, M., Meier, M., Vössing, T., Widder, S., Krammer, G., Sefton, M. A. & Herderich, M. J. (2008) From wine to pepper: rotundone, an obscure sesquiterpene, is a potent spicy aroma compound. *Journal of Agricultural and Food Chemistry, dx.doi.org/10.1021/jf800183k.*

Wurz, R. E. M., Kepner, R. E., & Webb, A. D. (1988) The biosynthesis of certain gamma-lactones from glutamic acid by film yeast activity on the surface of Flor Sherry. *American Journal of Enology and Viticulture, 39*, 234–238.

Yoshimoto, H., Fujiwara, D., Momma, T., Ito, C., Sone, H., Kaneko, Y., & Tamai, Y. (1998) Characterization of the ATF1 and Lg-ATF1 genes encoding alcohol acetyltransferases in the bottom fermenting yeast *Saccharomyces pastorianus*. *Journal of Fermentation and Bioengineering, 86*, 15–20.

Yunoki, K., Hirose, S., & Ohnishi, M. (2007) Ethyl esterification of long-chain unsaturated fatty acids derived from grape must by yeast during alcoholic fermentation. *Bioscience Biotechnology and Biochemistry, 71*, 3105–3109.

Zambonelli, C., Mutinelli, P., & Pacchetti, G. (1975) Biosynthesis of sulphur amino acids in *Saccharomyces cerevisiae*. I. Genetic analysis of leaky mutants of sulphite reductase. *Archives of Microbiology, 102*, 247–251.

Zeeman, W., Snyman, J. P., & van Wyk, C. J. (1982) The influence of yeast strain and malolactic fermentation on some volatile bouquet substances and on the quality of table wines. In A. D. Webb (Ed.), *Proceedings, University of California, davis Grape and Wine Symposium* (pp. 79–90). Davis, CA: University of California Press.

Zironi, R., Romano, P., Suzzi, G., Battistutta, F., & Comi, G. (1993) Volatile metabolites produced in wine by mixed and sequential cultures of *Hanseniaspora guilliermondii* or *Kloeckera apiculata* and *Saccharomyces cerevisiae. Biotechnology Letters, 15*, 235–238.

Zohre, D. E., & Erten, H. (2002) The influence of *Kloeckera apiculata* and *Candida pulcherrima* yeasts on wine fermentation. Process Biochemistry, *38*, 319–324.

Chapter 8E
Identification of Impact Odorants of Wines

Vicente Ferreira and Juan Cacho

Contents

8E.1 A General Framework for Wine Aroma Analysis

The aroma of wine is one of the most complex existing in nature for a number of reasons. The first is because there are many different wines showing quite different aromas, the second because the aromas of even a single wine change with time, while it is stored in the bottle and while it is waiting in the glass to be consumed. Finally, because in most cases wines do not have a simple characteristic aroma; rather they have a palette of subtle aromas which are very difficult to define and which surely are perceived differently by the different people.

Such sensorial complexity is, of course, caused by the chemical complexity of wine aroma as is detailed below.

1. To begin with, and in spite of the title of the chapter, most wines do not have genuine impact compounds; rather they contain a relatively large number of active-odorants that contribute to a larger or smaller extent to the different aroma nuances of wine. A genuine aroma impact compound is a molecule that is able to transmit entirely its sensory descriptors to a product, to the point that

V. Ferreira (✉)
Laboratory for Flavor Analysis and Enology, Department of Analytical Chemistry, Faculty of Sciences, University of Zaragoza. 50009 Zaragoza, Spain
e-mail: vferre@unizar.es

M.V. Moreno-Arribas, M.C. Polo (eds.), *Wine Chemistry and Biochemistry*,
DOI 10.1007/978-0-387-74118-5_15, © Springer Science+Business Media, LLC 2009

the odor of the product is closely related to the odor of such chemical (Belitz and Grosch 1988). In wine this only happens in some particular cases, such as Muscat. For convenience, however, we will retain the denomination of impact molecule to refer to a molecule which is a neat contributor to an aroma nuance. This fact has practical consequences for the researcher, since in many cases it is not possible to establish a clear and univocal link between a sensory descriptor and a single aroma molecule. Such a link will have to be established a posteriori in sensory studies working with the potentially relevant aroma molecules in model solutions or in wine.

2. A second question related to the previous one is the relatively large numbers of potential impact compounds in wine. The simultaneous presence of many different odor chemicals causes that the final perception will be the result of a complex brain processing in which some odors are integrated into a single perception; some act in a competitive or even destructive way (Atanasova et al., 2004), while others interact to form a new and different perception. A particular case is the presence in wine of whole sets of aroma chemicals members of a chemical homologous series and displaying similar odors. In this case the practical consequences are that the role of some chemicals should be considered as a part of acombination and, of course, that the final role can only be assessed via different sensory experiments.
3. The third question derives from the large diversity in physico-chemical properties of the different wine impact compounds. Some of them are quite hydrophobic and are released very easily from the wine matrix, while some others are quite hydrophilic and will be released with difficulty (Ferreira et al. 2006). The former are major constituents of the headspaces on a glass of wine, while the latter will reach the pituitary only when the level of liquid in the glass is very small or when the wine is swallowed. These differences in behavior, still not completely understood, may explain why different odors are perceived when the glass of wine is full or when it is nearly empty (Petka et al. 2006) and why sometimes there are marked differences between ortho and retronasal perceptions.
4. Finally, some of the most powerful wine aroma compounds take part in reversible interactions which evolve during wine aging and that can be reversed, at least in part, when the wine takes contact with air. These aspects have not yet been studied in depth, but it is well known that carbonyls form reversible associations with sulfur dioxide and that mercaptans take part in complex redox equilibria. These molecules involved in interactions or in redox equilibria are the most likely cause of the aromatic changes noted during the aging of wine or after the bottle is opened.

This aroma chemical complexity together with some relevant aspects of the wine as analytical and sensory matrix strongly constrains the way in which the study of wine impact molecules must be carried out. Some of the most important points are the following:

1. The first concerns the general composition of wine. Wine contains thousands of different molecules, and a number approaching one thousand is formed by

volatile molecules. Many of these volatile molecules have some aroma, but only a limited number of volatiles can be found in wines at concentrations high enough to be perceived. Obviously, these compounds are the ones which will have a major influence on the aroma characteristics of wine and, therefore, are the ones which will have to be analyzed and considered further in sensory studies. This is something that all researchers in wine chemistry should remember: there are so many volatiles that it is absolutely essential to screen the real odor-active molecules from all the volatiles present.

2. The second point is related to the simultaneous presence of odorants at g/L levels and of others that can be active at levels as low as ng/L. This means that although it makes sense to use a general screening procedure for detecting by olfactometry the potentially most relevant aroma molecules, it will not be possible to use a single isolation or preconcentration scheme to identify and further quantify the different aroma molecules. Rather, it will be necessary to have an array of chemical isolation and quantification procedures if a comprehensive aroma analysis is our objective.
3. The third and fourth influential factors on wine aroma chemistry arise from the problems caused by the matrix on both the isolation and preconcentration of molecules and on the sensory assessment of the role played by the impact odorants. The presence of major volatiles, such as ethanol and fusel alcohols, complicates the isolation of the other wine odorants, particularly of those present at low levels. This has a marked influence on the way in which the extracts for screening and for further qualitative or quantitative studies should be prepared. On the other hand, ethanol and the major fermentation volatiles have also a deep influence on the way in which the odor chemicals are released and perceived.

In conclusion, the study of the wine aroma chemicals and the understanding of the role they play in the different wine aroma nuances have to be structured into a numbers of steps strongly constrained by the previous considerations. Such steps will be the subject of thischapter. The first step is about the screening of aroma molecules, which will be carried out by using gas chromatography-olfactometry. The second will be the isolation and identification of odorants. The third is the quantitative determination, for which only a very brief outline will be given, and the fourth is about the sensory tools used to assess the sensory role played by the different odorants.

8E.2 Screening for Aroma Impact Molecules

Since all aroma molecules are more or less volatile, the technique that "a priori" is best suited to screen the odor active molecules from the rest of molecules is Gas Chromatography-Olfactometry (GC-O). This technique makes use of the human nose as detector for the compounds eluting out of the chromatographic column, typically a fused silica capillary column (Acree et al. 1984). There are several different approaches for GC-O differing in the way in which the olfactometric signals are recorded and processed, but the results most often rely most on the sample treatment used rather than on the type of signals recovered. Both aspects will be briefly discussed. A summary of the reports on the comprehensive study of the GC-O profiles is given in Table 8E.1.

Table 8E.1 A summary of the papers published in the last 10 years on the general GC-O profiling of wine

Reference	Type of wine/sample	Type of extract	GC-O strategy
Guth et al. (1997)	Gewürztraminer and Schereube	Successive extraction with ether	AEDA
Ferreira et al. (1998b)	Grenache	Continuous L-L extraction with Freon 11	AEDA
López et al. (1999)	Merlot, Cabernet Sauvignon, Grenache	Continuous L-L extraction with Freon 11	AEDA
Ong and Acree (1999)	Gewürztraminer	Sequential Freon 11 and ethyl acetate extraction	Charm
Kotseridis and Baumes (2000)	Cabernet Sauvignon and Merlot wines, juices and yeast extracts	Simple extraction with DCM XAD-2 extraction (musts)	AEDA
Escudero et al. (2000)	Oxidized white wines	Continuous L-L extraction with Freon 11	AEDA
Aznar et al. (2001)	Aged red Rioja wines	Solid Phase Extraction on XAD-4 resins	AEDA
Serot et al. (2001)	French and Romanian musts	Dichloromethane	NIF, OSME, AEDA
Ferreira et al. (2001a)	Spanish aged red wines	Solid Phase Extraction on XAD-4 resins	Quantitative AEDA
Ferreira et al. (2002a)	Grenache rosé wines	Solid Phase Extraction on XAD-4 resins	AEDA
Lopez et al. (2003b)	Canary white wines	Solid Phase Extraction on LiChrolut-EN resins	Posterior intensity
Lee and Noble (2003)	Chardonnay wines	Continuous L-L extraction with Freon 11	Frequency of detection
Marti et al. (2003)	Grenache red from Priorat	Head space SPME	AEDA
Ferreira ACS et al. (2003)	Port wine	Dichloromethane extraction	AEDA
López et al. (2004)	Hydrolyzed Grape Flavor precursors from Grenache and Tempranillo	Simple extraction with DCM XAD-2 extraction (musts)	AEDA
Cullere et al. (2004b)	Premium Spanish aged reds	Solid Phase Extraction on LiChrolut-EN resins	Posterior intensity

Table 8E.1 (continued)

Reference	Type of wine/sample	Type of extract	GC-O strategy
Buettner (2004)	Chardonnay	In-mouth Stir Bar Sorptive Extraction	Unspecific
Escudero et al. (2004)	Maccabeo white	Solid Phase Extraction on XAD-4 resins	AEDA
Fretz et al. (2005)	Petite Arvine	Dichloromethane extraction	OSME
Campo et al. (2005)	Spanish white monovarietals	Dynamic headspace with trapping on LiChrolut-EN	Posterior intensity
Fang and Qian (2005)	Pinot Noir (Oregon)	Extraction with pentane-diethyl ether	AEDA
Guarrera et al. (2005)	Sicilian Passito wines		OSME
Komes et al. (2006)	Croatian Riesling	Continuous L-L extraction with Freon 11	NIF
Gurbuz et al. (2006)	Merlot and Cabernet Sauvignon reds	Head space SPME	OSME
Campo et al. (2006b)	Madeira wines	Dynamic headspace with trapping on LiChrolut-EN	Posterior intensity
Bailly et al. (2006)	Sauternes wines	XAD-2 resins	AEDA
Petka et al. (2006)	Devin white wine	Dynamic headspace with trapping on LiChrolut-EN	Posterior intensity
Sarrazin et al. (2007)	Botrytized wines (Sauternes)	Dichloromethane extraction	AEDA
Gomez Miguez et al. (2007)	Zalema white wine	Dynamic headspace with trapping on LiChrolut-EN	Posterior intensity
Escudero et al. (2007)	Premium Spanish aged reds	Dynamic headspace with trapping on LiChrolut-EN	Posterior intensity

8E.2.1 Signal Recording and Processing in GC-Olfactometry

Generally speaking, the main purpose of the GC-O research is to list and rank the aroma compounds present in the foodstuff according to their potential importance in the food flavor. The ways in which these lists are built differ among the different GC-O techniques, which can be classified into three broad categories:

1. Based on determination of threshold concentration: Aroma Extract Dilution Analysis (AEDA) (Schieberle and Grosch 1987; Ullrich and Grosch 1987), and Charm analysis (Acree et al. 1984)
2. Based on the measurement of the frequency of citations (Ruth et al. 1995; Pollien et al. 1997)

3. Based on the assessment of intensity: OSME (McDaniel et al. 1990; Miranda-Lopez et al. 1992), cross-modality matching (Etiévant et al. 1999), and posterior intensity evaluation (Ferreira V et al. 2003b; van Ruth 2004)

8E.2.1.1 Techniques Based on Threshold

EDA is one of the most frequently used methods for the screening of flavor impact compounds, probably because of its simplicity. The theoretical background of the technique showing practical hints to obtain a reproducible and traceable signal have been clearly established (Ferreira et al. 2002b); however, most researchers keep on using the technique in an archaic and refutable way, surely because they think that the possible errors in the ranking of odorants will be corrected in the subsequent quantitative analysis. According to this technique, the flavor extract is sequentially diluted (following a rate R, where R is usually 2, 3, 5 or 10, but could be as high as 20) and each dilution is analyzed by GC-O by usually a small number of judges, often (most incorrectly) by a single one. The Flavor Dilution of an odorant corresponds to the maximum dilution at which that odorant can be perceived by at least one of the judges. Numerically, if the last dilution at which compound c was perceived was P (where P usually is 0, 1, 2,. . .n), its FD is R^P or, more correctly, $R^{P+0.5}$. When several judges are used in a study, usually (and incorrectly) the maximum FD is provided as the FD factor of that compound, although the result should be the geometric mean of the FD factors recorded for each judge (Ferreira et al. 2002b). Charm is similar in concept, but the recording of the signal is made with the help of a computer and the judge records not only whether there is an odorant, but the time that the odorant is perceived just by holding pressed one key of the keyboard. Consequently, in this case the signal is a kind of peak called Charm peak defined by its area and height. The height of the Charm peak is the same than the FD obtained in an AEDA experiment. A basic scheme of the techniques is shown in Fig. 8E.1.

These techniques provide a ranking of aroma compounds relating to the ratio concentration in the extract/odor threshold in air. Working properly (using a panel of judges and 1/10 dilutions) it is possible to obtain, with reasonable effort, confidence intervals for the results, which are most useful for comparative purposes, and to relate them to the concentration of the odorant in the extract.

8E.2.1.2 Techniques Based on Frequency of Citations

In these techniques an extract is injected in the GC-O system and the effluent is smelled by a group of judges. What is recorded is the number of people that are able to detect the odorants. The proportion of people able to detect an odorant is called the Nasal Impact Frequency (NIF), and if the time for which the sensation lasts is also recorded, the product NIF by time is called Surface of Nasal Impact frequency (SNIF). This technique obviously requires a group of judges and the experiment is carried out at a single dilution of the extract. The theoretical background and some interesting applications of the technique are well documented in the scientific literature (Pollien et al. 1999; Debonneville et al., 2002). The technique is more

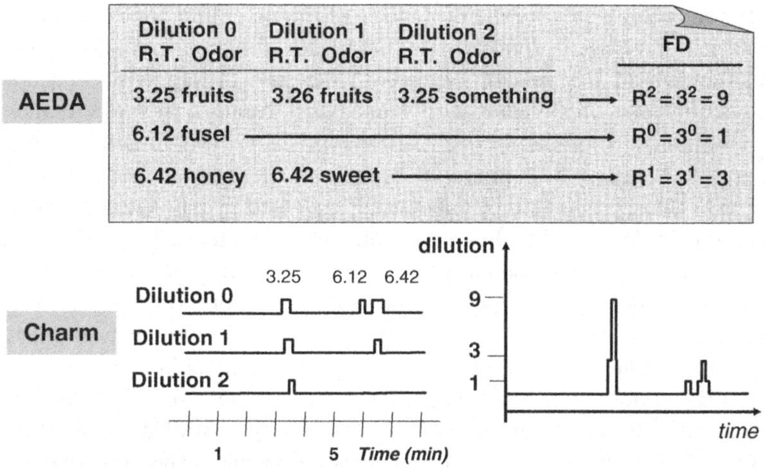

Fig. 8E.1 Scheme showing the basics of AEDA and Charm analysis. Different sequential dilutions of the sample extract are analyzed in the GC-O system. In AEDA the judge simply marks the retention times and the odor descriptions. The FD value is R^p, where p is the p-th dilution of the extract at which the odor was last detected and R is the dilution rate (3 in the figure). In Charm, he/she presses the space bar of the computer during the odor detection, and the outputs are combined to form the Charm chromatogram

reproducible than Charm analysis (Debonneville et al., 2002) or some posterior intensity methods (van Ruth, 2004). However, the dynamic range of the response is narrower in the case of NIF and this means that in the practice it is easier to discriminate concentration differences in posterior intensity methods (van Ruth, 2004).

8E.2.1.3 Techniques Based on the Measurement of Intensity

The first technique of this type was OSME (McDaniel et al. 1990; Miranda-Lopez et al. 1992) which was introduced at the beginning of the 1990s. In this technique the judge handles a knob connected to a variable resistance which is actuated following the intensity of the odor eluting out of the column. That is, the judge acts as a genuine GC detector. The output is something very similar to a chromatogram (aromagram, odorgram or osmegram) in which the peaks are related to the intensity (peak height) and intensity × time (peak area) of the odors presented in the sample. The technique is really appealing; however, in its original form the task imposed on the judges is extremely difficult and, consequently, the results are poorly reproducible and difficult to handle. It should be considered that the judges require some time to detect the odor, to estimate their intensity and to assess simultaneously its quality, and that a continuous monitoring requires a continuous and systematic feedback (da Silva et al. 1994). As a consequence, the technique has been simplified so that the judges do not try to follow closely the odor intensity of the effluent, but rather to detect the peak start and end and to provide an evaluation of the intensity. The peaks, then, look like simple triangles instead of genuine chromatographic peaks.

Two different alternatives for the evaluation of the magnitude of the odor intensity of the effluent have been proposed. The first is based on a psychophysical concept known as cross modality matching (Stevens 1975). According to this concept, there is a connection between the intensity of some basic stimulus, such as visual or olfactory stimulation, and some muscular responses, such as the contraction of the iris or the separation between the thumb and the index fingers (finger span) (Stevens and Stone 1959). This property is satisfactorily exploited in the so-called finger-span method (Guichard et al. 1995; Etiévant et al. 1999). The judge fits his/her index and thumb fingers in a device connected to a resistance which registers when and by how much the finger span is opened. The judge does not have to try to follow the intensity as in the initial version of OSME, but to give a simple measurement of the intensity of the olfactory stimulus as a whole. Therefore, although there is a time-intensity measurement (because, again, the time the finger span is actuated is also registered) only the maximum intensity (the measurement of the overall intensity of the stimulus) is recorded, at least in its original version. The procedure does not require a sophisticated assessors training (Bernet et al. 2002) and is easy to use. The finger span device is offered by a chromatographic company as an additional tool for its GC-olfactometric port (GC-Sniffer 9000. Product brochure 2008)

The second and easiest alternative is to ask the judges just to give a numerical measurement of the overall intensity of an odor eluted out of the column. Typically the panelists are asked to measure the overall intensity using a simple five, seven or nine point scale. The technique is very easy to implement and its reproducibility and quantitative ability have been well demonstrated (Le Guen et al. 2000; Serot et al. 2001; Ferreira V et al. 2003b); van Ruth 2004; Campo et al. 2005; Pet'ka et al. 2005).

8E.2.1.4 Overall Evaluation

In general, it must be said that the selection of the technique for recording and further processing the olfactometric signal is not critical provided that the experiment is carried out correctly. As can be deduced from the large differences in individual thresholds (Punter 1983; Cain and Gent 1991), and particularly from the documented existence of many odorants for which a part of the population shows specific anosmia (Amoore 1977), the use of a sensory panel is crucial. Our personal experience, however, suggests that the use of posterior intensity methods is most cost-effective. In these methods, as in all methods involving the use of a single dilution of the extract, it is important to choose adequately the concentration factor for such extract in order to ensure that only a reduced number of odorants reach odor intensities approaching the maximum of the scale.

8E.2.2 Preparation of Extracts for GC-Olfactometry

This is really a major and critical issue in GC-O, particularly in the study of wine and other alcoholic beverages and, in general, it can be said that there are better and worse alternatives, but that, unfortunately, there is not a perfect single solution. It must be taken into account that in the gas chromatograph all the volatiles injected are vaporized and chromatographied (provided that they do not decompose, react or interact irreversibly with the column) so that the extract should reflect as closely as possible the composition of the vapors reaching the pituitary during the olfaction process and not the composition in volatiles of the original sample.

This means that the classical approach in which the extract contains quantitatively all the volatiles present in the product is very far from the optimal solution. In the past, great effort was devoted to preparing total extracts "representative" enough of the original sample (Abbot et al. 1993; Moio et al. 1995; Priser et al. 1997; Bernet et al. 1999), since researchers were concerned mainly with the possibility of creating artifacts or loosing relevant aroma compounds during the preparation of the extract. However, all these techniques produce extracts whose composition is very far from those of the headspaces reaching the pituitary. The major problem with these extracts is that they ideally contain 100% of all the aroma molecules of wine and hence the GC-O experiment will detect 100% of all the aroma molecules originally present in the sample. However, during the normal olfaction (both ortho or retronasal) some aroma molecules are transported very efficiently from the liquid wine in the glass or in our mouth to the pituitary (the most volatile and nonpolar), while others are transported just in a negligible proportion (the least volatile and most polar) because they are strongly retained by the wine hydroalcoholic matrix. Consequently, the sensory importance of the least volatile and most polar compounds will be strongly overemphasized in the GC-O experiments, as it is schematically shown in Fig. 8E.2. It may be argued that such bias will be corrected by the quantitative determination and by the use of olfactory thresholds, but this involves a

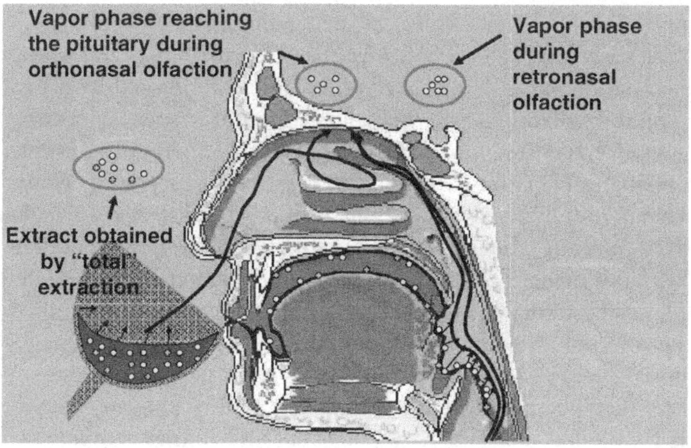

Fig. 8E.2 Schematic drawing showing the compositional differences between the extract obtained by a total extraction technique and the vapor phases reaching the pituitary during ortho or retronasal olfaction. The extract is enriched in polar volatiles that cannot reach the pituitary either by ortho or retronasal pathways lot of extra and in part, useless effort. In addition, wine extracts obtained following the total extraction approach are extremely complex with many odors and the olfactometric signals overlap, making very difficult the correct assignment of the identity of the odorant (Cullere et al. 2004b; Escudero et al. 2004). In spite of all this, most researchers keep on using these kinds of extracts for the GC-O screening of wine aromas due to their simplicity and to the existence of some reasonable doubts about the performance of headspace techniques.

The previous discussion leads to the conclusion that "a priori" headspace techniques should be preferred although, as will be discussed, there are some questions that as yet do not have a satisfactory answer. Here there are three different alternatives: static headspace sampling, dynamic headspace sampling and headspace solid phase microextraction sampling. Static headspace sampling has been used only rarely in the GC-O profiling of wines (Guth 1997a) and only as a complementary technique of a total extraction strategy. Surely this is because a large volume of headspace has to be injected to obtain clear olfactometric signals, and the presence of ethanol and water creates serious problems in the chromatography if such a large volume is injected. In addition, in classical static headspace there is no extract, so that, the GC-O experiment must be carried out on wine, which could cause problems arising from the stability and homogeneity of the samples if the experiment is going to take a long time.

The second possibility is the use of dynamic headspace sampling. This technique fulfils a priori the requirements for the preparation of extracts for screening by GC-O. However, the presence of ethanol in the headspaces decreases by factors as high as two orders of magnitude the capacity of the sorbents most frequently used for the trapping of volatiles, such as Tenax TA or Porapack Q resins, which helps to explain why the technique is only seldom used out of our laboratory (Le Fur et al. 2003). This limitation can be solved by increasing the size of the trap, by reducing the volume of sample, by using thermal desorption or by using a most efficient sorbent. The solution given by Le Fur et al. (2003) makes use of an automated purge and trap unit. Aroma volatiles are purged by a gentle stream of nitrogen on a small volume of wine and are further trapped in a cartridge filled with 140 mg of Tenax TA. The volatiles are finally thermally desorbed directly in the chromatographic column. In this strategy a new aliquot of wine is required each experiment and the analysis of their results suggests that the technique fails in transferring to the column some less-volatile but potentially important odorants, since the last odor detected corresponds to β-phenylethyl alcohol. Such limitation could be due to the difficult transference of those compounds from the trap to the column. The solution given in our laboratory makes use of a type of new generation sorbents, initially devised for the extraction of polar pesticides from water. These sorbents have demonstrated to have an amazing ability not only to extract aroma compounds from wine (Ferreira et al. 2004) but also to trap compounds present in effluents (Lopez et al. 2003a, 2007). In relation to Tenax TA, these sorbents can be up to 200 times more efficient, which makes it possible to use a relatively small amount of sorbent (400 mg) to collect efficiently all the volatiles purged out from 80 mL of wine heated at 37 °C (and diluted with synthetic saliva) by a large stream of nitrogen (100 mL/min) during a very long period of time (200 min). Desorption is carried out with the help of a solvent, which provides a single extract that can be used along the whole GC-O process. This strategy has been successfully used in the screening of very different wines (Campo et al. 2005, 2006a; Petka et al. 2006; Gomez-Miguez et al. 2007). It would not be fair, however, to say that this strategy represents a universal satisfactory solution: there are some evidences about difficulties in the elution of some mercaptans (Ferreira et al. 2007) and about a poor transfer of some polar compounds to the trap (Escudero et al., in preparation). However, the technique makes it possible to get quite easily simple extracts representing relatively well the composition of the vapors reaching the pituitary.

The third alternative is the use of a Solid Phase Microextraction fibre (SPME) to collect the volatiles in the headspace. The technique is clean, very easy to use and provides a good concentration of many volatiles on the headspaces of wine. Because of this, it has been the technique of choice in some recent works (Marti et al. 2003; Fan and Qian 2005; Gurbuz et al. 2006; Tat et al. 2007). Nevertheless, the use of this technique is not exempt from problems either. On the one hand the technique does not provide an extract, and on the other hand it is quite difficult to optimize and validate and therefore to assess the reliability of the results. It should be concluded that, although the technique is appealing, more research is needed in order to establish its advantages and drawbacks.

8E.3 Techniques for Further Isolation and Identification of Aroma Compounds

The results of the GC-O screening processes are hierarchical lists of odor zones ranked by their potential sensory importance attending to their FD factors, Charm values, Osme areas, NIF or SNIF values or odor intensities. At this time of the analysis, the single identity criteria are the nature of the odor and its retention time, most likely in a single GC column. The following step is to identify the odorant or odorants behind each of the most relevant odor zones. This is also a critical and sometimes difficult part of the work which should be carried out with rigor and perseverance. A recent paper gives some general guidelines that should be assumed by all authors (Molyneux and Schieberle 2007). A general scheme, with some specific comments from our personal experience working with wine odorants, is given below.

1. Standardization of retention times. This is the first step. Retention times of our odor zones have to be normalized by those of *n*-alkanes used as standards. To do that, the mixture of alkanes, which can be directly obtained from different chromatographic suppliers, is injected in exactly the same conditions in which the GC-O experiment was carried out. As nearly all the GC-O experiments use temperature gradients, the Retention Index or Linear Retention Index – not Kóvats Index according to the IUPAC (IUPAC 1997) – will be determined with the formula

$$I = 100 \times \left[\frac{t_{Ri} - t_{Rz}}{t_{R(z+1)} - t_{Rz}} + z \right]$$

where t_{Ri} is the Retention time of the odor zone i, t_{Rz} and $t_{R(z+1)}$ are the retention times of the *n*-alkanes eluting immediately before (z) and after (z+1) the odor zone, and z is the number of Carbon atoms in such *n*-alkane. It should be noted that if the extract is very concentrated, such as those obtained from total extraction, the retention times of some compounds eluting near of a major compound, such as isobutanol, isoamyl alcohol, ethyl lactate, diethyl succinate or β-phenylethyl alcohol, will be significantly shifted. In these cases it is advisable to spike the alkane solution to the extract. In the case an AEDA experiment has been carried out, the retention times of the higher dilutions are more accurate.

2. Getting retention data from our odor zones in a second chromatographic column with a phase of different polarity. Most typically the main column will be a Carbowax-20M type phase and the second one a 5% phenyl-dimethylpolysiloxane. In these last phases, many alcohols and particularly acids have problems of tailing and fronting, respectively, which complicates the correct identification of odor zones. This problem will be particularly acute, again, in concentrated extracts obtained by total extraction. In any case, particularly if the extracts have many odors, it will not be possible to establish a clear relation between the identities of the odor zones in the two columns. It should be noted that many relevant aroma compounds are really "difficult" molecules from the chromatographic point of view. This is particularly the case of aldehydes (especially methional and phenylacetaldehyde) mercaptans and some polar compounds such as Furaneol or Sotolon. These compounds are particularly sensitive to the existence of specific adsorption sites on the chromatographic inlets. The inertness of our columns is something that should be ensured before the GC-O experiment via the analysis of Grob-type mixtures (Grob et al. 1981).
3. Running the extract on a GC-MS system. The extract will be injected in a GC-MS system in scan mode using a column similar to the main column for the GC-O experiment. The *n*-alkane mixture will also be injected under the same conditions, and a screening will be carried out to obtain the MS spectrum and potential identification of those peaks eluting at approximately the Retention Index of the target odor zones.
4. Compiling data. Bringing together data of retention times and mass spectra to propose candidates for the different odor zones. At this stage it is convenient to do the search in "reverse" mode, i.e. to look specifically for those odorants that according to the literature are more likely to be present in the sample. There are several quite complete lists about the odorant composition of many wines in the literature (Aznar et al. 2001; Ferreira et al. 2001a) and a review (Ferreira et al. 2001b). Table 8E.1 presents a summary of some of the most recent references including the type of wine studied, the type of extract used and the olfactometric technique.
5. Confirmation of candidates. This will be done by the injection of the pure standards in the two columns in the GC-O system and in the main column in the

GC-MS system. The similarity of odors and odor intensities (at relatively equivalent levels) and odor thresholds (in AEDA or Charm experiments) must be used as an additional confirmatory criterion.

After this process, the identity of most of the "easy" odorants, i.e. those odorants present at concentrations large enough to produce clear peaks in the GC-MS system, will have been satisfactorily confirmed. Nevertheless, some of the most important wine impact aroma compounds belong to the class of difficult odorants and it will be nearly impossible to get a clear mass spectrum for such compounds in the original extract. For these compounds it will be necessary to obtain a more refined and concentrated fraction. The exact procedure to follow depends on different questions, some of which are detailed below:

a) The targeted odor zone is at the final part of the chromatogram. This means that the odorant is quite heavy and/or polar and that therefore, it is very difficult to concentrate in extracts obtained by headspace techniques (static or dynamic). In this case it should be advisable to get a total extract following some of the procedures recommended in the literature (see for instance Kotseridis and Baumes 2000; Ferreira et al. 2001a; Cullere et al. 2004a) and to apply further a fractionation of the extract. This can be carried out on silica (Kotseridis and Baumes 2000; Campo et al. 2006a), on reversed phase HPLC (Ferreira et al. 1999; Aznar et al. 2001), on normal phase HPLC (López et al. 1999), or even in a polymeric sorbent (Cullere et al. 2003). If the odorant is very polar a reversed phase fractionation on C18 or in a polymeric sorbent is the best option. The polarity of the odorant can be assessed from the difference between their Retention Index on polar and nonpolar stationary phases (Ferreira et al. 1998a): the higher the difference, the more polar the compound. In these cases even a direct isolation in the same solid phase extraction cartridge could be attempted (Ferreira V et al. 2003a).
b) The targeted odor zone corresponds to a very volatile compound. In this case the best isolation strategy is via headspace. A headspace SPME extraction with a Carboxen fibre is a good starting choice that should be complemented by the use of a chromatographic column with a very thick stationary phase or even by a Porous Layer Open Tubular (PLOT) column.
c) The targeted odor zone corresponds with a reasonable probability to an ultra-trace odorant belonging to the families of polyfunctional mercaptans or methoxypyrazines. In these cases, several procedures are described for the selective isolation of these odorants. See Tominaga et al. (1998), Schneider et al. (2003), Tominaga and Dubourdieu (2006), and Ferreira et al. (2007) for the selective isolation of polyfunctional mercaptans and Allen et al. (1994) and Sala et al. (2002) for those of methoxypyrazines. Some selective separation schemes for carbonyls have been also described in the literature (Ledauphin et al. 2006b), but they are extremely complicated, work intensive and some of the steps require harsh operating conditions under which some molecules could be formed (artifacts) or degraded.

d) The target odor zone corresponds to a nonpolar compound of average volatility. In this case a primary isolation by a dynamic headspace technique followed by a normal phase fractionation is a good choice.

8E.4 Quantitative Determination of Wine Active Odorants

Given the different concentration levels at which wine active odorants are found, it is absolutely necessary to use an array of analytical methods for their determination. Major volatiles ($C>0.1$ mg/L), such as fusel alcohols, fatty acids and their ethyl esters, can be directly determined by GC-FID using many different preconcentration techniques such as direct injection, simple solvent extraction, microextraction with a solvent, Stir Bar Sorptive Extraction (SBSE) dynamic headspace or SPME. Minor volatiles ($1<C<0.2$ mg/L) can also be determined by direct GC-MS analysis using a similar array of sample preparation techniques. Results obtained with the different sample preparation techniques differ in the range of analytes available with accuracy and in the level of automation of the method. From this last point of view, solventless fully automated sample preparation techniques such as SBSE, SPME, or its recently introduced competitor Solid Phase Dynamic Extraction (SPDE) present some major advantages, since they are very easy to use and are relatively cost-effective. From the point of view of the range of analytes available, however, they cannot compete today with the broad possibilities offered by Solid Phase Extraction. Headspace techniques fail in the analysis of some less volatile compounds, and SBSE at present, also has problems in properly extracting some polar compounds.

However, is not the analysis of minor wine volatiles that still presents difficulties. With the level of sensitivity and automation of the analytical techniques, the determination of many odorants at μg/L level is a simple analysis. The difficulties come when the analytes of interest cannot be easily determined using a single non-selective-preconcentration step. This will happen when the analytes are difficult to extract because they are very polar and/or not very volatile or when they are present at very low levels. The concentration level at which the analysis of an aroma compound becomes difficult is related to its polarity and to the quality of its mass spectrum. For instance, the analysis of 2,4,6-trichloroanisol (TCA) at, let's say, 20 ng/L is not a very difficult analysis, because this molecule is quite nonpolar (easily extractable, relative volatile) and has a mass spectrum with abundant high mass ions.In contrast, the analysis of methional or of sotolon at 1 μg/L is quite difficult because these compounds are very polar (difficult to extract, not very volatile) and their mass spectra lack powerful ions. For these difficult analytes, some of which are very important wine impact aromas, specific strategies must be developed:

- Carbonyls. The direct analysis of some relevant carbonyls such as methional, phenylacetaldehyde, isobutyraldehyde, isovaleraldehyde, 2-methylbutanal or (*E*)-2-nonenal, even if it can be carried out (da Costa et al. 2004), is not a very convenient technique. Different alternatives have been proposed, most of them making use of the formation of derivatives with pentafluorobenzyl hydroxylamine (Cullere et al. 2004a; Mateo-Vivaracho et al. 2006) or with 3-methylbenzothiazolidin-2-one hidrazone (Ledauphin et al. 2006a).

- Volatile sulfur compounds. Volatile sulfur compound are best determined in the headspaces using a sulfur selective detector. Different alternatives, based on direct headspace of headspace-SPME, have been developed (Rauhut et al. 1998; Mestres et al. 2000; López et al. 2007).
- Polyfunctional mercaptans. Polyfunctional mercaptans are some of the most powerful odorants in nature, and they must be quantified at extremely low levels. The methods developed for their determination make use either of a selective separation using *p*-hydroxy-mercurybenzoate (Tominaga et al. 1998; Tominaga and Dubourdieu 2006; Ferreira et al. 2007), of covalent chromatography (Schneider et al. 2003) or of the derivatization with pentaflurobenzyl bromide (Mateo-Vivaracho et al. 2006, 2007). A recent report using non-selective headspace isolation with SPME has also been published, but the limits of quantification are more than one order of magnitude above the odor thresholds (Fedrizzi et al. 2007).
- Polar trace compounds, such as sotolon and furaneol, are often directly determined by GC-MS of a polar extract obtained directly from wine (Cutzach et al. 1998; Camara et al. 2004; Moreno et al. 2005). However, this strategy only works when the compounds reach relatively high concentrations. For a determination at lower concentrations, a selective SPE isolation procedure has been proposed (Ferreira V et al. 2003a)
- Methoxypyrazines are compounds with very low detection thresholds which must be determined at very low levels. For these compounds, different selective isolation methods have been proposed (Allen et al. 1994; Sala et al. 2002). Some authors use a simple extraction (Kotseridis et al. 1999; Falcao et al. 2007) or an optimized headspace SPME procedure (Chapman et al. 2004; Prouteau et al. 2004) using in most cases isotopically-labelled internal standards to compensate for matrix effects. In spite of the claims of the authors, all these methods present some difficulties to accurately determine the compounds at the lowest levels at which they can be found. A recent report has presented an advanced method combining the preconcentration ability of headspace SPME with the selectivity of comprehensive GC (Ryan et al. 2005).

8E.5 Assessment of the Sensory Role Played by the Different Odorants

At this point in the chapter it is convenient to introduce a discussion about the role that the different wine odorants can play in the aroma of wine. This discussion is based on the experience gained in the last few years, which have shown that there are different and clearly identifiable roles in the way in which aroma compounds contribute to the formation of the different aroma nuances of wine.

1. Impact or highly active compounds, that are the compounds which can effectively transmit their specific (impact) or primary (highly active) aroma nuance to a given wine without the need of the support of more aroma chemicals. An example is linalool in Muscat wines.
2. Impact groups of compounds. These are families of compounds usually having similar chemical structures (chemical homologous series) and with quite close odor properties that can impart to the aroma of a wine the specific notes of the family. An example is the γ-lactones.
3. Subtle compounds or families. These are the compounds or groups of compounds which fail in transmitting their specific aroma nuances to the wine, but contribute decisively to the development in wine of some secondary-generic aroma nuance (for instance fruity, sweet), always with the need of more chemicals bearing a similarity in such odor notes. Compounds in categories 1 and 2, which do not reach enough concentration, or even if reaching it, they co-occur with many other powerful odorants (such as happens in complex wines), may fall into this category.
4. Compounds forming the base of wine aroma. These are the compounds, present in all wines at concentrations above their corresponding odor thresholds, which are no longer perceived as single entities because their aromas are fully integrated to form the complex concept of wine aroma. Within this group different roles can be found:

   a. Aroma enhancers
   b. Aroma depressors

The problem is that many wines, particularly the most complex, do not have clear impact compounds but rather families of compounds contributing to a given aroma nuance, and this fact must be considered for the design of the sensory experiments to assess the role of the different aromas.

A classical strategy makes use of odor thresholds calculated in matrix as similar as possible to the wine subject of the study. This strategy, for instance, demonstrated its usefulness for assessing the importance of wood constituents on wine aroma (Boidron et al. 1988) or for establishing the maximum limits beyond which some odorants exert a negative effect on wine aroma (Chatonnet et al. 1993). However, this strategy is quite limited in scope and can only be applied to those compounds that really play the role of impact compounds and to those cases in which is possible to find wines free from such substances. It is, therefore, very well suited to characterize taints and off-flavors.

A second more rigorous strategy is that proposed by Grosch (1993). This strategy first converts the concentration data into Odor Activity Values just normalizing the concentration by the corresponding odor threshold, calculated in a simple matrix similar to that of the product (in the case of wine the matrix could be a simple ethanolic solution with pH adjusted at 3.4). Second and most important, the odorants are mixed in a synthetic medium (this is called the model) to evaluate the degree of similarity between the model and the original product. And third, new models from which some of the odorants have been omitted are again prepared to evaluate the effect of the omission of the odorant in the model. Obviously, the most important odorants are those whose omission causes the highest impact. There are few examples of this technique applied to wine because it requires a complete quantification of all the aromas present in the sample (Guth 1997b; Ferreira et al. 2002a; Escudero et al. 2004). Furthermore, this technique fails when the sample does not contain really impact compounds, i.e. when the omission of the compounds (one at a time) does not have relevant sensory consequences, as was demonstrated in Escudero et al. (2004). Unfortunately this is something that is going to happen in many wines. The experience learned from that work makes it possible to extract some conclusions:

1. Compounds showing similar aromas should be considered together as members of a family, particularly in cases in which the compounds are formed along similar biochemical routes. For instance, there is no sense in trying to determine the impact of ethyl hexanoate, since this compound together with ethyl butyrate, ethyl octanoate and ethyl decanote forms a family of compounds produced along the same metabolic pathway and showing aromas which, together, our noses are not able to differentiate.
2. The impact of a compound cannot be predicted from its OAV value, but is the result of the interaction of the odorant with the other odorants present in the mixture. In general, experience has shown that impact compounds, whenever they exist, are odorants present at concentrations higher than the threshold and showing, usually, an odor quite different from the rest of odors present in the wine and demonstrating a personal character. For instance, out of all fruity esters of wine, only isoamyl acetate and phenylethyl acetate, both of which have clear specific and differentiable aromas, have been found to be able to act as impact compounds (Van Wyk et al. 1979; Tat et al. 2007). Similarly, the three mercaptans derived from cysteine, 4-methyl-4-mercaptopentan-2-one, 3-mercaptohexanol and 3-mercaptohexyl acetate, all of which have specific aromas, have been found to be key compounds in some wines (Darriet et al. 1995; Murat et al. 2001; Ferreira et al. 2002a; Campo et al. 2005).
3. Instead of making a big reconstitution experiment on a single wine, very good results can be obtained by studying simultaneously a group of 5–6 wines showing a wide range of variability in the most important sensory descriptors and trying to establish correlations between the scores of the main sensory descriptors of the wines and both the olfactometric scores and the chemical quantitative data of the most relevant odorants. Different types of easy sensory experiments, such as ranking tests, can be further carried out to confirm or reject the observed correlations. Examples of this kinds of work showing the effect of impact compound or of whole families of compounds can be found in Campo et al. (2005) and Escudero et al. (2007).

Acknowledgments This work has been supported by the Spanish MEC (projects AGL2004-06060/ALI and AGL2007-65139/ALI)

References

Abbot, N., Etiévant, P., Langlois, D., Lesschaeve, I., and Issanchou, S. (1993). Evaluation of the representativiness of the odor of beer extracts prior to analysis by GC Eluate Sniffing. *J. Agric. Food Chem., 41*, 777–780.

Acree, T.E., Barnard, J., and Cunningham, D.G. (1984). A procedure for the sensory analysis of gas chromatographic effluents. *Food Chem., 14*, 273–286.

Allen, M.S., Lacey, M.J., and Boyd, S. (1994). Determination of methoxypyrazines in red wines by stable isotope dilution gas chromatography-mass spectrometry. *J. Agric. Food Chem., 42*, 1734–1738.

Amoore, J.E. (1977). Specific anosmia and concept of primary odors. *Chem. Senses Flavor, 2*, 267–281.

Atanasova, B., Thomas-Danguin, T., Langlois, D., Nicklaus, S., and Etievant, P. (2004). Perceptual interactions between fruity and woody notes of wine. *Flav. Frag. J., 19*, 476–482.

Aznar, M., López, R., Cacho, J.F., and Ferreira, V. (2001). Identification and quantification of impact odorants of aged red wines from Rioja. GC-olfactometry, quantitative GC-MS, and odor evaluation of HPLC fractions. *J. Agric. Food Chem., 49*, 2924–2929.

Bailly, S., Jerkovic, V., Marchand-Bryaert, J., and Collin, S. (2006). Aroma extraction dilution analysis of Sauternes wines. Key role of polyfunctional thiols. *J. Agric. Food Chem., 54*, 7227–7234.

Belitz, H.D., and Grosch, W. (1988). Química de los Alimentos (Zaragoza, Ed. Acribia).

Bernet, C., Dirninger, N., Etievant, P., and Schaeffer, A. (1999). Evaluation of the odor representativeness of gewurztraminer wines extracts. *Sci. Alim., 19*, 701–709.

Bernet, C., Dirninger, N., Claudel, P., Etievant, P., and Schaeffer, A. (2002). Application of finger span cross modality matching method (FSCM) by naive assessors for olfactometric discrimination of gewurztraminer wines. *Lebens. Wiss. Technol., 35*, 244–253.

Boidron, J.N., Chatonnet, P., and Pons, M. (1988). Influence du bois sur certaines substances odorantes des vins. *Conn. Vigne Vin, 22*, 275–293.

Buettner, A. (2004). Investigation of potent odorants and afterodor development in two Chardonnay wines using the buccal odor screening system (BOSS). *J. Agric. Food Chem., 52*, 2339–2346.

Cain, W.S., and Gent, J.F. (1991). Olfactory sensitivity – reliability, generality, and association with aging. *J. Exp. Psychol. Human Percep. Perfor., 17*, 382–391.

Camara, J.S., Marques, J.C., Alves, M.A., and Ferreira, A.C.S. (2004). 3-hydroxy-4,5-dimethyl-2(5H)-furanone levels in fortified Madeira wines, Relationship to sugar content. *J. Agric. Food Chem., 52*, 6765–6769.

Campo, E., Ferreira, V., Escudero, A., and Cacho, J. (2005). Prediction of the wine sensory properties related to grape variety from dynamic-headspace gas chromatography-olfactometry data. *J. Agric. Food Chem., 53*, 5682–5690.

Campo, E., Cacho, J., and Ferreira, V. (2006a). Multidimensional chromatographic approach applied to the identification of novel aroma compounds in wine – Identification of ethyl cyclohexanoate, ethyl 2-hydroxy-3-methylbutyrate and ethyl 2-hydroxy-4-methylpentanoate. *J. Chromatogr. A., 1137*, 223–230.

Campo, E., Ferreira, V., Escudero, A., Marques, J.C., and Cacho, J. (2006b). Quantitative gas chromatography-olfactometry and chemical quantitative study of the aroma of four Madeira wines. *Anal. Chim. Acta, 563*, 180–187.

Chapman, D.M., Thorngate, J.H., Matthews, M.A., Guinard, J.X., and Ebeler, S.E. (2004). Yield effects on 2-methoxy-3-isobutylpyrazine concentration in Cabernet Sauvignon using a solid phase microextraction gas chromatography/mass spectrometry method. *J. Agric. Food Chem., 52*, 5431–5435.

Chatonnet, P., Dubourdieu, D., Boidron, J.N., and Lavigne, V. (1993). Synthesis of volatile phenols by Saccharomyces cerevisiae in wines. *J. Sci. Food Agric., 62*, 191–201.

Cullere, L., Aznar, M., Cacho, J., and Ferreira, V. (2003). Fast fractionation of complex organic extracts by normal-phase chromatography on a solid-phase extraction polymeric sorbent – Optimization of a method to fractionate wine flavor extracts. *J. Chromatogr. A., 1017*, 17–26.

Cullere, L., Cacho, J., and Ferreira, V. (2004a). Analysis for wine C5-C8 aldehydes through the determination of their O-(2,3,4,5,6-pentafluorobenzyl)oximes formed directly in the solid phase extraction cartridge. *Anal. Chim. Acta, 524*, 201–206.

Cullere, L., Escudero, A., Cacho, J., and Ferreira, V. (2004b). Gas chromatography-olfactometry and chemical quantitative study of the aroma of six premium quality Spanish aged red wines. *J. Agric. Food Chem., 52*, 1653–1660.

Cutzach, I., Chatonnet, P., and Dubourdieu, D. (1998). Study on sweet natural non Muscat wine aroma. I. Qualitative analysis of sweet natural wine aroma found during ageing. *J. Int. Sci. Vigne Vin, 32*, 99–110.

da Costa, M.S., Goncalves, C., Ferreira, A., Ibsen, C., de Pinho, P.G., and Ferreira, A.C.S. (2004). Further insights into the role of methional and phenylacetaldehyde in lager beer flavor stability. *J. Agric. Food Chem., 52*, 7911–7917.

da Silva, M.A.A.P., Lundahl, D.S., and McDaniel, M.R. (1994). The capability and psychophysics of osme, a new GC-olfactometry technique. In Trends in flavour research, H. Maarse and D.G. Van der heij, eds. (Elsevier science B.V.), pp. 191–209.

Darriet, P., Tominaga, T., Lavigne, V., Boidron, J.N., and Dubourdieu, D. (1995). Identification of a powerful aromatic component of Vitis vinifera L. var. Sauvignon wines: 4-mercapto-4-methylpentan-2-one. *Flav. Frag. J., 10*, 385–392.

Debonneville, C., Orsier, B., Flament, I., and Chaintreau, A. (2002). Improved hardware and software for quick gas chromatography-olfactometry using Charm and GC-"SNIF" analysis. *Anal. Chem., 74*, 2345–2351.

Escudero, A., Cacho, J., and Ferreira, V. (2000). Isolation and identification of odorants generated in wine during its oxidation: a gas chromatography-olfactometric study. *Eur. Food Res. Technol., 211*, 105–110.

Escudero, A., Gogorza, B., Melus, M.A., Ortin, N., Cacho, J., and Ferreira, V. (2004). Characterization of the aroma of a wine from Maccabeo. Key role played by compounds with low odor activity values. *J. Agric. Food Chem., 52*, 3516–3524.

Escudero, A., Campo, E., Farina, L., Cacho, J., and Ferreira, V. (2007). Analytical characterization of the aroma of five premium red wines. Insights into the role of odor families and the concept of fruitiness of wines. *J. Agric. Food Chem., 55*, 4501–4510.

Etiévant, P.X., Callement, G., Langlois, D., Issanchou, S., and Coquibus, N. (1999). Odor intensity evaluation in gas chromatography – olfactometry by finger span method. *J. Agric. Food Chem., 47*, 1673–1680.

Falcao, L.D., de Revel, G., Perello, M.C., Moutsiou, A., Zanus, M.C., and Bordignon-Luiz, M.T. (2007). A survey of seasonal temperatures and vineyard altitude influences on 2-methoxy-3-isobutylpyrazine, C-13-norisoprenoids, and the sensory profile of Brazilian Cabernet Sauvignon wines. *J. Agric. Food Chem., 55*, 3605–3612.

Fan, W.L., and Qian, M.C. (2005). Headspace solid phase microextraction and gas chromatography-olfactometry dilution analysis of young and aged Chinese "Yanghe Daqu" liquors. *J. Agric. Food Chem., 53*, 7931–7938.

Fang, Y., and Qian, M. (2005). Aroma compounds in Oregon Pinot Noir wine determined by aroma extract dilution analysis (AEDA). *Flav. Frag. J., 20*, 22–29.

Fedrizzi, B., Versini, G., Lavagnini, I., Nicolini, G., and Magno, F. (2007). Gas chromatography-mass spectrometry determination of 3-mercaptohexan-1-ol and 3-mercaptohexyl acetate in wine A comparison of headspace solid phase microextraction and solid phase extraction methods. *Anal. Chim. Acta, 596*, 291–297.

Ferreira, A.C.S., Barbe, J.C., and Bertrand, A. (2003). 3-hydroxy-4,5-dimethyl-2(5H)-furanone, A key odorant of the typical aroma of oxidative aged Port wine. *J. Agric. Food Chem., 51*, 4356–4363.

Ferreira, V., Ardanuy, M., López, R., and Cacho, J.F. (1998a). Relationship between flavor dilution values and odor unit values in hydroalcoholic solutions: Role of volatility and a practical rule for its estimation. *J. Agric. Food Chem., 46*, 4341–4346.

Ferreira, V., López, R., Escudero, A., and Cacho, J.F. (1998b). The aroma of Grenache red wine: Hierarchy and nature of its main odorants. *J. Sci. Food Agric., 77*, 259–267.

Ferreira, V., Hernandez-Orte, P., Escudero, A., López, R., and Cacho, J. (1999). Semipreparative reversed-phase liquid chromatographic fractionation of aroma extracts from wine and other alcoholic beverages. *J. Chromatogr. A., 864*, 77–88.

Ferreira, V., Aznar, M., López, R., and Cacho, J. (2001a). Quantitative gas chromatography-olfactometry carried out at different dilutions of an extract. Key differences in the odor profiles of four high-quality Spanish aged red wines. *J. Agric. Food Chem., 49*, 4818–4824.

Ferreira, V., López, R., and Aznar, M. (2001b). Olfactometry and aroma extract dilution analysis of wines. In analysis of taste and aroma, Volume 21, J. Jackson, ed. (Berlin: Springer-Verlag), pp. 89–122.

Ferreira, V., Ortin, N., Escudero, A., López, R., and Cacho, J. (2002a). Chemical characterization of the aroma of Grenache rose wines: Aroma extract dilution analysis, quantitative determination, and sensory reconstitution studies. *J. Agric. Food Chem., 50*, 4048–4054.

Ferreira, V., Pet'ka, J., and Aznar, M. (2002b). Aroma extract dilution analysis. Precision and optimal experimental design. *J. Agric. Food Chem., 50*, 1508–1514.

Ferreira, V., Jarauta, I., López, R., and Cacho, J. (2003a). Quantitative determination of sotolon, maltol and free furaneol in wine by solid-phase extraction and gas chromatography-ion-trap mass spectrometry. *J. Chromatogr. A., 1010*, 95–103.

Ferreira, V., Pet'ka, J., Aznar, M., and Cacho, J. (2003b). Quantitative gas chromatography-olfactometry. Analytical characteristics of a panel of judges using a simple quantitative scale as gas chromatography detector. *J. Chromatogr. A., 1002*, 169–178.

Ferreira, V., Jarauta, I., Ortega, L., and Cacho, J. (2004). Simple strategy for the optimization of solid-phase extraction procedures through the use of solid-liquid distribution coefficients Application to the determination of aliphatic lactones in wine. *J. Chromatogr. A.,* 1025, 147–156.

Ferreira, V., Pet'ka, A., and Cacho, J. (2006). Intensity and persistence profiles of flavor compounds in synthetic solutions. Simple model for explaining the intensity and persistence of their aftersmell. *J. Agric. Food Chem., 54*, 489–496.

Ferreira, V., Ortin, N., and Cacho, J.F. (2007). Optimization of a procedure for the selective isolation of some powerful aroma thiols – Development and validation of a quantitative method for their determination in wine. *J. Chromatogr. A, 1143*, 190–198.

Fretz, C., Kanel, S., Luisier, J.L., and Amado, R. (2005). Analysis of volatile components of Petite Arvine wine. *Eur. Food Res. Technol., 221*, 504–510.

Gomez-Miguez, M.J., Cacho, J.F., Ferreira, V., Vicario, I.M., and Heredia, F.J. (2007). Volatile components of Zalema white wines. *Food Chem., 100*, 1464–1473.

Grob, K., Grob, G., and Grob, K. (1981). Testing capillary gas-chromatographic columns. *J. Chromatogr., 219*, 13–20.

Grosch, W. (1993). Detection of potent odorants in foods by aroma extract dilution analysis. *Trends Food Sci. Technol., 4*, 68–73.

Guarrera, N., Campisi, S., and Asmundo, C.N. (2005). Identification of the odorants of two passito wines by gas chromatography-olfactometry and sensory analysis. *Am. J. Enol. Vitic., 56*, 394–399.

Guichard, H., Guichard, E., Langlois, D., Issanchou, S., and Abbott, N. (1995). GC sniffing analysis – olfactive intensity measurement by 2 methods. *Z. Lebensm. Unter. Forsch., 201*, 344–350.

Gurbuz, O., Rouseff, J.M., and Rouseff, R.L. (2006). Comparison of aroma volatiles in commercial Merlot and Cabernet Sauvignon wines using gas chromatography – Olfactometry and gas chromatography – Mass spectrometry. *J. Agric. Food Chem., 54*, 3990–3996.

Guth, H. (1997a). Identification of character impact odorants of different white wine varieties. *J. Agric. Food Chem., 45*, 3022–3026.

Guth, H. (1997b). Quantitation and sensory studies of character impact odorants of different white wine varieties. *J. Agric. Food Chem., 45*, 3027–3032.

GC-Sniffer 9000. Product brochure (2008).http://www.chromtech.de/index/php?CAT=635& SUBCAT=527& page=269.

IUPAC (1997). Compendium of chemical terminology, 2nd ed. http://www.iupac.org/goldbook/R05360.pdf.

Komes, D., Ulrich, D., and Lovric, T. (2006). Characterization of odor-active compounds in Croatian Rhine Riesling wine, subregion Zagorje. *Eur. Food Res. Technol., 222*, 1–7.

Kotseridis, Y., and Baumes, R. (2000). Identification of impact odorants in Bordeaux red grape juice, in the commercial yeast used for its fermentation, and in the produced wine. *J. Agric. Food Chem., 48*, 400–406.

Kotseridis, Y., Baumes, R.L., Bertrand, A., and Skouroumounis, G.K. (1999). Quantitative determination of 2-methoxy-3-isobutylpyrazine in red wines and grapes of Bordeaux using a stable isotope dilution assay. *J. Chromatogr. A., 841*, 229–237.

Le Fur, Y., Mercurio, V., Moio, L., Blanquet, J., and Meunier, J.M. (2003). A new approach to examine the relationships between sensory and gas chromatography-olfactometry data using generalized procrustes analysis applied to Six French Chardonnay wines. *J. Agric. Food Chem., 51*, 443–452.

Le Guen, S., Prost, C., and Demaimay, M. (2000). Critical comparison of three olfactometric methods for the identification of the most potent odorants in cooked mussels (Mytilus edulis). *J. Agric. Food Chem., 48*, 1307–1314.

Ledauphin, J., Barillier, D., and Beljean-Leymarie, M. (2006a). Gas chromatographic quantification of aliphatic aldehydes in freshly distilled Calvados and Cognac using 3-methylbenzothiazolin-2-one hydrazone as derivative agent. *J. Chromatogr. A., 1115*, 225–232.

Ledauphin, J., Basset, B., Cohen, S., Payot, T., and Barillier, D. (2006b). Identification of trace volatile compounds in freshly distilled Calvados and Cognac: Carbonyl and sulphur compounds. *J. Food Comp. Anal., 19*, 28–40.

Lee, S.J., and Noble, A.C. (2003). Characterization of odor-active compounds in Californian Chardonnay wines using GC-olfactometry and GC-mass spectrometry. *J. Agric. Food Chem., 51*, 8036–8044.

Lopez, P., Batlle, R., Nerin, C., Cacho, J., and Ferreira, V. (2007). Use of new generation poly(styrene-divinylbenzene) resins for gas-phase trapping-thermal desorption – Application to the retention of seven volatile organic compounds. *J. Chromatogr. A., 1139*, 36–44.

López, R., Ferreira, V., Hernandez, P., and Cacho, J.F. (1999). Identification of impact odorants of young red wines made with Merlot, Cabernet Sauvignon and Grenache grape varieties: a comparative study. *J. Sci. Food Agric., 79*, 1461–1467.

López, R., Ferreira, V., Culleré, L., Grasa, C., and Cacho, J. (2003a). Improvements for the analytical chemistry of flavours obtained from the use of new-generation polymeric sorbents. In Flavour Research at the Dawn of the Twenty-first Century, Proceedings of the 10th Weurman Flavour Research Symposium, J.L.L. Quéré and P.X. Etievant, eds. (Paris: Lavoisier, Ed. TecandDoc, 2003. ISBN 2-7430-0639-0), pp. pp. 662–665.

López, R., Ortin, N., Perez-Trujillo, J.P., Cacho, J., and Ferreira, V. (2003b). Impact odorants of different young white wines from the Canary Islands. *J. Agric. Food Chem., 51*, 3419–3425.

López, R., Ezpeleta, E., Sanchez, I., Cacho, J., and Ferreira, V. (2004). Analysis of the aroma intensities of volatile compounds released from mild acid hydrolysates of odourless precursors extracted from Tempranillo and Grenache grapes using gas chromatography-olfactometry. *Food Chem., 88*, 95–103.

López, R., Lapena, A.C., Cacho, J., and Ferreira, V. (2007). Quantitative determination of wine highly volatile sulfur compounds by using automated headspace solid-phase microextraction and gas chromatography-pulsed flame photometric detection – Critical study and optimization of a new procedure. *J. Chromatogr. A., 1143*, 8–15.

Marti, M.P., Mestres, M., Sala, C., Busto, O., and Guasch, J. (2003). Solid-phase microextraction and gas chromatography olfactometry analysis of successively diluted samples. A new approach of the aroma extract dilution analysis applied to the characterization of wine aroma. *J. Agric. Food Chem., 51*, 7861–7865.

Mateo-Vivaracho, L., Ferreira, V., and Cacho, J. (2006). Automated analysis of 2-methyl-3-furanthiol and 3-mercaptohexyl acetate at ng L-1 level by headspace solid-phase

microextracion with on-fibre derivatisation and gas chromatography-negative chemical ionization mass spectrometric determination. *J. Chromatogr. A., 1121*, 1–9.
Mateo-Vivaracho, L., Cacho, J., and Ferreira, V. (2007). Quantitative determination of wine polyfunctional mercaptans at nanogram per liter level by gas chromatography-negative ion mass spectrometric analysis of their pentafluorobenzyl derivatives. *J. Chromatogr. A., 1146*, 242–250.
McDaniel, M.R., Miranda-López, R., Walson, B.T., Micheals, N.J., and Libbey, L.M. (1990). Pinot Noir aroma: A sensory/gas chromatographic approach. In Flavors and Off-Flavors, G. Charalambous, ed. (Amsterdam: Elsevier), pp. 23–36.
Mestres, M., Busto, O., and Guasch, J. (2000). Analysis of organic sulfur compounds in wine aroma. *J. Chromatogr. A., 881*, 569–581.
Miranda-Lopez, R., Libbey, L.M., Watson, B.T., and McDaniel, M.R. (1992). Odor analysis of Pinot Noir wines from grapes of different maturities by a gas chromatography-olfactometry technique (osme). *J. Food Sci., 57*, 985–993 and 1019.
Moio, L., Chambellant, E., Lesschaeve, I., Issanchou, S., Schlich, P., and Etievant, P.X. (1995). Production of representative wine extracts for chemical and olfactory analysis. *It. J. Food Sci., 7*, 265–278.
Molyneux, R.J., and Schieberle, P. (2007). Compound identification: A Journal of agricultural and food chemistry perspective. *J. Agric. Food Chem., 55*, 4625–4629.
Moreno, J.A., Zea, L., Moyano, L., and Medina, M. (2005). Aroma compounds as markers of the changes in sherry wines subjected to biological ageing. *Food Control, 16*, 333–338.
Murat, M., Tominaga, T., and Dubourdieu, D. (2001). Mise en évidence de composés clefs dans l'arôme des vins rosés et clairets de Bordeaux. *J. Int. Sci. Vigne Vin, 35*, 99–105.
Ong, P.K.C., and Acree, T.E. (1999). Similarities in the aroma chemistry of Gewurztraminer variety wines and lychee (Litchi chinesis Sonn.) fruit. *J. Agric. Food Chem., 47*, 665–670.
Pet'ka, J., Ferreira, V., and Cacho, J. (2005). Posterior evaluation of odour intensity in gas chromatography-olfactometry: comparison of methods for calculation of panel intensity and their consequences. *Flavour Frag. J., 20*, 278–287.
Pet'ka, J., Ferreira, V., Gonzalez-Vinas, M.A., and Cacho, J. (2006). Sensory and chemical characterization of the aroma of a white wine made with Devin grapes. *J. Agric. Food Chem., 54*, 909–915.
Pollien, P., Ott, A., Montignon, F., Baumgartner, M., Munoz-Box, R., and Chaintreau, A. (1997). Hyphenated headspace Gas chromatography sniffing technique : screening of impact odorants and quantitative aromagram comparisons. *J. Agric. Food Chem., 45*, 2630–2637.
Pollien, P., Fay, L.B., Baumgartner, M., and Chaintreau, A. (1999). First attempt of odorant quantitation using gas chromatography-olfactometry. *Anal. Chem., 71*, 5391–5397.
Priser, C., Etievant, P.X., Nicklaus, S., and Brun, O. (1997). Representative champagne wine extracts for gas chromatography olfactometry analysis. *J. Agric. Food Chem., 45*, 3511–3514.
Prouteau, C., Schneider, R., Lucchese, Y., Nepveu, F., Renard, R., and Vaca-Garcia, C. (2004). Improving headspace-solid-phase microextraction of 3-isobutyl-2-methoxypyrazine by experimental design with regard to stable isotope dilution gas chromatography-mass spectrometric analysis of wine. *Anal. Chim. Acta, 513*, 223–227.
Punter, P.H. (1983). Measurement of human olfactory thresholds for several groups of structurally related compounds. *Chem. Senses, 7*, 215–235.
Rauhut, D., Kurbel, H., Macnamara, K., and Grossmann, M. (1998). Headspace GC-SCd Monitoring of Low Volatile Sulfur-Compounds During Fermentation and in Wine. *Analusis*, 26, 142–145.
Ruth, S.M.v., Roozen, J.P., and Posthumus, M.A. (1995). Instrumental and sensory evaluation of the flavour of dried French beans (Phaseolus vulgaris) influenced by storage conditions. *J. Sci. Food Agric., 69*, 393–401.
Ryan, D., Watkins, P., Smith, J., Allen, M., and Marriott, P. (2005). Analysis of methoxypyrazines in wine using headspace solid phase microextraction with isotope dilution and comprehensive two-dimensional gas chromatography. *J. Sep. Sci., 28*, 1075–1082.

Sala, C., Mestres, M., Marti, M.P., Busto, O., and Guasch, J. (2002). Headspace solid-phase microextraction analysis of 3-alkyl-2-methoxypyrazines in wines. *J. Chromatogr. A., 953*, 1–6.

Sarrazin, E., Dubourdieu, D., and Darriet, P. (2007). Characterization of key-aroma compounds of botrytized wines, influence of grape botrytization. *Food Chem., 103*, 536–545.

Schieberle, P., and Grosch, W. (1987). Evaluation of the flavour of wheat and rey bread crusts by aroma extract dilution analysis. *Z. Lebensm. Unters. Forsch., 185*, 111–113.

Schneider, R., Kotseridis, Y., Ray, J.L., Augier, C., and Baumes, R. (2003). Quantitative determination of sulfur-containing wine odorants at sub parts per billion levels. 2. Development and application of a stable isotope dilution assay. *J. Agric. Food Chem., 51*, 3243–3248.

Serot, T., Prost, C., Visna, L., and Burcea, M. (2001). Identification of the main odor-active compounds in musts from french and romanian hybrids by three olfactometric methods. *J. Agric. Food Chem., 49*, 1909–1914.

Stevens, S.S. (1975). Psychophysics. pp. 99–133, Wiley: New York.

Stevens, S.S., and Stone, G. (1959). Finger span: Ratio scale, category scale and JND scale. *J. Exp. Psychol., 57*, 91–95.

Tat, L., Comuzzo, P., Battistutta, F., and Zironi, R. (2007). Sweet-like off-flavor in aglianico del vulture wine: Ethyl phenylacetate as the mainly involved compound. *J. Agric. Food Chem., 55*, 5205–5212.

Tominaga, T., and Dubourdieu, D. (2006). A novel method for quantification of 2-methyl-3-furanthiol and 2-furanmethanethiol in wines made from Vitis vinifera grape varieties. *J. Agric. Food Chem., 54*, 29–33.

Tominaga, T., Murat, M.L., and Dubourdieu, D. (1998). Development of a method for analyzing the volatile thiols involved in the characteristic aroma of wines made from Vitis-Vinifera L. CV Sauvignon Blanc. *J. Agric. Food Chem., 46*, 1044–1048.

Ullrich, F., and Grosch, W. (1987). Identification of the most intense volatile flavour compounds formed during autoxidation of linoleic acid. *Z. Lebensm. Untersch. Forsch., 184*, 277–282.

van Ruth, S.M. (2004). Evaluation of two gas chromatography-olfactometry methods: the detection frequency and perceived intensity method. *J. Chromatogr. A., 1054*, 33–37.

Van Wyk, C.J., Augustyn, O.P.H., De Wet, P., and Joubert, W.A. (1979). Isoamyl acetate, a key fermentation volatile of wines of vitis vinifera cv. Pinotage. *Am. J. Enol. Vitic., 30*, 167–173.

Chapter 8F
Interactions Between Wine Matrix Macro-Components and Aroma Compounds

María Ángeles Pozo-Bayón and Gary Reineccius

Contents

8F.1 Introduction

Although there are many studies in the literature that have focused on the identification and quantification of wine aroma compounds, to understand fully wine aroma perception it is necessary to incorporate the study of the interactions between aroma components and non-volatile wine matrix macro-components. This phenomenon influences aroma volatility and solubility, and thus its release from wine. Aroma release ultimately influences aroma perception. Although some research has been devoted to the study of interactions of aroma compounds and non-volatile components of wine, the diversity and significant of these interactions have not been thoroughly considered. This chapter is devoted to a discussion of this topic.

8F.1.1 Partition Coefficient

Aroma perception of a food, e.g. wine, is strongly influenced by the way indigenous aroma molecules distribute between the gas and liquid phases. This distribution

M.Á. Pozo-Bayón (✉)
Instituto de Fermentaciones Industriales (CSIC), C/Juan de la Cierva 3, 28806 Madrid, Spain
e-mail: mdelpozo@ifi.csic.es

M.V. Moreno-Arribas, M.C. Polo (eds.), *Wine Chemistry and Biochemistry*,
DOI 10.1007/978-0-387-74118-5_16, © Springer Science+Business Media, LLC 2009

is determined by the vapour-liquid equilibrium (VLE) of each individual odorant, which is characterized by the partition coefficient (k_i) between the vapour phase and the liquid phase. This coefficient is expressed by the ratio between compound concentrations in the gas phase vs the liquid phase in the sample at equilibrium:

$$k_i = [C_i^{gas} \text{ in (mol or g)/m}^3]/[C_i^{liq} \text{ in (mol or g)/m}^3] \quad (8F.1)$$

The partition coefficient can be expressed in molar fraction, also called absolute volatility (Ki) or by the activity coefficient (γ_i). Both parameters are related as is follows:

$$K_i = y_i/x_i = \gamma_i * P^{\circ}_{i(T)}/P_T \quad (8F.2)$$

where x_i and y_i are the molar fractions in the liquid and gas phases, respectively, $P^{\circ}_{i(T)}$ is the vapour pressure of pure component i at a given temperature T (P_a), and P_T is the total pressure (Pa). Volatile compounds in food and beverages are usually near infinite dilution (concentrations lower than 10^{-4} mole fraction), and thus activity coefficient can be considered constant and is noted as γ_i^{∞}. The product $\gamma_i \times P^{\circ}_{i(T)}$ is Henry's constant.

Partitioning of volatile substances between the liquid and gas phases is mainly governed by aroma compound volatility and solubility. These physicochemical properties are expected to be influenced by wine constituents present in the medium, for instance polysaccharides, polyphenols, proteins among others. Consideration of the physicochemical interactions that occur between aroma compounds and wine constituents is necessary to understand the perception of wine aroma during consumption. The binding that occurs at a molecular level reflects changes at a macroscopic level of the thermodynamic equilibrium, such as volatility and solubility, or changes in kinetic phenomena. Thus, thermodynamic and dynamic approaches can be used to study the behaviour of aroma compounds in simple (model) or complex (foods) media.

8F.1.2 Methods to Measure Interactions Between Aroma Compounds and Wine Macro-Components

The methods employed to measure the interactions that occur between aroma compounds and other food or beverage constituents are frequently based on measuring changes in the vapour-liquid equilibrium when different macromolecules are present in the media. The determination of the gas-liquid partitioning with and without a food macromolecule is widely employed.

There are different experimental methods for determining the gas-liquid partition coefficients leading to the determination of activity coefficients at infinite dilution γ_i^{∞}. The most frequently used methods are dynamic and static headspace methods.

Dynamic headspace methods involving bubbling an inert gas through a dilute binary solution, e.g. exponential dilution methods, have been used to determine the activity coefficient at infinite dilution (γ^{∞}). For example, this method has been used by Langourieux and Crouzet (1997) to determine how wine polysaccharides modify the aroma vapour-liquid equilibrium and Dufour and Bayonove (1999b) have used this method to study the influence of polyphenols on it. Lubbers et al. (1994b) have used a *headspace stripping at equilibrium method* for a similar purpose. This approach requires an external calibration by injection of liquid standard solutions into the gas chromatograph.

Most of the *static headspace methods* determine the partition coefficient by quantifying volatile concentration above a sample by gas-chromatography. The *vapour phase calibration method (VPC)* uses an external vapour standard for calibration. One must assure that the pure component is completely vaporized before injection. A widely employed alternative is the *Liquid calibration static headspace (LC-SH) method* (Voilley et al. 1991; Nedjma 1997). A third approach uses HS-SPME. SPME may be used to determine partition coefficients if short sampling times are applied: the process must only *sample* the headspace and not disrupt the equilibrium (Jung and Ebeler 2003). This method has become very popular to study the effect of wine macromolecules on the liquid-vapor equilibrium, (Whiton and Zoecklein 2000; Escalona et al. 2002; Hartmann et al. 2002; Aronson and Ebeler 2004).

Some static headspace methods do not require an external calibration and are based on measurements performed at thermodynamic equilibrium between liquid and gas phase. In the *phase ratio variation method (PRV)* described by Ettre and Collaborators (1993), the partition coefficient calculation is based on the fact that the headspace concentration changes as a function of the phase volume ratio (gas and liquid phases), while the partition coefficient remains constant. This method has been recently applied to study the interactions between aroma compounds and macromolecules in different food systems (Savary et al. 2006, 2007) but so far not to the wine.

Athès et al. (2004) compared the data from three static headspace methodologies (VPC, PRV and LC-SH) for determining gas/liquid partition coefficients of two aroma compounds in hydroalcoholic, multicomponent solutions at "infinit" dilution. They found that PRV was a simpler method compared to VPC and LC-SH and that VPC and PRV were more accurate than LC-SH since errors due to gas leaks and adsorption in gastight syringes are avoided. They suggested that these issues could be responsible for significant bias (50% lower values) obtained when using the LC-SH method. Nevertheless, all three methods were able to find an effect of ethanol (up to 20%) on the release of aroma compounds from their model system (Fig. 8F.1).

Aznar et al. (2004) developed a static headspace APCI-MS methodology, employing ethanol such as make up gas to act as the proton transfer reagent ion. Using ethanol as the reagent gas avoided the problem of high levels of ethanol in the sample (wine, beer and other alcoholic beverages) relative to other volatile compounds. Using this technique they found that ethanol decreased the partition coefficients of most of the aroma compounds. The degree of reduction was related to

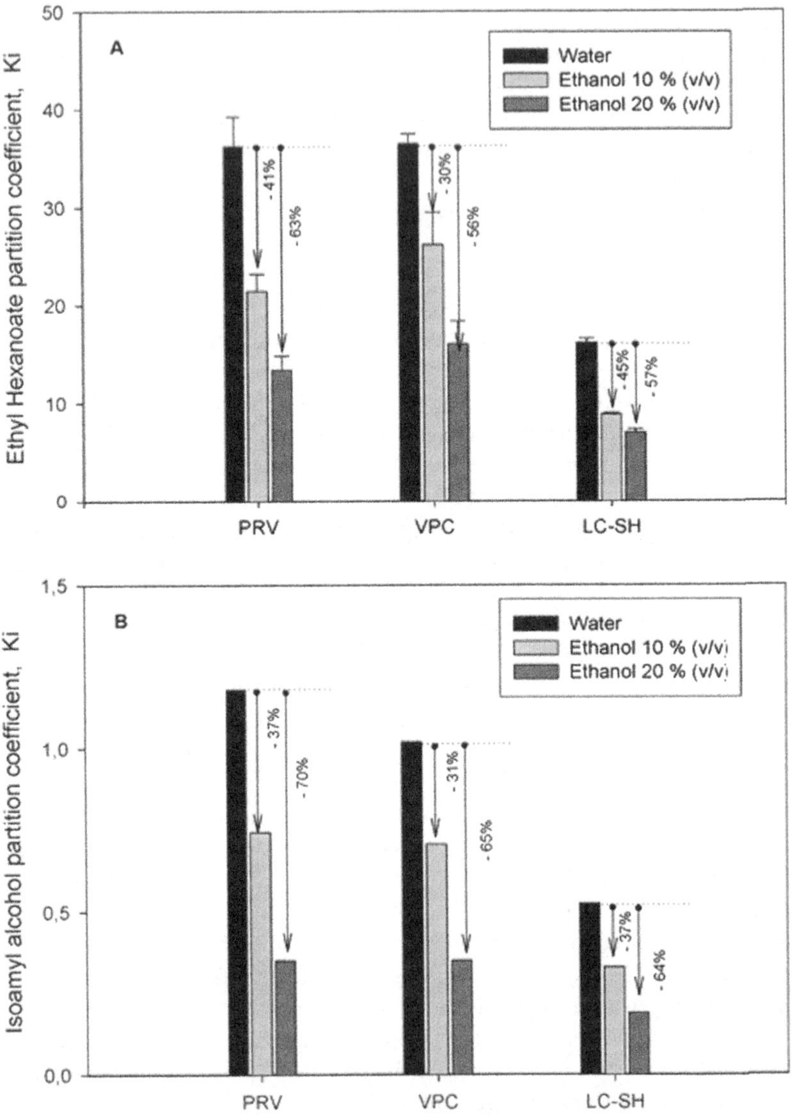

Fig. 8F.1 Ethyl hexanoate (**a**) and isoamyl alcohol (**b**) partition coefficients [K (mol/mol)/(mol/mol)] at 25 °C in water, in 10 vol.% ethanol, and in 20 vol.% ethanol, with three static headspace methods: PRV, VPC, and LC-SH (reprinted with permission from Athès et al. (2004) J Agric Food Chem 52:2021–2027. Copyright (2004) American Chemical Society)

the hydrophobicity of the aroma compound. The same group (Tsachaki et al. 2005) used this method in dynamic conditions that better correlated with the change in the perceived aroma profile during wine consumption.

Other methods to determine the interactions between aroma compounds and wine matrix components do not involve gas phase measurements. For example, the *equilibrium dialysis method* has been applied for determining interactions between yeast macromolecules and some wine aroma compounds (Lubbers et al. 1994a) and more recently to study the interaction of aroma compounds and catequins in aqueous solution (Jung and Ebeler 2003). While this method can be set up in different ways, a simple approach is to fill a dialysis cell (two chambers separated by a semipermeable membrane) with an aromatized liquid. A non-volatile component of wine can be added to one chamber of the cell and then the system allowed to come to equilibrium. If the added non-volatile component binds the aroma compound, the other chamber will be depleted by this binding. Quantification of this change in concentration permits calculating the quantity that is bound to the added substrate.

The above-mentioned methods can be used to demonstrate the existence of molecular interactions between aroma compounds and other wine macromolecules; nevertheless, they do not provide any insight into the nature of this interaction. Determining the nature of an interaction typically involves the use of spectroscopic methods; unfortunately, this methodology has not been extensively applied to studying wine flavour interactions.

Among the few works published determining mechanisms of flavour interactions in wines, the work of Dufour and Bayonove (1999b) is notable in that they studied the interactions between wine aroma compounds and polyphenols using exponential dilution analysis and H-1 NMR spectroscopy. They calculated the dissociation constants confirming the weak capacity of catequins to bind aroma compounds. Jung et al. (2000) also used NMR techniques to explore the mechanisms of interaction between some typical wine polyphenols (gallic acid and naringin) and two aroma compounds (ethyl hexanoate and 2-methylpyrazine).

8F.2 Interactions Between Aroma Compounds and Specific Wine Matrix Components

8F.2.1 Wine Matrix Composition

Wine is a complex mixture consisting of indigenous components and those obtained by chemical and biochemical transformations by wine microorganisms and/or during wine aging. Wine composition varies widely and is influenced by the grape (variety, quality) and by the winemaking conditions.

Many of the wine macro-components (e.g. carbohydrates, proteins, polyphenols), come from the skins and the pulp of grapes and from the cell walls of the yeast. Although this varies, the molecular weight of the majority of macromolecules is over 10,000 D and their final concentration ranges from 0.3 to 1 g/L (Voilley et al. 1991). Most macromolecules will be eliminated by clarification and stabilization treatments of the wine. Because of their interactions with wine aroma

Table 8F.1 β-Damascenone odour thresholds in ng/L. (reprinted with permission from Pineau et al. (2007) J Agric Food Chem 55:4103–4108. Copyright (2007) American Chemical Society)

Water/ethanol solution[a]	Model white wine[b]	Model red wine 1[c]	Model red wine 2[d]	Red wine
50	140	2100	850	7000

[a]Hydroalcoholic solution prepared with a mixture water:ethanol (88:12 v/v) + 4 g/L tartaric acid. pH was adjusted to 3.5 (0.5N KOH)

[b]Prepared after addition of charcoal to 1 L of Chardonnay wine to eliminate the β-damascenone

[c]Prepared after evaporation of 1.5 L of Merlot wine in rotavapor. The residue was diluted with a mixture water:ethanol (88:12 v/v)

[d]Prepared after evaporation of 1.5 L of Merlot wine in rotavapor until 2/3 of its volume and adjusting 180 mL of ethanol and milliQ water till obtain 1.5 L of wine

compounds, they could be responsible for undesirable flavour changes thereby producing wine of poor flavour (poor odour intensity and quality).

The impact of wine macromolecules on flavour quality/perception can be illustrated by some recent research. A term called Odour Activity Value (OAV) (ratio of concentration to threshold) is widely used in the field to predict the potential contribution of an aroma compound to overall product flavour. Pineau et al. (2007) have pointed out that wine matrix components have not been taken into account in the calculation of OAV of potent wine odorants. A consideration of interactions between aroma compounds and wine macromolecules has demonstrated that some components previously considered to be very important to wine aroma may not be as important as thought. For example, β-damascenone has a very low perception threshold in a pure hydroalcoholic solution as compared to reconstituted red wines where it is over 1000-fold higher (Table 8F.1). Thus, Pineau et al. (2007) suggested that aroma impact data for wines based on OAV calculated using sensory threshold values obtained in water or hydroalcoholic solutions may not be accurate, and could be highly overestimated. This type of error can result in academic and industry researchers focusing flavour research efforts on compounds that are of little importance and ignoring those aroma compounds of importance. It is highly desirable to understand how flavour interactions with the major components of wines influence flavour perception.

8F.2.2 Effect of Specific Components on Wine Aroma

8F.2.2.1 Ethanol

Ethanol is produced during yeast fermentation of grape sugars, and it is, after water, the major component of wines. Ethanol content is highly variable across wines depending on the sugar content of the must and on the winemaking technology, ranging between 10% and 15% for table wines to 18–21% for some young Porto and Sherry fortified wines (Hermosín 2003). It directly contributes to wine aroma and overall flavour since it is substantially above its perception threshold (from 0.1 to 100 ppm) (Bayonove et al. 2000) At lower concentrations than those found in wines (between 2% and 4%), it enhances the sour and sweet tastes, and also masks the perception of some aroma compounds, such as citral (Zamora et al. 2005). Because it can influence viscosity of the beverages, it could modify aroma release and thus, aroma perception (Nurgel and Pickering 2005).

Nevertheless, the most studied ethanol effect is related to its capacity to modify solution polarity, thus altering the gas-liquid partition coefficient. An increase in ethanol content has been shown to decrease the activity coefficients of many volatile compounds in wine because of an increase in solubility (Voilley et al. 1991). Hartmann et al. (2002) showed a decrease in the recovery of 3-alkyl-methoxy-pyrazynes extracted with a divinylbenzene/carboxen SPME fibre from wine model systems when the ethanol content increased from 0% to 20%. Similarly, Whiton and Zoecklein (2000) reported that a small increase in ethanol content (from 11% to 14%), in general, reduced the recovery of typical wine volatile compounds. Both of these studies suggest that increasing the alcohol content will reduce the release of volatile compounds from wines.

Aznar et al. (2004) used static headspace-APCI-MS to study the release of volatiles from water and hydroalcoholic systems (12 vol.%). They found a decrease in the headspace concentration of volatile compounds with an increase in the log P values (hydrophobicity values) until log P = 3. Nevertheless, for very non-polar compounds (log P > 3), they did not find this trend; this could be due to changes in hydrophobic interactions in the solution.

At higher ethanol concentrations (17–20 mL/100 mL), a decrease in volatility of ethyl esters and aldehydes has been found and this effect cannot be explained only by an increase in the solubility of the aroma compounds afforded by the added ethanol. This effect has been attributed to changes in the structure of the solution where ethanol molecules can aggregate at molar fractions above 0.05–0.06 (corresponding to 15–17 mL/100 mL of ethanol) creating hydrophobic areas (or ethanol clusters) able to retain other low water-soluble components (Conner 1994; Escalona et al. 1999). It has been found that the addition of wood extracts (in the case of model spirit solutions) increases this effect (reduction in activity coefficients) for some volatile compounds at ethanol strengths above 10 mL/100 mL (Conner et al. 1999). Nevertheless, Escalona et al. (2001) did not find the same effect after the addition of wood extracts to model wines. This author suggested that this discrepancy may have been due to differences in the wood extracts used in the studies. Those used in Conner's work were obtained in whisky aging, extracting fractions with higher ethanol solubility, while the extracts used in the wine study were obtained from wine aging and were mainly composed of more water soluble compounds that had no contribution to the formation of ethanol clusters.

There are few studies reporting on the effect of ethanol on the release of aroma compounds using dynamic methodologies. In one study (dynamic headspace analysis and APCI-MS), Tsachaki et al. (2005) observed that in aqueous systems (no ethanol) there was a rapid decrease in MS signal intensity until the rate of replenishment equalled the rate of loss from headspace purging. However, above ethanolic solutions, there was a similar initial rapid decrease followed by a "steady state"loss at much higher levels (at 50–90%) of the initial relative intensity depending on the volatile compound (Fig. 8F.2). In contrast to the aroma release effects noted under

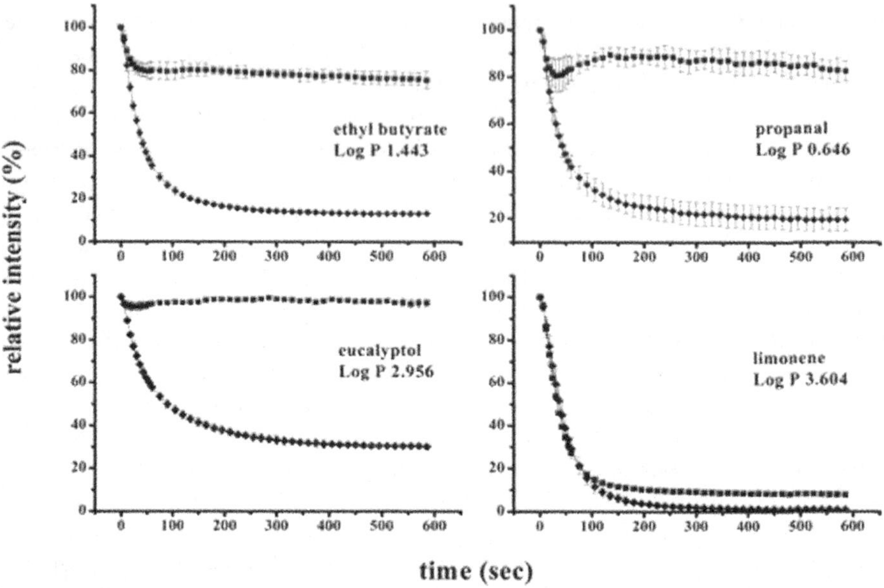

Fig. 8F.2 Dynamic headspace dilution profile of four volatiles in aqueous (♦) and ethanol (■) solutions (relative values). Each point is the mean of three replicates, *error bars* show standard deviation (reprinted with permission from Tsachaki et al. (2005) J Agric Food Chem 53: 8328–8333. Copyright (2005) American Chemical Society)

static conditions, the absolute volatile concentrations above ethanolic solutions were greater than those found above a water solution. This effect could not be completely explained by the log P value since some compounds with very different log P values showed very similar behaviours (Fig. 8F.2). Tsachaki et al. (2005) explained the effect observed during dynamic studies as a direct result of the properties of ethanol. Since ethanol is surface active, it will concentrate preferentially at the solution vapour-interface. When ethanol evaporates, some areas of the interface are depleted of ethanol, creating a surface tension gradient at the interface. Ethanol moves from the bulk phase to replenish the depleted surface areas carrying along an appreciable volume of underlying liquid (i.e. aroma compounds). This phenomenon is called the Marangoni effect (Spedding et al. 1993).

Concerning the impact of ethanol on aroma perception, Pet'ka et al. (2003) showed that ethanol at low concentrations (under 10%) could decrease aroma compound detection threshold. Nevertheless, Grosch (2001) observed that the less ethanol present in a complex wine model mixture, the greater the intensity of the fruity and floral odours. Although this effect could be easily explained by the increased partial pressure of the odorants with reduced ethanol concentration, they showed in GC-O (gas chromatography-olfactometry) experiments that ethanol strongly increased the odour threshold of wine volatiles. In fact the reduction in odour activity of the wine volatiles when ethanol was added was much larger than the reduction in their partial pressure.

More recently, Le Berre et al. (2007) observed reduced volatility of whisky lactone in hydroalcoholic solution compared to that in water, but not for isoamyl acetate. The same authors found a synergic effect of the woody note on the fruity odour in the aqueous solution, which disappeared with the addition of ethanol. They also observed that the woody note (for the highest woody odour concentrations) was masked by the fruity odour, in both aqueous and dilute alcohol solutions.

8F.2.2.2 Phenolic Compounds

Polyphenols, major non-volatile components in wine, have been reported to interact non-covalently with aroma compounds in solution. These interactions could affect the release of wine aroma compounds. Some of the changes in wine flavour resulting from using different grapes or wine making practices may in fact be due to the effect of these factors on the composition of the polyphenol fraction. For example, the removal of polyphenols through filtration or fining treatment and precipitation induced by increasing polymerization during wine aging has been suspected to produce flavour balance modifications (Voilley et al. 1991).

Early studies carried out by King and Solms (1982) documented interactions between phenolic compounds and aroma compounds in water systems. They suggested that hydrophobic interactions between aroma compounds and phenolic compounds increased solubility of aroma compounds thereby decreasing the activity coefficient of the aroma compounds.

Using exponential dilution analysis and an NMR technique, Dufour and Bayonove (1999) confirmed the existence of weak interactions between catequins and aroma compounds in model wine systems and they also agreed that mutual hydrophobicity was the driving force for this interaction. They also showed a different type of interaction depending on the type of polyphenols (catequin or tannin), and on the nature of the aroma compound.

There is substantial literature reporting the reaction of aldehydes (mainly acetaldehyde) with wine polyphenols such as flavanols and anthocyanidins (Fulcrand et al. 1996; Dallas et al. 1996; Saucier et al. 1997; Timberlake and Bridle 1976; Escribano-Bailón et al. 1996; Es-Safi et al. 1999, etc.). One of these reactions involves a Bayer acid-catalyzed condensation, giving rise to a condensation product composed of two flavanols, two anthocyanidins, or one flavanol and one anthocyanidin linked by an ethanolic bridge (formed from an aldehyde). A second type of reaction involving aldehydes is produced by adding the aldehyde to the anthocyanin molecule, forming an additional ring in the anthocyanin. All of these condensation products are directly related to the development of color and astringency during wine aging (Fulcrand et al. 1996; Atanasova et al. 2002; Mateus et al. 2002, etc.). This condensation mechanism could explain why Escalona et al. (2001) found that the flavanol (+)-catequin in hydroalcoholic solution (10–20 mL/100 mL) had little effect on the activity coefficient for ethyl hexanoate while octanal was significantly affected.

Recently Nonier et al. (2007) studied the reaction kinetics at different pH (3 and 3.5) of (+)-catequine and representative oak wood furans (furfuraldehyde [FA],

5-hydroxymethylfurfuraldehyde [HMF] and 5-methylfurfuraldehyde [5-OHFA]) and phenolic aldehydes (vanillin and syringaldehyde) in wine-like model solutions. The reactivity of both aldehyde families was very different: furanic aldehydes were more reactive than phenolic aldehydes (HMF was the most reactive, followed by FA, 5-OHFA, syringaldehyde and finally vanillin). Depending upon the aldehyde, the product formed was very different. Thus, the formation of these condensation products could play an important role in wine flavour development when aged in oak barrels.

It has been suggested that anthocyanins can form hydrogen bonds with some aroma compounds when they are present at high concentrations (Voilley et al. 1991). Dufour and Sauvaitre (2000) reported that at pH 3.5, malvin, the hemiacetal form of the anthocyanin 3,5 diglucoside, was primarily responsible for aroma interactions rather than the main anthocyanin, malvidin-3-glucoside. They also reported that benzaldehyde, furfural, and 2-isobutyl-3-methoxypyrazine did not contribute as copigments whereas all phenol-based flavour compounds (e.g. vanillin, syringaldehyde) led to copigmentation. Although they did not observe any visible effect on color by increasing the concentration of anthocyanins, the interactions of volatile compounds with the anthocyanins may affect the sensory characteristics of wines. Also during the aging of wine in oak casks, Escalona et al. (2002) have shown that some phenolic compounds extracted from the wood (ellagic tannins) could participate in polymerization reactions with aroma compounds.

Some NMR studies (Jung et al. 2000) have confirmed that interactions between some polyphenols (gallic acid and naringine) and some aroma compounds (e.g. ethyl hexanoate and 2-methylpyrazine) are due to π-π stacking of the gallic acid ring with the aromatic ring of a flavour compound. Hydrogen bonding between functional groups of a flavour compound and a polyphenol provide stability to the complex and help to orient the molecules in a specific conformation. Nevertheless, other studies performed with alkyl-methoxypyrazines in wine model systems supplemented with some of these polyphenols (e.g. gallic acid) did not shown any measurable effect (Hartmann et al. 2002), confirming the different effects of polyphenols depending on the type of aroma compound being considered.

Most of the studies on polyphenol/aroma compound interactions have been based on analytical determinations but there are few sensory studies documenting the effect of polyphenols on wine flavour. However, Aronson and Ebeler (2004) have shown that gallic acid (in 1% ethanol solution) significantly decreased the volatility of 2-methylpyrazine, while naringin at the same level had little effect. They also found that ethyl benzoate had little interaction with either polyphenol. One of the most valuable findings of this work was the agreement between the analytical and sensory data showing the effect of gallic acid and naringine interactions with 2-methylpyrazine.

8F.2.2.3 Polysaccharides and Proteins

Wine polysaccharides, ranging from 500 to 1500 mg/L (Will and Dietrich 1990), mainly come from grape primary cell walls, and from autolysis of micro-organisms such as yeast used in winemaking or *Botrytis cinerea,* a parasitic mould of the vine. Micro-organisms can release exocellular and cell wall polysaccharides during fermentation. *Sacharomyces cerevisiae* exocellular manoproteins have structural features in common with cell wall mannoproteins but have lower protein contents (Saulnier et al. 1991). This diversity of origins leads to wines differing in polysaccharides, i.e. different in composition and structure.

Dufour and Bayonove (1999a) reported two criteria for polysaccharide discrimination: acidity and protein content. Neutral peptic substances (type II arabinogalactans and arabinogalactans-proteins) represent 40% of the polysaccharides in wine and acidic pectic polysaccharides, (e.g. homogalacturonans and rhamnogalacturonans) account for 20% of them. Because of the difficulty in purifying wine polysaccharides, most of the studies on interactions between wine polysaccharides and aroma compounds have been carried out with exocellular and cell wall mannoproteins (thus mainly glycoproteins) of Saccharomyces (see effect of yeast and derivatives in the next section).

Nevertheless, Dufour and Bayonove (1999a) studied the effect of different wine polysaccharides isolated from wine, specifically arabinogalactan proteins (AGPs), monomeric and dimeric rhamnogalacturonans II (mRG-II, dRG-II) and mannoproteins (MPs), on the activity coefficients of some volatile compounds (isoamyl acetate, ethyl hexanoate, 1-hexanol, diacetyl). They found different effects depending on the type of polysaccharide and the nature of the aroma compound. They observed that the volatilities of isoamyl acetate and ethyl hexanoate were not affected by a range of polysaccharides at concentrations from 5 to 20 g/L. However, at higher concentrations, the volatilities of these two esters were decreased by protein rich polysaccharides and AGP0 (with the lower uronic acid content) and weakly salted out in the presence of the uronic acid rich fractions (AGP4). They did not find the effect in the presence of monomeric and dimeric rhamnogalacturonans. They also observed that the volatility of 1-hexanol in water was reduced in the order: AGP0>dRG-II>mRG-II>AGP4 while it was strongly salted out in presence of MP0 (mannoprotein rich in polysaccharides). Regarding diacetyl, its activity coefficient in water was not modified, increasing only at high concentration of AGP4.

Proteins are present in wines at very low and wide range of concentrations (between 30 and 269 mg/L) (Feuillat et al. 2000). Their concentration depends on the winemaking technology and grape type. Must and wine proteins have a molecular weigh between 25 and 35 kDa (Pueyo et al. 1993) and most are glycoproteins (Yokotsuka et al. 1991).

Other than studies on the role of proteins released by yeast during autolysis (mannoproteins) on wine aroma, little work has been reported on interactions of other proteins with aroma compounds. One study investigating such interactions was published by Druaux et al. (1995). They used synthetic wines and bovine serum albumin (BSA) as a model protein. This protein was found to bind δ-decalactone and there was greater binding when in water than in a model wine environment (pH 3.5 and 10% ethanol). To our knowledge this is the only study focused on elucidating the effect of proteins (others than mannoproteins) on the aroma release in wine or model wine.

8F.2.2.4 Macromolecules Derived from Wine Micro-Organisms

A major group of yeast macromolecules are the mannoproteins which represent ca. 35% of total wine polysaccharides. These mannoproteins can be classified into two main groups, those secreted into the wine by yeast during alcoholic fermentation and those released into wine due to yeast autolysis during aging on lees (Chalier et al. 2007). These macromolecules play an important role in winemaking; for example, they are involved in the improvement of tartaric acid stability, reduction of protein haze and the stabilization of wine colour and phenolic compounds among others (Feuillat 2003). However, since the addition of mannoproteins under EU law is only currently permitted for experimental trials (EU Regulation 1622/2000, art.41), other alternatives, i.e. ingredients that provide mannoproteins but are approved for use in wines (such as yeast derivatives, extracts and autolysates), are being used commercially to improve wine organoleptic characteristics. It must be considered that these alternatives bring in many other components (not just mannoproteins) which may have a negative effect on wine flavor. They may contribute undesirable aroma compounds, offer interactions that imbalance the wine, or result in new aroma compounds being formed during wine making. There is a risk in using these materials without adequate research on their overall effect.

Elaborating more on mannoproteins and their effects on wine aroma, Voilley et al. (1990) studied the binding of volatile compounds on a yeast-wall:bentonite mixture in finning experiments. The binding of β-ionone was higher (about 30%) than the three other volatile compounds studied (*n*-hexanol, ethyl hexanoate, isoamyl acetate). Lubbers et al. (1994a,b) found that these interactions were mainly hydrophobic in nature, although binding was dependent on the type of aroma compound and nature of the substrate.

Interactions between mannoproteins from yeast cell walls and aroma compounds have been studied by Langourieux and Crouzet (1997). They performed the experiments with crude mannoproteins extracts and observed no effect on the activity coefficient of isoamyl acetate, and a slight decrease on the activity coefficients of ethyl hexanoate and limonene. However, when they purified the mannoproteins or when they used a model glycopeptide, they did not observe any effect on limonene volatility. If the synthetic peptide was heat treated (50 °C), they observed a slight reduction on the activity coefficient of limonene. This was explained by an increase in the hydrophobicity of the glycopeptide after the thermal treatment.

Chalier et al. (2007), using mannoprotein at levels usually found in wines (150 mg/L), compared the effect of a whole mannoprotein extract (isolated from a synthetic medium subjected to alcoholic fermentation) to that of well characterized different mannoproteins fractions. From the four wine aroma compounds studied (isoamyl acetate, hexanol, ethyl hexanoate and β-ionone), all except isoamyl acetate showed a decrease in volatility (up to 80%) when mannoproteins were present (Fig. 8F.3). They suggested that both the glycosidic and the peptidic parts of these macromolecules may be responsible for the interaction. They also found that the interactions of the whole mannoprotein extract Vs. mannoprotein fractions were different, suggesting that the conformational and compositional structure of these

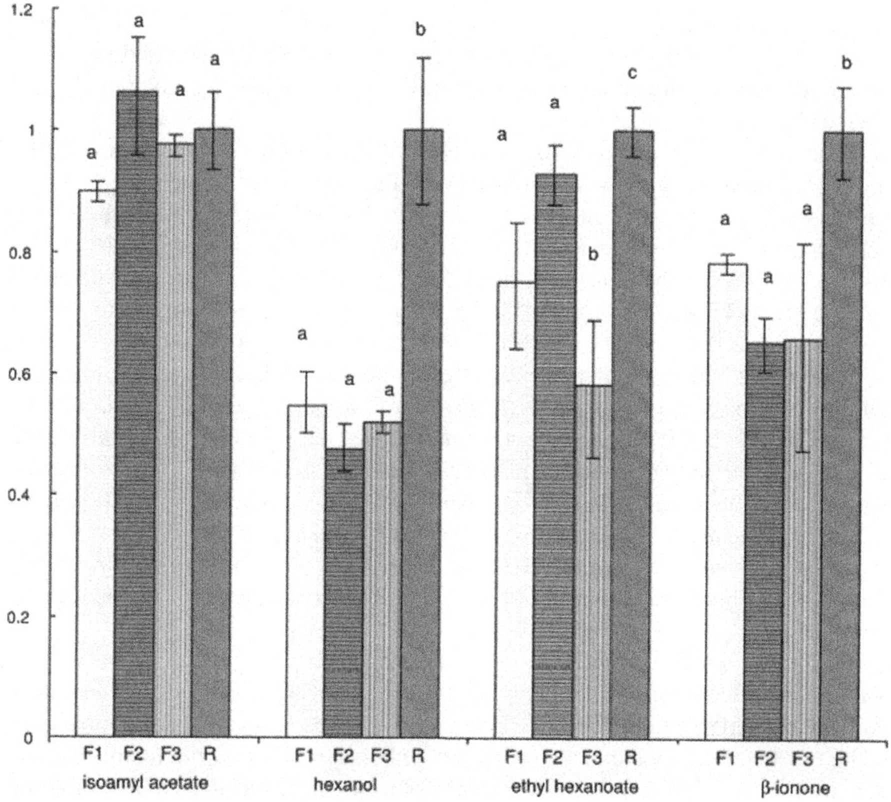

Fig. 8F.3 Effect of mannoproteins fractions (F1, F2 and F3) isolated from ICVD80 strain on aroma peak area in the headspace at 25 °C. Relative peak area is obtained by dividing the peak area of a volatile compound in the model solution added with mannoprotein fraction by the peak area in the model solution without mannoprotein. R represents relative peak area in the absence of mannoprotein. *Errors bars* represent standard deviation. Different labels (a–c) indicate that means significantly differ at $p < 0.05$ (based on Neuwman–Keuls test) (reprinted from Chalier et al. (2007) Food Chem 100:22–30. Copyright (2007), with permission from Elsevier)

macromolecules is important in the determining aroma interactions. Moreover they observed different effects depending on the yeast strain that produced the mannoproteins.

In another study on aroma interactions with industrial yeast derivatives (yeast extracts and autolysates), Comuzzo et al. (2006) also found that these macromolecules strongly modified wine aroma composition by either affecting the volatility of indigenous wine aroma compounds or by adding new aroma compounds, thereby modifying the original wine aroma profile. Both effects were related with the dosage of yeast derivatives. From a sensory point of view, they noted that the addition of yeast derivatives to wines with a strong varietal character was unadvisable, but for non-aromatic white wines, e.g. Pinot gris, the perception of a yeast-like note could be positive.

Since yeast lees may adsorb some aroma compounds responsible of off-flavours in wines (volatile phenols), these components have been also proposed such as a cost-effective and efficient approach to remove or to decrease organoleptic defects in wine (Chassagne et al. 2005).

In the case of wines aged in oak barrels with lees, aroma interactions with the lees could modify how the wine interacts with the wood. One would expect a competition between the wood and yeast derivatives for hydrophobic interactions with aroma compounds (Ramírez-Ramírez et al. 2004). One of the most recent studies in this field by Jimenez-Moreno and Ancín-Azpilicueta (2007) investigated the effect of lees extracted from wines on the binding of aroma compounds contributed by the wood. As one would expect, most of these *woody* aroma compounds decreased in the wines when in the presence of lees, independent of their hydrophobicity and concentration (same effect for 10 g/L or 50 g/L). This finding suggested the saturation of the binding sites at 10 g/L, and that the binding was likely due to non-hydrophobic interactions. A summary conclusion is that wines aged in oak barrels in the presence of lees could affect wine aroma giving wines with less woody, aromatic character.

Yeast cells can also influence wine flavour by binding aroma precursors such as glycoconjugated terpenes. Moio et al. (2004) have shown that this binding by yeast cells during fermentation could have important consequences on wine varietal aroma.

The interaction between aroma compounds and other wine micro-organisms (e.g. lactic acid bacteria) or with metabolites produced during malolactic fermentation has been studied to a limited extent. Interactions between polysaccharides produced by the most common wine lactic bacteria *(Oenoccocus oeni)* during malolactic fermentation have been shown to be responsible for the reduced volatility of some aroma compounds in wines (Boido et al. 2002). The possibility of direct interactions between the surface of the bacteria cells and aroma compounds should also be considered since this type of interaction has been found for other food lactic bacteria (Ly et al. 2008).

8F.2.2.5 Other Wine Components

Glycerol is also one of the most abundant components in wine. It is mainly produced during glycerol-pyruvic fermentation at the beginning of alcoholic fermentation. It contributes directly to wine flavour giving sweetness (Noble and Bursick 1984) and imparts some viscosity (Nurgel and Pickering 2005). Thus glycerol could directly affect flavour perception.

The effect of glycerol-flavour interactions in white wine has been studied by Lubbers et al. (2001). They found that glycerol at concentrations ranging from 5 to 50 mg/L did not modify the relative volatility of methyl-3-butanol, methyl-3-propanol, ethyl hexanoate, or methyl-3-butyl acetate. They also carried out sensory analysis, showing that the overall flavour of a model wine and white wine was not changed by the addition of glycerol, suggesting that the concentrations in wine are too low to have a significant effect on flavour perception.

In wines, De La Ossa and Galán (1986) studied the effect of adding different salts, such as $CaCl_2$in a concentration range between 0 and 9.2 mol/L on aroma compound volatility. They found a *salting out* effect for acetaldehyde, ethyl acetate, methanol and ethanol at salt concentrations from 0.5 to 2 mol/L, while other volatile compounds, such as propanol, and 1-butanol, were not affected. Interestingly, they did not find increased headspace concentrations of these volatiles as they increased the $CaCl_2$ levels above 2 mol/L. One has to question if salting out was actually the mechanism responsible for increased headspace concentrations of volatiles since 0.5 mol/L is relatively low for that mechanism to be effective and one would expect increased salting out with increased $CaCl_2$ concentrations (up to saturation).

It has been suggested that the addition of potassium bitartrate/tartaric acid to wines could induce a competition of these compounds for the free water involved in solvation enhancing ethanol aggregation. Although Escalona et al. (2001) did not find any effect of the addition of these compounds to model wine systems flavoured with ethyl hexanoate, they found a decrease in the activity coefficient of octanal. They explained this by the possible reactivity of the carbonyl group of the aldehyde with hydroxyl radicals.

In some winemaking practices, wine is aged in wood for a relatively long period of time. This practice is important to wine aroma because during this period the wine acquires an *aged bouquet,* gaining aroma compounds transferred from the wood. The major wood components are able to donate or accept electrons, giving wood an acid-base character (Escalona et al. 2001). Moreover, lignin has hydrophobic sites; thus wood can absorb aroma compounds from wine. Ramírez-Ramírez et al. (2001) and Chasagne et al. (2003) have reported the sorption capacity of oak wood for some aroma compounds, notably ethyl esters, under conditions simulating oak aging of wine. The composition of the wine itself can influence this absorption, e.g. the concentration of ethanol may change the solubility of some aroma compounds in the wine. Ramírez-Ramírez et al. (2004) observed good agreement between the hydrophobicity of a model aroma compound and the percentage of it absorbed by the wood. They also observed a reduction in the amount of volatiles absorbed by the wood as the alcohol concentration increased, e.g. for some esters the amount absorbed by wood in 10% ethanol media was half of the amount absorbed at 0%. They found the lowest absorption for most of the volatile compounds studied at 15% ethanol – the highest level studied. This fact could be due for the increasing in solubility for most of the aroma compounds. As expected, the type of aroma compound is also important in determining absorption by wood. For example, Hartmann et al. (2002) did not find any effect on the volatility of alkyl-methoxypyrazines (no interaction) when white oak sawdust was introduced to model wine systems.

8F.3 Conclusions and Future Trends

The effects of the main wine macro-components on wine aroma has been discussed in this chapter. Although these interactions have not received as much scientific attention as other aspects of wine flavour, it is clear that it is necessary to consider them since they may strongly influence the release of wine aroma components, thereby influencing flavour perception by the consumer.

In terms of research in this area, historically much more work has focused on the studying flavour release under equilibrium conditions as opposed to dynamic conditions. In the future, research should focus to a greater extent on flavour release under dynamic conditions which better represents aroma release during wine consumption.

Moreover, new winemaking practices, such as the use of yeast derivatives, the use of oak chips, the aging of wines on lees, and the use of specific micro-organisms to carry out the alcoholic and malolactic fermentation will require new studies, adding an understanding of all of these technologies to wine aroma release and ultimately flavour perception. The role of wine matrix compounds on wine aroma persistence, one of the most interesting sensory attributes of wines, has been largely neglected and should be the basis of additional research in the field.

References

Aronson, J., & Ebeler, S. (2004). Effect of polyphenol compounds on the headspace volatility of flavors. *Am. J. Enol. Vitic., 55*, 13–21.

Atanasova ,V., Fulcrand, H., Le Guernevé, C., Cheynier, V., & Moutonet, M. (2002). Structure of a new dimeric acetaldehyde malvidine-3-glucoside condensation product. *Tetrahedron Lett., 43*, 6151–6153.

Athes, V., Lillo, M.P.Y., Bernard, C., Perez-Correa, R., & Souchon I. (2004). Comparison of experimental methods for measuring infinite dilution volatilities of aroma compounds in water/ethanol mixtures *J. Agric. Food Chem., 52*, 2021–2027.

Aznar, M., Tsachaki, M., Linforth, R.S.T., Ferreira, V., & Taylor, A.J. (2004). Headspace analysis of volatile organic compounds from ethanolic systems by direct APCI-MS. *Int. J. Mass Spectrom., 239*, 17–25.

Bayonove, C., Baumes, R., Crouzet, J., & Günata, Z. (2000). In*: Enología: Fundamentos Científicos y Tecnológicos,* pp. 137–176. Coordinator: Claude Flanzy. AMV and Mundi-Prensa Editions, Madrid.

Boido, E., Lloret, A., Medina, K., Carrau, F., & Dellacassa, E. (2002). Effect of ss-glycosidase activity of *Oenococcus oeni* on the glycosylated flavor precursors of Tannat wine during malolactic fermentation. *J. Agric. Food Chem., 50*, 2344–2349.

Chalier, P., Angot, B., Delteil, D., Doco, T., & Gunata, Z. (2007). Interactions between aroma compounds and whole mannoprotein isolated from *Saccharomyces cerevisiae* strains. *Food Chem., 100*, 22–30.

Chasagne, D., Ramírez, G., Lubbers, S., Charpentier, C., Feuillat M., & Voilley, A. (2003). Sorption of aroma compounds to oak wood in wine ageing conditions. In: J.L. Le Quéré and P.X. Etiévant (Eds.), *Flavour research at the dawn of the twenty – first century* pp. 63–66. Cachan, France: Lavoisier

Chassagne, D., Guilloux-Benatier, M., Alexandre, H., & Voilley, A. (2005). Sorption of wine volatile phenols by yeast lees. *Food Chem., 91*, 39–44.

Comuzzo, P., Tat, L., Tonizzo, A., & Battistutta, F. (2006). Yeast derivatives (extracts and autolysates) in winemaking: Release of volatile compounds and effects on wine aroma volatility. *Food Chem., 99*, 217–230.

Conner, J.M., Paterson, A, & Piggott, J.R. (1994). Agglomeration of ethyl-esters in model spirit solution and malt whiskeys. *J. Sci. Food Agric., 66*, 45–53.

Conner, J.M., Paterson, A., & Piggott, J.R. (1999). Release of distillate flavour compounds in Scotch malt whisky. *J. Sci. Food Agric., 79*, 1015–1020.

Dallas, C., Ricardo-da-Silva, J.M., & Laureano, O. (1996). Products formed in model wine solutions involving anthocyanins, procyanidin B-2, and acetaldehyde. *J. Agric. Food Chem., 44*, 2402–2407.

De La Ossa, E.M, & Galán M.A. (1986). Salt effect on the vapor/liquid equilibrium of wine *Am. J. Enol. Vitic., 37*, 254–258.

Druaux, C., Lubbers, C., Lubbers, S., Charpentier, C., & Voilley, A. (1995). Effects of physicochemical parameters of a model wine on the binding of a model wine on the binding of gamma-decalactone on bovine serum-albumin. *Food Chem., 53*, 203–207.

Dufour, C., & Bayonove, C.L. (1999a). Influence of wine structurally different polysaccharides on the volatility of aroma substances in a model system. *J. Agric. Food Chem., 47*, 671–677.

Dufour, C., & Bayonove, C.L. (1999b). Interactions between wine polyphenols and aroma substances. An insight at the molecular level. *J. Agric. Food Chem., 47*, 678–684.

Dufour, C., & Sauvaitre I. (2000). Interactions between anthocyanins and aroma substances in a model system. Effect on the flavor of grape-derived beverages. *J. Agric. Food Chem., 48*, 1784–1788.

Escalona, H., Piggott, J.R., Conner, J.M., & Paterson, A. (1999). Effect of ethanol strength on the volatility of higher alcohols and aldehydes. *Ital J. Food Sci., 11*, 241–248.

Escalona, H., Homman-Ludiye, H., Piggott, J.R., & Paterson, A. (2001). Effect of potassium bitartrate, (+)-catechin and wood extracts on the volatility of ethyl hexanaote and octanal in ethanol/water solutions. *Food Sci. Tech., 34*, 76–80.

Escalona, H., Birkmyre, L., Piggott, J.R., & Paterson, A. (2002). Effect of maturation in small oak casks on the volatility of red wine aroma compounds *Anal. Chim. Acta., 458*, 45–54.

Escribano-Bailón, T., Dangles, O., & Brouillard, R. (1996). Coupling reactions between flavylium ions and catequin. *Phytochemistry, 41*, 1583–1592.

Es-Safi, N.E., Fulcrand, H., Cheynier, V., & Moutounet, M. (1999). Studies on the acetaldehyde-induced condensation of (–)-epicatechin and malvidin 3-O-glucoside in a model solution system. *J. Agric. Food Chem., 47*, 2096–2102.

Ettre, L.S., Welter, C., & Kolb, B. (1993). Determination of gas-liquid partition coefficients by automatic equilibrium headspace-gas chromatography utilizing the phase ratio variation method. *Chromatographia, 35*, 73–84.

Feuillat, M. (2003). Yeast Macromolecules: Origin, Composition and Enological Interest. *Am. J. Enol. Vitic., 54*, 211–213.

Feuillat, M., Charpentier, C., & Mauhean, A, (2000) In: *Enología : Fundamentos Científicos y Tecnológicos*, pp. 97–130. Coordinator: Claude Flanzy. AMV and Mundi-Prensa Editions, Madrid.

Fulcrand, H., Doco, T., EsSafi, N.E., Cheynier, V., & Moutounet M. (1996). Study of the acetaldehyde induced polymerisation of flavan-3-ols by liquid chromatography ion spray mass spectrometry. *J. Chromatogr. A, 752*, 85–91.

Grosch, W. (2001). Evaluation of the key odorants of foods by dilution experiments, aroma models and omission. *Chem. Senses, 26*, 533–545.

Hartmann, P.J., Mc Nair, H.M., & Zoecklein, W. (2002). Measurements of 3-alkyl-2-methoxypyrazine by Headspace Solid Phase Microextraction in spiked model wines. *Am. J. Enol. Vitic., 53*, 285–288.

Hermosín, I. (2003). Influence of Ethanol Content on the Extent of Copigmentation in a Cencibel Young Red Wine. *J. Agric. Food Chem., 51*, 4079–4083.

Jimenez-Moreno, N. & Ancín-Azpilicueta, C. (2007). Binding of oak volatile compounds by wine lees durimg simulation of wine ageing. *Food Sci. Tech., 40*, 619–624.

Jung, D.M., & Ebeler S.E. (2003). Headspace solid-phase microextraction method for the study of the volatility of selected flavor compounds. *J. Agric. Food Chem., 51*, 200–205.

Jung, D.M., de Ropp J. S., & Ebeler S.E. (2000). Study of interactions between food phenolics and aromatic flavors using one- and two-dimensional 1H NMR Spectroscopy. *J. Agric. Food Chem., 48*, 407–412.

King, B.M., & Solms, J. (1982). Interactions of volatile flavour compounds with propyl gallate and other phenols as compared with caffeine. *J. Agric. Food Chem., 30*, 838–840.

Langourieux, S., & Crouzet, J.C. (1997). Study of interactions between aroma compounds and glycopeptides by a model system. *J. Agric. Food Chem., 45*, 1873–1877.

Le Berre, E., Atanasova, B., Langlois, D., Etievant, P., & Thomas-Danguin, T. (2007). Impact of ethanol on the perception of wine odorant mixtures. *Food Qual. Prefer., 18*, 901–908

Lubbers, S., Charpentier, C., Feulliat M., & Voilley, A. (1994a). Influence of yeast walls on the behaviour of aroma compounds in a model wine. *Am. J. Enol. Vitic., 45*, 29–33,

Lubbers, S., Voilley, A., Feuillat, M., & Charpentier, C. (1994b). Influence of mannoproteins from yeast on the aroma intensity of a model wine. *Food Sci. Technol-Leb, 27*, 108–114

Lubbers, S., Verret, C., & Voilley, A. (2001). The effect of glycerol on the perceived aroma of a model wine and a white wine. *Food Sci. Technol-Leb, 34*, 262–265.

Ly, M.H., Covarrubias-Cervantes, M., Dury-Brun, C., Bordet S., Voilley, A., Le, T. M., Belin, J.M., Wache Y. (2008). Retention of aroma compounds by lactic acid bacteria in model food media. *Food Hydrocoll., 22*, 211–217.

Mateus, N., Silva, A.M.S., Rivas-Gonzalo, J.C., Santos-Buelga, C., & De Freitas, V.A.P. (2002). Identification of anthocyanin-flavanol pigments in red wines by NMR and mass spectrometry. *J. Agric. Food Chem., 50*, 2110–2116.

Moio, L., Ugliano, M., Gambuti, A., Genovese, A., & Piombino, P. (2004). Influence of clarification treatment on concentrations of selected free varietal aroma compounds and glycoconjugates in Falanghina (Vitis vinifera L.) must and wine.

Nedjma, M. (1997). Influence of complex media composition, Cognac's brandy, or Cognac, on the gas chromatography analysis of volatile sulfur compounds – Preliminary results of the matrix effect. *Am. J. Enol. Vitic., 48*, 333–338.

Noble, A.C., & Bursick, A.C. (1984) The contribution to glycerol to perceived viscosity and sweetness in white wine *Am. J. Enol. Vitic., 35*, 110–112.

Nonier, M.F., Vivas, N., de Gaulejac, N.V., Pianet, I., & Fouquet, E. (2007). A kinetic study of the reactions of (+)-catechin with aldehydes derived from toasted oak. *J. Sci. Food Agric., 87*, 2081–2091.

Nurgel, C., & Pickering G. (2005). Contribution of glycerol, ethanol and sugar to the perception of viscosity and density elicited by model white wines. *J. Texture Stud., 36*, 303–323.

Pet'ka, J., Cacho, J., & Ferreira, V. (2003). Comparison of flavor perception routes (orthonasal, bucal, retronasal and aftertaste) in a synthetic wine model and with GC-olfactometric data. Oral presentation. *Actualites Oenologiques. Bordeaux (Francia).*

Pineau, B., Barbe, J. B., Van Leeuwen, C., & Dubourdieu, D. (2007). Which Impact for β-damascenone on red wines aroma? *J. Agric. Food Chem., 55*, 4103–4108.

Pueyo, E., Dizy, M., & Polo, M.C. (1993). Varietal differenciation of musts and wines by means of protein fraction *Am. J. Enol. Vitic., 44*, 255–260

Ramírez-Ramírez, G., Lubbers, S., Charpentier, C., Feuillat, M., Voilley, A., & Chassagne, D. (2001). Aroma compound sorption by oak wood in a model wine. *J. Agric. Food Chem., 49*, 3893–3897.

Ramírez-Ramírez, G., Chassagne, D., Feuillat, M., Voilley, A., & Charpentier C. (2004). Effect of wine constituents on aroma compound sorption by oak wood in a model system. *Am J. Enol. Vitic., 55*, 22–26.

Saucier, C., Little, D., & Glories, Y. (1997). First evidence of acetaldehyde-flavanol condensation products in red wine. *Am. J. Enol. Vitic., 48*, 370–373.

Saulnier, L., Mercereau, T., & Vezhinet, E.F. (1991) Mannoproteins from flocculating and nonflocculating Saccharomyces cerevisiae yeasts. *J. Sci. Food Agric., 54*, 275–286.

Savary, G., Guichard, E., Doublier, J.L., & Cayot, N. (2006). Mixture of aroma compounds: Determination of partition coefficients in complex semi-solid matrices. *Food Res. Int., 39*, 372–379.

Savary, G., Lafarge, C., Doublier, J.L., & Cayot N. (2007) Distribution of aroma in a starch-polysaccharide composite gel. *Food Res. Int., 40*, 709–716.

Spedding, P.L., Grimshaw, J., & Ohare, K.D. (1993). Abnormal evaporation rate of ethanol from low concentration aqueous solutions. *Langmuir, 9*, 1408–1413.

Timberlake, C.F., & Bridle, P. (1976). Interactions between anthocyanins, phenolic compounds and acetaldehyde. *Am. J. Enol. Vitic., 27*, 97–105.

Tsachaki, M., Linforth, R.S.T., & Taylor A.J. (2005). Dynamic headspace analysis of the release of volatile organic compounds from ethanolic systems by direct APCI-MS. *J. Agric. Food Chem., 53*, 8328–8333.

Voilley, A., Lamer, C., Dubois, O., & Feuillat, M. (1990). Influence of macromolecules and treatments on the behaviour of aroma compounds in a model wine. *J. Agric. Food Chem., 38*, 248–251.

Voilley, A., Beghin, V., Charpentier, V., Charpentier, C., & Peyrond, D. (1991). Interactions between aroma substances and macromolecules in a model in a model wine. *Food Sci. Technol-Leb., 24*, 469–472

Whiton, R.S., & Zoecklein, B.W. (2000). Optimization of headspace solid-phase microextraction for analysis of wine aroma compounds. *Am. J. Enol. Vitic., 51*, 379–382.

Will, F., & Dietrich, H. (1990). Analysis of the monomer composition of wine polysacharydes. *Z. Lebensm. Unters For., 191*, 123–128.

Yokotsuka, K., Ebihara, T., & Sato, T. (1991). Comparison of soluble proteins in juice and wine from kushu grape, *J. Ferment Bioeng., 71*, 248–253

Zamora, M.C., Goldner, M.C., & Galmarini, M.V. (2006). Sourness-sweetness interactions in different media: White wine, ethanol and water. *J. Sens. Stud., 21*, 601–611.

Chapter 9
Phenolic Compounds

Chapter 9A
Anthocyanins and Anthocyanin-Derived Compounds

María Monagas and Begona Bartolomé

Contents

9A.1 General Considerations of Anthocyanins in Wine

Structural characteristics and occurrence. Anthocyanins are mainly located in the grape skins, with the exception of the teinturier varieties that also contain anthocyanins in the pulp. The anthocyanins identified in grape skins and wines from *Vitis vinifera* are the 3-*O*-monoglucosides and the 3-*O*-acylated monoglucosides of five main anthocyanidins – delphinidin, cyanidin, petunidin, peonidin and malvidin – which differ from each other by the number and position of the hydroxyl and methoxyl groups located in the B-ring of the molecule (Fig. 9A.1). Acylation occurs at the C-6 position of the glucose molecule by esterification with acetic, *p*-coumaric and caffeic acids (Mazza and Miniati 1993). Recently, the existence of anthocyanins acylated with lactic acid, originating in wine from grape anthocyanins, has been reported (Alcalde-Eon et al. 2006).

In the last few years, the use of mass spectrometry techniques (FAB-MS, ESI-MS/MS, MALDI-MS, HPLC/API-MS and HPLC/ESI-MS) has allowed

B. Bartolomé (✉)
Instituto de Fermentaciones Industriales, CSIC. Juan de la Cierva 3, 28006 Madrid, Spain.
e-mail: bartolome@ifi.csic.es

M.V. Moreno-Arribas, M.C. Polo (eds.), *Wine Chemistry and Biochemistry*,
DOI 10.1007/978-0-387-74118-5_17, © Springer Science+Business Media, LLC 2009

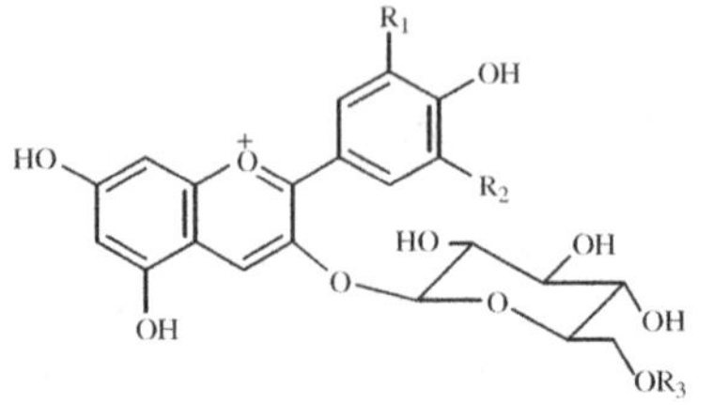

Anthocyanidin	R_1	R_2
Cyanidin	OH	H
Delphinidin	OH	OH
Peonidin	OCH_3	H
Petunidin	OCH_3	OH
Malvidin	OCH_3	OCH_3

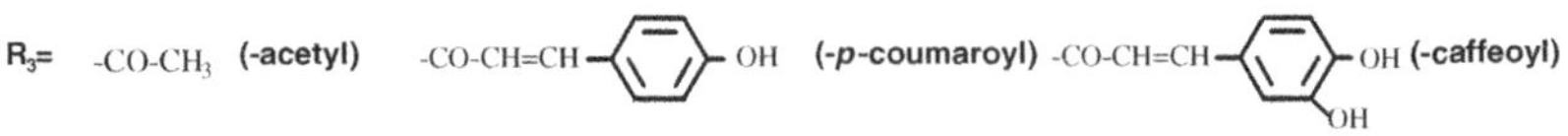

Fig. 9A.1 Chemical structures of anthocyanins

confirmation of the 3-*O*-glucosides, 3-*O*-acetylmonoglucosides and 3-*O*-*p*-coumaroylmonoglucosides of delphinidin, cyanidin, petunidin, peonidin and malvidin as well as the 3-*O*-caffeoylmonoglucosides of malvidin and peonidin in grapes and wines (Baldi et al. 1995; Giusti et al. 1999; Revilla et al. 1999; Wang and Sporns 1999; Favretto and Flamini 2000; Monagas et al. 2003; Wang et al. 2003b; Núñez et al. 2004). Recently, the 3-*O*-caffeoylmonoglucosides of cyanidin, delphinidin and petunidin have also been identified in grape skin (Vidal et al. 2004) and wine (Alcalde-Eon et al. 2006) fractions. The occurrence of the *cis* isomer of malvidin-3-(6-*p*-coumaroyl)-glucoside has also been confirmed in grapes and wine from *Vitis vinifera* (Monagas et al. 2003; Núñez et al. 2004). Later on, Alcalde-Eon et al. (2006) also detected in red wines the *cis* isomer of delphinidin, cyanidin- and petunidin-3-(6-*p*-coumaroyl)-glucosides together with the possible *cis* isomer of malvidin-3-(6-caffeoyl)-glucoside.

Although originally thought to be present only in grapes from non *V. vinifera* spp., the use of modern and more sensitive analytical techniques has allowed confirmation of the occurrence of anthocyanidin-3,5-diglucosides in *V. vinifera* grape skin extracts (Baldi et al. 1995; Vidal et al. 2004) and wines (Heier et al. 2002; Alcalde-Eon et al. 2006). Recently, the presence of 3,7-diglucosides has also been proposed (Alcalde-Eon et al. 2006). Finally, Vidal et al. (2004) has confirmed for the first time the existence of anthocyanin oligomers up to trimers in grape skin extracts. Oligomers were possibly linked by either an A-type (carbon-carbon and ether bonds) or B-type (carbon-carbon bond) linkages.

Varietal characteristics. The distribution and concentration of grape anthocyanins depends on the cultivar, maturity, climatic conditions, production area and fruit yield (Roggero et al. 1986a; González-San José et al. 1990; Vivas et al. 2001). In general, malvidin is the major anthocyanidin in red grape varieties, representing up to 90% in Grenache and less than 50% in Sangiovese. However, the amount of acylated anthocyanins is largely influenced by the grape variety and they could be absent from some varieties such as Pinot noir (Ribéreau-Gayón et al. 2000).

The anthocyanin profile has been used as chemotaxonomy criteria to establish differences between grape varieties. Relationships between the individual or the total concentration of the different anthocyanidins have been proposed for varietal characterization (Roggero et al. 1986b; Ortega Meder et al. 1994; Baldi et al. 1993; Roggero et al. 1988; Mattivi et al. 2006). These relationships are related to the enzymatic activities of flavonoid-3'-hydroxylase and *o*-dihydroxyphenol-*O*-methyltransferase (Ortega Meder et al. 1994; Roggero et al. 1986b). Another widely used variety classification is based on the presence and relative abundance of anthocyanins acylated with acetic and *p*-coumaric acids (Ortega Meder et al. 1994; Marx et al. 2000; Núñez et al. 2004), which is related to the grape's acetyl and cinnamoyl transferase activities (Roggero et al. 1988). The grape variety determines the production of each of these enzymes since they are direct expressions of the genome.

The wine anthocyanin composition depends on the original grape profile but also on the extraction and winemaking techniques employed. Maceration, which allows the diffusion of anthocyanins and other phenolic compounds from the solid part of the grape to the must, can occur before fermentation, as in the case of thermovinification, or during the alcoholic fermentation using crushed (traditional vinification) or whole (carbonic maceration) grapes. After reaching a maximum level after a few days of fermentation, the concentration of anthocyanins decreases as a consequence of their adsorption on yeast cell walls (Sect. 9A.4), precipitation in the form of colloidal material together with tartaric salts and elimination during filtration and fining. Hydrolysis reactions (i.e., enzymatic deglycosilation; Sect. 9A.3), as well as condensation reactions with other phenols (Sect. 9A.2) during winemaking also modify the anthocyanin composition of wines. Despite all these changes, the wine anthocyanin profile has also been used as chemotaxonomy criteria to establish differences between grape varieties, vineyard localization and yield, vintages and winemaking techniques (Etiévant and Schilich 1988; González-San José et al. 1990; Almela, et al. 1996; Arozarena et al. 2000).

Anthocyanin reactivity. Four different anthocyanin structures exist in equilibrium in acidic or neutral medium: the flavylium cation (red), the quinoidal base (blue), the hemiketal or carbinol pseudo-base (colorless) and the chalcone (colorless) (Brouillard 1982) (Fig. 9A.2). At wine pH ($\approxeq$3.5), the equilibrium is largely displaced towards to the colorless hemiketal form. Depending on pH, anthocyanins can act as electrophiles in the flavylium form through their C-2 and C-4 positions (C-ring), or as nucleophiles in the hemiketal form through their C-6 and C-8 positions (A-ring). However, the low content of flavylium cations present at wine pH does not seem to limit the progress of chemical reactions requiring this specie, as it will be discussed later (Sect. 9A.2).

Anthocyanin bleaching in wine can occur by the nucleophilic addition of either water at C-2 (Cheminat and Brouillard 1986) or bisulfite (an antifungal and antioxidant normally used during winemaking) at the C-4 position of the flavylium cation (Berke et al. 1998) (Fig. 9A.2). However, wine anthocyanins and thus color, can be stabilized either by copigmentation or through their conversion into more stable pigments by different condensation reactions that occur during the winemaking process. Copigmentation consists of the hydrophobic interaction of the polarizable planar nuclei of the colored form of anthocyanins (flavylium cation and quinoidal base) with another molecule or copigment (intermolecular copigmentation) or with

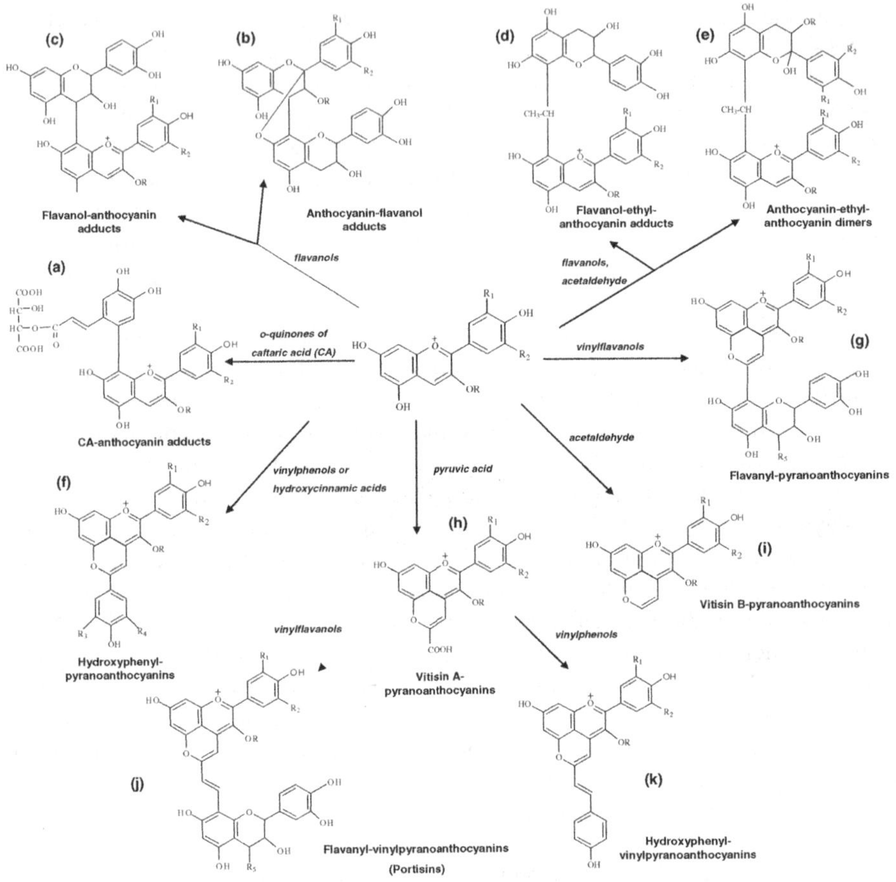

Fig. 9A.2 Anthocyanin equilibrium forms and bleaching reaction with bisulfite

an aromatic residue linked to the pigment (intramolecular copigmentation). Through this interaction, the nucleophilic attack of water at C-2 is partially reduced. The color properties of anthocyanins in wine are described in detail in Chapter 9B.

9A.2 Chemical Reactions of Anthocyanins During Winemaking: Main Anthocyanin-Derived Compounds

Main chemical reactions involving anthocyanins during winemaking include reactions with enzymatically generated *o*-quinones, direct and acetaldehyde-mediated anthocyanin-flavanol and anthocyanin-anthocyanin condensation reactions and reactions leading to pyranoanthocyanin formation (Fig. 9A.3). For each of these reactions, the precursors, mechanism, products obtained in model solution and identified in wine, are provided. Evidence of factors affecting these reactions, such as the concentration of the reactants, pH, temperature, metal ions and oxygen, among others, are also reported. Since flavanols and other phenolic compounds are also involved in many of these reactions, they are also described in Chapters 9B and 9C. The structure/organoleptic properties (color and taste) relationship of the corresponding reaction products is covered in Chapter 4.49D.

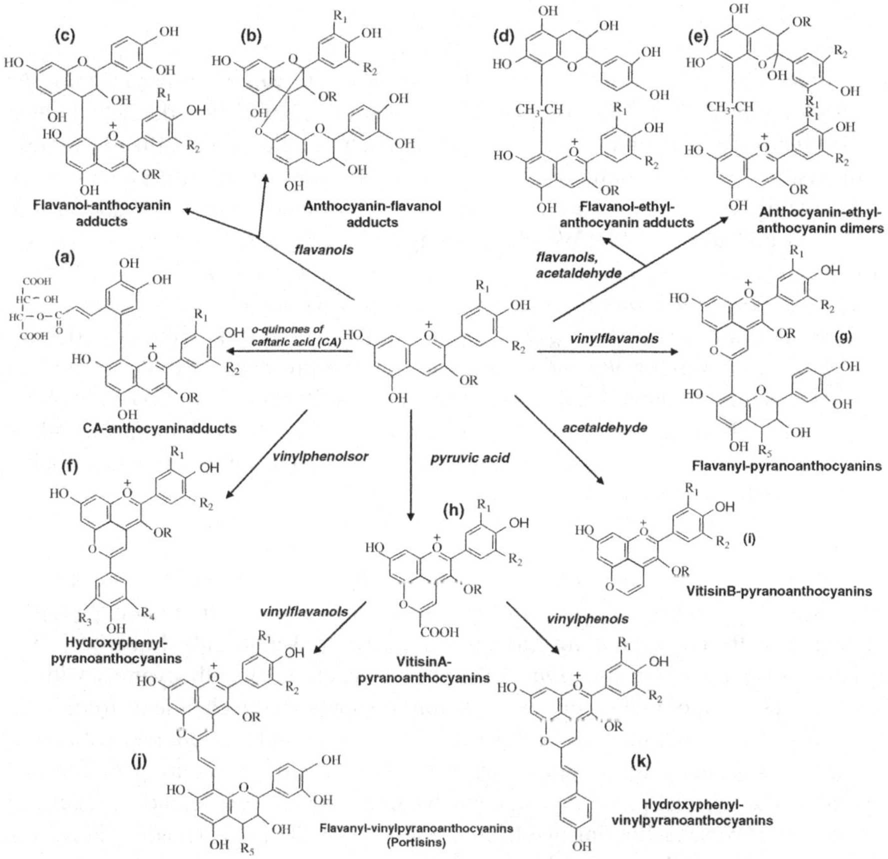

Fig. 9A.3 Main chemical reactions of anthocyanins during winemaking

9A.2.1 Reactions of Anthocyanins with Enzymatically Generated o-Quinones

Precursors. Precursors for this reaction are anthocyanins and *o*-quinones of caffeoyltartaric (caftaric acid) and *p*-coumaroyltartaric (cutaric acid) acids generated by enzymatic oxidation by grape polyphenoloxidase (PPO) (Singleton et al. 1985). PPO first adds an OH group to monophenols and then oxidizes the resulting *o*-diphenol to *o*-quinone.

Mechanism of reaction. Under oxidative conditions and for grapes containing high levels of hydroxycinnamic acids and low of levels of glutathione (a compound that easily reacts with *o*-quinones, preventing oxidation of phenolic compounds in wines) (Sarni-Manchado et al. 1995), some phenolics are able to react with the *o*-quinones of caftaric acid (Cheynier et al. 1986) (see also Chapter 9B).

Two different mechanisms have been postulated in order to explain the reactions of anthocyanins with these *o*-quinonic species (Cheynier et al. 1997; Sarni-Manchado et al. 1997):

a) *Coupled oxidation of anthocyanins to very unstable quinones regenerating the initial caftaric acid.* The first evidence of this mechanism was described when the glycosides of cyanidin, pelargonidin and delphinidin were oxidized by enzymatically generated *o*-quinones of catechol, (+)-catechin and chlorogenic acid, producing discoloration of the anthocyanin solution (Peng and Markakis 1963; Pifferi and Cultrera 1974; Wesche-Ebeling and Montgomery 1990).
b) *Condensation reactions of anthocyanin in its hemiketal form (nucleophilic) to the o-quinone (electrophilic) producing a colorless adduct which is then equilibrated to the corresponding flavylium form* (Fig. 9A.3a). In solutions containing malvidin-3-glucoside and caftaric acid in the presence of PPO, adducts of caftaric acid *o*-quinones with anthocyanins in hemiketal (colorless) or flavylium (red) forms have been observed (Sarni-Manchado et al. 1997). Although no NMR characterization of these adducts has been provided, the anthocyanin is probably linked to the quinone by its C8 or C6 positions.

Evidence in wine. The caftaric-malvidin-3-glucoside condensation product in flavylium form has been detected in red wine after the oxidation of Gamay grapes, together with the corresponding derivative of peonidin-3-glucoside (Labarbe 2000).

Factors affecting the reaction. The type of reaction of anthocyanins with *o*-quinones (coupled oxidation or condensation) depends on the chemical structure of the anthocyanin. *o*-Diphenolic anthocyanins (cyanidin, delphinidin and petunidin-3-glucosides) usually react rapidly by coupled oxidation, whereas peonidin and malvidin-3-glucosides participate in condensation reactions with quinonic species at a slower rate (Wesche-Ebeling and Montgomery 1990; Sarni-Manchado et al. 1997). As expected, the nucleophilic addition of anthocyanins on *o*-quinones, which requires the hemiketal form of the anthocyanin, was faster at pH 3.4 than at pH 1.7 (Sarni-Manchado et al. 1997). In relation to the effect of oxygen, Cheynier et al. (1997) concluded that in red vinification under oxidative conditions, the enzymatic oxidation of tartaric esters of hydroxycinnamic acids and the coupled oxidation of anthocyanins (delphinidin-3-glucoside) with enzymatically generated caftaric acid *o*-quinones, could take place but only resulting in relatively low losses of these compounds.

9A.2.2 Anthocyanin-Flavanol and Anthocyanin-Anthocyanin Direct Condensation Reactions

Precursors. The precursors of these reactions are anthocyanins and flavanols, acting both as electrophiles and nucleophiles (see also Chapter 9B).

Mechanism. Two different mechanisms have been proposed for this type of reaction leading to the anthocyanin-flavanol (A-F) and to the flavanol-anthocyanin (F-A) products (Fig. 9A.3):

a) *Anthocyanin-flavanol (A-F) direct condensation reaction.* The mechanism of reaction proposed for the A-F condensation reaction starts with the nucleophilic attack of the C-8 o C-6 position of the flavanol to the electrophilic C-4 position of the anthocyanin in the form of flavylium cation, giving rise either to a flavene, that can be oxidized to the corresponding flavylium cation and finally dehydrated to a xanthylium salt of yellow color (Jurd and Somers 1970; Somers 1971; Baranowski and Nagel 1983; Liao et al. 1992; Santos-Buelga et al. 1999), or to a colorless bicyclic condensation product (Fig. 9A.3b) first characterized in model solutions containing malvidin-3,5-diglucoside and (+)-catechin (Bishop and Nagel 1984).
b) *Flavanol-anthocyanin direct condensation reaction (F-A).* In the case of F-A condensation reaction, carbocations generated from the acid-catalyzed interflavanic bond cleavage typical of the chemistry of procyanidins (Haslam 1980) (Chapter 9B), act as electrophilic agents and react with the nucleophilic C-6 or C-8 position of the anthocyanin in its hydrated hemiketal form, giving rise to a colorless dimer that could be dehydrated to the corresponding red flavylium form (Fig. 9A.3c).

Evidence in wine. The products resulting from both mechanisms have been detected in red wine fractions with the aid of thiolysis and mass spectrometry (Remy et al. 2000). For F-A condensation products, the thiolysis results indicated that malvidin-3-glucoside was linked through its C-6 or C-8 position to an (epi)catechin moiety, generating the pigment (epi)catechin-malvidin-3-glucoside in flavylium form (MW=781). The presence of this compound has been extensively reported in wine (Vivar-Quintana et al. 2002; Monagas et al. 2003; Alcalde-Eon et al. 2004; Alcalde-Eon et al. 2006) but its structure has been only recently confirmed as an F-A product using ESI-MS, model solutions and a synthesized adduct identified as catechin-(4α $\rightarrow$ 8)-malvidin-3-glucoside by NMR (Salas et al. 2004a,b, 2005a,b). F-A products of the entire series of anthocyanidin-3-glucosides and their acylated derivates with (epi)catechin and (epi)gallocatechin have been also detected in wine fractions (Alcalde-Eon et al. 2006; Boido et al. 2006).

In the case of A-F condensation products, the work of Remy et al. (2000) confirmed that the anthocyanin was linked through its C-4 position to the (epi)catechin moeity forming colorless malvidin-3-glucoside-(epi)catechin adducts (MW=782) that were first tentatively identified as adducts in flavene form (Santos-Buelga et al. 1999) in model solutions containing catechin and malvidin-3-glucoside. However, the conversion of the flavene to the flavylium xanthylium previously described in the literature (Somers 1971) could not be confirmed. Later on, the formation of these A-F colorless adducts was investigated by HPLC/ESI-MS and thiolysis in model solutions containing malvidin-3-glucoside, (–)-epicatechin and/or

(+)-catechin and ethanol (Remy-Tanneau et al. 2003). Based on their resistance to thiolysis, it was postulated that the adducts were of bicyclic nature containing both C-C and ether type bonds (structure similar to A-type procyanidins), confirming the mechanism first proposed by Bishop and Nagel (1984). The major dimer was identified by NMR as malvidin-3-glucoside-(C2-*O*-C7, C4-C8)-epicatechin (Remy-Tanneau et al. 2003) and recently detected again in wine fractions obtained by high speed countercurrent chromatography (HSCC) (Salas et al. 2005a). According to Remy et al. (2000) both types of products, A-F and F-A, seem to be more associated with oligomeric procyanidins than with the polymeric ones. However, Hayasaka and Kennedy (2003) have detected molecular masses of A-F and F-A polymers in wine up to the level of octamers. Evidence of (epi)cat$_n$-anthocyanins and anthocyanin-*O*-(epi)cat$_n$ ($n \geq 2$) has also been obtained by thiolysis of HSCC wine fractions (Salas et al. 2005a).

Anthocyanin-anthocyanin dimers have also been detected in red wine fractions (Vidal et al. 2004; Salas et al. 2005a; Alcalde-Eon et al. 2007). Both, the flavene (malvidin-3-glucoside[flavene]-(C4-C8)-malvidin-3-glucoside[A^+]) and the bicyclic structure (malvidin-3-glucoside[flavane]-(C2-*O*-C7, C4-C8)-malvidin-3-glucoside [A^+]) have been proposed for these compounds. Other dimers containing malvidin-3-glucoside and either delphinidin, cyaniding, peonidin and petunidin -3-glucosides have also been detected in wine fractions (Salas et al. 2005a; Alcalde-Eon et al. 2007). Recently, F-A-A^+ oligomers consisting of (epi)catechin or (epi)gallocatechin and a dimeric anthocyanin have been detected in red wine fractions (Alcalde-Eon et al. 2007).

Factors affecting the reaction. Salas et al. (2003) have studied the influence of pH (2.0 and 3.8) on the progress of the reaction between the procyanidin dimer B2-3′-*O*-gallate (Ec-EcG) and malvidin-3-glucoside. Recently, similar experiments have been performed with a flavanol monomer ((–)-epicatechin) instead of B2-3′-*O*-gallate in the pH range 2.0-6.0 (Dueñas et al. 2006). The nucleophilic addition of epicatechin onto the flavylium cation of malvidin-3-glucoside occurred at all pH values leading to the malvidin-3-glucoside-epicatechin adduct in the flavene form. However, the conversion of this flavene into further products differed according to the pH. At pH 2 it proceeded to a colorless A-type dimer whereas at higher pH values (3.2–6.0), it was converted into new xanthylium pigments (Dueñas et al. 2006). Nevertheless, the precursor of these compounds, the intermediate flavylium form of the A-F product, could not be detected in this case as in previous experiments carried out with procyanidin dimer B2-3′-*O*-gallate at pH 3.8 (Salas et al. 2003) or with a flavanol trimer at pH 3.0 at high temperature (50 °C) (Malien-Aubert et al. 2002), indicating the role of the flavanol mDP in the stabilization of this specie. However, there is one case where a stable A-F dimer in the flavylium form was synthesized from a flavanol monomer (Escribano-Bailón et al. 1996), but a synthetic flavylium with no OH group at C-5 position was used instead of a common anthocyanin. Finally, when the experiments were performed at pH ≥4.0, (–)-epicatechin was oxidized to *o*-quinones. The nucleophilic addition of the anthocyanin in its hemiketal form onto this specie was promoted, leading to a new series of F-A adducts.

Studies in model solution containing an anthocyanin and flavanol oligomers (up to tetramers) at pH 3 carried out at 50 °C demonstrated that temperature is another factor that affects the progress of direct condensation reactions (Malien-Aubert et al. 2002). At acidic pH and high temperature, the anthocyanin is in equilibrium with the colorless chalcone. Although breakage of the flavanol C-C bond occurred under these conditions, the formation of the chalcone impeded the synthesis of F-A products and only A-F adducts were formed (Malien-Aubert et al. 2002).

9A.2.3 Anthocyanin-Flavanol and Anthocyanin-Anthocyanin Aldehyde-Mediated Condensation Reactions

Precursors. Acetaldehyde is a yeast metabolite present in wine as a result of the alcoholic fermentation. Acetaldehyde is also gradually generated during wine aging from the oxidation of ethanol in the presence of polyphenols (Wildenradt and Singleton 1974). Other aldehydes such as formaldehyde, propionaldehyde, isovaleraldehyde, benzaldehyde, isobutyraldehyde and 2-methylbutyraldehyde are also minor constituent of wines, but important components of spirits used for the fortification of Port wines (Pissarra et al. 2005b). Furfural, 5-hydroxymethylfurfural and vanillin are compounds resulting from oak toasting, could also be found in wines submitted to barrel aging. Other compounds bearing an aldehyde moiety, such as glyoxylic acid (COOH-CHO), can be generated from the oxidation of tartaric acid by traces of iron or copper ions (Oszmianski et al. 1996; Clark and Scollary 2002; Es-Safi et al. 2003).

Mechanism of reaction. The mechanism for aldehyde-mediated condensation reactions was first proposed by Timberlake and Bridle (1976). The aldehyde, in the form of carbocation, reacts with the flavanol (or tannin) at position C-6 or C-8 of the phloroglucinol ring (Chapter 9B). After dehydratation, this flavanol-aldehyde adduct gives rise to a new carbocation that attacks the anthocyanin. The resulting compound is stabilized by deprotonation forming a quinoidal base of violet color. The proposed structure involves the flavanol and the anthocyanin linked by an ethyl bridge -(CH-CH_3) (Fig. 9A.3d). The flavanol is attached to ethyl bridge by its C-8 or C-6 positions and the anthocyanin by its C-8 position resulting in the C8-(CH-CH_3)-C8 and in the C6-(CH-CH_3)-C8 isomers. In according to this, model solutions containing malvidin-3-glucoside and either (+)-catechin or (–)-epicatechin in the presence of acetaldehyde generally give rise to two major products (Roggero et al. 1987; Bakker et al. 1993; Rivas-Gonzalo et al. 1995). Due to the existence of an asymmetric carbon in the ethyl bridge, two diastereoisomers differing on the stereochemistry (*R* or *S*) of the methine carbon are also formed for each of the regioisomers (Rivas-Gonzalo et al. 1995; Escribano-Bailón et al. 1996). Lee et al. (2004) have recently reported the NMR data of ethyl-linked malvidin-3-glucoside-(epi)catechin pigments.

Ethyl-linked products usually undergo polycondensation; the initial products evolve to new pigments with high degree of polymerization that finally precipitate

(Bakker et al. 1993; García-Viguera et al. 1994; Escribano-Bailón et al. 1996; Es-Safi et al. 1999). Studies in model solutions using HPLC/ESI-MS have revealed the detection of trimeric and tetrameric pigments linked by an ethyl bridge being composed of various flavanol units but no more than two anthocyanin moeties, since only one of the A-ring positions of the latter (C-8) was believed to be involved in the polymerization process, while two positions (C-6 and C-8) could be used in the case of the flavanol (Es-Safi et al. 1999). Contrarily to these findings, Atanasova et al. (2002b) demonstrated that the C-6 position of anthocyanins was also reactive, although to a lesser extent than the C-8 position since the polymerization of anthocyanins in the absence of flavanols was also induced by acetaldehyde, giving rise to dimers, trimers and tetramers composed of ethyl-linked anthocyanins units in different structural forms (flavylium, hemiketal and quinoidal) (Fig. 9A.3e). Both C-6 and C-8 positions seemed reactive in the hemiketal form whereas in the case of the flavylium form only C-8 position was reactive and therefore ceased the polymerization process.

Model solutions containing malvidin-3-glucoside and flavanols (monomers and dimers) in the presence of other aldehydes, such as isovaleraldehyde, benzaldehyde, propionaldehyde, isobutyraldehyde, formaldehyde, 2-methylbutyraldehyde, vanillin, furfural and hydroxymethylfurfural also resulted in anthocyanin-aryl/alkyl-flavanol pigments through the same mechanism described above for acetaldehyde (Pissarra et al. 2003; Sousa et al. 2007). The NMR characterization of malvidin-3-glucoside-catechin aryl/alkyl-linked pigments have recently been described (Pissarra et al. 2004, 2005a; Sousa et al. 2007). Similar pigments have also been reported in model solutions containing furfural and hydroxymethylfurfural as aldehydes (Es-Safi et al. 2000).

Colored adducts in which the anthocyanin and the flavanol are linked by a carboxy-methine bridge, have also been identified, in addition to colorless carboxy-methine-linked catechin dimers (Chapter 9B), in model solutions containing (+)-catechin, malvidin-3-glucoside and glyoxylic acid (Es-Safi et al. 2003).

Evidence in wine. The occurrence of ethyl-linked condensation products in wine is well documented. The adducts malvidin-3-glucoside-(CH-CH_3)-catechin (Revilla et al. 1999; Vivar-Quintana et al. 1999, 2002; Atanasova et al. 2002a; Heier et al. 2002; Mateus et al. 2002a; Alcalde-Eon et al. 2004, 2006; Monagas et al. 2003; Wang et al. 2003a; Salas et al. 2005a; Boido et al. 2006), malvidin-3-(6-*p*-coumaroyl)-glucoside-(CH-CH_3)-catechin (Mateus et al. 2003a; Alcalde-Eon et al. 2004, 2006; Monagas et al. 2003; Boido et al. 2006) and malvidin-3-glucoside[AOH]-(CH-CH_3)-malvidin-3-glucoside[A^+] (Atanasova et al. 2002b; Salas et al. 2005a) were first confirmed both in wine and in wine fractions by ESI-MS. Ethyl-(epi)catechin adducts of the glucosides and acylated-glucosides of the entire series of anthocyanidin-3-glucosides (cyanidin, delphinidin, peonidin, petunidin and malvidin) as well as a ethyl-procyanidin dimer of malvidin-3-glucoside have been later described in red wines (Atanasova et al. 2002a; Heier et al. 2002; Alcalde-Eon et al. 2004, 2006; Salas et al. 2005a). The occurrence of ethyl-linked adducts of (epi)gallocatechin with the acetyl-glucoside of malvidin as well as with the different anthocyanidin-glucosides has also been reported in wine

(Alcalde-Eon et al. 2004, 2006; Boido et al. 2006). Finally, evidence of (epi)cat$_n$-ethyl-anthocyanins adducts ($n \geq 2$) has been recently obtained by thiolysis of HSCC wine fractions (Salas et al. 2005a). However, to date the occurrence of condensation products between anthocyanins and aldehydes other than acetaldehyde has not been confirmed in wines.

Factors affecting the reaction. The anthocyanin-flavanol acetaldehyde-mediated condensation reaction follows a first-order kinetic in relation to the disappearance of the anthocyanin but does not follow any simple reaction order regarding the appearance of polymers, probable due to their eventual precipitation (Baranowski and Nagel 1983). The reaction rate is higher in the presence of oxygen and at acidic pH, since the formation of acetaldehyde and its protonated form are favoured under these conditions, respectively (García-Viguera et al. 1994; Rivas-Gonzalo et al. 1995; Atanasova et al. 2002a). Temperature also affects the evolution and accumulation of the new formed pigments. At low temperatures (15 °C), the polymers are slowly accumulated and are more stable in relation to their degradation and precipitation (Baranowski and Nagel 1983; Rivas-Gonzalo et al. 1995).

The rate of the reaction between procyanidin B2 and different antocyanidin-3-glucosides in the presence of acetaldehyde, increases in the following order: malvidin-3-glucoside<cyanidin-3-glucoside<peonidin-3-glucoside (Dallas et al. 1996b). On the other hand, the degree of polymerization of the flavanol is also important for the reaction rate. In the presence of acetaldehyde, the reaction with malvidin-3-glucoside increases in the following order: (+)-catechin<(–)-epicatechin <procyanidin B3<procyanidin B2-3'-*O*-gallate<procyanidin B2<procyanidin B1 <procyanidin C1 (Dallas et al. 1996a).

9A.2.4 Reactions Leading to Pyranoanthocyanin Formation

9A.2.4.1 Anthocyanin-Vinylphenol/Anthocyanin-Hydroxycinnamic Acid Condensation Reactions: Hydroxyphenyl-Pyranoanthocyanins

Precursors. Both hydroxycinnamic acids and 4-vinylphenols can lead to the formation of hydroxyphenyl-pyranoanthocyanins. The main hydroxycinnamic acids present in wines are *p*-coumaric, caffeic, ferulic and sinapic acids. 4-Vinylphenol and 4-vinylguaiacol are volatile phenols associated with off flavors in wine (Etiévant 1981) and arise from the decarboxylation of *p*-coumaric and ferulic acid, respectively, via the yeast cinnamate decarboxylase (CD) (Chatonnet et al. 1993).

Mechanism of reaction. The mechanism of the reaction between anthocyanins and 4-vinylphenols was first proposed by Fulcrand et al. (1996). Hydroxyphenyl-pyranoanthocyanins result from the cycloaddition of the ethylenic bond of the 4-vinylphenol molecule at positions C-4 and C-5 of the anthocyanin followed by an oxidation process, resulting in a pyrane ring (Fulcrand et al. 1996) for which this type of compound receives the name of pyranoanthocyanins (Fig. 9A.3f). The vinylphenol addition at C-4 protects the anthocyanin of being hydrated.

Schwarz et al. (2003b) proposed a different mechanism for the formation of hydroxyphenyl-pyranoanthocyanins in red wines, including the free hydroxycinnamic acid and the anthocyanin without enzymatic support (Fig. 9A.3f). The C-2 (nucleophilic) position of caffeic acid is initially linked to the C-4 (electrophilic) position of malvidin-3-glucoside giving rise to an electron deficient intermediate (carbenium ion) which is stabilized with the aromatic ring substituents (electron donors) of the cinnamic acid moiety. The intermediate carbenium ion can be intramolecularly trapped by the anthocyanin-OH group at C-5 forming a pyrane ring. The final product is then formed by oxidation and decarboxylation of the intermediate. Consequently, only cinnamic acids with electron-donor substituents, like *p*-coumaric, ferulic, caffeic and sinapic acids, could be involved in the reaction in order to stabilize the carbenium ion. Through this new mechanism it was possible to explain why caffeic and sinapic acids, compounds for which there is no evidence for an enzymatic decarboxylation by *Saccharomyces cerevisiae* CD into their respective 4-vinylphenols, can also give rise to hydroxyphenyl-pyranoanthocyanins (Chatonnet et al. 1993).

Evidence in wine. Hydroxyphenyl-pyranoanthocyanins, firstly detected and isolated from polymeric membranes employed for red wine microfiltration (Cameira dos Santos et al. 1996), have been studied by UV-visible, mass and NMR spectrometry (Fulcrand et al. 1996). This study confirmed the presence of phenyl-pyrano derivatives of malvidin-3-glucoside and malvidin-3-(6-*p*-coumaroyl)-glucoside (Fulcrand et al. 1996). The corresponding phenyl-pyrano of the entire series of anthocyanidin-3-glucosides and their *p*-coumaroyl derivatives (with the exception of cyanidin), malvidin-3-(6-acetyl)-glucoside and malvidin-3-(6-caffeoyl)-glucoside, have been also reported in wines (Asenstorfer et al. 2001; Hayasaka and Asenstorfer 2002; Mateus et al. 2003a; Alcalde-Eon et al. 2004, 2006; Monagas et al. 2003; Wang et al. 2003a; Pozo-Bayón et al. 2004; Boido et al. 2006).

Using a combination of mass spectrometry techniques (nano-ESI-MS/MS), Hayasaka and Asenstorfer (2002) later identified a new series of hydroxyphenyl-pyranoanthocyanins in red wine fractions, including the catechyl-, guaiacyl- and siringyl-pyrano derivatives of malvidin-3-glucoside. Catechyl-pyranomalvidin-3-glucoside (also called pinotin A) has been isolated from *Vitis vinifera* cv Pinotage red wine and characterized it by HPLC-ESI/MS and NMR (Schwarz et al. 2003a). The presence of catechyl-pyrano derivatives of the entire series of anthocyanidin-3-glucoside (with the exception of cyanidin), malvidin-3-(6-acetyl)-glucoside and delphinidin, petunidin and malvidin-3-(6-*p*-coumaroyl)-glucosides, as well as the guaiacyl-pyrano derivatives of malvidin-3-glucoside and its acylated forms have been later confirmed in wines (Alcalde-Eon et al. 2004, 2006; Monagas et al. 2003; Wang et al. 2003a; Pozo-Bayón et al. 2004; Boido et al. 2006).

Factors affecting the reaction. Schwarz, Hofmann, and Winterhalter (2004) have investigated the factors influencing the formation of Pinotin A and its correlation with wine age. It was found that the formation of this pigment was more dependent on the concentration of caffeic acid than in that of malvidin-3-glucoside. Although an exponential increase of the concentration of Pinotin A was observed with prolonged aging time, the most rapid synthesis was observed when malvidin-3-glucoside was degraded to a larger extent (2.5–4 year old wines) due to its participation in other chemical reactions. This led to an increase of the ratio of caffeic acid/malvidin-3-glucoside since the concentration of caffeic acid remained very stable during the storage time. However, a minimum concentration of malvidin-3-glucoside (5–10 mg/L) was found to be needed to maintain the reaction rate high enough to compensate for its simultaneous incorporation into other products. Recently, Rentzsch et al. (2007) also confirmed that Pinotin A formation in Grenache red wines only takes place after a prolonged aging time, whereas hydroxyphenyl-pyranoanthocyanins derived from *p*-coumaric and ferulic acids were produced by the enzymatic pathway during fermentation and by chemical synthesis during the aging process.

9A.2.4.2 Anthocyanin-Vinylflavanol Condensation Reactions: Flavanyl-Pyranoanthocyanins

Precursors. The precursors for this reaction are anthocyanins, flavanols or flavanols containing a vinyl residue at C-8 (i.e., 8-vinylflavanols). 8-Vinylflavanols could arise from the cleavage of flavanol-ethyl-flavanol oligomers or from the dehydration of the flavanol-ethanol adduct formed after the attack of aldehyde cation to the flavanol (Chapter 9B). Saucier et al. (1997) have supported evidence for this precursor when detecting an ion corresponding to vinyl-catechin from the fragmentation of ethyl-linked catechin dimers under ESI-MS in positive or negative mode.

Mechanism of reaction. The mechanism proposed for the formation of flavanyl-pyranoanthocyanins (Francia-Aricha et al. 1997; Mateus et al. 2003a,b) (Fig. 9A.3g), is similar to that described above by Fulcrand et al. (1996) for the formation of hydroxyphenyl-pyranoanthocyanins (Sect. 9A.2.4.1; Fig. 9A.3f).

Evidence in wine. Flavanyl-pyranoanthocyanins derived from the reaction between malvidin-3-glucoside and (+)-catechin, (–)-epicatechin and procyanidin B2, first reported in model solutions containing malvidin-3-glucoside, acetaldehyde and the respective flavan-3-ol (Francia-Aricha et al. 1997), have also been identified in wines (Mateus et al. 2003a,b; Atanasova et al. 2002a; Alcalde-Eon et al. 2004, 2006; Mateus et al. 2003a; Monagas et al. 2003; Wang et al. 2003a; Boido et al. 2006). Complete NMR characterization has been provided for the flavanyl-pyrano derivatives of malvidin-3-glucoside and its *p*-coumaroyl ester with (+)-catechin, (–)-epicatechin and procyanidin B3 and for derivatives of malvidin-3-(6-*p*-coumaroyl)-glucoside with (+)-catechin, (–)-epicatechin and procyanidin B1 isolated from Port wines (Mateus et al. 2002b, 2003a). Asenstorfer et al. (2001) have identified a large family of flavanyl pyranoanthocyanins of oligomeric nature including the –di-, tri- and tetra-catechin-pyrano derivatives of malvidin-3-glucoside and its acylated forms in a commercial grape marc. Atanasova et al. (2002a) also demonstrated, via thiolysis, that both monomeric flavanols and proanthocyanidins could be involved in the formation of anthocyanin-vinylflavanol adducts.

He et al. (2006a) isolated and quantified oligomeric flavanyl-pyranomalvidin-3-glucosides [(+)-catechin, (–)-epicatechin and procyanidin B3] and flavanyl-pyrano-malvidin-3-(6-*p*-coumaroyl)-glucosides [(+)-catechin, (–)-epicatechin and procyanidin B1] from Port wine from different ages (3-, 4- and 6-year-old wines). In general, the profile of flavanyl-pyranoanthocyanin pigments in the different wines was similar to that of their precursor in grapes. The concentration of malvidin-3-glucoside-based flavanyl-pyranoanthocyanins was higher than those of malvidin-3-(6-*p*-coumaroyl)-based pigments. Pyranoanthocyanin-procyanidin adducts were more abundant than the corresponding monomeric ones and among these latter forms, pyranoanthocyanin-(–)-epicatechin were higher that the corresponding (+)-catechin derivatives.

Factors affecting the reaction. To date there has been only one study in relation to the evolution of these pigments during wine aging. According to He et al. (2006a), the concentration of flavanyl-pyranoanthocyanin could be affected during wine aging. An increase in the concentration of (+)-catechin and (–)-epicatechin derivatives of pyranomalvidin-3-glucoside was observed from 3 to 6 years of aging, probably due to the increase of the vinyl(epi)catechin precursors from ethyl-linked (epi)catechin oligomers or to the hydrolysis of the corresponding acylated pyranoanthocyanins.

9A.2.4.3 Condensation Reaction Between Anthocyanins and Enolizable Aldehydes and Ketones: Carboxy-Pyranoanthocyanins and Related Compounds

Precursors. Precursors for this reaction are compounds exhibiting keto-enol tautomerism. These compounds are usually secondary metabolites derived from the glycolysis cycle of yeast metabolism during fermentation. Pyruvic acid is one of the main precursor compounds involved in this type of reaction. During yeast fermentation it is decarboxylated to acetaldehyde and then reduced to ethanol. Acetone, acetoin (3-hydroxybutan-2-one), oxalacetic acid, acetoacetic acid and diacetyl, among others, are also secondary metabolites likely to participate in this kind of condensation reaction with anthocyanins.

Mechanism of reaction. The adduct of malvidin-3-glucoside with pyruvic acid, also known as vitisin A(Fig. 9A.3h), was firstly detected in fortified red wines (Bakker et al. 1997) and in a grape marc (Fulcrand et al. 1998) and further isolated and characterized by NMR (Bakker et al. 1997; Fulcrand et al. 1998). According to Fulcrand et al. (1998), the reaction between pyruvic acid and grape anthocyanins occurs through a series of steps similar to those previously described for the hydroxyphenyl-pyranoanthocyanins (Sect. 9A.2.4.1; Fig. 9A.3f). Later studies performed by NMR (Mateus et al. 2001b) and mass spectrometry (Asenstorfer et al. 2001; Hayasaka and Asenstorfer 2002) have confirmed the structure proposed by Fulcrand et al. (1998). This mechanism is extended to the condensation reaction between anthocyanins and other enolizable precursors found in wine (Benabdeljalil et al. 2000).

Evidence in wine. The carboxy-pyranoanthocyanins have been extensively studied in wines. The derivatives of malvidin-3-glucoside and of its *p*-coumaroyl and acetyl esters, were first identified both in red wines (Revilla et al. 1999; Vivar-Quintana et al. 1999; Asenstorfer et al. 2001; Hayasaka and Asenstorfer 2002; Atanasova et al. 2002a; Heier et al. 2002; Alcalde-Eon et al. 2004, 2006; Monagas et al. 2003; Morata et al. 2003a; Pozo-Bayón et al. 2004; Boido et al. 2006) and in Port wines (Bakker et al. 1997; Romero and Bakker 2000b; Mateus et al. 2003a).

Similarly, the entire series of anthocyanidin-3-glucosides and most of their *p*-coumaroyl and acetyl esters have also been reported (Atanasova et al. 2002a; Heier et al. 2002; Mateus et al. 2003a; Alcalde-Eon et al. 2004, 2006; Wang et al. 2003a; Boido et al. 2006).

Pyranoanthocyanins arising from the condensation between anthocyanins and acetaldehyde have also been identified in wines. The corresponding derivatives of malvidin-3-glucoside and its acetyl ester, also known as vitisin B (Fig. 9A.3i) and acetylvitisin B, respectively, firstly isolated from Port wines (Bakker and Timberlake 1997), were later identified in a synthetic medium fermented by yeasts (Benabdeljalil et al. 2000), as well as in red wines and in red wine fractions (Revilla et al. 1999; Vivar-Quintana et al. 1999, 2002; Atanasova et al. 2002a; Hayasaka and Asenstorfer 2002; Heier et al. 2002; Mateus et al. 2002b; Monagas et al. 2003; Wang et al. 2003a; Alcalde-Eon et al. 2006; Boido et al. 2006). The corresponding B-type vitisins of numerous glucosides (delphinidin, petunidin and peonidin-3-glucosides) and acylated-glucosides (peonidin-3-(6-acetyl)-glucosides and malvidin-3-(6-*p*-coumaroyl)-glucosides) of anthocyanidins have been identified in wine (Heier et al. 2002; Alcalde-Eon et al. 2006; Boido et al. 2006).

Other pyranoanthocyanins bearing a methyl moiety, resulting from the reaction between anthocyanins and acetone (Benabdeljalil et al. 2000; Lu and Foo 2001; Hayasaka and Asenstorfer 2002; Alcalde-Eon et al. 2006) or with acetoacetic acid (He et al. 2006a) have also been reported in wine. Recently, complete characterization by UV-visible spectroscopy, NMR and mass spectrometry has been provided for the methyl-linked pyranomalvidin-3-glucoside and its *p*-coumaroyl ester (He et al. 2006b). Finally, pyranoanthocyanins resulting from the reaction of anthocyanins and vinylalcohol have also been described in red wines (Hayasaka and Asenstorfer 2002).

Factors affecting the reaction. The extent of the reactions between anthocyanins and pyruvic acid in model solutions follows a first order kinetic with respect to the anthocyanin disappearance. This reaction is affected by several factors, such as: anthocyanin composition, pH, pyruvic acid concentration, temperature and acetaldehyde concentration. The maximum formation took place at pH 2.7–3.0 due to requirement of the anthocyanin flavylium form, at high pyruvic acid concentration, at low storage temperature (10–15 °C) and in the absence of acetaldehyde (Romero and Bakker 1999a,b, 2000a,b).

Recent studies indicated that fermentation was the most important stage for the production of malvidin-3-glucoside-pyruvate (Asenstorfer et al. 2003). Maximum production of this compound in *Vitis vinifera* cv Shiraz musts, occurred in the period corresponding to 20–85% of glucose utilization, coinciding with the maximum concentration of both precursors, malvidin-3-glucoside and pyruvic acid. Morata et al. (2003a) have reported that the yeast strain used in the alcoholic fermentation also affected the production of malvidin-3-glucoside-pyruvate, and that the concentration of the pigment was in direct relation with the production of pyruvic acid by the yeast. Moreover, the content of SO_2 in must was also shown to influence the production of malvidin-3-glucoside-pyruvate since SO_2 regulates the concentration of pyruvic acid through the formation of a weak bisulfite addition compound

(Asenstorfer et al. 2003). Similarly, malolactic fermentation can also affect the production of the pigment since lactic acid bacteria have the capacity of using pyruvic acid (Asenstorfer et al. 2003). No synthesis and/or losses of the pigment were found at low SO_2 concentration and occurrence of malolactic fermentation, while a maximum production was achieved under the opposite conditions.

The evolution of anthocyanin-pyruvic acid adducts during wine aging seems to follow different patterns in function of the fermentation and aging conditions employed during the winemaking process. For instance, in wines treated with pectolitic enzymes, Revilla and González-San José (2001) found an increase in the concentration of malvidin-3-glucoside pyruvate during the first six months of aging in bottle, followed by a slight decrease. Similarly, Atanasova et al. (2002a) found an increase in these pigments both in non-oxygenated and oxygenated wines during a seven-month period, concluding that the anthocyanin-pyruvic acid condensation reaction was not influenced by oxygen. On the other hand, Mateus et al. (2001) reported losses (9–18%) of malvidin-derived pyruvic acid adducts in Port wines during 38 months in bottles, although losses were higher (70%) in wines stored in oak barrels (oxidative conditions) during the same period of aging. Finally, Pérez-Magariño and González-San José (2004) reported an increment of these derivatives in red wines during the first eight months of storage in oak barrels.

9A.2.4.4 Pyranoanthocyanins as Precursors of Other Anthocyanin-Derived Pigments: Vinylpyranoanthocyanin Pigments

Precursors. Carboxy-pyranoanthocyanins (Sect. 9A.2.4.3), 8-vinylflavanols and vinylphenols are the precursors for this reaction. The origin of 8-vinylflavanols and vinylphenols in wine has been already described in Sects. 9A.2.4.2 and 9A.2.4.1, respectively.

Mechanism of reaction. Flavanyl-vinylpyranoanthocyanin pigments, also called "portisins", were first isolated from aged Port wine (Mateus et al. 2003b). Their characterization by ESI/MS and NMR revealed a structure comprised of a pyranoanthocyanin linked to a flavanol by a vinyl bridge (Mateus et al. 2004) (Fig. 9A.3j). The proposed mechanism suggests that carboxy-pyranoanthocyanins, derived from the condensation reaction between anthocyanins and pyruvic acid, react at its C-10 position with the vinyl group of an 8-vinylflavanol derivative. The last step of the synthesis involves the loss of the formic group followed by oxidation giving rise to a pigment of blue color. This mechanism could be also extrapolated to the formation of phenyl-vinylpyranoanthocyanin pigments formed from the reaction between carboxy-pyranoanthocyanins and vinylphenols (Mateus et al. 2006; Oliveira et al. 2007) (Fig. 9A.3k).

Evidence in wine. The procyanidin dimer-vinylpyranomalvidin-3-glucoside and its *p*-coumaroyl ester have been isolated from Port wine and completely characterized by ESI/MS and NMR (Mateus et al. 2003b). Later, the catechin-vinylpyrano derivatives of petunidin, peonidin and malvidin-3-glucosides, malvidin-3-(6-acetyl)-glucoside and peonidin and malvidin-3-(6-*p*-coumaroyl)-glucosides, have also been identified in Port wine fractions (Mateus et al. 2005). Similarly, the pigment phenylvinylpyranomalvidin-3-glucoside has been recently detected in aged red wine (Mateus et al. 2006).

Factors affecting the reaction. Factors affecting the formation vinylpyranoanthocyanins must comprised those previously described for their precursor compounds.

9A.3 Enzymatic Deglycosilation of Anthocyanins During Winemaking

The β-glucosidase activity of certain strains of *Saccharomyces cerevisiae* wine yeast, which has a positive effect in wine aroma releasing the volatile aglycone of terpenol glycosides (Palmeri and Spagna 2007 and references therein), is also responsible for breakdown of the glucosidic bond of the anthocyanidin-3-glucosides (anthocyanin-β-glucosidase or anthocyanase activity). The released anthocyanidin is very unstable and is rapidly degraded leading to losses in wine color (Mazza and Miniati 1993).

There is very little information in relation to the effects of yeast β-glucosidase activity on grape and wine anthocyanins. Although *S. cerevisiae*, the main wine yeast, is not a good producer of β-glucosidase, other non-*Saccharomyces* wine yeasts (*Brettanomyces, Candida, Debaromyces, Dekkera, Hanseniaspora, Hansenula, Kloeckera, Pichia, Rhodotorula, Saccharomycodes, Schizosaccharomyces, Torulaspora,* among others) are considered potential sources of this enzyme. Sánchez-Torres et al. (1998) reported that the action of a *Candida molischiana* β-glucosidase activity resulted in a 50% color loss during the fermentation time. More recently, Manzanares et al. (2000) studied the glucosidase activity in 53 yeast strains belonging to the genera *Candida, Dekkera, Hanseniaspora, Metschnikowia, Pichia, Rhodotorula, Schizosaccharomyces* and *Zygosaccharomyces*, mainly isolated from grape and wine. Yeast strains showing the largest β-glucosidase activity belonged to the genera *Candida, Hanseniaspora* and*Pichia* and exhibited cell wall bond activity. The anthocyanin-β-glucosidase activity resulted in 30–70% color loss among yeast strains considered as best producers of the enzyme. To date, no structural dependent effect on anthocyanin-β-glucosidase activity has been reported.

9A.4 Adsorption of Anthocyanins on Yeast Cell Walls

The main structural constituents of *Saccharomyces cerevisiae* yeast cell wall are glucans and mannans with a minor proportion of chitin (Walker 1998). Mannoproteins are located in the outer layer of the yeast cell wall and determine most of the surface properties of the wall. Vasserot et al. (1997) studied the capacity of yeast lees to adsorb anthocyanins in an attempt to reduce the detrimental effects of charcoal on the color of red musts and wines. Experiments based on model wine solutions revealed that yeast lees possess a greater affinity for anthocyanins than charcoal but that the adsorption mechanism is based on weak and reversible interactions. The adsorption capacity of yeast cell was limited by the partition equilibrium between the anthocyanins to be adsorbed and the fraction remaining free in solution and at the same time was affected by the temperature, anthocyanin initial content and concentration of yeast lees employed. The amount of anthocyanin adsorbed was also influenced by the ethanol concentration, SO_2, pH and pigment structure (Vasserot et al. 1997). Anthocyanins were absorbed in proportion to their polarity as follows: delphinidin>cyanidin>petunidin>peonidin>malvidin. In contrast, Mazauric and Salmon (2005), by analyzing the remnant polyphenols in the medium in experiments carried out in model wine solution, recently found that the adsorption of anthocyanins on yeast lees was unrelated to their polarity. Desorption experiments indicated that the acetyl derivatives remained more adsorbed than the remaining ones (Mazauric and Salmon 2006). These findings were partly in accordance with the results obtained in a recent study performed in model solutions with fresh yeast (Medina et al. 2005). Morata et al. (2003, 2005) demonstrated that the yeast strain largely influenced the anthocyanin adsorption capacity of yeast cell walls during fermentation of *V. vinifera* Cabernet Sauvignon and Graciano musts. However, in contrast to the results obtained by Vasserot et al. (1997) on yeast lees, the less polar acylated anthocyanins, in particular the cinnamoyl derivatives were more strongly adsorbed than the more polar non-acylated ones on the cell walls of fermenting yeasts. Discrepancies between these studies could be attributed to the use of model solutions vs must and to different physiological status of the yeast used in the experiments.

Acknowledgments The authors are grateful to the AGL2006-04514 project (Ministerio de Educación y Ciencia) for financial support of this work.

References

Alcalde-Eon, C., Escribano-Bailón, M.T., Santos-Buelga, C., & Rivas-Gonzalo, J.C. (2004). Separation of pyranoanthocyanins from red wine by column chromatography. *Anal. Chim. Acta 513*, 305–318.

Alcalde-Eon, C., Escribano-Bailón, M.T., Santos-Buelga, C., & Rivas-Gonzalo, J.C. (2006). Changes in the detailed pigment composition of red wine during maturity and ageing. A comprehensive study. *Anal. Chim. Acta 563*, 238–254.

Alcalde-Eon, C., Escribano-Bailón, M.T., Santos-Buelga, C., & Rivas-Gonzalo, J.C. (2007). Identification of dimeric anthocyanins and new oligomeric pigments in red wine by jeans of HPLC-DAD-ESI-MS^n. *J. Mass Spect., 42*, 735–748.

Almela, L., Javaloy, S., Fernández-López, J.A., & López-Roca, J.M. (1996). Varietal classification of young red wines in terms of chemical and colour parameters.*J. Sci. Food Agric., 70*, 173–180.

Arozarena, I., Casp, A., Marín, R., & Navarro, M. (2000). Multivariate differentiation of Spanish red wines according to region and variety. *J. Sci. Food Agric. 70*, 1909–1917.

Asenstorfer, R.E., Hayasaka, Y., & Jones, G.P. (2001). Isolation and structures of oligomeric wine pigments by bisulfite-mediated ion-exchange chromatography. *J. Agric. Food Chem. 4*, 5957–5963.

Asenstorfer, R.E., Markides, A.J., Iland, P.G., & Jones, G.P. (2003). Formation of vitisin A during red wine vinification and maturation. *Aus. J. Grape Wine Res. 9*, 40–46.

Atanasova, V., Fulcrand, H., Cheynier, V., & Moutounet, M. (2002a). Effect of oxygenation on polyphenol changes occuring in the course of wine-making. *Anal. Chim. Acta 458*, 15–27.

Atanasova, V., Fulcrand, H., Le Guerneve, C., Cheynier, V., & Moutounet, M. (2002b). Structure of new dimeric acetaldehyde malvidin-3-glucoside condensation product. *Tetrahedron Lett. 43*, 6151–6153.

Bakker, J., & Timberlake, C.F. (1997). Isolation, identification and characterization of new color-stable anthocyanins occuring in some red wines. *J. Agric. Food Chem. 45*, 35–43.

Bakker, J., Picinelli, A., & Bridle, P. (1993). Model wine solutions: colour and composition changes during ageing.*Vitis, 32*, 111–118.

Bakker, J., Bridle, P., Honda, T., Kuwano, H., Saito, N., Terahara, N., & Timberlake, C. (1997). Identification of an anthocyanin occuring in some red wines. *Phytochemistry 44*, 1375–1382.

Baldi, A., Romani, A., Mulinacci, N., & Vincieri, F.F. (1993). Composés phenoliques dans les cépages de Toscane de *Vitis vinifera* L. *J. Int. Sci. Vigne Vin 27*, 201–215.

Baldi, A., Romani, A., Mulinacci, N., Vincieri, F., & Caseta, B. (1995). HPLC/MS application to anthocyanins of *Vitis vinifera* L. *J. Agric. Food Chem., 43*, 2104–2109.

Baranowski, E.S., & Nagel, C.W. (1983). Kinetics of malvidin-3-glucoside condensation in wine model systems. *J. Food Sci., 48*, 419–429.

Benabdeljalil, C., Cheynier, V., Fulcrand, H., Hakiki, A., Mosaddak, M., & Moutounet, M. (2000). Mise en évidence de nouveaux pigments formés par réaction des anthocyanes avec des métabolites de levure. *Sci. Aliment. 20*, 203–220.

Berke, B., Chèze, C., Vercauteren, J., & Deffieux, G. (1998). Bisulfite addition to anthocyanins : revisited structures of colourless adducts. *Tetrahedron Lett. 39*, 5771–5774.

Bishop, P.D., & Nagel, C.W. (1984). Characterization of the condensation product of malvidin 3,5-diglucoside and catechin. *J. Agric. Food Chem., 32*, 1022–1026.

Boido, E., Alcalde-Eon, C., Carrau, F., Dellacassa, E., & Rivas-Gonzalo, J.C. (2006). Aging effect on the pigment composition and color of Vitis vinifera L. cv Tannat wines. Contribution of the main pigment families to wine color. *J. Agric. Food Chem. 54*, 6692–6704.

Brouillard, R. (1982). Chemical structure of anthocyanins. In P. Markakis (Ed.), *Anthocyanins as Food Colors* (pp. 1–38). New York: Academic Press.

Cameira dos Santos, P.J., Briollouet, J.Y., Cheynier, V., & Moutounet, M. (1996). Detection and partial characterization of new anthocyanin derived pigments in wines. *J. Sci. Food Agric. 70*, 204–208.

Chatonnet, P., Dubourdieu, D., Boidron, J.-N., & Lavigne, V. (1993). Synthesis of volatile phenols by *Saccharomyces cerevisiae* in wines. *J. Sci. Food Agric. 62*, 191–202.

Cheminat, A., & Brouillard, R. (1986). NMR investigation of 3-O(β-D-glucosyl)malvidin structural transformation in aqueous solutions. *Tetrahedron Lett. 27*, 4457.

Cheynier, V., Trousdale, E.K., Singleton, V.L., Salgues, M.J., & Wylde, R. (1986). Characterization of 2-S-glutathionylcaftaric acid and its hydrolysis in relation to grape wines. *J. Agric. Food Chem. 34*, 217–221.

Cheynier, V., Souquet, J.M., Kontek, A., & Moutounet, M. (1997). Estimation of the oxidative changes in phenolic compounds of Carignane during winemaking. *Am. J. Enol. Vitic., 48*, 225–228.

Clark, A.C., & Scollary, G.R. (2002). Copper(II)-mediated oxidation of (+)-catechin in a model white wine system. *Au. J. Grape Wine Res. 8*, 186–195.

Dallas, C., Ricardo da Silva, J.M., & Laureano, O. (1996a). Interactions of oligomeric procyanidins in model wine solutions containing malvidin-3-glucoside and acetaldehyde. *J. Sci. Food Agric. 70*, 493–500.

Dallas, C., Ricardo da Silva, J.M., & Laureano, O. (1996b). Products formed in model wine solutions involving anthocyanins, procyanidin B2 and acetaldehyde. *J. Sci. Food Agric., 44*, 2402–2407.

Dueñas, M., Fulcrand, H., & Cheynier, V. (2006). Formation of anthocyanin-flavanol adducts in model solutions. *Anal. Chim. Acta, 563*, 15–25.

Escribano-Bailón, T., Dangles, O., & Brouillard, R. (1996). Coupling reactions between flavylium ions and catechin. *Phytochemistry 41*, 1583–1592.

Es-Safi, N., Fulcrand, H., Cheynier, V., & Moutounet, M. (1999). Studies on the acetaldehyde-induced condensation of (-)-epicatechin and malvidin 3-*O*-glucoside in a model solution system.*J. Agric. Food Chem. 47*, 2096–2102.

Es-Safi, N.E., Cheynier, V., & Moutounet, M. (2000). Study of the reactions between (+)-catechin and furfural derivaties in the presence or absence of anthocyanins and their implication in food color change.*J. Agric. Food Chem., 48*, 5946.

Es-Safi, N.E., Cheynier, V., & Moutounet, M. (2003). Implication of phenolic reactions in food organoleptic properties. *J. Food Comp. Anal. 16*, 535–553.

Etiévant, P.X. (1981). Volatil phenols determination in wines. *J. Agric. Food Chem. 29*, 65–67.

Etiévant, P., & Schilich, P. (1988). Varietal and geographic classification of french red wines in terms of pigments and flavanoid compounds. *J. Sci. Food Agric. 42*, 39–54.

Favretto, D., & Flamini, R. (2000). Application of electrospray ionization mass spectrometry to the study of grape anthocyanins. *Am. J. Enol. Vitic. 51*, 55–64.

Francia-Aricha, E.M., Guerra, M.T., Rivas-Gonzalo, J.C., & Santos-Buelga, V. (1997). New anthocyanin pigments formed after condensation with flavanols. *J. Agric. Food Chem., 45*, 2262–2266.

Fulcrand, H., Cameira Dos Santos, P.J., Sarni-Manchado, P., Cheynier, V., & FabreBonvin, J. (1996). Structure of new anthocyanin-derived wine pigments. *J. Chem. Soc. Perkin Trans., 1*, 735–739.

Fulcrand, H., Benabdeljalil, C, Rigaud, J., Cheynier, V., & Moutounet, M. (1998). A new class of wine pigments generated by reaction between pyruvic acid and grape anthocyanins. *Phytochemistry 47*, 1401–1407.

García-Viguera, C., Bridle, P., & Bakker, J. (1994). The effect of pH on the formation of coloured compounds in model solutions containing anthocyanins, catechin and acetaldehyde. *Vitis, 33*, 37–40.

Giusti, M.M., Rodríguez-Saona, L.E., Griffin, D., & Wrolstad, R.E. (1999). Electrospray and tandem mass spectrometry as tools for anthocyanin characterization. *J. Agric. Food Chem. 47*, 4657–4664.

González-San José, M., Santa María, G., & Diez, C. (1990). Anthocyanins as parameters for differentiating wines by grape variety, wine-growing region and wine-making methods. *J. Food Comp. Anal., 3*, 54–66.

Haslam, E. (1980). In Vino veritas: oligomeric procyanidins and the ageing of red wines. *Phytochemistry, 16*, 1625–1670.

Hayasaka, Y., & Asenstorfer, R.E. (2002). Screening for potencial pigments derived from anthocyanins in red wine using nanoelectrospray tandem mass spectrometry.*J. Agric. Food Chem. 50*, 756–761.

Hayasaka, Y., & Kennedy, J.A. (2003). Mass spectrometric evidence for the formation of pigmented polymers in red wine. *Aus. J. Grape Wine Res. 9*, 210–220.

He, J., Santos-Buelga, C., Silva A.M.S., Mateus N., & De Freitas, V. (2006a). Isolation and quatification of oligomeric pyranoanthocyanin-flavanol pigments from red wines by combination of column chromatographic techniques. *J. Chromatog. A 1134*, 215–225.

He, J., Santos-Buelga, C., Silva A.M.S., Mateus N., & De Freitas, V. (2006b). Isolation and structural characterization of new anthocyanin-derived yellow pigments in aged red wines. *J. Agric. Food Chem. 54*, 9598–9603.

Heier, A., Blaas, W., Droβ, A., & Wittkowski, R. (2002). Anthocyanin analysis by HPLC/ESI-MS. *Am. J. Enol. Vitic. 53*, 78–86.

Jurd, L., & Somers, T.C. (1970). The formation of xanthylium salts from proanthocyanidins. *Phytochemistry, 9*, 419–427.

Labarbe, B. (2000). Le potentiel phénolique de la grappe de *Vitis vinifera* var. Gamay noir et son devenir en vinification beaujolaise (Doctoral dissertation ENSA-M, UMI, UMII. Montpellier, France, 2000).

Lee, D.F., Swinny, E.E., & Jones, G.P. (2004). NMR identification of ethyl-linked anthocyanin-flavanol pigments formed in model wine ferments. *Tetrahedron Lett., 45*, 1671–1674.

Liao, H., Cai, Y., & Haslam, E. (1992). Polyphenols interactions. Anthocyanins: copigmentation and colour changes in young red wines. *J. Sci. Food Agric. 59*, 299–305.

Lu, Y., & Foo, Y. (2001). Unusual anthocyanin reaction with acetone leading to pyranoanthocyanin formation. *Tetrahedron Lett. 42*, 1371–1373.

Malien-Aubert, C., Dangles, O., & Amito, M.J. (2002). Influence of procyanidins on the color stability of eonin solutions. *J. Agric. Food Chem. 50*, 3299–3305.

Manzanares, P., Rojas, V., Genovés, S., & Vallés, S. (2000). A preliminary search for anthocyanin-β-D-glucosidase activity in non-*Saccharomyces* wine yeast. *Int. J. Food Sci. Technol. 35*, 95–103.

Marx, R., Holbach, B., & Otteneder, H. 2000. Determination of nine characteristic Anthocyanins in wine by HPLC. *F.V. OIV* No. 1104.

Mateus, N., & De Freitas, V. (2001). Evolution and stability of anthocyanin-derived pigments during port wine aging. *J. Agric. Food Chem. 49*, 5217–5222.

Mateus, N., Silva M.S., A., Vercauteren, J., & De Freitas, V. (2001). Occurrence of anthocyanin-derived pigments in red wines. *J. Agric. Food Chem., 49*, 4836–4840.

Mateus, N., De Pascual-Teresa, S., Rivas-Gonzalo, J.C., Santos-Buelga, C., & De Freitas, V. (2002a). Structural diversity of anthocyanin-derived pigments in port wines. *Food Chem. 76*, 335–342.

Mateus, N., Silva M.S., A., Santos-Buelga, C., Rivas-Gonzalo, J.C., & De Freitas, V. (2002b). Identification of anthocyanin-flavanol pigments in red wines by NMR and mass spectrometry. *J. Agric. Food Chem. 50*, 2110–2116.

Mateus, N., Carvalho, E., Carvalho, A.R.F., Melo, A., González-Paramás, A.M., Santos-Buelga, C., Silva, A.M.S., & De Freitas, V. (2003a). Isolation and structural characterization of new acylated anthocyanin-vinyl-flavanol pigments occuring in aging red wines. *J. Agric. Food Chem., 51*, 277–282.

Mateus, N., Silva, A.M.S., Rivas-Gonzalo, J.C., Santos-Buelga, C., & De Freitas, V. (2003b). A new class of blue anthocyanin-derived pigments isolated from red wines. *J. Agric. Food Chem. 51*, 1919–1923.

Mateus, N., Oliveira, J., Santos-Buelga, C., Silva, A.M.S., & De Freitas, V. (2004). NMR structural characterization of a new vinylpyranoanthocyanin-catechin pigment (a portisin). *Tetrahedron Lett., 45*, 3455–3457.

Mateus, N., Oliveira, J., González-Paramás, A.M., Santos-Buelga, C., & De Freitas, V. (2005). Screening of portisins (vinylpyranoanthocyanin pigments) in port wine by LC/DAD-MS. *Food Sci. Technol. Inter. 5*, 353–358.

Mateus, N., Oliveira, J., Pisarra, J., González-Paramás, A.M., Rivas-Gonzalo, J.C., Santos-Buelga, C., Silva, A.M.S., & De Freitas, V. (2006). A new vinylpyranoanthocyanin pigment occurring in aged red wine. *Food Chem. 97*, 689–695.

Mattivi, F., Guzzon, R., Vrhovsek, U., Stefanini, M., & Velasco, R. (2006). Metabolite profiling of grape: flavonols and anthocyanins. *J. Agric. Food Chem., 54*, 7692–7702.

Mazauric, J.P., & Salmon, J.M. (2005). Interactions between yeast lees and wine polyphenols during simulation of wine aging: I. Analysis of remnant polyphenolic compounds in the resulting wines. *J. Agric. Food Chem. 53*, 5647–5653.

Mazauric, J.P., & Salmon, J.M. (2006). Interactions between yeast lees and wine polyphenols during simulation of wine aging: I. Analysis of desorbed polyphenol compounds from yeast lees. *J. Agric. Food Chem. 54*, 3876–3881.

Mazza, G., & Miniati, E. (1993). *Anthocyanins in Fruits, Vegetables and Grains*. USA: CRC Press.

Medina, K., Boido, E., Dellacassa, E., & Carrau, F. (2005). Yeast interaction with anthocyanins during red wine fermentation. *Am. J. Enol. Vitic. 56*, 104–109.

Monagas, M., Núñez, V., Bartolomé, B., & Gómez-Cordoves, C. (2003). Anthocyanin-derived pigments in Graciano, Tempranillo and Cabernet Sauvignon wines produced in Spain. *Am. J. Enol. Vitic. 54*, 163–169.

Morata, A., Gómez-Cordovés, M.C., Colomo, B., & Suárez, J.A. (2003a). Pyruvic acid and acetaldehyde production by different strains of *Saccharomyces cerevisiae* with vitisin A and B formation in red wines. *J. Agric. Food Chem. 51*, 7402–7409.

Morata, A., Gómez-Cordovés, M.C., Suberviola, J., Bartolomé, B., Colomo, B., & Suárez, J.A. (2003b). Adsorption of anthocyanins by yeast cell walls during the fermentation of red wines. *J. Agric. Food Chem., 51*, 4084–4088.

Morata, A., Gómez-Cordovés, M.C., Colomo, B., & Suárez, J.A. (2005). Cell wall anthocyanin adsorption by different *Saccharomyces* strains during the fermentation of *Vitis vinifera* L. cv Graciano grapes. *Eur. Food Res. Technol., 220*, 341–346.

Núñez, V., Monagas, M., Gómez-Cordovés, C., & Bartolomé, B. (2004). *Vitis vinifera* L. cv. Graciano grapes characterized by its anthocyanin profile. *Postharvest Biol. Technol. 31*, 69–79.

Oliveira, J., De Freitas, V., Silva, A.M.S., & Mateus N. (2007). Reaction between hydroxycinnamic acids and anthocyanin-pyruvic adducts yielding new portisins. *J. Agric. Food Chem. 55*, 6349–6359.

Ortega Meder, M.D., Rivas Gonzalo, J.C., Vicente, J.L., & SantosBuelga, C. (1994). Differentiation of grapes according to the skin anthocyanin composition. *RECTA 34*, 409–426.

Oszmianski, J., Cheynier, V., & Moutounet, M. (1996). Iron-catalized oxidation of (+)-catechin in model systems. *J. Agric. Food Chem., 44*, 1712–1715.

Palmeri, R., & Spagna, G. (2007). Beta-glucosidase in cellular and acellular form for winemaking application. *Enz. Microbial Technol., 30*, 382–389.

Peng, C.Y., & Markakis, P. (1963). Effect of phenolase on anthocyanins. *Nature 199*, 596–597.

Pérez-Magariño, S., & González-San José, M.L. (2004). Evolution of flavanols, anthocyanins and their derivatives during the aging of red wines elaborated from grapes harvested at different stages of ripening. *J. Agric. Food Chem. 52*, 1181–1189.

Pifferi, P.G., & Cultrera, R. (1974). Enzymatic degradation of anthocyanins: the role of sweet cherry polyphenoloxidase. *J Food Sci. 39*, 786–791.

Pissarra, J., Mateus, N., Rivas-Gonzalo, J., Santos-Buelga, C., & De Freitas, V. (2003). Reaction between malvidin-3-glucoside and (+)-catechin in model solutions containing different aldehydes. *J. Food Sci. 68*, 476–481.

Pissarra, J., Lourenço, S., González-Paramás, A.M., Mateus, N., Santos-Buelga, C., Silva, A.M.S., & De Freitas, V. (2004). Structural characterization of new malvidin-3-glucoside-catechin aryl/alkyl pigments. *J. Agric. Food Chem. 52*, 5519–5526.

Pissarra, J., Lourenço, S., González-Paramás, A.M., Mateus, N., Santos-Buelga, C., Silva, A.M.S., & De Freitas, V. (2005a). Isolation and structural characterization of new anthocyanin-alkyl-catechin pigments. *Food Chem. 90*, 81–87.

Pissarra, J., Sandra, L., Machado J.M., Mateus, N., Guimaraens, D., & De Freitas, V. (2005b). Contribution and importance of wine spirit to the port wine final quality- initial approach. *J. Sci. Food Agric. 85*, 1091–1097.

Pozo-Bayón, M.A., Monagas, M., Polo, M.C., & Gómez-Cordovés, C. (2004). Occurrence of pyranoanthocyanins in sparkling wines manufactured with red grape varieties. *J. Agric. Food Chem. 52*, 1300–1006.

Remy, S., Fulcrand, H., Labarbe. B., Cheynier, V., & Moutounet, M. (2000). First confirmation in red wine of products resulting from direct anthocyanin-tannin reactions. *J. Sci. Food Agric. 80*, 745–751.

Remy-Tanneau, S., Le Guernevé, C., Meudec, E., & Cheynier, V. (2003). Characterization of a colorless anthocyanin-flavan-3-ol dimer containing both carbon-carbon and ether interflavanoid linkages by NMR and mass spectrometry. *J. Agric. Food Chem. 51*, 3592–3597.

Rentzsch, M., Schwarz, M., Winterhalter, P., & Hermosín-Guitiérrez, I. (2007). Formation of hydroxyphenyl-pyranoanthocyanins in Grenache wines: precursor levels and evolution during aging. *J. Agric. Food Chem. 55*, 4883–4888.

Revilla, I., & González-San José, M.L. (2001). Evolution during the storage of red wines treated with pectolytic enzymes: New anthocyanin pigment formation. *J. Wine Res. 12*, 183–197.

Revilla, I., Pérez-Magariño, S., González-San José, M.L., & Beltrán, S. (1999). Identification of anthocyanin derivaties in grape skin extracts and red wines by liquid chromatography with diode array and mass spectrometric detection. *J. Chromatog. A 847*, 83–90.

Ribéreau-Gayón, P., Glories, Y., Maujean, A., & Dubourdieu, D. (2000). Hanbook of Enology. Vol. 2. The Chemistry of Wine Stabilization and Treatments. England: John Wiley and Sons, Inc.

Rivas-Gonzalo, J.C., Bravo-Haro, S., & Santos-Buelga, C. (1995). Detection of compounds formed through the reaction of malvidin 3-monoglucoside and catechin in the presence of acetaldehyde. *J. Agric. Food Chem., 43*, 1444–1449.

Roggero, J.P., Coen, S., & Larice, J.L. (1986a). Etude comparative de la composition anthocyanique des cépages. Essai de classification. *Bull. Liaison Groupe Polyphénols 13*, 380–388.

Roggero, J.P., Coen, S., & Ragonet, B. (1986b). High performance liquid chromatography survey on changes in pigment content in repening grapes of Syrah. An approach to anthocyanin metabolism. *Am. J. Enol. Vitic. 37*, 77–83.

Roggero, J.P., Coen, S., Archier, O., & Rocheville-Divorne, C. (1987). Etude par C.L.H.P. de la réaction glucoside de malvidine.acétaldéhyde-composé phénolique. *Conn. Vigne Vin 21*, 163–168.

Roggero, J.P., Larice, J.L., Rocheville-Divore, C., Archier, P., & Coen, S., 1988. Composition anthocyanique des cépages. II. Essai de classificaton sur trois ans par analyse en composantes principales et par analyse factorielle discriminante. *Rev. Française d'Oenologie 112*, 41–48.

Romero, C., & Bakker, J. (1999a). Interactions between grape anthocyanins and pyruvic acid, with effect of pH and acid concentration on anthocyanin composition and color in model solutions. *J. Agric. Food Chem., 47*, 3130–3139.

Romero, C., & Bakker, J. (1999b). Effect of storage temperature and pyruvate on kinetics of anthocyanin degradation, vitisin A derivative formation and color characteristics of model solutions. *J. Agric. Food Chem., 48*, 2135–2141.

Romero, C., & Bakker, J. (2000a). Effect of acetaldehyde and several acids on the formation of vitisin A in model wine anthocyanin and colour evolution. *Int. J. Food Sci. Technol., 35*, 129–140.

Romero, C., & Bakker, J. (2000b). Anthocyanin and colour evolution during maturation of four port wines: effect of pyruvic acid addition. *J. Sci. Food Agric., 81*, 252–260.

Salas, E., Fulcrand, H., Meude, E., & Cheynier, V. (2003). Reaction of anthocyanin and tannins in model solutions. *J. Agric. Food Chem. 51*, 7951–7961.

Salas, E., Atanasova, V., Poncet-Legrand, C., Meudec, E., Mazauric, J.P., & Cheynier, V. (2004a). Demonstration of the occurrence of flavanol-anthocyanin adducts in wine and in model solutions. *Anal. Chim. Acta 513*, 325–332.

Salas, E., Le Guernevé, C., Fulcrand, H., Poncet-Legrand, C., & Cheynier, V. (2004b). Structure determination and colour properties of a new directly linked flavanol-anthocyanin dimer. *Tetrahedron Lett. 45*, 8725–8729.

Salas, E., Dueñas, M., Schwarz, M., Winterhalter, P., Cheynier, V., & Fulcrand, H. (2005a). Characterization of pigmets from different high speed countercurrent chromatography wine fractions. *J. Agric. Food Chem. 53*, 4536–4546.

Salas, E., Fulcrand, H., Poncet-LeGrand, C., Meudec, E., Köhler, N., Winterhalter, P., & Cheynier, V. (2005b). Isolation of flavanol-anthocyanin adducts by countercurrent chromatography. *J. Chromatog. Sci. 43*, 488–493.

Sánchez-Torres, P., González-Candelas, L., & Ramo, D. (1998). Heterologous expresión of a Candida molischiana anthocyanin-β-glucosidase in a wine yeast strain. *J. Agric. Food Chem. 46*, 354–360.

Santos-Buelga, C., Francia-Aricha, E.M., De Pascual-Teresa, S., & Rivas-Gonzalo, J.C. (1999). Contribution to the identification of the pigments responsible for the browning of anthocyanin-flavanol solutions. *Eur. Food Res. Technol. 209*, 411–415.

Sarni-Manchado, P., Fulcrand, H., Souillol, V., Souquet, J.M., & Cheynier, V. (1995). Mechanisms of anthocyanin degradation in grape must-like model solutions. *J. Sci. Food Agric., 69*, 385–391.

Sarni-Manchado, P., Cheynier, V., & Moutounet, M. (1997). Reaction of enzimatically generated quinones with malvidin-3-glucoside. *Phytochemistry 45*, 1365.

Saucier, C., Guerra, C., Pianet, I., Laguerres, M., & Glories, Y. (1997). (+)-Catechin-acetaldehyde condensation products in relation to wine-ageing. *Phytochemistry 46*, 229–234.

Schwarz, M., Jerz, G., & Winterhalter, P. (2003a). Isolation and structure of Pinotin A, a new anthocyanin derivative from Pinotage wine. *Vitis 42*, 105–106.

Schwarz, M., Wabnitz, T.C., &Winterhalter, P. (2003b). Pathway leading to the formation of anthocyanin-vinylphenol adducts and related pigments in red wines. *J. Agric. Food Chem. 51*, 3682–3687.

Schwarz, M., Hofmann, G., & Winterhalter, P. (2004). Investigations on anthocyanins in wines from Vitis vinifera cv. pinotage: Factors influencing the formation of pinotin a and its correlation with wine age. *J. Agric. Food Chem., 44*, 498–504.

Singleton, V.L., Salgues, M., Zaya, J., & Trousdale, E. (1985). Caftaric acid dissapearence and conversion to products of enzymatic oxidation in grape must and wines.*Am. J. Enol. Vitic., 36*, 50–56.

Somers, T.C. (1971). The phenolic nature of wine pigments. *Phytochemistry, 10*, 2175–2186.

Sousa, C., Mateus, N., Silva, A.M.S., González-Paramás, A.M., Santos-Buelga, C., & De Freitas, V. (2007). Structural and chromatic characterization of a new malvidin-3-glucoside-vanillyl-catechin pigment. *Food Chem. 102*, 1344–1351.

Timberlake, C.F., & Bridle, P. (1976). Interactions between anthocyanins, phenolic compounds and acetaldehyde and their significance in red wines. *Am. J. Enol. Vitic. 27*, 97–105.

Vasserot, Y., Caillet, S., & Maujean, A. (1997). Study of anthocyanin adsorption by yeast lees. Effect of some physicochemical parameters. *Am. J. Enol. Vitic. 48*, 433–437.

Vidal, S., Meudec, E., Cheynier, V., Skouroumounis, G., & Hayasaka, Y. (2004). Mass spectrometric evidence for the existence of oligomeric anthocyanins in grape skins. *J. Agric. Food Chem. 52*, 7144–7151.

Vivar-Quintana, A.M., Santos-Buelga, C., Francia-Aricha, E., & Rivas-Gonzalo, J.C. (1999). Formation of anthocyanin-derived pigments in experimental red wines. *Food Sci. Technol. Inter. 5*, 347–352.

Vivar-Quintana, A.M., Santos-Buelga, C., & Rivas-Gonzalo, J.C. (2002). Anthocyanin-derived pigments and colour of red wines. *Anal. Chim. Acta 458*, 147–155.

Vivas, N.G., Nonier, M.-F., Guerra, C., & Vivas, N. (2001). Anthocyanin in grape skins during maturation of *Vitis vinifera* L. cv. Caberent Sauvignon and Merlot Noir from different bordeaux terroirs. *J. Int. Sci. Vigne Vin 35*, 149–156.

Walker, G. (1998). *Yeast physiology and biotechnology*. UK: Wiley.

Wang, J., & Sporns, P. (1999). Analysis of anthocyanins in red wine and fruit juice using MALDI-MS. *J. Agric. Food Chem., 47*, 2009–2015.

Wang, H., Race, E.J., & Shrikhande, A.J. (2003a). Anthocyanin transformation in Cabernet Sauvignon wine during aging. *J. Agric. Food Chem. 51*, 7989–7994.

Wang, H., Race, E.J., & Shrikhande, A.J. (2003b). Characterization of anthocyanins in grape juices by ion trap liquid chromatography-mass spectrometry. *J. Agric. Food Chem. 51*, 1836–1844.

Wesche-Ebeling, P, & Montgomery, M.W. (1990). Strawberry polyphenoloxidase: its role in anthocyanin degradation. *J. Food Sci., 55*, 731–734, 745.

Wildenradt, H.L., & Singleton V.L. (1974). Production of aldehydes as a result of oxidation of polyphenolic compounds and its relation to wine aging. *J. Agric. Food Chem. 25*, 119–126.

Chapter 9B
Flavanols, Flavonols and Dihydroflavonols

Nancy Terrier, Céline Poncet-Legrand, and Véronique Cheynier

Contents

9B.1 Structure and Occurrence in Grape

9B.1.1 General Structural Characteristics

Flavanols, flavonols, and dihydroflavonols, like the anthocyanins presented in Chapter 9A, belong to the flavonoid family. These compounds are phenolic compounds that share a common C6-C3-C6 skeleton consisting of two phenolic rings (named A and B) linked together by a heterocyclic pyran ring (C-ring) as shown in Fig. 9B.1. Among them, several classes can be distinguished on the basis of

V. Cheynier (✉)
INRA, UMR Sciences pour l'Oenologie, 2, place Viala, F-34060
Montpellier, France
e-mail: cheynier@supagro.inra.fr

M.V. Moreno-Arribas, M.C. Polo (eds.), *Wine Chemistry and Biochemistry*,
DOI 10.1007/978-0-387-74118-5_18,

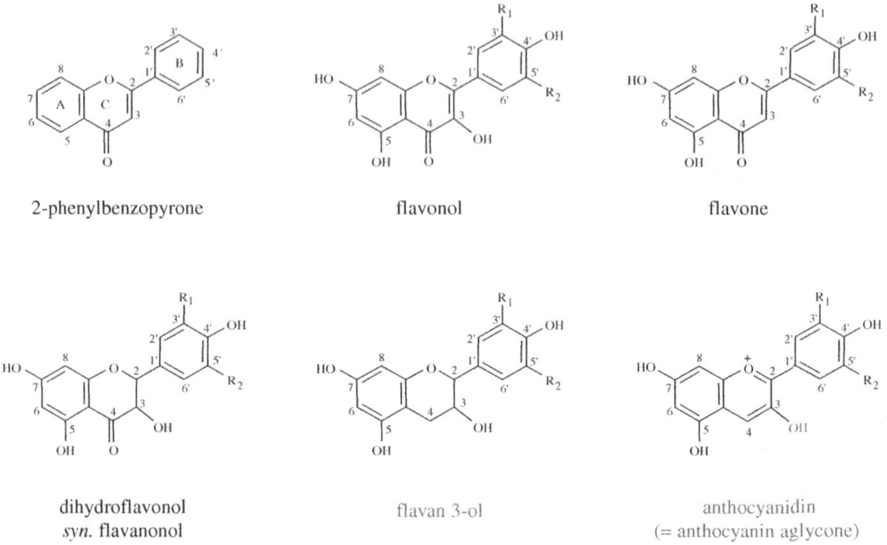

Fig. 9B.1 Chemical structures of flavonoids R1, R2 = H, OH, OCH3

the oxidation state of the C-ring. Flavonoids *sensu stricto* are based on the 2-phenylbenzopyrone structure characterized by a double link between carbons C2 and C3 and a ketone group in C4. They are represented in grapes by flavonols (that also have a hydroxyl group in C3), while flavones are also present in vine leaves. Flavonoids *sensu largo* comprise additional classes which do not show these characteristic features. Among them, anthocyanins, flavanols and dihydroflavonols are encountered in grapes. The first two groups are particularly abundant in grape and wine and essential to wine quality. Indeed, anthocyanins are the red pigments of grapes and responsible for the colour of red wines whereas flavanols contribute to taste (especially astringency and bitterness) and are also involved in the development of oxidative browning, haze and precipitates, as described in Chapter 9D.

Within each flavonoid class, further diversity results from modifications of their three ring skeleton, including

- hydroxylation
- methylation of some of the phenolic hydroxyls
- glycosylation: as illustrated in the case of anthocyanins, flavonoids can be glycosylated by various sugars (monomeric or oligomeric), usually by C-O-C glycosidic linkages. In some plant species, C-glycosylation is also encountered
- acylation of the alcoholic hydroxyl groups
- polymerisation, especially in the flavanol family

Grape flavonoids are hydroxylated on their C5 and C7 carbons, so that their A-ring is a phloroglucinol ring, and, on the B-ring, in the 4′ position. They can also be hydroxylated in 3′ or 3′ and 5′ positions and substituted by glycosidic or acyl groups on the alcoholic OH group in position 3.

9B.1.2 Flavanols

9B.1.2.1 Structure and Localisation

Grape flavanols, more accurately called flavan 3-ols as they are hydroxylated in the 3 position, are found as monomers but also as oligomers and polymers.

The major flavan 3-ol monomers in grape are (+)-catechin and its isomer, (–)-epicatechin, and, to a lesser extent, the gallic ester of (–)-epicatechin, (–)-epicatechin 3-gallate (Su and Singleton 1969) (Fig. 9B.2). Gallocatechin (Piretti et al. 1976; Czochanska et al. 1979b) has also been reported in *Vitis vinifera* and catechin 3-gallate (Lee and Jaworski 1987) and gallocatechin 3-gallate (Lee and Jaworski 1990) have been detected in some non-*Vinifera* varieties.

Flavanol oligomers and polymers are also called condensed tannins or proanthocyanidins. The term tannin refers to their capacity to interact or react with proteins and precipitate them out. When heated under acidic conditions, these molecules release red anthocyanidin pigments, hence the term proanthocyanidins. The term leucoanthocyanidin, also referring to this particular property, is sometimes encountered in the literature. However, this should be restricted to another group of compounds, flavan 3,4-diols, which are intermediates in the biosynthetic pathway leading to flavanols and anthocyanins (Stafford and Lester 1984; Nakajima et al. 2001; Abrahams et al. 2003) but have never been isolated from grapes, presumably due to their instability.

In B-type proanthocyanidins (Fig. 9B.2), the flavanol constitutive units are linked by C4-C8 and/or C4-C6 bonds, opening the possibility for branched structures.

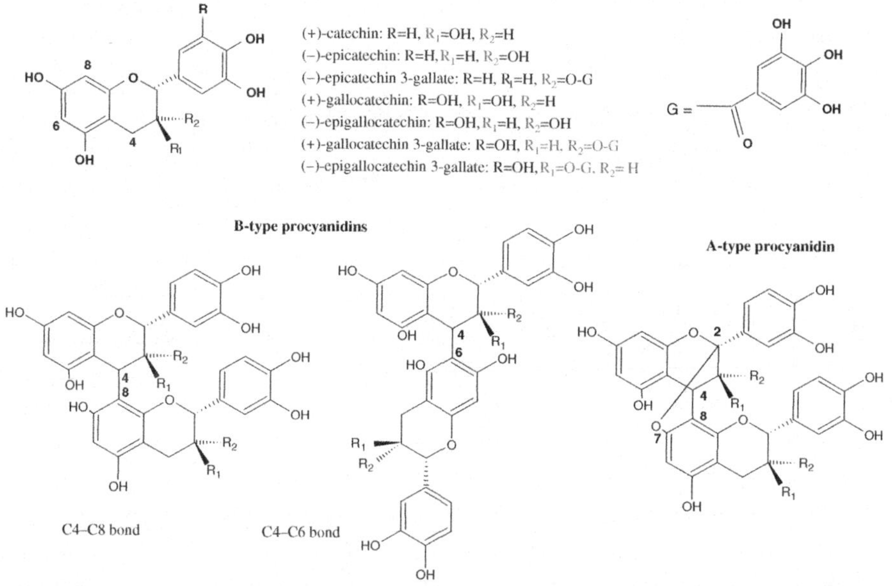

Fig. 9B.2 Structures of flavanol monomers and dimers

Double linkages, with C2-O-C7 or C2-O-C5 bond in addition to the C4-C6 or C4-C8 bond, give rise to A-type proanthocyanidins. Depending on the nature of the anthocyanidin released (see Chapter 9A), several groups of proanthocyanidins can be distinguished. Grape and wine proanthocyanidins belong to two of these groups, namely procyanidins and prodelphinidins that release cyanidin and delphinidin, respectively (Fig. 9B.3). Besides, mass signals that may correspond to methoxylated flavanol dimers have been detected in wine (Cooper and Marshall 2001) but their identification has not been confirmed. Finally, some of the constitutive units can be substituted (e.g. galloylated).

Flavanols are present in various vine plant tissues, including wood (Boukharta et al. 1988), leaves (Bogs et al. 2005; Tesnière et al. 2006), stems (Souquet et al. 2000), and fruit. Within the grape berry, they are particularly abundant in seeds and skins.

Table 9B.1 lists the flavanol dimers and trimers found in grape berries and the methods used for identification. Procyanidin dimers and trimers were first identified in seeds but they are also present in skins and stems with different distributions (Ricardo da Silva et al. 1991a) and trace amounts of B1 through B4 have been detected in pulp (Bourzeix et al. 1986). However, analysis of grape extracts by acid degradation and ^{13}C NMR established the presence of prodelphinidins along with the expected procyanidins (Czochanska et al. 1980). In agreement with this finding,

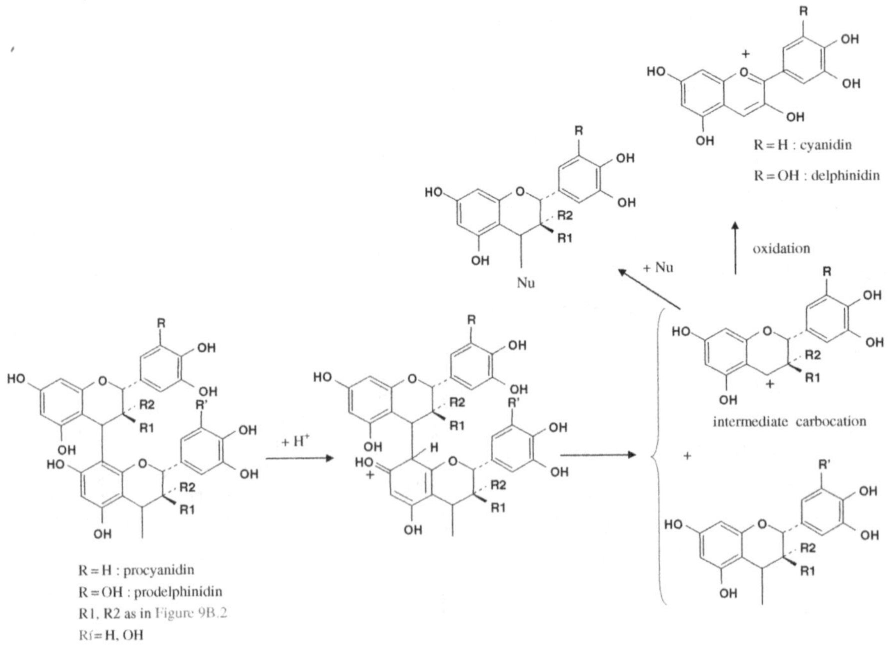

Fig. 9B.3 Acid-catalysed cleavage of proanthocyanidin interflavanic linkages and subsequent reactions, namely oxidation to anthocyanidin and nucleophilic (Nu) addition

Table 9B.1 Flavanol oligomers, flavonols and dihydroflavonols identified in grapes and wines, methods of detection

	Source	Reference	Detection
Flavanol dimers			
(–)-Epicatechin-(4β-8)-(+)-catechin (B1)	Berries	Weinges and Piretti (1971)	LC -IR, -^{1}H NMR, -MS
(–)-Epicatechin-(4α-8)-(–)-epicatechin (B2)	Berries	Weinges and Piretti (1971)	LC -IR, -^{1}H NMR, -MS
(+)-Catechin-(4α-8)-(+)-catechin (B3)	Berries	Weinges and Piretti (1971)	LC -IR, -^{1}H NMR, -MS
(+)-Catechin-(4α-8)-(–)-epicatechin (B4)	Berries	Weinges and Piretti (1971)	LC -IR, -^{1}H NMR, -MS
(–)-Epicatechin-(4α-6)-(–)-epicatechin (B5)	Seeds	Ricardo da Silva et al. (1991c)	HPLC, TLC, –^{1}H NMR, FAB-MS after enzymatic hydrolysis, acid hydrolysis, partial acid-catalysed degradation with phloroglucinol or phenylmethanethiol
(+)-Catechin-(4α-6)-(+)-catechin (B6)	Seeds	Ricardo da Silva et al. (1991c)	HPLC, TLC, -^{1}H NMR, FAB-MS after enzymatic hydrolysis, acid hydrolysis, partial acid-catalysed degradation with phloroglucinol or phenylmethanethiol
(–)-Epicatechin-(4β-6)-(+)-catechin (B7)	Seeds	Ricardo da Silva et al. (1991c)	HPLC, TLC, -^{1}H NMR, FAB-MS after enzymatic hydrolysis, acid hydrolysis, partial acid-catalysed degradation with phloroglucinol or phenylmethanethiol
(+)-Catechin-(4α-6)-(–)-epicatechin (B8)	Seeds	Ricardo da Silva et al. (1991c)	HPLC, TLC, -^{1}H NMR, FAB-MS after enzymatic hydrolysis, acid hydrolysis, partial acid-catalysed degradation with phloroglucinol or phenylmethanethiol
(–)-Epigallocatechin-(+)-catechin	Wine	Remy (1999)	HPLC-DAD, LC-MS
(–)-Epicatechin-(+)-gallocatechin	Wine	Remy (1999)	HPLC-DAD, LC-MS
(–)-Epicatechin-(+)-epigallocatechin	Wine	Remy (1999)	HPLC-DAD, LC-MS
(–)-Epicatechin 3-gallate-(4β-8)-(+)-catechin (B1 3-gallate)	Seeds	Ricardo da Silva et al. (1991c)	HPLC, TLC, NMR, FAB-MS after enzymatic hydrolysis, acid hydrolysis, partial acid-catalysed degradation with phloroglucinol or phenylmethanethiol

Table 9B.1 (continued)

	Source	Reference	Detection
(–)-Epicatechin 3-gallate-(4β-8)-(–)-epicatechin (B2 3 gallate)	Seeds	Ricardo da Silva et al. (1991c)	HPLC, TLC, NMR, FAB-MS after enzymatic hydrolysis, acid hydrolysis, partial acid-catalysed degradation with phloroglucinol or phenylmethanethiol
(–)-Epicatechin-(4β-8)-(–)-epicatechin 3-gallate (B2 3' gallate)	Berries	Czochanska et al. (1979a)	PLC, -MS, -^{1}H NMR
(–)-Epicatechin-(4β-8)-(–)-epicatechin 3-gallate (B2 3' gallate)	Stem	Boukharta et al. (1988)	LC-,UV spectroscopy, -^{1}H NMR after enzymatic hydrolysis, acid hydrolysis, or with α-thiol
(+)-Catechin-(4α-8)-(–)-epicatechin 3-gallate (B4 3 gallate)	Seeds	Ricardo da Silva et al. (1991c)	HPLC, TLC, NMR, FAB-MS after enzymatic hydrolysis, acid hydrolysis, partial acid-catalysed degradation with phloroglucinol or phenylmethanethiol
(–)-Epicatechin 3-gallate-(4β-8)-(–)-epicatechin 3-gallate (B2 3, 3' digallate)	Seeds	Ricardo da Silva et al. (1991c)	HPLC, TLC, NMR, FAB-MS after enzymatic hydrolysis, acid hydrolysis, partial acid-catalysed degradation with phloroglucinol or phenylmethanethiol
Flavanol trimers			
(–)-Epicatechin-(4β-8)- (–)-epicatechin-(4β-8)-(–)- epicatechin (C1)	Wine	Lea et al. (1979)	PC, TLC after toluene-thiol hydrolysis
(–)-Epicatechin-(4β-8)-(–)-epicatechin-(4β-8)-(+)-catechin	Wine	Lea et al. (1979)	PC, TLC after toluene-thiol hydrolysis
(–)-Epicatechin-(4β-8)-(–)-epicatechin-(4β-6)-(+)-catechin	Seeds	Ricardo da Silva et al. (1991c)	HPLC, TLC, – ^{1}H NMR, FAB-MS after enzymatic hydrolysis, acid hydrolysis, partial acid-catalysed degradation with phloroglucinol or phenylmethanethiol
(–)-Epicatechin-(4β-6)-(–)-epicatechin-(4β-8)-(–)-epicatechin	Seeds	Ricardo da Silva et al. (1991c)	HPLC, TLC, NMR, FAB-MS after enzymatic hydrolysis, acid hydrolysis, partial acid-catalysed degradation with phloroglucinol or phenylmethanethiol
(–)-Epicatechin-(4β-8)-(–)-epicatechin-(4β-6)-(–)-epicatechin	Seeds	Ricardo da Silva et al. (1991c)	HPLC, TLC, NMR, FAB-MS after enzymatic hydrolysis, acid hydrolysis, partial acid-catalysed degradation with phloroglucinol or phenylmethanethiol
(–)-Epicatechin-(4β-6)-(–)-epicatechin-(4β-8)-(+)-catechin	Seeds	Ricardo da Silva et al. (1991c)	HPLC, TLC, NMR, FAB-MS after enzymatic hydrolysis, acid hydrolysis, partial acid-catalysed degradation with phloroglucinol or phenylmethanethiol

Table 9B.1 (continued)

	Source	Reference	Detection
(–)-Epicatechin-(4β-8)-(–)-epicatechin 3-gallate-(4β-8)-(+)- catechin	Seeds	Ricardo da Silva et al. (1991c)	HPLC, TLC, NMR, FAB-MS after enzymatic hydrolysis, acid hydrolysis, partial acid-catalysed degradation with phloroglucinol or phenylmethanethiol
Flavonols			
Quercetin, (R1=OH, R2=H) 3-glucoside	Skin	Ribéreau-Gayon (1964)	PC- Fluorescence, -UV spectroscopy, hydrolysis
Quercetin 3-glucuronide	Skin	Ribéreau-Gayon (1964)	PC- Fluorescence, -UV spectroscopy, hydrolysis
Quercetin 3-glucosylgalactoside	Skin	Cheynier and Rigaud (1986)	HPLC-UV spectrometry, -MS, -^{1}H NMR, -TLC after hydrolysis
Quercetin 3-glucosylxyloside	Skin	Cheynier and Rigaud (1986)	HPLC-UV spectrometry, -MS, -^{1}H NMR, -TLC after hydrolysis
Quercetin 3-rhamnosylglucoside	Skin	Cantos et al. (2002)	HPLC-DAD, -MS-MS
Quercetin 3-rhamnosylglucoside	Leaves	Hmamouchi et al. (1996)	HPLC-UV spectrometry, -MS, -^{1}H and ^{13}C NMR, -TLC after hydrolysis
Kampferol (R1=R2=H) 3-glucoside	Skin	Ribéreau-Gayon (1964)	PC- Fluorescence, -UV spectroscopy, hydrolysis
Kampferol 3-glucuronide	Skin	Cheynier and Rigaud (1986)	HPLC-UV spectrometry, -MS, -^{1}H NMR, -TLC after hydrolysis
Kampferol 3-galactoside	Skin	Cheynier and Rigaud (1986)	HPLC-UV spectrometry, -MS, -^{1}H NMR, -TLC after hydrolysis
Kampferol 3-glucosylarabinoside	Skin	Cheynier and Rigaud (1986)	HPLC-UV spectrometry, -MS, -^{1}H NMR, -TLC after hydrolysis
Myricetin (R1=R2=OH) 3-glucoside	Skin	Ribéreau-Gayon (1964)	PC- Fluorescence, -UV spectroscopy, hydrolysis
Myricetin 3-glucuronide	Skin	Cheynier and Rigaud (1986)	HPLC-UV spectrometry, -MS, -^{1}H NMR, -TLC after hydrolysis

Table 9B.1 continued

	Source	Reference	Detection
Isorhamnetin (R1=OCH_3, R2=H) 3-glucoside	Skin	Cheynier and Rigaud (1986)	HPLC-UV spectrometry, -MS, -^{1}H NMR, -TLC after hydrolysis
Luteolin (R1=R2=H, O-β gluc at C7) 7-glucoside	Leaves	Hmamouchi et al. (1996)	HPLC-UV spectrometry, -MS, -^{1}H NMR, -TLC after hydrolysis
Apigenin (R1=OH R2=H, O-β gluc at C7) 7-glucoside	Leaves	Hmamouchi et al. (1996)	HPLC-UV spectrometry, -MS, -^{1}H NMR, -TLC after hydrolysis
Laricetin (R1=OCH_3, R2=OH) 3-glucoside and 3-galactoside	Skin	Mattivi et al. (2006)	HPLC-DAD, -MS before and after hydrolysis
Syringetin (R1=R2=OCH_3) 3-glucoside and 3-galactoside	Skin	Mattivi et al. (2006)	HPLC –DAD, -MS before and after hydrolysis
Dihydroflavonols			
Dihydroquercetin (R1=OH, R2=H) 3-rhamnoside (astilbin)	Skin	Trousdale and Singleton (1983)	LC-^{1}H NMR
Dihydroquercetin, (R1=OH, R2=H) 3-rhamnoside (astilbin)	Stem	Souquet et al. (2000)	LC-MS, ^{1}H NMR
Dihydroquercetin 3-galactoside	Skin	Masa et al. (2007)	HPLC-DAD, TLC, PC
Dihydroquercetin 3-glucoside	Skin	Masa et al. (2007)	HPLC-DAD, TLC, PC
Dihydrokampferol (R1=R2=H) 3-rhamnoside (engeletin)	Skin	Trousdale and Singleton (1983)	LC-^{1}H NMR
Dihydrokampferol 3-rhamnoside (engeletin)	Stem	Souquet et al. (2000)	LC-MS, – ^{1}H NMR

additional dimers containing gallocatechin and epigallocatechin units, either in the lower or in the upper position have been detected in wine (Fulcrand et al. 1999; de Pascual-Teresa et al. 2000).

The reversed-phase HPLC separation methods used for monomer and oligomer analysis become poorly resolutive as the molecular weight increases, owing to the large number of possible isomers distributed along the chromatographic profile and smaller amounts of each individual compound. Thus, detection of species beyond the tetramer by using this technique is almost impossible, especially in grape extracts which contain proanthocyanidins based on several constitutive units. Nevertheless, the prevalence of larger oligomers and polymers has been demonstrated in numerous plant species including grapes (Czochanska et al., 1980).

Alternative methods have been developed to characterize proanthocyanidins polymers. Some separation according to molecular weight can be achieved by adsorption chromatography on Sephadex (Lea and Timberlake 1974) or Fractogel (Derdelinckx and Jerumanis 1984; Ricardo da Silva et al. 1991c) columns or by gel permeation chromatography (Bae et al. 1994; Le Bourvellec et al. 2006; Yanagida et al. 1999) but the resolution is rather poor especially for polymers. Normal phase HPLC on silica (Rigaud et al. 1993; Yanagida et al. 2000) or diol (Kelm et al. 2006) columns improve separation of the polymeric proanthocyanidins and was successfully applied to grape seed (Prieur et al. 1994; Rigaud et al. 1993) and grape skin (Souquet et al. 1996) extracts. The elution profiles showed that grape seed proanthocyanidins consist of oligomers and lower molecular weight polymers while skin proanthocyanidins are polymeric material eluted as a large hump at the end of the chromatogram. However, no relationship between the retention time and the chain length could be established when comparing both extracts, due to their different compositions (Cheynier et al. 1999b). Proanthocyanidin polymers can be analysed by HPLC after acid catalysed cleavage in the presence of a nucleophilic agent. In these methods, the intermediate carbocations released from the upper units (initially substituted in C4) after cleavage of the interflavanic bonds react with the nucleophile (usually phloroglucinol or toluene-α-thiol) to yield an adduct while the other units are released as such (Fig. 9B.3). HPLC analysis of the reaction products thus gives access to the total amount of proanthocyanidins (calculated as the sum of released units), the nature and proportions of each constitutive unit and the mean degree of polymerisation (mDP).

Application of thiolysis to grape proanthocyanidins showed that those extracted from seeds are partly galloylated procyanidins (Prieur et al. 1994), based on epicatechin, catechin and epicatechin units, with degrees of polymerisation ranging from 1 to 16 in the fractions analysed. Those from skins (Souquet et al. 1996) consist of both procyanidins and prodelphinidins with mDP around 30 (up to 80 in the polymeric fractions). They contained epicatechin and epigallocatechin as their major constitutive units and very low levels of galloylated units (a few percent). Proanthocyanidins from stems (Souquet et al. 2000) and pulp (Mané et al. 2007b; Souquet et al. 2006) are also mixed procyanidin/prodelphinidin polymers with lower proportions of epigallocatechin units and higher levels of galloylation than skin proanthocyanidins. The mean degrees of polymerization calculated for pulp proanthocyanidins (about 20) are also intermediate between those in seeds and skins in all analysed varieties (Mané et al. 2007b; Souquet et al. 2006). The flavanol concentration is higher in seeds than in skins and pulp contain only small amounts. However, the contribution of skins to the entire berry content may exceed that of seeds in some varieties (Mané et al. 2007a).

9B.1.2.2 Variability in Grape

Flavanol composition is developmental stage-, genetic- and growth condition-dependent. Skin flavanols are synthesised mainly during a few weeks after flowering (Kennedy et al. 2001; Downey et al. 2003a). The skin flavanol pool is considered as almost constant during ripening when expressed on a per berry basis, on a quantitative and qualitative point of view (Fournand et al. 2006). In contrast, other authors have described a decrease in the content of flavanol monomers and proanthocyanidins after veraison (Downey et al. 2003a; Kennedy et al. 2002), and a concomitant increase (Kennedy et al. 2001) or decrease (Downey et al. 2003a) of proanthocyanidin mDP. In seeds, accumulation of flavanols is a bit delayed when compared to skin, and maximal concentration is reached a few weeks after veraison (Bogs et al. 2005; Downey et al. 2003a). The concentration of flavanol monomers then decreases sharply (Downey et al. 2003a; Romeyer et al. 1986) while slight accumulation of procyanidin oligomers is observed (Romeyer et al. 1986). The total amount of proanthocyanidins that can be extracted by the usual solvents decreases over the same period. However, the additional units measured by subjecting the residue to acid catalysed cleavage compensate for this loss, suggesting that polymeric flavanols have been strongly absorbed on the plant cell wall material rather than degraded (Downey et al. 2003a).

Flavanol composition, like anthocyanin profiles (Roggero et al 1988; Mazza and Miniati 1993), is greatly affected by genetics. Reported values for skin mDP ranged from 16 (Muscat de Hambourg; Souquet et al. 2006) to 86 (Cabernet-Sauvignon; Monagas et al. 2003) and for seed from 3 (Pinot noir; Mané et al. 2007b) to 10 (Maccabeo; Souquet et al. 2006). It should however be emphasized that extreme values reported by different authors are also related to variations in the analysis protocols. For instance, some authors separate oligomers from polymers whereas others do not. The percentage of B-ring trihydroxylation of subunits from skin proanthocyanidins is also a variable parameter, with values as high as 31% for Cabernet-Sauvignon (Monagas et al. 2003) but of only 3% in Maccabeo (Souquet et al. 2006). Percentage of galloylation of skin and seed flavanols seems to be independent and ranged from 1.1 to 6.5 in skin and from 9.5 to 20.6 in seeds (Mané et al. 2007b; Souquet et al. 2006; Monagas et al. 2003). Comparison of total amount is more delicate since values are expressed differently among publications: per gram of tissue (dry or fresh) weight, per gram of entire berry, per berry. Values found in literature for total amount of skin tannins is between 1.76g/kg of berries for Mourvedre and almost 3.15 for Muscat de Hambourg (Downey et al. 2003a; Cortell et al. 2005; Fournand et al. 2006; Souquet et al. 2006; Mané et al. 2007b).This concentration seems independent of anthocyanin concentration even if the number of varieties studied is statistically rather small. Contribution of the seeds to the total composition of the berry is also cultivar dependant; Maccabeo berries contain two times less proanthocyanidin in the seed than in the skin (Souquet et al. 2006) whereas seed contribution to the total flavanol pool of Pinot berries is greater than 60% (Mané et al. 2007b).

This effect can be modulated by environmental factors. Significant increases were found in skin proanthocyanidin content, proportion of (–)-epigallocatechin, and average DP in berries from zones with a low vine vigor (Cortell et al. 2005). In reaction to sun exposure, skin proanthocyanidin content tends to increase, particularly trihydroxylated subunits and mDP is enhanced (Downey et al. 2004; Cortell and Kennedy 2006). Shaded fruits reached a lower maximum in proanthocyanidin content than sun-exposed ones but the contents at harvest were similar. Most authors agree that water stress had only slight effects on tannin composition (Ojeda et al. 2002; Kennedy et al. 2002; Castellarin et al. 2006). Seed flavanol composition seems hardly affected by environmental factors.

9B.1.3 Flavonols and Dihydroflavonols

9B.1.3.1 Structure and Localisation

Flavonols, which play a protective role against UV radiations, are found in grape skins and in leaves. Some flavonols were also detected in pulp (Pereira et al. 2006), but none in the seeds (Rodriguez Montealegre et al. 2006).

The major flavonols in grape are 3-glycosides of quercetin that is dihydroxylated on the B-ring (Fig. 9B.1, R1=OH, R2=H), and especially its 3-glucoside and 3-glucuronide (Ribéreau-Gayon 1964; Wulf and Nagel 1976; Cheynier and Rigaud 1986; Downey et al. 2003b; Price et al. 1995). Other flavonols and dihydroflavonols (syn. flavanonols) were also identified in different parts of the plant as listed in Table 9B.1.

9B.1.3.2 Variability in Grape

The amount of flavonols in the berries also depends on the developmental stage, genetic and environmental factors. Flavonol biosynthesis occurs at flowering and again after veraison. Small amounts can be detected at the green stage but synthesis mainly occurs during ripening and a constant increase in flavonol content per berry is measured (Downey et al. 2003b).

Each cultivar possesses a specific flavonol and dihydroflavonol profile (quantitative and qualitative aspect) that could be used for taxonomic characterization (Mattivi et al. 2006; Masa et al. 2007). It was believed that white cultivars did not contain any methylated flavonol until Rodriguez Montealegre et al. (2006) detected small amounts of isorhamnetin glucoside in their skins. Based on a survey of 91 *Vitis vinifera* cultivars, Mattivi et al. (2006) concluded that isorhamnetin derivatives are present in small amounts in white cultivars whereas derivatives of myricetin, laricitrin and syringetin appear to be specific of red cultivars. However, myricetin was detected in some white Muscadine cultivars (*Vitis rotundifolia* sp.) (Talcott and

Lee 2002). The amount of flavonols ranged between 2 and 30mg/kg of berry for white skinned cultivars and 4 and 78mg/kg of berry for black skinned cultivars (Mattivi et al. 2006). Rodriguez Montealegre et al. (2006) agreed that red skinned cultivars contain more flavonols but they detected higher concentrations, up to 170mg/kg for Viognier berries or 200mg/kg for Shiraz berries.

Flavonol and especially quercetin levels in the skins of Pinot noir grapes showed a dramatic response to sun exposure (Price et al. 1995). The effect of bunch shading on flavonol accumulation was confirmed in Merlot (Spayd et al. 2002) and Syrah (Downey et al. 2004) while temperature had little or no effect. This suggests that biosynthesis of flavonols is light-induced, accordingly with the role of flavonols as UV-protectant (Cortell and Kennedy 2006; Downey et al. 2004; Spayd et al. 2002). This effect was maximum when shading was applied a few weeks before flowering, almost preventing any flavonol synthesis. When the treatment was applied after flowering, the amount of berry flavonols was 8–10 times lower than in the control berries (Cortell and Kennedy 2006; Downey et al. 2004). No detailed impact on the different grape flavonols in skins is available. The flavonol content in pulp also discriminated between shaded and light exposed berries, kaempferol and quercetin derivatives being more abundant in the pulp of sun exposed berries, and myricetin derivatives, along with other unidentified flavonols, in that of shaded berries (Pereira et al. 2006).

9B.2 Extraction into the Wine

Wine flavonoid composition depends not only on the grape composition but also on their extraction and subsequent reactions during the wine-making and aging process. Thus, white wines obtained by direct pressing with minimum skin contact contain mostly the flavonoids originating from pulp. In rosé and red wine technology, the procedures used to extract the anthocyanin pigments from the skins also result in increased extraction of other flavonoids from skins, seeds and eventually stems or leaves if present in the fermentation tank. Extraction continues until the wine is separated from the solid residue (marc or pomace) by racking or pressing. Its kinetics depend on the solubility of the compounds and on their accessibility within the berry tissues, which can be modulated by physiological factors such as the maturation stage. It is further influenced by other technological factors, including the concentration of alcohol and of sulfur dioxide in the liquid phase, the temperature and the extent of must homogenization. Consequently, the wine flavonoid composition is influenced by the duration of pre- and post-fermentation maceration phases and by treatments enhancing cell wall or berry degradation (e.g. use of pectinolytic enzymes).

9B.2.1 White Wines

Quercetin 3-glucuronide was the only flavonol detected in free run juices and wines (Alonso et al. 1986; Betes-Saura et al. 1996), along with trace amounts of kampferol

3-glucoside in Riesling wine (Baderschneider and Winterhalter, 2001). Its average concentration was found to be 0.5mg/L and 0.25mg/L, in free run juices and wines, respectively and 0.4mg/L in Cava sparkling wines. Champagne wines made from Pinot noir and Chardonnay contained quercetin aglycone and trace amounts of astilbin and engeletin (Chamkha et al. 2003) which have also been reported in other white wines (Trousdale and Singleton 1983). Finally, leaf contamination of the grape crush may result in increased flavonol concentration in wines (Somers and Ziemelis 1985).

Flavanol monomers and oligomers have been found in small amounts (a few mg/L) in white wines made without maceration (Cheynier et al. 1989b; Betes-Saura et al. 1996; Chamkha et al. 2003; Ricardo da Silva et al. 1993). Delays between harvest and pressing, especially if sulfur dioxide is added to prevent oxidation, as well as thorough pressing, result in increased concentrations of flavonoids in white musts and wines (Yokotstuka 1990; Somers and Pocock 1991). Skin contact before fermentation is sometimes used in white wine making to favour the extraction of volatile compounds and increase wine varietal character. This practice also resulted in an increase of flavanol concentration in wine (Cheynier et al. 1989b; Ricardo da Silva et al. 1993). Procyanidin B1 was the major dimer and galloylated dimers were present in very low amounts, suggesting that flavanols in white wine do not originate from seeds.

To our knowledge, proanthocyanidin polymers have not been analysed in white wines. In a recent study performed in our laboratory, no flavanol derivatives could be detected in Champagne wines after thiolysis (Mané 2007) although the pulp of all three Champagne cultivars contained about 20mg of proanthocyanidins per kg of berries (Mané et al. 2007b). This can be due to adsorption of the higher polymers on the grape cell wall material, as described for apple (Renard et al. 2001) or to oxidation during pressing of the must. Indeed, the role of enzymatic oxidation, catalyzed by the grape polyphenoloxidase (PPO) during obtention of white musts is well documented. Flavanol monomers are rather poor substrates for PPO and proanthocyanidins cannot be oxidized by the enzyme, presumably due to steric hindrance. However, all these compounds are readily oxidized by the quinones resulting from enzymatic oxidation of caffeoyl tartaric acid, the major substrate of PPO in grape (Cheynier et al. 1988; Cheynier and Ricardo Da Silva 1991). Increasing the level of oxygen exposure before fermentation resulted in much lower amounts of flavanols in wine, confirming the role of oxidation (Cheynier et al. 1989b; Ricardo da Silva et al. 1993).

9B.2.2 Red Wines

Anthocyanin content reaches a maximum early in fermentation (Nagel and Wulf 1979) whereas tannin extraction continues throughout pomace contact (Singleton and Draper 1964). Monitoring of red must flavonoid composition during maceration showed that the extraction of flavonols and of proanthocyanidins from skins roughly parallels that of anthocyanins while that of seed flavanols is slower (Cheynier et al. 1997b; Morel-Salmi et al. 2006). The initial rate of flavonoid extraction from skins was not influenced by the alcohol content, although the final concentration reached in water was lower than in 6.5% or 13% ethanol whereas extraction of flavonoids from seeds increased with concentration of ethanol (Canals et al. 2005). Whether this is due to their chemical structure, seed procyanidins being less hydrophilic than other flavonoids, or to the different characteristics of both plant tissues is unknown.

Consequently, pre-fermentation maceration at low temperature increases the level of anthocyanins and flavanols from pulp and skins while post-fermentation maceration increases that of proanthocyanidins, as a result of enhanced extraction from seeds (Cheynier et al. 2006).

Treatments favouring contact between solid and liquid phases such as pumping over or punching down are traditionally used to enhance extraction. Alternative processes have been proposed more recently. These include physical processes such as thermovinification and flash release (in which the grapes are heated quickly at high temperature ($> 95^\circ C$) and then placed under vacuum, to produce instant vaporization and cooling) and the use of pectinolytic enzymes.

Flash release and thermovinification greatly accelerated extraction of all flavonoids from Grenache, Carignan or Mourvedre grapes and can be used to produce polyphenol-enriched juices (Morel-Salmi et al. 2006). Maceration after flash release treatment further increased extraction of flavonoids and especially of flavanols. After maceration, the concentration of flavanols in flash release treated wines was much higher than in the control wine while that of flavonols and anthocyanins was hardly modified. The use of pectinolytic enzymes results in increased juice yield (Berg 1959; Ough and Crowell 1979) along with increased browning of white wines (Berg 1959) and faster colour development (Ough et al. 1975), enhanced extraction of phenolic compounds and colour in red wines (Fernandez-Zurbano et al. 1999).

Numerous protocols have been proposed to estimate flavonoid extractibility from grape and relate it to ripeness and crop quality. Such approaches have shown that anthocyanin extraction increases as they accumulate in the berries (Canals et al. 2005; Fournand et al. 2006) but that their extraction yield is stable (Fournand et al. 2006). The rate of proanthocyanidin extraction from skins as well as the quantity extracted remain constant throughout berry development. Extracted proanthocyanidins showed lower mDP and lower rate of galloylation than those remaining in the marc. Although the extraction yield did not change, this selectivity was somewhat lower in riper grapes. Extraction of flavonols from shaded grapes appeared more efficient than from sun exposed grapes, as the concentration of flavonols in the simulated maceration was only decreased 2.5-fold (vs 8-fold in the grape skins) (Cortell and Kennedy 2006). The extraction rate of skin proanthocyanidins was higher from sun exposed than from shaded grapes but the concentrations in both extracts were similar (Cortell and Kennedy 2006).

Nevertheless, the proposed protocols do not simulate flavonoid extraction during the maceration and fermentation process, as they do not reproduce the effect of gradual alcohol accumulation, temperature increase and duration of the maceration phase.

The concentrations of anthocyanins and proanthocyanidins present in red Grenache wines at the end of fermentation represented about 30% and 50%, respectively, of their amounts in grape (Morel-Salmi et al. 2006). Extraction of the pomace allowed to recover most of the proanthocyanidins but hardly increased the yield of anthocyanins. This indicates that a major proportion of anthocyanins have been converted to other species and/or been irreversibly adsorbed on the solid material during fermentation, in agreement with earlier studies reporting a drop in anthocyanin concentration after the initial increase (Nagel and Wulf 1979; Gao et al. 1997). Monitoring of flavonoid composition in wines made by pressing immediately after flash release and fermentation in the liquid phase demonstrated that anthocyanins, flavonols and proanthocyanidins undergo rapid changes (about 50% loss for flavonols and anthocyanins, 40% loss for proanthocyanidins after five days of fermentation), while flavanol monomers are not affected (Morel-Salmi et al. 2006). This provides good evidence that anthocyanin and tannin reactions that have been reported to take place slowly during aging actually start very early in the wine-making process.

9B.3 Reactions in Wine

Changes in flavonoid composition taking place during wine making and aging involve both enzymatic and chemical processes. The former processes, due to grape, yeast or fungi and exogenous enzymes occur mostly in to the early stages while chemical reactions continue during aging.

9B.3.1 Flavonoid Reactivity

Reactions of flavonoids are primarily due to the reactivity of the phenolic rings. On one hand, the resonance between the free electron pair of a phenolic oxygen and the benzene ring enhances electron delocalisation and confers the position adjacent to the hydroxyl a partial negative charge and thus a nucleophilic character (showing an excess of electrons and thus prone to react with electrophiles, showing an electron deficiency). Such nucleophilic sites are encountered on the phloroglucinol A-ring of flavonoids, in C6 and C8 (for the numbering, see Fig. 9B.1), due to their meta hydroxyl substitution pattern. On the other hand, the acidity of the phenolic hydroxyl groups leads to formation of phenate ions and subsequent oxidation to a semiquinone radical, or in the case of *o*-diphenolic groups as often encountered in the B-ring, to an *o*-quinone. The latter is an electrophilic species and thus prone to suffer nucleophilic addition. Other examples of electrophilic species include the anthocyanin flavylium cations and the carbocations resulting from acid-catalysed cleavage of proanthocyanidin interflavanic linkages.

9B.3.2 Enzymatic Processes

9B.3.2.1 Enzymatic Oxidation

The major grape enzymes involved in flavonoid degradation is polyphenoloxidase (PPO) that catalyses *o*-hydroxylation of monophenols (Fig. 9B.4(1)) and oxidation of *o*-diphenols to the corresponding *o*-quinones (Fig. 9B.4(2)). Its action is particularly important in white wine technology as extensive decompartmentation of the enzyme and its substrates (phenolic compounds and oxygen) takes place at pressing. Catechins and flavonol aglycones are rather poor substrates for grape PPO compared to caffeoyltartaric acid. Glycosylated flavonoids, epicatechin gallate and proanthocyanidins cannot be oxidized directly by grape PPO, presumably due to steric hindrance. However, they can react with the enzymically generated caffeoyltartaric acid quinone through coupled oxidation (Fig. 9B.3) and nucleophilic addition reactions (Fig. 9B.4) (Cheynier and Van Hulst 1988; Cheynier et al. 1988, 1989a, 1995; Cheynier and Ricardo Da Silva 1991), as described in Chapter II.2.3 for anthocyanins. In addition, disproportionation of an *o*-quinone and an *o*-diphenol can yield two semiquinone radicals (Fig. 9B.5) which can then undergo radical coupling (Fig. 9B.6). Enzymatic oxidation can also be catalysed by other enzymes such as laccase arising from *Botrytis cine*rea (grey mold or noble rot) and peroxidases which accept a wider range of substrates.

9B.3.2.2 Enzymatic Hydrolysis

Other enzymes degrading flavonoids include various hydrolases, originating from microorganisms or from pectolytic preparations added during wine-making. Some yeast strains show glycosidase activities, including β-glucosidase which is active on flavono and anthocyanin 3-glucosides. Tannase activity (tannin acyl hydrolase, EC 3.1.1.20) catalysing hydrolysis of galloyl esters has been observed in numerous fungi, including *B. cinerea*, and in lactic acid bacteria (Matthews et al. 2006). The use of exogenous enzymes is becoming more and more popular to help clarification, increase press yield, extraction, and release of aroma compounds from their glucosidic precursors. These enzyme preparations are pectinases and α-glucanases. They may show side activities such as cinnamate esterase, tannase and β-glucosidase. Hydrolysis of flavonol glycosides has been reported to result in haze development as the insoluble flavonol aglycones precipitated out (Somers and Ziemelis 1985).

9B.3.3 Chemical Reactions

As stated above, flavonoids react as nucleophiles through their C8 and C6 positions. Acid-catalysed cleavage of the interflavanic linkages of proanthocyanidins (cf. Fig. 9B.3) also takes place spontaneously in wine, yielding an intermediate electrophile. A third group of reactions involves the *o*-diphenolic B-ring which can be oxidized to electrophilic quinones. Several mechanisms arise from these reactivities.

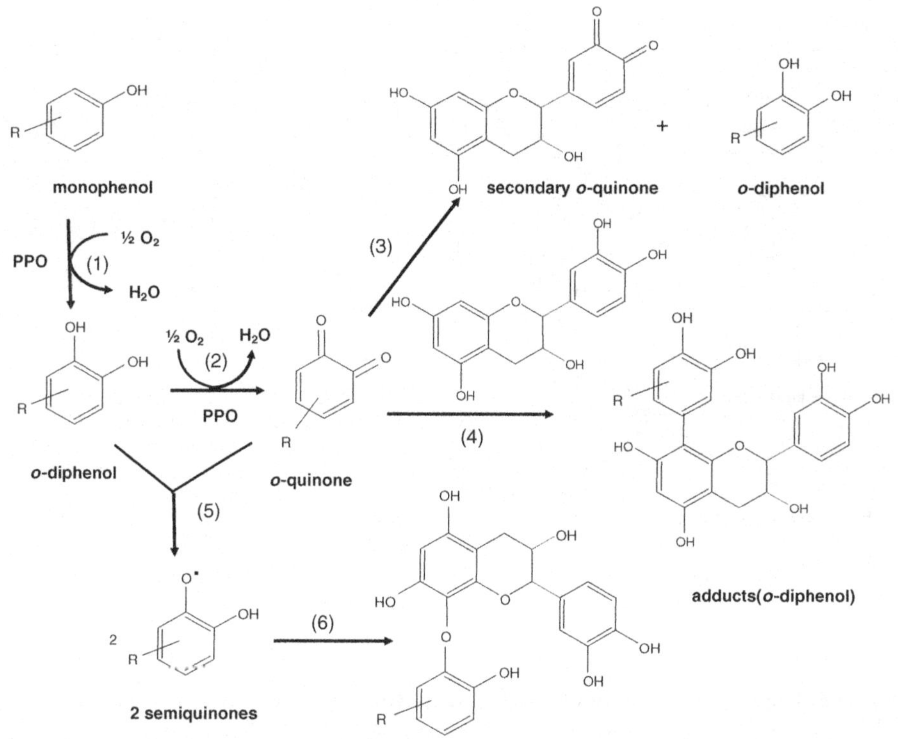

Fig. 9B.4 Enzymatic formation and reactions of quinones

Those involving both flavonol and anthocyanins have been described in Chapter 9A and will not be detailed here.

9B.3.3.1 Reactions Based on Acid-Catalysed Cleavage of Proanthocyanidins

Precursors

The precursors of these reactions are, on one hand, proanthocyanidins and, on the other hand, any kind of flavonoid that can act as a nucleophile. The latter include flavonols, dihydroflavonols, flavanol monomers, proanthocyanidins, and anthocyanins under their hemiketal form (for anthocyanin reactivity, see Chapter 9A).

Reaction Mechanism

The reaction starts with acid catalysed cleavage of a proanthocyanidin interflavanic linkage (Figs. 9B.3 and 9B.5(1)). The intermediate carbocation thus generated then undergoes nucleophilic addition. When the nucleophile is another flavanol (Fig. 9B.5(2a)), the product is a new proanthocyanidin molecule. As a result of this

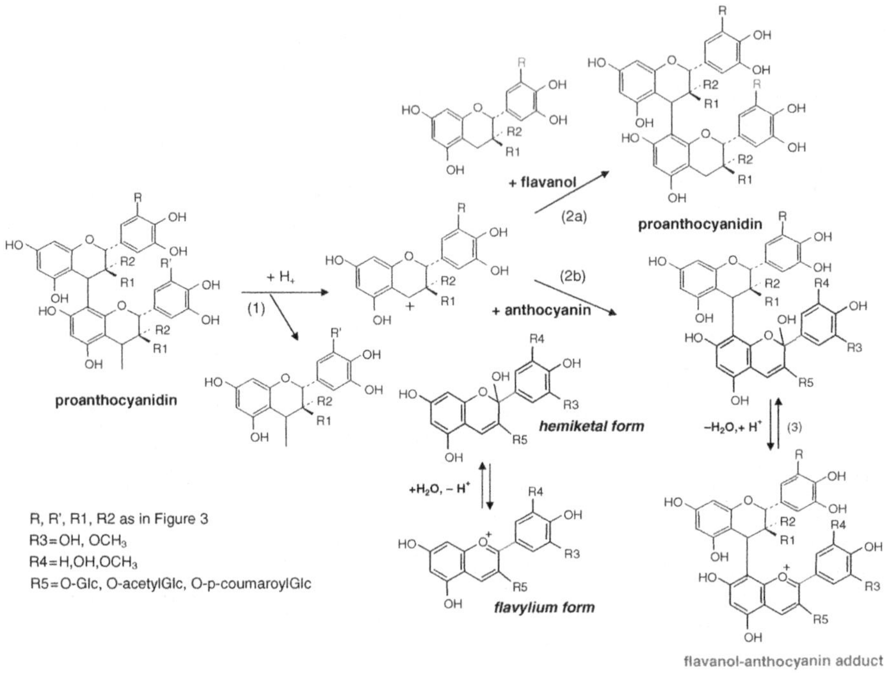

Fig. 9B.5 Polymerisation reactions based on acid-catalysed cleavage of proanthocyanidins

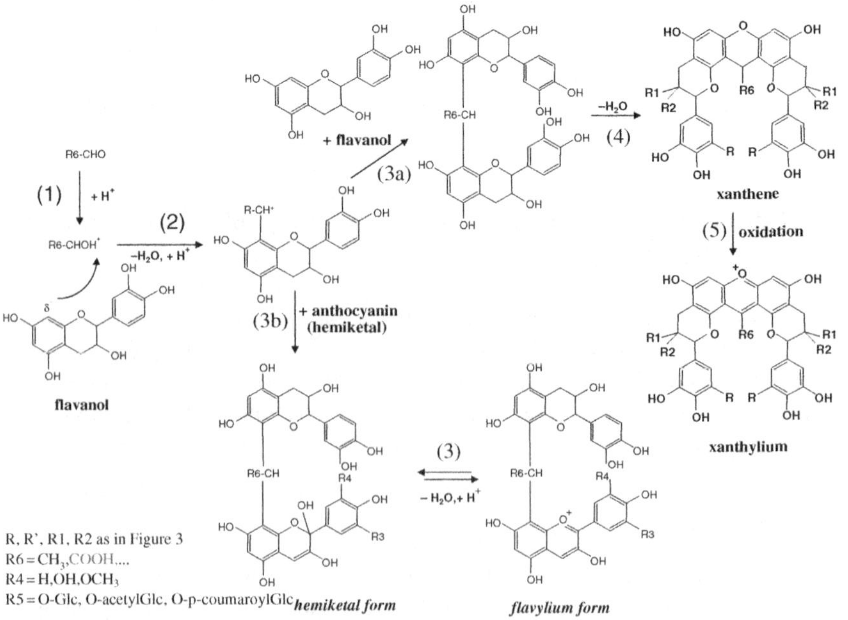

Fig. 9B.6 Condensation reactions with aldehydes

process, first described by Haslam (1980), the average DP of proanthocyanidins can be modified, either increased if the nucleophilic flavanols are polymers, or decreased if they are low molecular weight compounds. In the presence of excess amounts of monomers, side reactions of the carbocations leading to unknown species are much reduced and polymers are gradually replaced by oligomers (Vidal et al. 2002). Anthocyanins (under their hemiketal form) and presumably flavonols can replace the flavanol as a nucleophile. In the case of anthocyanins, the product is a colorless flavanol-anthocyanin hemiketal adduct (Fig. 9B.5(2b)) which then can give rise to the corresponding flavylium adduct, through protonation and dehydration (Fig. 9B.5(3)) (Ribéreau-Gayon 1982). Another mechanism involving the flavene form of the anthocyanin and subsequent oxidation of the resulting adduct to the flavylium pigment has also been proposed (Jurd 1969). Recent mass spectrometry studies have confirmed the former pathway (Salas et al. 2003) and established the presence of flavanol-anthocyanin adducts in wine (Salas et al. 2004, 2005).

Evidence in Wine

Evidence of such adducts in wine fractions has been provided, as detailed in Chapter 9A. These include F-A^+ (Alcalde-Eon et al. 2006; Boido et al. 2006) and F-A-A^+ (Alcalde-Eon et al. 2006) adducts based on different flavanol and anthocyanin units and $(F)_n$-A^+ adducts deriving from different flavanols monomers and oligomers (Hayasaka and Kennedy 2003). Proanthocyanidins arising from these reactions cannot be distinguished from those extracted from grapes. However, detection of F-A^+ adducts without prior fractionation (Morel-Salmi et al. 2006) confirmed the occurrence of the acid-catalyzed interflavanic bond breaking process in wines.

Factors Affecting the Reaction

The first step of the reaction is acid-catalyzed and thus largely determined by pH. When used for analytical purposes in reactions such as thiolysis or phloroglucinolysis, the reaction is performed under strongly acidic conditions. However, it takes place spontaneously at pH values such as encountered in wines. For instance, conversion of proanthocyanidin polymers to dimers was demonstrated at pH 3.2 (Vidal et al. 2002). When procyanidin dimer B2-3′-*O*-gallate (Ec-EcG) was incubated at pH 2, $(Ec)_n$-EcG oligomers (with $n = 1, 2, 3$) were formed. These oligomers that result from successive cleavage of the interflavanic bond and addition of B2-3′-*O*-gallate onto epicatechin carbocation were not detected at pH 3.8. In model systems containing B2-3′-*O*-gallate and malvidin 3-glucoside, flavanol anthocyanin adducts were formed at pH 2 but not at pH 3.8 (Salas et al. 2003), meaning that the acid catalyzed cleavage rather than the proportion of anthocyanin in the hemiketal form (decreasing with pH) was the limiting factor. However, such adducts were detected in a series of red wines (Morel-Salmi et al. 2006) showing pH values in the range 3.59–3.82 (Doco et al. 2007).

9B.3.3.2 Reactions Involving Aldehydes

Precursors

Various aldehydes are encountered in wine. The most abundant is acetaldehyde which is both a product of yeast metabolism and an oxidation product of ethanol. Glyoxylic acid, resulting from oxidation of tartaric acid, especially catalyzed by metal ions (Fe, Cu) or ascorbic acid, can also be present. Other aldehydes reported to participate in these reactions include furfural and 5-hydroxymethylfurfural that are degradation products of sugar and can be extracted from barrels (Es-Safi et al. 2000), vanillin which also results from oak toasting, isovaleraldehyde, benzaldehyde, propionaldehyde, isobutyraldehyde, formaldehyde and 2-methylbutyraldehyde which are present in the spirits used to produce fortified wines (Pissara et al. 2003).

Reaction Mechanism

The reaction mechanisms, first proposed by Timberlake and co-workers (Timberlake and Bridle 1976), was recently established by mass spectrometry (Fulcrand et al. 1996). It starts with protonation of the aldehyde (Fig. 9B.6(1)), yielding an intermediate carbocation which then suffers nucleophilic addition from the A-ring of the flavonoid. The resulting adduct, through protonation and dehydration steps, gives rise to another carbocation (Fig. 9B.6(2)) which reacts with a second flavonoid molecule (Fig. 9B.6(3a,b)). As both the C6 and C8 positions are reactive, the reaction can continue through the remaining free nucleophilic sites, leading to polymerisation. When several nucleophiles are present, they combine randomly to form a large variety of oligomers and polymers (Es-Safi et al. 1999a). Polymerisation was first established in the case of flavanol monomers and acetaldehyde. When both anthocyanins and flavanols are present, the latter react more readily with acetaldehyde and anthocyanins were initially thought to terminate the polymerisation chain. However, the detection of methylmethine-linked anthocyanin oligomers resulting from this process ruled out this hypothesis (Atanasova et al 2002). The hydration constants of methylmethine-linked malvidin 3-glucoside dimer (Atanasova et al 2002) and of flavanol-methylmethine-malvidin 3-glucoside (Duenas et al. 2006b) were determined. This indicated that one of the units of the anthocyanin dimer is under the hemiketal form and can be involved in further polymerisation while the anthocyanin in the flavanol adduct is predominantly under the flavylium form and thus less prone to react as a nucleophile.

Similar processes have been observed with other aldehydes, such as glyoxylic acid. However, the carboxymethine-linked oligomers resulting from reaction with glyoxylic acid proceeded to xanthylium salts rather than to larger polymers (Es-Safi et al. 1999b). The postulated pathway involves dehydration and cyclisation of the carboxymethine dimer (Fig. 9B.6(4)) followed by oxidation of the resulting xanthene (Fig. 9B.6(5)) that was also detected in the medium. Formation of xanthylium salts was also shown in the case of furfural and hydroxymethylfurfural (Es-Safi et al. 2000).

Evidence in Wine

A number of flavanol-methylmethine-anthocyanins have been detected in wine (see Chapter 9A for extensive list) as well as methylmethine-linked flavanol dimers (Cheynier et al. 1997a; Saucier et al. 1997b). More recently, a method to determine methyl-methine linkages has been proposed (Drinkine et al. 2007a), showing that they represented less than 4% of bonds between flavanol units in wine although this proportion increased with aging (Drinkine et al. 2007b). The other aldehyde derivatives have not yet been detected in wine.

Factors Affecting the Reaction

The rate of condensation reactions depends on the concentration of the precursors involved, and especially on that of the aldehyde, and on its protonation rate, which is primarily determined by pH. On the other hand, acidic conditions also favour cleavage of the products. The methylmethine bridges are particularly susceptible to acid-catalysed cleavage which leads to rearrangement and further polymerisation (Es-Safi et al. 1999a; Escribano-Bailon et al. 2001). This also yields vinylflavanol structures which add to anthocyanins to form flavanyl-pyranoanthocyanins (Cheynier et al. 1999a; Mateus et al. 2002) and flavanylvinylpyranoanthocyanins (portisins) (Mateus et al. 2003) as described in Chapter 9A.

Acetaldehyde is generated during alcoholic fermentation but its concentration can be much increased as a result of oxygen exposure in the wine-making or aging process. Metal ions such as iron (Oszmianski et al. 1996) or copper (Clark and Scollary 2002) much enhance browning of catechin in wine-like medium containing tartaric acid due to the formation of xanthylium pigments (Es-Safi et al. 1999b). Metals were first postulated to catalyse oxidation of tartaric acid to glyoxylic acid but it was latter demonstrated that copper(II) also accelerated the bridging of two (+)-catechin units by glyoxylic acid (Clark et al. 2003). Ascorbic acid, after an initial antioxidant phase, also increased formation of xanthylium salts, as it crossed-over to pro-oxidant (Bradshaw et al. 2003). Catechin reaction was faster with glyoxylic acid than with acetaldehyde and even faster when both aldehydes were present (Drinkine et al. 2005). Under the conditions used in this experiment, methylmethine and carboxymethine bridged dimers as well as mixed polymers were observed but no xanthylium salt could be detected, either because of the rather short incubation time used or because no metal catalyst was present. However, products arising from glyoxylic acid condensation have never been detected in wine, possibly because its formation rate from tartaric acid is the limiting factor.

9B.3.3.3 Reactions Involving Quinones

Precursors

These reactions involve, on one hand, the nucleophilic A-rings of flavonoids such as flavanols, and on the other hand, electrophilic quinones arising from enzymatic or chemical oxidation of phenolic compounds.

Reaction Mechanism

Quinones are electrophilic species which can thus add various nucleophiles, including flavonoids. They are also powerful oxidants which can oxidize *o*-diphenolic compounds such as *o*-diphenolic flavonoids to secondary quinones as described above. Addition of catechin to its quinone generated by enzymatic oxidation with polyphenoloxidase (Guyot et al. 1996b) or peroxidase (Weinges and Muller 1972) and by autoxidation (Hathway and Seakins 1957; Young et al. 1987) yield B-type dehydrodicatechins in which the catechin moieties are linked through a biphenyl bond between the A-ring of one (initially acting as the nucleophile) and the B-ring of the other (initially present as the quinone). These colorless species further oxidize to yellow A-type dehydrodicatechins. Additional dehydrodicatechins linked through biphenylether bonds and thus presumably resulting from radical coupling of two catechin semiquinones were also formed after enzymatic oxidation, especially at lower pH values which stabilise the semiquinones radicals (Guyot et al. 1996b). When other oxidizable phenolics such as phenolic acids are present, formation of codimers is also observed (Fulcrand et al. 2006; Richard-Forget et al. 1992). Depending on the relative redox potentials of the phenol/quinone couples, the flavanol can act as the nucleophile or as the electrophile in these reactions. In particular, in the presence of malvidin 3-glucoside which cannot be oxidized to an *o*-quinone, oxidation products were formed by addition of the anthocyanin onto the catechin quinones (Duenas et al. 2006a).

Evidence in Wine

To our knowledge, no oxidation product of flavanols or flavonols has been yet detected in wine, the only oxidation products evidenced so far being Grape Reaction Product (GRP, i.e. 2, *S*-glutathionyl caftaric acid) and anthocyanin caftaric acid adducts arising from addition of glutathione and of anthocyanins, respectively, onto caftaric acid quinone. However, the role of flavanols in non-enzymic browning of white wines is well documented (Simpson 1982; Cheynier et al. 1989b). Whether this results from autoxidation reactions, from formation of xanthylium salts through condensation reactions with aldehydes, or from other unknown processes remains to be established.

Factors Affecting the Reaction

Enzymatic reactions occur early in the wine making process unless laccase is present as a result of grape fungal contamination. They depend primarily on the available oxygen, on the presence of reductants such as ascorbic and sulfites, and on the ratio of glutathione to caftaric acid. Caftaric acid quinone is the major product of enzymatic oxidation in grape musts. Reductants reduce the quinone back to caftaric acid while glutathione traps it as GRP. When both are depleted, the excess quinones, if any, react with flavonoids through coupled oxidation or nucleophilic addition reactions. Indeed, aeration of musts induced losses of flavanols and compensated for their increased extraction following skin contact in white wine making (Cheynier et al. 1989b).

Flavonoid autoxidation in wine is a slow process but its rate increases with pH. For instance, products arising from addition of malvidin 3-glucoside onto epicatechin quinone were observed only at pH 4 and above (Duenas et al. 2006a). Oxidation of flavanols and formation of B-type and A-type dehydrodicatechins was observed in the wine pH range (Oszmianski et al. 1996). When catalysts such as metal ions are present, oxidation of tartaric acid and subsequent formation of xanthylium pigments compete with autoxidation reactions.

9B.3.3.4 Reactions with Other Electrophiles

Precursors

Other electrophiles include anthocyanin flavylium cations. The intermediate cation generated, in mildly acidic conditions, from vescalagin, an ellagitannin present in wines after barrel aging or addition of oak chips and tannin extracts, has also been shown to participate to such reactions (Quideau et al. 2003).

Reaction Mechanism

Again this involves nucleophilic substitution of a flavonoid at its nucleophilic C8 or C6 centre on an electrophilic center, which may be the C4 of an anthocyanin flavylium or the carbocation generated by protonation and dehydration of vescalagin.

As described in Chapter 9A, nucleophilic addition of the flavanol onto the C4 position of the flavylium ion generates an anthocyanin flavanol (A-F) flavene adduct which can either oxidize to the flavylium pigment (Jurd 1967, 1969) or rearrange to another colorless structure in which the anthocyanin and flavanol units are linked through an additional 2-O-7 ether bond as found in A-type proanthocyanidins (Jurd and Waiss 1965; Bishop and Nagel 1984; Remy-Tanneau et al. 2003). Further reactions of the flavylium salt have also been reported to yield yellow xanthylium salts (Jurd 1967, 1969; Jurd and Somers 1970; Somers 1971; Timberlake and Bridle 1976; Baranowski and Nagel 1983; Liao et al. 1992; Santos-Buelga et al. 1995). However, the postulated structures have never been confirmed. Detection of other xanthylium salts resulting from degradation of the intermediate flavylium adduct suggests that the latter is rather unstable (Duenas et al. 2006a).

Addition of catechin or epicatechin onto vescalagin yield complex tannin structures called acutissimin and epiacutissimin (Quideau et al. 2003) that have been isolated earlier from *Quercus acutissima.* It is worth noting that the isomer of vescalagin (i.e. castalagin) fails to undergo this reaction.

Evidence in Wine

A-type bound anthocyanin-flavanols were detected in wine as dimers and as larger oligomers (anthocyanin-(epi)cat$_n$, with $n > 1$) (Remy et al. 2000; Salas et al. 2005). The presence of such oligomers ($n = 1$ through 7) was confirmed by mass spectrometry (Hayasaka and Kennedy 2003).

Factors Affecting the Reaction

Nucleophilic addition of flavanols onto malvidin 3-glucoside was observed in a wide range of pH values (2–6). However, in the case of proanthocyanidins, flavanol-anthocyanin adducts resulting from acid catalysed cleavage of the interflavanic bonds were the predominant species formed at pH 2. The intermediate A-F flavene proceeded to different products according to the pH value and substrates. At pH 2, in the case of epicatechin it yielded the A-type A-F dimer while at pH 3 and above, it oxidized to the flavylium (Duenas et al. 2006a). Such flavylium adducts formed from proanthocyanidins appeared rather stable (Malien-Aubert et al. 2002; Salas et al. 2003). That formed from epicatechin could not be detected but was converted to xanthylium salts through a mechanism involving opening and cleavage of the anthocyanin C-ring (Duenas et al. 2006a).

9B.4 Interactions with Other Grape and Wine Constituents

Flavonoids are prone to interact between them and with other wine constituents such as proteins or polysaccharides. Associations of anthocyanins (i.e. self association and copigmentation) modify tint and enhance the intensity and stability of colour. Aggregation of flavanols as well as their interactions with proteins is responsible for haze development in wines and other beverages such as beer. Moreover, astringency perception results from their interaction with salivary proteins, as detailed in Chapter 9B. Interactions of flavonoids with proteins or polysaccharides lead to colloidal phenomena that may impede the efficiency of clarification and stabilization treatments but are also taken advantage of in processes such as fining and addition of protecting colloids. In addition, various technological problems are related to adsorption of flavonoids on surfaces. In particular, adsorption on plant cell walls limits extraction of flavonoids into the must and wine while adsorption on tank surfaces and filtration membranes result in difficult cleaning and/or clogging of the equipment.

9B.4.1 Interaction Processes

Interactions involving flavonoids are based on several phenomena, all deriving from electrostatic interactions:

- hydrogen bonding, which is a non covalent bond arising mainly from permanent dipole – permanent dipole interactions (a particular type of Keesom interactions).
- london interactions, also called dispersion forces, occurring between two induced dipoles. These interactions are further strengthened by the partial desolvation of the hydrophobic surfaces coming into contact and release of some water molecules from the solvation shell into the bulk phase (hydrophobic effect).

These forces are rather weak when compared to covalent binding, but their range of action is larger.

In addition to these non-covalent and reversible bonds, covalent binding through nucleophilic addition or radical coupling can occur, in particular as a result of oxidation reactions. Copigmentation has thus been proposed to be the first step leading to formation of anthocyanin-flavanol adducts in red wine (Brouillard and Dangles 1994). Besides, quinones arising from oxidation of phenolic compounds can suffer nucleophilic addition from the SH or NH2 groups of proteins (Pierpoint 1969).

9B.4.2 Flavonoid Interactions

9B.4.2.1 Copigmentation

Among flavonoid interaction processes, copigmentation has been extensively studied; for a review see Brouillard and Dangles (1993).

Actors

It involves, on one hand an anthocyanin under its flavylium or quinonoidal base form, and on the other hand another planar hydrophobic structure that can be another anthocyanin unit (self-association), or another colorless species (copigment), covalently bound (intramolecular copigmentation) or not (intermolecular copigmentation) to the anthocyanin pigment.

Interaction Mechanisms

Copigmentation is driven by hydrophobic vertical stacking between the anthocyanin and the copigment to form π-π complexes from which water is excluded. The flavylium cation as well as the quinonoidal base are planar hydrophobic structures and can be involved in such complexes whereas the hemiketal form cannot. The association thus results in displacement of the anthocyanin hydration equilibrium from the colorless hemiketal to the red flavylium form that can be easily measured by spectrophotometry.

Evidence in Wine

Copigmentation in wine is estimated from the effect of dilution on the absorbance values in the visible range, at a given pH and ethanol content (Boulton R 2001). Indeed, decreases in the absorbance values (reflecting the concentration of flavylium ions) higher than the dilution factor are attributed to disruption of the copigmentation complexes. On this basis, copigmentation has been reported to contribute 30–50% of the color of young red wines.

Factors Affecting the Interaction

Copigmentation, involving a shift from the colorless hemiketal to the flavylium, is especially important in mildly acidic conditions such as the wine pH range where hemiketal forms are prevalent. Association constants determined by spectrophotometry (89M-1; for epicatechin (Malien-Aubert et al. 2002), 101M-1 for catechin (Mirabel et al. 1999b) indicate that flavanol monomers are poor copigments of malvidin 3-glucoside compared to flavonols ($>10^3$/M; Malien-Aubert et al. 2002). Association constants could not be accurately determined in the case of proanthocyanidins which were not available as pure compounds, but appeared even lower (4/M for a fraction of procyanidin oligomers (Mirabel et al. 1999b).

Surprisingly, copigmentation was enhanced in the presence of 12% ethanol (Mirabel et al. 1999b), reaching 157/M for epicatechin and 68/M for the procyanidin fraction.

9B.4.2.2 Self Association and Aggregation

Flavanol self-association has been demonstrated by means of NMR (Dufour and Bayonove 1999; Mirabel et al. 1999a) and mass spectrometry (Sarni-Manchado and Cheynier 2002) and their aggregation has been studied by means of dynamic light scattering (Poncet-Legrand et al. 2003; Riou et al. 2002; Saucier et al. 1997a) and cryo-transmission electron microscopy (Poncet-Legrand et al. 2003).

Actors

Self association of catechin and epicatechin was revealed by NMR (Dufour and Bayonove 1999) but did not lead to aggregation (Poncet-Legrand et al. 2003). In contrast, epicatechin gallate and, to a lesser extent, epigallocatechin gallate (Poncet-Legrand et al. 2003), as well as procyanidins (Riou et al. 2002) and methyl-methine linked catechin oligomers resulting from acetaldehyde condensation (Saucier et al. 1997a) aggregate into metastable colloidal particles in wine-like ethanolic solutions. Aggregation and precipitation of flavonols is restricted to aglycones which exhibit lower solubility than their glycosides.

Interaction Mechanisms

Van der Waals interactions between similar entities in a polar solvant are attractive. The formation of hydrogen bonds between the solvant and the solute ensures its solubility. As flavanol aggregates are not charged and ionic interactions are not significantly involved, the large incidence of ionic strength indicates that hydrophobic effect is the major driving force (Poncet-Legrand et al. 2003).

Wine often exhibits turbidity due to the presence of micro-organisms, cell debris, potassium hydrogen tartrate crystals and other insoluble material. Flavonol aglycones have been shown to be responsible for the formation of haze and deposits in white wines (Somers and Ziemelis 1985). In red wines, the presence of colloidal size-range particles was shown by light scattering experiments after centrifugation (Vernhet et al. 2003). Phenolic compounds and especially proanthocyanidins are involved in the formation of protein haze (Waters et al. 1995) and are major components of precipitates and aggregates adsorbed on tank material (Vernhet et al. 1999a, 1999b) or filtration membranes (Vernhet and Moutounet 2002). However, these particles also contain other material such as proteins, polysaccharides or potassium hydrogen tartrate so that self-aggregation of phenolic compounds in wine and its role in the aggregation processes cannot be easily determined.

Factors Affecting the Interaction

The structure of the molecule itself affects the interaction mechanisms. In addition to the molecular formula, external parameters such as flavonoid concentration and medium composition play an important part. Self-aggregation was observed with galloylated monomers and proanthocyanidin fractions, but not with catechin or epicatechin. Thus, flavanol aggregation seems to require the presence of at least three phenolic rings (or two *o*-diphenolic rings) in the molecule as this enables it to establish bridges with other polyphenols (Baxter et al. 1997a). Aggregation of procyanidins first increased with mDP up to 5 for non-galloylated procyanidin fractions and to DP 10 for galloylated procyanidins from grape seeds and then decreased for larger polymers, suggesting that higher molecular weight procyanidins can adopt a conformation that increases their solubility. The gallic acid ring favours self-association, as evidenced by NMR (Baxter et al. 1997a), but this was not confirmed in the case of oligomeric fractions. Scattered intensity, aggregate size and polydispersity indexes increased with the flavanol concentration. Size and polydispersity indexes also increased with ionic strength and decreased when the ethanol content was raised. No aggregation was observed at 20% ethanol for any of the fractions up to 5g/L (Poncet-Legrand et al. 2003). Self association constants recorded for epicatechin were also five times weaker in 10% ethanol (Dufour and Bayonove 1999) than in water (Baxter et al. 1997a), which is in agreement with the proposed hydrophobic interaction mechanism.

9B.4.3 Flavonoid Interactions with Other Macromolecules

9B.4.3.1 Interactions with Proteins

Interactions between tannins and proteins have been extensively studied (Hagerman 1989; Haslam and Lilley 1988; Haslam et al. 1992), owing to their role in haze formation, astringency perception, and nutritional and anti-nutritional effects resulting from inhibition of various enzymes and reduction of dietary protein digestion. Other effects include reduced adsorption of β-casein at the air-liquid interface in the presence of epigallocatechin gallate with potential consequences on foam properties (Sausse et al. 2003).

Actors

Flavonoid protein complexation shows little specificity. However, lower molecular weight flavonoids (i.e. flavonols, non galloylated flavanol monomers) display moderate affinity for proteins and do not form aggregates. Similarly, although all proteins interact with tannins, proline rich structures such as encountered in proteins most commonly used as fining agents (e.g. gelatin, casein) or in salivary proline rich proteins (PRP) involved in astringency perception, are particularly prone to interact with tannins. Binding of flavonols and flavones to some proteins such as serumalbumine which is involved in their transport in plasma is well documented (Boulton et al. 1998; Dufour and Dangles 2005). Flavonols have also been shown to adsorb on polyvinylpolypyrrolidone (PVPP) (Laborde et al. 2006) but they have not been reported to interact with wine proteins.

Interaction Mechanisms

Interactions between flavonoids and proteins rely upon both Van der Waals-London interactions and hydrogen bonding (Oh et al. 1980; Luck et al. 1994; Murray et al. 1994; Charlton et al. 1996). A study performed by using NMR indicated stacking of the phenolic rings with the proline residues in proline sequences and stabilisation of the complexes through hydrogen bonding between the H acceptor site of the adjacent peptide bond and the hydrogen atom of the phenolic hydroxyl (Murray et al. 1994). More recently, an isothermal titration calorimetry (ITC) experiment showed that the interaction of flavanols with poly-L-proline involves both entropic (associated to hydrophobic effect and conformational changes) and enthalpic (attributed to hydrogen bonding) phenomena (Poncet-Legrand et al. 2007). The latter is prevalent in the case of flavanol monomers and the former in that of polymers. Interaction does not necessarily lead to precipitation. Flavonoids can form soluble complexes with peptides and proteins, as shown by NMR (Baxter et al. 1997b; Charlton et al. 2002b; Hemingway et al. 1999), mass spectrometry (Sarni-Manchado and Cheynier 2002) or fluorimetry (Dufour and Dangles 2005). Aggregation of flavanols with casein (Jobstl et al. 2006), poly-l-proline

(Poncet-Legrand et al. 2006) and human salivary PRP (Pascal et al. 2007) occurs in three stages:

- ligand binding and saturation of binding sites together with folding of the protein, in the case of intrinsically unstructured proteins such as salivary PRP (Pascal et al. 2007) or casein (Jobstl et al. 2006)
- formation of rather small and homogenous protein-flavonoid aggregates
- further aggregation and precipitation

Adsorption of flavonols on PVPP also involves Van der Waals interactions with associated hydrophobic effect and hydrogen bonding (Laborde et al. 2006).

Evidence in Wine

Protein haze due to interactions of flavanols with proteins and peptides is well documented in beer (Outtrup 1989) but also takes place in wine. Pathogenesis related proteins which are synthesized by vine following pathogen attacks seem to be particularly involved in these processes (Waters et al. 1996). Protein precipitation also occurs as a result of fining treatments that consist in adding exogenous proteins to precipitate tannins out in order to stabilise the wine and reduce its astringency. The proteins most commonly used for red wine fining are gelatins, albumins and caseins. Plant proteins from lupine or wheat have recently been tested as alternatives to gelatin (Maury et al. 2003). All of them, as well as salivary proteins (Sarni-Manchado and Cheynier 2002) selectively precipitate higher molecular weight flavanols (Ricardo da Silva et al. 1991b; Maury et al. 2001; Sarni-Manchado et al. 1999), which also exhibit higher astringency (Vidal et al. 2003). However, only a small proportion of flavonoids was recovered in the pellet and the wine composition was not significantly modified by the treatment (Maury et al. 2003). The loss of astringency observed after fining may thus be partly due to the inclusion of flavanols in soluble complexes.

Factors Affecting the Interaction

Interactions and formation of insoluble complexes with proteins increase with the number of phenolic rings, and especially of *o*-diphenolic rings and thus with polymerisation and galloylation (Haslam and Lilley 1988; McManus et al. 1985). The intensity of mass spectrometry signals corresponding to soluble peptide flavanol complexes increased from the monomers to the dimers and with galloylation (Sarni-Manchado and Cheynier 2002). ITC experiments failed to detect any interaction of poly-L-proline with catechin or epicatechin while association constants of 3.7×10^4/M and 8.1×10^4/M were determined for epigallocatechin gallate and epicatechin gallate, respectively. That of an oligomeric procyanidin fraction from grape seeds was even higher (3.4×10^5/M) confirming that the affinity of flavanols for proteins increases with their chain length. Furthermore, interaction of flavanol monomers with proteins follows the same order as their partition coefficients between octanol and water, meaning that it increases with hydrophobicity of the phenolic compound (Poncet-Legrand et al. 2007). Phenolic oxidation, generating polymeric species, resulted in enhanced protein interactions as evidenced by higher inhibition of enzymes (Guyot et al. 1996a), or changes in casein adsorption properties at the air/liquid interface (Sausse et al. 2003). As mentioned above, higher molecular weight flavanols are also selectively precipitated out by proteins. Moreover, within gelatins (Maury et al. 2001) or glutens (Maury et al. 2003), smaller molecular weight proteins appeared more selective than larger ones. The interaction mechanism also depends on protein concentration. At low concentration, it occurs in three stages as the polyphenol/protein ratio increases, as described above: saturation of the interaction sites, formation of metastable colloids, and aggregation leading to haze. At high protein concentration, direct bridging occurs, resulting in lower aggregation and turbidity thresholds. Interactions of flavonols with proteins (Dufour and Dangles 2005) as well as their adsorption on PVPP is much more efficient with aglycones as the sugar residue on the glycosides weakens the driving forces (Laborde et al. 2006). Finally, other parameters such as the solvent characteristics, the presence of other solutes and the temperature influence protein/flavanol association and the properties of resulting complexes. Thus the affinity between tannins and PRPs is lower at higher temperatures (Charlton et al. 2002a). The presence of polysaccharides prevents coprecipitation of tannins and proteins (Luck et al. 1994; Cheynier et al. 2006). Ionic strength and pH affect proteins solubility. Precipitation of tannin protein complexes is highest at the protein pHi as electrostatic repulsions are minimal (Calderon et al. 1968; Perez-Maldonado et al. 1995; Charlton et al. 2002a; Kawamoto and Nakatsubo 1997). The effect of ionic strength and ethanol content on the interactions of epigallocatechin gallate with a salivary PRP was investigated (Pascal et al. 2006). Increasing the ionic strength with sodium chloride or tartrate ions resulted in an increased stability of the aggregates, meaning that aggregation was not driven by repulsive electrostatic interactions. In 12% ethanol, the protein was not fully dissolved and aggregation with epigallocatechin gallate required much higher concentrations of the latter, confirming the role of hydrophobic interactions.

9B.4.3.2 Interactions with Polysaccharides

Actors

Major wine polysaccharides, including mannoproteins originating from yeasts and plant cell wall constituents (e.g. arabinogalactan proteins (AGP) and rhamnogalacturonan II (RGII)), have been shown to interact with flavanols (Riou et al. 2002). Besides, arabic gum (a mixture of arabinogalactans and arabinogalactan proteins) can be added as a protecting colloid to limit or prevent aggregation, flocculation and precipitation of tannins and tannin-protein complexes (Pellerin and Cabanis 1998).

However, at higher doses, it can lead to instability and haze development (Saucier et al. 1996; Siebert et al. 1996).

Interaction Mechanisms

Light scattering studies have shown that interactions of mannoproteins and of some arabinogalactan proteins with procyanidins prevent aggregation of the latter and result in small and stable particles. Adding other polysaccharides such as RGII monomer had no effect while RGII dimer increased aggregation and led to precipitation (Riou et al. 2002). The efficiency of mannoproteins as particle stabilizers decreased as their molecular weight increased, suggesting that the mechanism involved is steric stabilisation (Poncet-Legrand et al. 2007). The stabilising effect increased with ionic strength, ruling out an electrostatic stabilisation mechanism. Polysaccharides were also shown to limit precipitation of tannin protein complexes (Cheynier et al. 2006; Haslam et al. 1992; Luck et al. 1994). This was attributed to formation of soluble ternary protein-polysaccharide-flavonoid complexes, again mediated by hydrogen bonding and hydrophobic effect (McManus et al. 1985).

Evidence in Wine

Among wine polysaccharides, mannoproteins play an important role in protein haze stabilisation (Waters et al. 1994; Dupin et al. 2000). Gelatin fining of a wine phenolic extract in wine-like solution resulted in a much higher precipitation rate than when the same treatment was applied on the original wine. After addition of wine polysaccharides at the concentration normally encountered in wines, precipitation was reduced back to the level measured in wine, confirming the stabilizing effect of polysaccharides (Cheynier et al. 2006).

Factors Affecting the Interaction

The size and conformation of both the polyphenol and the polysaccharide are important. The presence of hydrophobic cavities such as encountered in cyclodextrins favours the interactions with phenolic molecules of appropriate shape or mobility (Smith et al. 1994). Polyphenols bind to dextran gels such as Sephadex used in chromatography, with affinity increasing with their molecular weight (Lea and Timberlake 1974), but do not interact with dextran oligomers (Williamson et al. 1995). The effect of polysaccharides also depends on the medium composition. Thus the effect of mannoproteins on procyanidin aggregation was stronger at lower ethanol concentration, i.e. under conditions in which procyanidins were in poor solvent and polysaccharides in good solvent. At high ionic strength, all mannoprotein fractions, including those of higher molecular weight, efficiently stabilized polyphenols.

9B.4.4 Flavonoid Adsorption on Solid Material

9B.4.4.1 Adsorption on Plant and Yeast Cell Walls

Actors

Plant and yeast cell walls consist mostly of polysaccharides along with smaller amounts of proteic material. Adsorption of flavanols on isolated plant cell walls (Renard et al. 2001) and on yeast lees (Mazauric and Salmon 2005, 2006) has been demonstrated. The latter also retained anthocyanins, as stated in Chapter 9A.

Mechanisms

Adsorption of proanthocyanidins on cell walls is driven by low energy bonds as described above for polysaccharides. The apparent affinity constants were higher with polysaccharides showing hydrophobic domains such as cross-linked pectins and xyloglucans than for cellulose (Le Bourvellec et al. 2005). Adsorption increased greatly with concentration, suggesting that stacking of polyphenols takes place once they are bound with the cell wall material (Renard et al. 2001). Cooperative mechanisms were also postulated to explain why lower molecular weight procyanidins were more easily desorbed (Renard et al. 2001).

Evidence in Wine

Adsorption of flavonoids on plant and yeast cell walls has been reported to contribute to poor retention of flavanols in wine (Sun et al. 1999) as well as losses of colouring matter (Vasserot et al. 1997). However, all these studies have been performed using model systems which might not reproduce the complexity of real wines. Moreover, adsorption on solid plant material may also impede extraction into the wine. In red wine-making, large proportions of flavonoids and especially of proanthocyanidins are not retained in the wine but recovered in the pomace after pressing (Morel-Salmi et al. 2006). Various technologies, including physical treatments such as thermovinification, must freezing or flash release and treatment with pectolytic enzymes have become of common practice in red wine making. They are reported to enhance the release of phenolic compounds into the must, as a result of cell or vacuole membrane damage or enhanced desorption of polyphenol from the cell wall material. Higher extraction rates of all phenolics were observed after flash release or thermotreatment (Morel-Salmi et al. 2006). The efficiency of pectolytic enzymes appears more variable: the amount of total phenolic compounds, and especially of derived pigments in red wines was enhanced in some cases (Pardo et al. 1999; Bautista-Ortin et al. 2005) and not modified in others (Bautista-Ortin et al. 2005; Doco et al. 2007). More recent studies have shown that the increase in colour is associated to enhanced proanthocyanidin content and higher conversion rate of anthocyanins to derived pigments (Ducasse et al. 2007).

Factors Affecting the Interaction

Phenolic acids and flavanol monomers do not absorb on plant cell walls while adsorption of procyanidins increases with their degree of polymerisation (Renard et al. 2001). Selective adsorption of molecules showing higher proportions of galloylated units and of catechin was also demonstrated (Le Bourvellec et al. 2004). Similarly, higher molecular weight proanthocyanidins were preferentially adsorbed on yeast lees (Mazauric and Salmon 2006). Adsorption of procyanidins on plant cell walls was not affected by pH. It increased with ionic strength and decreased as the temperature increased and in the presence of ethanol (Le Bourvellec et al. 2004), like procyanidin self-aggregation (Poncet-Legrand et al. 2003). Partial desorption from isolated plant cell wall material could be achieved by rinsing with buffer and was more efficient in the presence of dissolved polysaccharides (Renard et al. 2001). In contrast, desorption from yeast lees was extremely difficult, indicating stronger adsorption (Mazauric and Salmon 2006). This is possibly due to the presence of larger amounts of non polysaccharidic components such as proteins. Indeed proanthocyanidins were also show to be difficult to release from the pellets recovered after protein fining (Maury et al. 2003).

9B.4.4.2 Adsorption on Winery Equipment

Actors

Formation of tartrate crystals is a major source of instability in wines and treatments such as cold stabilisation, electrodialysis and ion-exchange are usually performed to prevent it from occurring in the bottled wine (Vernhet et al. 1999a). Polysaccharides and polyphenols, as well as yeast cells, were shown to be associated to tartrate crystals in the deposits formed after cold stabilisation both in white wines (Vernhet et al. 1999a) and in red wines (Vernhet et al. 1999b). Wine polysaccharides and polyphenols are also involved in the fouling of organic membranes during microfiltration of red wine (Vernhet and Moutounet 2002).

Mechanisms

Yeast cells represented at least 20% of the tartrate deposits in red wines. Scanning electron microscopy observations revealed that they adhere first on the stainless steel and suggest that they act as primary nucleation germs for crystallization (Vernhet and Moutounet 2002). The differences in shapes and composition of crystals formed in white and red wines may be attributed to higher adsorption of organic material in the latter (Vernhet and Moutounet 2002) as this is known to block the crystal growth (Rodriguez-Clemente and Correa-Gorospe 1988). Arabinogalactan protein and mannoproteins are the major polysaccharides in crystals and seem to take part in the reduced growth rate (Vernhet and Moutounet 2002). Since they had almost no effect on tartrate crystallization in model ethanolic solutions (Gerbaud and Gabas 1997), this may be related to their interaction with procyanidins (Riou et al. 2002).

The mechanisms of microfiltration membrane fouling were investigated in wine and model solutions (Vernhet and Moutounet 2002). The sharp decline observed in microfiltration fluxes within the first minutes of the process could not be attributed to adsorption alone. They can be explained by a two step mechanisms involving first interaction of the wine constituents with the membrane, quickly followed by their aggregation at the pore entrance (Vernhet and Moutounet 2002).

Factors Affecting the Interaction

Crystal appearance and growth are slower in red wines than in white wines and also differ within red wines. Arabinogalactan-proteins and mannoproteins were the major polysaccharides in the precipitates while rhamnogalaturonan II could not be detected. The average degree of polymerisation of proanthocyanidins in the deposit was higher that that of wine proanthocyanidins, indicating that polymers were selectively associated with the tartrate crystals. A preferential association of apolar flavonols was similarly observed, presumably as their lower solubility favours adsorption on surfaces.

Polyphenol adsorption under static conditions increased with the polarity of the membrane and the ability of its surface to act as hydrogen acceptor in hydrogen bonding which strengthens the interaction (Vernhet and Moutounet 2002). Polysaccharide adsorption was negligible in static conditions and decreased as the polarity of the membrane surface increased (Vernhet et al. 1997). However, polysaccharides played a major role in the fouling process, presumably due to the formation of colloidal aggregates with procyanidins. Indeed, fouling appeared largely determined by the pore size distribution of the membranes (Vernhet and Moutounet 2002). Moreover, the performance of the membranes and of back-pulsing operations in restoring the flux is related to the ratio of fine (colloidal) particles to large particles (e.g. yeast cells) as the former can penetrate into the membrane pores and produce irreversible internal fouling while the latter develop external fouling which is mostly reversible (Boissier et al. 2008; Vernhet et al. 2003).

References

Abrahams, S., Lee, E., Walker, A. R., Tanner, G. J., Larkin, P. J., & Ashton, A. R. (2003). The Arabidopsis TDS4 gene encodes leucoanthocyanidin dioxygenase (LDOX) and is essential for proanthocyanidin synthesis and vacuole development.*Plant J., 35*, 624–636.

Alcalde-Eon, C., Escribano-Bailon, M., Santos-Buelga, C., & Rivas-Gonzalo, J. (2006). Changes in the detailed pigment composition of red wine during maturity and ageing – A comprehensive study. *Anal. Chim. Acta, 563*, 238–254.

Alcalde-Eon, C., Escribano-Bailon, M., Santos-Buelga, C., & Rivas-Gonzalo, J. (2007). Identification of dimeric anthocyanins and new oligomeric pigments in red wine by means of HPLC-DAD-ESI/MSn. *J. Mass Spectrom., 42*, 735–748.

Alonso, E., Estrella, I., & Revilla, E. (1986). Presence of quercetin-3-O-glucuronoside in spanish table wines. *J. Sci. Food Agric., 37*, 1118–1120.

Atanasova, V., Fulcrand, H., Le Guerneve, C., Cheynier, V., & Moutounet, M. (2002a). Structure of a new dimeric acetaldehyde malvidin 3-glucoside condensation product. *Tetrahedron Lett., 43*, 6151–6153.

Atanasova, V., Fulcrand, H., Le Guernevé, C., Dangles, O., & Cheynier, V. (2002b). First evidence of acetaldehyde-induced anthocyanin polymerisation. Polyphenol Communications 2002. Marrakech, pp. 417–418.

Baderschneider, B., & Winterhalter, P. (2001). Isolation and characterization of novel benzoates, cinnamates, flavonoids, and lignans from Riesling wine and screening for antioxidant activity. *J. Agric. Food Chem., 49*, 2788–2798.

Bae, Y. S., Foo, L. Y., & Karchesy, J. J. (1994). GPC of natural procyanidin oligomers and polymers.*Holzforschung, 48*, 4–6.

Baranowski, J. D., & Nagel, C. W. (1983). Kinetic of malvidin-3-glucoside condensation in wine model systems.*J. Food Sci., 48*, 419–421.

Bautista-Ortin, A. B., Martinez-Cutillas, A., Ros-Garcia, J. M., Lopez-Roca, J. M., & Gomez-Plaza, E. (2005). Improving colour extraction and stability in red wines: the use of maceration enzymes and enological tannins.*Int. J. Food Sci. Technol., 40*, 867–878.

Baxter, N. J., Lilley, T. H., Haslam, E., & Williamson, M. P. (1997a). Multiple interactions between polyphenols and a salivary proline-rich protein repeat result in complexation and precipitation. *Biochemistry, 36*, 5566–5577.

Baxter, N. J., Lilley, T. H., Haslam, E., & Williamson, M. P. (1997b). Multiple interactions between polyphenols and a salivary proline-rich protein repeat results in complexation and precipitation. *Biochemistry, 36*, 5566–5577.

Berg, H. W. (1959). The effects of several fungal pectic enzyme preparations on grape musts and wines.*Am. J. Enol. Vitic., 10*, 130–134.

Betes-Saura, C., Andres-Lacueva, C., & Lamuela-Raventos, R. M. (1996). Phenolics in white free run juices and wines from Penedes by High-performance liquid chromatography: Changes during vinification. *J. Agric. Food Chem., 44*, 3040–3046.

Bishop, P. B., & Nagel, C. W. (1984). Characterization of the condensation product of malvidin 3,5-diglucoside and catechin.*J. Agric. Food Chem., 32*, 1022–1026.

Bogs, J., Downey, M., Harvey, J., Ashton, A., Tanner, G., & Robinson, S. (2005). Proanthocyanidin synthesis and expression of genes encoding leuanthocyanidin reductase and anthocyanidin reductase in developing grape berries and grapevine leaves. *Plant Physiol., 139*, 652–663.

Boido, E., Alcalde-Eon, C., Carrau, F., Dellacassa, E., & Rivas-Gonzalo, J. C. (2006). Aging effect on the pigment composition and color of Vitis vinifera L. Cv. Tannat wines. Contribution of the main pigment families to wine color. *J. Agric. Food Chem., 54*, 6692–6704.

Boissier, B., Lutin, F., Moutounet, M., & Vernhet, A. (2008). Particles deposition during the cross-flow microfiltration of red wines -incidence of the hydrodynamic conditions and of the yeast to fines ratio. *Chem. Engin. Process.*, 47(3) 276–286

Boukharta, M., Girardin, M., & Metche, M. (1988). Procyanidines galloylées du sarment de vigne (Vitis vinifera) separation et identification par chromatographie liquide haute performance et chromatographie en phase gazeuse.*J. Chromatogr., 455*, 406–409.

Boulton, D., Walle, U., & Walle, T. (1998). Extensive binding of the bioflavonoid quercetin to human plasma proteins.*J. Pharmacy Pharmacol., 50*, 243–249.

Boulton, R. (2001). The copigmentation of anthocyanins and its role in the color of red wine: a critical review.*Am. J. Enol. Vitic., 52*, 67–87.

Bourzeix, M., Weyland, D., & Heredia, N. (1986). Etude des catéchines et des procyanidols de la grappe de raisin, du vin et d'autres dérivés de la vigne. *Bull. OIV,* 669–670, 1171–1253.

Bradshaw, M. P., Cheynier, V., Scollary, G. R., & Prenzler, P. D. (2003). Defining the ascorbic acid crossover from anti-oxidant to pro-oxidant in a model wine matrix containing (+)-catechin. *J. Agric. Food Chem., 51*, 4126–4132.

Brouillard, R., & Dangles, O. (1993). Flavonoids and flower colour in Harborne, J. B. (Ed), The flavonoids. Advances in research since 1986, Chapman and Hall, pp. 565–588.

Brouillard, R., & Dangles, O. (1994). Anthocyanin molecular interactions : the first step in the formation of new pigments during wine aging. *Food Chem., 51*, 365–371.

Calderon, P., VanBuren, J., & Robinson, W. (1968). Factors influencing the formation of precipitates and hazes by gelatin and condensed and hydrolyzable tannins. *J. Agric. Food Chem., 16*, 479–482.

Canals, R., Llaudy, M., Valls, J., Canals, J., & Zamora, F. (2005). Influence of ethanol concentration on the extraction of color and phenolic compounds from the skins and seeds of Tempranillo grapes at different stages of ripening. *J. Agric. Food Chem., 53*, 4019–4025.

Cantos, E., Espin, J., & Tomas-Barberan, F. (2002). Varietal differences among the polyphenol profiles of seven table grape cultivars studied by LC-DAD-MS-MS. *J. Agric. Food Chem., 50*, 5691–5696.

Castellarin, S., Di Gaspero, G., Marconi, R., Nonis, A., Peterlunger, E., Paillard, S., Adam-Blondon, A., & Testolin, R. (2006). Colour variation in red grapevines (Vitis vinifera L.): genomic organisation, expression of flavonoid 3'-hydroxylase, flavonoid 3',5'-hydroxylase genes and related metabolite profiling of red cyanidin-/blue delphinidin based anthocyanins in berry skin. *BMC Genomics, 7*, 12.

Chamkha, M., Cathala, B., Cheynier, V., & Douillard, R. (2003). Phenolic composition of champagnes from Chardonnay and Pinot Noir vintages. *J. Agric. Food Chem., 51*, 3179–3184.

Charlton, A. J., Baxter, N. J., Lilley, T. H., Haslam, E., McDonald, C. J., & Williamson, M. P. (1996). Tannin interactions with a full-length human salivary proline-rich protein display a stronger affinity than with proline-rich repeats. *FEBS Lett., 382*, 289–292.

Charlton, A., Baxter, N. J., Khan, M. L., Moir, A. J. G., Haslam, E., Davis, A. P., & Williamson, M. P. (2002a). Polyphenol/peptide binding and precipitation. *J. Agric. Food Chem., 50*, 1593–1601.

Charlton, A. J., Haslam, E., & Williamson, M. P. (2002b). Multiple conformations of the proline-rich-protein/epigallocatechin gallate complex determined by time-averaged nuclear overhauser effects.*J. Am. Chem. Soc., 124*, 9899–9905.

Cheynier, V., & Ricardo Da Silva, J. M. (1991). Oxidation of grape procyanidins in model solutions containing trans-caffeoyl tartaric acid and grape polyphenoloxidase. *J. Agric. Food Chem., 39*, 1047–1049.

Cheynier, V., & Rigaud, J. (1986). HPLC separation and characterization of flavonols in the skins of Vitis vinifera var. Cinsault. *Am. J. Enol. Vitic., 37*, 248–252.

Cheynier, V., & Van Hulst, M. W. (1988). Oxidation of trans-caftaric acid and 2-S-glutathionyl caftaric acid in model solutions. *J. Agric. Food Chem., 36*, 10–15.

Cheynier, V., Osse, C., & Rigaud, J. (1988). Oxidation of grape juice phenolic compounds in model solutions. *J. Food Sci., 53*, 1729–1732.

Cheynier, V., Basire, N., & Rigaud, J. (1989a). Mechanism of trans-caffeoyl tartaric acid and catechin oxidation in model solutions containing grape polyphenoloxidase. *J. Agric. Food Chem., 37*, 1069–1071.

Cheynier, V., Rigaud, J., Souquet, J. M., Barillère, J. M., & Moutounet, M. (1989b). Effect of pomace contact and hyperoxidation on the phenolic composition and quality of Grenache and Chardonnay wines.*Am. J. Enol. Vitic., 40*, 36–42.

Cheynier, V., Fulcrand, H., Guyot, S., Oszmianski, J., & Moutounet, M. (1995). Reactions of enzymically generated quinones in relation to browning in grape musts and wines in Lee, C. Y., & Whitaker, J. R. (Eds), Enzymatic browning and its prevention in foods, American Chemical Society, pp. 130–143.

Cheynier, V., Fulcrand, H., Sarni, P., & Moutounet, M. (1997a). Reactivity of phenolic compounds in wine: Diversity of mechanisms and resulting products. *In Vino analytica scientia.* Bordeaux, pp. 143–154.

Cheynier, V., Prieur, C., Guyot, S., Rigaud, J., & Moutounet, M. (1997b). The structures of tannins in grapes and wines and their interactions with proteins in Watkins, T. R. (Ed), Wine. Nutritional and therapeutic benefits, American Chemical Society, pp. 81–93.

Cheynier, V., Es-Safi, N.-E., & Fulcrand, H. (1999a). Structure and colour properties of anthocyanins and related pigments.*International Congress on Pigments in Food and Technology*. Sevilla (Spain), pp. 23–35.

Cheynier, V., Souquet, J.-M., Roux, E. L., Guyot, S., & Rigaud, J. (1999b). Size separation of condensed tannins by normal-phase high-performance liquid chromatography in Packer, L. (Ed), *Methods Enzymol.*, Volume 299. Oxidants and antioxidants. Part A., Academic Press, pp. 178–184.

Cheynier, V., Dueñas-Paton, M., Salas, E., Maury, C., Souquet, J.-M., Sarni-Manchado, P., & Fulcrand, H. (2006). Structure and properties of wine pigments and tannins. *Am. J. Enol. Vitic.*, *57*, 298–305.

Clark, A. C., & Scollary, G. R. (2002). Copper(II)-mediated oxidation of (+)-catechin in a model white wine system.*Aust. J. Grape Wine Res.*, *8*, 186–195.

Clark, A. C., Prenzler, P. D., & Scollary, G. R. (2003). The role of copper(II) in the bridging reactions of (+)-catechin by glyoxylic acid in a model white wine. *J. Agric. Food Chem.*, *51*, 6204–6210.

Cooper, J. J., & Marshall, A. G. (2001). Electrospray ionisation Fourier transform mass spectrometric analysis of wine. *J. Agric. Food Chem.*, *49*, 5710–5718.

Cortell, J., & Kennedy, J. A. (2006). Effect of shading on accumulation of flavonoid compounds in (Vitis vinifera L.) Pinot Noir fruit and extraction in a model system. *J. Agric. Food Chem.*, *54*, 8510–8520.

Cortell, J. M., Halbleib, M., Gallagher, A. V., Righetti, T. L., & Kennedy, J. A. (2005). Influence of vine vigor on grape (Vitis vinifera L. Cv. Pinot Noir) and wine proanthocyanidins. *J. Agric. Food Chem.*, *53*, 5798–5808.

Czochanska, Z., Foo, L., & Porter, L. (1979a). Compositional changes in lower molecular weight flavans during grape maturation.*Phytochemistry*, *18*, 1819–1822.

Czochanska, Z., Foo, L. Y., Newman, R. H., Porter, L. J., Thomas, W. A., & Jones, W. T. (1979b). Direct proof of a homogeneous polyflavan-3-ol structure for polymeric proanthocyanidins. *J. Chem. Soc. Chem. Comm.*, *8*, 375–377.

Czochanska, Z., Foo, L. Y., Newman, R. H., & Porter, J. L. (1980). Polymeric proanthocyanidins. Stereochemistry, structural units and molecular weight. *J. Chem. Soc, Perkin Trans. I*, 2278–2286.

de Pascual-Teresa, S., Rivas-Gonzalo, J. C., & Santos-Buelga, C. (2000). Prodelphinidins and related flavanols in wine. *Int. J. Food Sci. Technol.*, *35*, 33–40.

Derdelinckx, G., & Jerumanis, J. (1984). Separation of malt hop proanthocyanidins on Fractogel TSK HW-40 (S). *J. Chromatogr.*, *285*, 231–234.

Doco, T., Williams, P., & Cheynier, V. (2007). Effect of flash release and pectinolytic enzyme treatments on wine polysaccharide composition. *J. Agric. Food Chem.*, *55*, 6643–6649.

Downey, M., Harvey, J., & Robinson, S. (2003a). Analysis of tannins in seeds and skins of Shiraz grapes throughout berry development. *Austr. J. Grape Wine Res.*, *9*, 15–27.

Downey, M., Harvey, J., & Robinson, S. (2003b). Synthesis of flavonols and expression of flavonol synthase genes in the developing grape berries of Shiraz and Chardonnay *(Vitis vinifera L.). Austr. J Grape Wine Res.*, *9*, 110–121.

Downey, M. O., Harvey, J. S., & Robinson, S. P. (2004). The effect of bunch shading on berry development and flavonoid accumulation in Shiraz grapes. *Aust. J. Grape Wine Res.*, *10*, 55–73.

Drinkine, J., Glories, Y., & Saucier, C. (2005). (+)catechin-aldehyde condensations: competition between acetaldehyde and glyoxylic acid. *J. Agric. Food Chem.*, *53*, 7552–7558.

Drinkine, J., Lopes, P., Kennedy, J., Teissedre, P., & Saucier, C. (2007a). Analysis of ethylidene-bridged flavan-3-ols in wine. *J. Agric. Food Chem.*, *55*, 1109–1116.

Drinkine, J., Lopes, P., Kennedy, J., Teissedre, P., & Saucier, C. (2007b). Ethylidene-bridged flavan-3-ols in red wine and correlation with wine age. *J. Agric. Food Chem.*, *55*, 6292–6299.

Ducasse, M.-A., Souquet, J.-M., Fulcrand, H., & Cheynier, V. (2007). Impact des traitements enzymatiques sur la composition phénoliques des vins rouges. 8th Symposium International d'Œnologie. Bordeaux, France.

Duenas, M., Fulcrand, H., & Cheynier, V. (2006a). Formation of anthocyanin-flavanol adducts in model solutions. *Anal. Chim. Acta., 563*, 15–25.

Duenas, M., Salas, E., Cheynier, V., Dangles, O., & Fulcrand, H. (2006b). UV-Visible spectroscopic investigation of the 8-8-methylmethine catechin-malvidin 3-glucoside pigments in aqueous solution : structural transformations and molecular complexation with chlorogenic acid. *J. Agric. Food Chem., 54*, 189–196.

Dufour, C., & Bayonove, C. (1999). Interactions between wine polyphenols and aroma substances. An insight at the molecular level. *J. Agric. Food Chem., 47*, 678–684.

Dufour, C., & Dangles, O. (2005). Flavonoid-serum albumin complexation: determination of binding constants and binding sites by fluorescence spectroscopy. *Biochim. Biophys. Acta*-General Subjects 1721, 164–173.

Dupin, I. V. S., McKinnon, B. M., Ryan, C., Boulay, M., Markides, A. J., Jones, G. P., Williams, P. J., & Waters, E. J. (2000). Saccharomyces cerevisiae mannoproteins that protect wine from protein haze: their release during fermentation and lees contact and a proposal for their mechanism of action.*J. Agric. Food Chem., 48*, 3098–3105.

Escribano-Bailon, T., Alvarez-Garcia, M., Rivas-Gonzalo, J. C., Heredia, F. J., & Santos-Buelga, C. (2001). Color and stability of pigments derived from the acetaldehyde-mediated condensation between malvidin-3-O-glucoside and (+)-catechin. *J. Agric. Food Chem., 49*, 1213–1217.

Es-Safi, N., Fulcrand, H., Cheynier, V., & Moutounet, M. (1999a). Competition between (+)-catechin and (–)-epicatechin in acetaldehyde-induced polymerization of flavanols. *J. Agric. Food Chem., 47*, 2088–2095.

Es-Safi, N. E., Guerneve, C. L., Fulcrand, H., Cheynier, V., & Moutounet, M. (1999b). New polyphenolic compounds with xanthylium skeletons formed through reaction between (+)-catechin and glyoxylic acid. *J. Agric. Food Chem., 47*, 5211–5217.

Es-Safi, N. E., Cheynier, V., & Moutounet, M. (2000). Study of the reactions between (+)-catechin and furfural derivatives in the presence or absence of anthocyanins and their implication in food color change. *J. Agric. Food Chem., 48*, 5946–5954.

Fernandez-Zurbano, P., Ferreira, V., Pena, C., Escudero, A., & Cacho, J. (1999). Effects of maceration time and pectolytic enzymes added during maceration on the phenolic composition of must. *J. Food Sci. Technol. Internat., 5*, 319–325.

Fournand, D., Vicens, A., Sidhoum, L., Souquet, J.-M., Moutounet, M., & Cheynier, V. (2006). Accumulation and extractability of grape skin tannins and anthocyanins at different advanced physiological stages. *J. Agric. Food Chem., 54*, 7331–7338.

Fulcrand, H., Doco, T., Es-Safi, N., Cheynier, V., & Moutounet, M. (1996). Study of the acetaldehyde induced polymerisation of flavan-3-ols by liquid chromatography ion spray mass spectrometry. *J. Chromatogr., 752*, 85–91.

Fulcrand, H., Remy, S., Souquet, J.-M., Cheynier, V., & Moutounet, M. (1999). Study of wine tannin oligomers by on-line liquid chromatography electrospray ionisation mass spectrometry. *J. Agric. Food Chem., 47*, 1023–1028.

Fulcrand, H., Morel-Salmi, C., Poncet-Legrand, C., Vernhet, A., & Cheynier, V. (2006). Tannins: From reactions to complex supramolecular structures.*Austr. J. Grape Wine Res.* Adélaïde, Australie, pp. 12–17.

Gao, L., Girard, B., Mazza, G., & Reynolds, A. G. (1997). Changes in anthocyanins and color characteristics of Pinot Noir wines during different vinification processes. *J. Agric. Food Chem., 45*, 2003–2008.

Gerbaud, V., & Gabas, N. (1997). Influence of wine polysaccharides and polyphenols on the crystallization of potassium hydrogen tartrate. *J. Int. Sci. Vigne Vin*, 31, 65–83.

Guyot, S., Pellerin, P., Brillouet, J., Moutounet, M., & Cheynier, V. (1996a). Inhibition of b-glucosidase (Amygdalae Dulces) by (+)-catechin oxidation products and procyanidin dimers. *Biosci., Biotech. Biochem., 60*, 1131–1135.

Guyot, S., Vercauteren, J., & Cheynier, V. (1996b). Colourless and yellow dimers resulting from (+)-catechin oxidative coupling catalysed by grape polyphenoloxidase.*Phytochemistry, 42*, 1279–1288.

Hagerman, A. E. (1989). Chemistry of tannin-protein complexation in Hemingway, R. W., & Karchesy, J. J. (Eds), Chemistry and significance of condensed tannins, Plenum Press, pp. 323–331.

Haslam, E. (1980). In vino veritas: oligomeric procyanidins and the ageing of red wines. *Phytochemistry, 19*, 2577–2582.

Haslam, E., & Lilley, T. H. (1988). Natural astringency in foodstuffs. A molecular interpretation. *Crit. Rev. Food Sci. Nutr., 27*, 1–40.

Haslam, E., Lilley, T. H., Warminski, E., Liao, H., Cai, Y., Martin, R., Gaffney, S. H., Goulding, P. N., & Luck, G. (1992). Polyphenol complexation. A study in molecular recognition. in Ho, C.-T., Lee, C. Y., & Huang, M.-T. (Eds), Phenolic compounds in food and their effects on health, American Chemical Society, pp. 8–50.

Hathway, D. E., & Seakins, J. W. T. (1957). Autoxidation of polyphenols. Part III. Autoxidation in neutral aqueous solutions of flavans related to catechin.*J. Chem. Soc., 300*, 1562–1566.

Hayasaka, Y., & Kennedy, J. A. (2003). Mass spectrometric evidence for the formation of pigmented polymers in red wine. *Austr. J. Grape Wine Res., 9*, 210–220.

Hemingway, R. W., Steynberg, P. J., Steynberg, J. P., & Hatano, T. (1999). NMR studies on the conformation of polyflavanoids and their association with proteins in Argyropoulos, D. S. (Ed), Advances in lignocellulosics characterization, TAPPI Press, pp. 157–178.

Hmamouchi, M., Es-Safi, N., Lahrichi, M., Fruchier, A., & Essassi, E. M. (1996). Flavones and flavonols in leaves of some Moroccan Vitis vinifera cultivars. *J. Agric. Food Chem., 47*, 186–192.

Jobstl, E., Howse, J. R., Fairclough, J. P. A., & Williamson, M. P. (2006). Noncovalent cross-linking of casein by epigallocatechin gallate characterized by single molecule force microscopy. *J. Agric. Food Chem., 54*, 4077–4081.

Jurd, L. (1967). Anthocyanidins and related compounds-XI. Catechin-flavylium salt condensation reactions. *Tetrahedron, 23*, 1057–1064.

Jurd, L. (1969). Review of polyphenol condensation reactions and their possible occurrence in the aging of wines.*Am. J. Enol. Vitic., 20*, 195–197.

Jurd, L., & Somers, T. C. (1970). The formation of xanthylium salts from proanthocyanidins. *Phytochemistry, 9*, 419–427.

Jurd, L., & Waiss, A. C. (1965). Anthocyanins and related compounds-VI Flavylium salt-phloroglucinol condensation product. *Tetrahedron, 21*, 1471–1483.

Kawamoto, H., & Nakatsubo, F. (1997). Effects of environmental factors on two-stage tannin-protein co-precipitation. *Phytochemistry, 46*, 479–483.

Kelm, M. A., Johnson, J. C., Robbins, R. J., Hammerstone, J. F., & Schmitz, H. H. (2006). High-Performance Liquid Chromatography Separation and Purification of Cacao (Theobroma cacao L.) Procyanidins According to Degree of Polymerization Using a Diol Stationary Phase. *J. Agric. Food Chem., 54*, 1571–1576.

Kennedy, J. A., Hayasaka, Y., Vidal, S., Waters, E. J., & Jones, G. P. (2001). Composition of grape skin proanthocyanidins at different stages of berry development. *J. Agric. Food Chem., 49*, 5348–5355.

Kennedy, J. A., Matthews, M. A., & Waterhouse, A. L. (2002). Effect of maturity and vine water status on grape skin and wine flavonoids. *Am. J. Enol. Vitic., 53*, 268–274.

Laborde, B., Moine-Ledoux, V., Richard, T., Saucier, C., Dubourdieu, D., & Monti, J.-P. (2006). PVPP-Polyphenol complexes: A molecular approach. *J. Agric. Food Chem., 54*, 4383–4389.

Le Bourvellec, C., Guyot, S., & Renard, C. M. G. C. (2004). Non-covalent interaction between procyanidins and apple cell wall material. Part I – Effect of some environmental parameters. *Biochim. Biophys. Acta, 1672*, 192–202.

Le Bourvellec, C., Bouchet, B., & Renard, C. M. G. C. (2005). Non-covalent interaction between procyanidins and apple cell wall material. Part III: Study on model polysaccharides. *Biochim. Biophys. Acta, 1725*, 10–18.

Le Bourvellec, C., Picot, M., & Renard, C. (2006). Size-exclusion chromatography of procyanidins: Comparison between apple and grape procyanidins and application to the characterization of fractions of high degrees of polymerization.*Anal. Chim. Acta, 563*, 33–43.

Lea, A. G. H., & Timberlake, C. F. (1974). The phenolics of ciders.1.Procyanidins. *J. Sci. Food Agric., 25*, 1537–1545.

Lea, A. G. H., Bridle, P., Timberlake, C. F., & Singleton, V. L. (1979). The procyanidins of white grapes and wines. *Am. J. Enol. Vitic., 30*, 289–300.

Lee, C. Y., & Jaworski, A. (1987). Phenolic compounds in white grapes grown in New York. *Am. J. Enol. Vitic., 38*, 277–281.

Lee, C. Y., & Jaworski, A. W. (1990). Identification of some phenolics in white grapes. *Am. J. Enol. Vitic., 41*, 87–89.

Liao, H., Cai, Y., & Haslam, E. (1992). Polyphenol interactions. Anthocyanins: co-pigmentation and colour changes in red wines. *J. Sci. Food Agric., 59*, 299–305.

Luck, G., Liao, H., Murray, N. J., Grimmer, H. R., Warminski, E. E., Willamson, M. P., Lilley, T. H., & Haslam, E. (1994). Polyphenols, astringency and prolin-rich proteins. *Phytochemistry, 37*, 357–371.

Malien-Aubert, C., Dangles, O., & Amiot, M.-J. (2002). Influence of procyanidins on the color stability of oenin solutions.*J. Agric. Food Chem., 50*, 3299–3305.

Mané, C. (2007). Phénomènes oxydants et composés phénoliques dans les vins blancs de Champagne: développements méthodologiques pour l'analyse des polymères, Formation Doctorale Sciences des Aliments, Montpellier Supagro, p. 279.

Mané, C., Sommerer, N., Yalcin, T., Cheynier, V., Cole, R. B., & Fulcrand, H. (2007a). Assessment of the molecular weight distribution of tannin fractions through MALDI-TOF MS analysis of protein-tannin complexes. *Anal. Chem., 79*, 2239–2248.

Mané, C., Souquet, J. M., Olle, D., Verries, C., Veran, F., Mazerolles, G., Cheynier, V., & Fulcrand, H. (2007b). Optimization of simultaneous flavanol, phenolic acid, and anthocyanin extraction from grapes using an experimental design: Application to the characterization of champagne grape varieties. *J. Agric. Food Chem., 55*, 7224–7233.

Masa, A., Vilanova, M., & Pomar, F. (2007). Varietal differences among the flavonoid profile of white grape cultivars studied by high performance liquid chromatography. *J. Chromatogr. A 1164*, 291–297.

Mateus, N., de Pascual-Teresa, S., Rivas-Gonzalo, J., Santos-Buelga, C., & De Freitas, V. (2002). Structural diversity of anthocyanin-derived pigments in port wines. *Food Chem., 76*, 335–342.

Mateus, N., Silva, A. M. S., Rivas-Gonzalo, J. C., Santos-Buelga, C., & De Freitas, V. (2003). A new class of blue anthocyanin-derived pigments isolated from red wines.*J. Agric. Food Chem., 51*, 1919–1923.

Matthews, A., Grbin, P. R., & Jiranek, V. (2006). A survey of lactic acid bacteria for enzymes of interest to oenology.*Austr. J. Grape Wine Res., 12*, 235–244.

Mattivi, F., Guzzon, R., Vrhovsek, U., Stefanini, M., & Velasco, R. (2006). Metabolite Profiling of Grape: Flavonols and Anthocyanins. *J. Agric. Food Chem., 54*, 7692–7702.

Maury, C., Sarni-Manchado, P., Lefebvre, S., Cheynier, V., & Moutonet, M. (2001). Influence of fining with different molecular weight gelatins on proanthocyanidin composition and perception of wines. *Am. J. Enol. Vitic., 52*, 140–145.

Maury, C., Sarni-Manchado, P., Lefebvre, S., Cheynier, V., & Moutounet, M. (2003). Influence of fining with plant proteins on proanthocyanidin composition of red wines. *Am. J. Enol. Vitic., 54*, 105–111.

Mazauric, J.-P., & Salmon, J.-M. (2005). Interactions between yeast lees and wine polyphenols during simulation of wine aging: I. Analysis of remnant polyphenolic compounds in the resulting wines.*J. Agric. Food Chem., 53*, 5647–5653.

Mazauric, J.-P., & Salmon, J.-M. (2006). Interactions between yeast lees and wine polyphenols during simulation of wine aging. II. Analysis of desorbed polyphenol compounds from yeast lees. *J. Agric. Food Chem., 54*, 3876–3881.

Mazza, G., & Miniati, E. (1993). Grapes in Mazza, G., & Miniati, E. (Eds), Anthocyanins in fruits, vegetables and grains, CRC Press, pp. 149–199.

McManus, J. P., Davis, K. G., Beart, J. E., Galffney, S. H., Lilley, T. H., & Haslam, E. (1985). Polyphenol interactions. Part 1. Introduction; some observations on the reversible complexation of polyphenols with proteins and polysaccharides. *J. Chem. Soc. Perkin Trans, II*, 1429–1438.

Mirabel, M., Glories, Y., Pianet, I., & Dufourc, E. J. (1999a). Towards high resolution H-1 NMR spectra of tannin colloidal aggregates.*J. Chimie Physique Physico-Chimie Biologique, 96*, 1629–1634.

Mirabel, M., Saucier, C., Guerra, C., & Glories, T. (1999b). Copigmentation in model wine solutions: Occurrence and relation to wine aging. *Am. J. Enol. Vitic., 50*, 211–218.

Monagas, M., Gómez-Cordovés, C., Bartolomé, B., Laureano, O., & Silva, J. M. R. D. (2003). Monomeric, oligomeric, and polymeric flavan-3-ol composition of wines and grapes from Vitis vinifera L. Cv. Graciano, Tempranillo, and Cabernet Sauvignon. *J. Agric. Food Chem., 51*, 6475–6481.

Morel-Salmi, C., Souquet, J. M., Bes, M., & Cheynier, V. (2006). The effect of flash release treatment on phenolic extraction and wine composition. *J. Agric. Food Chem.,54*, 4270–4276.

Murray, N. J., Williamson, M. P., Lilley, T. H., & Haslam, E. (1994). Study of the interaction between salivary proline-rich proteins and a polyphenol by 1H-NMR spectroscopy. *Eur. J. Biochem., 219*, 923–935.

Nagel, C. W., & Wulf, L. W. (1979). Changes in the anthocyanins, flavonoids and hydroxycinnamic acid esters during fermentation and aging of merlot and cabernet sauvignon. *Am. J. Enol. Vitic., 30*, 111–116.

Nakajima, J. J., Tanaka, Y., Yamazaki, M., & Saito, K. (2001). Reaction mechanism from leucoanthocyanidin to anthocyanin-3-glucoside, a key reaction for coloring in anthocyanin biosynthesis. *J. Biol. Chem., 276*, 25797–25803.

Oh, H. I., Hoff, J. E., Armstrong, G. S., & Haff, L. A. (1980). Hydrophobic interactions in tannin-protein complexes. *J. Agric. Food Chem., 28*, 394–398.

Ojeda, H., Andary, C., Kraeva, E., Carbonneau, A., & Deloire, A. (2002). Influence of pre- and post-véraison water deficit on synthesis and concentration of skin phenolic compounds during berry growth of Vitis vinifera cv. Shiraz. *Am. J. Enol. Vitic., 53*, 261–267.

Oszmianski, J., Cheynier, C., & Moutounet, M. (1996). Iron-catalyzed oxidation of (+)-catechin in wine-like model solutions. *J. Agric. Food Chem., 44*, 1972–1975.

Ough, C., & Crowell, E. (1979). Pectic-enzyme treatment of white grapes: temperature, variety and skin-contact time factors.*Am. J. Enol. Vitic., 30*, 22–27.

Ough, C. S., Noble, A. C., & Temple, D. (1975). Pectic Enzyme Effects on Red Grapes.*Am. J. Enol. Vitic., 26*, 195–200.

Outtrup, H. (1989). Haze active peptides in beer. E.B.C. Congress, pp. 609–616.

Pardo, F., Salinas, M. R., Alonso, G. L., Navarro, G., & Huerta, M. D. (1999). Effect of diverse enzyme preparations on the extraction and evolution of phenolic compounds in red wines. *Food Chem., 67*, 135–142.

Pascal, C., Poncet-Legrand, C., Sarni-Manchado, P., Cheynier, V., & Vernhet, A. (2006). Effect of ionic strength, tartaric acid and ethanol on the interactions between flavan-3-ols and salivary proline rich proteins. Macromolecules and Secondary metabolites in Grapevine and Wines. Reims.

Pascal, C., Poncet-Legrand, C., Imberty, A., Gautier, C., Sarni-Manchado, P., Cheynier, V., & Vernhet, A. (2007). Interactions between a non glycosylated human proline-rich protein and flavan-3-ols are affected by protein concentration and polyphenol/protein ratio. *J. Agric. Food Chem., 55*, 4895–4901.

Pellerin, P., & Cabanis, J.-C. (1998). Les glucides in Flanzy, C. (Ed), Oenologie, Lavoisier Tec & Doc, pp. 40–92.

Pereira, G. E., Gaudillere, J.-P., Pieri, P., Hilbert, G., Maucourt, M., Deborde, C., Moing, A., & Rolin, D. (2006). Microclimate Influence on Mineral and Metabolic Profiles of Grape Berries. *J. Agric. Food Chem., 54*, 6765–6775.

Perez-Maldonado, R. A., Norton, B. W., & Kerven, G. L. (1995). Factors affecting in vitro formation of tannin -protein complexes. J. Sci. Food Agric., *69*, 291–298.

Pierpoint, W. S. (1969). o-quinones formed in plant extracts : their reactions with amino acids and peptides. *Biochem. J. 112*, 609–617.

Piretti, M. V., Ghedini, M., & Serrazanetti, G. (1976). Isolation and identification of the polyphenolic and terpenoid constituents of Vitis vinifera. v. Trebbiano variety. *Annali di Chimica, 66*, 429–437.

Pissara, J., Mateus, N., Rivas-Gonzalo, J., Santos-Buelga, C., & De Freitas, V. (2003). Reaction between malvidin 3-glucoside and (+)-catechin in model solutions containing different aldehydes. *J. Food Sci., 68*, 476–481.

Poncet-Legrand, C., Cartalade, D., Putaux, J.-L., Cheynier, V., & Vernhet, A. (2003). Flavan-3-ol aggregation in model ethanolic solutions: incidence of polyphenol structure, concentration ethanol content and ionic strength.*Langmuir, 19*, 10563–10572.

Poncet-Legrand, C., Edelmann, A., Putaux, J.-L., Cartalade, D., Sarni-Manchado, P., & Vernhet, A. (2006). Poly(L-proline) interactions with flavan-3-ols units: Influence of the molecular structure and the polyphenol/protein ratio. *Food Hydrocolloids, 20*, 687–697.

Poncet-Legrand, C., Gautier, C., Cheynier, V., & Imberty, A. (2007). Interactions between flavan-3-ols and poly(L-proline) studied by isothermal titration calorimetry: Effect of the tannin structure.*J. Agric. Food Chem., 55*, 9235–9240.

Price, S. F., Breen, P. J., Vallado, M., & Watson, B. T. (1995). Cluster sun exposure and quercetin in pinot noir grapes and wine. *Am. J. Enol. Vitic., 46*, 187–194.

Prieur, C., Rigaud, J., Cheynier, V., & Moutounet, M. (1994). Oligomeric and polymeric procyanidins from grape seeds.*Phytochemistry, 36*, 781–784.

Quideau, S., Jourdes, M., Saucier, C., Glories, Y., Pardon, P., & Baudry, C. (2003). DNA topoisomerase inhibitor acutissimin A and other flavano-ellagitannins in red wine. *Angew. Chem. Int. Ed., 42*, 6012–6014.

Remy, S. (1999). Caractérisation des composés phénoliques polymériques des vins rouges, INAPG, pp. 199.

Remy, S., Fulcrand, H., Labarbe, B., Cheynier, V., & Moutounet, M. (2000). First confirmation in red wine of products resulting from direct anthocyanin-tannin reactions. *J. Sci. Food Agric., 80*, 745–751.

Remy-Tanneau, S., Guerneve, C. L., Meudec, E., & Cheynier, V. (2003). Characterization of a colorless anthocyanin-flavan-3-ol dimer containing both carbon-carbon and ether interflavanoid linkages by NMR and mass spectrometries.*J. Agric. Food Chem., 51*, 3592–3597.

Renard, C. M. G. C., Baron, A., Guyot, S., & Drilleau, J. (2001). Interactions between apple cell walls and native apple polyphenols: quantification and some consequences. *Int. J. Biol. Macromolecules, 29*, 115–125.

Ribéreau-Gayon, P. (1964). Les compo*sés phénoliques du raisin et du vin II. Les flavonosides et les anthocyanosides. Ann. Physiol. Vég., 6*, 211–242.

Ribéreau-Gayon, P. (1982). The anthocyanins of grapes and wines in Markakis, P. (Ed), Anthocyanins as food colors, Academic Press, pp. 209–244.

Ricardo da Silva, J. M., Bourzeix, M., Cheynier, V., & Moutounet, M. (1991a). Procyanidin composition of Chardonnay, Mauzac and Grenache blanc grapes. *Vitis, 30*, 245–252.

Ricardo da Silva, J. M., Cheynier, V., Souquet, J.-M., Moutounet, M., Cabanis, J.-C., & Bourzeix, M. (1991b). Interaction of grape seed procyanidins with various proteins in relation to wine fining. *J. Sci. Food Agric., 57*, 111–125.

Ricardo da Silva, J. M., Rigaud, J., Cheynier, V., Cheminat, A., & Moutounet, M. (1991c). Procyanidin dimers and trimers from grape seeds.*Phytochemistry, 30*, 1259–1264.

Ricardo da Silva, J. M., Cheynier, V., Samson, A., & Bourzeix, M. (1993). Effect of pomace contact, carbonic maceration and hyperoxidation on the procyanidin composition of Grenache blanc wines. *Am. J. Enol. Vitic., 44*, 168–172.

Richard-Forget, F., Rouet-Mayer, M.-A., Goupy, P. M., Philippon, J., & Nicolas, J. J. (1992). Oxidation of chlorogenic acid, catechins, and 4-methylcatechol in model solutions by apple polyphenol oxidase. *J. Agric. Food Chem., 40*, 2114–2122.

Rigaud, J., Escribano-Bailon, M. T., Prieur, C., Souquet, J.-M., & Cheynier, V. (1993). Normal-phase high-performance liquid chromatographic separation of procyanidins from cacao beans and grape seeds.*J. Chromatogr. A, 654*, 255–260.

Riou, V., Vernhet, A., Doco, T., & Moutounet, M. (2002). Aggregation of grape seed tannins in model – effect of wine polysaccharides.*Food Hydrocolloids, 16*, 17–23.

Rodriguez Montealegre, R., Romero Peces, R., Chacon Vozmediano, J. L., Martinez Gascuena, J., & Garcia Romero, E. (2006). Phenolic compounds in skins and seeds of ten grape Vitis vinifera varieties grown in warm climates.*J. Food Compos. Anal., 19*, 687–693.

Rodriguez-Clemente, R., & Correa-Gorospe, I. (1988). Structural, morphological and kinetic aspects of potassium hydrogen tartrate precipitation from wines and ethanolic solutions. *Am. J. Enol. Vitic., 30*, 169–179.

Roggero, J. P., Larice, J. L., Rocheville Divorne, C., Archier, P., & Coen, S. (1988). Composition anthocyanique des cépages. I : Essai de classification par analyse en composantes principales et par analyse factorielle discriminante.*Rev. F. Oenol., 112*, 41–48.

Romeyer, F., Macheix, J. J., & Sapis, J. C. (1986). Changes and importance of oligomeric procyanidins during maturation of grape seeds. *Phytochemistry, 25*, 219–221.

Salas, E., Fulcrand, H., Meudec, E., & Cheynier, V. (2003). Reactions of anthocyanins and tannins in model solutions.*J. Agric. Food Chem., 51*, 7951–7961.

Salas, E., Atanasova, V., Poncet-Legrand, C., Meudec, E., Mazauric, J., & Cheynier, V. (2004). Demonstration of the occurrence of flavanol-anthocyanin adducts in wine and in model solutions. *Anal. Chim. Acta, 513*, 325–332.

Salas, E., Dueñas, M., Schwarz, M., Winterhalter, P., Cheynier, V., & Fulcrand, H. (2005). Characterization of pigments from different high speed countercurrent chromatography wine fractions.*J. Agric. Food Chem., 53*, 4536–4546.

Santos-Buelga, C., Bravo-Haro, S., & Rivas-Gonzalo, J. (1995). Interactions between catechin and malvidin-3-monoglucoside in model solutions. *Z. Lebens.-Unter.Forsch.,* 201.

Sarni-Manchado, P., & Cheynier, V. (2002). Study of noncovalent complexation between catechin derivatives and peptide by electrospray ionization-mass spectrometry (ESI-MS). *J. Mass Spectrom., 37*, 609–616.

Sarni-Manchado, P., Deleris, A., Avallone, S., Cheynier, V., & Moutounet, M. (1999). Analysis and characterization of wine condensed tannins precipitated by protein used as fining agent in enology. *Am. J. Enol. Vitic., 50*, 81–86.

Saucier, C., Bourgeaois, G., Vitry, C., Roux, D., & Glories, Y. (1997a). Characterization of (+)-catechin-acetaldehyde polymers: A model for colloidal state of wine polyphenols.*J. Agric. Food Chem., 45*, 1045–1049.

Saucier, C., Guerra, C., Pianet, I., Laguerre, M., & Glories, Y. (1997b). (+)catechin-acetaldehyde condensation products in relation to wine ageing.*Phytochemistry, 46*, 229–234.

Saucier, C., Roux, D., & Glories, Y. O., Symp. Int. Oenol., 5th Meeting; Tec & Doc – Lavoisier: Paris, 1996. (1996). Stabilité colloïdale de polymères catéchiques: influence des polysaccharides in Lonvaud-Funel, A. (Ed), Oenologie95, Lavoisier, Tec et Doc, pp. 395–400.

Sausse, P., Aguie-Beghin, V., & Douillard, R. (2003). Effects of epigallocatechin gallate on beta-casein adsorption at the air/water interface.*Langmuir, 19*, 737–743.

Siebert, K. J., Carrasco, A., & Lynn, P. Y. (1996). Formation of protein-polyphenol haze in beverages.*J. Agric. Food Chem., 44*, 1997–2005.

Simpson, R. F. (1982). Factors affecting oxidative browning of white wine.*Vitis, 21*, 233–239.

Singleton, V. L., & Draper, D. (1964). The transfer of polyphenolic compounds from grape seeds into wines.*Am. J. Enol. Vitic., 15*, 34–40.

Smith, V. K., Ndou, T. T., & Warner, I. M. (1994). Spectroscopic study of the interaction of catechin with a-, b- and g- cyclodextrins.*J. Physical Chem., 98*, 8627–8631.

Somers, T. C. (1971). The polymeric nature of wine pigments. *Phytochemistry, 10*, 2175–2186.

Somers, T. C., & Pocock, K. F. (1991). Phenolic assessment of white musts: varietal differences in free-run juices and pressings. *Vitis, 30*, 189–201.

Somers, T. C., & Ziemelis, G. (1985). Flavonol haze in white wines. *Vitis, 24*, 43–50.

Souquet, J.-M., Cheynier, V., Brossaud, F., & Moutounet, M. (1996). Polymeric proanthocyanidins from grape skins. *Phytochemistry, 43*, 509–512.

Souquet, J.-M., Labarbe, B., Le Guernevé, C., Cheynier, V., & Moutounet, M. (2000). Phenolic composition of grape stems.*J. Agric. Food Chem., 48*, 1076–1080.

Souquet, J.-M., Veran, F., Mané, C., & Cheynier, V. (2006). Optimization of extraction conditions on phenolic yields from the different parts of grape clusters – Quantitative distribution of their proanthocyanidins. XXIII International Conference on Polyphenols. Winnipeg, Manitoba, Canada.

Spayd, S., Tarara, J., Mee, D., & Ferguson, J. (2002). Separation of Sunlight and Temperature Effects on the Composition of Vitis vinifera cv. Merlot Berries.*Am. J. Enol. Vitic., 53*, 171–182.

Stafford, H., & Lester, H. (1984). Flavan-3-ol biosynthesis; the conversion of (+)-dihydroquercetin and flavan-3,4-cis-diol (leucocyanidin) to (+) catechin by reductases extracted from cell suspension cultures of Douglas fir. *Plant Physiol., 76*, 184–186.

Su, C., & Singleton, V. (1969). Identification of three flavan-3-ols from grapes. *Phytochemistry, 8*, 1553–1558.

Sun, B. S., Pinto, T., M.C. Leandro, Ricardo-Da-Silva, J. M., & Spranger, M. I. (1999). Transfer of catechins and proanthocyanidins from solid parts of the grape cluster into wine. *Am. J. Enol. Vitic., 50*, 179–184.

Talcott, S., & Lee, J. (2002). Ellagic acid and flavonoid antioxidant content of Muscadine wine and juice. *J. Agric. Food Chem., 50*, 3186–3192.

Tesnière, C., Torregrosa, L., Pradal, M., Souquet, J.-M., Gilles, C., Dos Santos, K., Chatelet, P., & Günata, Z. (2006). Effects of genetic manipulation of alcohol dehydrogenase levels on the response to stress and the synthesis of secondary metabolites in grapevine leaves.*J. Exp. Bot., 57*, 91–99.

Timberlake, C. F., & Bridle, P. (1976). Interactions between anthocyanins, phenolic compounds, and acetaldehyde and their significance in red wines. *Am. J. Enol. Vitic., 27*, 97–105.

Trousdale, E., & Singleton, V. L. (1983). Astilbin and engeletin in grapes and wines. *Phytochemistry, 22*, 619–620.

Vasserot, Y., Caillet, S., & Maujean, A. (1997). Study of anthocyanin adsorption by yeast lees. AEffect of some physicochemical parameters. *Am. J. Enol. Vitic., 48*, 433–437.

Vernhet, A., & Moutounet, M. (2002). Fouling of organic microfiltration membranes by wine constituents : importance, relative impact of wine polysaccharides and polyphenols and incidence of membrane properties.*J. Membrane Sci., 201*, 101–122.

Vernhet, A., Bellon-Fontaine, M. N., Brillouet, J.-M., Roesink, E., & Moutounet, M. (1997). Wetting properties of microfiltration membrane: determination by means of the capillary rise technique and incidence on the adsorption of wine polysaccharide and tannins. *J. Membrane Sci. 128*, 163–174.

Vernhet, Cartalade, D., & Moutounet, M. (2003). Contribution to the understanding of fouling build-up during microfiltration of wines. *J. Membr. Sci.*e *211*, 357–370.

Vernhet, A., Dupré, K., Boulangé L., Cheynier V., Pellerin P., & Moutounet M. (1999a). Composition of tartrate precipitates deposited on stainless steel tanks during the cold stabilization of wines. Part I : white wines. *Am. J. Enol. Vitic., 50*, 391–397.

Vernhet, A., Dupré, K., Boulangé L., Cheynier V., Pellerin P., & Moutounet M. (1999b). Composition of tartrate precipitates deposited on stainless steel tanks during the cold stabilization of wines. Part II: red wines. *Am. J. Enol. Vitic., 50*, 398–403.

Vidal, S., Cartalade, D., Souquet, J., Fulcrand, H., & Cheynier, V. (2002). Changes in proanthocyanidin chain-length in wine-like model solutions. *J. Agric Food Chem., 50*, 2261–2266.

Vidal, S., Francis, L., Guyot, S., Marnet, N., Kwiatkowski, M., Gawel, R., Cheynier, V., & Waters, E. J. (2003). The mouth-feel properties of grape and apple proanthocyanidins in a wine-like medium.*J. Sci. Food Agric., 83*, 564–573.

Waters, E., Pellerin, P., & Brillouet, J. (1994). A Saccharomyces mannoprotein that protects wine from protein haze.*Carbohydrate Polymers, 23*, 185–191.

Waters, E. J., Peng, Z., Pocock, K. F., & Williams, P. J. (1995). Proteins in white wine, I: procyanidin occurrence in soluble protein hazes and its relationships to protein instability. *Austr. J. Grape Wine Res.*, *1*, 86–93.

Waters, E. J., Shirley, N. J., & Williams, P. J. (1996). Nuisance proteins of wine are grape pathogenesis-related proteins. *J. Agric. Food Chem.*, *44*, 3–5.

Weinges, K., & Muller, O. (1972). Uber die enzymatische oxydative Kupplung der nat rlichen Polyhydroxyflavane. *Chemiker Zeitung*, 96, 96.

Weinges, K., & Piretti, M. V. (1971). Isolierung des C30H26O12-procyanidins B1 aus Wientrauben.*Liebigs Ann. Chem.*, *748*, 218–220.

Williamson, M. P., Trevitt, C., & Noble, J. M. (1995). NMR studies of dextran oligomer interactions with model polyphenols. *Carbohydrate Research, 266*, 229–235.

Wulf, L. W., & Nagel, C. W. (1976). Analysis of phenolic acids and flavonoids by high-pressure liquid chromatography. *J. Chromatogr.*, *116*, 271–279.

Yanagida, A., Kanda, T., Shoji, T., Ohnishi-Kameyama, M., & Nagata, T. (1999). Fractionation of apple procyanidins by size-exclusion chromatography.*J. Chromatogr. A.*, *855*, 181–190.

Yanagida, A., Kanda, T., Takahashi, T., Kamimura, A., Hamazono, T., & Honda, S. (2000). Fractionation of apple procyanidins according to their degree of polymerization by normal-phase high-performance liquid chromatography.*J. Chromatogr. A.*, *890*, 251–259.

Yokotstuka, K. (1990). Effect of press design and pressing pressures on grape juice components. *J. Ferment. Bioeng.*, *70*, 15–21.

Young, D. A., Young, E., Roux, D. G., Brandt, E. V., & Ferreira, D. (1987). Synthesis of condensed tannins. Part 19. Phenol oxidative coupling of (+)-catechin and (+)-mesquitol. Conformation of Bis (+)-catechins.*J. Chem. Soc. Perkin Trans. I*, 2345–2351.

Chapter 9C
Non-flavonoid Phenolic Compounds

Michael Rentzsch, Andrea Wilkens, and Peter Winterhalter

Contents

The non-flavonoid phenolic constituents in wine are divided into hydroxybenzoic acids and hydroxycinnamic acids, volatile phenols, stilbenes and miscellaneous compounds (e.g. lignans and coumarins). Although non-colored, the non-flavonoid constituents are known to enhance and stabilize the color of red wines by intra- and intermolecular reactions. They furthermore contribute to wine flavor (volatile phenolic acids) and some of them (e.g. resveratrol) exhibit potent biological activities.

9C.1 Phenolic Acids

9C.1.1 Hydroxybenzoic Acids (HBA)

Being derived from benzoic acid, hydroxybenzoic acids are characterized by a C6-C1 skeleton (Fig. 9C.1). The most common derivatives found in wine are gallic acid, gentisic acid, *p*-hydroxybenzoic acid, protocatechuic acid, syringic acid, salicylic acid, and vanillic acid. In wine the different hydroxybenzoic acids can be mainly found in their free form (Drawert et al. 1974; Fernandéz de Simon et al. 1992; Garcia-Viguera and Bridle 1995; Güntert et al. 1986; Monagas et al. 2005a,b; Peña-Neira et al. 2000; Pozo-Bayón et al. 2003; Salagoïty-Auguste and Bertrand 1984;

P. Winterhalter (✉)
Institut für Lebensmittelchemie, TU Braunschweig, Schleinitzstrasse 20, D-38106 Braunschweig, Germany
e-mail: p.winterhalter@tu-bs.de

M.V. Moreno-Arribas, M.C. Polo (eds.), *Wine Chemistry and Biochemistry*,
DOI 10.1007/978-0-387-74118-5_19,

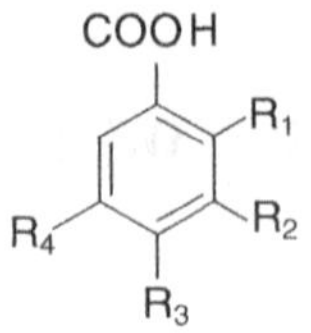

Hydroxybenzoic acids	R_1	R_2	R_3	R_4	MW
Gallic acid	H	OH	OH	OH	170
Gentisic acid	OH	H	H	OH	154
p-Hydroxybenzoic acid	H	H	OH	H	138
Protocatechuic acid	H	OH	OH	H	154
Salicylic acid	OH	H	H	H	138
Syringic acid	H	OCH_3	OH	OCH_3	198
Vanillic acid	H	OCH_3	OH	H	168

Fig. 9C.1 Structures of hydroxybenzoic acids in wine

Vanhoenacker et al. 2001). Gallic acid is one of the HBA with the highest concentrations in wine. It not only originates from the grape itself but is also formed by hydrolysis of hydrolyzable and condensed tannins, i.e. the gallic acid esters of flavan-3-ols.

The levels of HBA in wine show great variability depending on grape variety and growing conditions. Pozo-Bayón et al. (2003) reported values between 0.3 and 1.3 mg/L for gallic acid in Spanish sparkling wines from white and red grapes, respectively. For protocatechuic acid the values were between 0.5 and 0.93 mg/L, while the concentrations for *p*-hydroxycinnamic acid were in the range of 0–0.22 mg/L. Peña-Neira et al. (2000) detected concentrations of gallic acid in wine from La Rioja with up to 2.29 mg/L while Sladkovský et al. (2004) reported concentrations of 4.8 mg/L in tawny port. In a study that evaluated the effect of certain yeast strains during malolactic fermentation on non-flavonoid polyphenols in red wine from La Rioja (Hernández et al. 2007) the authors report a gallic acid concentration of up to 41.6 mg/L. However, in comparison to the group of hydroxycinnamic acids, the overall concentration of HBAs in wine is relatively low.

Among the derivates of HBAs, further compounds have been identified. Güntert et al. (1986) identified ethyl esters of vanillic acid and *p*-hydroxybenzoic acid, and methyl esters of vanillic acid and protocatechuic acid. Ethyl esters of protocatechuic acid and vanillic acid, as well as the glucose ester of vanillic acid, were isolated from a German Riesling wine (Baderschneider and Winterhalter 2001). Analytically, HBA are mostly determined as trimethylsilane derivatives by using gas chromatographic analysis. A good overview on analytical methods as well as mass spectrometric data is given by Monagas et al. (2005b).

9C.1.2 Hydroxycinnamic Acids (HCA)

Hydroxycinnamic acids possess a C6-C3 skeleton and formally belong to the group of phenylpropanoids. The different compounds present in wine are mainly derived from the hydroxycinnamic acids caffeic acid, *p*-coumaric acid, ferulic acid, and sinapic acid (Fig. 9C.2). These derivatives can be present in *cis*- and *trans*-configured forms, while the *trans* forms are more stable and therefore more prevalent. In wine HCA are present in low amounts in their free form, while the depside forms, i.e. esters of l-(+)-tartaric acid, are predominant. The ubiquitous chlorogenic acids, esters of HCA and quinic acid, cannot be found in wine but are replaced by the tartaric acid esters instead (Ong and Nagel 1978; Singleton et al. 1978; Somers et al. 1987).

Among the hydroxycinnamic acids, caftaric acid predominates (up to 50% of total hydroxycinnamic acids). Other important substances are the tartaric esters of *p*-coumaric acid and ferulic acid, and the *trans-p*-coumaric glucoside (Somers et al. 1987). The concentration levels of hydroxycinnamic acid derivatives in wine depend on many factors like grape variety, growing conditions, climate, etc. It is

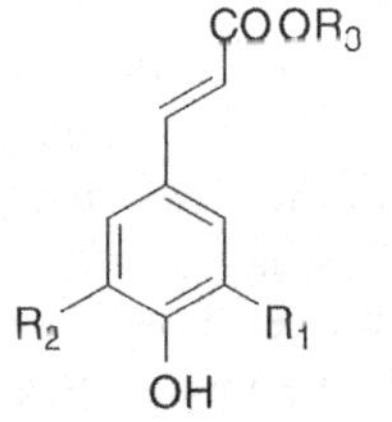

Hydroxycinnamic acids	R_1	R_2	R_3	MW
Caffeic acid	OH	H	H	180
Caftaric acid	OH	H	Tartaric acid	312
p-Coumaric acid	H	H	H	164
p-Coutaric acid	H	H	Tartaric acid	296
Ferulic acid	OCH_3	H	H	194
Fertaric acid	OCH_3	H	Tartaric acid	326
Sinapic acid	OCH_3	OCH_3	H	224

Fig. 9C.2 Structures of hydroxycinnamic acids in wine

therefore not surprising to find great differences in the concentrations published for different wines. In general, concentrations around 100 mg/L are described. For *p*-coutaric and fertaric acid the values are around 55 mg/L and 16 mg/L, respectively (Andres-Lacueva et al. 1996; Baldi et al. 1993; Boursiquot et al. 1986; Cheynier et al. 1989; Herrick and Nagel 1985; Herrmann 1989; Lao et al. 1996; Lee and Jaworski 1989; Okamura and Watanabe 1981; Ricardo-da-Silva et al. 1993; Singleton et al. 1978; Singleton et al. 1986). While in juices and young wines the content of free hydroxycinnamic acids is very low, an increase can be observed during storage. The tartaric esters are hydrolyzed and give rise to free forms of HCAs (Andres-Lacueva et al. 1996; Betés-Saura et al. 1996; Somers et al. 1987; Würdig and Woller 1989).

Betés-Saura et al. (1996) reported a loss of 27% of hydroxycinnamates during vinification in white wines from Penedès (46.76 mg/L in juices vs 34.07 mg/L in wine). With a share of around 73% of total phenolics, hydroxycinnamates were the most important group of phenolics in these white wines. Among them *trans*-caftaric acid was the major phenol with values between 10 and 13 mg/L. Okamura and Watanabe (1981) gave an overview of the concentrations of caftaric, coutaric, caffeic, and coumaric acid in commercial white wines. Average values were 23.0, 5.0, 0.9, and 0.7 mg/L for Semillon wines (Bordeaux), 29.0, 10.0, 1.7, and 0.9 mg/L in Chardonnay (California), 112.0, 53.0, 3.1, and 2.1 mg/L for Koshu wine (Japan), and 51.0, 13.0, 4.1, and 2.9 mg/L for Riesling wine (Germany).

There are also reports of numerous derivatives of HCAs occurring during aging. Apart from tartaric esters, ethyl esters of caffeic acid and coumaric acid, as well as ethyl esters and diethyl esters of caftaric acid were detected and characterized (Baderschneider and Winterhalter 2001; Somers et al. 1987). Also described were the 4-*O*-glucosides of ferulic acid and coumaric acid, the glucosides of caffeic acid, coumaric acid, ferulic acid, and sinapic acid (Baderschneider and Winterhalter 2001; Cooper and Marshall 2001; Monagas et al. 2005a). Cheynier et al. (1986) demonstrated that the reaction product of caftaric acid and glutathione, 2-*S*-glutathionylcaftaric acid, is the major phenolic product formed by enzymatic oxidation. This colorless product that is also called "grape reaction product (GRP)" is considered as a measure to estimate the oxidation and browning potential of grape musts. High GPR levels were reported to correlate with a lower sensitivity to browning reactions (Rigaud et al. 1991).

Additionally, hydroxycinnamates are important constituents of acylated anthocyanins. In red wine, the 6-*O*-coumaroylglucosides and caffeoylglucosides are common constituents (Mazza and Miniati 1993). In the authenticity control of red wine the relation of acetylated and coumaroylated anthocyanins can be used for the assessment of the grape variety (Holbach et al. 2001; Otteneder et al. 2002).

9C.1.2.1 Reactions of Hydroxycinnamic Acids with Anthocyanins

Hydroxycinnamic acids take part in numerous reactions that occur during winemaking and wine aging. They are important compounds in oxidation processes of wine. During storage oxygen reacts over a coupled oxidation with vicinal di- and trihydroxyphenols, like caffeic acid, to produce the corresponding chinones (Singleton 1987; Wildenradt and Singleton 1974) The subsequent reactions result in the oxidation of ethanol to acetaldehyde, which has been shown to take part in condensation reactions between the wine polyphenols, such as anthocyanins and flavan-3-ols (Dallas et al. 1996a, b; Es-Safi et al. 1999; Fulcrand et al. 1996b; Rivas-Gonzalo et al. 1995; Santos-Buelga et al. 1995) forming a great variety of new, partly ethyl-linked, pigments (Dallas et al. 1996a, b; Es-Safi et al. 1999; Fulcrand et al. 1996b; Rivas-Gonzalo et al. 1995; Santos-Buelga et al. 1995; Timberlake and Bridle 1976). The oxidation of hydroxycinnamates also contributes to the browning of white wines during aging (Cheynier et al. 1990; Cilliers and Singleton 1990). The browning effect seemed to correlate with the oxidation of caffeic acid although it was shown that the effect of catechin was far greater than that of hydroxycinnamic acids (Okamura and Watanabe 1981).

The color of red wine is also strongly influenced by the presence of hydroxycinnamic acids. They play an important role in the phenomenon of copigmentation which describes the color intensification of anthocyanin solutions by the presence of other phenolic compounds (Baranac et al. 1996; Darias-Martín et al. 2002; Dimitric Markovic et al. 2000, 2005; Miniati et al. 1992). The intensification and the observed bathochromic shift of color are due to two main mechanisms, the intermolecular and the intramolecular copigmentation. Intermolecular copigmentation describes the stabilization of flavylium cations of anthocyanins by copigments over π-complexes between different molecules. Intramolecular copigmentation can occur for example in the case of anthocyanins acylated with hydroxycinnamic acids (sandwich type) (Figueiredo et al. 1999). A summary of copigmentation effects in red wine is given by Boulton (2001).

Another important reaction of hydroxycinnamic acids with anthocyanins is the formation of pyranoanthocyanins (Rentzsch et al. 2007b). These pigments are formed by direct reaction of hydroxycinnamic acids and their corresponding vinylphenols with anthocyanins present in red wine (Fig. 9C.3) (Fulcrand et al. 1996a; Hayasaka and Asenstorfer 2002; Håkansson et al. 2003; Sarni-Manchado et al. 1996; Schwarz et al. 2003). While in young wines the reaction of the vinylphenols is predominating, due to an enzymatic decarboxylation of coumaric and ferulic acid, during storage the share of direct reaction products of caffeic acid becomes more important (Rentzsch et al. 2007a; Schwarz et al. 2004). It was shown that caffeic and sinapic acid are not decarboxylated to their vinylphenols by yeast activity (Chatonnet et al. 1993); instead the formation of pyranoanthocyanis takes place over the direct reaction of caffeic acid, sinapic acid and anthocyanins. Schwarz et al. (2003) described the pathway of formation of hydroxyphenyl-pyranoanthocyanins over direct reaction with hydroxycinnamates. The stabilization of an intermediated carbenium ion structure by electron donating substituents on the aromatic ring of hydroxycinnamic acids is essential for the formation of these new pigments. Among the hydroxycinnamic acids, sinapic acid showed the fasted reaction followed by caffeic acid and ferulic acid. Coumaric acid exhibited the slowest reactivity. However, due to the high concentration of coumaric acid and caffeic acid in red wine, the

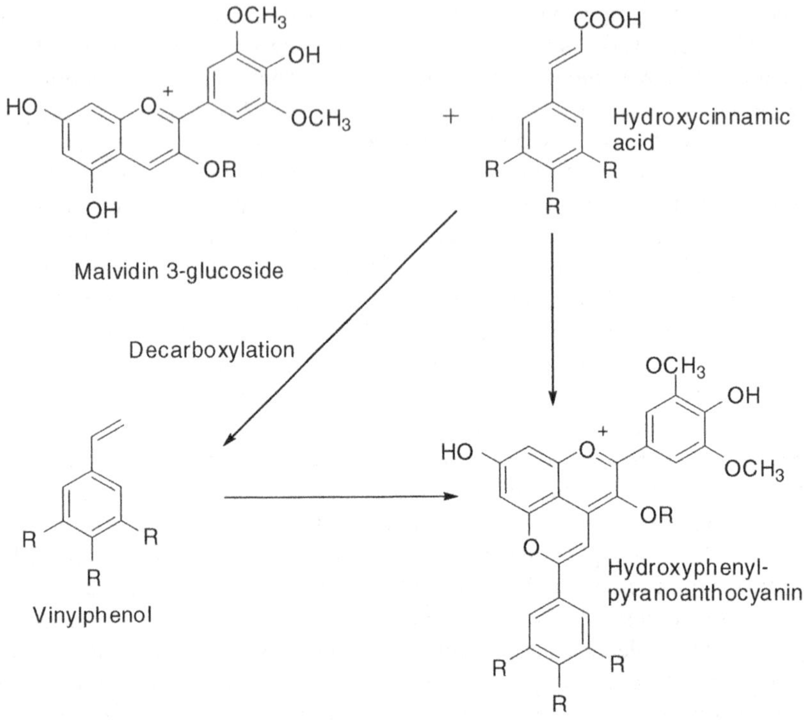

Fig. 9C.3 Reactions of hydroxycinnamic acids and their corresponding vinylphenols hydroxyphenyl-pyranoanthocyanins

hydroxyphenyl-pyranoanthocyanins resulting from these precursors are predominant. An overview on pyranoanthocyanins is given by Rentzsch et al. (2007b).

9C.2 Volatile Phenols

The volatile phenols possess the lowest concentration among the phenolic compounds in wine. However, due to their odor activity they have a great influence on the sensory characteristics of wine. Two sources for volatile phenols can be distinguished. One is the enzymatic formation from precursors present in wine and the other origin is due to migration from wood during barrel maturation. Among the volatile phenols, the vinyl and ethyl phenols play the most important role. Emerging from the decarboxylation of hydroxycinnamic acids, these compounds are responsible for off-flavors of wines. Vinylphenols exhibit unpleasant odors. With thresholds of between 420 μg/L for a 10/1 mixture of 4-vinlyphenol and 4-vinylguaiacol in white wine and 720 μg/L for a 1/1 mixture of the ethyl-phenols in red wine, these compounds can easily spoil the wine. As reported by Chatonnet et al. (1989), the yeast *Saccharomyces cerevisiae* can only decarboxylate coumaric and ferulic acid.

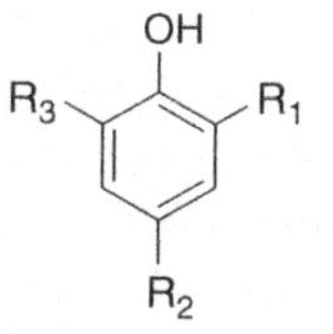

Phenols	R_1	R_2	R_3	MW
Ethyl phenol	H	C_2H_5	H	122
Vinyl phenol	H	C_2H_3	H	120
Guaiacol	OCH_3	OH	H	124
Methyl guaiacol	OCH_3	CH_3	H	122
Ethyl guaiacol	OCH_3	C_2H_5	H	136
Vinyl guaiacol	OCH_3	C_2H_3	H	134
Propyl guaiacol	OCH_3	C_3H_7	H	150
Allyl guaiacol	OCH_3	C_3H_5	H	148
Syringol	OCH_3	H	OCH_3	154
Methyl syringol	OCH_3	CH_3	OCH_3	152

Fig. 9C.4 Structures of volatile phenols in wine

More potent yeasts responsible for spoilages of wines with volatile phenols are of *Brettanomyces/Dekkerra* type that can produce the vinyl phenols and ethyl phenols of hydroxyl cinnamic acids. Ribéreau-Gayon et al. (2000) give an overview on this topic. Another source of volatile phenols is the oak wood of barrel and oak chips. These phenols are characterized by smoked and toasted aromas. Important volatile phenols in wine are depicted in Fig. 9C.4.

9C.3 Stilbenes

Stilbenes are a subclass of phenolic compounds naturally occurring in various families of plants, but grapes and wine are considered the most important dietary sources of these substances (Guebailia et al. 2006). Stilbenes can be biosynthesized by grapevines as a defence response to stress, such as microbial infection and UV irradiation, and they are transferred during the winemaking process into the must and wine. Due to their antioxidative, anticarcinogenic and antimutagenic potency, stilbenes are considered to play a central role in the human diet (Buiarelli et al. 2007).

Stilbenoids are derived from cinnamic acid and three acetate units from malonyl coenzym A. The first part of the biosynthesis is in common to flavonoids. The two biosynthetic routes are diverging at the point of cyclization of a styryl-3,5,7-triketoheptanoic acid. A C-acylation produces a chalcone and subsequent modifications lead to the flavonoids. An aldol condensation of the same intermediate polyketide produces a stilbene-2-carboxylic acid that is unstable and constitutes a range of structures known as stilbenoids. Figure 9C.5 shows an overview of the biosynthetic pathway (Gorham 1995).

One of the most relevant and extensively studied stilbene is *trans*-resveratrol (3,5,4'-trihydroxystilbene), a phytoalexin produced by grapevines in response to fungal infection, particularly *Botrytis cinerea.* Synthesis of resveratrol in grape berries is mainly located in the skin cells and is absent or low in the fruit flesh. In nature, resveratrol exists in two isomeric forms (*cis*- and *trans*-configured) in the free as well as in ß-glucoconjugated form. The 3-O-ß-D-glucosides of *cis*- and *trans*-resveratrol cis-and trans-configured are called piceids. The chemical structures of resveratrol and further stilbenes are depicted in Fig. 9C.6.

Stilbenes can also occur in oligomeric and polymeric forms, so-called viniferins. They are induced by oxidative polymerization of the monomer resveratrol through the activity of a peroxidase (Jean-Denis et al. 2006). Figure 9C.7 shows the biosynthesis of *trans*-ε-viniferin.

Numerous stilbenes such as ε-viniferin (Landrault et al. 2002) and δ-viniferin, two resveratrol dehydrodimers (Vitrac et al. 2005), the resveratrol dimer pallidol (Vitrac et al. 2001), α-viniferin, a trimer of resveratrol (Pryce and Langcake 1977)

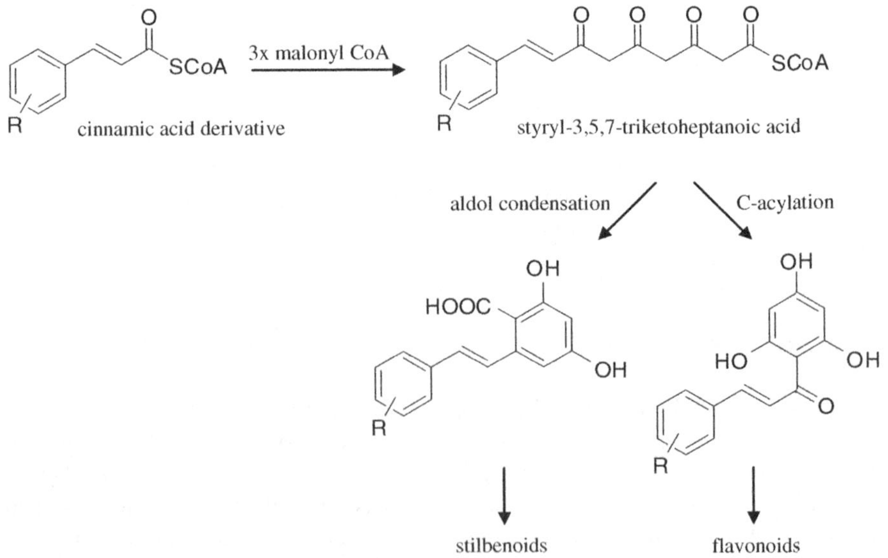

Fig. 9C.5 Common biosynthetic pathway of stilbenoids and flavonoids

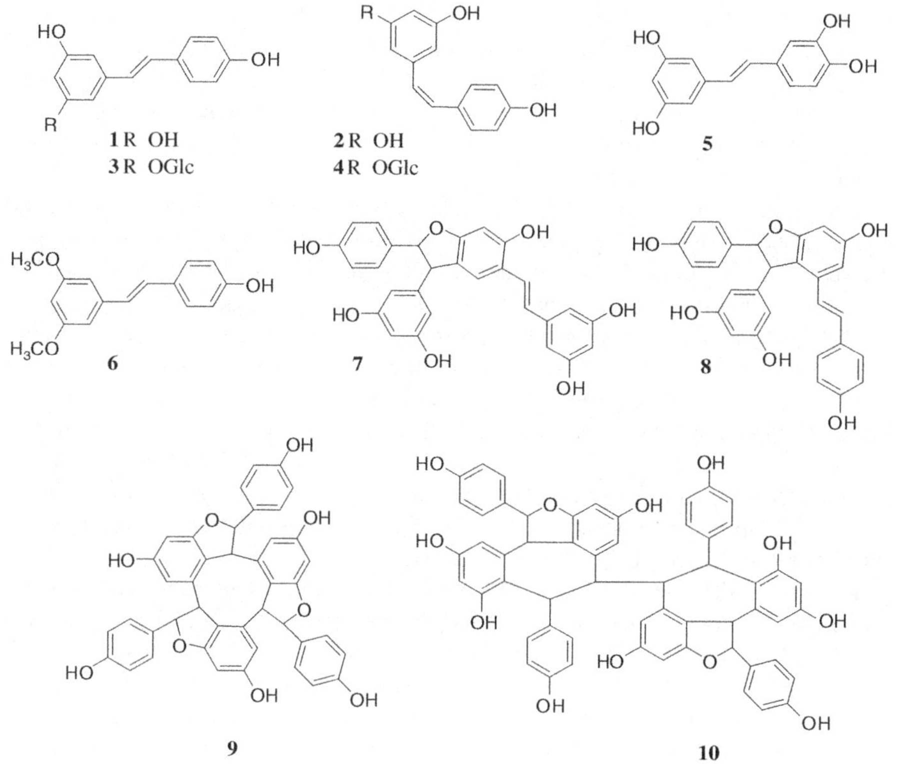

Fig. 9C.6 Chemical structures of stilbenes: *trans*-resveratrol (1), *cis*-resveratrol (2), *trans*-piceid (3), *cis*-piceid (4), *trans*-piceatannol (5), *trans*-pterostilbene (6), *trans*-δ-viniferin (7), *trans*- ε-viniferin (8), α-viniferin (9), hopeaphenol (10)

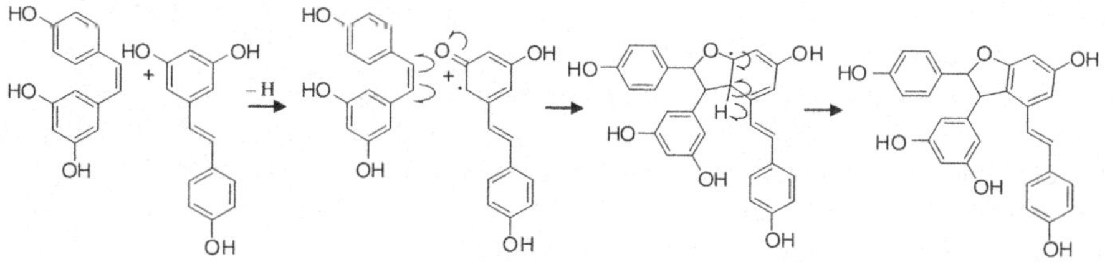

Fig. 9C.7 Biosynthesis of *trans*-ε-viniferin

and the resveratrol tetramer hopeaphenol (Guebailia et al. 2006) have been identified in grapevine and wine. In addition, some oligomeric glucosides like *cis*- and *trans*-ε-viniferin diglucosides and pallidol-3-3″-diglucoside could be isolated and characterized from Riesling wine (Baderschneider and Winterhalter 2000).

The content of stilbenes in wine varies considerably and depends on several factors including climate, grape variety, fungal infections (Perrone et al. 2007; Jeandet et al. 1995), UV light, heavy metal ions (Püssa et al. 2006) and enological methods

(Cantos et al. 2003; Gambuti et al. 2004). It is also influenced by yeast enzymatic activities, in particular those of isomerase and glucosidase (Jeandet et al. 1994). Equally, activities of lactic acid bacteria, which are responsible for malolactic fermentation (Hernández et al. 2007), can also affect stilbene content in wine (Poussier et al. 2003). Aging of wine appears to have no important influence on the concentration of stilbenes (Jeandet et al. 1995).

Red wines usually contain higher stilbene concentrations than rosé or white wines. This depends on the more prolonged skin contact of the must during fermentation and the high phenolic content of red grape cultivars (Perrone et al. 2007).

Free *trans-* and *cis*-resveratrols are present in a concentration range of 0.2–13 mg/L in red wines and 0.1–0.8 mg/L in white wines. Consistently, high concentrations of *trans*-resveratrol are obtained in wines from Pinot noir. On the one hand, Cabernet Sauvignon wines cover a wide range of concentrations, with relatively high values in those from cool-climate countries such as Ontario and the Bordeaux region of France, whereas, on the other hand, wines from warmer climates like California, South America, and Australia tend to have much lower concentrations. In comparison to wine, grape juice offers a content of *trans*-resveratrol in a range of 0.09–0.18 mg/L (Stecher et al. 2001).

For piceid, the resveratrol-3-*O*-glucoside, concentrations are reported to be in a range of 0.3–9 mg/L in red and 0.1–2.2 mg/L in white wines (Goldberg et al. 1995, 1996a, b; Lamuela-Raventós et al. 1995; Sato et al. 1997; Naugler et al. 2007; Romero-Pérez et al. 1996). In Portuguese red wines piceids were even determined in concentrations up to 68 mg/L (Ribeiro de Lima et al. 1999). In comparison to wine, grapes were found to contain mainly *trans*-resveratrol glucoside in concentrations ranging from 1.5 to 7.3 μg/g (Burns et al. 2002).

In a survey of commercial wines from the South of France, levels of pallidol and viniferin have been reported. Viniferin was found to be present in red and botrytized sweet white wines with levels between 0.1 and 1.63 mg/L. Pallidol was not found in dry and sweet white wines but only in wines made by maceration with stems, with levels between 0.38 and 2.22 mg/L (Landrault et al. 2002).

In addition, Guebailia et al. (2006) have investigated the concentration of hopeaphenol in North African wines. Ksarwine presented the highest concentration of hopeaphenol (3.8 mg/L), followed by Muscat (3.06 mg/L), Guerrouane (2.68 mg/L), Merlot (2.1 mg/L), Cabernet Sauvignon (1.48 mg/L), Sidi-Brahim (0.61 mg/L), Amjad (0.34 mg/L), and Gris d'Algérie (0.3 mg/L).

Since Siemann and Creasy (1992) described the presence of *trans*-resveratrol in wines, many different methods have been developed to determine this compound and its derivatives (Romero-Pérez et al. 1996). Various methods use RP18-HPLC and gradient elution with UV detection (Lamuela-Raventós et al. 1995; Naugler et al. 2007; Ribeiro de Lima et al. 1999; Vitrac et al. 2005).

Trans-configured stilbenes such as *trans*-resveratrol, *trans*-pterostilbene and *trans*-ε-viniferin show two characteristic bands corresponding to high absorbances from 308 to 336 nm and from 281 to 313 nm. Cis-configured stilbenes like *cis*-resveratrol, *cis*-pterostilbene and *cis*-ε-viniferin exhibit a UV maximum at 285nm (Jeandet et al. 1997).

In addition, stilbenes are fluorescent compounds which are easily detected by fluorometry. For resveratrol, fluorescence detection is highly selective and even twice as sensitive as UV detection (Stecher et al. 2001). Due to this potential, methods by using HPLC coupled with absorbance and fluorescence detection were developed (Jeandet et al. 1997; Vitrac et al. 2002). Otherwise identification of stilbenes in wine by HPLC-DAD can be limited by coelution of two or more compounds. Therefore application of modern mass-spectrometry techniques is important to confirm the structure of stilbenes and to detect novel compounds in wine (Monagas et al. 2005b; Buiarelli et al. 2007; Careri et al. 2004; Jean-Denis et al. 2006; Kammerer et al. 2004; Mark et al. 2005; Püssa et al. 2006; Stecher et al. 2001). Table 9C.1 shows an overview on stilbenoid molecular ions and fragments identified in wine and grapevine by mass spectrometry.

It is widely accepted that polyphenols in wine are responsible for beneficial health effects (Sun et al. 2006). Particularly *trans*-resveratrol has been intensively studied and marked biological activities with regard to the prevention of cardiovascular disease and cancer have been reported (Ito et al. 2003). Other stilbenes also have properties similar to those of *trans*-resveratrol. Therefore, monitoring new stilbene derivatives in wine is of particular relevance (Guebailia et al. 2006).

9C.4 Miscellaneous Compounds

In addition to phenolic acids, volatile phenols, and stilbenes, other non-flavonoid phenolic compounds are also known to occur in wine. Marinos et al. (1992) were able to identify the lignans isolariciresinol-4′-*O*-ß-D-glucopyranoside and seco-isolariciresinol-ß-D-glucoside in an Australian Riesling wine as first representatives of a new category of wine phenols. By applying preparative all-liquid chromatographic techniques (i.e. countercurrent chromatography), Baderschneider and Winterhalter (2001) succeeded in isolating and fully characterizing nine additional lignans and neolignans from a German Riesling. Their structures are depicted in Fig. 9C.8. Nurmi et al. (2003) reported lignan concentrations in red wines in the range of 0.8-1.4 mg/L, with isolariciresinol being the main compound. Until now, it has not been clearly distinguished as to which extent these compounds are genuine grape constituents or rather formed during aging of wines in oak barrels. Clearly oak-wood derived are the coumarins which can be considered as cinnamic acid derivatives. The coumarins umbelliferone, 4-methyl-umbelliferone, esculin, and scopoletin have been extracted from oak wood (Puech and Moutounet 1988), and scopoletin (7-hydroxy-6-methoxycoumarin) has been reported as marker for the storage of wine in oak barrels (Tricard et al. 1987). More recently, oak-derived ellagtannins were reported to react with various nucleophilic wine constituents, such as catechin, epicatechin, anthocyanins, glutathione, and ethanol during barrel aging, giving rise to a formation of condensation products including ß-1- *O*-ethylvescalin and the potent antitumor agent Acutissimin A (Quideau et al. 2003, 2005; Saucier et al. 2006).

Table 9C.1 Stilbenoids identified in wine and grapevine by mass spectrometry

Compounds	$[M\text{-}H]^-$	Fragment *m/z*	Technique	Sample	Reference
Monomers					
Resveratrol	227 $[M\text{-}H]^-$	185, 183, 159, 157, 143	HPLC-ESI-MS/MS	Grapevine	Püssa et al. (2006)
Piceatannol	243 $[M\text{-}H]^-$	225, 201, 200, 199, 175	HPLC-ESI-MS/MS	Grapevine	Püssa et al. (2006)
Astringin	405 $[M\text{-}H]^-$	243, 173, 159	HPLC-TIS-MS/MS	Red wine	Buiarelli et al. (2007)
Piceid	389 $[M\text{-}H]^-$	305, 289, 227, 175	HPLC-ESI-MS/MS	Grapevine	Püssa et al. (2006)
Dimers					
ε-Viniferin	453$[M\text{-}H]^-$	435, 411, 369, 359, 347	HPLC-ESI-MS/MS	Grapevine	Püssa et al. (2006)
Pallidol	455 $[M+H]^+$		FAB-MS (+)	Red wine	Vitrac et al. (2001)
Pallidol-3-*O*-glucoside	634 $[M + NH_4]^+$		DCI-MS	White wine	Baderschneider and Winterhalter (2000)
Pallidol-3-3″-O-glucoside	777$[M\text{-}H]^-$	615	HPLC-ESI-MS/MS	White wine	Baderschneider and Winterhalter (2000)
ε-Viniferin diglucoside	777	615	HPLC-ESI-MS/MS	White wine	Baderschneider and Winterhalter (2000)

ESI-MS/MS: Electro spray ionization-tandem mass spectrometry; TIS-MS/MS: Turbo Ion Spray interface-tandem mass spectrometry; FAB-MS: Fast atom bombardment-mass spectrometry; DCI-MS: Desorption chemical ionization-mass spectrometry

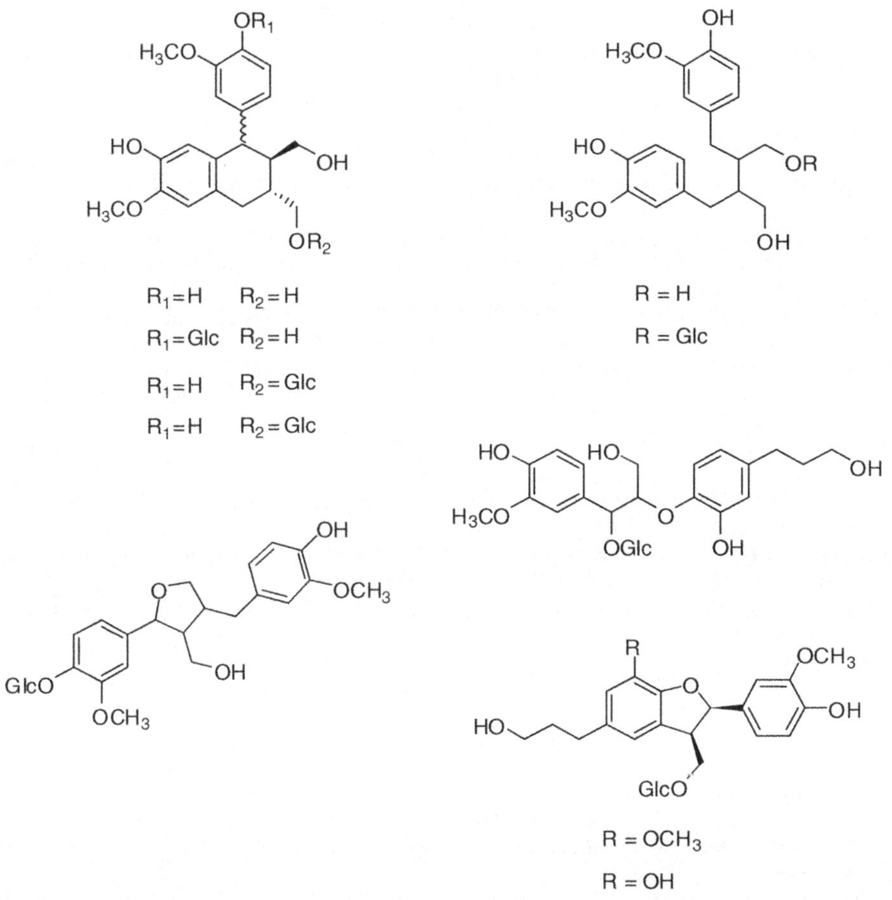

Fig. 9C.8 Structures of lignans and neolignans isolated from a German Riesling wine

References

Andres-Lacueva, C., Lamuela-Raventós, R. M., Buxadera, S., Torre-Bonat, M. C. (1996). Polyphenol content in sparkling wines (Cava) at different aging period. In J. Vercauteren, C. Chèze, M.C. Dumon, J.F. Weber (Eds.), *Polyphenols communications 96/XVIIIth International Conference on Polyphenols* (Vol. 2, pp. 305–306). Bordeaux: Secrétariat du Groupe Polyphénols.

Baderschneider, B., & Winterhalter, P. (2000). Isolation and characterization of novel stilbene derivatives from Riesling wine. *J. Agric. Food Chem.*, *48*, 2681–2686.

Baderschneider, B., & Winterhalter, P. (2001). Isolation and characterization of novel benzoates, cinnamates, flavonoids, and lignans from Riesling wine and screening for antioxidant activity. *J. Agric. Food Chem.*, *49*, 2788–2798.

Baldi, A., Romani, A., Mulinacci, N., & Vincieri, F. F. (1993). Composés phenoliques dans les cépages de Toscane de *Vitis vinifera* L. *J. Int. Sci. Vigne Vin*, *27*, 201–215.

Baranac, J. M., Petranovic, N. A., & Dimitric-Markovic, J. M. (1996). Spectrophotometric study of anthocyan copigmentation reactions. *J. Agric. Food Chem.*, *44*, 1333–1336.

Betés-Saura, C., Andrés-Lacueva, C., & Lamuela-Raventós, R. M. (1996). Phenolics in white free run juices and wines from Penedès by High-Performance Liquid Chromatography: Changes during vinification. *J. Agric. Food Chem., 44*, 3040–3046.

Boulton, R. (2001). The copigmentation of anthocyanins and its role in the color of red wine: A critical review. *Am. J. Enol. Vitic., 52*, 67–87.

Boursiquot, J. M., Sapis, J. C., & Macheix, J. J. (1986). Les esters hydroxycinnamiques chez le genre *Vitis*. Essais d'application taxonomique premiers résultats. *C.R. Acad. Sci. Serie Iii- Life Sciences, 302*, 177–180.

Buiarelli, F., Coccioli, F., Jasionowska, R., Merolle, M., & Terracciano, A. (2007). Analysis of some stilbenes in Italian wines by liquid chromatography/tandem mass spectrometry. *Rap. Commun. Mass Spectrom., 21*, 2955–2964.

Burns, J., Yokota, T., Ashihara, H., Lean, M. E. J., & Crozier, A. (2002). Plant foods and herbal sources of resveratrol. *J. Agric. Food Chem., 50*, 3337–3340.

Cantos, E., Espín, J. C., Fernandez, M. J., Oliva, J., & Tomas-Barberan, F. A. (2003). Postharvest UV-C-irradiated grapes as a potential source for producing stilbene-enriched red wines. *J. Agric. Food Chem., 51*, 1208–1214.

Careri, M., Corradini, C., Elviri, L., Nicoletti, I., & Zagnoni, I. (2004). Liquid Chromatography-Electrospray Tandem Mass Spectrometry of *cis*-resveratrol and *trans*-resveratrol: Development, validation, and application of the method to red wine, grape, and winemaking byproducts. *J. Agric. Food Chem., 52*, 6868–6874.

Chatonnet, P., Dubourdieu, D., & Boidron, J. N. (1989). Effects of certain factors on yeast phenolic acid decarboxylation. *Conn. Vigne* Vin., *23*, 59–62.

Chatonnet, P., Dubourdieu, D., Boidron, J.-N., & Lavigne, V. (1993). Synthesis of volatile phenols by *Saccharomyces cerevisiae* in wines. *J. Sci. Food Agric., 62*, 191–202.

Cheynier, V. F., Trousdale, E. K., Singleton, V. L., Salgues, M. J., & Wylde, R. (1986). Characterization of 2-*S*-glutathionylcaftaric acid and its hydrolysis in relation to grape wines. *J. Agric. Food Chem., 34*, 217–221.

Cheynier, V., Rigaud, J., Souquet, J. M., Barillere, J. M., & Moutounet, M. (1989). Effect of pomace contact and hyperoxidation on the phenolic composition and quality of Grenache and Chardonnay wines. *Am. J. Enol. Vitic., 40*, 36–42.

Cheynier, V., Rigaud, J., Souquet, J. M., Duprat, F., & Moutounet, M. (1990). Must Browning in relation to the behavior of phenolic compounds during oxidation. *Am. J. Enol. Vitic., 41*, 346–349.

Cilliers, J. J. L., & Singleton, V. L. (1990). Caffeic Acid Autoxidation and the Effects of Thiols. *J. Agric. Food Chem., 38*, 1789–1796.

Cooper, H. J., & Marshall, A. G. (2001). Electrospray Ionization Fourier Transform Mass Spectrometric Analysis of Wine. *J. Agric. Food Chem., 49*, 5710–5718.

Dallas, C., Ricardo-da-Silva, J. M., & Laureano, O. (1996a). Interactions of oligomeric procyanidins in model wine solutions containing malvidin-3-glucoside and acetaldehyde. *J. Sci. Food Agric., 70*, 493–500.

Dallas, C., Ricardo-da-Silva, J. M., & Laureano, O. (1996b). Products formed in model wine solutions involving anthocyanins, procyanidin B_2, and acetaldehyde. *J. Agric. Food Chem., 44*, 2402–2407.

Darias-Martín, J., Martín-Luis, B., Carrillo-López, M., Lamuela-Raventós, R., Díaz-Romero, C., & Boulton, R. (2002). Effect of caffeic acid on the color of red wine. *J. Agric. Food Chem., 50*, 2062–2067.

Dimitric Markovic, J. M. D., Petranovic, N. A., & Baranac, J. M. (2000). A Spectrophotometric study of the copigmentation of malvin with caffeic and ferulic acids. *J. Agric. Food Chem., 48*, 5530–5536.

Dimitric Markovic, J. M. D., Petranovic, N. A., & Baranac, J. M. (2005). The copigmentation effect of sinapic acid on malvin: A spectroscopic investigation on colour enhancement. *J. Photochem. Photobiol. B: Biol., 78*, 223–228.

Drawert, F., Schreier, P., & Scherer, W. (1974). Gaschromatographisch-massenspektrometrische Untersuchung flüchtiger Inhaltsstoffe des Weines III. Säuren des Weinaromas. *Z. Lebensm. Unters. Forsch., 155*, 342–347.

Es-Safi, N.-E., Fulcrand, H., Cheynier, V., & Moutounet, M. (1999). Studies on acetaldehyde-induced condensation of (-)-epicatechin and malvidin 3-*O*-glucoside in a model solution system. *J. Agric. Food Chem., 47*, 2096–2102.

Fernandéz de Simon, B., Pérez-Ilzarbe, J., Hernández, T., Gómez-Cordovés, C., & Estrella, I. (1992). Importance of phenolic compounds for the characterization of fruit juices. *J. Agric. Food Chem., 40*, 1531–1535.

Figueiredo, P., George, F., Tatsuzawa, F., Toki, K., Saito, N., & Brouillard, R. (1999). New features of intramolecular copigmentation by acylated anthocyanins. *Phytochemistry, 51*, 125–132.

Fulcrand, H., Cameira dos Santos, P.-J., Sarni-Manchado, P., Cheynier, V., & Favre-Bonvin, J. (1996a). Structure of new anthocyanin-derived wine pigments. *J. Chem. Soc. Perkin Trans. 1: Organic and Bio-Organic Chemistry*, 735–739.

Fulcrand, H., Doco, T., Es-Safi, N.-E., Cheynier, V., & Moutounet, M. (1996b). Study of the acetaldehyde induced polymerisation of flavan-3-ols by liquid chromatography-ion spray mass spectrometry. *J. Chromatogr. A, 752*, 85–91.

Gambuti, A., Strollo, D., Ugliano, M., Lecce, L., & Moio, L. (2004). *trans*-Resveratrol, quercetin, (+)-catechin, and (-)-epicatechin content in south italian monovarietal wines: Relationship with maceration time and marc pressing during winemaking. *J. Agric. Food Chem., 52*, 5747–5751.

Garcia-Viguera, C., & Bridle, P. (1995). Analysis of non-coloured phenolic compounds in red wines. A comparison of High-Performance Liquid Chromatography and Capillary Zone Electrophoresis. *Food Chem., 54*, 349–352.

Goldberg, D. M., Yan, J., Ng, E., Diamandis, E. P., Karumanchiri, A., Soleas, G., & Waterhouse, A. L. (1995). A global survey of trans-resveratrol concentrations in commercial wines. *Am. J. Enol. Vitic., 46*, 159–165.

Goldberg, D. M., Tsang, E., Karumanchiri, A., Diamandis, E. P., Soleas, G., & Ng, E. (1996a). Method to assay the concentrations of phenolic constituents of biological interest in wines. *Anal. Chem., 68*, 1688–1694.

Goldberg, D. M., Ng, E., Karumanchiri, A., Diamandis, E. P., & Soleas, G. J. (1996b). Resveratrol glucosides are important components of commercial wines. *Am. J. Enol. Vitic., 47*, 415–420.

Gorham, J. (1995). *The Biochemistry of the Stilbenoids*. London: Chapman & Hall.

Guebailia, H. A., Chira, K., Richard, T., Mabrouk, T., Furiga, A., Vitrac, X., Monti, J. P., Delaunay, J. C., & Mérillon, J. M. (2006). Hopeaphenol: The first resveratrol tetramer in wines from North Africa. *J. Agric. Food Chem., 54*, 9559–9564.

Güntert, M., Rapp, A., Takeoka, G. R., & Jennings, W. (1986). HRGC and HRGC-MS Applied to wine constituents of lower volatility. *Z. Lebens. Unter. Forsch., 182*, 200–204.

Håkansson, A. E., Pardon, K., Hayasaka, Y., de Sa, M., & Herderich, M. (2003). Structures and colour properties of new red wine pigments. *Tetrahedron Lett., 44*, 4887–4891.

Hayasaka, Y., & Asenstorfer, R. E. (2002). Screening for potential pigments derived from anthocyanins in red wine using nanoelectrospray tandem mass spectrometry. *J. Agric. Food Chem., 50*, 756–761.

Hernández, T., Estrella, I., Pérez-Gordo, M., Alegría, E. G., Tenorio, C., Ruiz-Larrrea, F., & Moreno-Arribas, M. V. (2007). Contribution of malolactic fermentation by *Oenococcus oeni* and *Lactobacillus plantarum* to the changes in the nonanthocyanin polyphenolic composition of red wine. *J. Agric. Food Chem., 55*, 5260–5266.

Herrick, I. W., & Nagel, C. W. (1985). The caffeoyl tartrate content of white Riesling wines from California, Washington, and Alsace. *Am. J. Enol. Vitic., 36*, 95–97.

Herrmann, K. (1989). Occurrence and content of hydroxycinnamic and hydroxybenzoic acid compounds in foods. *Crit. Rev. Food Sci. Nutr., 28*, 315–347.

Holbach, B., Marx, R., & Zimmer, M. (2001). Bedeutung der Shikimisäure und des Anthocyanspektrums für die Charakterisierung von Rebsorten. *Lebensmittelchemie, 55*, 32–34.

Ito, T., Akao, Y., Yi, H., Ohguchi, K., Matsumoto, K., Tanaka, T., Iinuma, M., & Nozawa, Y. (2003). Antitumor effect of resveratrol oligomers against human cancer cell lines and the molecular mechanism of apoptosis induced by vaticanol C. *Carcinogesis, 24*, 1489–1497.

Jean-Denis, J. B., Pezet, R., & Tabacchi, R. (2006). Rapid analysis of stilbenes and derivatives from downy mildew-infected grapevine leaves by Liquid Chromatography-Atmospheric Pressure Photoionisation Mass Spectrometry. *J. Chromatogr. A, 1112*, 263–268.

Jeandet, P., Bessis, R., Sbaghi, M., & Meunier, P. (1994). Occurrence of a resveratrol b-D-glucoside in wine: Preliminary studies. *Vitis, 33*, 183–184.

Jeandet, P., Bessis, R., Sbaghi, M., Meunier, P., & Trollat, P. (1995). Resveratrol content of wines of different ages: Relationship with fungal disease pressure in the vineyard. *Am. J. Enol. Vitic., 46*, 1–4.

Jeandet, P., Breuil, A. C., Adrian, M., Weston, L. A., Debord, S., Meunier, P., Maume, G., & Bessis, R. (1997). HPLC analysis of grapevine phytoalexins coupling Photodiode Array Detection and Fluorometry. *Anal. Chem., 69*, 5172–5177.

Kammerer, D., Claus, A., Carle, R., & Schieber, A. (2004). Polyphenol screening of pomace from red and white grape varieties (*Vitis vinifera* L.) by HPLC-DAD-MS/MS. *J. Agric. Food Chem., 52*, 4360–4367.

Lamuela-Raventós, R. M., Romero-Pérez, A. I., Waterhouse, A. L., & de la Torre-Boronat, M. C. (1995). Direct HPLC analysis of *cis*- and *trans*-resveratrol and piceid isomers in Spanish red *Vitis vinifera* wines. *J. Agric. Food Chem., 43*, 281–283.

Landrault, N., Larronde, F., Delaunay, J.-C., Castagnino, C., Vercauteren, J., Mérillon, J.-M., Gasc, F., Cros, G., & Teissédre, P.-L. (2002). Levels of stilbene oligomers and astilbin in French varietal wines and in grapes during Noble Rot development. *J. Agric. Food Chem., 50*, 2046–2052.

Lao, C., López-Tamames, E., Lamuela-Raventós, R. M., Buxaderas, S., Torre-Bonat, M. C. (1996). Effect of grape pectic enzyme treatment on phenolics of white musts and wines. In J. Vercauteren, C. Chèze, M. C. Dumon, J. F. Weber (Eds.), *Polyphenols communications 96/XVIIth International Conference on Polyphenols* (Vol. 2, pp. 303–304). Bordeaux: Secrétariat du Groupe Polyphénols.

Lee, C. Y., Jaworski, A. (1989). Major phenolic compounds in ripening white grapes. *Am. J. Enol. Vitic., 40*, 43–46.

Marinos, V. A., Tate, M. E., Williams, P. J. (1992). Lignan and phenylpropanoid glycerol glucosides in wine. *Phytochemistry, 31*, 4307–4312.

Mark, L., Nikfardjam, M. S. P., Avar, P., & Ohmacht, R. (2005). A validated HPLC method for the quantitative analysis of *trans*-resveratrol and *trans*-piceid in Hungarian wines. *J. Chromatogr. Sci., 43*, 445–449.

Mazza, G., Miniati, E. (1993). *Anthocyanins in fruits, vegetables, and grains*. Boca Raton: CRC Press.

Miniati, E., Damiani, P., & Mazza, G. (1992). Copigmentation and self-association of anthocyanins in food model systems. *It. J. Food Sci., 4*, 109–116.

Monagas, M., Bartolomé, B., & Gómez-Cordovés, C. (2005a). Updated knowledge about the presence of phenolic compounds in wine. *Cr. Rev. Food Sci. Nutr., 45*, 85–118.

Monagas, M., Suárez, R., Gómez-Cordovés, C., & Bartolomé, B. (2005b). Simultaneous determination of nonanthocyanin phenolic compounds in red wines by HPLC-DAD/ESI-MS. *Am. J. Enol. Vitic., 56*, 139–147.

Naugler, C., McCallum, J. L., Klassen, G., & Strommer, J. (2007). Concentrations of *trans*-resveratrol and related stilbenes in Nova Scotia wines. *Am. J. Enol. Vitic., 58*, 117–119.

Nurmi, T., Heinonen, S., Mazur, W., Deyama, T., Nishibe, S., & Adlercreutz, H. (2003). Lignans in selected wines. *Food Chem., 83*, 303–309.

Okamura, S., & Watanabe, M. (1981). Determination of phenolic cinnamates in white wine and their effect on wine quality. *Agric. Biol. Chem., 45*, 2063–2070.

Ong, B. Y., & Nagel, C. W. (1978). High-Pressure Liquid-Chromatographic analysis of hydroxycinnamic acid - tartaric acid esters and their glucose esters in *Vitis vinifera*. *J. Chromatogr., 157*, 345–355.

Otteneder, H., Holbach, B., Marx, R., & Zimmer, M. (2002). Rebsortenbestimmung in Rotwein anhand der Anthocyanspektren. *Mitt. Klosterneuburg, 52*, 187–194.

Peña-Neira, A., Hernández, T., García-Vallejo, C., Estrella, I., & Suarez, J. A. (2000). A Survey of phenolic compounds in Spanish wines of different geographical origin. *Eur. Food Res. Technol., 210*, 445–448.

Perrone, G., Nicoletti, I., Pascale, M., De Rossi, A., De Girolamo, A., & Visconti, A. (2007). Positive correlation between high levels of ochratoxin A and resveratrol-related compounds in red wines. *J. Agric. Food Chem., 55*, 6807–6812.

Poussier, M., Guilloux-Benatier, M., Torres, M., Heras, E., & Adrian, M. (2003). Influence of different maceration techniques and microbial enzymatic activities on wine stilbene content. *Am. J. Enol. Vitic., 54*, 261–266.

Pozo-Bayón, M. A., Hernández, M. T., Martín-Álvarez, P. J., & Polo, M. C. (2003). Study of low molecular weight phenolic compounds during the aging of sparkling wines manufactured with red and white grape varieties. *J. Agric. Food Chem., 51*, 2089–2095.

Pryce, R. J., & Langcake, P. (1977). α-Viniferin: An antifungal resveratrol trimer from grapevines. *Phytochemistry, 16*, 1452–1454.

Puech, J. L., & Moutounet, M. (1988). Liquid chromatographic determination of scopoletin in hydroalcoholic extract of oak wood and in matured distilled alcoholic beverages. *J. Assoc. Off. Anal. Chem., 71*, 512–514.

Püssa, T., Floren, J., Kuldepp, P., & Raal, A. (2006). Survey of grapevine *Vitis vinifera* stem polyphenols by Liquid Chromatography-Diode Array Detection-Tandem Mass Spectrometry. *J. Agric. Food Chem., 54*, 7488–7494.

Quideau, S., Jourdes, M., Saucier, C., Glories, Y., Pardon, P., & Baudry, C. (2003). DNA Topoisomerase Inhibitor Acutissimin A and Other Flavano-Ellagitannins in Red Wine. *Angew. Chem. Int. Ed., 42*, 6012–6014.

Quideau, S., Jourdes, M., Lefeuvre, D., Montaudon, D., Saucier, C., Glories, Y., Pardon, P., & Pourquier, P. (2005). The chemistry of wine polyphenolic *C*-glycosidic ellagitannins targeting human topoisomerase II. *Chem. Eur. J., 11*, 6503–6513.

Rentzsch, M., Schwarz, M., Winterhalter, P., & Hermosin-Gutiérrez, I. (2007a). Formation of hydroxyphenyl-pyranoanthocyanins in Grenache wines: precursor levels and evolution during aging. *J. Agric. Food Chem., 55*, 4883–4888.

Rentzsch, M., Schwarz, M., & Winterhalter, P. (2007b). Pyranoanthocyanins – an overview on structures, occurrence, and pathways of formation. *Trends Food Sci. Technol., 18*, 526–534.

Ribeiro de Lima, M. T., Waffo-Téguo, P., Teissédre, P. L., Pujolas, A., Vercauteren, J., Cabanis, J. C., & Mérillon, J. M. (1999). Determination of stilbenes (*trans*-astringin, *cis*- and *trans*-piceid, and *cis*- and *trans*-resveratrol) in Portuguese wines. *J. Agric. Food Chem., 47*, 2666–2670.

Ribéreau-Gayon, P., Glories, Y., Maujean, A., & Dubourdieu, D. (2000). *Handbook of Enology. Volume 2: The chemistry of wine: stabilization and treatments*. Chichester: Wiley

Ricardo-da-Silva, J. M., Cheynier, V., Samsom, A., & Bourzeix, M. (1993). Effect of pomace contact, carbonic maceration, and hyperoxidation on the procyanidin composition of Grenache Blanc wines. *Am. J. Enol. Vitic., 44*, 168–172.

Rigaud, J., Cheynier, V., Souquet, J. M., & Moutounet, M. (1991). Influence of must composition on phenolic oxidation kinetics. *J. Sci. Food Agric., 57*, 55–63.

Rivas-Gonzalo, J. C., Bravo-Haro, S., & Santos-Buelga, C. (1995). Detection of compounds formed through the reaction of malvidin 3-monoglucoside and catechin in the presence of acetaldehyde. *J. Agric. Food Chem., 43*, 1444–1449.

Romero-Pérez, A. I., Lamuela-Raventós, R. M., Waterhouse, A. L., & de la Torre-Boronat, M. C. (1996). Levels of *cis*- and *trans*-resveratrol and their glucosides in white and rosé *Vitis vinifera* wines from Spain. *J. Agric. Food Chem., 44*, 2124–2128.

Salagoïty-Auguste, M. H., & Bertrand, A. (1984). Wine phenolics - analysis of low molecular weight components by High-Performance Liquid-Chromatography. *J. Sci. Food Agric., 35*, 1241–1247.

Santos-Buelga, C., Bravo-Haro, S., & Rivas-Gonzalo, J. C. (1995). Interactions between catechin and malvidin-3-monoglucoside in model solutions. *Z. Lebens. Unters. Forsch., 201*, 269–274.

Sarni-Manchado, P., Fulcrand, H., Souquet, J.-M., Cheynier, V., & Moutounet, M. (1996). Stability and color of unreported wine anthocyanin-derived pigments. *J. Food Sci., 61*, 938–941.

Sato, M., Suzuki, Y., Okuda, T., & Yokotsuka, K. (1997). Contents of resveratrol, piceid, and their isomers in commercially available wines made from grapes cultivated in Japan. *Biosci. Biotechnol. Biochem., 61*, 1800–1805.

Saucier, C., Jourdes, M., Glories, Y., & Quideau, S. (2006). Extraction, detection, and quantification of flavano-ellagtannins and ethylvescalagin in a Bordeaux red wine aged in oak barrels. *J. Agric. Food Chem., 54*, 7349–7354.

Schwarz, M., Wabnitz, T. C., & Winterhalter, P. (2003). Pathway leading to the formation of anthocyanin-vinylphenol adducts and related pigments in red wines. *J. Agric. Food Chem., 51*, 3682–3687.

Schwarz, M., Hofmann, G., & Winterhalter, P. (2004). Investigations on anthocyanins in wines from *Vitis vinifera* cv. Pinotage: Factors influencing the formation of pinotin A and its correlation with wine age. *J. Agric. Food Chem., 52*, 498–504.

Siemann, E. H., & Creasy, L. L. (1992). Concentration of the phytoalexin resveratrol in wine. *Am. J. Enol. Vitic., 43*, 49–52.

Singleton, V. L. (1987). Oxygen with phenols and related reactions in musts, wines, and model systems: observations and practical implications. *Am. J. Enol. Vitic., 38*, 69–77.

Singleton, V. L., Timberlake, C. F., & Lea, A. G. H. (1978). The phenolic cinnamates of white grapes and wine. *J. Sci. Food Agric., 29*, 403–410.

Singleton, V. L., Zaya, J., & Trousdale, E. K. (1986). Caftaric and coutaric acids in fruit of *Vitis*. *Phytochemistry, 25*, 2127–2133.

Sladkovský, R., Solich, P., & Urbánek, M. (2004). High-Performance Liquid Chromatography determination of phenolic components in wine using off-line Isotachophoretic pretreatment. *J. Chromatogr., A, 1040*, 179–184.

Somers, T., Vérette, E., & Pocock, K. (1987). Hydroxycinnamate esters of *Vitis vinifera*: Changes during white vinification, and effects of exogenous enzymic hydrolysis. *J. Sci. Food Agric., 40*, 67–78.

Stecher, G., Huck, C. W., Popp, M., & Bonn, G. K. (2001). Determination of flavonoids and stilbenes in red wine and related biological products by HPLC and HPLC-ESI-MS-MS. *Fresenius J. Anal. Chem., 371*, 73–80.

Sun, B., Ribes, A. M., Leandro, M. C., Belchior, A. P., & Spranger, M. I. (2006). Stilbenes: Quantitative extraction from grape skins, contribution of grape solids to wine and variation during wine maturation. *Anal. Chim. Acta, 563*, 382–390.

Timberlake, C. F., & Bridle, P. (1976). Interactions between anthocyanins, phenolic compounds, and acetaldehyde and their significance in red wines. *Am. J. Enol. Vitic., 27*, 97–105.

Tricard, C., Salagoity, M. H., & Sudraud, P. (1987). Scopoletin: A marker of storage in oak barrels. *Conn. Vigne Vin, 21*, 33–41.

Vanhoenacker, G., De Villiers, A., Lazou, K., De Keukeleire, D., & Sandra, P. (2001). Comparison of High-Performance Liquid Chromatography – Mass Spectroscopy and Capillary Electrophoresis – Mass Spectroscopy for the analysis of phenolic compounds in diethyl ether extracts of red wines. *Chromatographia, 54*, 309–315.

Vitrac, X., Castagnino, C., Waffo-Téguo, P., Delaunay, J.-C., Vercauteren, J., Monti, J.-P., Deffieux, G., & Mérillon, J.-M. (2001). Polyphenols newly extracted in red wine from Southwestern France by centrifugal partition chromatography. *J. Agric. Food Chem., 49*, 5934–5938.

Vitrac, X., Monti, J.-P., Vercauteren, J., Deffieux, G., & Mérillon, J.-M. (2002). Direct liquid chromatographic analysis of resveratrol derivatives and flavanonols in wines with absorbance and fluorescence detection. *Anal. Chim. Acta, 458*, 103–110.

Vitrac, X., Bornet, A., Vanderlinde, R., Valls, J., Richard, T., Delaunay, J.-C., Mérillon, J.-M., & Teissédre, P.-L. (2005). Determination of stilbenes (δ-viniferin, *trans*-astringin, *trans*-piceid, *cis*- and *trans*-resveratrol, ε-viniferin) in Brazilian wines. *J. Agric. Food Chem., 53*, 5664–5669.

Wildenradt, H. L., & Singleton, V. L. (1974). The production of aldehydes as a result of oxidation of polyphenolic compounds and its relation to wine aging. *Am. J. Enol. Vitic., 25*, 119–126.

Würdig, G., & Woller, R. (1989). *Chemie des Weines.* Stuttgart: Eugen Ulmer

Chapter 9D
Influence of Phenolics on Wine Organoleptic Properties

Celestino Santos-Buelga and Victor de Freitas

Contents

9D.1 Introduction

In his reference paper about the polymeric nature of wine pigments published in 1971, Somers stated "*the chemistry of wine phenolics, which entirely responsible for wine colour and tannin character, is still largely a matter of speculation*". Although relevant advances in the knowledge of the structures of wine pigments and tannins and the reactions taking place during aging have been made since then, that statement still applies. Many gaps still exist regarding the precise nature of the compounds and mechanisms that determine red wine color expression and astringency perception and their changes during wine life. These aspects have not been fully elucidated and their control continues basically as an empirical matter and still constitutes "*a major challenge in oenological research*" (Somers 1971).

C. Santos-Buelga (✉)
Unidad de Nutrición y Bromatología, Facultad de Farmacia, Universidad de Salamanca, Campus Miguel de Unamuno s/n, E-37007 Salamanca, Spain
e-mail: csb@usal.es

M.V. Moreno-Arribas, M.C. Polo (eds.), *Wine Chemistry and Biochemistry*,
DOI 10.1007/978-0-387-74118-5_20,

Among polyphenolic compounds, two types of flavonoids, the anthocyanins and flavanols (i.e., catechins, proanthocyanidins, condensed tannins), are particularly relevant to the quality of red wines, as they are key compounds for color definition and astringency. Other flavonoids such as flavonols may have some influence on color and bitterness, although they are present in red wines in much lower amounts. Phenolic acids and hydrolysable tannins, released from barrel wood, may also have an influence on wine taste and color, and hydroxycinnamoyl derivatives from grape must are involved in the oxidative browning of white wines together with flavanols. Besides, some of these perceptions may be modified by other sensory characteristics (e.g. sourness, sweetness) related to other wine components (Preys et al. 2006).

Wine phenolic composition depends on the original grape and on oenological practices and storage conditions. Anthocyanins and tannins are located in the solid parts of the cluster from where they are released during winemaking. Further, they undergo various enzymatic and chemical reactions as the wine is made and aged. The new compounds formed often exhibit sensory properties different from those of their precursors, hence modifying the quality of the wines (Cheynier et al. 2006).

In this chapter we will try to revise what is known about relationships between phenolic composition and sensory properties of red wine, namely color and astringency, although it is quite possible that in the end more questions than answers will be found.

9D.2 The Color of Red Wines

9D.2.1 Origin and Components

The initial purple-red color of young red wines arises from the anthocyanins extracted from grape skins, while during aging this color shifts to a more reddish brown hue, mainly due to progressive structural changes of anthocyanins. These changes occur through different mechanisms that have been revised by Monagas and Bartolome in a previous chapter of this book. Although these chemical transformations and the corresponding wine color changes have been largely studied and demonstrated to occur during wine life, the interpretation of wine color in terms of phenolics continues to be a pending matter, as it is affected by the numerous factors that influence wine phenolic composition, the own reactive nature of phenolic compounds, the anthocyanin equilibria, and the external and internal conditions of the wines (temperature, oxygen access, pH, acetaldehyde, or SO_2 content) (Monagas et al. 2006).

The anthocyanins are structurally dependent on the conditions and composition of the media where they are dissolved and suffer interactions among them and with other compounds that influence their structural equilibria and modify their color. Anthocyanins are usually represented as their red flavylium cation, but in aqueous media this form undergoes rapid proton transfer reactions, leading to blue quinonoidal bases, and hydration, generating colorless hemiketals in equilibrium with chalcone structures. The proportion of each form is determined by the pH

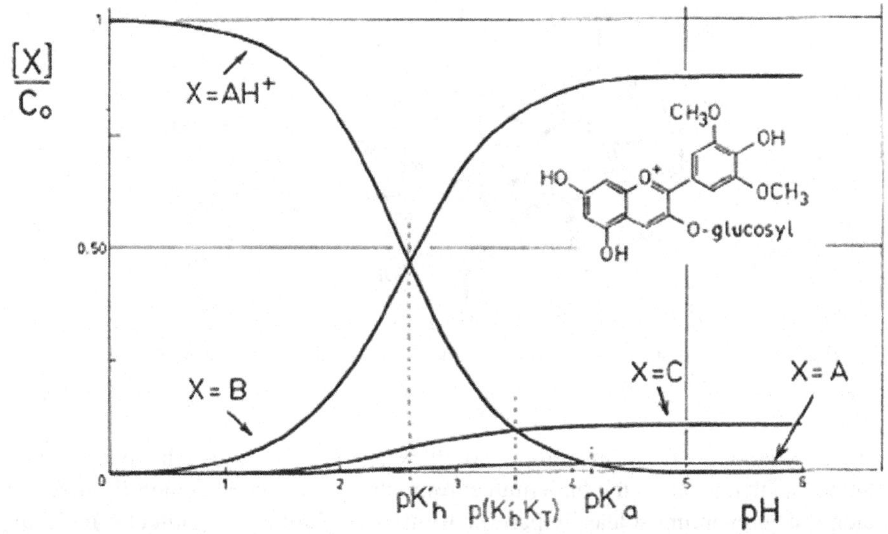

Fig. 9D.1 Equilibrium distribution at 25 °C among structural forms of malvidin 3-glucoside as a function of pH. AH+, flavylium cation; A, quinonoidal base; B, hemiketal; C, chalcone (from Brouillard 1982, with permission from Elsevier)

value, and the flavylium ions predominate only in very acidic solutions (Brouillard et al. 1977). Weakly acidic aqueous or hydroalcoholic solutions of pure anthocyanins show a poor color or are practically colorless depending on the pH value, due to the predominant presence of colorless hydrated hemiketal forms (Fig. 9D.1). In red wines the usual presence of sulfites, which form adducts with the anthocyanin flavylium in the same way as water to give colorless products, may provide further discoloration. Even in these circumstances, red wines continue showing an intense red color, indicating that pigment stabilizing mechanisms exist that provide a sufficient presence of anthocyanin chromophores to ensure the definition and the maintenance of the color even over years. It is generally assumed that such stabilization is achieved through two major mechanisms: (1) the non-covalent associations of anthocyanin chromophores with other wine compounds, process called copigmentation, and (2) the progressive substitution of grape anthocyanins by more stable pigments (Cheynier et al. 2006). The first process would have particular importance in young red wines, whereas the second one mainly participates in the color of aged red wines.

9D.2.2 Copigmentation

9D.2.2.1 Influence in the Color of Red Wines

The copigmentation phenomenon consists of hydrophobic interactions (π-π stacking) between the planar polarizable nuclei of the colored forms of the anthocyanins (i.e., flavylium ion and quinonoidal forms) with other organic molecules

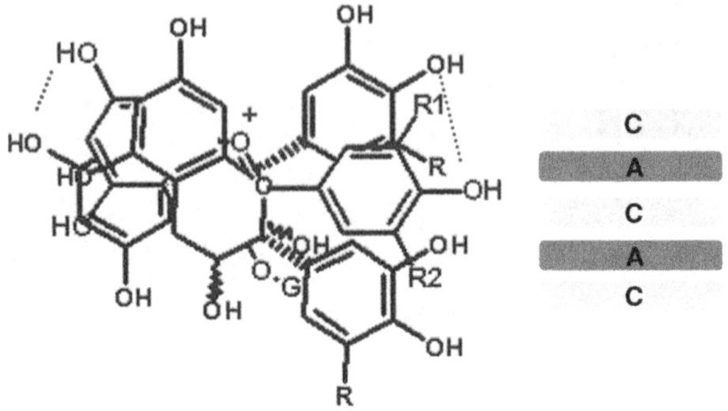

Fig. 9D.2 Vertical π–π stacking between an anthocyanin (A) and a copigment (C)

(i.e., copigments). Copigmentation complexes adopt a sandwich configuration (Fig. 9D.2) that protects the flavylium chromophore from the nucleophilic attack of water, thus preventing at least in part the formation of colorless hemiketal and chalcone forms. The final result is that the anthocyanin solutions show a more intense color that theoretically could be expected according to the pH value of the media. Occasionally a bathochromic effect can also be produced due to the proton transfer equilibrium between the flavylium cation and the quinonoidal base and/or preferent association between quinonoidal forms and copigments. Depending on the type of anthocyanins and copigments and their concentrations there will be variations in color hue and intensity. Thus at the same time color stabilization and variation can be obtained (Brouillard et al. 2003).

It is well known that the copigmentation phenomenon occurs in plant tissues offering an explanation for the large color variations in flowers and fruits; excellent reviews on the subject have been published by Goto and Kondo (1991) and Brouillard and Dangles (1993). However, the influence of copigmentation in the color of red wines is subjected to more controversy, due to the lower anthocyanin concentrations existing in wine compared to those in plant vacuoles and the dissociating effect of the ethanol on the copigmentation complexes (Dangles and Brouillard 1992). Despite this, it is assumed that it must also constitute a relevant process for the definition of the color in young red wines (Boulton 2001).

Many of the studies about copigmentation have been carried out in model systems and they are not always strictly applicable to wine. A method for the estimation of the copigmentation effect in red wines, based on the comparison of the absorbance at 520 nm before and after disruption of the copigmentation complexes by dilution with a wine-like solution, was proposed by Boulton (1996). That method has the inconvenient that only takes into account the variation at λ_{max} of the flavylium ion in the visible region, thus ruling out the modifications that the copigmentation induces in other regions of the visible spectrum and that logically have also qualitative and quantitative influence on wine color. Gonnet (1998) proposed another approach based on the colorimetric analysis in the CIELAB color space that consider the spectral changes over the complete range of visible wavelengths.

This latter methodology has hardly been used to evaluate the strength of the copigmentation effect, although it has been employed to characterise the color of red wines.

By using Boulton's method diverse authors (e.g., Darias-Martin et al. 2007; Hermosin, 2003; Levengood and Boulton 2004; Schwarz et al. 2005) have concluded that copigmentation could account for up to half of the observed color of young red wines, depending on the levels of colorless phenolic compounds or cofactors present, which could even have more importance than anthocyanins to define the color in those wines.

During wine life a decrease is produced in the concentrations of both anthocyanins and cofactors in the wines, so that the extension of the copigmentation also decreases up to having no influence on the color of aged wines. The results obtained about the contribution of the copigmentation to the color of red wines differ among studies. Hermosin et al. (2005), in assays with Spanish wines of the varieties Cencibel (*syn* Tempranillo), Cabernet Sauvignon and Syrah, found that copigmentation accounted for 32–45% of the color of red wines recently elaborated to decrease to 20–34% after three months and be practically negligible after nine months (0–5%). In studies carried out with wines made with grape varieties of the Canary Islands (Listan Negro and Negramoll), Darias-Martin et al. (2007) calculated that copigmentation contributed 22% and 19% of color in red wines of one and two years of age, respectively. Similar results were obtained by Lorenzo et al. (2005) that estimated that around 18% of copigmentation still existed in their wines after nine months of oak aging. The differences in the results among authors might be explained by the variations in wine composition, as determined by grape characteristics and the winemaking processes, which provoke differences in the type and levels of anthocyanins and cofactors that affect the extension and stability of the copigmentation process. Further, the conditions of storage and aging, as well as the scale of the vinifications should also play a determinant role on the results obtained.

The influence of the ethanol on the copigmentation was studied by Hermosin (2003). In assays with de-alcoholised wines further reconstituted found that as ethanol content increased the percentage of copigmentation diminished following a quadratic relation; with ethanol contents between 12% and 14% the wines maintained about three quarters of the copigmentation observed at 0% ethanol. This observation confirms the dissociating role of ethanol on copigmentation complexes but also shows that copigmentation still occurs in red wines. In model solutions, the increase in ethanol in the medium has been seen to induce a bathochromic shift in maximum wavelength in the visible region of the spectrum in relation to water (Brouillard et al. 1991). However, this effect was hardly observed in the reconstituted wines, when only a slight bathochromic shift was observed in wines with ethanol content higher than 20–22% (Hermosin 2003), suggesting that only some contribution of this effect to the blue hues of red wines containing higher ethanol levels (e.g., fortified red wines) could be expected.

Copigmentation may not only affect color definition of red wines but also influence its stability. Chemical reactions that take place in wines, like oxidation and polymerization, are likely to be related with the free concentrations of phenolic substrates, since part of them is involved in the copigmentation slower reaction rates and evolution might be expected in wines with greater copigmentation (Boulton 2001). On the other hand, some authors have suggested that the copigmentation could act as a first stage in the formation of new pigments and determine both the type of compounds formed and their levels, thereby affecting the color of aged red wines (Brouillard et al. 2003; Brouillard and Dangles 1994). Likewise, copigmentation could also influence the extraction of pigments from the grape and their retention in the must during winemaking (Boulton 2001). Some of these points will be revised below.

Different types of copigmentation can be distinguished: *intermolecular copigmentation* when anthocyanins associate with other molecules, *self-association* when it involves anthocyanins themselves, and *intramolecular copigmentation* when the anthocyanin chromophore interacts with other residues of its own molecule. This latter type of copigmentation is restricted to anthocyanins which are acylated by phenolic acids linked to the anthocyanidin through a suitable spacer, so that it allows the molecule to fold in such a way that the aromatic acyl group(s) can interact with the flavylium nucleus and protect it from hydration (Dangles et al. 1993). *Vitis vinifera* anthocyanins have only one sugar moiety and at most one hydroxycinnamoyl residue (i.e., *p*-coumaric acid or less frequently caffeic acid). Although the possibility of some intramolecular copigmentation cannot be ruled out, they are not as flexible as more complex flower anthocyanins having at least two sugar molecules between the anthocyanidin and the acyl moiety allowing a suitable folding of the structure. Furthermore, the contents of acylated anthocyanins are reduced in most grape varieties and are even worse represented in red wines. In these circumstances, it does not seem that the intramolecular copigmentation can constitute an important mechanism for the enhancement of the color in red wines, although acylated anthocyanins may be more involved in intermolecular copigmentation than non-acylated ones (Boulton 2001).

A particular case of copigmentation is self-association, consisting of a positive deviation from Beer's law that occurs on increasing the concentration of anthocyanins in the medium. Although it is accepted that this phenomenon contributes to the color expression in flowers (Hoshino 1991; Hoshino et al. 1981), its participation in the color of red wine is matter of discussion. Self-association at acidic pH was found to occur between the flavylium cation itself and with its *Z*-chalcone (Houbiers et al. 1998). The interaction with *Z*-chalcone forms cannot be expected to explain self-association in red wines, given the low existing concentrations of these form that should exist in wine. However, it might be possible the interaction among flavylium forms of different anthocyanins that can act as cofactors for each other. Somers and Evans (1979) considered that self-association could be responsible for much of the non-Beer's law behaviour in red wines, whereas Boulton (2001), based on studies carried out by circular dichroism, concluded that self-association was not much relevant to the enhancement of color in young red wines, but intermolecular copigmentation between anthocyanins and different phenolic compounds would be mainly responsible for the nonlinear color deviations observed. Assays carried out in our laboratory with anthocyanins in a wine concentration range (50–600 mg/L) concluded that self-association could be responsible between 8% and up to 60% of the absorbance increase at 520 nm observed in wine-like solutions, depending on the type and concentration of the anthocyanins involved (González-Manzano et al. 2008b). Support to the existence of self-association in wines was also be provided by the observations of di Stefano et al. (2005) that found that young red wines deprived of the classical cofactors continued to deviate from the Beer's law. This situation should change in the presence of copigments. In model assays carried out with wine anthocyanins in the presence of different mixtures of flavanols, it was observed that self-association still contributes in some extent to the increase the absorbance in the solutions (Gonzalez-Manzano 2007). Nevertheless, it is necessary to take into account that flavanols are poor anthocyanin copigments and that in wines more efficient cofactors also exist, such as flavonols or hydroxycinnamic acids, which could favour intermolecular copigmentation with regard to self-association. Further studies would be thus necessary to assess the actual effect of self-association in the color of red wines.

Another process that influences the color of the anthocyanins and that is sometimes included as copigmentation is metal complexation. Metal cations like iron, aluminium or magnesium can form complexes with anthocyanins that have free *o*-hydroxyl groups in ring B. The metal selectively links with the quinonoidal forms of the anthocyanin, modifying anthocyanin equilibria towards these structures and provoking a change of the color to more violet hues (Dangles et al. 1994). Nevertheless, metal complexation should not play a relevant role in wine color since malvidin 3-glucoside, the majority anthocyanin in red wines, cannot form this type of complex; in addition, low levels in metals should exist in wine to avoid *casses* (Riberéau-Gayón et al. 2000). At present, there is no evidence that the levels of metal cations in wine are at all correlated with color expression (Boulton 2001).

9D.2.2.2 Type of Copigments

Different types of compounds have been evaluated as potential anthocyanin copigments, including alkaloids, amino acids, nucleotides, carbohydrates or phenolic compounds (Asen et al. 1972; Brouillard et al. 1991; Mistry et al. 1991). Many of these compounds can be found in wine, although some of them should not be expected to have a relevant contribution to the copigmentation phenomenon in red wine, due to their low levels or their comparatively poor ability to act as anthocyanin copigments. Thus, sugars (i.e., fructose and glucose) have been found to produce no change or slightly increase the absorbance of anthocyanin solutions even at high concentrations (50% in water) (Lewis et al. 1995). Similar observations were made for polysaccharides like pectins (Lewis et al. 1995) or mannoproteins. Guadalupe et al. (2007) observed that wines elaborated with addition of commercial mannoproteins or inoculation of must with yeast over-expressing mannoproteins had lower values of color intensity than controls at the end of fermentation and that these differences increased during malolactic fermentation and aging; this observation would question the usual belief that polysaccharides may protect wine color.

Benzoic acids (gallic acid, protocatechuic acid, etc.), which can be transferred from the grape, released from barrels wood or derived from anthocyanin breakdown, are considered poor anthocyanin copigments compared with other phenolic compounds (Brouillard et al. 1991; Eiro and Heinonen 2002). By contrast, some hydrolysable tannins could be potentially good copigments (Mistry et al. 1991), but their presence in wine is restricted to wood-aged wines, in which the concentration of anthocyanins is already reduced and copigmentation would not play a relevant role on the color. Furthermore, although notable variations may exist depending on the type and age of the barrel, the contents of this type of tannins in wine are usually very low. Thus, concentrations of a few mg/L were found for ellagic acid and ellagitannins after aging in oak barrels of different origins (Fernandez de Simon et al. 2003; Perez-Prieto et al. 2003), French oak and new barrels releasing more ellagitannins than American oak.

Among red wine components, flavonoids (in particular, flavonols and flavanols) and hydroxycinnamoyl derivatives appear as the more promising compounds to act as anthocyanin copigments. In assays carried out in model solutions, flavonols behave as powerful anthocyanin copigments (Asen et al. 1972; Davies and Mazza 1993; Baranac et al. 1996, 1997a, 1997b), whereas flavan-3-ols are comparatively poor cofactors with hydroxycinnamic acids showing intermediate behaviors (Brouillard et al. 1991; Gomez-Miguez et al. 2006). However, studies conducted directly in red wines are scarce, so that the actual influence of all these compounds on their color must rather be speculated.

The levels of the different compounds able to participate in the copigmentation are extremely variable in red wines, depending on the characteristics of the grape and the winemaking process. In general, it is assumed that the contents of flavonols and hydroxycinnamoyl derivatives are much lower to those of flavanols, which, as either monomers (catechins) or polymers (i.e., condensed tannins, proanthocyanidins), would constitute major phenolic components in red wines (Cheynier et al. 2003). Flavonols are found in grape as glycosides (mainly in the form of quercetin 3-glucuronide, quercetin 3-glucoside and myricetin 3-glucoside), which are transferred to wine during maceration. In the course of wine life minor amounts of the aglycones are released as a result of the cleavage of their glycosidic linkages. Contents of total flavonols in young red wines range from traces to values around 100 mg/L, depending on the grape variety and winemaking procedures (Hermosin et al. 2005; Hollman and Arst 2000; McDonald et al. 1998); however, a rapid decrease in their levels is produced, thereby after nine months between 50% and 80% of their original content is lost (Hermosin et al. 2005). Hydroxycinnamic acids (caffeic, *p*-coumaric, and ferulic acids) appear in wine as a consequence of the hydrolysis of grape hydroxycinnamoyltartaric esters (i.e., caftaric, coutaric and fertaric acids). Further, they can also be released from the barrel wood and minor amounts are also produced by cleavage of the acylated anthocyanins. Thus, contrary to flavonols, their levels could increase during red wine life. Concentrations of total hydroxycinnamic derivatives in red wines seem to be in the same order of magnitude as flavonols (some tens of mg/L) (Clifford 2000; Cheynier et al. 2003). Flavanols are the most abundant phenolic compounds in red wines, with contents that may reach some hundreds of mg/L (Santos-Buelga and Scalbert 2000) or be even in the level of g/L when considered the badly characterized and quantified proanthocyanidin polymers (Cheynier et al. 2003). Thus, in spite of being in general worse copigments than flavonols or hydroxycinnamates, it is assumed that flavanols should play a role in copigmentation and color of red wine. Indeed, the importance of tannins to provide support to color and ensure its stabilization during aging in red wines is a fact empirically assumed in enology.

The comparatively smaller ability of flavanols to act as copigments is attributed to their non-planar structure that impedes a closer approach to the anthocyanin and reduces the potential surface area available for hydrophobic stacking. However, important differences can exist among flavanols regarding their efficiency as anthocyanin copigments. Thus, regarding flavanol monomers, epicatechin is a better copigment than catechin (Brouillard et al. 1991; Liao et al. 1992; Mirabel et al. 1999), which is explained by the preferential quasi-equatorial conformation of its ring B that allows both aromatic nuclei to be approximately coplanar and participate in the π-π stacking with the flavyilum or quinonoidal forms of the anthocyanins (Liao et al. 1992). On the other hand, procyanidin dimers with C4-C6 interflavanic linkages seem to be better copigments than their respective C4-C8 dimers (Berke and de Freitas 2007), possibly due to the fact that they possess a more flexible and open conformational structure that allows them better interaction with the anthocyanin (Fig. 9D.3) (de Freitas et al. 1998). Moreover, galloylation at C-3 of the catechin units improves the ability of flavan-3-ols to act as copigment (Berke and de Freitas 2005; Liao et al. 1992). The substantial enhancement of copigmentation in the case the dimer B2-3′-*O* gallate compared to its analogue dimer B2 could be explained by the presence of the well exposed planar π-π systems forming a pocket into which the anthocyanin may intercalate, thereby offering an important interaction (Fig. 9D.4) (Berke and de Freitas 2005).

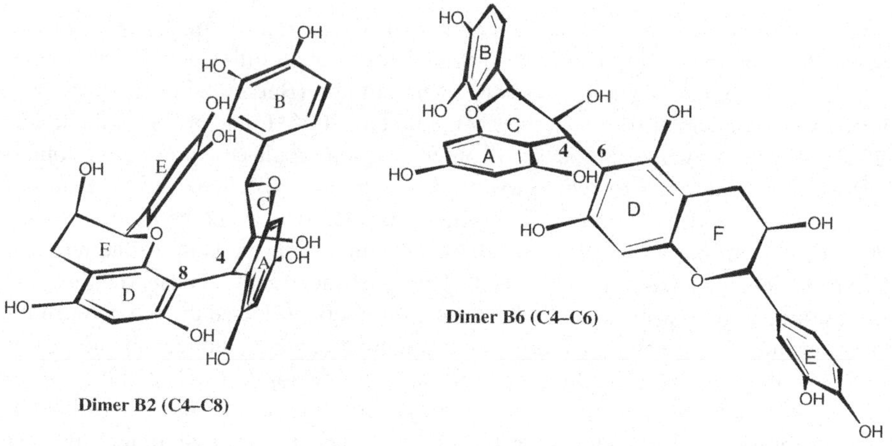

Fig. 9D.3 Representations of the preferred conformations of procyanidin dimers B2 (epicatechin-4,8-epicatechin) and B6 (epicatechin-4,6-epicatechin) (adapted from de Freitas et al. 1998)

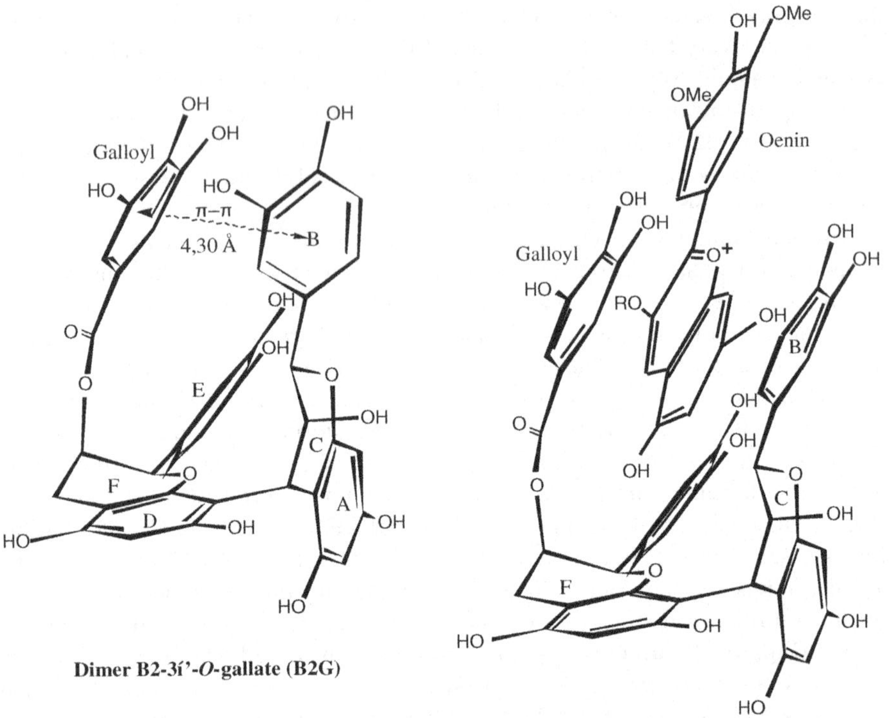

Dimer B2-3í'-*O*-gallate (B2G)

B2G/oenin sandwich-type complex

Fig. 9D.4 Preferred conformation of dimer B2-3′- *O*-gallate (B2G) and suggested conformational arrangements of the oenin intercalated between the galloyl ester group and catechol ring B of B2G (adapted from Berke and de Freitas 2005)

The studies about copigmentation ability of flavanols have mostly been carried out with monomers and dimers, but little is known about efficiency of compounds with greater degree of polymerisation. Some authors (Berke and de Freitas 2007; Escribano-Bailon et al. 1999; Gomez-Miguez et al. 2006) found procyanidin dimers B3 and B2 to be worse anthocyanin copigments than their constituting monomers (catechin and epicatechin), which was explained by the existence of conformational restraints imposed by the increase in the molecular size, whereas Malien-Aubert et al. (2002) did not observe great differences among flavanols up to the tetramer regarding their efficiency as copigments. However, these authors observed that with the passage of time procyanidin trimers and tetramers protected the red color in the solutions more efficiently than monomers and dimers, which induced the formation of xanthylium pigments leading to a yellowing in the solutions. The ability of the flavanols to maintain the color of the anthocyanin solutions was also determined by their different stability. Thus, the procyanidin B3 was more susceptible to thermal degradation than B2 and produced a faster alteration in the color (Malien-Aubert et al. 2002). More recently, in studies carried out with condensed tannin analogues consisting of ethyl-bridged oligomers containing up to six catechins units obtained by acetaldehyde-mediated condensation, it was found that dimers and trimers were more effective as malvidin 3-glucoside copigments than monomers and compounds with higher degree of polymerisation (Gonzalez-Manzano et al. 2008a). All in all, more studies are required to decide about the influence of the polymerisation degree on the copigmentation effect.

The greater or lesser ability of the distinct compounds to act as anthocyanin copigments does not ensure an adequate maintenance of the color. In assays carried out in model solutions, it was observed that, although flavonols are good copigments, they might not provide a suitable preservation of the color with the passage of time. Flavonols induce a pure copigmentation effect, which decreases as their concentration in the medium declines, being apparently replaced by other cofactors, such as hydroxycinnamic acids and flavanols, which are able to react with the anthocyanins to yield new pigments that help the color be kept (Hermosin et al. 2005; Gomez-Miguez et al. 2006).

Pre-fermentation addition of different copigments during winemaking has been tried by some authors as a strategy to improve wine color. Darias-Martin et al. (2001) in experimental red wines made with authoctonous grapes of the Canary Islands (i.e., '*Listan Negro*' and '*Negramoll*') observed enhancements of the color (measured as the increase in the absorbance at 520 nm) of 13% and 60% in the wines added with catechin and caffeic acid (120 mg/L each), respectively, in relation to the controls. However, the wines added with catechin lost their red color more rapidly than the control wines, whereas those added with caffeic acid still showed 23% more color than the controls after 210 days of storage. In a further assay, the same group found that the enhancement in the color was greater when the amount of caffeic acid added increased (in the range 120–960 mg/L), reaching up to 111% of increase in the absorbance at 520 nm; this effect is more noticeable in wines made with cultivars less rich in color (Darias-Martín et al. 2002).

Bloomfield et al. (2003), in wines made adding caffeic and *p*-coumaric acids to '*Cabernet Sauvignon*' and '*Pinot noir*' grapes after crushing, also found an enhancement of the color due to copigmentation, with *p*-coumaric acid being more effective than caffeic acid. Nevertheless, the fact that these authors observed a hypsochromic shift in the wavelength of maximum absorption in the visible region of the wines spectra indicated that not only a copigmentation effect was produced but also that the presence of newly-formed pyranthocyanins pigments should be accounting for the color.

In wines made with pre-fermentative addition of rutin, caffeic acid or *p*-coumaric acid to '*Cabernet Sauvignon*' and '*Tempranillo*' grapes, Schwarz et al. (2005) observed that only the addition of rutin was always accompanied by a significant hyperchromic shift at 520 nm (9% for '*Tempranillo*' and 35% for '*Cabernet Sauvignon*' wines) explained by a copigmentation effect. However, the addition of hydroxycinnamic acids had variable effects depending on structure and the grape cultivar. A hyperchromic effect was only found when *p*-coumaric acid was added to '*Cabernet Sauvignon*' wine (22% increase of *A*520), whereas the pre-fermentation addition of caffeic acid always resulted in hypochromic effects, contrary to the results previously obtained by other authors. In further analyses carried out after four months of storage, the percentage of color due to copigmentation decreased in all the wines, and after nine months no color attributable to copigmentation was noticed.

Assays were also made with the addition of grape seeds (supplementary amounts of 60 g/L) during fermentation, so as to provide supplementary sources for the extraction of proanthocyanidins in '*Garnacha*', '*Tempranillo*' and '*Vranac*' red wines (Kovac et al. 1992; Kovac et al. 1995). In all cases, the wines obtained showed greater contents of total phenolics and free anthocyanins, as well as an increase in their color intensity and a stabilization of wine color through time. However, these benefits were not produced when higher amounts of seeds were added (120 g/L), which was explained by an increased retention of the coloring matter by pomace. The enrichment in seeds during wine making was also seen to lead to wines with a greater concentration of total anthocyanins in red wines obtained by microvinification by Canals et al. (2008), although that increase was slight and not statistically significant. Nevertheless, the amounts of seeds added by these authors (up to 384 g added to 12 kg of grapes) were lower than those used in the studies of Kovac and coworkers. Similar results were also obtained in wines subjected to delestage, a practice to strength maceration (Canals et al. 2008).

An interesting observation made by all these authors was that the initial increase in the color of the wines made with pre-fermentative addition of cofactors of either nature (i.e., flavonols, hydroxycinnamic acids or flavanols) was not only explained by a copigmentation effect but also by an enhanced anthocyanin extraction from grapes. This seems to support the views of Boulton (2001) that color extraction and retention in the wine is not only depending on the concentrations of pigments in the berry, but also on the levels of cofactors. Hence the final contents of anthocyanins in the wine would be fixed once saturation has been reached for the major cofactors, the extent of this equilibrium being predetermined by the composition of the grape. Thus, according to this author the factors controlling the solubility and retention of pigments in young wines would be more important than contacting methods in determining wine color. The incorporation of extra copigments, either by external addition or co-vinification of different grape varieties contributing additional cofactors, would shift the extraction equilibrium out causing more pigment to move from the skins into the wine. Further, the increase in the anthocyanin concentration in the must would induce more copigments to be extracted from the grapes; thus the levels of other phenolics in the wines would also be increased. Experimental support to this view has been contributed by Lorenzo et al. (2005) in studies where '*Monastrell*' grapes were co-fermented in the presence of '*Cabernet Sauvignon*' and '*Merlot*' grapes.

The increased phenolic extraction provided by the copigmentation should have implications not only on the color but also on other sensory properties, as well as on the redox characteristics of the wines, as many of these compounds show recognized antioxidant and oxygen scavenger capacities. It is known that flavanols display different degrees of astringency and bitterness. According to Boulton (2001), their rates of binding to saliva proteins and receptors in the mouth could be related to their free concentrations rather than to their total concentrations. Therefore, although the anthocyanins themselves may not contribute to taste, they could somehow modulate astringency and bitterness in wines perception through the retention of flavanols and other phenolics in the copigmentation complexes, thus reducing the levels of their free forms. In spite of the knowledge accumulated over the last few decades there are still many gaps to fill about copigmentation and its implications on color and other sensory properties of the wines. Among them, it is necessary to know more about the efficiency as copigments of the different grape components, which anthocyanins are better candidates to participate in the process, which are the optimal total and relative concentrations of anthocyanins and cofactors to provide a suitable color and, in summary, which is the more suitable phenolic profile of grapes and wine to favor a stronger copigmentation and a good and more stable color.

9D.2.3 Changes During Wine Maturation and Aging

9D.2.3.1 Nature of the Pigment Material

It is very well known that during wine life a shift in the color of red wines is produced from the purple-red hue of young red wines to orange-brick hues typical of the aged ones. This change is explained by the progressive displacement of grape anthocyanins by more stable pigment structures that are less sensitive to the pH value and sulfur dioxide bleaching than anthocyanins, thus being able to express their color in wine conditions at which the anthocyanins are mostly colorless (Vivar-Quintana et al. 2002). These pigments have been classically designed as "polymeric pigments" that are associated with the non-dialysable coloring material that remains in the aqueous phase after extraction of the red wine by isoamyl alcohol (Somers 1971). They were long believed to result from reactions involving anthocyanins and/or tannins, either direct or mediated by acetaldehyde (Haslam 1980; Jurd 1969; Somers 1971; Timberlake and Bridle 1976), and over the last two decades many of these pigments have been positively identified in wines. A comprehensive relation can be found in the article of Alcalde-Eon et al. (2006) where more than one hundred derived pigments detected in red wine are listed. However, all of the identified pigments are rather relatively small molecules, thus raising doubts about the actual polymeric nature of the wine pigments.

A method for the fractionation by column chromatography of the coloring material of red wines in three fractions corresponding to anthocyanin monomers, and the so-called "red polymers" and "brown polymers" was proposed by Bourzeix et al. (1980), who roughly estimated the molecular weight of the red polymers in 560, meaning that they would not consist of more than two flavonoid units. No indication about the size or nature of the brown polymers was made. It is usually assumed that polymeric pigments are responsible for the "humps" observed in the RP-HPLC chromatograms obtained for matured and aged red wines. Fractionation of red wines to separate and collect the hump material in order to characterize it further has been tried by some authors (Mateus et al. 2002a; Remy et al 2000; Salas et al. 2005;

Vivar-Quintana et al. 2002). Using thiolysis, mean polymerization degrees between 2.9 and 10.7 have been obtained for the compounds present in those humps (Remy et al 2000; Salas et al. 2005), although not only pigments but also proanthocyanidins and colorless compounds present in the same extracts must account for those results. According to our knowledge, the largest anthocyanin-tannin oligomers detected in red wines correspond to colorless adducts consisting of one anthocyanin moiety linked to a procyanidin trimer (Salas et al. 2005), although the possibility of larger adducts, either colored or not, is hypothesized by the same group. Nonetheless, as far as we know, no actual polymeric pigments have been characterized in red wines, which does not mean that their presence should be discarded, as they may instable and/or escape the usual techniques for separation and analysis. On the other hand, the existence of various centres of asymmetry in the derived pigments that contain tannin moieties makes possible the occurrence of many isomers whose difficult chromatographic separation could also be accounting for the humps observed in the HPLC profiles of matured and aged red wines.

In assays carried out by our group, it was found that when 'hump' fractions previously separated from red wines were further analyzed by RP-HPLC after changing their pH to neutrality or by simple diluting them, they suffered a notably decrease in their size at the same time that peaks of monomeric anthocyanins (mostly malvidin 3-glucoside) not observed in the original hump fraction appeared in the chromatograms (non-published results). This observation allows speculating that, although the existence of large pigment structures cannot be ruled out, the 'polymeric pigment material' of red wines could be constituted, at least in part, by anthocyanins retained physically rather than chemically into tannin clusters. This had already been postulated by Somers (1966), who found that wine tannin material separated by gel filtration on Sephadex gels contained pigment molecules that were easily released by mild acid treatment, indicating that weak linkages were involved in the retention of anthocyanins into the tannin matrix.

The pyranoanthocyanins in their diverse forms are the derived pigments more usually found in red wines. They have been estimated to constitute about 70% of the total derived pigments present in a two-year-old red wine (Alcalde-Eon et al. 2006) and account for up to 50% of overall pigment material in a five-year-old wine (Boido et al. 2006). Anthocyanin-ethyl-flavanol derivatives and, more recently, the products of the direct condensation between anthocyanins and tannins are also commonly detected in red wines, but their concentrations are lower than those of pyranoanthocyanins and their relative contribution decrease in older wines (Alcalde-Eon et al. 2006). When evaluated by the peaks observed in the HPLC chromatograms, the levels of all these pigments are apparently too low to explain the color of the wines. The fact that they are totally or partially resistant to the discoloring effect of the low acidity and the presence SO_2 and, therefore, able to express their color in wine conditions, contrary to the anthocyanins, might in part explain this contradiction (Vivar-Quintana et al. 2002). In addition, the sum of the contribution of the minor amounts of the diverse existing pigments and possible synergisms among them might also account for some of the color expression. Despite this, it is assumed that a significant part of color expressed by aged red wines is still not satisfactorily explained and even in the case of the identified pigment families the actual contribution of each of them is not yet well established. Thus, further studies are required to elucidate the actual contribution of each of them to the definition of the color of red wines.

9D.2.3.2 Pyranoanthocyanins

As previously revised in this same book, this type of pigments contains an additional pyran ring attached to positions 4 and 5 of the anthocyanin structure. Three families of pyranoanthocyanins have been described according to the nature of the substituents on that pyran ring as influenced by the type of precursors involved in their formation, either products of the microbial metabolism (e.g., pyruvic acid, acetaldehyde or vinylphenols), compounds extracted from grape or released by the barrels (e.g., hydroxycinnamic acids) or products derived from chemical reactions taking place in wine (e.g., vinyl-flavanols). Depending on the type of precursors, pyranoanthocyanins can be formed at different stages of the wine life. Those resulting from products of the metabolism of yeasts, like the so-called vitisins A and B, first described by Bakker and coworkers (Bakker et al. 1997; Bakker and Timberlake, 1997), can be formed in early events of winemaking (Asenstorfer et al. 2003). However, at that stage high anthocyanin concentrations still exist in wine and, therefore, pyranoanthocyanins are not expected to have a relevant impact in wine color. Hydroxyphenyl-pyranoanthocyanins, whose formation involve hydroxycinnamic acids (Schwarz and Winterhalter 2004) or vinylphenols (Fulcrand et al. 1996), and flavan-pyranoanthocyanins, derived from vinyl-flavanols released from the cleavage of ethyl-linked oligomers (Francia-Aricha et al. 1997; Mateus et al. 2002b), would appear in further stages of the wine life (Schwarz and Winterhalter 2004) and could have a determining influence in the color of aged red wines. The absorption spectra of all these pigments in the visible region is characterised by a maximum wavelength hypsochromically shifted (20–40 nm) with regard to that of the anthocyanins (Fig. 9D.5). Pyranoanthocyanins are more resistant against sulfites and pH-induced color loss (Bakker et al. 1997). This is explained by the substitution at position C4 of the anthocyanin that confer them protection against the nucleophilic attack of SO_2 and water. Although pyranoanthocyanins could likely undergo similar reactions at C2, the presence of the additional aromatic pyran ring (as well as other conjugated aromatic rings in the case of hydroxyphenyl and vinyl-flavanol derivatives) contribute to the delocalisation of the positive charge and help to stabilize the colored flavylium cation (Hakansson et al. 2003). This is illustrated in Fig. 9D.6, where the influence of the pH value and the addition of SO_2 on the spectra of cyanidin 3-glucoside and a pyranoanthocyanin is shown.

The color of pyranoanthocyanins is more orange than that of anthocyanins, which leads one to suppose that they can contribute to the red tile and orange hues characteristics of aged wines. Similar molar extinction coefficients values for malvidin 3-glucoside (1.6×10^4 L/mol/cm) and its corresponding carboxypyranoanthocyanin (i.e., vitisin A; 1.3×10^4 L/mol/cm) were determined by Mateus and

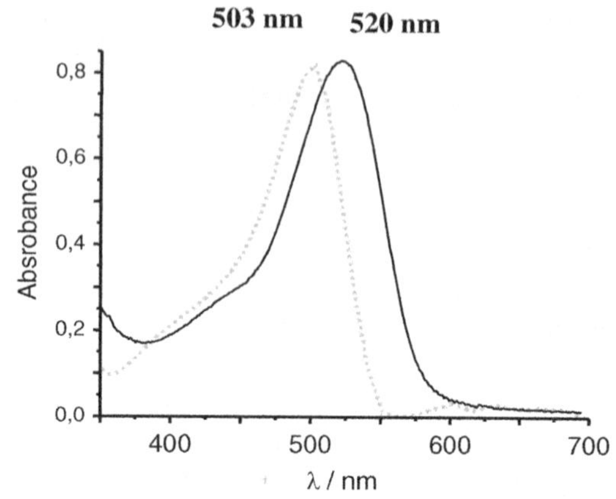

Fig. 9D.5 Absorption spectra in the visible region of malvidin 3-glucoside (Mv3glc) and a pyranoanthocyanin (pyranoMv3glc-catechin)

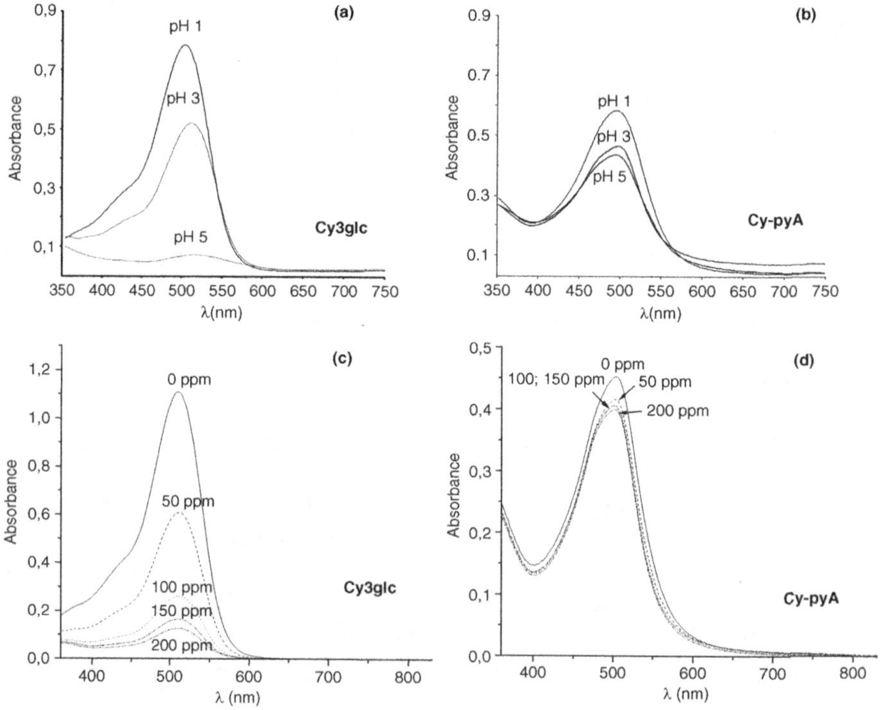

Fig. 9D.6 Absorption spectra in the visible region of cyanidin 3-glucoside (Cy3glc) and carboxypyranoCy3glc (Cy-py A) as a function of pH and concentration of SO_2 (adapted from Oliveira et al. 2006a)

de Freitas (2001) when measured at their maximum wavelengths in methanol/HCl 0.01%. Nevertheless, at wine pH conditions the difference between the extinction of both pigments must be higher in favor of the pyranoanthocyanins, due to its greater resistance to pH bleaching. This was clearly shown by Hakansson et al. (2003), who found that at pH 1.5 the vinylsyringol and vinylcatechol adducts of malvidin 3-glucoside (pinotin A) possessed about one-third of the molar extinction coefficient of the anthocyanin (27,600). However, in model wine (pH 3.6) closer extinctions existed for the three pigments (7,100, 10,000 and 6,200 for malvidin 3-glucoside and its vinylcatechol and vinylsyringol adducts at their maximum wavelengths in the visible region, respectively). This suggests that, for a same concentration and at wine pH, the color expression of these pigments would be rather similar and, thus, in wine their contribution to the color would mostly be depending on their concentrations (Rentzsch et al. 2007). However, Schwarz and coworkers found the visual detection limits for malvidin 3-glucoside, pinotin A (Schwarz and Winterhalter 2004) and vitisin A (Schwarz et al. 2003) at wine pH to be 0.14, 0.03, and 0.07 mg/mL, respectively, pointing out that for the same concentration pyranoanthocyanins would show higher color expression than anthocyanins.

Estimations about the actual contribution of pyranoanthocyanins to the color of red wines differ according to the authors, which may be due to the types of wines studied and the way used for pigment calculation. Alcalde-Eon et al. (2006) found that 70% of the pigments in a two-year old red wine were still anthocyanins, and that pyranoanthocyanins represented about 15%. If one assumes the observations of Schwarz et al. (2003) that pyranoanthocyanins have a visual impact 2–4 times greater than anthocyanins, and taking into account that the percentage of anthocyanins expected to be in colored forms at wine pH would be around 15% (Brouillard 1982), this would mean that a relevant part of the color of those wines would be determined by the pyranoanthocyanins. There is nothing to say in older wines in which the pyranoanthocyanins could be the most abundant pigments representing up to 50% of total pigments, as determined by Boido et al. (2006) in a 64-months old wine sample. Monagas et al. (2006) using chromatic and polynomial regression analyses to evaluate the influence of the different pigment families in the color of red wines up to 26 months of bottling concluded that both anthocyanins and pyranoanthocyanins were involved in the definition of the color. Among these, the pyruvic acid adducts (i.e., carboxypyranoanthocyanins) were those that provided the best model for predicting most of the color parameters. These views do not seem to be shared by Schwarz et al. (2003) that estimated that the contribution of carboxypyranoanthocyanins (vitisins A) to the color in aged red wines would be below 5%. This sounds logical since the formation of those pigments in the wine takes place in early stages to further decline (Alcalde-Eon et al. 2006).

A particular type of pyranoanthocyanins derivatives are the vinylpyranoanthocyanin-flavanol pigments, first described by Mateus et al. (2003) and also called Portisins (Mateus et al 2004b), which were isolated from Port wines. These pigments have particular chromatic features. Their absorption spectrum in the visible region is largely shifted towards higher wavelengths (Fig. 9D.7) and they show a blue color, likely due to the extended conjugation of their π electrons, which confers

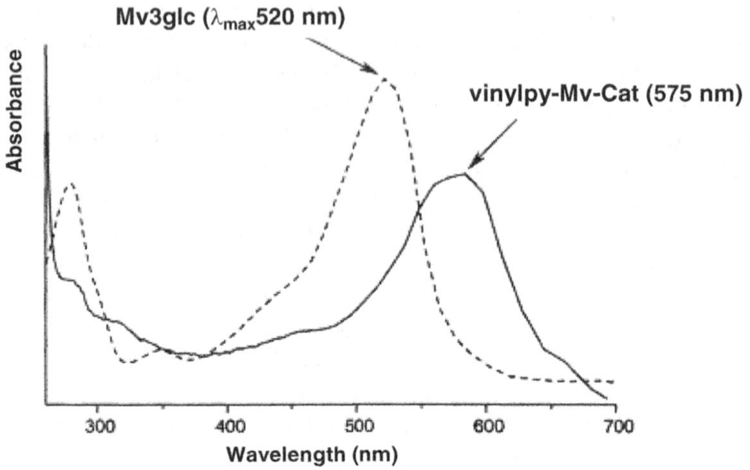

Fig. 9D.7 UV-visible spectra of malvidin 3-glucoside (Mv3glc) and vinylpyrano-Mv3glc-catechin (vinylpy-Mv-Cat) (adapted from Mateus et al. 2003)

a higher stability to their molecules. These pigments also have increased resistance to sulfites and pH-induced discoloration (Oliveira et al. 2006b) and their formation involves the reaction between pyranoanthocyanins and vinyl-flavanols, thus constituting a further step in the evolution of wine pigments, in which anthocyanins are no longer the main precursor. Although these blue pigments were only detected in very small quantities in fortified wines, they present unique spectroscopic features that may somehow contribute to the changing color of aged wines.

9D.2.3.3 Anthocyanin-ethyl-flavanol Pigments

These pigments were described by Timberlake and Bridle (1976) as resulting from the condensation between anthocyanins and flavanol mediated by acetaldehyde. They have absorption spectra with maximum wavelengths in the visible region about 15 nm bathochromically shifted with regard to that of the parent anthocyanins and show a more violet hue in wine-like solutions. Despite having the positions C2 and C4 of the anthocyanin moiety free, these pigments are partially resistant to pH-induced discoloration and sulfite bleaching (Fig. 9D.8), which is explained by their folded spatial conformation where the flavylium and catechin nuclei form a cavity that could accommodate part of a second molecule of adduct. This would favour the formation of non-covalent dimers where two pigments are stacked on one another in a way that protects their flavylium nuclei from the nucleophilic attack of water and SO_2 (Escribano-Bailon et al. 1996). A hydration constant of 4.17 was calculated for catechin-ethyl-malvidin 3-glucoside, higher than its proton transfer

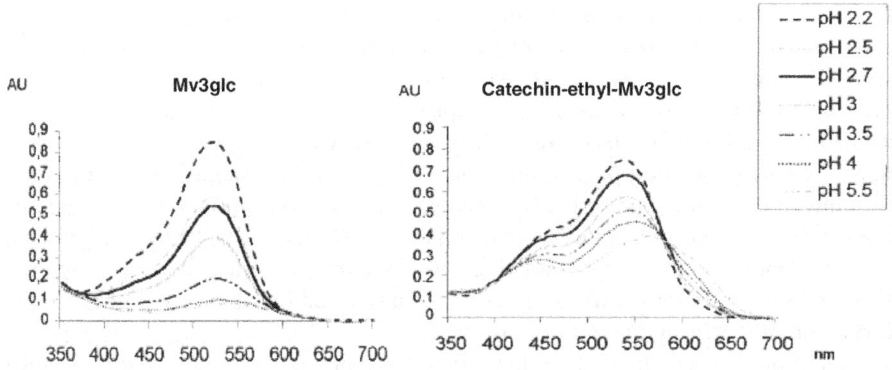

Fig. 9D.8 Absorption spectra in the visible region of malvidin 3-glucoside (Mv3glc) and catechin-ethyl-Mv3glc as a function of pH (adapted from Escribano-Bailon et al. 2001)

constant (3.44), meaning that it is mostly present in wine as the blue quinonoidal base turning more blue as pH increases (Cheynier et al. 2006). This increased tendency of the anthocyanin-ethyl-flavanol pigments to stabilize by deprotonation to their quinonoidal bases had already been noticed by Timberlake and Bridle (1976). The greater bathochromic displacement in maximum wavelength observed in their absorption spectra with pH increase and the presence of an isosbestic point around 580 nm (Fig. 9D.8) is proof of the shift of the equilibria in that direction.

The extinction of catechin-ethyl-malvidin 3-glucoside was calculated to be 17,100/M at pH 0.5 in 10% ethanol by Escribano-Bailon et al. (2001) slightly lower than that of malvidin 3-glucoside (20,200/M). In spite of their smaller extinction as compared with anthocyanins, the increased resistance to pH-induced and sulfite discoloration makes these pigments express more color than free anthocyanins in wine conditions. However, the pigments of this family do not seem to have a relevant direct contribution to the color of red wines in a great extent, as they are usually present in very low amounts. According to Alcalde-Eon et al. (2006) and Boido et al. (2006) they would not constitute more than 1% of total pigments either in young or aged red wines and their relative contribution to pigment composition would even decrease with aging. This scarce presence may be explained by their instability, which would also be the reason why only pigments constituted by an anthocyanin residue linked to a flavanol monomer or dimer are usually found in red wines. When sufficient acetaldehyde is available and a relatively low pH value exists, they rapidly progress to larger structures by the incorporation of new ethyl-flavanol units until a critical size is reached, after which they precipitate (Es-Safi et al. 1999; Rivas-Gonzalo et al. 1995). In addition, soluble ethyl-linked pigments are quite instable and suffer an easy cleavage at the level of the ethyl bridge releasing the anthocyanin moiety and ethyl-flavanols (Escribano-Bailon et al. 2001) that can either react with anthocyanins (or flavanols) to give new ethyl-bridged derivates or alternatively the more stable flavan-pyranoanthocyanins (Mateus et al. 2002b). Important amounts of acetaldehyde are formed during the alcoholic fermentation, in which the formation of ethyl-linked polymers can be expected to contribute to the anthocyanin losses by precipitation. In further stages of wine life some amounts of acetaldehyde can also be produced during malolactic fermentation as well as from ethanol by coupled oxidation of polyphenols (Wildenradt and Singleton 1974). In these stages less acetaldehyde is available and the conditions are less favorable for the formation of condensed pigments, so that no great amounts of them are expected to be formed. Nevertheless, although these pigments are not expected to contribute directly to the color of red wines, they could play an important indirect role on the color changes due to their contribution to anthocyanin losses by precipitation in early stages of winemaking and further involvement in the formation of flavan-pyranoanthocyanins.

In addition to acetaldehyde, other aldehydes can also exist in wines (e.g., propionaldehyde, isovaleraldehyde, isobutyraldehyde, benzaldehyde or vanillinaldehyde), released from the wood of the barrels used in the aging process or as a result of the addition of wine spirit in the case of fortified wines like Port wine. These compounds can also be involved in the formation of alkyl/aryl adducts between anthocyanins and flavanols similar to those induced by acetaldehyde (Pissarra et al. 2003, 2004, 2005; Sousa et al. 2007). The UV-visible spectra of all these pigments display a bathochromic shift in their λ_{max} in relation to that of the parent anthocyanins, resulting in a "blueing" effect of different magnitude depending on the aldehyde. Thus, the bathochromic shift seems to be higher for branched aldehydes, like benzaldehyde or isobutyraldehyde, than for non-branched ones (e.g., acetaldehyde, isovaleraldehyde or propionaldehyde). The steric hindrance promoted by the more branched substituents probably favors a more π-electron conjugated conformation leading to a greater displacement of the λ_{max} in the visible region (Pissarra et al. 2003). This pigment family could play a significant role in the color of Port wines, especially during the initial stages of aging. The content and profile of aldehydes in the wine spirit added would determine the type of aryl/alkyl adducts formed, with different chromatic characteristics, thus having an influence on color definition (Pissarra et al. 2004). The possibility of some aldehydes extracted during oak aging being involved in the formation of pigments cannot be discarded either. Quite recently a pigment from the condensation between malvidin 3-glucoside and catechin mediated by vanillin aldehyde has been characterised (Sousa et al. 2007). This aldehyde is reported to occur in oak aged wines as released from wood (Escalona et al. 2002).

9D.2.3.4 Anthocyanin-flavanol Condensed Pigments

The formation in red wines of anthocyanin-flavanol adducts had been early hypothesised by Jurd (1967, 1969). Two types of adducts seem possible, (1) F-A derivatives resulting from the nucleophilic addition of the hemiketal form of an anthocyanin through their C-8 or C-6 positions at C-4 of a carbocation resulting from the cleavage of a procyanidin, and (2) A-F derivatives from the electrophilic substitution of the anthocyanin flavylium form (C-4) by a flavanol (C-8 or C-6). The former

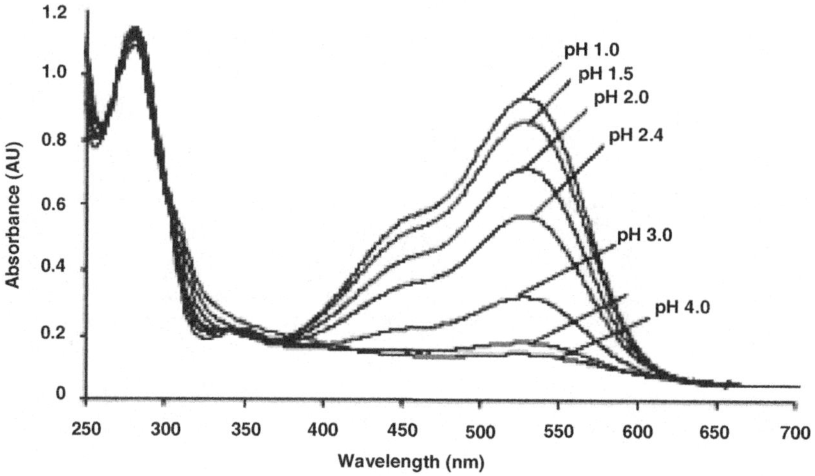

Fig. 9D.9 Absorption spectra in the visible region of a catechin-Mv3glc direct condensation dimer as a function of pH (adapted from Salas et al. 2004b with permission from Elsevier)

could oxidize to the corresponding F-A^+ pigments, whereas the latter seem to yield preferentially doubly-linked A-F colorless structures (Bishop and Nagel 1984; Remy-Tanneau et al. 2003) or rearrange towards yellow xanthylium ions (Jurd and Somers 1970; Dueñas et al. 2006; Liao et al. 1992). Recent evidence has been obtained regarding the presence in wines of colorless A-F adducts (Remy et al 2000) and F-A^+ pigments (Alcalde-Eon et al. 2006, 2007; Salas et al. 2004a, 2005; Vivar-Quintana et al. 1999). The latter show absorption spectra and color characteristics similar to anthocyanins being subjected to color modifications by the effect of pH (Fig. 9D.9), and they can also be supposed to be sensible to sulfite bleaching (Salas et al. 2004b). A hydration constant of 2.64 has been calculated for the dimer catechin-malvidin 3-glucoside similar to that of malvidin 3-glucoside (Cheynier et al. 2006). Even if their contribution to the pigment composition in red wines seems to increase slightly during aging, their levels remain low (Alcalde-Eon et al. 2006; Boido et al. 2006), which linked to their pH-induced discoloration lead to suppose that they do not contribute in an important extent to the color of red wines.

9D.2.3.5 Other Pigments

Other pigment families have been shown to occur in red wine, like anthocyanin oligomers (Salas et al. 2005; Vidal et al. 2004a), caftaric acid-anthocyanin adducts (Sarni-Manchado et al. 1997) or catechin-pyrylium derived pigments (i.e., oaklins) resulting from the reaction between catechin and wood aldehydes, like coniferaldehyde or sinapaldehyde (de Freitas et al. 2004; Sousa et al. 2005). These

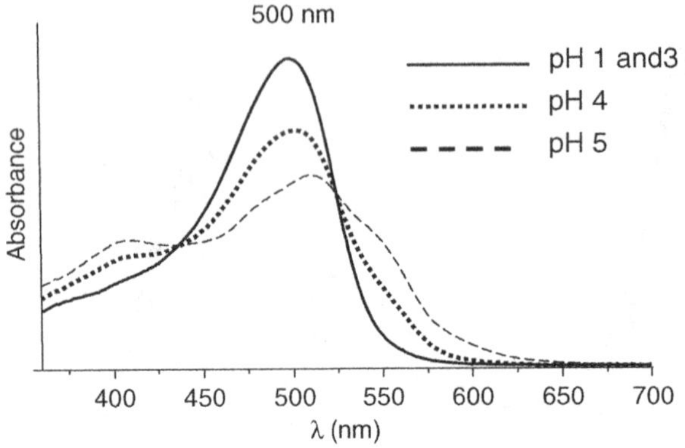

Fig. 9D.10 Absorption spectra in the visible region of a catechin-pyrylium derived pigment derived from the reaction between catechin and sinapaldehyde as a function of pH

latter pigments show a π-electron conjugation analogous to anthocyanins and have similar absorption spectra with maximum absorptivity around 480–520 nm, although their color is less influenced by pH than anthocyanins in the range of values existing in wines (Fig. 9D.10). Nonetheless, the levels of all these types of pigments in red wines seem to be very low and their importance for color definition is unknown. In the particular case of oaklins, some contribution to the color of red wines aged in oak barrels might be expected but further studies are still required to conclude about it.

9D.3 Astringency

9D.3.1 Astringency Perception

Phenolic compounds are a very complex group of natural compounds with large structural diversity that contribute directly to the flavor, namely to the bitterness and astringency perceived in a variety of food and beverages such as unripe fruits, wines, teas and beers. While bitterness is a taste felt by specific receptors in the tongue, astringency is usually defined as the array of tactile sensations felt around the mouth transduced by the trigeminal nerve (Green 1993). Astringent primary descriptors include dryness (lack of lubrication or moisture resulting in friction between oral surfaces), roughness (harsh texture of the oral cavity marked by edges and projections that are felt when oral surfaces contact with each other) and constriction (feeling of puckering and contraction felt in the mouth lips and cheeks) (Clifford 1997; Lee and Lawless 1991; Gawel et al. 2001).

It is usual amongst experienced tasters to employ general and subjective terms to define astringency or sub-qualities of astringency. To categorize the vocabulary,

Gawel et al. (2000) have proposed an interesting hierarchically structured vocabulary of mouth-feel sensations, presented as a "Mouth-feel Wheel", to assist tasters in their interpretation of wine astringency. Nevertheless, this new vocabulary seems to be too extensive, which hampers their implementation in a common language.

In contrast to other foods, astringency in red wine is one of the major attributes that the most exigent consumers of red wines learn to like. The levels of astringency in high quality red wines should be balanced: low levels of astringency could lead to the wine being flat and uninteresting, and an excess could mask other wine characteristics and make it harsh and dry (Gawel 1998).

Astringency is often the last sensation felt since it is a dynamic process that requires a period of time to develop. Lesschaeve and Noble (2005) evaluated the time-intensity curves of the astringency perceived from a red wine and observed that its intensity reaches a maximum 6–8 s after wine ingestion. Moreover, they also observed that the astringency intensity is build-up when the samples are tasted: the maximal intensity of astringency increased with repeating sipping of red wine at 25-s intervals (Lesschaeve and Noble 2005). Nevertheless, the perception capacity of astringency and bitterness change between individuals depending on intrinsic factors, such as the salivary protein composition and the flow-rate of saliva. Although the physicochemical mechanism of astringency is not completely understood, it is widely accepted that astringency results from the interaction between tannins and salivary proteins, resulting in the formation of protein-tannin aggregates in the mouth (Haslam 1998a; de Wijk and Prinz 2005). Some experimental evidence for this hypothesis was provided by Kallithraka et al. (1998) who observed a loss in saliva proteins after the tasting of astringent tannin-rich solutions and red wines, as a result of their complexation with phenolic compounds. Other authors have established a linear correlation between the astringency of tannin-rich solutions, perceived by a taste panel, and the capacity of those solutions to precipitate proteins (Llaudy et al. 2004; Monteleone et al. 2004; Troszynska et al. 2006).

There are many other factors that influence astringency in wine such as pH, sweetness, viscosity and levels of ethanol. The role of ethanol, the second most important component in wines, in wine taste is not totally clear. Fischer & Noble (1994) observed only a slight decrease of astringency of a white wine when its alcohol level rose from 8 vol.% to 14 vol.% while bitterness increases considerably. According to Gawel (1998) and our own point-of-view, the presence of other substances (carbohydrates, glycerol, etc.) in complex beverages such as wine and beer could be responsible for the perception of subtle sub-descriptors of astringency, such as soft, grainy, harsh, green, chalky, that are not perceived in tannin model solutions.

9D.3.2 Interactions Between Proteins and Tannins

The term "tannin" is often carelessly employed to identify some phenolic compounds as a result of the structural ambiguity of that wide group of compounds.

One of the most important properties of tannins, and why tannins are recognised, is their ability to interact with proteins. In fact, the term "tannin" has been typically employed to designate the substances of vegetable origin able of transform fresh skin into leather, thereby interacting and precipitating proteins of the animal skin. Perhaps the more practical, functional and understandable definition of "tannin" is that proposed many years ago by Bate-Smith & Swain (1962) as "*water soluble phenolic compounds having molecular weights between 500 and 3000 and, besides giving the usual phenolic reactions, having special properties such as the ability to precipitate alkaloids, gelatin and other proteins*". Nowadays, with the relatively recent advances in the chromatographic and structural identification techniques of tannins, it is generally accepted that the molecular weight range proposed by these authors is very narrow and that tannin structures might reach many thousands of Dalton.

The general consensus is that proteins and tannins interact via hydrogen bonding and hydrophobic effects (Fig. 9D.11). Ionic interactions do not seem to be important at neutral or slightly acidic pH, because under these conditions phenolic compounds are not supposed to be charged as they are weak acids (pKa 9–10) (Oh et al. 1980; Vernhet et al. 1996). Hydrogen bonding can be established between

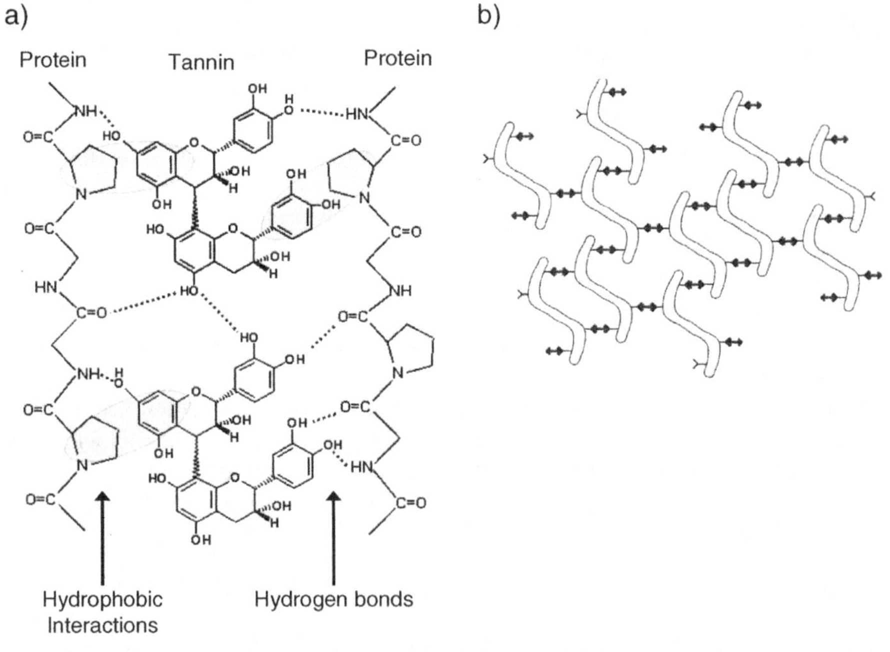

Fig. 9D.11 Schematic interaction between tannins and proteins: **a** main driving forces between phenolic rings (cross-linkers) of tannins and the amide groups and apolar side chains of amino acids such as proline; **b** protein-tannin aggregates: the *grey* "Ss" represent proteins with a number of tannin binding places, and the *black arrows* represent tannins with protein binding sites the hydroxyl groups of phenolic compounds and carbonyl and amide groups of proteins. Hydrophobic interaction can occur between the benzenic ring of phenolic compounds and the apolar side chains of amino acids such as leucine, lysine or proline in proteins (Hatano and Hemingway 1996; Jobstl et al. 2006; Oh et al. 1980; Wroblewski et al. 2001).

The proline residue is generally considered to be a good binding site as it provides a flat, rigid, hydrophobic surface that is favourable to interactions with other planar hydrophobic surfaces such as benzenic rings. Effectively, NMR studies performed by the Haslam research team in the University of Sheffield have shown the existence of hydrophobic stacking between the benzenic ring of several phenolic compounds and the apolar face of the proline ring of peptides rich in this amino acid (Baxter et al. 1997; Charlton et al. 2002a; Murray et al. 1994). On the other hand, a carbonyl group of tertiary amides, as it occurs in a protein with proline residues, is a hydrogen acceptor more efficient than a carbonyl in a secondary or primary amide. Hence, hydrogen bonding involving the carbonyl group of proline is predictable, concomitantly with hydrophobic interactions. Although studies performed by other authors suggest that the driving force of protein-tannin interactions are mainly hydrogen bonding (Frazier et al. 2006; Simon et al. 2003), it seems more prudent to accept that both type of interactions are present, strengthening each other, depending on the protein and tannin structure and medium conditions (Artz et al. 1987; Hagerman et al. 1998; Oh et al. 1980).

Protein-tannin interactions are importantly affected by pH (Calderon et al. 1968; Naczk et al. 1996). Apparently there is higher protein precipitation at pH values close to the protein isoelectric point (pI) where protein-protein electrostatic repulsions are minimized (Hagerman and Butler 1981): proteins with acidic pI have higher affinities to complex with tannins at lower pH, whereas basic proteins aggregate preferentially at higher pH (Yan and Bennick 1995). An interesting study by NMR performed by Charlton et al. (2002b) with EGCG and a basic PRP fragment of 19 amino acids, showed that pH greatly affects precipitation of aggregates but does not change the protein-tannin binding affinities. These authors suggested that protein-tannin particles are in a colloidal state, stabilized by repulsions of proteins of the same charge. When this charge is minimized (at pH=pI), there is no repulsion and particles aggregate together and precipitate. Therefore, although the initial binding is due to specific protein-tannin interactions, the aggregation process seems to be due mainly to surface charge effects.

The affinity of tannins to bind proteins is favored by their ability to work as multidentate ligands (cross-linking) in which one tannin is able to bind to more than one protein at one time (Fig. 9D.11b) or to bind to more than one point in the same protein (Charlton et al. 2002a; Siebert et al. 1996). These associations between tannin and protein could result in aggregates that precipitate depending on the ratio of protein/tannin and also the concentration of protein (Frazier et al. 2003; Poncet-Legrand et al. 2006).

Besides the relative concentrations, those interactions are affected by the structures of both tannin and protein.

9D.3.3 Tannin Structures and Their Influence in Wine Astringency

Tannins are ubiquitous compounds widespread in the plant kingdom, and are classically divided into hydrolysable tannins and condensed tannins according to their chemical structure. Condensed tannins are abundant in the different parts of the grape (seeds, skins and stems) and are extracted to wine during winemaking. Hydrolysable tannins are absent in grapes, and it is generally accepted that they pass to wine from oak during aging in barrels (Clifford and Scalbert 2000). However, as already referred to above, only a few mg/L of hydrolysable tannins have been reported in red wines (Fernandez de Simon et al. 2003; Perez-Prieto et al. 2003).

Hydrolysable tannins are composed of polyols (such as glucose and quinic acid) linked to at least one gallic acid (gallotannins) or one hexahydroxydiphenic acid (ellagic tannins). Condensed tannins (also known as proanthocyanidins) are polymers of flavan-3-ols. Flavan-3-ol units are usually linked through C-C interflavanol bonds established between the C4 of one flavan-3-ol unit and the C8 or C6 of another unit. There are also tannins with an additional ether linkage between the C2 of the upper unit and the oxygen-bearing C7 or C5 of the lower unit, in addition to the usual C4-C8 or C4-C6 interflavanol bond. However, only traces of that type of tannins have been reported in grape and wine (Krueger et al. 2000; Vivas de Gaulejac 2001).

Condensed tannins can be found in grape with different degrees of polymerization with two or more units of flavan-3-ols, reaching as high as 80 units of flavanol. Degrees of polymerization (DP) between 3–83 and 2–16 have been reported in grape skins and grape seeds, respectively (Souquet et al. 1996). Red wine might contain up to 4 g/L or even more of condensed tannins and their degree of polymerization change continuously during aging as a result of the chemical transformations.

Mouth-feel properties of tannins-rich solutions depend on tannin structure and concentration (Noble 1990; Hagerman et al. 1998; Wroblewski et al. 2001). The perception of the astringency of condensed tannins increases with tannin size and degree of galloylation (Vidal et al. 2003a) as does their ability to complex with proteins probably because they have more interaction sites (Bacon and Rhodes 1998; de Freitas and Mateus 2001; Baxter et al. 1997). However, the influence of the bigger structures of condensed tannins in astringency is controversial. Some authors believe that these complex structures (mean DP >7) are not enough soluble to be astringent (Lea 1990), while others have worked with highly polymerized structures of proanthocyanidins (mean DP up to 70) and have observed that they were soluble in water-alcohol solution and highly astringent (Cheynier et al. 2006; Vidal et al. 2003a).

Although they were still able to bind proteins efficiently, the bigger structures of condensed tannins have been shown to have more difficulty to bind proteins, which has been attributed to the decrease in their flexibility (de Freitas and Mateus 2002; Siebert et al. 1996). Effectively, it has been reported that more flexible condensed tannins have better ability to bind proteins because they are more efficient cross linkers (Lea 1992; de Freitas and Mateus 2001; Goldstein and Swain 1963; Frazier et al. 2003; Okuda et al. 1985).

9D.3.4 Protein Structures

The interaction between tannins and proteins can be affected by the size of the protein (Hagerman and Butler 1981; Maury et al. 2003), its charge (Hagerman and Butler 1981), the presence of side chains (Maury et al. 2003), and its conformation (Wroblewski et al. 2001).

As already mentioned, the presence of proline is apparently a common characteristic of proteins with high binding affinities towards tannins (Hagerman and Butler 1981). Apart from being a binding site, proline residues are also useful for maintaining the peptide in an extended conformation, thereby providing a bigger surface of protein to binding (Baxter et al. 1997).

Saliva is produced by two groups of glands: parotid and submandibular/sublingual. Submandibular/sublingual saliva is produced in non-stimulated conditions (between meals) and the major organic components are mucins, proteins that have lubricatory properties. Parotid saliva is secreted in stimulated conditions and is mainly composed of 30% of α-amilase, and 70% of Proline-Rich Proteins (PRPs) with high levels of proline (35–40%) (Dodds et al. 2005). PRPs can be divided into three major groups: acidic, basic and glycosylated proteins (Kauffman and Keller 1979). All salivary PRPs show affinity towards dietary tannins (Bacon and Rhodes 1998) and are easily precipitated (Bacon and Rhodes 1998; Yan and Bennick 1995; Kallithraka et al. 1998). Basic PRPs seem to be more effective in precipitating tannins when compared to acidic and glycosilated PRPs (Lu and Bennick 1998). A group of eleven proteins (IB1-IB7, IB8a, b and c and IB9) can be found among the family of basic PRPs having very similar sequences and displaying repetitive patterns, in which some sequences of amino acids can be found several times in the same protein and in different proteins (Kauffman et al. 1991). It has been proposed that the presence of multiple repeated regions rich in proline could provide sites favourable for tannin binding, as well as flexible hinges on the protein. This latter feature allows the protein to fold and "wrap around" the tannin, thereby increasing the association by cooperative intermolecular interactions (Charlton et al. 1996).

Previous works have shown that a basic PRP (IB8c) binds to condensed tannins much more effectively than α-amylase (de Freitas and Mateus 2002). This can be explained by the 3D structure of proteins: α-amylase is a globular protein, and IB8c is likely to adopt an extended random coil conformation, which would allow the protein to offer more contact sites to interact with tannins. However, α-amylase seems to be more specific and selective than PRPs in the aggregation with samples containing different amounts of procyanidins (Mateus et al. 2004c).

The astringency sensation is felt differently by different tasters (Gawel et al. 2001) probably due to differences in individuals' saliva, namely its protein composition (Lesschaeve and Noble 2005; Horne et al. 2002). Recently, a study concerning the characterization of basic PRPs in thirteen normal adults has demonstrated that the protein IB9 was only detected in nine adults saliva, while IB7 was not detected at all (Messana et al. 2004). The salivary flow rate was also described as a factor that varies between subjects and can affect astringency perception (Lesschaeve and Noble 2005; Horne et al. 2002), with which it seems to be inversely related (Lesschaeve and Noble 2005; Condelli et al. 2006). Low-flow subjects felt astringency later and more intensively for a longer time than did high-flow subjects (Lesschaeve and Noble 2005). This effect could be explained by the increase of the saliva pH and the protein concentration decrease for higher saliva flow.

9D.3.5 Changes in Tannin Composition and Astringency During Wine Aging

It is generally considered that wine astringency intensity is higher in young wine and decreases during aging "softening" the wine. After being extracted from grapes, phenolic compounds – and namely flavanols – involved in astringency perception and anthocyanins responsible for wine colour undergo chemical reactions yielding new simple phenolic compounds and more complex tannin-like structures. The latter could include in their structures anthocyanins or other chromophore groups ('polymeric pigments'). Different pathways have been proposed to explain some of those chemical transformations that occur during wine aging, some of them revised in previous chapter of this book.

It has been known for a long time that the oxidative polymerization of flavanols could directly involve oxygen (catalyzed by traces of metal ions) via quinones, or be mediated by acetaldehyde mainly arising from ethanol oxidation (Somers 1971; Timberlake and Bridle 1976; Wildenradt and Singleton 1974; Rivas-Gonzalo et al. 1995). The role of acetaldehyde in the polyphenol transformation has been extensively studied by different research groups and is well documented. Acetaldehyde-mediated condensation between flavanols themselves (Peleg et al. 1999) or with anthocyanins leads to ethyl-bridged oligomers (Timberlake and Bridle 1976) and also to flavan-pyranoanthocyanins (Francia-Aricha et al. 1997; Mateus et al. 2003). Thus, acetaldehyde formed in situ from ethanol oxidation should play an important role in the transformation and polymerization (insolubilization) of water-soluble proanthocyanidins, causing the loss of astringency. In wines there are other aldehydes such as furfural and benzaldehyde that could be involved in such reactions, as it was demonstrated in wine-like model solutions (Es-Safi et al. 2002; Nonier et al. 2006; Pissarra et al. 2004). On the other hand, wine tannin-like polyphenolic compounds could derive from direct reactions between flavanols and anthocyanins yielding anthocyanin-flavanol and flavanol-anthocyanin adducts (Remy et al 2000; Salas et al. 2005). Some of these tannin-like polyphenolic compounds have been suggested to be less bitter and astringent than the original precursors (Vidal et al. 2003b, 2004a). As a result of this whole field of polycondensations, tannins become consecutively larger and the biggest structure are presumed to precipitate forming an insoluble deposit, as it was often observed specially in red wines that age in barrel or bottle (Haslam 1980; Singleton and Noble 1976).

An alternative point of view is that some tannins (proanthocyanidins) might be depolymerised by acid catalysis, thereby becoming smaller and reducing astringency

(Cheynier et al. 2006). In conclusion, and attending to all the information available concerning this matter, it is more prudent to accept that all the phenomena referred above are involved altogether in the evolution of the mouth-feel sensation of wine during aging.

At present the structures of the wine-derived-tannins are practically unknown and a great part of the structures identified so far have been only demonstrated in wine model studies. Hence, the influence, or contribution, of wine tannins to astringency is far from being ascertained. For that reason, it is crucial to make efforts to clarify those structures and the mechanisms involved in their formation. Moreover, the role of oxygen is not fully understood and it is important to know how to deal with it during winemaking to control oxidation, and therefore improve the wine taste characteristics.

9D.3.6 Polysaccharides and Astringency

Research work performed in the end of the last century by Haslam and co-workers showed that carbohydrates inhibit the protein-tannin interactions in solution (Luck et al. 1994; Ozawa et al. 1987). More recently, similar findings have been reported by de Freitas and co-workers using nephelometric techniques (Fig. 9D.12) for different carbohydrates and protein-tannin systems (Carvalho et al. 2006a, b; de Freitas et al. 2003; Mateus et al 2004a).

This inhibition of the interactions between tannins and salivary protein by carbohydrates has been proposed to contribute to the loss of astringency during ripening of some fruits (Luck et al. 1994; Ozawa et al. 1987; Taira et al. 1997).

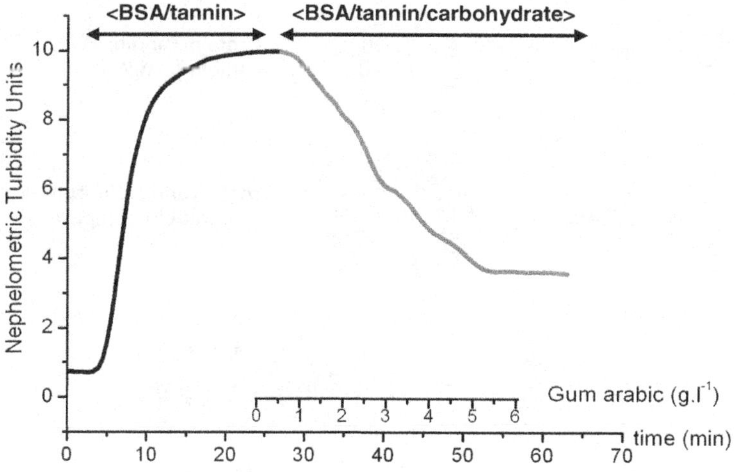

Fig. 9D.12 Flow nephelometric analysis of the inhibitory effect of gum arabic on aggregate formation between procyanidins (tetramers–digallate and pentamers–gallate, 0.034 g/L) and BSA (0.33 g/L) (adapted from Carvalho et al. 2006a)

In fact, the cellular structure of fruits softens during ripening as a result of the enzymatic depolymerization of structural carbohydrates in cell walls (pectins, hemicellulose, cellulose) releasing soluble fragments (Fry 1995; Huber 1983; Brummell and Harpster 2001; Prasanna et al. 2007). The increase in soluble pectin fragments could inhibit salivary protein-tannin interactions in the mouth leading to a decreased astringent response. Some other mechanisms have been proposed to explain this phenomenon. The decrease of astringency could also result from the changes of polyphenolic composition during ripening (Haslam 1998b; Goldstein and Swain 1963). Some authors have shown that the concentration of flavan-3-ols and namely procyanidins in grapes decrease during maturity (de Freitas et al. 2001; Czochanska et al. 1979; Kennedy et al. 2000). It has also been demonstrated that during grape ripening, tannins bind to the internal face of tonoplasts, and also to the cell wall polysaccharides through osidic bonds (Amrani Joutei et al. 1994). Other findings showed that acetaldehyde formed in situ during ripening of some fruits, such as persimmon, induces the polymerization (insolubilization) of water-soluble proanthocyanidins, decreasing astringency (Tanaka et al. 1994).

Regarding the influence of polysaccharides, two mechanisms have been proposed to explain their influence on protein-tannin interactions in solution (Haslam 1998b; Luck et al. 1994) (Fig. 9D.13):

(i) Polyelectrolyte polysaccharides could form ternary complexes with the protein-tannin aggregates, enhancing its solubility in aqueous medium
(ii) Polysaccharides could encapsulate tannins interfering with their ability to bind and precipitate proteins

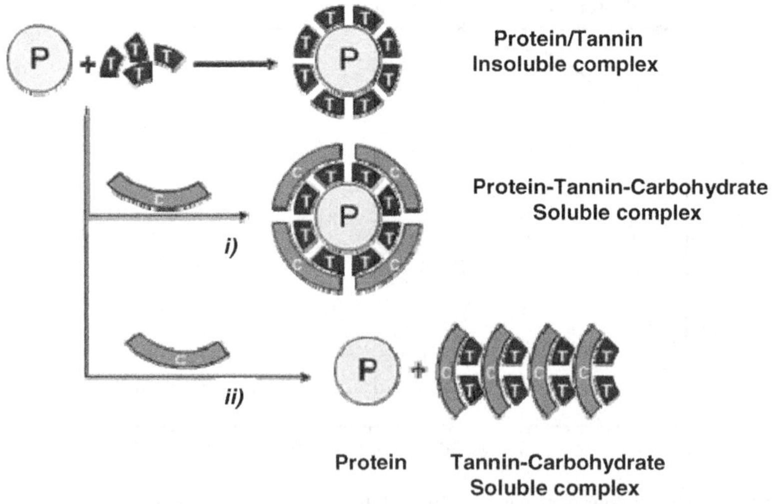

Fig. 9D.13 Mechanism of carbohydrate inhibition of protein tannin interactions (Mateus et al. 2004) (adapted from Mateus et al. 2004a)

In general, the effectiveness of carbohydrates to prevent protein-tannin aggregation increase with their ionic character (Luck et al. 1994; Carvalho et al. 2006b; de Freitas et al. 2003): neutral carbohydrates practically do not affect aggregation.

Ionic polysaccharides are present in wine, and are expected to affect mouth feel properties and mainly astringency acting on the interaction between tannins and salivary proteins. The major wine polysaccharides are arabinogalactan proteins (AGP) and rhamnogalacturonan II (RGII), pectic polysaccharides which originate from grape cell walls, and mannoproteins (MP) produced by yeast during wine fermentation (Doco et al. 2000). Vidal et al. (2004c) have demonstrated that procyanidin astringency decreases in the presence of RGII, while MPs and AGPs decrease bitterness (Vidal et al. 2004c). Escot et al. (2001) have demonstrated that the wine structure was modified by the addition of mannoproteins and that this addition reduces astringency of red wine due to a higher tannin/mannoprotein condensation level.

Recently, Carvalho et al. (2006b) studied the influence of wine polysaccharides (AGP, RGII and MP) on salivary protein-tannin interactions. The results showed that the most acidic fractions of AGPs and MPs have the ability to inhibit the formation of aggregates between condensed tannins and two different salivary proteins (α-amylase and IB8c). The concentrations tested are below to those present in wine which means that they could have an influence in wine astringency.

On the other hand, it has also been found that polysaccharides present in wines interfere with the self-aggregation of proanthocyanidins (Riou et al. 2002). The change in the proanthocyanidin colloidal state in wines could also affect their ability to complex with salivary proteins and thereby their sensory properties.

9D.3.7 Experimental Studies of the Interactions Between Proteins and Tannins, and Astringency

Beside sensorial analysis, several other experimental techniques have been used to study the interaction between tannin and proteins, a characteristic that is involved in astringency. The development of new techniques that only require small amounts of compounds, the elucidation of new polyphenol structures and proteins engaged in astringency perception, and the improvement of the preparative techniques to isolate these compounds have contributed importantly to the advances in this field. These techniques include NMR, microcalorimetry, enzyme inhibition, microscopy techniques, radioactivity measurements, HPLC, fluorescence, electrophoresis, ESI-MS, infrared spectroscopy, marker utilization, molecular modeling and light scattering measuring techniques such as dynamic light scattering (DLS) and nephelometry (Chapon 1993; de Freitas and Mateus 2002; Edelmann and Lendl 2002; Hagerman and Butler 1980; Sarni-Manchado and Cheynier 2002; Beart et al. 1985; Kandra et al. 2004; Papadopoulou et al. 2005; Soares et al. 2007; Condelli et al. 2006; Kallithraka et al. 1998; Siebert et al. 1996; Simon et al. 2003; Wroblewski et al. 2001).

Some studies have been made relating specifically to astringency. Some of these studies have been focused directly on interactions between tannins and salivary proteins, and in the changes in saliva protein composition after interaction with tannins. Other studies have correlated the sensorial astringency with protein-tannin interactions using several proteins as a model such as mucin, ovalbumin, gelatin, BSA and salivary proteins. In fact, the astringency felt when sampling different tannin solutions can be correlated with the ability of the same tannins to precipitate proteins.

Apart from salivary proteins, other proteins have been used in the tannin-protein interaction studies due to some characteristics that make them similar to PRPs, like casein, gelatin, polyproline (Jobstl et al. 2004; Calderon et al. 1968; Luck et al. 1994; Poncet-Legrand et al. 2006; Siebert et al. 1996). Although it is not a protein, the polymer polyvinylpolypyrrolidone as also been used in these studies (Hagerman and Butler 1981; Laborde et al. 2006). Recently, an electronic tongue based on protein-tannin interactions has been developed to measure astringency (Edelmann and Lendl 2002). Despite the unquestionable importance of all these works to understand the interaction between tannins and proteins, extrapolation to the real context of wine sensory should be done with care.

References

Alcalde-Eon, C., Escribano-Bailon, M. T., Santos-Buelga, C., & Rivas-Gonzalo, J. C. (2006). Changes in the detailed pigment composition of red wine during maturity and ageing – A comprehensive study. *Anal. Chim. Acta, 563*, 238–254.

Alcalde-Eon, C., Escribano-Bailón, M. T., Santos-Buelga, C., & Rivas-Gonzalo, J. C. (2007). Identification of dimeric anthocyanins and new oligomeric pigments in red wine by means of HPLC-DAD-ESI/MS^n *J. Mass Spectr., 42*, 735–748.

Amrani Joutei, K., Glories Y., & Mercier M. (1994). Localisation des tanins dans la pellicule de baie de raisin. *Vitis, 33*, 133–138.

Artz, W. E., Bishop, P. D., Dunker, A. K., Schanus, E. G., & Swanson, B. G. (1987). Interaction of synthetic proanthocyanidin dimer and trimer with bovine serum-albumin and purified bean globulin fraction G-1. *J. Agric. Food Chem., 35*, 417–421.

Asen, S., Stewart, R. N., & Norris, K. H. (1972). Co-pigmentation of anthocyanins in plant tissues and its effects on color. *Phytochemistry, 11*, 1139–1144.

Asenstorfer, R. E., Markides, A. J., Iland, P. G., & Jones, G. P. (2003). Formation of vitisin A during red wine vinification and maturation. *Aust. J. Grape Wine Res., 9*, 40–46.

Bacon, J. R., & Rhodes, M. J. C. (1998). Development of a competition assay for the evaluation of the binding of human parotid salivary proteins to dietary complex phenols and tannins using a peroxidase-labeled tannin. *J. Agric. Food Chem., 46*, 5083–5088.

Bakker, J., & Timberlake, C. F. (1997). Isolation, identification and characterization of new color-stable anthocyanins occuring in some red wines. *J. Agric. Food Chem., 46*, 35–43.

Bakker, J., Bridle, P., Honda, T., Kuwano, H., Saito, N., Terahara, N., & Timberlake, C. F. (1997). Identification of an anthocyanin occuring in some red wines. *Phytochemistry, 44*, 1375–1382.

Baranac, J. M., Petranovic, N. A., & Dimitric-Markovic, J. M. (1996). Spectrophotometric study of anthocyan copigmentation reactions. *J. Agric. Food Chem., 44*, 1333–1336.

Baranac, J. M., Petranovic, N. A., & Dimitric-Markovic, J. M. (1997a). Spectrophotometric study of anthocyan copigmentation reactions.2. Malvin and the nonglycosidized flavone quercetin. *J. Agric. Food Chem., 45*, 1694–1697.

Baranac, J. M., Petranovic, N. A., & Dimitric-Markovic, J. M. (1997b). Spectrophotometric study of anthocyan copigmentation reactions.3. Malvin and the nonglycosidized flavone morin. *J. Agric. Food Chem., 45*, 1698–1700.

Bate-Smith, E. C., & Swain, T. (1962). Flavonoid Compounds. In H. S. Mason & A. M. Florkin (Ed.) *Comparative Biochemistry* (pp. 755–809). New York: Academic Press.

Baxter, N. J., Lilley, T. H., Haslam, E., & Williamson, M. P. (1997). Multiple interactions between polyphenols and a salivary proline-rich protein repeat result in complexation and precipitation. *Biochemistry, 36*, 5566–5577.

Beart, J. E., Lilley, T. H., & Haslam, E. (1985). Plant Polyphenols - Secondary Metabolism and Chemical Defense – Some Observations. *Phytochemistry, 24*, 33–38.

Berke, B., & de Freitas, V. A. P. (2005). Influence of procyanidin structures on their ability to complex with oenin. *Food Chem., 90*, 453–460.

Berke, B., & de Freitas, V. A. P. (2007). A colorimetric study of oenin copigmented by procyanidins. *J. Sci. Food Agric., 87*, 260–265.

Bishop, P. D., & Nagel, C. W. (1984). Characterization of the condensation product of malvidin 3,5-diglucoside and catechin. *J. Agric. Food Chem., 32*, 1022–1026.

Bloomfield, D. G., Heatherbell, D. A., & Nikfardjam, M. S. P. (2003). Effect of *p*-coumaric acid on the color in red wine. *Mitt. Klosterneuburg, 53*, 195–198.

Boido, E., Alcalde-Eon, C., Carrau, F., Dellacassa, E., & Rivas-Gonzalo, J. C. (2006). Aging effect on the pigment composition and color of Vitis vinifera L. cv Tannat wines. Contribution of the main pigment families to wine color. *J. Agric. Food Chem., 54*, 6692–6704.

Boulton, R. (1996). A method for the assessment of copigmentation in red wines. Presented at the *Forty-seventh Annual Meeting of the American Society of Enology and Viticulture*, Reno NV, June 1996.

Boulton, R. (2001). The copigmentation of anthocyanins and its role in the color of red wine: A critical review. *Am. J. Enol. Vitic., 52*, 67–84.

Bourzeix, M., Heredia, N., Estrella, M. I., Puech, J. L., & Fartsov, K. (1980). Estimation quantitative de la matière colorante rouge des moûts concentrés et des vins. *Bull. Liaison-Groupe Polyphenols, 9*, 131–142.

Brouillard, R. (1982). Chemical structure of anthocyanins. In P. Markakis (Ed.), *Anthocyanins as Food Colors* (pp. 1–38). New York: Academic Press.

Brouillard, R., & Dangles, O. (1993). Flavonoids and flower colour. In J. B. Harborne (Ed.) *The Flavonoids. Advances in research since 1986* (pp. 565–588). London: Chapman & Hall.

Brouillard, R., & Dangles, O. (1994). Anthocyanin molecular-interactions - the first step in the formation of new pigments during wine aging. *Food Chem., 51*, 365–371.

Brouillard, R., Delaporte, B., & Dubois, J. E. (1977). Chemistry of anthocyanins pigments 2. Kinetic and thermodinamic study of proton transfer, hydration, and tautomeric reactions of malvidin-3-glucoside. *J. Am. Chem. Soc., 99*, 8461–8468.

Brouillard, R., Wigand, M. C., Dangles, O., & Cheminat, A. (1991). pH and solvent effects on the copigmentation reaction of malvin with polyphenols, purine and pyrimidine-derivatives. *J. Chem. Soc. Perkin Trans., 2*, 1235–1241.

Brouillard, R., Chassaing, S., & Fougerousse, A. (2003). Why are grape/fresh wine anthocyanins so simple and why is it that red wine color lasts so long? *Phytochemistry, 64*, 1179–1186.

Brummell, D. A., & Harpster, M. H. (2001). Cell wall metabolism in fruit softening and quality and its manipulation in transgenic plants. *Plant Mol. Biol., 47*, 311–340.

Calderon, P., Vanburen, J., & Robinson, W. B. (1968). Factors influencing formation of precipitates and hazes by gelatin and condensed and hydrolysable tannins. *J. Agric. Food Chem., 16*, 479–482.

Canals, R., Llaudy, M. C., Canals, J. M., & Zamora, F. (2008). Influence of the elimination and addition of seeds on the colour, phenolic composition and astringency of red wine. *Eur. Food Res. Tech., 226*, 1183–1190.

Carvalho, E., Povoas, M., Mateus, N., & de Freitas, V. A. P. (2006a). Application of flow nephelometry to the analysis of carbohydrate influence on protein-tannin interactions. *J. Sci. Food Agric., 86*, 891–896.

Carvalho, E., Mateus, N., Plet, B., Pianet, I., Dufourc, E., & dde Freitas, V. A. P. (2006b). Influence of wine pectic polysaccharides on the interactions between condensed tannins and salivary proteins. *J. Agric. Food Chem., 54*, 8936–8944

Chapon, L., (1993). Nephelometry as a Method for Studying the Relations between Polyphenols and Proteins. *J. Inst. Brew., 99*, 49–56.

Charlton, A. J., Baxter, N. J., Lilley, T. H., Haslam, E., McDonald, C. J., & Williamson, M. P. (1996). Tannin interactions with a full-length human salivary proline-rich protein display a stronger affinity than with single proline-rich repeats. *FEBS Letters, 382*, 289–292.

Charlton, A. J., Haslam, E., & Williamson, M. P. (2002a). Multiple conformations of the proline-rich protein/epigallocatechin gallate complex determined by time-averaged nuclear Overhauser effects. *J. Am. Chem. Soc., 124*, 9899–9905.

Charlton, A. J., Baxter, N. J., Khan, M. L., Moir, A. J. G., Haslam, E., Davies, A. P., & Williamson, M. P. (2002b). Polyphenol/peptide binding and precipitation. *J. Agric. Food Chem., 50*, 1593–1601.

Cheynier, V., Moutounet, M., & Sarni-Manchado, P. (2003). Los compuestos fenólicos. In C. Flanzy (Ed.), *Enologia: fundamentos cientificos y tecnologicos* (pp. 114–136). Madrid: Ediciones Mundi-Prensa.

Cheynier, V., Dueñas, M., Salas, E., Maury, C., Souquet, J. M., Sarni-Manchado, P., & Fulcrand, H. (2006). Structure and properties of wine pigments and tannins. *Am. J. Enol. Vitic., 57*, 298–305.

Clifford, M. N. (1997). Astringrency. In F. A. Tomás-Barberán & R. J. Robins (Ed.), *Phytochemistry of Fruits and Vegetables: proceedings of the phytochemical society of Europe* (pp.87–107). Oxford: Clarendon Press.

Clifford, M. N. (2000). Anthocyanins – nature, occurrence and dietary burden. *J. Sci. Food Agric., 80*, 1063–1072.

Clifford, M. N., & Scalbert, A. (2000). Ellagitannins – nature, occurrence and dietary burden. *J. Sci. Food Agric., 80*, 1118–1125.

Condelli, N., Dinnella, C., Cerone, A., Monteleone, E., & Bertuccioli, M. (2006). Prediction of perceived astringency induced by phenolic compounds II: criteria for panel selection and preliminary application on wine samples. *Food Qual. Pref., 17*, 96–107.

Czochanska, Z., Foo L. Y., & Porter L. J. (1979). Compositional changes in lower molecular weight flavans during grape maturation. *Phytochemistry, 18*, 1819–1822.

Dangles, O., & Brouillard, R. (1992). Polyphenol Interactions – the copigmentation case - thermodynamic data from temperature-variation and relaxation kinetics – medium effect. *Can. J. Chem., 70*, 2174–2189.

Dangles, O., Saito, N., & Brouillard, R. (1993). Kinetic and thermodynamic control of flavilium hydration in the pelargonidin–cinnamic acid complexation. Origin of the extraordinary flower colour diversity of *Pharbitis* nil. *J. Am. Chem. Soc., 115*, 3125–3132.

Dangles, O., Elhabiri, M., & Brouillard, R. (1994). Kinetic and thermodynamic investigation of the aluminum anthocyanin complexation in aqueous-solution. *J. Chem. Soc. Perkin Trans., 2*, 2587–2596.

Darias-Martín, J., Carrillo, M., Díaz, E., & Boulton R. B. (2001). Enhancement of red wine colour by pre-fermentation addition of copigments. *Food Chem., 73*, 217–220.

Darias-Martín, J., Martin-Luis, B., Carrillo-Lopez, M., Lamuela-Raventos, R., Díaz-Romero, C., & Boulton R. B. (2002). Effect of caffeic acid on the color of red wine. *J. Agric. Food Chem., 50*, 2062–2067.

Darias-Martin, J., Carrillo-Lopez, M., Echavarri, J. F., & Diaz-Romero, C. (2007). The magnitude of copigmentation in the colour of aged red wines made in the Canary Islands. *Eur. Food Res. Tech., 224*, 643–648.

Davies, A. J., & Mazza, G. (1993). Copigmentation of simple and acylated anthocyanins with colorless phenolic-compounds. *J. Agric. Food Chem., 41*, 716–720.
de Freitas, V. A. P., & Mateus, N. (2001). Structural features of procyanidin interactions with salivary proteins. *J. Agric. Food Chem., 49*, 940–945.
de Freitas, V. A. P., & Mateus, N. (2002). Nephelometric study of salivary protein-tannin aggregates. *J. Sci. Food Agric., 82*, 113–119.
de Freitas, V. A. P., Glories, Y., Bourgeois, G., & Vitry, C. (1998). Characterisation of oligomeric and polymeric procyanidins from grape seeds by liquid secondary ion mass spectrometry. *Phytochemistry, 49*, 1435–1441.
de Freitas, V. A. P., Glories Y., & Augustin, M. (2001). Developmental changes of procyanidins in grapes of red Vitis vinifera varieties and their composition in respective wines. *Am. J. Enol. Vitic., 51*, 397–403.
de Freitas, V. A. P., Carvalho, E., & Mateus, N. (2003). Study of the Influence of carbohydrates on the Protein-Tannin aggregation by nephelometry. *Food Chem., 81*, 503–509.
de Freitas, V., Sousa, C., Silva, A., Santos-Buelga, C., & Mateus, N. (2004). Synthesis of a new catechin-pyrylium derived pigment. *Tetrahedron Lett., 45*, 9349–9352.
de Wijk, R. A., & Prinz, J. F. (2005). The role of friction in perceived oral texture. *Food Qual. Pref., 16*, 121–129.
di Stefano, R., Gentilini, N., & Panero, L. (2005). Experimental observations about copigmentation phenomenon. *Riv. Vit. Enol., 58*, 35–50.
Doco, T., Williams, P., Moutounet, M., & Pellerin, P. (2000). Les polysaccharides du vin. *Bull. O.I.V., 73*, 785–792.
Dodds, M. W. J., Johnson, D. A., & Yeh, C. K. (2005). Health benefits of saliva: a review. *J. Den., 33*, 223–233.
Dueñas, M., Fulcrand, H., & Cheynier, V. (2006). Formation of anthocyanin-flavanol adducts in model solutions. *Anal. Chim. Acta, 563*, 15–25.
Edelmann, A., & Lendl, B. (2002). Toward the optical tongue: Flow-through sensing of tannin-protein interactions based on FTIR spectroscopy. *J. Am. Chem. Soc., 124*, 14741–14747.
Eiro, M. J., & Heinonen, M. (2002). Anthocyanin color behavior and stability during storage: Effect of intermolecular copigmentation. *J. Agric. Food Chem., 50*, 7461–7466.
Escalona, H., Birkmyre, L., Piggott, J. R., & Paterson, A. (2002). Effect of maturation in small oak casks on the volatility of red wine aroma compounds. *Anal. Chim. Acta, 458*, 45–54.
Escot, S., Feuillat, M., Dulau, L., & Charpentier, C. (2001). Release of polysaccharides by yeasts and the influence of released polysaccharides on colour stability and wine astringency. *Aust. J. Grape Wine Res., 7*, 153–159.
Escribano-Bailon, T., Dangles, O., & Brouillard, R. (1996). Coupling reactions between flavylium ions and catechin. *Phytochemistry, 41*, 1583–1592.
Escribano-Bailon, M. T., Santos-Buelga, C., Francia-Aricha, E. M., Rivas-Gonzalo, J. C., & Heredia, F. J. (1999). Flavanol-anthocyanin and colour quality. In *Proceedings of 1st International Congress Pigment in Food Technology* (pp. 363–367). Sevilla, Spain.
Escribano-Bailon, T., Alvarez-Garcia, M., Rivas-Gonzalo, J. C., Heredia, F. J., & Santos-Buelga, C. (2001). Color and stability of pigments derived from the acetaldehyde-mediated condensation between malvidin 3-O-glucoside and (+)-catechin. *J. Agric. Food Chem., 49*, 1213–1217.
Es-Safi, N., Fulcrand, H., Cheynier, V., & Moutounet, M. (1999). Studies on the acetaldehyde-induced condensation of (-)-epicatechin and malvidin 3-*O*-glucoside in a model solution system.*J. Agric. Food Chem., 47*, 2096–2102.
Es-Safi, N.-E., Cheynier, V., & Moutounet, M. (2002). Interactions between cyanidin 3-O-glucoside and furfural derivatives and their impact on food color changes. *J. Agric. Food Chem., 50*, 5586–5595.
Fernandez de Simon, B., Hernandez, T., Cadahia, E., Dueñas, M., & Estrella, I. (2003). Phenolic compounds in a Spanish red wine aged in barrels made of Spanish, French and American oak wood. *Eur. Food Res. Tech., 216*, 150–156.

Fischer, U., & Noble A. C. (1994). The effect of ethanol, catechin concentration, and pH on sournss and bittherness of wine. *Am. J. Enol. Vitic., 45*, 6–10.

Francia-Aricha, E. M., Guerra, M. T., Rivas-Gonzalo, J. C., & Santos-Buelga, C. (1997). New anthocyanin pigments formed after condensation with flavanols. *J. Agric. Food Chem., 45*, 2262–2266.

Frazier, R. A., Papadopoulou, A., Mueller-Harvey, I., Kissoon, D., & Green, R. J. (2003). Probing protein-tannin interactions by isothermal titration microcalorimetry. *J. Agric. Food Chem., 51*, 5189–5195.

Frazier, R. A., Papadopoulou, A., & Green, R. J. (2006). Isothermal titration calorimetry study of epicatechin binding to serum albumin. *J. Pharm. Biomed. Anal., 41*, 1602–1605.

Fry, S. C. (1995). Polysaccharide-modifying enzymes in the plant-cell wall. *Ann. Rev. Plant Physiol. Plant Mol. Biol., 46*, 497–520.

Fulcrand, H., Cameira Dos Santos, P. J., Sarni-Manchado, P., Cheynier, V., & FabreBonvin, J. (1996). Structure of new anthocyanin-derived wine pigments. *J. Chem. Soc. Perkin Trans., 1*, 735–739.

Gawel, R. (1998). Red wine astringency: a review. *Aust. J. Grape Wine Res., 4*, 74–95.

Gawel, R., Oberholster A., & Francis, I. L. (2000). A "Mouth-feel Wheel": terminology for communicating the mouth-feel characteristics of red wine. *Aust. J. Grape Wine Res., 6*, 203–207.

Gawel, R., Iland, P. G., & Francis, I. L. (2001). Characterizing the astringency of red wine: a case study. *Food Qual. Pref., 12*, 83–94.

Goldstein, J. L., & Swain, T. (1963). Changes in tannins in ripening fruits. *Phytochemistry, 2*, 371–383.

Gomez-Miguez, M., Gonzalez-Manzano, S., Escribano-Bailon, M. T., Heredia, F. J., & Santos-Buelga, C. (2006). Influence of different phenolic copigments on the color of malvidin 3-glucoside. *J. Agric. Food Chem., 54*, 5422–5429.

Gonnet, J. F. (1998). Colour effects of co-pigmentation of anthocyanins revisited – 1. A colorimetric definition using the CIELAB scale. *Food Chem., 63*, 409–415.

Gonzalez-Manzano, S. (2007). Anthocyanins and flavanols in grape and wine. Influence of the composition in the processes of copigmentation and colour stability. *Ph.D. dissertation*, University of Salamanca (Spain).

Gonzalez-Manzano, S., Mateus, N., de Freitas, V. A. P., & Santos-Buelga, C. (2008a). Influence of the degree of polymerisation in the ability of catechins to act as anthocyanin copigments *228*, 83–92.

González-Manzano, S., Santos-Buelga, C., Dueñas, M., Rivas-Gonzalo, J. C., & Escribano-Bailón, M. T. (2007b). Colour implications of self-association processes of wine anthocyanins.*Eur. Food Res. Tech.* (in press). DOI 10.1007/s00217-007-0560-9.

Goto, T., & Kondo, T. (1991). Structure and molecular stacking of anthocyanins – flower color variation. *Angewadte Cemie International English Edition, 30*, 17–33.

Green, B. G. (1993). Oral Astringency – A tactile component of flavor. *Acta Psych., 84*, 119–125.

Guadalupe, Z., Palacios, A., & Ayestaran, B. (2007). Maceration enzymes and mannoproteins: a possible strategy to increase colloidal stability and color extraction in red wines.*J. Agric. Food Chem., 55*, 4854–4862.

Hagerman, A. E., & Butler, L. G. (1980). Determination of protein in tannin-protein Precipitates. *J. Agric. Food Chem., 28*, 944–947.

Hagerman, A. E., & Butler, L. G. (1981). The specificity of proanthocyanidin-protein interactions. *J. Biol. Chem., 256*, 4494–4497.

Hagerman, A. E., Rice, M. E., & Ritchard, N. T. (1998). Mechanisms of protein precipitation for two tannins, pentagalloyl glucose and epicatechin(16) (4 -> 8) catechin (procyanidin). *J. Agric. Food Chem., 46*, 2590–2595.

Hakansson, E., Pardon, K., Hayasaka, Y., de Sa, M., & Herderich, M. (2003). Structures and colour properties of new red wine pigments. *Tetrahedron Lett., 44*, 4887–4891.

Haslam, E. (1980). In vino veritas: oligomeric procyanidins and the ageing of red wines. *Phytochemistry, 16*, 1625–1670.

Haslam, E. (1998a). Taste, bitterness and astringency. In E. Haslam (Ed.), *Pratical polyphenolics: from structure to molecular recognition and physiological action* (pp. 178–225). Cambridge: University Press.
Haslam, E. (1998b). Maturation – Changes in astringency. In E. Haslam (Ed.), *Pratical polyphenolics: from structure to molecular recognition and physiological action* (pp. 226–261). Cambridge: University Press.
Hatano, T., & Hemingway, R. W. (1996). Association of (+)-catechin and catechin-(4 alpha->8)-catechin with oligopeptides. *Chem. Comm., 22*, 2537–2538.
Hermosin, I. (2003). Influence of ethanol content on the extent of copigmentation in a Cencibel Young Red Wine.*J. Agric. Food Chem., 51*, 4079–4083.
Hermosin, I., Sanchez-Palomo, E., & Vicario-Espinosa, A. (2005). Phenolic composition and magnitude of copigmentation in young and shortly aged red wines made from cultivars, Cabernet Sauvignon, Cencibel, and Syrah. *Food Chem., 92*, 269–283.
Hollman, P. C. H., & Arst, I. C. W. (2000). Flavonols, flavones and flavanols – nature, occurrence and dietary burden. *J. Sci. Food Agric., 80*, 1081–1092.
Horne, J., Hayes, J., & Lawless, H. T. (2002). Turbidity as a measure of salivary protein reactions with astringent substances. *Chem. Sens., 27*, 653–659.
Hoshino, T. (1991). An approximate estimate of self-association constants the self-stacking conformation of malvin quinoidal bases studied by 1H NMR. *Phytochemistry, 30*, 2049–2055.
Hoshino, T., Matsumoto, U., Harada, N., & Goto, T. (1981). Chiral excitation coupled stacking of anthocyanins: Interpretation of the origin of anomalous CD induced by anthocyanin association. *Tetrahedron Lett., 22*, 3621–3624.
Houbiers, C., Lima, J. C., Maçanita, A. L., & Santos, H. (1998). Color stabilization of malvidin 3-glucoside: Self-aggregation of the flavylium cation and copigmentation with the *Z*-chalcone form. *J. Phy. Chem. B, 102*, 3578–3585.
Huber, D. J. (1983). Polyuronide Degradation and Hemicellulose Modifications in Ripening Tomato Fruit. *J. Am. Soc. Hort. Sci., 108*, 405–409.
Jobstl, E., O'Connell, J., Fairclough, J. P. A., & Williamson, M. P. (2004). Molecular model for astringency produced by polyphenol/protein interactions. *Biomacromolecules, 5*, 942–949.
Jobstl, E., Howse, J. R., Fairclough, J. P. A., & Williamson, M. P. (2006). Noncovalent cross-linking of casein by epigallocatechin gallate characterized by single molecule force microscopy. *J. Agric. Food Chem., 54*, 4077–4081.
Jurd, L. (1967). Catechin-flavylium salt condesation reactions. *Tetrahedron, 23*, 1057–1064.
Jurd, L. (1969). Review of polyphenol condensation reactions and their possible occurrence in the aging of wines. *Am. J. Enol. Vitic., 20*, 191–195.
Jurd, L., & Somers, T. C. (1970). The formation of xanthylium salts from proanthocyanidins. *Phytochemistry, 9*, 419–427.
Kallithraka, S., Bakker, J., & Clifford, M. N. (1998). Evidence that salivary proteins are involved in astringency. *J. Sens. Stud., 13*, 29–43.
Kandra, L., Gyemant, G., Zajacz, A., & Batta, G. (2004). Inhibitory effects of tannin on human salivary alpha-amylase. *Biochem. Bioph. Res. Comm., 319*, 1265–1271.
Kauffman, D. L, & Keller, P. J. (1979). Basic proline-rich proteins in human-parotid saliva from a single subject. *Arch. Oral Biol., 24*, 249–256.
Kauffman, D. L., Bennick, A., Blum, M., & Keller, P. J. (1991). Basic proline-rich proteins from human parotid-saliva – Relationships of the covalent structures of 10 proteins from a single individual. *Biochemistry, 30*, 3351–3356.
Kennedy, J. A., Matthews M. A., & Waterhouse A. L. (2000). Changes in grape seed polyphenols during fruit ripening. *Phytochemistry, 55*, 77–85,
Kovac, V., Alonso, E., Bourzeix, M., & Revilla, E. (1992). Effect of several enological practices on the content of catechins and proanthocyanidins of red wines. *J. Agric. Food Chem., 40*, 1953–1957.
Kovac, V., Alonso, E., & Revilla, E. (1995). The effect of adding supplementary quantities of seeds during fermentation on the phenolic composition of wines. *Am. J. Enol. Vitic., 46*, 363–367.

Krueger, C. G., Dopke, N. C., Treichel, P. M., Folts, J., & Reed, J. D. (2000). Matrix-assisted laser desorption/ionization time-of-flight mass spectrometry of polygalloyl polyflavan-3-ols in grape seed extract. *J. Agric. Food Chem., 48*, 1663–1667.

Laborde, B., Moine-Ledoux, V., Richard, T., Saucier, C., Dubourdieu, D., & Monti, J.-P. (2006). PVPP-polyphenol complexes: A molecular approach. *J. Agric. Food Chem., 54*, 4383–4389.

Lea, A. G. H. (1990). Bitterness and astringency: the procyanidines of fermented apple ciders. In R. L. Rouseff (Ed.), *Bitterness in Foods and Beverages. Developments in Food Science 25* (pp. 123–143). Amsterdam: Elsevier.

Lea, A. G. H. (1992). Flavor, color, and stability in fruit products: The effect of polyphenols. In R. W. Hemingway & P. E. Laks (Ed.), *Plant Polyphenols* (pp. 827–837). New York: Plenum Press.

Lee, C. B., & Lawless, H. T. (1991). Time-course of astringent sensations. *Chem. Senses, 16*, 225–238.

Lesschaeve, I., & Noble, A. C. (2005). 1 Polyphenols: factors influencing their sensory properties and their effects on food and beverage preferences. *Am. J. Clin. Nut., 81*, 330S–335S.

Levengood, J., & Boulton, R. (2004). The variation in the color due to copigmentation in young Cabernet Sauvignon wines. In A. L. Waterhouse & J. A. Kennedy (Eds.), *Red wine color: revealing the mysteries* (pp. 35–52). Washington DC: American Chemical Society.

Lewis, C. E., Walker, J. R. L., & Lancaster, J. E. (1995). Effect of polysaccharides on the color of anthocyanins. *Food Chem., 54*, 315–319.

Liao, H., Cai, Y., & Haslam, E. (1992). Polyphenols interactions. Anthocyanins: copigmentation and colour changes in young red wines. *J. Sci. Food Agric., 59*, 299–305.

Llaudy, M. C., Canals, R., Canals, J. M., Rozes, N., Arola, L., & Zamora, F. (2004). New method for evaluating astringency in red wine. *J. Agric. Food Chem., 52*, 742–746.

Lorenzo, C., Pardo, F., Zalacain, A., Alonso, G. L., & Salinas M. R. (2005). Effect of Red Grapes Co-winemaking in Polyphenols and Color of Wines. *J. Agric. Food Chem., 53*, 7609–7616.

Lu, Y., & Bennick, A. (1998). Interaction of tannin with human salivary proline-rich proteins. *Arch. Oral Biol., 43*, 717–728.

Luck, G., Liao, H., Murray, N. J., Grimmer, H. R., Warminski, E. E., Williamson, M. P., Lilley, T. H., & Haslam, E. (1994). Polyphenols, astringency and proline-rich proteins. *Phytochemistry, 37*, 357–371.

Malien-Aubert, C., Dangles, O., & Amiot, M. J. (2002). Influence of procyanidins on the color stability of eonin solutions. *J. Agric. Food Chem., 50*, 3299–3305.

Mateus, N., & de Freitas, V. A. P. (2001). Evolution and stability of anthocyanin-derived pigments during port wine aging. *J. Agric. Food Chem., 49*, 5217–5222.

Mateus, N., de Pascual-Teresa, S., Rivas-Gonzalo, J. C., Santos-Buelga, C., & de Freitas, V. A. P. (2002a). Structural diversity of anthocyanin-derived pigments in port wines. *Food Chem., 76*, 335–342.

Mateus, N., Silva, A. M. S., Santos-Buelga, C., Rivas-Gonzalo, J. C., & de Freitas, V. A. P. (2002b). Identification of anthocyanin-flavanol pigments in red wines by NMR and mass spectrometry. *J. Agric. Food Chem., 50*, 2110–2116.

Mateus, N., Silva, A. M. S., Rivas-Gonzalo, J. C., Santos-Buelga, C., & de Freitas, V. A. P. (2003). A new class of blue anthocyanin-derived pigments isolated from red wines. *J. Agric. Food Chem., 51*, 1919–1923.

Mateus, N., Carvalho, E., Luis, C., & de Freitas, V. A. P. (2004a). Influence of the tannin structure towards the disruption effect of carbohydrates on protein-tannin aggregates. *Anal. Chim. Acta, 513*, 135–140.

Mateus, N., Oliveira, J., Santos-Buelga, C., Silva, A. M. S., & de Freitas, V. A. P. (2004b). NMR structural characterization of a new vinylpyranoanthocyanin-catechin pigment (a portisin). *Tetrahedron Lett., 45*, 3455–3457.

Mateus, N., Pinto, R., Ruão, P., & de Freitas, V. A. P. (2004c). Influence of the addition of grape seed procyanidins to Port wines in the resulting reactivity with human salivary proteins. *Food Chem., 84*, 195–200.

Maury, C., Sarni-Manchado, P., Lefebvre, S., Cheynier, V., & Moutounet, M. (2003). Influence of fining with plant proteins on proanthocyanidin composition of red wines. *Am. J. Enol. Vitic.*, *54*, 105–111.

McDonald, M. S., Hughes, M., Burns, J., Lean, M. E. J., Matthews, D., & Crozier, A. (1998). Survey of the free and conjugated myricetin and quercetin content of red wines of different geographical origins. *J. Agric. Food Chem.*, *46*, 368–375.

Messana, I., Cabras, T., Inzitari, R., Lupi, A., Zuppi, C., Olmi, C., Fadda, M. B., Cordaro, M., Giardina, B., & Castagnola, M. (2004). Characterization of the human salivary basic proline-rich protein complex by a proteomic approach. *J. Proteome Res.*, *3*, 792–800.

Mirabel, M., Saucier, C., Guerra, C., & Glories, Y. (1999). Copigmentation in model wine solutions: Ocurrence and relation to wine ageing. *Am. J. Enol. Vitic.*, *50*, 211–218.

Mirabel, E. (1991). Polyphenol Interactions.5. Anthocyanin Copigmentation. *J. Chem. Soc. Perkin Trans.*, *2*, 1287–1296.

Monagas, M., Martin-Alvarez, P. J., Bartolome, B., & Gomez-Cordoves, C. (2006). Statistical interpretation of the color parameters of red wines in function of their phenolic composition during aging in bottle. *Eur. Food Res. Tech.*, *222*, 702–709.

Monteleone, E., Condelli, N., Dinnella, C., & Bertuccioli, M. (2004). Prediction of perceived astringency induced by phenolic compounds. *Food Qual. Pref.*, *15*, 761–769.

Murray, N. J., Williamson, M. P., Lilley, T. H., & Haslam, E. (1994). Study of the interaction between salivary proline-rich proteins and a polyphenol by 1H-NMR Spectroscopy. *Eur. J. Biochem.*, *219*, 923–935.

Naczk, M., Oickle, D., Pink, D., & Shahidi, F. (1996). Protein precipitating capacity of crude canola tannins: Effect of pH, tannin, and protein concentrations. *J. Agric. Food Chem.*, *44*, 2144–2148.

Noble, A. (1990). Bitterness and astringency in wine. In R. Rousseff (Ed.), *Bitterness in Foods and Beverages* (pp. 145–158). Amsterdam: Elsevier.

Nonier, M. F., Pianet, I., Laguerre, M., Vivas, N., & Vivas de Gaulejac, N. (2006). Condensation products derived from flavan-3-ol oak wood aldehydes reaction. 1. Structural investigation. *Anal. Chim. Acta*, *563*, 76–83.

Oh, H. I., Hoff, J. E., Armstrong, G. S., & Haff, L. A. (1980). Hydrophobic interaction in tannin-protein complexes. *J. Agric. Food Chem.*, *28*, 394–398.

Okuda, T., Mori K., & Hatano T. (1985). Relationship of the structure of tannins to the binding activities with hemoglobin and methylene blue.*Chem. Pharm. Bull.*, *33*, 1424–1433.

Oliveira, J., Fernandes, V., Miranda, C., Santos-Buelga, C., Silva, A. M. S., de Freitas, V. A. P., & Mateus, N. (2006a). Color properties of four cyanidin-pyruvic acid adducts. *J. Agric. Food Chem.*, *54*, 6894–6903.

Oliveira, J., Santos-Buelga, C., Silva, A. M. S., de Freitas, V. A. P., & Mateus, N. (2006b). Chromatic and structural features of blue anthocyanin-derived pigments present in Port wine. *Anal. Chim. Acta*, *563*, 2–9.

Ozawa, T., Lilley, T. H., & Haslam, E. (1987). Polyphenol interactions - astringency and the loss of astringency in ripening fruit. *Phytochemistry*, *26*, 2937–2942.

Papadopoulou, A., Green, R. J., & Frazier, R. A. (2005). Interaction of flavonoids with bovine serum albumin: A fluorescence quenching study. *J. Agric. Food Chem.*, *53*, 158–163.

Peleg, H., Karine Gacon, K., Schlich, P., & Noble, A. C. (1999). Bitterness and astringency of flavan-3-ol monomers, dimers and trimers. *J. Sci. Food Agric.*, *79*, 1123–1128.

Perez-Prieto, L. J., de la Hera-Orts, M. L., Lopez-Roca, J. M., Fernandez-Fernandez, J. I., & Gomez-Plaza, E. (2003). Oak-matured wines: influence of the characteristics of the barrel on wine colour and sensory characteristics. *J. Sci. Food Agric.*, *83*, 1445–1450.

Pissarra, J., Mateus, N., Rivas-Gonzalo, J., Santos-Buelga, C., & de Freitas, V. A. P. (2003). Reaction between malvidin-3-glucoside and (+)-catechin in model solutions containing different aldehydes. *J. Food Sci.*, *68*, 476–481.

Pissarra, J., Lourenço, S., González-Paramás, A. M., Mateus, N., Santos-Buelga, C., Silva, A. M. S., & de Freitas, V. A. P. (2004). Structural characterization of new malvidin-3-glucoside-catechin aryl/alkyl pigments. *J. Agric. Food Chem.*, *52*, 5519–5526.

Pissarra, J., Lourenço, S., González-Paramás, A. M., Mateus, N., Santos-Buelga, C., Silva, A. M. S., & de Freitas, V. A. P. (2005). Isolation and structural characterization of new anthocyanin-alkyl-catechin pigments. *Food Chem., 90*, 81–87.

Poncet-Legrand, C., Edelmann, A., Putaux, J. L., Cartalade, D., Sarni-Manchado, P., & Vernhet, A. (2006). Poly(L-proline) interactions with flavan-3-ols units: Influence of the molecular structure and the polyphenol/protein ratio. *Food Hydrocoll., 20*, 687–697.

Prasanna, V., Prabha, T. N., & Tharanathan, R. N. (2007). Fruit ripening phenomena – An overview. *Crit. Rev. Food Sci. Nut., 47*, 1–19.

Preys, S., Mazerolles, G., Courcoux, P., Samson, A., Fischer, U., Hanafi, M., Bertrand, D., & Cheynier, V. (2006). Relationship between polyphenolic compostion and some sensory properties in red wines using multiway analyses. *Anal. Chim. Acta, 563*, 126–136.

Remy, S., Fulcrand, H., Labarbe, B., Cheynier, V., & Moutounet, M. (2000). First confirmation in red wine of products resulting from direct anthocyanin-tannin reactions. *J. Sci. Food Agric., 80*, 745–751.

Remy-Tanneau, S., Le Guernevé, C., Meudec, E., & Cheynier, V. (2003). Characterization of a colorless anthocyanin-flavan-3-ol dimer containing both carbon-carbon and ether interflavanoid linkages by NMR and mass spectrometry. *J. Agric. Food Chem., 51*, 3592–3597.

Rentzsch. M., Schwarz, M., Winterhalter, P., & Hermosin, I., (2007). Formation of hydroxyphenyk-pyranoanthocyanins in Grenache wines: precursor levels and evolution during aging. *J. Agric. Food Chem., 55*, 4883–4888.

Ribéreau-Gayón, P., Glories, Y., Maujean, A., & Dubourdieu, D. (2000). *Handbook of Enology. Vol. 2. The chemistry of wine stabilization and treatments.* London: John Wiley and Sons, Inc.

Riou, V., Vernhet, A., Doco, T., & Moutounet, M. (2002). Aggregation of grape seed tannins in model wine – effect of wine polysaccharides. *Food Hydrocol., 16*, 17–23.

Rivas-Gonzalo, J. C., Bravo-Haro, S., & Santos-Buelga, C. (1995). Detection of compounds formed through the reaction of malvidin 3-monoglucoside and catechin in the presence of acetaldehyde. *J. Agric. Food Chem., 43*, 1444–1449.

Salas, E., Atanasova, V., Poncet-Legrand, C., Meudec, E., Mazauric, J. P., & Cheynier, V. (2004a). Demonstration of the occurrence of flavanol-anthocyanin adducts in wine and in model solutions. *Anal. Chim. Acta, 513*, 325–332.

Salas, E., Le Guerneve, C., Fulcrand, H., Poncet-Legrand, C., & Cheynier, V. (2004b). Structure determination and colour properties of a new directly linked flavanol-anthocyanin dimer. *Tetrahedron Lett., 45*, 8725–8729.

Salas, E., Dueñas, M., Schwarz, M., Winterhalter, P., Cheynier, V., & Fulcrand, H. (2005). Characterization of pigments from different high speed countercurrent chromatography wine fractions. *J. Agric. Food Chem., 53*, 4536–4546.

Santos-Buelga, C., & Scalbert, A. (2000). Proanthocyanidins and tannin-like compounds – nature, occurrence, dietary intake and effects on nutrition and health. *J. Sci. Food Agric., 80*, 1094–1117.

Sarni-Manchado, P., & Cheynier, V. (2002). Study of non-covalent complexation between catechin derivatives and peptides by electrospray ionization mass spectrometry. *J. Mass Spectr., 37*, 609–616.

Sarni-Manchado, P., Cheynier, V., & Moutounet, M. (1997). Reaction of enzimatically generated quinones with malvidin-3-glucoside. *Phytochemistry, 45*, 1365.

Schwarz, M., Jerz, G., & Winterhalter, P. (2003). Isolation and structure of Pinotin A, a new anthocyanin derivative from Pinotage wine. *Vitis 42*, 105–106.

Schwarz, M., & Winterhalter, P. (2004). Novel aged anthocyanins from Pinotage wines: isolation, characterization, and pathway of formation. In A. L. Waterhouse & J. A. Kennedy (Eds.), *Red wine color: revealing the mysteries* (pp. 179–197). Washington DC.

Siebert, K. J., Troukhanova, N. V., & Lynn, P. Y. (1996). Nature of polyphenol-protein interactions. *J. Agric. Food Chem., 44*, 80–85.

Simon, C., Barathieu, K., Laguerre, M., Schmitter, J. M., Fouquet, E., Pianet, I., & Dufourc, E. J. (2003). Three-dimensional structure and dynamics of wine tannin-saliva protein complexes. A multitechnique approach. *Biochemistry, 42*, 10385–10395.

Singleton, V. L., & Noble A. C. (1976). Wine flavor and phenolic substances. In C. O. Chichester (Ed.), *Advance in Food Research, Suppl. 3* (pp. 47–70). Washington: American Chemical Society.

Soares, S., Mateus, N., & de Freitas, V. A. P. (2007). Interaction of different polyphenols with bovine serum albumin (BSA) and human α-amylase (HSA) by quenching fluorescence. *J. Agric. Food Chem., 55*, 6726–6735.

Somers, T. C. (1966). Wine tannins – isolation of condensed flavonoid pigments by gel-filtration. *Nature, 209*, 368–370.

Somers, T. C. (1971). The phenolic nature of wine pigments. *Phytochemistry, 10*, 2175–2186.

Somers, T. C., & Evans, M. E. (1979). Grape pigment phenomena - interpretation of major color losses during vinification. *J. Sci. Food Agric., 30*, 623–633.

Souquet, J. M., Cheynier, V., Brossaud, F., & Moutounet, M. (1996). Polymeric proanthocyanidins from grape skins. *Phytochemistry, 43*, 509–512.

Sousa, C., Mateus, N., Perez-Alonso, J., Santos-Buelga, C., & de Freitas, V. A P. (2005). Preliminary study of oaklins, a new class of brick-red catechin-pyrylium pigments resulting from the reaction between catechin and wood aldehydes. *J. Agric. Food Chem., 53*, 9249–9256.

Sousa, C., Mateus, N., Silva, A. M. S., González-Paramás, A.,M., Santos-Buelga, C., & de Freitas, V. (2007). Structural and chromatic characterization of a new malvidin-3-glucoside-vanillyl-catechin pigment. *Food Chem., 102*, 1344–1351.

Taira, S., Ono, M., & Matsumoto, N. (1997). Reduction of persimmon astringency by complex formation between pectin and tannins. *Postharvest Biol. Technol., 12*, 265–271.

Tanaka, T., Takahashi, R., Houno, I., & Nonaka G.-I. (1994). Chemical evidence for the de-aastringency (insolubilization of tannins) of persimmon fruit. *J. Chem. Soc. Perkin Trans., 1*, 3013–3022.

Timberlake, C. F., & Bridle, P. (1976). Interactions between anthocyanins, phenolic compounds, and acetaldehyde and their significance in red wines. *Am. J. Enol. Vitic., 27*, 97–105.

Troszynska, A., Amarowicz, R., Lamparski, G., Wolejszo, A., & Barylko-Pikielna, N. (2006). Investigation of astringency of extracts obtained from selected tannins-rich legume seeds. *Food Qual. Pref., 17*, 31–35.

Vernhet, A., Pellerin, P., Prieur, C., Osmianski, J., & Moutounet, M. (1996). Charge properties of some grape and wine polysaccharide and polyphenolic fractions. *Am. J. Enol. Vitic., 47*, 25–30.

Vidal, S., Francis, L., Guyot, S., Marnet, N., Kwiatkowski, M., Gawel, R., Cheynier, V., & Waters E. (2003a). The mouth-feel properties of grape and apple proanthocyanidins in a wine-like medium. *J. Sci. Food Agric., 83*, 564–573.

Vidal, S., Cheynier, V, Waters E, & Noble A. C. (2003b). Effect of tannin composition and wine carbohydrates on astringency and bitterness. In H. Meiselman (Ed.), *5th Pangborn Sensory Science Symposium*. Boston: Elsevier.

Vidal, S., Francis, L., Noble, A., Kwiatkowski, M., Cheynier, V., & Waters, E. (2004a). Taste and mouth-feel properties of different types of tannin-like polyphenolic compounds and anthocyanins in wine. *Anal. Chim. Acta, 513*, 57–65.

Vidal, S., Meudec, E., Cheynier, V., Skouroumounis, G., & Hayasaka, Y. (2004b). Mass spectrometric evidence for the existence of oligomeric anthocyanins in grape skins. *J. Agric. Food Chem., 52*, 7144–7151.

Vidal, S., Courcoux, P., Francis, L., Kwiatkowski, M., Gawel, R., Williams, P., Waters, E., & Cheynier, V. (2004c). Use of an experimental design approach for evaluation of key wine components on mouth-feel perception. *Food Qual. Pref., 15*, 209–217.

Vivar-Quintana, A. M., Santos-Buelga, C., Francia-Aricha, E., & Rivas-Gonzalo, J. C. (1999). Formation of anthocyanin-derived pigments in experimental red wines. *Food Sci. Technol. Inter., 5*, 347–352.

Vivar-Quintana, A. M., Santos-Buelga, C., & Rivas-Gonzalo, J. C. (2002). Anthocyanin-derived pigments and colour of red wines. *Anal. Chim. Acta, 458*, 147–155.

Vivas de Gaulejac, N., Vivas, N., Absalon, C., & Nonier, M. F. (2001). Identification of procyanidin A2 in grape and wine of Vitis vinifera L. cv Merlot Noir and Carbernet-Sauvignon. *J. Int. Sci. Vigne Vin, 35*, 51–56.

Wildenradt, H. L., & Singleton V. L. (1974). Production of aldehydes as a result of oxidation of polyphenolic compounds and its relation to wine aging. *J. Agric. Food Chem., 25*, 119–126.

Wroblewski, K., Muhandiram, R., Chakrabartty, A., & Bennick, A. (2001). The molecular interaction of human salivary histatins with polyphenolic compounds. *Eur. J. Biochem., 268*, 4384–4397.

Yan, Q. Y., & Bennick, A. (1995). Identification of histatins as tannin-binding proteins in human saliva. *Biochem. J., 311*, 341–347.

Chapter 9E
Health-Promoting Effects of Wine Phenolics

Alberto Dávalos and Miguel A. Lasunción

Contents

The first systematic observations concerning the health-promoting effects of wine consumption were obtained within the MONICA project (WHO 1989), a worldwide monitoring system for cardiovascular disease organized by the World Health Organization. Despite the similar prevalence of risk factors such as hypercholesterolemia, hypertension, diabetes and high intake of saturated fats in different western industrialized nations, deaths from coronary heart disease are much lower in France than in the United Kingdom and USA. This has been attributed to the higher consumption of alcohol in France, particularly to the regular drinking of red wine (Renaud and de Lorgeril 1992; St Leger et al. 1979). As consumption of ethanol can itself produce beneficial effects similar to those attributed to wine consumption, such as an increase in high density lipoproteins (HDL) and a decrease in platelet aggregation (Mukamal et al. 2005; Sesso 2001), the relative contribution of alcohol to the effects of wine are unclear, and efforts have been made to identify specific components of red wine that account for the reduction of coronary heart disease in the context of the "French paradox." Polyphenols from red wine have been reported to exert potent antioxidant effects that prevent low-density lipoprotein (LDL) oxidation (Frankel et al. 1993; Fuhrman et al. 1995; Nigdikar et al. 1998), and despite some reports of the absence of an association (Sesso et al. 2003), they are serious candidates to explain the protective effects of vegetable and fruit consumption against cancer and cardiovascular diseases (Arts and Hollman 2005; Mink

A. Dávalos (✉)
Servicio de Bioquímica-Investigación, Hospital Ramón y Cajal, Ctra. de Colmenar, km 9, E-28034, Instituto de Salud Carlos III, Madrid, Spain
e-mail: adavalos.hrc@salud.madrid.org

M.V. Moreno-Arribas, M.C. Polo (eds.), *Wine Chemistry and Biochemistry*,
DOI 10.1007/978-0-387-74118-5_21,

et al. 2007). In addition to antioxidant activity, long before the French paradox, wine polyphenols had been described to elicit other biological effects including vasorelaxation, inhibition of platelet aggregation, inhibition of cell proliferation, migration, and angiogenesis, and effects on lipid metabolism. Furthermore, consumption of certain wine polyphenols mimics caloric restriction to extend longevity in some organisms (Iijima et al. 2002; Stoclet et al. 2004; Vita 2005; Wood et al. 2004). However, epidemiological observations indicate that the health benefits of wine are noted only when consumption is moderate, because excessive alcohol intake is of itself detrimental (Pitsavos et al. 2005; Sasaki 2000).

9E.1 Bioavailability

The biological effects of wine consumption depend on the bioavailability of the different polyphenols it contains. The first indirect evidence of their bioavailability came from the observation of increased plasma antioxidant capacity and reduced susceptibility of LDL to ex vivo oxidation after consumption of red wine polyphenols (Fuhrman et al. 1995; Maxwell et al. 1994; Nigdikar et al. 1998; Serafini et al. 1998). Direct evidence of the bioavailability of wine polyphenols was obtained by measuring their concentration in plasma or urine after the ingestion of wine or wine polyphenols (Bell et al. 2000; Bub et al. 2001; Donovan et al. 1999). Different groups of polyphenols from red and white wine have been reported to be absorbed and metabolized, including anthocyanins (Bub et al. 2001; Frank et al. 2003), flavonols (de Vries et al. 2001), stilbenes (Vitaglione et al. 2005), phenolic acids (Caccetta et al. 2000; Simonetti et al. 2001), and a variety of other polyphenols from different sources (Scalbert and Williamson 2000). In general, bioavailability of polyphenols varies widely depending on the dietary sources and the forms they contain (Manach et al. 2005). Based on ingestion of 50 mg of aglycone equivalent, plasma concentration of total metabolites has been calculated to range from 0 to 4μmol/L (Manach et al. 2005), with isoflavones and gallic acid being better absorbed and proanthocyanidins and anthocyanins less bioavailable.

Major factors determining the bioavailability and metabolic fate of polyphenols in the organism include the chemical structure, the amount of polyphenol ingested, the food matrix, and dietary factors (Bitsch et al. 2004; Goldberg et al. 2003; Nemeth et al. 2003; Scalbert and Williamson 2000). The first step in the absorption and metabolism of dietary flavonoid glycosides involves deglycosylation in the small intestine by lactase phlorizin hydrolase (LPH) and the cytosolic β-glucosidase (CBG) (Day et al. 2000; Nemeth et al. 2003). LPH is a membrane-bound β-glucosidase that is primarily responsible for the hydrolysis of milk lactose and is exposed on the luminal surface of enterocytes, whereas CBG is a broad-specificity β-glucosidase and is located intracellularly (de Graaf et al. 2001; Mantei et al. 1988). Considerable evidence is available supporting models of absorption of dietary flavonoid glycosides and even of the 3-β-glucoside of *trans*-resveratrol. In these models, polyphenol glycosides reach the small intestine, where they may be hydrolyzed by LPH in the lumen or transported into the enterocyte by glucose transporters such as SGLT-1 and then hydrolyzed by CBG, releasing the aglycone (Ader et al. 2001; Henry-Vitrac et al. 2006; Nemeth et al. 2003; Shimizu et al. 2000). Then, the aglycone can be further metabolized into conjugates, diffuse passively to the circulation, or be excreted from the enterocyte back to the lumen through ATP-binding cassette (ABC) transporters (Nemeth et al. 2003; Henry-Vitrac et al. 2006). Very little is known about the absorption and metabolism of hydroxycinnamic and hydroxybenzoic acids. Bioavailability studies from different sources suggest that chlorogenic acid may be absorbed mainly in the colon but also in the stomach after hydrolysis by microbial esterases (Lafay et al. 2006; Nardini et al. 2002; Olthof et al. 2003), while ferulic acid may be absorbed in the small intestine (Bourne et al. 2000; Manach et al. 2005; Silberberg et al. 2006). Anthocyanins are reported to be absorbed and eliminated very rapidly, and their bioavailability is very low (Frank T et al. 2003; Bitsch et al. 2004; Manach et al. 2005). The stability of anthocyanins in the gastrointestinal tract is variable and identification of their different metabolites should be considered in order to evaluate the true degree of bioavailability (Borges et al. 2007; Manach et al. 2005; McDougall et al. 2005). Most polyphenols are excreted in urine and their metabolites exhibit different half-lives. Although the human tissue distribution has recently been described (Henning et al. 2006), tissue accumulation has not been yet demonstrated. However, plasma accumulation with repeated ingestion is expected at least with metabolites exhibiting longer half-lives (Manach et al. 2005).

9E.2 Antioxidant Properties and Vascular Effects

Reactive oxygen species (ROS) have been reported to act via different molecular pathways to play important roles in diverse pathological processes associated with aging, including cardiovascular diseases, certain types of cancer, hypertension, inflammation, neurological disorders, diabetes, and chronic kidney disease (Valko et al. 2007). They are generated as a result of normal cell metabolism. However, different external agents (cytokines, toxins, drugs, radiation, etc.) can also trigger ROS production. This can be largely counteracted by antioxidant defense systems, which can be enzymatic (superoxide dismutase, catalase, glutathione peroxidase), non-enzymatic (glutathione, paraoxonase), or dietary (antioxidant vitamins A, C, and E, and polyphenols). Increased ROS levels may damage lipids, proteins, and DNA, and can also trigger a stress signal that activates different redox-sensitive signaling pathways that can have either damaging or potentially protective effects (Finkel and Holbrook 2000). In general, it has been described that phenolic compounds are secondary antioxidants included in the category of free radical terminators. Phenolic antioxidants are excellent hydrogen or electron donors and their phenoxy radical intermediates are relatively stable due to resonance delocalization of unpaired electrons around the aromatic ring and lack of suitable sites for attack by molecular oxygen (Shahidi et al. 1992). Consumption of red wine (Maxwell et al. 1994; Serafini et al. 1998; Whitehead et al. 1995) and red grape juice (Castilla et al. 2006) have been shown to increase plasma/serum antioxidant activity and to protect LDL from oxidation evaluated both by the ex vivo susceptibility of the LDL particles to oxidation induced by Cu^{2+} or other oxidant agents and by analysis of oxidized LDL in plasma/serum (Avellone et al. 2006; Covas et al. 2003; di Bari et al. 2003; Frankel et al. 1993; Castilla et al. 2006). Consequently, the potent antioxidant activity of phenolic compounds in red wine has been proposed as an explanation for the French paradox (de Whalley et al. 1990). In vivo inhibition of LDL oxidation is only a minor part of the real complex effect exhibited by polyphenols in different processes. The recent in vivo demonstration that polyphenols from supplementation, in this case tea polyphenols, effectively reach tissue cells (Henning et al. 2006) provides new insight into the effect of wine polyphenols and polyphenols from other sources, complementing the extensive data on their effects at the cellular level, evaluated to date only in in vitro models. Free radical quenching is only a simplistic explanation of how polyphenols could exert their effects. The action of polyphenols on different pathways is complex and is still not well understood.

The mitogen-activated protein kinases (MAPKs) are a family of serine/threonine kinases that play an important role in the transduction of signals that coordinate the cellular response to different cellular environmental stimuli. The pathway ultimately modulates gene transcription in the nucleus and mediates various physiological and pathological changes in cell function (Roux and Blenis 2004; Huang et al. 2004). JNK/SAPK, p38, and ERK are the major proteins in the MAPK family. Several approaches have been used to evaluate the role of MAPKs in modulating the beneficial effects of polyphenols. Atherosclerosis progression is characterized by endothelial dysfunction. Independently of their antioxidant activity, which may increase the bioavailability of nitric oxide (NO), wine polyphenols improve endothelial function both via an increased expression of endothelial nitric oxide synthase (eNOS) (Leikert et al. 2002; Wallerath et al. 2002) and an activation of eNOS that is dependent upon the activation of phosphatidylinositol 3-kinase (PI3K)-Akt pathway through the activation of p38MAPK alpha, resulting in an increased formation of NO (Anter et al. 2004; Ndiaye et al. 2005; Stoclet et al. 2004). Synthesis of endothelin-1, an endothelium-derived peptide with potent vasoconstrictor and proliferative properties, have also been reported to be inhibited by wine procyanidins (Corder et al., 2001, 2006). The mechanism involved in this effect has yet to be fully elucidated; however, interfering with the ERK1/2 pathway through attenuation of ROS formation may be involved (Liu et al. 2003).

Advanced plaque formation is accelerated by formation of new blood vessels, proliferation and migration of vascular smooth muscle cells (VSMCs), and increased degradation of extracellular matrix predominantly by matrix metalloproteinases (MMP). Red wine polyphenols inhibit VSMC migration induced by platelet-derived grow factor via inhibition of PI3K activity and the p38 MAPK pathway; this effect is mediated by inhibition of phosphorylation of MKK3/6, a kinase upstream of p38MAPK (Iijima et al. 2002). Expression of major proangiogenic factors, vascular endothelial growth factor (VEGF), and MMP-2 has been shown to be reduced by wine polyphenols (Baron-Menguy et al. 2007; El Bedoui et al. 2005; Oak et al. 2003, 2004). VEGF expression was inhibited by preventing the redox-sensitive activation of the p38 MAPK pathway (Oak et al. 2003), while MMP activity was reduced via direct inhibition of MT1-MMP activity, the physiological activator of pro-MMP-2 (Oak et al. 2004; El Bedoui et al. 2005; Zhen et al. 2006). Polyphenol-mediated inhibition of metastasis-specific MMPs in tumor cells has been associated with the inhibition of activation of MAPK and nuclear factor-κB (NF-κB) pathways (Vayalil et al. 2004).

NF-κB is a transcription factor that plays a key role in various physiological processes and human diseases (Karin and Greten 2005; Hoffmann et al. 2006). It consists of different hetero- and homodimer proteins (RelA (p65), RelB, c-Rel, p50, and p52), with the RelA:p50 heterodimer acting as the primary mediator of NF-κB transcriptional activity. Four IκB proteins have been described in the NF-κB signaling module: IκBα, -β, and -ε, and nfκb2/p100 (Basak et al. 2007; Hoffmann et al. 2006). Different stimuli that activate NF-κB mediate the phosphorylation of IκBα through several IκB kinases (IKKs), and in this way IκBα becomes a target for ubiquitination and subsequent proteasomal degradation. IκBα degradation leads to the release from the cytosol of retained NF-κB dimers, which then migrate to the nucleus to exert transcriptional activity (Gilmore 2006; Hoffmann et al. 2006; Karin and Ben-Neriah 2000). NF-κB controls the expression of molecules implicated in most inflammatory processes, including pro-inflammatory cytokines and chemokines (IL-1, 2, 6, 8, TNFα, MCP-1, etc.), adhesion molecules (E-selectin, ICAM, VCAM, etc.), and several other proteins (Hoffmann et al. 2006). Data from in vitro models have shown that different polyphenols reduce NF-κB DNA binding activity induced by different stimuli (Nam 2006; Schubert et al. 2002; Terra et al. 2007). Our group and others have demonstrated that supplementation with a source of polyphenols such as grape juice or red wine reduces serum/plasma levels of MCP-1, VCAM, and ICAM (Blanco-Colio et al. 2007; Castilla et al. 2006; Estruch et al. 2004). The mechanism involved in this effect may be through the NF-κB pathway. In vivo, ingestion of polyphenols from different sources, including red wine, has been shown to effectively decrease NF-κB activation in mononuclear cells (Blanco-Colio et al. 2000, 2007; Davis et al. 2001). Concerning the mechanism underlying these effects, it has been reported that certain polyphenols, but not common dietary antioxidants, inhibit IKK activity induced by different stimuli and p65 phosphorylation (Ukil et al. 2006; Wheeler et al. 2004; Yang F et al. 2001). Inhibition of MAPK-mediated activation of NF-κB has also been proposed as the mechanism responsible for the effects of certain polyphenols (Hou et al. 2007; Lee et al. 2006).

9E.3 NADPH Oxidase

Overproduction of superoxide ($\bullet O_2^-$) has been implicated in the pathogenesis of various cardiovascular diseases. The main sources of human superoxide include the nicotinamide adenine dinucleotide phosphate oxidase (NADPH oxidase) enzyme complex, cyclooxygenase, mitochondrial oxidases, xanthine oxidase, and nitric oxide synthase. The NADPH oxidase enzyme system is considered a major source of superoxide in vascular cells, phagocytic polymorphonuclear neutrophils, monocytes, and platelets (Bedard and Krause 2007). Phagocyte NADPH oxidase is comprised of the membrane-bound catalytic core flavocytochrome b_{558}, formed by the subunit NOX2/gp91phox and p22phox, and the cytosolic components p47phox, p67phox, p40phox, the small G-protein Rac, and rap1A. Once cytosolic components are activated, they assemble with the flavocytochrome and share the capacity to transport electrons across the plasma membrane to generate superoxide and other downstream ROS. Homologs of the cytochrome subunits—NOX1, 3, 4, 5, DUOX1, and DUOX2—have been characterized. Nox1 and Nox4 are abundantly expressed in VSMCs and endothelial cells, respectively, but the distribution of the different members of the family is markedly different (Bedard and Krause 2007). In contrast to abundant phagocytic production of superoxide for host defense, vascular and other non-phagocytic cells produce superoxide, but to a significantly lesser extent. Increased NADPH oxidase activity contributes to a large number of diseases, in particular cardiovascular diseases and neurodegeneration (Cave et al. 2006; Infanger et al. 2006). NADPH oxidase overexpression and superoxide production have been shown to correlate with the severity of atherosclerosis (Sorescu et al. 2002), plaque stability (Azumi et al. 2002), oxidative stress in coronary artery disease (Guzik et al. 2000), plasma metalloproteinase-9 levels (Zalba et al. 2007), and circulating oxidized LDL (Carnevale et al. 2007; Fortuño et al. 2006), supporting a role for this enzyme in the pathogenesis of atherosclerosis (Azumi et al. 2002; Guzik et al. 2006). Thus, it might be suggested that reduction of NADPH-oxidase production and levels of systemic superoxide would reduce vascular disease associated with oxidative stress.

In vitro studies revealed that polyphenols reduce expression of the NADPH oxidase subunits p22phox and p67phox in endothelial cells (Xu et al. 2004; Ying et al. 2003), and NADPH oxidase-dependent platelet recruitment via the inhibition of protein kinase C (PKC) (Pignatelli et al. 2006a). Activation of PKC triggers the respiratory burst of phagocytes liberating large amounts of superoxide. Inhibition of PKC-induced activation by collagen was achieved by the synergistic effect of quercetin and catechin, but was unaffected by single polyphenols (Pignatelli et al. 2006b). In animal models, polyphenols prevented angiotensin II or deoxycorticosterone acetate salt-induced expression of p22phox or p47phox associated with hypertension (Jimenez et al. 2007; Sanchez et al. 2007; Sarr et al. 2006) and prevented the increased expression of NOX2 in fructose-fed rats (Al-Awwadi et al. 2005). Red wine polyphenols prevent angiotensin II-induced hypertension and endothelial dysfunction in rats by reducing the expression of p22phox and Nox-1 (Sarr et al. 2006). The mechanism through which polyphenols reduce NADPH oxidase expression and activity is not well understood; however, the redox-sensitive NFκB pathway may be involved. TNFα-induced activation of NFκB regulates the transcriptional activation of p47phox, p67phox, and NOX2 (Gauss et al. 2007), and NFκB sites have been described in the p22phox gene promoter (Moreno et al. 2003).

We have recently observed that in hemodialysis patients dietary supplementation with concentrated red grape juice rich in polyphenols reduces the phagocytic

NADPH oxidase activity in circulating neutrophils (unpublished observations). NADPH oxidase-dependent superoxide production appears to be abnormally high in mononuclear cells from these patients (Fortuño et al. 2005). The absence of a commercial pharmacological treatment to reduce or inhibit systemic NADPH oxidase enzyme complex highlights the importance of our results, which demonstrate for the first time that the oxidative stress produced by systemic NADPH oxidase may be counteracted by supplementation with polyphenols. Our in vitro studies suggest that the mechanism involved in this effect is the reduction of p22phox, p47phox, and NOX expression (unpublished observations).

9E.4 Effects on Lipid Metabolism

More than 93% of the body's cholesterol is located within the cells, where it plays various structural and metabolic roles. The rest circulates in the plasma within the lipoprotein particles. Given the link between plasma cholesterol level and cardiovascular diseases, the regulation of cholesterol homeostasis is of great interest. Cholesterol homeostasis depends on a highly regulated balance between the absorption of dietary cholesterol, de novo cholesterol biosynthesis, and biliary clearance and excretion. Familial hypercholesterolemia is one of the most common disorders of cell cholesterol homeostasis and is caused by a mutation in the gene encoding the LDL receptor. Plasma LDL binds to the receptor and is taken into cells to supply cholesterol for vital cellular functions. Defects in the LDL receptor lead to an increase in the plasma concentration of LDL-cholesterol, which may deposit in arteries and promote atherosclerosis. On the other hand, excess intracellular cholesterol induces cytotoxicity and apoptosis (Feng et al. 2003). ABC transporters are a large family of proteins that transport different molecules across the cell membrane (Kaminski et al. 2006). Several ABC transporters are involved in cholesterol efflux from cells. ABCA1 and ABCG1 are implicated in the reversal of cholesterol transport to Apo A-I rich particles, and ABCG5 and ABCG8 are implicated in the efflux of plant sterols to the intestinal lumen from enterocytes, as well as in bile acid excretion in the liver (Oram and Vaughan 2006). Mutations in either ABCG5 or ABCG8 produce the rare autosomal recessive disease called sitosterolemia, resulting in hyperabsorption of plant sterols (Berge et al. 2000). Mutations in ABCA1 lead to Tangier disease, characterized by cholesterol accumulation in macrophages and reduction of circulating mature HDL (Clee et al. 2000). In cultured macrophages, anthocyanins and various polyphenols induce cholesterol efflux and ABCA1 expression (Xia et al. 2005). These effects have been reported to be due, at least in part, to the activation of liver x receptor α (LXRα) and associated changes in gene expression (Sevov et al. 2006; Xia et al. 2005; 2007). The nuclear receptor LXRα is activated by oxysterols and is involved in regulating the metabolism of cholesterols and bile acids. It limits cholesterol accumulation through the expression of the cholesterol efflux genes ABCA1, ABCG1, ABCG5, and ABCG8. LXR-α also serves as a molecular link between cholesterol metabolism and inflammation (Tontonoz and Mangelsdorf 2003). Interestingly, anthocyanins have been reported to modify cholesterol distribution in lipid rafts in endothelial cells, possibly by activating ABCA1-mediated cholesterol efflux, which prevents the recruitment of TRAF-2 to lipid rafts and thus inhibits the CD40-CD40L inflammatory pathway associated with NFκB (Xia et al. 2007).

Cholesterol homeostasis is characterized by a feedback regulation achieved through the membrane-bound sterol regulatory element-binding protein (SREBP) family of transcription factors (Brown et al. 2000). SREBPs are synthesized as inactive precursor proteins and anchored to the endoplasmic reticulum (ER), where they interact with the SREBP-cleavage-activating protein (Scap). Insigs proteins are resident in the ER and interact with Scap, retaining the SREBP/Scap complex in the ER while Scap binds cholesterol. Scap possesses a sterol-sensing domain. In response to low cellular cholesterol levels, a conformational change in the Scap protein is induced and Scap is released by Insig. It then escorts the SREBPs to the golgi, where they are processed by two membrane-associated proteases (S1P and S2) that cleave at a site in the lumen (S1P) and a site within the first transmembrane domain of SREBPs, liberating the active SREBP fragment (Brown et al. 2000). The active fragments translocate to the nucleus, where they bind to the promoter of the different genes regulated through the SREBPs (Goldstein et al. 2006). Target genes regulated by SREBPs include the LDL receptor, 3-hydroxy-3-methylglutaryl coenzyme A reductase (the rate limiting enzyme in de novo cholesterol biosynthesis), several genes involved in the synthesis and metabolism of sterols, and embryonic signaling proteins. Lowering plasma cholesterol by reducing intestinal absorption, raising LDL receptor expression, reducing de novo biosynthesis, increasing reverse cholesterol transport, or promoting cholesterol elimination through the biliary system is of great interest to prevent atherosclerosis.

The first reported beneficial effects of wine intake on lipid metabolism were those concerning its antioxidant activity against LDL oxidation. Oxidative modification of LDL in the arterial wall plays a key role in the pathogenesis of atherosclerosis (Witztum and Steinberg 1991). Polyphenols from grapes have been demonstrated to inhibit both in vitro and ex-vivo LDL oxidation and lipid peroxidation, after ingestion of different wine or grape juice types, for different periods of time or in different amounts (Abu-Amsha et al. 2001; Aviram and Fuhrman 1998; Castilla et al. 2006; Frankel et al. 1993; Hayek et al. 1997; Ivanov et al. 2001; Nigdikar et al. 1998; O'Byrne et al. 2002; Pignatelli et al. 2006b; Stein et al. 1999). A plausible mechanism to explain this effect is through sparing of endogenous alpha-tocopherol content of LDL, thereby avoiding LDL oxidation (Deckert et al. 2002; Frank et al. 2006), and preservation or increase of paraoxonase activity (Fuhrman and Aviram 2002; Gouedard et al. 2004). Human paraoxonases are serum HDL-associated enzymes that can hydrolyze and reduce lipid peroxides in lipoproteins and arterial cells. Consumption of wine polyphenols and other sources of polyphenols reduces total plasma cholesterol and/or LDL cholesterol, both in healthy subjects (Cartron et al. 2003; Castilla et al. 2006; Zern et al. 2005) and different groups with varying health status, including hypercholesterolemic subjects

(Baba et al. 2007a; Jung et al. 2003), post-menopausal women (Zern et al. 2005), and hemodialysis patients (Castilla et al. 2006). Increased HDL cholesterol caused by polyphenols has also been reported in healthy individuals and patients (Baba et al. 2007b; Castilla et al. 2006; Covas et al. 2006; Mursu et al. 2004). In vitro models have provided evidence of other mechanisms that may explain these effects (Fig. 9E.1). A reduced synthesis and secretion of apolipoprotein B (ApoB) has been observed in both CaCo-2 enterocytes and HepG2 hepatocytes following treatment with polyphenols (Pal et al. 2005; Vidal et al. 2005; Takechi et al. 2004); such an effect could attenuate the synthesis and secretion of proatherogenic intestinal lipoproteins. Inhibition of microsomal transfer protein, diacylglycerol acyltransferase,

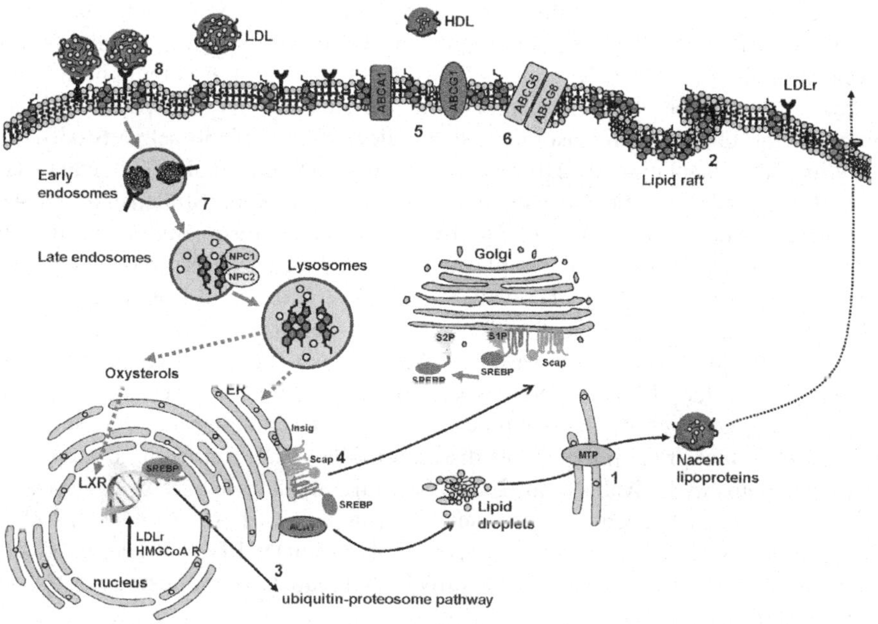

Fig. 9E.1 Effects of polyphenols on cholesterol homeostasis. Low-density lipoprotein (LDL) is taken up by the LDL receptor (LDLr) and transported through a complex vesicular pathway, from early endosomes to late endosomes and lysosomes. From there, cholesterol can enter the plasma membrane or reach the endoplasmic reticulum (ER), where cholesterol-sensing machinery regulates cholesterol homeostasis through the sterol regulatory element-binding protein (SREBP) pathway. Low levels of free cholesterol in the ER activate the SREBP pathway, leading to increased transcription of target genes including the LDLr. Excess free cholesterol is stored as cholesteryl ester in lipid droplets through the action of acyl coenzyme A cholesterol acyltransferase (ACAT). Microsomal triglyceride transfer protein (MTP) is necessary for lipoprotein assembly. Intracellular excess of free cholesterol activates cholesterol efflux by ATP-binding cassette (ABC) transporter proteins. Polyphenols have been reported to inhibit MTP (1), change cholesterol distribution in lipid rafts (2), inhibit the ubiquitin-proteasome pathway (3), enhance SREBP processing (4), increase liver x receptor-mediated expression of ABC efflux proteins (5), increase sterol elimination through bile acid excretion (6), disrupt endocytic LDL trafficking (7), and increase LDLr expression and activity (8)

and acyl CoA:cholesterol acyltransferase are the main mechanisms postulated for the reduced Apo B secretion (Allister et al. 2005; Borradaile et al. 2003; Casaschi et al. 2002; Wilcox et al. 2001). In animal models, it has also been shown that grape procyanidins increase cholesterol 7α-hydroxylase (CYP7A1), an observation which suggests increased cholesterol elimination via bile acids (Del Bas et al. 2005). CYP7A1 is a liver-specific enzyme that catalyzes the 7α-hydroxylation of cholesterol, the rate limiting step in the pathway responsible for the conversion of cholesterol into bile acids. Other sources of polyphenols have also been shown to increase bile acids, cholesterol, or steroid excretion (Sembries et al. 2006; Yang and Koo 2000). A possible mechanism explaining this effect is through the orphan nuclear receptor small heterodimer partner (Del Bas et al. 2005).

Evidence from in vitro and in vivo animal models suggests that the increased LDL receptor expression and the decrease in cholesterol biosynthesis are the main factors contributing to the polyphenol-induced reduction of plasma LDL-cholesterol levels (Bursill et al. 2007; Kim et al. 2004; Pal et al. 2003;). It is thought that the increased expression of LDL receptors is regulated by the SREBP pathway. Inhibition of the ubiquitin/proteasome-mediated degradation of the active SREBP-2, resulting in up-regulation of LDL receptor expression (Kuhn et al. 2004), enhanced SREBP processing via the SREBP cleavage activating protein (Mullen et al. 2004), and increases in the conversion of SREBP-1 to its active form (Bursill and Roach 2006), has been proposed as a possible mechanism. We have observed that polyphenols from grape juice reduce the availability of free cholesterol from endocytosed LDL in the endoplasmic reticulum for regulatory purposes (Dávalos et al. 2006). Thus, SREBP processing is increased and SRE-regulated target genes are induced. As a result, surface LDL receptor expression is increased and this increases the clearance of plasma cholesterol, thus decreasing plasma cholesterol levels. Our findings indicate that grape polyphenols disrupt or delay LDL trafficking through the endocytic pathway (Dávalos et al. 2006). Several polyphenols have been shown to be transported by LDL particles (Lamuela-Raventos et al. 1999; Natella et al. 2007). Whether increased concentrations of polyphenols within the LDL particles or effects on proteins responsible for cholesterol trafficking disrupt LDL trafficking has not yet been established.

9E.5 Life Span

Caloric restriction, defined as a reduction in calorie intake below usual ad libitum intake without malnutrition, has been widely demonstrated to extend life span in different species from yeast to mammals and delays several age-associated biochemical, physiological, and behavioral changes, thus indicating a broad relationship between energy intake and aging (Fontana and Klein 2007; Masoro 2000; Sohal and Weindruch 1996). Sirtuins are a family of NAD^+-dependent deacetylases conserved from *Escherichia coli* to humans and are known to mediate gene silencing at telomeres, DNA repair, and ribosomal DNA recombination (Guarente 1999; Imai et al. 2000). Caloric restriction has been shown to extend life span through activation of the silencing information regulator 2 (Sir2) gene in *Saccharomyces cerevisiae* (Lin et al. 2000) and *Caenorhabditis elegans* (Tissenbaum and Guarente 2001). The human homolog of Sir2, SIRT1, is a p53 deacetylase that promotes cell survival by negatively regulating the p53 tumor suppressor (Luo et al. 2001; Vaziri et al. 2001). Inducing SIRT1 expression by caloric restriction promotes long-term survival of mammalian cells (Cohen et al. 2004). Recently it has been shown that SIRT1 controls the gluconeogenesis/glycolytic pathway in the liver in response to fasting signals through the transcriptional coactivator of PGC-1α (Rodgers et al. 2005), suggesting a role for SIRT1 in energy metabolism, diabetes, and life span. Thus, the use of molecules that trigger SIRT1 activity, mimicking the effect of caloric restriction, would allow individuals to eat normally while the body responds as though food were in short supply. Different plant polyphenols have been found to stimulate SIRT1 activity; these include quercetin, piceatannol, and resveratrol (de Boer et al. 2006; Howitz et al. 2003). These and other compounds were found to have a chemical structure similar to that of the *trans* stilbene ring. Thus, the *trans* stilbene resveratrol showed the highest stimulation of SIRT1 activity (Howitz et al. 2003). Resveratrol has since been demonstrated to effectively extend life span in different species, including *S. cerevisiae* (Howitz et al. 2003), *C. elegans*, and the fruit fly *Drosophila melanogaster* (Wood et al. 2004), an effect that is dependent on the Sir2 gene. In mammals, resveratrol has been shown to exert a protective effect in several disease models, including cancer and diabetes (Garvin et al. 2006; Su et al. 2006). The SIRT1 pathway also improves health and survival in mice on high-calorie diets (Baur et al. 2006) and protects against neurodegeneration in mouse models of Alzheimer's disease and amyotrophic lateral sclerosis (Kim et al. 2007). The amount of resveratrol in grapes and red wine, however, is relatively low and depends on several strain-related and environmental factors, which together with the low bioavailability of resveratrol in mammals, its low solubility, and sensitivity to light and oxidation (Yang et al. 2007), may limit the achievement of target plasma concentrations from normal food ingestion. Also, although it has been shown that polyphenols protect against several pathological processes associated with aging and have remarkable effects mimicking caloric restriction in vitro and in animal models, it is not known whether resveratrol or other SIRT1 activators can extend life span in humans. Further proof that a drug extends life span would require difficult, extended clinical trials. Nevertheless, this promising effect of resveratrol along with the other cardioprotective effects of wine polyphenols support the contention that moderate wine consumption is health promoting.

The beneficial effects of wine may not be attributable to a single polyphenol but rather to the complex mixture of polyphenols it contains. There is evidence that antioxidant properties underlie most of the effects of wine, but at the same time, certain effects cannot be mimicked by common dietary antioxidants. Although polyphenols are not a *panacea*, those present in red grape and its derivatives appear to exert beneficial effects on health.

Acknowledgments The work was supported by grants from the Instituto Nacional de Investigación y Tecnología Agraria y Alimentaria (VIN03-027), Fondos FEDER, and Ministerio de Educación y Ciencia (AGL2004-07075), Spain. We thank Gema de la Peña for data revision and Dr. Iain Patten for editorial advice.

References

Abu-Amsha Caccetta, R., Burke, V., Mori, T. A., Beilin, L. J., Puddey, I. B., & Croft, K. D. (2001). Red wine polyphenols, in the absence of alcohol, reduce lipid peroxidative stress in smoking subjects. *Free Rad. Biol. Med., 30*, 636–642.

Ader, P., Block, M., Pietzsch, S., & Wolffram, S. (2001). Interaction of quercetin glucosides with the intestinal sodium/glucose co-transporter (SGLT-1). *Cancer Lett., 162*, 175–180.

Al-Awwadi, N. A., Araiz, C., Bornet, A., Delbosc, S., Cristol, J. P., Linck, N., Azay, J., Teissedre, P. L., & Cros, G. (2005). Extracts enriched in different polyphenolic families normalize increased cardiac NADPH oxidase expression while having differential effects on insulin resistance, hypertension, and cardiac hypertrophy in high-fructose-fed rats. *J. Agric. Food Chem., 53*, 151–157.

Allister, E. M., Borradaile, N. M., Edwards, J. Y., & Huff, M. W. (2005). Inhibition of microsomal triglyceride transfer protein expression and apolipoprotein B100 secretion by the citrus flavonoid naringenin and by insulin involves activation of the mitogen-activated protein kinase pathway in hepatocytes. *Diabetes, 54*, 1676–1683.

Anter, E., Thomas, S. R., Schulz, E., Shapira, O. M., Vita, J. A., & Keaney, J. F. Jr. (2004). Activation of endothelial nitric-oxide synthase by the p38 MAPK in response to black tea polyphenols. *J. Biol. Chem., 279*, 46637–46643.

Arts, I. C., & Hollman, P. C. (2005). Polyphenols and disease risk in epidemiologic studies. *Am. J. Clin. Nutr., 81*, 317S–325S.

Avellone, G., Di Garbo, V., Campisi, D., De Simone, R., Raneli, G., Scaglione, R., & Licata, G. (2006). Effects of moderate Sicilian red wine consumption on inflammatory biomarkers of atherosclerosis. *Eur. J. Clin. Nutr., 60*, 41–47.

Aviram, M., & Fuhrman, B. (1998). Polyphenolic flavonoids inhibit macrophage-mediated oxidation of LDL and attenuate atherogenesis. *Atherosclerosis, 137*, S45–S50.

Azumi, H., Inoue, N., Ohashi, Y., Terashima, M., Mori, T., Fujita, H., Awano, K., Kobayashi, K., Maeda, K., Hata, K., Shinke, T., Kobayashi, S., Hirata, K., Kawashima, S., Itabe, H., Hayashi, Y., Imajoh-Ohmi, S., Itoh, H., & Yokoyama, M. (2002). Superoxide generation in directional coronary atherectomy specimens of patients with angina pectoris: important role of NAD(P)H oxidase. *Arterioscler. Thromb. Vasc. Biol., 22*, 1838–1844.

Baba, S., Natsume, M., Yasuda, A., Nakamura, Y., Tamura, T., Osakabe, N., Kanegae, M., & Kondo, K. (2007a). Plasma LDL and HDL cholesterol and oxidized LDL concentrations are altered in normo- and hypercholesterolemic humans after intake of different levels of cocoa powder. *J. Nutr., 137*, 1436–1441.

Baba, S., Osakabe, N., Kato, Y., Natsume, M., Yasuda, A., Kido, T., Fukuda, K., Muto, Y., & Kondo, K. (2007b). Continuous intake of polyphenolic compounds containing cocoa powder reduces LDL oxidative susceptibility and has beneficial effects on plasma HDL-cholesterol concentrations in humans. *Am. J. Clin. Nutr., 85*, 709–717.

Baron-Menguy, C., Bocquet, A., Guihot, A. L., Chappard, D., Amiot, M. J., Andriantsitohaina, R., Loufrani, L., & Henrion, D. (2007). Effects of red wine polyphenols on postischemic neovascularization model in rats: low doses are proangiogenic, high doses anti-angiogenic. *FASEB J., 13*, 3511–3521.

Basak, S., Kim, H., Kearns, J. D., Tergaonkar, V., O'Dea, E., Werner, S. L., Benedict, C. A., Ware, C. F., Ghosh, G., Verma, I. M., & Hoffmann, A. (2007). A fourth IkappaB protein within the NF-kappaB signaling module. *Cell, 128*, 369–381.

Baur, J. A., Pearson, K. J., Price, N. L., Jamieson, H. A., Lerin, C., Kalra, A., Prabhu, V. V., Allard, J. S., Lopez-Lluch, G., Lewis, K., Pistell, P. J., Poosala, S., Becker, K. G., Boss, O., Gwinn, D., Wang, M., Ramaswamy, S., Fishbein, K. W., Spencer, R. G., Lakatta, E. G., Le Couteur, D., Shaw, R. J., Navas, P., Puigserver, P., Ingram, D. K., de Cabo, R., & Sinclair, D. A. (2006). Resveratrol improves health and survival of mice on a high-calorie diet. *Nature, 444*, 337–342.

Bedard, K., & Krause, K. H. (2007). The NOX family of ROS-generating NADPH oxidases: physiology and pathophysiology. *Physiol. Rev., 87*, 245–313.

Bell, J. R., Donovan, J. L., Wong, R., Waterhouse, A. L., German, J. B., Walzem, R. L., & Kasim-Karakas, S. E. (2000). (+)-Catechin in human plasma after ingestion of a single serving of reconstituted red wine. *Am. J. Clin. Nutr., 71*, 103–108.

Berge, K. E., Tian, H., Graf, G. A., Yu, L., Grishin, N. V., Schultz, J., Kwiterovich, P., Shan, B., Barnes, R., & Hobbs, H. H. (2000). Accumulation of dietary cholesterol in sitosterolemia caused by mutations in adjacent ABC transporters. *Science, 290*, 1771–1775.

Bitsch, R., Netzel, M., Frank, T., Strass, G., & Bitsch, I. (2004). Bioavailability and Biokinetics of Anthocyanins From Red Grape Juice and Red Wine. *J. Biomed. Biotechnol., 5*, 293–298.

Blanco-Colio, L. M., Valderrama, M., Alvarez-Sala, L. A., Bustos, C., Ortego, M., Hernandez-Presa, M. A., Cancelas, P., Gomez-Gerique, J., Millan, J., & Egido, J. (2000). Red wine intake prevents nuclear factor-kappaB activation in peripheral blood mononuclear cells of healthy volunteers during postprandial lipemia. *Circulation, 102*, 1020–1026.

Blanco-Colio, L. M., Muñoz-García, B., Martín-Ventura, J. L., Alvarez-Sala, L. A., Castilla, M., Bustamante, A., Lamuela-Raventós, R. M., Gómez-Gerique, J., Fernández-Cruz, A., Millán, J., & Egido, J. (2007). Ethanol beverages containing polyphenols decrease nuclear factor kappa-B activation in mononuclear cells and circulating MCP-1 concentrations in healthy volunteers during a fat-enriched diet. *Atherosclerosis, 192*, 335–341.

Borges, G., Roowi, S., Rouanet, J. M., Duthie, G. G., Lean, M. E., & Crozier, A. (2007). The bioavailability and absorption of anthocyanins: towards a better understanding. *Mol. Nutr. Food Res., 51*, 702–713.

Borradaile, N. M., de Dreu, L. E., Barrett, P. H., Behrsin, C. D., & Huff, M. W. (2003). Hepatocyte apoB-containing lipoprotein secretion is decreased by the grapefruit flavonoid, naringenin, via inhibition of MTP-mediated microsomal triglyceride accumulation. *Biochemistry, 42*, 1283–1291.

Bourne, L., Paganga, G., Baxter, D., Hughes, P., & Rice-Evans, C. (2000). Absorption of ferulic acid from low-alcohol beer. *Free Rad. Res., 32*, 273–280.

Brown, M. S., Ye, J., Rawson, R. B., & Goldstein, J. L. (2000). Regulated intramembrane proteolysis: a control mechanism conserved from bacteria to humans. *Cell, 100*, 391–398.

Bub, A., Watzl, B., Heeb, D., Rechkemmer, G., & Briviba, K. (2001). Malvidin-3-glucoside bioavailability in humans after ingestion of red wine, dealcoholized red wine and red grape juice. *Eur. J. Nutr., 40*, 113–120.

Bursill, C. A., & Roach, P. D. (2006). Modulation of cholesterol metabolism by the green tea polyphenol (-)-epigallocatechin gallate in cultured human liver (HepG2) cells. *J. Agric. Food Chem., 54*, 1621–1626.

Bursill, C. A., Abbey, M., & Roach, P. D. (2007). A green tea extract lowers plasma cholesterol by inhibiting cholesterol synthesis and upregulating the LDL receptor in the cholesterol-fed rabbit. *Atherosclerosis, 193*, 86–93.

Caccetta, R. A., Croft, K. D., Beilin, L. J., & Puddey, I. B. (2000). Ingestion of red wine significantly increases plasma phenolic acid concentrations but does not acutely affect ex vivo lipoprotein oxidizability. *Am. J. Clin. Nutr., 71*, 67–74.

Carnevale, R., Pignatelli, P., Lenti, L., Buchetti, B., Sanguigni, V., Di Santo, S., & Violi, F. (2007). LDL are oxidatively modified by platelets via GP91(phox) and accumulate in human monocytes. *FASEB J., 21*, 927–934.

Cartron, E., Fouret, G., Carbonneau, M. A., Lauret, C., Michel, F., Monnier, L., Descomps, B., & Leger, C. L. (2003). Red-wine beneficial long-term effect on lipids but not on antioxidant characteristics in plasma in a study comparing three types of wine–description of two O-methylated derivatives of gallic acid in humans. *Free Rad. Res., 37*, 1021–1035.

Casaschi, A., Wang, Q., Dang, K., Richards, A., & Theriault, A. (2002). Intestinal apolipoprotein B secretion is inhibited by the flavonoid quercetin: potential role of microsomal triglyceride transfer protein and diacylglycerol acyltransferase. *Lipids, 37*, 647–652.

Castilla, P., Echarri, R., Dávalos, A., Cerrato, F., Ortega, H., Teruel, J. L., Lucas, M. F., Gómez-Coronado, D., Ortuno, J., & Lasunción, M. A. (2006). Concentrated red grape juice exerts antioxidant, hypolipidemic, and antiinflammatory effects in both hemodialysis patients and healthy subjects. *Am. J. Clin. Nutr., 84*, 252–262.

Cave, A. C., Brewer, A. C., Narayanapanicker, A., Ray, R., Grieve, D. J., Walker, S., & Shah, A. M. (2006). NADPH oxidases in cardiovascular health and disease. *Antiox. Redox Signal., 8*, 691–728.

Clee, S. M., Kastelein, J. J., van Dam, M., Marcil, M., Roomp, K., Zwarts, K. Y., Collins, J. A., Roelants, R., Tamasawa, N., Stulc, T., Suda, T., Ceska, R., Boucher, B., Rondeau, C., DeSouich, C., Brooks-Wilson, A., Molhuizen, H. O., Frohlich, J., Genest, J. Jr., & Hayden, M. R. (2000). Age and residual cholesterol efflux affect HDL cholesterol levels and coronary artery disease in ABCA1 heterozygotes. *J. Clin. Inves., 106*, 1263–1270.

Cohen, H. Y., Miller, C., Bitterman, K. J., Wall, N. R., Hekking, B., Kessler, B., Howitz, K. T., Gorospe, M., de Cabo, R., & Sinclair, D. A. (2004). Calorie restriction promotes mammalian cell survival by inducing the SIRT1 deacetylase. *Science, 305*, 390–392.

Corder, R., Douthwaite, J. A., Lees, D. M., Khan, N. Q., Viseu Dos Santos, A. C., Wood, E. G., & Carrier, M. J. (2001). Endothelin-1 synthesis reduced by red wine. *Nature, 414*, 863–864.

Corder, R., Mullen, W., Khan, N. Q., Marks, S. C., Wood, E. G., Carrier, M. J., & Crozier, A. (2006). Oenology: red wine procyanidins and vascular health. *Nature, 444*, 566.

Covas, M. I., Konstantinidou, V., Mysytaki, E., Fito, M., Weinbrenner, T., De La Torre, R., Farre-Albadalejo, M., & Lamuela-Raventos, R. (2003). Postprandial effects of wine consumption on lipids and oxidative stress biomarkers. *Drugs Exp. Clin. Res., 29*, 217–223.

Covas, M. I., Nyyssonen, K., Poulsen, H. E., Kaikkonen, J., Zunft, H. J., Kiesewetter, H., Gaddi, A., de la Torre, R., Mursu, J., Baumler, H., Nascetti, S., Salonen, J. T., Fito, M., Virtanen, J., Marrugat, J., & EUROLIVE Study Group. (2006). The effect of polyphenols in olive oil on heart disease risk factors: a randomized trial. *Ann. Intern. Med., 145*, 333–341.

Dávalos, A., Fernández-Hernando, C., Cerrato, F., Martínez-Botas, J., Gómez-Coronado, D., Gómez-Cordoves, C., & Lasunción. M. A. (2006). Red grape juice polyphenols alter cholesterol homeostasis and increase LDL-receptor activity in human cells in vitro. *J. Nutr., 136*, 1766–1773.

Davis, J. N., Kucuk, O., Djuric, Z., & Sarkar, F. H. (2001). Soy isoflavone supplementation in healthy men prevents NF-kappa B activation by TNF-alpha in blood lymphocytes. *Free Rad. Biol. Med., 30*, 1293–1302.

Day, A. J., Canada, F. J., Diaz, J. C., Kroon, P. A., Mclauchlan, R., Faulds, C. B., Plumb, G. W., Morgan, M. R., & Williamson, G. (2000). Dietary flavonoid and isoflavone glycosides are hydrolysed by the lactase site of lactase phlorizin hydrolase. *FEBS Letters, 468*, 166–170.

De Boer, V. C., de Goffau, M. C., Arts, I. C., Hollman, P. C., & Keijer, J. (2006). SIRT1 stimulation by polyphenols is affected by their stability and metabolism. *Mech. Ageing Dev., 127*, 618–627.

De Graaf, M., van Veen, I. C., van der Meulen-Muileman, I. H., Gerritsen, W. R., Pinedo, H. M., & Haisma, H. J. (2001). Cloning and characterization of human liver cytosolic beta-glycosidase. *Biochem. J., 356*, 907–910.

De Vries, J. H., Hollman, P. C., van Amersfoort, I., Olthof, M. R., & Katan, M. B. (2001). Red wine is a poor source of bioavailable flavonols in men. *J. Nutr., 131*, 745–748.

De Whalley, C. V., Rankin, S. M., Hoult, J. R., Jessup, W., & Leake, D. S. (1990). Flavonoids inhibit the oxidative modification of low density lipoproteins by macrophages. *Biochem. Pharmacol., 39*, 1743–1750.

Deckert, V., Desrumaux, C., Athias, A., Duverneuil, L., Palleau, V., Gambert, P., Masson, D., & Lagrost, L. (2002). Prevention of LDL alpha-tocopherol consumption, cholesterol oxidation, and vascular endothelium dysfunction by polyphenolic compounds from red wine. *Atherosclerosis, 165*, 41–50.

Del Bas, J. M., Fernandez-Larrea, J., Blay, M., Ardevol, A., Salvado, M. J., Arola, L., & Blade, C. (2005). Grape seed procyanidins improve atherosclerotic risk index and induce liver CYP7A1 and SHP expression in healthy rats. *FASEB J., 19*, 479–481.

Di Bari, M., Zacchei, S., Kritchevsky, S. B., Anichini, M., Cesaretti, S., Chiarlone, M., Masotti, G., & Marchionni, N. (2003). Anti-oxidized LDL antibodies and wine consumption: a population-based epidemiological study in Dicomano, Italy. *Ann. Epidemiol., 13*, 189–195.

Donovan, J. L., Bell, J. R., Kasim-Karakas, S., German, J. B., Walzem, R. L., Hansen, R. J., & Waterhouse, A. L. (1999). Catechin is present as metabolites in human plasma after consumption of red wine. *J. Nutr., 129*, 1662–1668.

El Bedoui, J., Oak, M. H., Anglard, P., & Schini-Kerth, V. B. (2005). Catechins prevent vascular smooth muscle cell invasion by inhibiting MT1-MMP activity and MMP-2 expression. *Cardiovasc. Res., 67*, 317–325.

Estruch, R., Sacanella, E., Badia, E., Antunez, E., Nicolas, J. M., Fernandez-Sola, J., Rotilio, D., de Gaetano, G., Rubin, E., & Urbano-Marquez, A. (2004). Different effects of red wine and gin consumption on inflammatory biomarkers of atherosclerosis: a prospective randomized crossover trial. Effects of wine on inflammatory markers. *Atherosclerosis, 175*, 117–123.

Feng, B., Yao, P. M., Li, Y., Devlin, C. M., Zhang, D., Harding, H. P., Sweeney, M., Rong, J. X., Kuriakose, G., Fisher, E. A., Marks, A. R., Ron, D., & Tabas, I. (2003). The endoplasmic reticulum is the site of cholesterol-induced cytotoxicity in macrophages. *Nat. Cell Biol., 5*, 781–792.

Finkel, T., & Holbrook, N. J. (2000). Oxidants, oxidative stress and the biology of ageing. *Nature, 408*, 239–247.

Fontana, L., & Klein, S. (2007). Aging, adiposity, and calorie restriction. *JAMA, 297*, 986–994.

Fortuño, A., Beloqui, O., San Jose, G., Moreno, M. U., Zalba, G., & Diez, J. (2005). Increased phagocytic nicotinamide adenine dinucleotide phosphate oxidase-dependent superoxide production in patients with early chronic kidney disease. *Kidney Inter. Suppl., 99*, S71–S75.

Fortuño, A., San José, G., Moreno, M. U., Beloqui, O., Diez, J., & Zalba, G. (2006). Phagocytic NADPH oxidase overactivity underlies oxidative stress in metabolic syndrome. *Diabetes, 55*, 209–215.

Frank, J., Budck, A., Lundh, T., Parker, R. S., Swanson, J. E., Lourenco, C. F., Gago, B., Laranjinha, J., Vessby, B., & Kamal-Eldin, A. (2006). Dietary flavonoids with a catechol structure increase alpha-tocopherol in rats and protect the vitamin from oxidation in vitro. *J. Lipid Res., 47*, 2718–2725.

Frank, T., Netzel, M., Strass, G., Bitsch, R., & Bitsch, I. (2003). Bioavailability of anthocyanidin-3-glucosides following consumption of red wine and red grape juice. *Can. J. Physiol. Pharmacol., 81*, 423–435.

Frankel, E. N., Kanner, J., German, J. B., Parks, E., & Kinsella, J. E. (1993). Inhibition of oxidation of human low-density lipoprotein by phenolic substances in red wine. *The Lancet, 341*, 454–457.

Fuhrman, B., & Aviram, M. (2002). Preservation of paraoxonase activity by wine flavonoids: possible role in protection of LDL from lipid peroxidation. *Ann. N.Y. Acad. Sci., 957*, 321–324.

Fuhrman, B., Lavy, A., & Aviram, M. (1995). Consumption of red wine with meals reduces the susceptibility of human plasma and low-density lipoprotein to lipid peroxidation. *Am. J. Clin. Nutr., 61*, 549–554.

Garvin, S., Ollinger, K., & Dabrosin, C. (2006). Resveratrol induces apoptosis and inhibits angiogenesis in human breast cancer xenografts in vivo. *Cancer Lett., 231*, 113–122.

Gauss, K. A., Nelson-Overton, L. K., Siemsen, D. W., Gao, Y., Deleo, F. R., & Quinn, M. T. (2007). Role of NF-kappaB in transcriptional regulation of the phagocyte NADPH oxidase by tumor necrosis factor-alpha. *J. Leukocite Biol., 82*, 729–741.

Gilmore, T. D. (2006). Introduction to NF-kappaB: players, pathways, perspectives. *Oncogene, 25*, 6680–6684.

Goldberg, D. M., Yan, J., & Soleas, G. J. (2003). Absorption of three wine-related polyphenols in three different matrices by healthy subjects. *Clin. Biochem., 36*, 79–87.

Goldstein, J. L., DeBose-Boyd, R. A., & Brown, M. S. (2006). Protein sensors for membrane sterols. *Cell, 124*, 35–46.

Gouedard, C., Barouki, R., & Morel, Y. (2004). Dietary polyphenols increase paraoxonase 1 gene expression by an aryl hydrocarbon receptor-dependent mechanism. *Mol. Cell. Biol., 24*, 5209–5222.

Guarente, L. (1999). Diverse and dynamic functions of the Sir silencing complex. *Nat. Genet., 23*, 281–285.

Guzik, T. J., West, N. E., Black, E., McDonald, D., Ratnatunga, C., Pillai, R., & Channon, K. M. (2000). Vascular superoxide production by NAD(P)H oxidase: association with endothelial dysfunction and clinical risk factors. *Circ. Res., 86*, E85–E90.

Guzik, T. J., Sadowski, J., Guzik, B., Jopek, A., Kapelak, B., Przybylowski, P., Wierzbicki, K., Korbut, R., Harrison, D. G., & Channon, K. M. (2006). Coronary artery superoxide production and nox isoform expression in human coronary artery disease. *Arterioscler. Thromb. Vasc. Biol., 26*, 333–339.

Hayek, T., Fuhrman, B., Vaya, J., Rosenblat, M., Belinky, P., Coleman, R., Elis, A., & Aviram, M. (1997). Reduced progression of atherosclerosis in apolipoprotein E-deficient mice following consumption of red wine, or its polyphenols quercetin or catechin, is associated with reduced susceptibility of LDL to oxidation and aggregation. *Arterioscler. Thromb. Vasc. Biol., 17*, 2744–2752.

Henning, S. M., Aronson, W., Niu, Y., Conde, F., Lee, N. H., Seeram, N. P., Lee, R. P., Lu, J., Harris, D. M., Moro, A., Hong, J., Pak-Shan, L., Barnard, R. J., Ziaee, H. G., Csathy, G., Go, V. L., Wang, H., & Heber, D. (2006). Tea polyphenols and theaflavins are present in prostate tissue of humans and mice after green and black tea consumption. *J. Nutr., 136*, 1839–1843.

Henry-Vitrac, C., Desmoulière, A., Girard, D., Mérillon, J. M., & Krisa, S. (2006). Transport, deglycosylation, and metabolism of trans-piceid by small intestinal epithelial cells. *Eur. J. Nutr., 45*, 376–382.

Hoffmann, A., Natoli, G., & Ghosh, G. (2006). Transcriptional regulation via the NF-kappaB signaling module. *Oncogene, 25*, 6706–6716.

Hou, D. X., Masuzaki, S., Hashimoto, F., Uto, T., Tanigawa, S., Fujii, M., & Sakata, Y. (2007). Green tea proanthocyanidins inhibit cyclooxygenase-2 expression in LPS-activated mouse macrophages: molecular mechanisms and structure-activity relationship. *Arch. Biochem. Biophys., 460*, 67–74.

Howitz, K. T., Bitterman, K. J., Cohen, H. Y., Lamming, D. W., Lavu, S., Wood, J. G., Zipkin, R. E., Chung, P., Kisielewski, A., Zhang, L. L., Scherer, B., & Sinclair, D. A. (2003). Small molecule activators of sirtuins extend Saccharomyces cerevisiae lifespan. *Nature, 425*, 191–196.

Huang, C., Jacobson, K., & Schaller, M. D. (2004). MAP kinases and cell migration. *J. Cell Sci., 117*, 4619–4628.

Iijima, K., Yoshizumi, M., Hashimoto, M., Akishita, M., Kozaki, K., Ako, J., Watanabe, T., Ohike, Y., Son, B., Yu, J., Nakahara, K., & Ouchi, Y. (2002). Red wine polyphenols inhibit vascular smooth muscle cell migration through two distinct signaling pathways. *Circulation, 105*, 2404–2410.

Imai, S., Armstrong, C. M., Kaeberlein, M., & Guarente, L. (2000). Transcriptional silencing and longevity protein Sir2 is an NAD-dependent histone deacetylase. *Nature, 403*, 795–800.

Infanger, D. W., Sharma, R. V., & Davisson, R. L. (2006). NADPH oxidases of the brain: distribution, regulation, and function. *Antiox. Redox Signal., 8*, 1583–1596.

Ivanov, V., Carr, A. C., & Frei, B. (2001). Red wine antioxidants bind to human lipoproteins and protect them from metal ion-dependent and -independent oxidation. *J. Agric. Food Chem., 49*, 4442–4449.

Jimenez, R., Lopez-Sepulveda, R., Kadmiri, M., Romero, M., Vera, R., Sanchez, M., Vargas, F., O'Valle, F., Zarzuelo, A., Duenas, M., Santos-Buelga, C., & Duarte, J. (2007). Polyphenols restore endothelial function in DOCA-salt hypertension: role of endothelin-1 and NADPH oxidase. *Free Rad. Biol. Med., 43*, 462–473.

Jung, U. J., Kim, H. J., Lee, J. S., Lee, M. K., Kim, H. O., Park, E. J., Kim, H. K., Jeong, T. S., & Choi, M. S. (2003). Naringin supplementation lowers plasma lipids and enhances erythrocyte antioxidant enzyme activities in hypercholesterolemic subjects. *Clin. Nutr., 22*, 561–568.

Kaminski, W. E., Piehler, A., & Wenzel, J. J. (2006). ABC A-subfamily transporters: structure, function and disease. *Biochim. Biophis. Acta, 1762*, 510–524.

Karin, M., & Ben-Neriah, Y. (2000). Phosphorylation meets ubiquitination: the control of NF-[kappa]B activity. *Ann. Rev. Immunol., 18*, 621–663.

Karin, M., & Greten, F. R. (2005). NF-kappaB: linking inflammation and immunity to cancer development and progression. *Nat. Rev. Immunol., 5*, 749–759.

Kim, D., Nguyen, M. D., Dobbin, M. M., Fischer, A., Sananbenesi, F., Rodgers, J. T., Delalle, I., Baur, J. A., Sui, G., Armour, S. M., Puigserver, P., Sinclair, D. A., & Tsai, L. H. (2007). SIRT1 deacetylase protects against neurodegeneration in models for Alzheimer's disease and amyotrophic lateral sclerosis. *EMBO J., 26*, 3169–3179.

Kim, H. J., Oh, G. T., Park, Y. B., Lee, M. K., Seo, H. J., & Choi, M. S. (2004). Naringin alters the cholesterol biosynthesis and antioxidant enzyme activities in LDL receptor-knockout mice under cholesterol fed condition. *Life Sci., 74*, 1621–1634.

Kuhn, D. J., Burns, A. C., Kazi, A., & Dou, Q. P. (2004). Direct inhibition of the ubiquitin-proteasome pathway by ester bond-containing green tea polyphenols is associated with increased expression of sterol regulatory element-binding protein 2 and LDL receptor. *Biochim. Biophis. Acta, 1682*, 1–10.

Lafay, S., Morand, C., Manach, C., Besson, C., & Scalbert, A. (2006). Absorption and metabolism of caffeic acid and chlorogenic acid in the small intestine of rats. *British J. Nutr., 96*, 39–46.

Lamuela-Raventos, R. M., Covas, M. I., Fito, M., Marrugat, J., & de La Torre-Boronat, M. C. (1999). Detection of dietary antioxidant phenolic compounds in human LDL. *Clin. Chem., 45*, 1870–1872.

Lee, S. O., Jeong, Y. J., Yu, M. H., Lee, J. W., Hwangbo, M. H., Kim, C. H., & Lee, I. S. (2006). Wogonin suppresses TNF-alpha-induced MMP-9 expression by blocking the NF-kappaB activation via MAPK signaling pathways in human aortic smooth muscle cells. *Biochem. Biophys. Res. Comm., 351*, 118–125.

Leikert, J. F., Rathel, T. R., Wohlfart, P., Cheynier, V., Vollmar, A. M., & Dirsch, V. M. (2002). Red wine polyphenols enhance endothelial nitric oxide synthase expression and subsequent nitric oxide release from endothelial cells. *Circulation, 106*, 1614–1617.

Lin, S. J., Defossez, P. A., & Guarente, L. (2000). Requirement of NAD and SIR2 for life-span extension by calorie restriction in Saccharomyces cerevisiae. *Science, 289*, 2126–2128.

Liu, J. C., Chen, J. J., Chan, P., Cheng, C. F., & Cheng, T. H. (2003). Inhibition of cyclic strain-induced endothelin-1 gene expression by resveratrol. *Hypertension, 42*, 1198–1205.

Luo, J., Nikolaev, A. Y., Imai, S., Chen, D., Su, F., Shiloh, A., Guarente, L., & Gu, W. (2001). Negative control of p53 by Sir2alpha promotes cell survival under stress. *Cell, 107*, 137–148.

Manach, C., Williamson, G., Morand, C., Scalbert, A., & Remesy, C. (2005). Bioavailability and bioefficacy of polyphenols in humans. I. Review of 97 bioavailability studies. *Am. J. Clin. Nutr., 81*, 230S–242S.

Mantei, N., Villa, M., Enzler, T., Wacker, H., Boll, W., James, P., Hunziker, W., & Semenza, G. (1988). Complete primary structure of human and rabbit lactase-phlorizin hydrolase: implications for biosynthesis, membrane anchoring and evolution of the enzyme. *EMBO J., 7*, 2705–2713.

Masoro, E. J. (2000). Caloric restriction and aging: an update. *Exp. Gerontol., 35*, 299–305.

Maxwell, S., Cruickshank, A., & Thorpe, G. (1994). Red wine and antioxidant activity in serum. *The Lancet*, 344, 193–194.

McDougall, G, J., Fyffe, S., Dobson, P., & Stewart, D. (2005). Anthocyanins from red wine – their stability under simulated gastrointestinal digestion. *Phytochemistry, 66*, 2540–2548.

Mink, P. J., Scrafford, C. G., Barraj, L. M., Harnack, L., Hong, C. P., Nettleton, J. A., & Jacobs, D. R. Jr. (2007). Flavonoid intake and cardiovascular disease mortality: a prospective study in postmenopausal women. *Am. J. Clin. Nutr., 85*, 895–909.

Moreno, M. U., San Jose, G., Orbe, J., Paramo, J. A., Beloqui, O., Diez, J., & Zalba, G. (2003). Preliminary characterisation of the promoter of the human p22(phox) gene: identification of a new polymorphism associated with hypertension. *FEBS Letters, 542*, 27–31.

Mukamal, K. J., Massaro, J. M., Ault, K. A., Mittleman, M. A., Sutherland, P. A., Lipinska, I., Levy, D., D'Agostino, R. B., & Tofler, G. H. (2005). Alcohol consumption and platelet activation and aggregation among women and men: the Framingham Offspring Study. *Alcohol. Clin. Exp. Res., 29*, 1906–1912.

Mullen, E., Brown, R. M., Osborne, T. F., & Shay, N. F. (2004). Soy isoflavones affect sterol regulatory element binding proteins (SREBPs) and SREBP-regulated genes in HepG2 cells. *J. Nutr., 134*, 2942–2947.

Mursu, J., Voutilainen, S., Nurmi, T., Rissanen, T. H., Virtanen, J. K., Kaikkonen, J., Nyyssönen, K., & Salonen, J. T. (2004). Dark chocolate consumption increases HDL cholesterol concentration and chocolate fatty acids may inhibit lipid peroxidation in healthy humans. *Free Rad. Biol. Med., 37*, 1351–1359.

Nam, N. H. (2006). Naturally occurring NF-kappaB inhibitors. *Mini Reviews in Medicinal Chemistry, 6*, 945–951.

Nardini, M., Cirillo, E., Natella, F., & Scaccini, C. (2002). Absorption of phenolic acids in humans after coffee consumption. *J. Agric. Food Chem., 50*, 5735–5741.

Natella, F., Nardini, M., Belelli, F., & Scaccini, C. (2007). Coffee drinking induces incorporation of phenolic acids into LDL and increases the resistance of LDL to ex vivo oxidation in humans. *Am. J. Clin. Nutr., 86*, 604–609.

Ndiaye, M., Chataigneau, M., Lobysheva, I., Chataigneau, T., & Schini-Kerth, V. B. (2005). Red wine polyphenol-induced, endothelium-dependent NO-mediated relaxation is due to the redox-sensitive PI3-kinase/Akt-dependent phosphorylation of endothelial NO-synthase in the isolated porcine coronary artery. *FASEB J., 19*, 455–457.

Nemeth, K., Plumb, G. W., Berrin, J. G., Juge, N., Jacob, R., Naim, H. Y., Williamson, G., Swallow, D. M., & Kroon, P. A. (2003). Deglycosylation by small intestinal epithelial cell beta-glucosidases is a critical step in the absorption and metabolism of dietary flavonoid glycosides in humans. *Eur. J. Nutr., 42*, 29–42.

Nigdikar, S. V., Williams, N. R., Griffin, B. A., & Howard, A. N. (1998). Consumption of red wine polyphenols reduces the susceptibility of low-density lipoproteins to oxidation in vivo. *Am. J. Clin. Nutr., 68*, 258–265.

Oak, M. H., Chataigneau, M., Keravis, T., Chataigneau, T., Beretz, A., Andriantsitohaina, R., Stoclet, J. C., Chang, S. J., & Schini-Kerth, V. B. (2003). Red wine polyphenolic compounds inhibit vascular endothelial growth factor expression in vascular smooth muscle cells by preventing the activation of the p38 mitogen-activated protein kinase pathway. *Arterioscler. Thromb. Vasc. Biol., 23*, 1001–1007.

Oak, M. H., El Bedoui, J., Anglard, P., & Schini-Kerth, V. B. (2004). Red wine polyphenolic compounds strongly inhibit pro-matrix metalloproteinase-2 expression and its activation in response to thrombin via direct inhibition of membrane type 1-matrix metalloproteinase in vascular smooth muscle cells. *Circulation, 110*, 1861–1867.

O'Byrne, D. J., Devaraj, S., Grundy, S. M., & Jialal, I. (2002). Comparison of the antioxidant effects of Concord grape juice flavonoids alpha-tocopherol on markers of oxidative stress in healthy adults. *Am. J. Clin. Nutr., 76*, 1367–1374.

Olthof, M. R., Hollman, P. C., Buijsman, M. N., van Amelsvoort, J. M., & Katan, M. B. (2003). Chlorogenic acid, quercetin-3-rutinoside and black tea phenols are extensively metabolized in humans. *J. Nutr., 133*, 1806–1814.

Oram, J. F., & Vaughan, A. M. (2006). ATP-Binding cassette cholesterol transporters and cardiovascular disease. *Circ. Res., 99*, 1031–1043.

Pal, S., Ho, N., Santos, C., Dubois, P., Mamo, J., Croft, K., & Allister, E. (2003). Red wine polyphenolics increase LDL receptor expression and activity and suppress the secretion of ApoB100 from human HepG2 cells. *J. Nutr., 133*, 700–706.

Pal, S., Ho, S. S., & Takechi, R. (2005). Red wine polyphenolics suppress the secretion of ApoB48 from human intestinal CaCo-2 cells. *J. Agric. Food Chem., 53*, 2767–2772.
Pignatelli, P., Di Santo, S., Buchetti, B., Sanguigni, V., Brunelli, A., & Violi, F. (2006a). Polyphenols enhance platelet nitric oxide by inhibiting protein kinase C-dependent NADPH oxidase activation: effect on platelet recruitment. *FASEB J., 20*, 1082–1089.
Pignatelli, P., Ghiselli, A., Buchetti, B., Carnevale, R., Natella, F., Germano, G., Fimognari, F., Di Santo, S., Lenti, L., & Violi, F. (2006b). Polyphenols synergistically inhibit oxidative stress in subjects given red and white wine. *Atherosclerosis, 188*, 77–83.
Pitsavos, C., Makrilakis, K., Panagiotakos, D. B., Chrysohoou, C., Ioannidis, I., Dimosthenopoulos, C., Stefanadis, C., & Katsilambros, N. (2005). The J-shape effect of alcohol intake on the risk of developing acute coronary syndromes in diabetic subjects: the CARDIO2000 II Study. *Diabet. Med., 22*, 243–248.
Renaud, S., & de Lorgeril, M. (1992). Wine, alcohol, platelets, and the French paradox for coronary heart disease. *The Lancet, 339*, 1523–1526.
Rodgers, J. T., Lerin, C., Haas, W., Gygi, S. P., Spiegelman, B. M., & Puigserver, P. (2005). Nutrient control of glucose homeostasis through a complex of PGC-1alpha and SIRT1. *Nature, 434*, 113–138.
Roux, P. P., & Blenis, J. (2004). ERK and p38 MAPK-activated protein kinases: a family of protein kinases with diverse biological functions. *Microbiol. Mol. Biol. Rev., 68*, 320–344.
Sanchez, M., Lodi, F., Vera, R., Villar, I. C., Cogolludo, A., Jimenez, R., Moreno, L., Romero, M., Tamargo, J., Perez-Vizcaino, F., & Duarte, J. (2007). Quercetin and isorhamnetin prevent endothelial dysfunction, superoxide production, and overexpression of p47phox induced by angiotensin II in rat aorta. *J. Nutr., 137*, 910–915.
Sarr, M., Chataigneau, M., Martins, S., Schott, C., El Bedoui, J., Oak, M. H., Muller, B., Chataigneau, T., & Schini-Kerth, V. B. (2006). Red wine polyphenols prevent angiotensin II-induced hypertension and endothelial dysfunction in rats: role of NADPH oxidase. *Cardiovasc. Res., 71*, 794–802.
Sasaki, S. (2000). Alcohol and its relation to all-cause and cardiovascular mortality. *Acta Cardiol., 55*, 151–156.
Scalbert, A., & Williamson, G. (2000). Dietary intake and bioavailability of polyphenols. *J. Nutr., 130*, 2073S–2085S.
Schubert, S. Y., Neeman, I., & Resnick, N. (2002). A novel mechanism for the inhibition of NF-kappaB activation in vascular endothelial cells by natural antioxidants. *FASEB J., 16*, 1931–1933.
Sembries, S., Dongowski, G., Mehrländer, K., Will, F., & Dietrich, H. (2006). Physiological effects of extraction juices from apple, grape, and red beet pomaces in rats. *J. Agric. Food Chem., 54*, 10269–10280.
Serafini, M., Maiani, G., & Ferro-Luzzi, A. (1998). Alcohol-free red wine enhances plasma antioxidant capacity in humans. *J. Nutr., 128*, 1003–1007.
Sesso, H. D. (2001). Alcohol and cardiovascular health: recent findings. *Am. J. Cardiovasc. Drug.: drugs, devices, and other interventions, 1*, 167–172.
Sesso, H. D., Gaziano, J. M., Liu, S., & Buring, J. E. (2003). Flavonoid intake and the risk of cardiovascular disease in women. *Am. J. Clin. Nutr., 77*, 1400–1408.
Sevov, M., Elfineh, L., & Cavelier, L. B. (2006). Resveratrol regulates the expression of LXR-alpha in human macrophages. *Biochem. Biophys. Res. Comm., 348*, 1047–1054.
Shahidi, F., Janitha, P. K., & Wanasundara, P. D. (1992). Phenolic antioxidants. *Crit. Rev. Food Sci. Nutr., 32*, 67–103.
Shimizu, M., Kobayashi, Y., Suzuki, M., Satsu, H., & Miyamoto, Y. (2000). Regulation of intestinal glucose transport by tea catechins. *Biofactors, 13*, 61–65.
Silberberg, M., Morand, C., Mathevon, T., Besson, C., Manach, C., Scalbert, A., & Remesy, C. (2006). The bioavailability of polyphenols is highly governed by the capacity of the intestine and of the liver to secrete conjugated metabolites. *Eur. J. Nutr., 45*, 88–96.

Simonetti, P., Gardana, C., & Pietta, P. (2001). Plasma levels of caffeic acid and antioxidant status after red wine intake. *J. Agric. Food Chem., 49*, 5964–5968.

Sohal, R. S., & Weindruch, R. (1996). Oxidative stress, caloric restriction, and aging. *Science, 273*, 59–63.

Sorescu, D., Weiss, D., Lassegue, B., Clempus, R. E., Szocs, K., Sorescu, G. P., Valppu, L., Quinn, M. T., Lambeth, J. D., Vega, J. D., Taylor, W. R., & Griendling, K. K. (2002). Superoxide production and expression of nox family proteins in human atherosclerosis. *Circulation, 105*, 1429–1435.

St Leger, A. S., Cochrane, A. L., & Moore, F. (1979). Factors associated with cardiac mortality in developed countries with particular reference to the consumption of wine. *The Lancet, 1*, 1017–1020.

Stein, J. H., Keevil, J. G., Wiebe, D. A., Aeschlimann, S., & Folts, J. D. (1999). Purple grape juice improves endothelial function and reduces the susceptibility of LDL cholesterol to oxidation in patients with coronary artery disease. *Circulation, 100*, 1050–1055.

Stoclet, J. C., Chataigneau, T., Ndiaye, M., Oak, M. H., El Bedoui, J., & Chataigneau, M., (2004). Schini-Kerth, V. B. Vascular protection by dietary polyphenols. *Eur. J. Pharmacol., 500*, 299–313.

Su, H. C., Hung, L. M., & Chen, J. K. (2006). Resveratrol, a red wine antioxidant, possesses an insulin-like effect in streptozotocin-induced diabetic rats. *Am. J. Physiol. Endocrinol. Metab., 290*, E1339–E1346.

Takechi, R., Hiramatsu, N., Mamo, J. C., & Pal, S. (2004). Red wine polyphenolics suppress the secretion and the synthesis of Apo B48 from human intestinal CaCo-2 cells. *Biofactors, 22*, 181–183.

Terra, X., Valls, J., Vitrac, X., Merrillon, J. M., Arola, L., Ardevol, A., Blade, C., Fernandez-Larrea, J., Pujadas, G., Salvado, J., & Blay, M. (2007). Grape-seed procyanidins act as antiinflammatory agents in endotoxin-stimulated RAW 264.7 macrophages by inhibiting NFkB signaling pathway. *J. Agric. Food Chem., 55*, 4357–4365.

Tissenbaum, H. A., & Guarente, L. (2001). Increased dosage of a sir-2 gene extends lifespan in Caenorhabditis elegans. *Nature, 410*, 227–230.

Tontonoz, P., & Mangelsdorf, D. J. (2003). Liver X receptor signaling pathways in cardiovascular disease. *Mol. Endocrinol., 17*, 985–993.

Ukil, A., Maity, S., & Das, P. K. (2006). Protection from experimental colitis by theaflavin-3,3'-digallate correlates with inhibition of IKK and NF-kappaB activation. *Br. J. Pharmacol., 149*, 121–131.

Valko, M., Leibfritz, D., Moncol, J., Cronin, M. T., Mazur, M., & Telser, J. (2007). Free radicals and antioxidants in normal physiological functions and human disease. *Int. J. Biochem. Cell Biol., 39*, 44–84.

Vayalil, P. K., Mittal, A., & Katiyar, S. K. (2004). Proanthocyanidins from grape seeds inhibit expression of matrix metalloproteinases in human prostate carcinoma cells, which is associated with the inhibition of activation of MAPK and NF kappa B. *Carcinogenesis, 25*, 987–995.

Vaziri, H., Dessain, S. K., Ng Eaton, E., Imai, S. I., Frye, R. A., Pandita, T. K., Guarente, L., & Weinberg, R. A. (2001). hSIR2(SIRT1) functions as an NAD-dependent p53 deacetylase. *Cell, 107*, 149–159.

Vidal, R., Hernandez-Vallejo, S., Pauquai, T., Texier, O., Rousset, M., Chambaz, J., Demignot, S., & Lacorte, J. M. (2005). Apple procyanidins decrease cholesterol esterification and lipoprotein secretion in Caco-2/TC7 enterocytes. *J. Lipid Res., 46*, 258–268.

Vita, J. A. (2005). Polyphenols and cardiovascular disease: effects on endothelial and platelet function. *Am. J. Clin. Nutr., 81*, 292S–297S.

Vitaglione, P., Sforza, S., Galaverna, G., Ghidini, C., Caporaso, N., Vescovi, P. P., Fogliano, V., & Marchelli, R. (2005). Bioavailability of trans-resveratrol from red wine in humans. *Mol. Nutr. Food Res., 49*, 495–504.

Wallerath, T., Deckert, G., Ternes, T., Anderson, H., Li, H., Witte, K., & Forstermann, U. (2002). Resveratrol, a polyphenolic phytoalexin present in red wine, enhances expression and activity of endothelial nitric oxide synthase. *Circulation, 106*, 1652–1658.

Wheeler, D. S., Catravas, J. D., Odoms, K., Denenberg, A., Malhotra, V., & Wong, H. R. (2004). Epigallocatechin-3-gallate, a green tea-derived polyphenol, inhibits IL-1 beta-dependent proinflammatory signal transduction in cultured respiratory epithelial cells. *J. Nutr., 134*, 1039–1044.

Whitehead, T. P., Robinson, D., Allaway, S., Syms, J., & Hale, A. (1995). Effect of red wine ingestion on the antioxidant capacity of serum. *Clin. Chem., 41*, 32–35.

Wilcox, L. J., Borradaile, N. M., de Dreu, L. E., & Huff, M. W. (2001). Secretion of hepatocyte apoB is inhibited by the flavonoids, naringenin and hesperetin, via reduced activity and expression of ACAT2 and MTP. *J. Lipid Res., 42*, 725–734.

Witztum, J. L., & Steinberg, D. (1991). Role of oxidized low density lipoprotein in atherogenesis. *J. Clin. Inves., 88*, 1785–1792.

Wood, J. G., Rogina, B., Lavu, S., Howitz, K., Helfand, S. L., Tatar, M., & Sinclair, D. A. (2004). Sirtuin activators mimic caloric restriction and delay ageing in metazoans. *Nature, 430*, 686–689.

World Health Organisation. (1989). World health statistics annual. Geneve, World Health Organisation.

Xia, M., Hou, M., Zhu, H., Ma, J., Tang, Z., Wang, Q., Li, Y., Chi, D., Yu, X., Zhao, T., Han, P., Xia, X., & Ling, W. (2005). Anthocyanins induce cholesterol efflux from mouse peritoneal macrophages: the role of the peroxisome proliferator-activated receptor gamma-liver X receptor alpha-ABCA1 pathway. *J. Biol. Chem., 280*, 36792–36801.

Xia, M., Ling, W., Zhu, H., Wang, Q., Ma, J., Hou, M., Tang, Z., Li, L., & Ye, Q. (2007). Anthocyanin prevents CD40-activated proinflammatory signaling in endothelial cells by regulating cholesterol distribution. *Arterioscler. Thromb. Vasc. Biol., 27*, 519–524.

Xu, J. W., Ikeda, K., & Yamori, Y. (2004). Genistein inhibits expressions of NADPH oxidase p22phox and angiotensin II type 1 receptor in aortic endothelial cells from stroke-prone spontaneously hypertensive rats. *Hypertens. Res., 27*, 675–683.

Yang, F., Oz, H. S., Barve, S., de Villiers, W. J., McClain, C. J., & Varilek, G. W. (2001). The green tea polyphenol (-)-epigallocatechin-3-gallate blocks nuclear factor-kappa B activation by inhibiting I kappa B kinase activity in the intestinal epithelial cell line IEC-6. *Molecular Pharmacology, 60*, 528–533.

Yang, H., Baur, J. A., Chen, A., Miller, C., Adams, J. K., Kisielewski, A., Howitz, K. T., Zipkin, R. E., & Sinclair, D. A. (2007). Design and synthesis of compounds that extend yeast replicative lifespan. *Aging Cell, 6*, 35–43.

Yang, T. T., & Koo, M. W. (2000). Chinese green tea lowers cholesterol level through an increase in fecal lipid excretion. *Life Sci.s, 66*, 411–423.

Ying, C. J., Xu, J. W., Ikeda, K., Takahashi, K., Nara, Y., & Yamori, Y. (2003). Tea polyphenols regulate nicotinamide adenine dinucleotide phosphate oxidase subunit expression and ameliorate angiotensin II-induced hyperpermeability in endothelial cells. *Hypertens. Res., 26*, 823–828.

Zalba, G., Fortuño, A., Orbe, J., San Jose, G., Moreno, M. U., Belzunce, M., Rodriguez, J. A., Beloqui, O., Paramo, J. A., & Diez, J. (2007). Phagocytic NADPH oxidase-dependent superoxide production stimulates matrix metalloproteinase-9: implications for human atherosclerosis. *Arterioscler. Thromb. Vasc. Biol., 27*, 587–593.

Zern, T. L., Wood, R. J., Greene, C., West, K. L., Liu, Y., Aggarwal, D., Shachter, N. S., & Fernandez, M. L. (2005). Grape polyphenols exert a cardioprotective effect in pre- and postmenopausal women by lowering plasma lipids and reducing oxidative stress. *J. Nutr., 135*, 1911–1917.

Zhen, M. C., Huang, X. H., Wang, Q., Sun, K., Liu, Y. J., Li, W., Zhang, L. J., Cao, L. Q., & Chen, X. L. (2006). Green tea polyphenol epigallocatechin-3-gallate suppresses rat hepatic stellate cell invasion by inhibition of MMP-2 expression and its activation. *Acta Pharmacol. Sin., 27*, 1600–1607.

Part III
Spoilage of Wines

Chapter 10
Aromatic Spoilage of Wines by Raw Materials and Enological Products

Alain Bertrand and Angel Anocibar Beloqui

Contents

For wine to earn its reputation and merit its price, it must fulfil certain quality standards but, above all, must be free from defects. These latter factors may be inherent in the raw materials used and result in spoilage of the wine, may be derived from the products used to treat the vines, or be caused by accidental contamination of the grapes. The wine can be contaminated in any stage of its manufacture, even in the bottles (Blaise and Bertrand 1998).

A. Bertrand (✉)
Regents professor. Université V. Segalen Bordeaux 2. Faculté d'œnologie, 351 Cours de la Libération, F33405 Talence, France
e-mail: alain.bertrand@oenologie.u-bordeaux2.fr

M.V. Moreno-Arribas, M.C. Polo (eds.), *Wine Chemistry and Biochemistry*,
DOI 10.1007/978-0-387-74118-5_22, © Springer Science+Business Media, LLC 2009

10.1 Contaminations of Technological Origin

10.1.1 Hydrocarbonates and Derivatives

It is not recommended to resurface tarmacked roads in the weeks prior to the grape harvest to protect wine quality, since this process can produce strong tar-like or naphthalene flavors and aromas. Any contamination by diesel-oil type hydrocarbons can render wines undrinkable.

Oil leaks from the hydraulic jack of the harvesting machines and presses can also contaminate the grapes (Fig. 10.1).

Another source of contamination corresponds to fluid leaks from the refrigerating apparatus inside the tanks, producing excess propan-1,2-diol. This compound, naturally present in wines in tens of mg/L, does not produce an aromatic defect, but does give a slightly sugary flavor. Maximum levels should be limited to 150 mg/L in still wines and 300 mg/L in sparkling wines, according to the OIV, and contaminated wines must be destroyed. It is strongly recommended not to use glycol and/or diethylene glycol as coolants owing to their toxic nature.

10.1.2 Phthalates

Phthalates are plasticizers that can migrate from the materials in contact with the wine. Although they do not cause pronounced changes in flavor they can reduce the limpidity and freshness of wine.

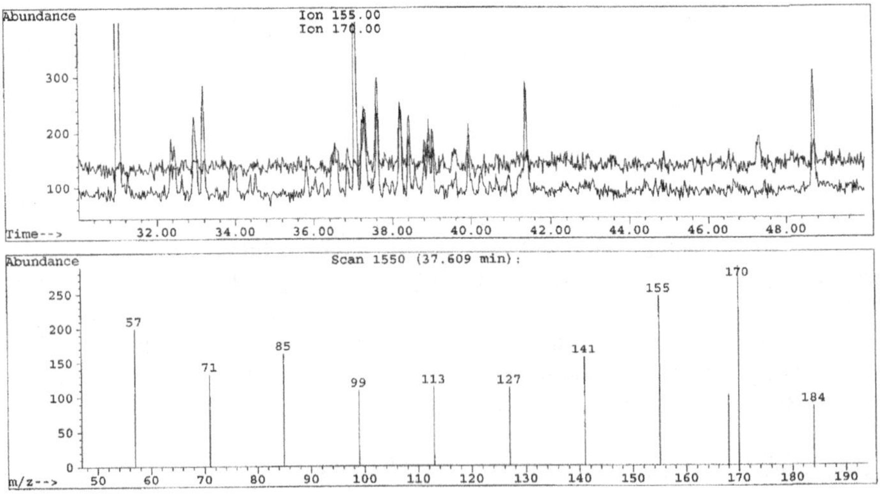

Fig. 10.1 Chromatogram of wine extract mildly contaminated with diesel oil. The peaks of ions 155 and 170 are typical of this type of contamination (naphthalene derivatives present in all the hydrocarbons)

In an important study on alcoholic drinks presented to the OIV, Bertoli et al. (2004) studied different phthalates (dimethyl, diethyl, diisobutyl, dibutyl, benzyl, butyl, bi 2-ethylhexyl, dioctyl). They detected the constant presence of these compounds in alcoholic drinks due to environmental contamination and recommend that special care be taken when choosing contact materials (for pipes, tanks, etc.). In any case, the levels present in wines rarely exceed tens of μg/L; for dibutyl phthalate they are lower than 10 μg/L and for the other phthalates are often below the detection limit.

10.2 Defects Caused by Residues and Wine Materials

10.2.1 Bitter Almond Flavor

Wines stored in vats lined with epoxy resins can present unusually high levels of benzoic aldehyde (several mg/L). Benzylic alcohol is both a plasticizer and diluent of these resins. Its conversion into benzaldehyde can be due (Blaise 1986) to the action of an exocellular enzyme of *Botrytis cinerea* called Alcohol Benzylic Oxidase (ABO, E.C. 1.1.3.7.) responsible for this oxidation process (Blaise and Brun 1986).

10.2.2 Styrene or Vinylbenzene Flavor

Styrene flavor is one of the commonest defects caused by the polyester "plastic material" of the tanks used to store the wines (Brun et al. 1982), especially in plastic vats. It gives off a plastic or burnt rubber aroma and has a perception threshold of around 100 μg/L.

10.2.3 Defects Derived from the Epoxy Resins

The solvents or diluents applied to line the tanks can affect wine quality even at very low doses, of the order of a few μg/L. These should be completely removed before the tanks are filled with wine.

These often correspond to aromatic hydrocarbons (xylene, toluene, ethylbenzenes, etc.) and ketones (methyl-ethylketone, methyl-isobutylketone, methyl-propylketone, etc.).

There are also other alterations of the same origin that produce other flavor defects: – a very strong phenol flavor produced by the resin monomers or impurities (cresols, phenol, 2-ethylphenol, 3-ethylphenol, *m*-terbutylphenol (Fig. 10.2), 2,4-dichlorophenol, etc.).

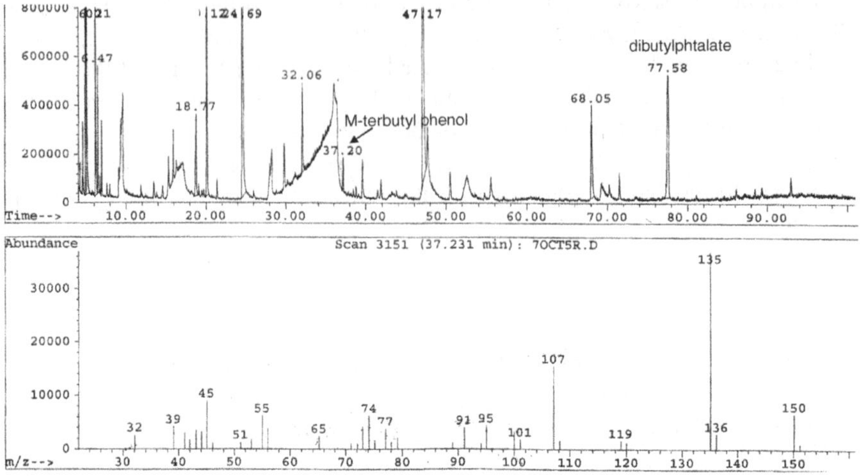

Fig. 10.2 Chromatogram of a wine contaminated by the release of *m*-terbutylphenol and dibutyl phthalate from the epoxidic lining of a tank

10.3 Defects Caused by Some Preservatives

10.3.1 Geranium Aroma

The addition of sorbic acid to a wine should always be carried out in the presence of free SO_2, since bacteria that are not inhibited by this acid can break it down to 2-ethoxyhexa-3,5-diene (Lacoste 1973; Crowell and Guymon 1975; Radler 1976). The perception threshold of this ether in wines is $1 \times 10-7$ g/L (Würdig 1977).

10.3.2 Mustard Aroma

Allyl isothiocyanate (CH_2=$CHCH_2N$=C=S) has been studied as a contaminant in wine from discs treated with this substance. It is used at the surface of tanks to prevent the development of yeast velum, which is prohibited in most countries. We did not find allyl isothiocyanate in any of the wines studied.

10.4 Exogenous Sulphur Compounds

10.4.1 Compounds of Technological Origin (Rubbery Flavor)

10.4.1.1 Benzothiazole

Benzothiazole is a sulfurous compound naturally present in wines at levels below 7 μg/L (Keck 1989). At higher concentrations, the benzothiazole may be derived from contamination by the rubber pipes.

10.4.1.2 Cyclohexyl Isothiocyanate

Anocibar Beloqui (1998) demonstrated that a wine with a solvent or rubbery aroma came from a batch of barrels that had been filled with a new hose. The wine from the first filled barrel only had a few more μg/L of benzothiazole than the wine from the other barrels and, in any case, the concentrations were lower than 15 μg/L. These values are below the perception threshold for this substance in a synthetic solution (50 μg/L) and well below the perception threshold in red wine (200 μg/L) reported by Lavigne (1996).

Analysis of a sample from the hose pipe indeed showed the presence of benzothiazole and when the chromatogram of this sample was compared to that of the wine sample contaminated with this piping, there was a peak common to both of them. The mass spectra of the compounds found in the wine (Fig. 10.3) and of the sample from the hose were also similar. In both cases, from the library of reference spectra this compound has been identified as cyclohexyl isothiocyanate: C_6H_{11}–N=C=S and is a contaminant produced in the manufacture of rubber materials (Cocheo et al. 1983).

10.4.2 Flavors Caused by Pesticide Degradation

For many years it was assumed that the residues of sulphurous pesticides caused the "reducing" defects in wines (Burkhardt 1978).

Cantarelli et al. (1964) have shown that yeasts can use ethylene bi-thiocarbamates and produce sulphurous compounds such as H_2S and CS_2. Marshall (1977) described the degradation mechanism of dithiocarbamates to ethylenediamine and CS_2 or to ethylenethiouramide that would be transformed into ethylenethiourea and into CS_2.

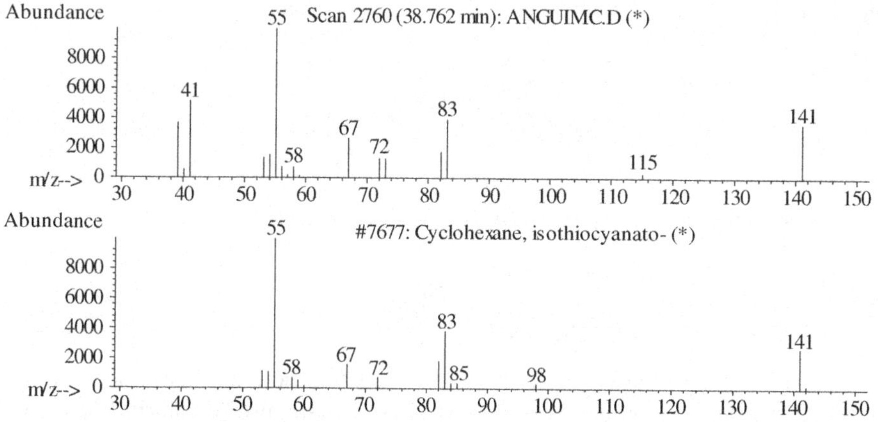

Fig. 10.3 Mass spectra of a compound found in a contaminated wine extract (*top*) compared with that from the spectra library NBS75K.l. This compound is a contaminant produced during the manufacture of cork materials (Cocheo et al. 1983)

Several works focus on the release of sulphurous compounds into wine, from dithiocarbamates. Maujean (1989) also show that these molecules, and especially "thirame", a fungicide with antibotyritis action, can degrade into isothiocyanate and H_2S, ethylenediamine and CS_2, or into ethylenethiourea and CS_2.

10.4.2.1 Dimethyl Disulphide Derived from "Acephate"

At wine pH, "acephate" or "orthene", organothiophosphorus pesticide is slowly hydrolyzed to "metamidophos" first and, later, into methanethiol that can be oxidised to dimethyl sulphide. The rate of hydrolysis is faster at lower pH and when the temperature exceeds 20 °C the rate increases rapidly (Rauhut 1990).

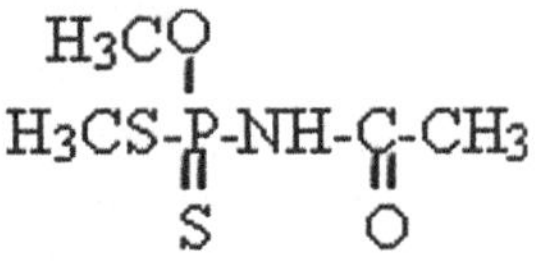

Acéphate

On the other hand, methanethiol can be oxidised to dimethyl disulphide (that has a perception threshold in the air seven times weaker than hydrogen sulphide.

$$2\ CH_3SH \rightarrow CH_3S - SCH_3 + 2H$$

10.4.2.2 Residues from Methomyl Degradation

If the residues of "méthomyl", an insecticide commonly used in viticulture, are present in the must they can lead to the formation of methanethiol, ethanethiol, diethyl sulphide and diethyl disulphide. If these compounds are present in wine they can give it "cooked cauliflower" or "wet wool" aromas.

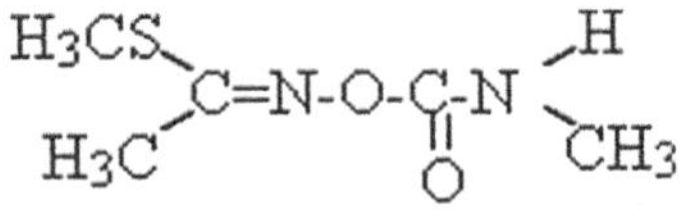

Méthomyl

10.4.2.3 Degradation of "Thirame"

Head space analysis shows that, in addition to CS_2, there are also traces of H_2S which, according to Maujean et al. (1993), are due to impurities entering during the preparation. This is in accordance with the research carried out by Cantagrel et al. (1994).

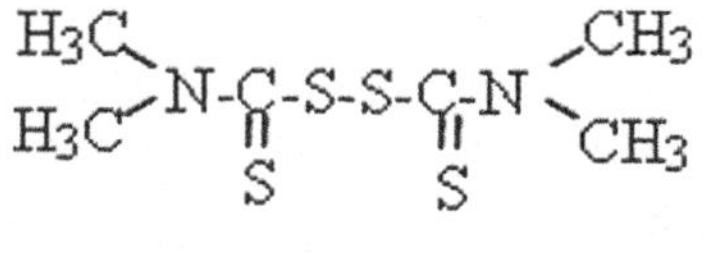

Thirame

In the literature, a series of compounds are mentioned as possibly derived from the breakdown of "thirame", including:

- Dimethyldithiocarbamate (Jacob 1961)
- Dimethyldithiocarbamate, CS_2, DMS and COS (Schmitt 1987)
- Isothiocyanate and H_2S or ethylenediamine and CS_2 or ethylenethiouramide or ethylenethiourea or ethylenediamine and COS and CS_2 (Marshall 1977; Maujean 1989)

This passes through an intermediate, which usually has the formula $C_5H_9NS_3$, but which does not appear owing to the action of the yeasts (Anocibar Beloqui 1998). This could be one of the two formulae presented (Fig. 10.4).

Analysis of wines from different châteaux and different harvests showed that there are always high doses of carbon disulphide when this molecule is present, although a correlation was not established (Fig. 10.5).

The sensorial analysis shows that, for CS_2, the descriptors used by the tasters are "sulphurous", "rubbery", "burnt" and "reduced".

The perception threshold of CS_2in water is around 20 μg/L, a little lower than that recorded by Spedding and Raut (1982) in white wines (38 μg/L). Tasters can recognise this substance at concentrations of around 150 μg/L. However, at lower levels, the aromatic tone of the wine changes and it tends to mask some fruity aromas, while it can enhance the mouldy aroma or even that of volatile phenols and the wine can be described as lacking limpidity.

Of all the wines studied, a large number contain concentrations of CS_2 far higher than those we recorded, which could contribute to wine spoilage. Moreover, carbon disulphide can evolve in the presence of alcohol into ethyl monothiocarbamate, although we did not detect this compound. We have conducted experiments to

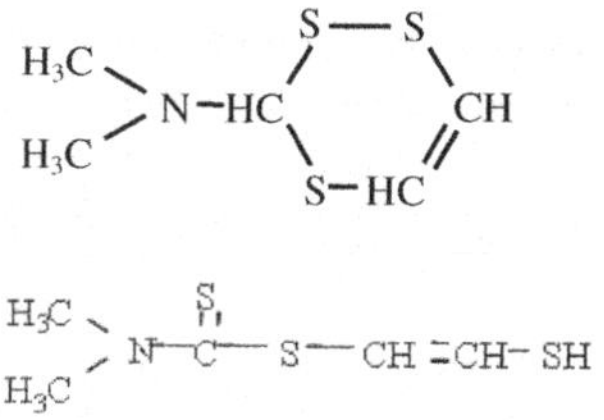

Fig. 10.4 Supposed molecular structures corresponding to the degradation products of thirame by yeast

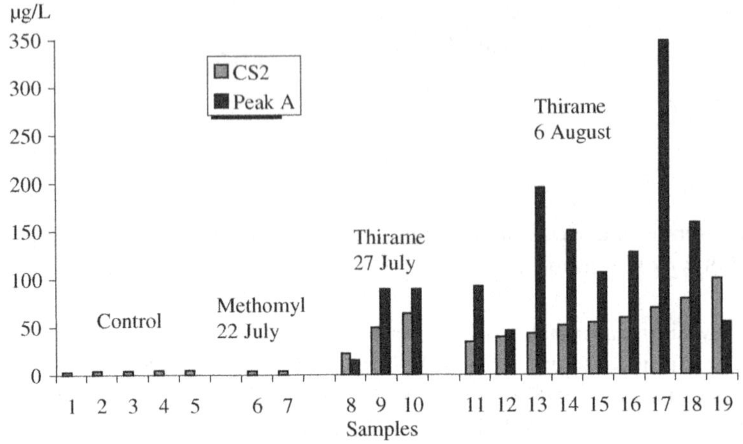

Fig. 10.5 Influence of phytosanitary vine treatments on the amount of residues in different Beaujolais wines. Peak A: $C_5H_9NS_3$

remove CS_2 from wine. Treatment with copper sulphate, which gives good results with compounds containing a thiol group, do not work with CS_2. The only treatment that has proven quite useful is to bubble nitrogen through it.

10.4.2.4 Products Derived from the Sulphur Used in Vine Treatment

Sulphur is a phytosanitary product that has been used for some time in the treatment of grapevine powdery mildew (*Uncinula Necator*). It can be used in powder form to stop the spread of this disease (sublimable or sulphur flower), generally in large doses, or dissolved (soluble sulphur).

There are few data in the literature about the amount of elemental sulphur residue in the vine. Thomas et al. (1993) found sulphur concentrations in the grapes always at concentrations lower than 3 µg/g of grape, equivalent to 3.4 mg/L of sulphur in the musts for vine treatments of 13 kg/ha. If the treatments are carried out before ripening, as they should be, the concentrations found as residue are between 0.9 and 1.7 µg/L.

At least part of the hydrogen sulphide, H_2S, was formed in the wine by the reduction of elemental sulphur (Maw 1965; Wainwright 1971; Eschenbruch 1974; Schütz and Kunkee 1977; Wenzel et al. 1980; Wenzel and Dittrich 1978, 1983; Stratford and Rose 1985; Colagrande et al. 1988; Thomas et al. 1993). Below, we show the mechanism of this reaction starting with elemental sulphur:

$$2\,R - SH + S \quad \rightarrow R - S - S - R + H_2S$$

Thomas et al. (1993) added from 0 to 3.4 mg/L of sulphur to musts that are fermented using different yeast strains. They did not find any correlation between residual sulphur and the amount of H_2S formed. Other authors, however, found a dose of residual sulphur higher than 1 mg/L. The amounts of H_2S formed exceed

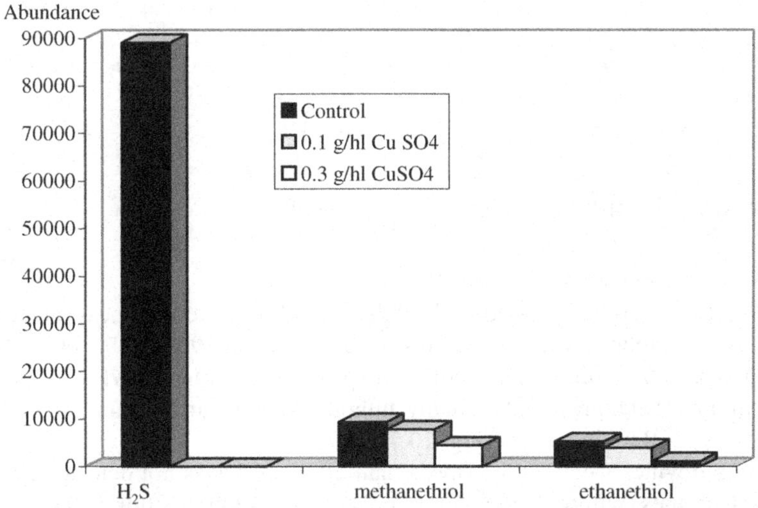

Fig. 10.6 Influence of treatments with copper sulphate on the amount of sulphur compounds found

the threshold reported by Wenzel and colleagues (Wenzel et al. 1980; Wenzel and Dittrich 1983). The removal of hydrogen sulphide should be carried out quickly after alcoholic fermentation because this molecule is highly reactive. One treatment with copper sulphate is highly effective (Fig. 10.6) although much less so for the mercaptanes, in which case it is necessary to use silver salts.

The perception threshold of H_2S is difficult to establish. Wenzel et al. (1980) found it to be between 50 and 80 μg/L in wine and in a synthetic solution it is around 0.8 μg/L (Lavigne 1996). The aromatic tone given by this compound is of rotten eggs. However, according to Dittrich and Staudenmayer (1970), it can contribute positively to the aromas of young wines if present in amounts of 20–30 μg/L.

Several authors have reported a wide range of concentrations for this compound in wines (Dittrich and Staudenmayer 1970; Wenzel et al. 1980; Lavigne 1996). In wines with reduction aroma defects, it is present in concentrations between 0.8 and over 80 μg/L.

In conclusion, the way to prevent these defects lies in a good strategy to treat powdery mildew.

10.5 Evidence for Compounds Derived from H_2S

10.5.1 Ethanethiol

Hydrogen sulphide is a highly reactive substance and can react with most chemical compounds. With carbon dioxide it reacts to produce carbonyl sulphide, a highly toxic but also highly volatile compound (with a boiling point of −50°C) (Cantagrel et al. 1994).

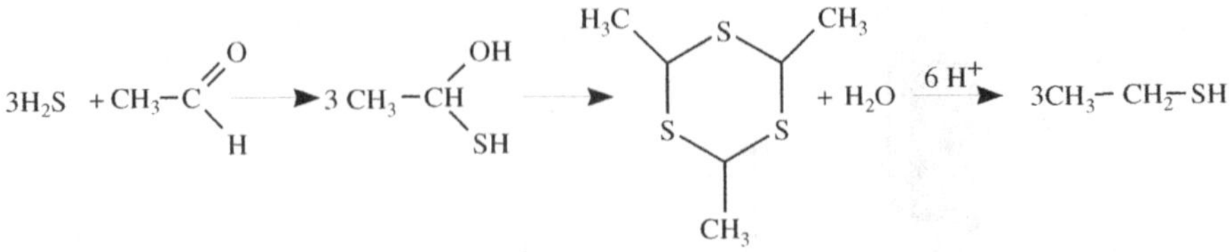

Fig. 10.7 Reaction of H_2S with ethanal to produce ethanethiol

Mercaptanes can be oxidised to disulphides and the latter can be reduced to the respective mercaptanes with the addition of SO_2 (Tanner 1969; Bobet et al. 1990).

The reaction between H_2S and ethanal to produce 2,4,6-trimethyl-*S*-trithiane and ethanethiol has been proposed by many authors (Neuberg and Grauer 1952; Wilson and Katz 1972; Rankine 1963) (Fig. 10.7).

In the red wines we analysed (more than 200), COS was not detected in any of the healthy wines (without defects) nor in the wines with reduction defects.

Ethanethiol has a very weak perception threshold (1.1 μg/L) according to Goniak and Noble (1987), and has never been detected in wines without reduction defects. In some of the wines with a very high H_2S concentration (estimated at around 20 μg/L), the ethanethiol concentrations are very low (lower than 2 μg/L). Also, concentrations in some white wines are higher than those found in red wines (Lavigne 1996). The very different chemical composition of a white wine compared to a red wine (especially in relation to the presence of phenolic compounds) could explain the difference in results. On the other hand, no author has demonstrated the formation of ethanethiol from H_2S, ethanol or ethanal in a red wine.

10.5.2 Trithiolanes

Other reaction mechanisms have also been proposed between ethanal and H_2S (Werkhoff et al. 1991). H_2S reacts with ethanal through 1-hydroxy-ethanethiol (1-sulphanylethanol) forming ethane-1,1-dithiol, a basic compound in many production cycles of other sulphur compounds (Zhang et al. 1988). This compound has been isolated and identified in yeast extracts by Werkhoff et al. (1989). These authors claim that the aroma it produces depends on its concentration, and is similar to that of "onion", "leek", "cabbage" or "peanut". In the presence of oxygen, ethane-1,1-dithiol can be oxidised to produce *cis-/trans*-3,6-dimethyl-1,2,4,5-tetrathiane that has rubbery aromas (Werkhoff et al. 1989). After oxidation/cyclization, *cis-/trans*-3,5-dimethyl-1,2,4-trithiolane is produced that can give the wine "meaty", "spicy" and "garlic" aromas. These compounds were first detected in roast and boiled chicken, pork and beef (Brinkmann et al. 1972). The reaction between 1,1-ethanedithiol and H_2S forms *cis-/trans*-4,7-dimethyl-1,2,3,5,6-pentathiepane that produces meaty aromas (Chen and Ho 1986). The reactions involved and cited by the above authors are described (Fig. 10.8).

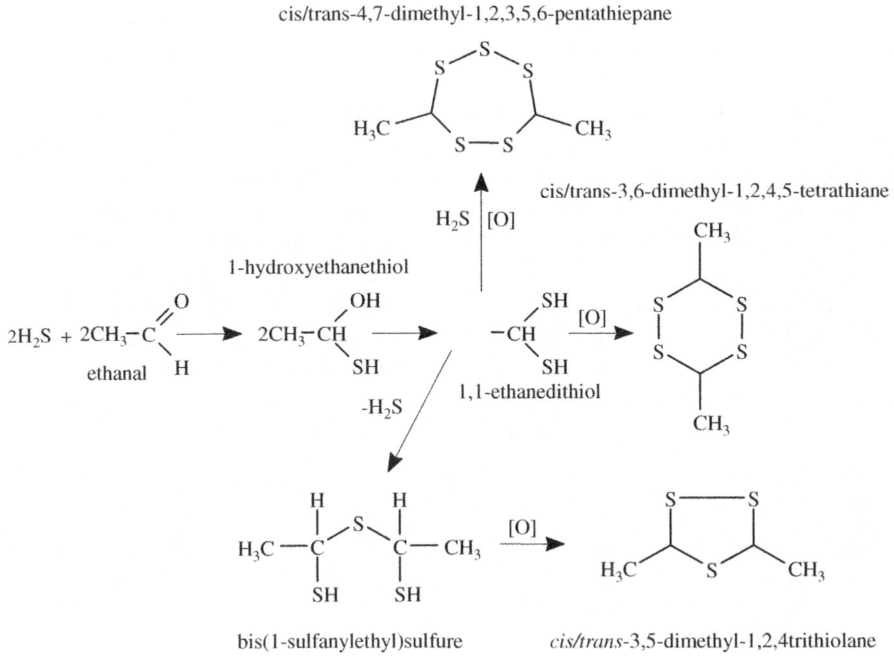

Fig. 10.8 Sulphur compounds formed in the reaction of H_2S with ethanal (according to different authors)

These compounds, produced in the reaction of H_2S and ethanal, have been detected in a synthetic solution and in a white wine by Rauhut and Dittrich (1993). These authors made H_2S at concentrations from 1 to 5 mg/L react with ethanal at variables concentrations from 100 to 500 mg/L. This reaction produces *cis-/trans*-3,6-dimethyl-1,2,4,5-tetrathiane, the *cis-/trans*-4,7-dimethyl-1,2,3,5,6-pentathiepane and the *cis-/trans*-3,5-dimethyl-1,2,4-trithiolane and the precursor of these compounds, 1,1-ethanedithiol. However, the same reaction with copper does not prevent formation of these compounds. In contrast, the reaction reported by some authors between H_2S and ethanal to form ethanethiol, does not take place.

We did not detect any of the reaction products between H_2S and ethanal in any of the red wines studied, not even in the wines with a strong concentration of H_2S.

We carried out some experiments in an attempt to explain this phenomenon. The first consisted in carrying out the reaction in a synthetic solution, as Rauhut and Dittrich (1993) has done previously and, indeed, we too detected the sulphur compounds described by the authors.

Then 3 mg of H_2S were added to a red wine containing 75 mg/L of ethanal (a higher value than that normally detected in red wines). The dose of free SO_2 remained at 30 mg/L, but after 10 days none of the compounds described were detected. There are several possible reasons for this. The first is that H_2S reacts with other red wine compounds, such as phenolic compounds (De Freitas 1995).

The second is that H_2S cannot react with ethanal because it is bound to SO_2in the bisulphitic combination or to phenolic compounds.

Two experiments were performed to try to determine which of these reactions was occurring. After macerating the red wine grapes for one day the coloured must obtained was seeded with yeasts (100 mg/L of LSA). Also, a commercial white must was seeded with the same dose of yeasts. When fermentation started, 5 mg/L of H_2S were added and the analysis was carried out one week after finishing alcoholic fermentation. In both cases, the formation of *cis-/trans*-3,6-dimethyl-1,2,4,5-tetrathiane and of *cis-/trans*-3,5-dimethyl-1,2,4-trithiolane was detected. The mass spectra are shown in Figs. 10.9 and 10.10. By contrast, in this experiment *cis-/trans*-4,7-dimethyl-1,2,3,5,6, pentathiepane and 1,1-ethanedithiol did not appear.

Isomers of these compounds have very similar retention times in these conditions (Fig. 10.11). With a polar column, the *cis-/trans*-3,5-dimethyl-1,2,4-trithiolane exit first, followed by *cis-/trans*-3,6-dimethyl-1,2,4,5-tetrathiane.

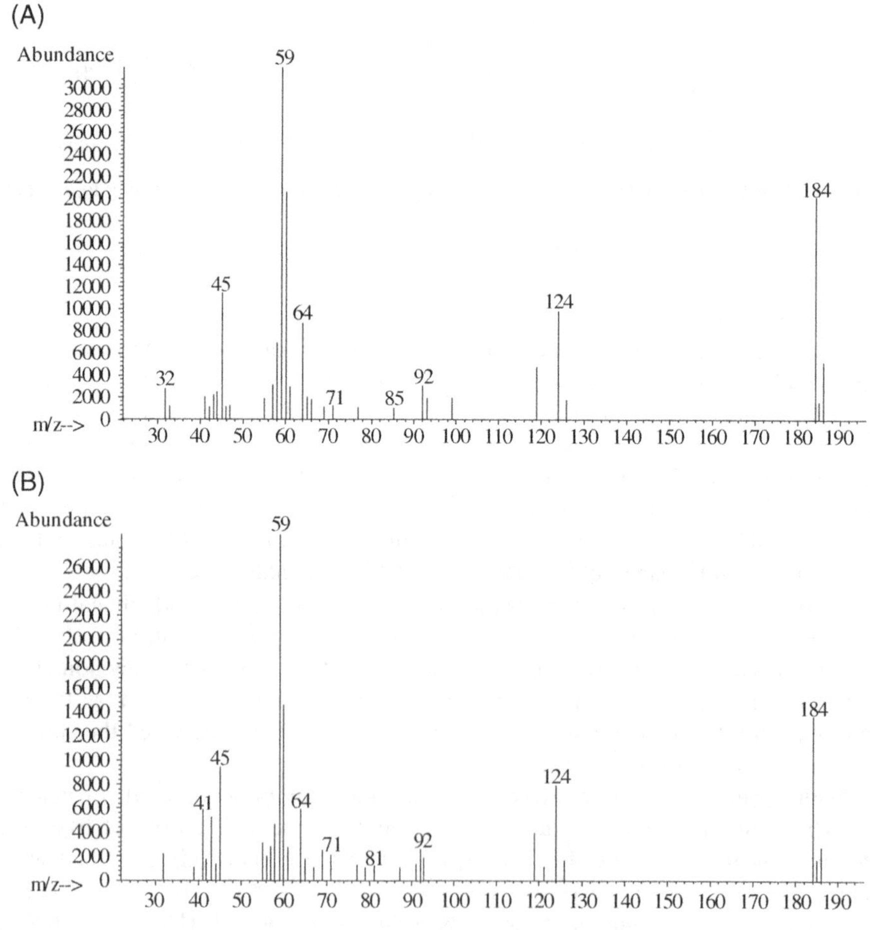

Fig. 10.9 Mass spectra of *cis-/trans*-3,6-dimethyl-1,2,4,5-tetrathiane

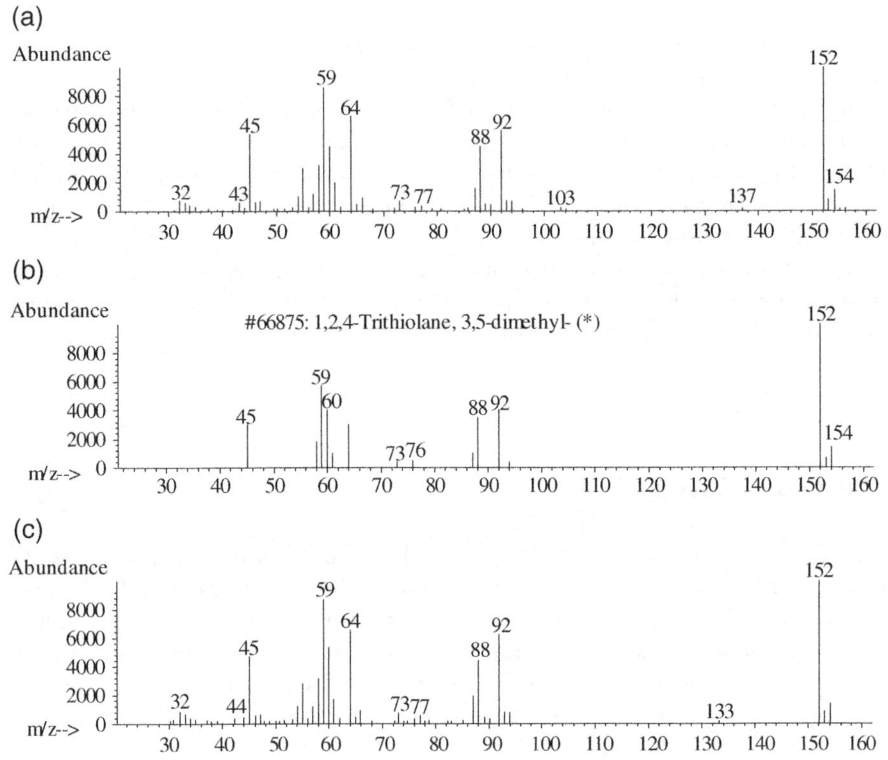

Fig. 10.10 Mass spectra of *cis-/trans*-3,5-dimethyl-1,2,4-trithiolane. At the centre is the mass spectra of this compound found in the library NBS75K.L

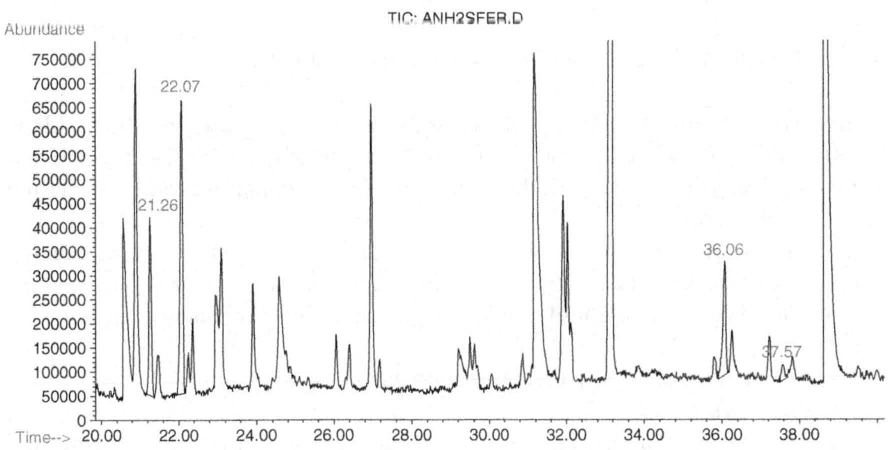

Fig. 10.11 Chromatogram of wine extract fermented with H_2S. The first peaks with retention times recorded are for the compounds *cis-/trans*-3,5-dimethyl-1,2,4-trithiolane, and after *cis-/trans*-3,6-dimethyl-1,2,4,5-tetrathiane

These compounds were not found in the wines studied. Nonetheless, some were detected in a grape marc spirit with a clear reducing defect in which we detected the presence of *cis-/trans*-3,5-dimethyl-1,2,4-trithiolane and *cis-/trans*-3,6-dimethyl-1,2,4,5-tetrathiane, although neither *cis-/trans*-4,7-dimethyl-1,2,3,5,6-pentatiepane nor 1,1-ethanedithiol were detected. These compounds could possibly come from the base wine used for the distillation, although it is prohibited to use SO_2 in wines used for this purpose. The hypothesis according to which ethanal is free to react with H_2S is, therefore, reinforced.

In conclusion, high doses of hydrogen sulphide in wines caused by the late treatment of vines with sulphur or some pesticides can cause serious defects in the grape marc. Moreover, it is almost impossible to remove the products from the compounds formed.

10.6 Cork Flavor

Cork flavor, or the olfactory defect caused by moulds, is the commonest of the undesirable flavors found in wine. In most cases, the cork is responsible for tainting wine with 2,4,6-trichloroanisol (TCA) (Tanner et al. 1981). However, barrelled wine may also be contaminated with 2,3,4,6-tetrachloroanisol (TeCA) that can be present in the damp atmosphere of wine cellars in the presence of woods treated by polychlorophenols. TCA has a slightly different aroma than TeCA but the two contaminants are often confused. Another possibility is that the cork itself may be contaminated by other corks (TCA) or by TeCA.

Wines often have this defect and it can affect up to 5% of bottles.

10.6.1 Origin of the 2,4,6-Trichloroanisol (TCA)

When a wine is tainted with cork flavor and this effect is caused by TCA, this is almost always caused by the cork. This can be verified by analysing the wine and also the central third of the cork that cannot have been contaminated by either the wine or by the exterior.

Consumers reject a wine containing more than 10 ng/L of TCA. Wine experts, however, can detect this defect from levels of 5 ng/L and can even detect its negative effects at levels of around 2 ng/L when comparing with an uncontaminated control wine.

Two different cases of cork contamination can be observed.

when the cork material is the source of contamination, very high values can be recorded such as 1500 ng/g. These high values are almost always recorded when the cork has high concentrations of trichlorophenol (TCP) (Table 10.1).

Several studies have shown that the main cause of this is treating the corks with chlorine or bleach solutions. It has also been demonstrated that when a wine is

Table 10.1 Example of contaminated corks (ng/g)

	TCA[a]	TCP	TeCA	TeCP	PCA	PCP
Cork no. 1	543	319	8	3	48	5
Cork no. 2	1190	1146	3	5	13	6

[a] TCA, 2,4,6-trichloroanisol; TeCA, 2,3,4,6-tetrachloroanisol; PCA, pentachloroanisol; TCP, 2,4,6-trichlorophenol; TeCP, 2,3,4,6-tetrachlorophenol; PCP, pentachlorophenol

Table 10.2 Examples of corks contaminated by the breakdown of wood treatment products (ng/g of cork)

	TCA[a]	TCP	TeCA	TeCP	PCA	PCP
Cork no. 1	1	0	379	8	4350	17
Cork no. 2	3	0.3	46	0.3	160	16

[a] TCA, 2,4,6-trichloroanisol; TeCA, 2,3,4,6-tetrachloroanisol; PCA, pentachloroanisol; TCP, 2,4,6-trichlorophenol; TeCP, 2,3,4,6-tetrachlorofenol; PCP, pentachlorophenol

strongly contaminated by cork TCA, all the bottles stopped with corks from the same batch are also contaminated to some degree.

On the other hand, there are also corks that are simply contaminated by storage in an atmosphere contaminated by the breakdown products of the polychlorophenols used to treat the pinewood of the palettes or shelving where the bottles are stored (Bertrand and Barrios 1994). This contamination is easy to detect because TCA is always present at lower levels than TeCA and PCA (Table 10.2).

If the contamination has taken place after the bottles were corked during storage on the palettes then the defect will be on the external part of the cork.

10.6.2 Origin of the 2,3,4,6-Tetrachloroanisol

Apart from the case reported above, this defect usually affects wine in the barrel. The perception threshold for TeCA is around 15 ng/L for experts and 35 ng/L for experienced consumers. The flavor defect is described as powdery, dry and musty.

Pentachlorophenol (PCP) and its sodium derivative (Na PCP) have been used for some time to prevent the pinewood from taking on a bluish color. In addition, it is an excellent pesticide and, also, seems to function as an insecticide. It often contains 2,3,4,6-tetrachlorophenol as an impurity and, less frequently, 2,4,6-TCP. At the surface of the wood, at a depth of 1 or 2 mm, the levels of PCP may exceed 1 mg/g (Table 10.3).

In the damp atmosphere of the wine cellars, moulds can break down PCP and become detoxified as pentachloroanisol, which is volatile but has little aroma. In contrast, tetrachloroanisol, even at very weak doses can contaminate the wine during decanting or bottling (Table 10.4).

Table 10.3 Examples of wood shavings from boxes or wooden structures treated with polychlorophenols (μg/g of wood)

	TCA[a]	TCP	TeCA	TeCP	PCA	PCP
Box no. 1	0.008	1	0.16	0.05	1	26
Box no. 2	0.001	0.001	0.12	2.7	1.8	1100
Box no. 4	0.001	0.001	0.10	2.6	0.86	415
Box no. 5	0.070	1	1.6	1.3	8	1000
Powder	0.001	0.001	0.01	2	0.3	120

[a] TCA, 2,4,6-trichloroanisol; TeCA, 2,3,4,6-tetrachloroanisol; PCA, pentachloroanisol; TCP, 2,4,6-trichlorophenol; TeCP, 2,3,4,6-tetrachlorophenol; PCP, pentachlorophenol

Table 10.4 Examples of wines contaminated by the breakdown of wood treatment products (ng/L of wine)

	TCA[a]	TCP	TeCA	TeCP	PCA	PCP
Wine no. 1	0	6	28	8	73	5
Wine no. 2	2	11	68	74	485	718
Wine no. 4	1	7	33	3	136	15

[a] TCA, 2,4,6-trichloroanisol; TeCA, 2,3,4,6-tetrachloroanisol; PCA, pentachloroanisol; TCP, 2,4,6-trichlorophenol; TeCP, 2,3,4,6-tetrachlorophenol; PCP, pentachlorophenol

10.6.3 2-Methoxy-3,5-Dimethylpyrrhazine

The compound 2-methoxy-3,5-dimethylpyrrhazine has a musty aroma and its perception threshold is around 3 ng/L (Sefton and Simpson 2005). This molecule is also found in wine-making equipment in poor condition and can be very difficult to eliminate.

10.7 Other Defects

In a sample of defective wines, most in the barrel, half of them contained TCA or TeCA at clearly perceptible doses. Wines may also have defects described as "earthy", "weedy", "wild mushrooms", "wet cardboard" and considerable research has centred on this. Lee and Simpson (1993) analysed them and identifieds compounds such as 2,4-dichloro-6-methylanisol, chlorated cresols, oct-1-en-3-one, oct-1-en-3-ol, *cis*-octa-1,5-dien-3-one, *cis*-octa-1,5-dien-3-ol, guayacol, 2-methylisoborneol and geosmine. Rapp (1992) also detected other substances in the corks such as: 6-chlorovainilline, 4-chloro guayacol, 4,5-dichloro guayacol and veratrol.

References

Anocibar Beloqui, A., (1998). Contribution à l'étude des composés volatils des vins rouges. (Doctoral dissertation N 611, University V. Segalen, Bordeaux II, France, 1982).

Bertoli, M., Caussanel, J. L., & Médina, B. (2004). Dosage des phtalates dans les boissons alcooliques. FV OIV, 1234.

Bertrand, A., & Barrios, M. L. (1994). Contamination des bouchons par les produits de traitement des palettes de stockage des bouteilles. *Rev. Fr. Oenol., 145*, 9–17.

Blaise, A. (1986). Les résines époxydiques et leur utilisation dans le domaine vinicole. Migration de l'alcool benzylique et goût d'amande amère. (Doctoral dissertation, University of Montpellier I, 1986).

Blaise, A., & Bertrand, A. (1998). Altérations organoleptiques des vins. In C. Flanzy (Ed.), *Œnologie, fondements scientifiques et technologiques* (pp. 1183–1211). Tec Doc, Paris

Blaise, A., & Brun, S. (1986). Un phénomène enzymatique à l'origine du goût d'amande amère dans les vins. *C.R. Acad. Agric. Fr., 72*(2), 73–77.

Bobet, R. A., Noble, A. C., & Boulton, R. B. (1990). Kinetics of the ethanethiol and diethyl disulfide interconversion in wine-like solutions. *J. Agric. Food. Chem., 38*, 449–452.

Brinkmann H. W., Copier H., De Leuw J. J. M., & Tjan S. B. (1972). Components contribution to beef flavour: Analysis of the headspace volatiles of beef broth. *J. Agric. Food Chem., 20*, 177–181.

Brun, S., Giffone, M., & Chabas, J. (1982). Cuves vinicoles en résines polyesters. In S. Brun (Ed.), *Les matériaux en contact avec les aliments* (pp. 279–282). Tec Doc, Lavoisier, Paris.

Burkhardt, R. (1978). Verschiedene geruchswirksame Verbindungen des Schwefels in mosten und Weinen. *Dtsch. Weinbau, 28*, 1281–1283.

Cantagrel, R., Lurton, L., Desache, F., Vidal, J. P., Maujean, A., & Medjma M. (1994). Incidence des produits de traitement utilisés au vignoble sur l'apparition de défauts olfactifs dans les eaux-de-vie. F.V OIV 991, 22311/130495.

Cantarelli, D., Tarufi, F., & Martini A. (1964). Chemical and microbiological surveys on the effects of dithiocarbamate fungicides on wine-making. *J. Sci. Food Agric., 15*, 186–196.

Chen, C. C., & Ho, C. T. (1986). Identification of sulfurous compounds of Shiitake Mushroom (*Lentinus edodes, Lange Sing*). *J. Agric. Food Chem., 34*, 830–833.

Cocheo V., Bellomo, M. L., & Bombi, G. G. (1983). Rubber manufacture: sampling and identification of volatile polluants. *Am. Ind. Hig. Assoc. J., 44*, 521–527.

Colagrande, O., Mazzoleni, V., & Silva, A. (1988). Génèse des odeurs et des goûts anormaux des vins. In *Symposium sur les odeurs et goûts anormaux des vins* (pp. 29–39), Alexandria.

Crowell, E. A., & Guymon, J. F. (1975). Wine constituents arising from sorbic acid addition, and identification of 2-ethoxyhexa-3,5-diene as source of geranium-like off-odor. *Am. J. Enol. Vitic., 2*, 97–102.

De Freitas, V. (1995). Recherches sur les tanins condensés: Application à l'étude des structures et propriétés des procyaanidines du raisn et du vin. (Doctoral Dissertation, University of Bordeaux II, France, 1995).

Dittrich, H. H., & Staudenmayer, T. (1970). Über die Zusammenhänge zwischen der Sulfit-Bildung und Schwefelwasserstoff-Bildung bei Saccharomyces cerevisiae. *Zentralbl. Bakteriol. Parasitenkd. Infektionshr. Hyg., II, 124*, 113–118.

Eschenbruch, R. (1974). Sulfite and sulfide formation during wine makinf. -a review. *Am. J. Enol. Vitic., 25*, 157–161.

Goniak, O. J., & Noble, A. C. (1987). Sensory study of selected volatile sulfur compounds in white wine.*Am. J. Enol. Vitic., 38*, 223–227.

Jacob, L. (1961). Veränderungen von Thiuramen bei der alkoholischen Gärung. *Wein-Wiss., 16*, 97–104.

Keck, S. (1989). Untersuchungen zur Bedeutung flüchtiger phenolischer, schwefelhaltiger und stickstoffhaltiger Verbindungen für unerwünschte. Aromanoten des Weines mittels Gaschromatographie/Massenspektrometrie. (Doctoral dissertation, University of Karlsruhe, Germany).

Lacoste, J. (1973). Utilisation du sorbate de potassium. *Rev. Fr. Oenol., 50*, 31.

Lavigne, V. (1996). Recherche sur les composés soufrés volatils formés par la levure au cours de la vinification et l'élevage des vins blancs secs. (Doctoral dissertation, University of Bordeaux II, France).

Lee, T. H., & Simpson, R. F. (1993). Microbiology and chemistry of cork taints in wine. In G. H. Fleet (Ed.), *Wine Microbiology and Biotechnology*. Chur, Switzerland: Harwood Academic Publishers.

Marshall, W. (1977). Thermal decomposition of ethylenbisdithiocarbamate fungicide to ethylenthiourea in aqueous media.*J. Agric. Food Chem., 25*, 357–361.

Maujean, A. (1989). Goûts anormaux. *Rev. Oenol.*, N°53 (novembre), 31–32.

Maujean, A., Medina, M., Vidal, J. P. Cantagrel, R., & Lurton, L. (1993). Etude du comportement chimique et thermique des produits soufrés dans les vins et les eaux-de-vie. In R. Cantagrel (Ed.), *Elaboration et connaissance des spiritueux* (pp. 110–122). Ed BNIC, Diffusion Technique et Documentation Lavoisier, Paris.

Maw, G. A. (1965). The role of sulfur in yeast growth and in brewing. *Wallerstein Laboratory Communications,* XXVIII, N° 95, April, 49–70.

Neuberg, C., & Grauer, A. (1952). Biosynthese von Mercaptanen. Hopp-Seyler's. *Z. Physiol. Chem., 289*, 253–256.

Radler, F. (1976). Dégradation de l'acide sorbique par les bactéries. *Bull. l'O.I.V.*, 545–546, 629–635.

Rankine, B. C. (1963). Nature, origin and prevention of hydrogen sulphide aroma in wines. *J. Sci. Food Agric., 14*, 79–91.

Rapp, A. (1992). Foreign and undesirable flavours in wine. *In*: Donéche B. *Les acquisitions récentes en chromatographie du vin-Applications à l'analyse sensorielle des vins.* Tec Doc Lavoisier, Paris, 151–173.

Rauhut, D. (1990). Trace analysis of sulfurous off-flavours in wine caused by extremely volatile S-containing metabolites of pesticides e.g. Orthene. In *Actualités Oenologiques 89, Comptes Rendus du 4 Symposium International d'Oenologie*, Bordeaux, 15–17 June 1989, Paris Dunod, pp. 482–487.

Rauhut, D., & Dittrich, H. H. (1993). Sulfur compounds and their influence on wine quality. *Schwefelverbindungen und ihr Einflub auf die Weinqualität*, 214–218.

Schmitt, A. (1987). Untersuchung des chemischen Verhaltrens von Tetramethyl-thiuramdisulfid in Lebensmitteln. (Doctoral dissertation, University of Karlsruhe, Germany, 1987).

Schütz, M., & Kunkee, R. E. (1977). Formation of hydrogen sulfide from elemental sulfur during fermentation by wine yeast.*Am. J. Enol. Vitic., 28*, 137–144.

Sefton, M. A, & Simpson, R. F. (2005). Compounds causing cork taint and the factors affecting their transfer from natural cork closures to wine – a review. *Austr. J. Grape Wine Res., 11*, 226–240.

Spedding, D. J., & Raut, P., (1982). The influence of dimethyl sulphide and carbon disulphide in the bouquet of wine. *Vitis, 21*, 240–246.

Stratford, M., & Rose, A. H. (1985). Hydrogen sulphide production from sulphite by *Saccharomyces cerevisiae. J. Gen. Microbiol., 131*, 1417–1424.

Tanner, H. (1969). Der Weinböckser, Entstehung und Beseitigung. *Schweiz. Z. Obst-Weinbau, 105*, 252–258.

Tanner, H., Zannier, C., & Buser, H. R. (1981). 2, 4, 6-Trichloranisol : eine dominierende Komponente des Korkgeschmackes. *Schweiz. Z. Obst- Weinbau, 117*, 97–103.

Thomas, C. S., Boulton, R. B., Silacci, M. W., & Gubler, W. D. (1993). The effect of elemental sulfur, yeast strain, and fermentation medium on hydrogen sulfide production during fermentation. *Am. J. Enol. Vitic., 44*, 211–216.

Wainwright, T. (1971). Production of H_2S by yeast : role of nutrients. *J. App. Bacteriol., 34*, 161–171.

Wenzel, K., & Dittrich, H. H. (1978). Zur Beeinflussung der Schwefelwasserstoff-Bildung der Hele durch Trub, Stickstoffgehalt, molekularen Schwefel und Kupfer bei der Vergärung von Traubenmost. *Wein Wiss., 33*, 200–213.

Wenzel, K., & Dittrich, H. H. (1983). Hydrogen sulfide formation by yeast during wine making. In *3rd. International Mycology Congress*, Tokyo – Japan, Abstract 339.

Wenzel, K., Dittrich, H. H., Seyffard, H. P., & Bohner, J. (1980). Schwefelrückstände auf Trauben und im Most und ihr Einflub auf die H_2S-Bildung. *Wein Wiss., 35*, 414–420.

Werkhoff, P., Brüning, J., Emberger, R., Matthias, G., Köpsel, M., & Kuhn, W. (1989). Studies on volatile sulphur-containing flavour components in yeast extrat. In *11 th International Congress of Essential Oils, Fragances and Flavours*. New Dehli. *Proceedings, 4*, 215–243.

Werkhoff, P., Bretschneider, W., Emberger, R., Guntert, M., Hopp R., & Kopsel M. (1991). Recent developments in sulfur flavour chemistry of yeast extracts. *Chem. Mikrobiol. Technol. Lebensm., 13*, 30–57.

Wilson, R. A., & Katz, I. (1972). Review of literature on chicken flavor and report of isolation of several new chicken flavor components from aqueous cooked chicken broth. *J. Agric. Food Chem., 20*, 741–747.

Würdig, G. (1977). Problèmes de la conservation des vins doux. In: 57éme *Assemblée générale de l'OIV, Nyons-Changins, Suisse*. O.I.V., Paris, 21–29.

Zhang, Y., Chien, M., & Ho, C. T. (1988). Comparaison of the volatile compounds obtained from thermal degradation of cysteine and glutathione. *J. Agric. Food Chem., 36*, 992–996.

Chapter 11
Wine Spoilage by Fungal Metabolites

Manuel Malfeito-Ferreira, André Barata, and Virgilio Loureiro

Contents

11.1 Introduction

This chapter is devoted to the description of toxic and spoiling secondary metabolites produced by moulds and yeasts in grapes and wines. The toxic compounds affect human health being a food safety issue while spoilage concerns wine organoleptical quality being, therefore, a technological matter. Amongst several compounds with detrimental effects on human health reported as occurring in wine – lead, pesticides, ethyl carbamate, biogenic amines, and more recently ochratoxin A (OTA) – only the last is of fungal origin and the most preoccupying of them. The occurrence

Virgilio Loureiro (✉)
Tapada da Ajuda, 1349-017 Lisbon, Portugal
e-mail: vloureiro@isa.utl.pt

M.V. Moreno-Arribas, M.C. Polo (eds.), *Wine Chemistry and Biochemistry*,
DOI 10.1007/978-0-387-74118-5_23,

of OTA in grapes and wines has a special meaning in enology, because wine has been known, since remote antiquity, as "the safest beverage". In fact, the enormous success of wine (and beer) for millennia in the Mediterranean basin is due to the bad sanitary quality of water, frequently contaminated by salmonella or other faecal pathogens around urban agglomerates. In the second half of the nineteenth century, Louis Pasteur stated that "Wine is the most hygienic of the beverages", implying that any pathogenic bacteria had capacity to grow or survive in wine. The occurrence of OTA in wine changes the paradigm of "the safest beverage", thus imposing a food safety concern to the wine industry.

Microorganisms have long been known as one of the main sources of wine spoilage. Earlier enology treatises mention the spoiling activities of lactic and acetic acid bacteria and yeasts, concerns being preferentially directed to the deleterious effects of acetic and lactic acid bacteria (Ribéreau-Gayon et al. 2006). More recently, yeasts have become a source of serious trouble for winemakers, especially after the confirmation of *Dekkera/Brettanomyces bruxellensis* as the producers of unwanted levels of 4-ethylphenol (4-EP) leading to phenolic or "horse-sweat" taints. These yeast species and lactic acid bacteria are also responsible for the production of tetrahydropyridines (THP), connected with the taint described as "mousiness".

This chapter outlines the most recent advances in the awareness of the problems raised by ochratoxin A, volatile phenols and tetrahydropyridines. We will focus on the factors leading to their occurrence, the relevance to wine quality and safety and the respective preventive measures suitable under vineyard and winery practices.

11.2 Ochratoxin A

Mycotoxins are metabolites produced by fungi and affecting human health, causing mycotoxicoses. These diseases have been known for a long time but only after the early 1960s have they been appropriately studied (Bennett and Klich 2003; van Egmond and Schothorst 2007). The first mycotoxin to be studied was aflatoxin. Since then the number of known mycotoxins have increased to 300 or 400 compounds of variable toxicity and occurrence. The main food commodities affected are cereals used for direct human consumption, for processing, or as animal feeds and dried fruits (van Egmond and Schothorst 2007). Some mycotoxins have been mentioned in grape products (patulin, aflatoxins, trichothecenes) but OTA is the main toxin of concern in the wine industry (Hocking et al. 2007). The occurrence of OTA in wine was reported for the first time in 1995 (Zimmerli and Dick 1996) and much data has been generated since then for wine, grape juice and raisins, especially in the last seven years.

11.2.1 Chemical Structure

Ochratoxin A was originally described as a metabolite of the mould *Aspergillus ochraceus* in a laboratory screening for toxinogenic fungi (van der Merwe et al. 1965).

It is the most abundant and hence the most commonly detected member of the family of ochratoxins produced as secondary metabolites by moulds. Apart from ochratoxin α (OTα), the ochratoxins comprise a polyketide-derived dihydroisocoumarin moiety linked via the 7-carboxy group to l-β-phenylalanine by an amide bond. Ochratoxins consist of ochratoxin A (OTA), its methyl ester and its ethyl ester, also known as ochratoxin C (OTC), 4-hydroxyochratoxin A (4-OH OTA), ochratoxin B (OTB) and its methyl and ethyl esters and ochratoxin α (OTα), where the phenylalanine moiety is missing (Fig. 11.1). All of them behave like weak organic acids and the differences in the chemical structures have marked effects on their respective toxic potentials, OTA being the most toxic of the group (Ringot et al. 2006). The presence of chlorine in the OTA structure makes it unique among mycotoxins (Murphy et al. 2006). The empirical formula is $C_{20}H_{18}O_6NCl$ and the molecular weight is 403.82. The IUPAC developed formula of OTA is l-phenylalanine-*N*-[(5-chloro-3,4-dihydro-8-hydroxy-3-methyl-1-oxo-1*H*-2-benzopyran-7-yl)carbonyl]-(*R*)-isocoumarin (Ringot et al. 2006). The present knowledge of the biosynthetic pathway of OTA, which has not yet been fully established, has been reviewed by Ringot et al. (2006).

11.2.2 Toxicity and Public Health Safety

OTA is a mycotoxin considered to be a possible carcinogen (Class 2B) for humans (IARC 1993) and it has been shown to be nephrotoxic, hepatotoxic, teratogenic, carcinogenic and immunotoxic to several species of animals and to cause kidney and liver tumours in mice and rats (JECFA 2001). In humans it is accumulated in body tissues because it appears to be slowly eliminated, but the effects are not well established (Ringot et al. 2006). Ruminant animals, such as cows and ewes, are generally resistant to the effects of OTA due to its hydrolysis to non-toxic metabolites by protozoa in the stomachs before absorption into the blood (Kiessling et al. 1984). Recent studies on the toxic mechanisms are focused on the OTA ability to disturb cellular signalling and regulation and to modulate physiological signals and thereby

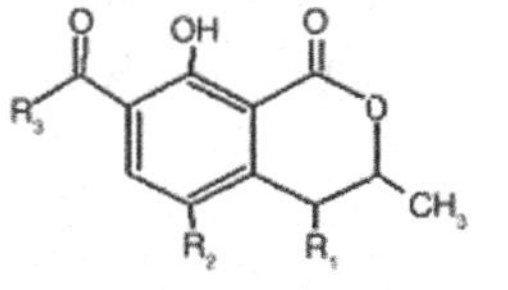

Ochratoxins	R1	R2	R3
OTA	H	Cl	-NH-CH(COOH)-CH_2- Phenyl
OTB	H	H	-NH-CH(COOH)-CH_2- Phenyl
OTC	H	Cl	-NH-CH($COOC_2H_5$)-CH_2-Phenyl
4-hydroxyochratoxin A	OH	Cl	-NH-CH(COOH)-C_2H- Phenyl
OTα	H	Cl	-OH

Fig. 11.1 Chemical structure of ochratoxins (reprinted from Ringot et al. 2006, permission to be obtained)

to influence cells viability and proliferation, but its modes of toxicity remain "a continuing enigma" (see review of Ringot et al. 2006).

Given their toxicity, mycotoxins are subjected to regulations determining the maximum allowable levels. As a rule food products have mycotoxin levels lower than the limits (Jorgensen 2005) but about 40% of the notifications received in 2005 by an European rapid alert system for food and feed were related to risks to human health by mycotoxins (van Egmond and Schothorst 2007). These authors further mentioned that almost 90% of these notifications were related to aflatoxin in nuts and nut products imported to the EU. Cases related to OTA in wine were not referred to in this report. In addition, an indirect measure of the toxicity and risk to human health may be given by the standards of international legislation or guidelines of advisory boards. Concerning OTA there are no regulations in the USA or in Codex Alimentarius Commission, in contrast to patulin, fumonisin, aflatoxin and deoxynivalenol (DON) (Murphy et al. 2006), and so it seems admissible that these toxins represent a bigger threat than OTA.

At European Union level the Council Regulation (EEC) 315/93 of 8th February 1993 provided the legal framework for establishing maximum levels for food contaminants at Community level. In 1995, the European Commission (EC) initiated the activity SCOOP (scientific cooperation on questions related to food), which included a project to provide data on the occurrence of OTA in food commodities on the European market and on the dietary exposure to OTA in the EU member states (Jorgensen 2005). As a consequence, many data on the occurrence of OTA in human food and human blood plasma have become available since 1995. After the first SCOOP report, known as SCOOP-1 (European Commission 1997) a second SCOOP task was performed, extended to other commodities and processed foods, including wine and other grape products, to evaluate if the additional studies changed the conclusions of first SCOOP report. Not surprisingly, due to the detection of OTA in wine, much occurrence data has been produced since then, not only for wine but also for dried vine fruits (currants, raisins, and sultanas) and grape juice, particularly after 2000. The most relevant conclusion from those data was that the overall mean level of OTA in wine was 0.36 μg/kg (mean of 1470 samples), representing the second source, after cereals, to the OTA exposure in the European diet (European Commission 2002), raising relevant concern and electing OTA as a threat to the European wine industry. In particular, JECFA (2001) calculated that the human OTA exposure was of 58%, 21%, 7%, 5% and 3% of total OTA intake for cereals, wine, grape juice, coffee and pork, respectively (Murphy et al. 2006). However, the calculation of these figures was based on the controversial assumption that the mean intake is represented by the arithmetic mean value. If the median value was used, the value of OTA in red wine would have been only 0.02 μg/kg and the contribution of wine consumption for OTA intake-rate would drop to 2% (Otteneder and Majerus 2000). The maximum allowable limits for OTA in several food products have been established recently in the EU, being the adult's strictest value applied to wine (Table 11.1). In addition, a provisional tolerable weekly intake (PTWI) of 100–120 ng/kg body weight (bw) is advised (JECFA 2001) that was set based on a safety factor of 450 related with the renal function deterioration of pigs,

Table 11.1 European Union and Swiss regulations for ochratoxin A (μg/kg)

Product	Maximum allowable concentration	Reference
Unprocessed cereals	5	Commission Regulation (2006)
All products derived from unprocessed cereals, including processed cereal products and cereals intended for direct human consumption with the exception of foodstuffs	3	
Dried vine fruit (currants, raisins and sultanas)	10	
Roasted coffee beans and ground roasted coffee, excluding soluble coffee	5	
Soluble coffee (instant coffee)	10	
Wine (including sparkling wine, excluding liqueur wine and wine with an alcoholic strength of not less than 15 vol.%) and fruit wine	2	
Aromatised wine, aromatised wine-based drinks and aromatised wine-product cocktails	2	
Grape juice, concentrated grape juice as reconstituted, grape nectar, grape must and concentrated grape must as reconstituted, intended for direct human consumption	2	
Processed cereal-based foods and baby foods for infants and young children	0.5	
Dietary foods for special medical purposes intended specifically for infants	0.5	
Coffee products (Switzerland)	8	Taniwaki (2006)

for which the lowest observed effect level (LOEL) was 0.008 mg/kg bw per day (Bakker and Pieters 2002). Current levels of weekly exposure to OTA in adults of EU member states vary between 15 and 60 ng/kg bw which is, at most, half quantity of the PTWY (CONTAM 2006).

Further evaluation of OTA occurrence in wines worldwide has revealed that variable proportions of wines are contaminated but only a rather small number has levels over the maximum allowable limit of 2.0 μg/kg (Ng et al. 2004; Mateo et al. 2007). Higher concentrations are more frequent in wines produced from dried grapes and in raisins, but are less consumed than table wines (Burdaspal and Legarda 2007; Valero et al. 2007b). Burdaspal and Legarda (2007) showed that sweet wines only contributed with 3.1–3.8% (regular consumers) or 0.3–0.4% (whole adult population) to the PTWI. Therefore, wines do not seem to have a significant contribution to human exposure to OTA.

11.2.3 Occurrence

Ochratoxin A is among the most important mycotoxins affecting foods being described in food commodities together with aflatoxin, deoxyvalenol, T-2 toxins and trichotecenes, although the relative weight of each other in public health has not been mentioned (see reviews of Frisvad et al. 2006 and Richard 2007). As in other mycotoxins, OTA can contaminate a wide variety of food commodities as a result of fungal infection in crops, in the field during growth, at harvest, in storage, and in shipment under favourable environmental conditions, especially when the foods are not properly dried. Then, if used in animal's feed it may contaminate their products (Murphy et al. 2006). OTA occurs mainly in the storage of cereals and grains under conditions that favour mould growth and toxin production. Therefore it is associated with cereal and cereal products, soy products, coffee and cocoa products (CAST 2003; Jorgensen 2005; Frisvad et al. 2006). Meat and meat products such as salami and hams, hard cheeses, spices and beer have also been related with OTA (Jorgensen 2005; Frisvad et al. 2006). Dried fruits (dates, plums, apricots, figs, sultanas) are typical affected commodities (Iamanaka et al. 2006). OTA may also occur in house dust and other airborne particulates (Richard 2007).

Concerning the wine industry, OTA has been detected in wines, grape juice, vinegar and raisins (Jorgensen 2005; Varga and Kozakiewicz 2006). These products were the last to be associated with OTA but studies in wines have greatly increased since the first report in 1995, being the OTA related food commodity most studied in the last decade (Jorgensen 2005), demonstrating the effort put into the solution of the problem by the wine sector. Table 11.2 lists the results of recent relatively large surveys worldwide demonstrating, as already mentioned, that wines with levels higher than the maximum allowable concentration of 2.0 μg/L are not frequent. As a rule, the OTA concentration increases from white and rosé to red wines, from cold to warmer regions, and dessert or sweet wines have higher mean levels (Varga and Kozakiewicz 2006 and Table 11.2). We are not aware of large surveys of OTA levels in grapes but raisins and sultanas have higher mean OTA levels than fresh grapes (Varga and Kozakiewicz 2006; Iamanaka et al. 2006).

11.2.4 Ochratoxin A Production on Grapes

11.2.4.1 OTA Producing Species

Ochratoxin A of food commodities is produced by a small number of fungal species in the genera *Penicillium*, *Aspergillus* and the *Aspergillus* teleomorphs *Petromyces* and *Neopetromyces* (Frisvad and Samson 2000; Frisvad et al. 2004).

The main species that occur in grapes and, consequently, in grape juices, raisins, wine and wine derivatives belong to the so-called black aspergilli, taxonomically included in the *Aspergillus* section *Nigri*. Unfortunately, the taxonomy of this section is not completely known, creating many difficulties on the identification of strains, originating a proliferation of taxa, including species, subspecies, and varieties (for a discussion see Samson et al. 2004 and Frisvad et al. 2006). Based on phenotypic comparisons of a broad collection of black aspergilli, Samson et al. (2004) considered 15 species provisionally accepted in *Aspergillus* section *Nigri*, four of those producing OTA and only two occurring on grapes, raisins and in

Table 11.2 Occurrence and concentration of OTA (μg/L) in wines obtained in recent large scale surveys

Wine type	Number of samples	Percentage of positive samples	Percentage of samples > 2μg/L	Range	Mean	Median	Reference
White	60	25	0	<0.010–1.36	0.108	0.01	Otteneder and Majerus (2000)
Rosé	55	40	–	<0.010–2.38	0.119	0.01	
Red	305	54	–	<0.010–3.31	0.201	0.02	
Red (North Italy)	17	70	0	<0.010–0.54	0.12	0.05	Perrone et al. (2007)
Red (Central Italy)	46	59	0	<0.010–0.80	0.07	0.02	
Red (South Italy)	49	100	18	0.02–4.93	1.36	1.03	
Sweet[a]	290	96.9	6.8	<0.010–4.63	0.499	0.14	Burdaspal and Legarda (2007)
Sweet (A, B, CI)[b]	20	0	0	<0.024	<0.024	–	Valero et al. (2007b)
Sweet (CII)	26	54	15.3	<0.024–27.79	2.01	–	
Sweet (CIII)	75	60	26.7	<0.024–15.62	1.71	–	

[a]Maximum allowable levels for sweet and fortified wines are not legislated (see Table 11.1)

[b]European regions of wine production based on production conditions, soil, region and climate (Valero et al. 2007b)

wine – *A. carbonarius* and to a lesser extent *A. niger*. Recently, a new species of section *Nigri* isolated from grapes and OTA non-producer – *A. ibericus* – was proposed (Serra et al. 2006), being likely that new species can enlarge that section in the near future. In fact, with recent work on the molecular characterization of a Southern Europe population of black aspergilli isolated from grapes it was concluded that these represent a complex of species, where some of them are peculiar to grapes (Perrone et al. 2006b). Another study performed with strains isolated from grapes in Italy indicated that *A. tubingensis* is able to produce OTA and that, together with *A. carbonarius* and *A. niger*, it may be responsible for the OTA contamination of Italian wines (Perrone et al. 2006a). Some references in the literature also describe the occurrence of the yellowish *A. ochraceus* (belonging to the *Aspergillus* section *Circumdati*) and the blue-green *Penicillium* species on grapes (Frisvad et al. 2004), but its importance seems to be minor when compared with black aspergilli.

The evaluation of OTA production by fungi on infected grapes is essential to establish the real producing ability by the different species because the results of tests on synthetic culture media are not always coincident with in vivo determinations (Bellí et al. 2007). Overall, most *A. carbonarius* strains have the ability to produce OTA in grapes whereas the proportion is lower in the other toxigenic species (Perrone et al. 2006a, 2006b).

From a scientific point of view, precise strain identification of OTA fungi producers is important to establish the phylogenetic relationships among species, to recognize the mycota of foodstuffs, and to understand the peculiarities and ecological needs of the species. However, from a viticultural and enological point of view, it will be much more important to establish the environmental conditions that are ideal for OTA fungi producers on grapes than their identification.

11.2.4.2 Factors Affecting Fungal Growth and OTA Production

The primary sources of *A. carbonarius* and *A. niger* are soil, bunch remnants or vine trash on soil vineyards, which are transported by wind from soil onto berry surfaces (Leong et al. 2006a; Hocking et al. 2007). Generally, the colonisation of grape bunches by black aspergilli and other fungi occurs when berry skin damage allows the entry into fruit tissues, where the low pH and high sugar content under aerobic conditions provide a competitive advantage for moulds. However, fungal invasion may occur without visible symptoms (Bellí et al. 2007). As a rule, the competition among contaminant microorganisms is more favourable to *Botrytis cinerea*, the agent responsible for common grey rot. However, other rot processes, such as sour rot and brown rot, can occur together with black aspergilli. The population dynamics of fungi outside or inside grape berries with skin damage is still poorly known, as it is difficult to establish which environmental conditions promote the dominance of each fungal species. Nevertheless, it has been shown that high temperatures (30 °C) and high relative humidity, between 80% and 100%, give rise to higher amounts of OTA produced by *A. carbonarius* on grapes (Bellí et al. 2007), suggesting that such conditions give competitive advantages to the black aspergilli population. Medina (2007) also showed that *A. carbonarius* growth was favoured by high a_w (0.98) and temperature (28 °C), whereas OTA production was increased at mild temperature (20 °C) and 0.96–0.98 a_w. In addition, Astoreca et al. (2007) found that optimum conditions of a_w and temperature are more restrictive for OTA production than for *Aspergillus* growth. The influence of relative humidity, rain or high a_w is less important than the influence of high temperature (Bellí et al. 2007; Medina 2007), explaining the occurrence of these fungi in hot and dry climate regions (Hocking et al. 2007). The influence of a_w appears to be more important in dried grapes, where black sultanas ($a_w = 0.629$) showed levels of OTA higher than 10 μg/kg when white sultanas ($a_w = 0.567$) did not (Iamanaka et al. 2006).

The frequency of the occurrence of OTA producing species on grapes is then essentially limited to conditions of high humidity and temperature, typical of subtropical and Mediterranean climates. In fact, OTA has been detected on grapes produced in France, South America, Spain, Italy, Portugal, Greece and Australia (as revised by Leong et al. 2006a), that have some wine regions with climatic conditions favouring black aspergilli species. In general, *A. carbonarius* is highly dominant, particularly in warmer regions, because their black spores are resistant to UV light and sun-drying (Leong et al. 2006a). However, in colder regions such as Germany, Northern Hungary, the Czech Republic or Northern parts of Portugal, France and Italy, black aspergilli have not been isolated from grape berries in spite of the presence of OTA in wines (for references see Blesa et al. 2006 and Varga and Kozakiewicz 2006) suggesting that other species, mainly *Penicillium*, adapted to cool temperatures, should be involved.

Zimmerli and Dick (1996) were the first authors to show that OTA content in southern wine-growing regions are higher than those of wines from northern areas, results also supported by the extensive work of Otteneder and Majerus (2000). More recently, OTA has been detected on grapes produced in many wine countries, where the highest amounts of OTA detected in each survey is generally correlated to vines growing in the warmest regions of each country (Blesa et al. 2006 and Table 11.2). However, the above-mentioned assumptions were not observed in Australia and South Africa, where no correlation was found between OTA incidence and wine region climate (Leong et al. 2006a). This fact was probably due to the rather low detected levels making these correlations imprecise. Further results in Australia showed apparently lower incidences in the cooler climate of Tasmania (Hocking et al. 2007). Canadian wines and grape juices have OTA levels comparable with data from cold climate wine growing areas in Europe (Ng et al. 2004 and Table 11.2). The fact that sweet wines from colder regions do not show high levels of OTA in contrast to those from warmer climates (Valero et al. 2007b) indicates that the main factor determining the OTA concentration in wines is the contamination of grapes by toxinogenic moulds.

11.2.5 Prevention and OTA Production Control on Grapes

Given the ubiquity of black aspergilli in vineyards of warm regions, all agents involved in the wine industry must learn how to live with them, minimising the

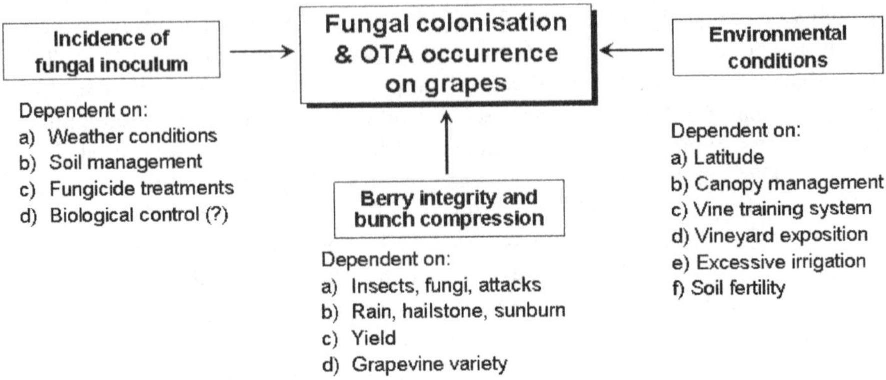

Fig. 11.2 Factors affecting grape colonisation by OTA producer fungi

effect of all factors – biotic and abiotic – that contribute to their infection and growth on grape berries (see Fig. 11.2). Tillage operations contribute to the spreading of fungal spores and should be minimised when possible (Leong et al. 2006a). Concerning fungicide treatments, although some compounds have shown its efficiency, others seem to stimulate OTA production (Magan 2006) and so the most significant function of agrichemicals is to avoid grape damage by phytopathogenic moulds or insects which provide an easy entry for spore infection. Recently, a Code of Good Viticultural Practices was recommended by the Office International de la Vigne et du Vin (OIV 2005) to be put in use mainly in critical areas of *A. carbonarius* occurrence. Prevention is especially important for dried grapes, either for direct consumption or for dessert wines, given their higher levels of OTA. The main measures to consider were the following, mainly directed to avoid berry damage:

- To avoid all the cropping practices that lead to an excessive vigour of vine plants and an exaggerated increase in the yield, which make bunches more compact and thereby susceptible to berry splitting (Mínguez et al. 2004; Leong et al. 2006a);
- To avoid the use of cultivars with thin skinned berries and too compact bunches (Bellí et al. 2007);
- To carry out a plant sanitary program with efficient agrichemicals against moulds, particularly powdery mildew and grey rot, and insects, mainly *Lobesia* spp., using the adequate dose and timing, and making sure that the active ingredients reach all parts of the bunch as well as penetrate its interior (Mínguez et al. 2004; Valero et al. 2007a);
- To control berry splitting due to rain just before harvest (JECFA 2001).

11.2.6 OTA Control Strategies in Wine

The factors affecting the levels of OTA in wines are systematised in Fig. 11.3. The presence of OTA in wines results exclusively from grape contamination. Thus, the

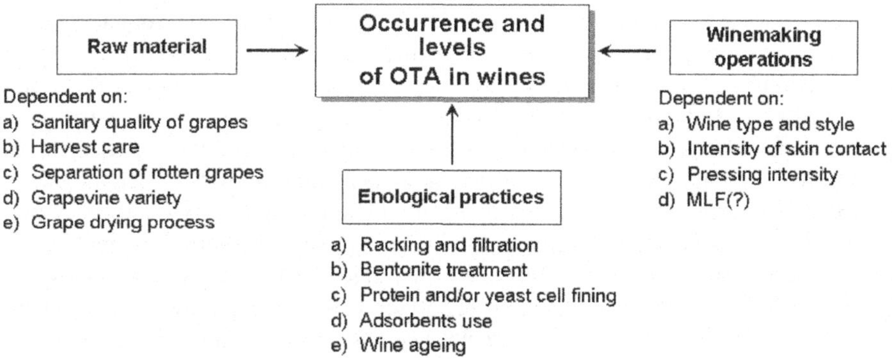

Fig. 11.3 Factors affecting the levels of OTA in wines

main measure to prevent the entry of grapes with high levels of OTA in wineries is to reject the rotten bunches and particularly those presenting brown or black rot. This is difficult to implement during harvest because rot may not be easily visible due to its development in the inner part of the bunch. In fact, bunch selection is only effective in small wineries, where the harvest is by hand and the personnel may be well trained. For large cooperatives or company wineries it is not so easy to separate the rot bunches, except if grape payment can be differentiated according to its health status. OTA determination is not an easy task at winery level (Cigič et al. 2006 and references cited therein). Rapid diagnostic tools are being developed but have not yet reached industrial laboratories (Magan 2006). This drawback can, however, be overtaken by the use of instruments that provide indicators of grape composition and health status, like the modern and costly FTIR (Fourier Transformed Infrared) spectroscopy. This method gives a measure of the grape rot that may be an indicator of the presence of OTA because gluconic acid and glycerol are correlated with *Botrytis*, sour rot and *Aspergillus* attack (Mínguez et al. 2004). When mechanical harvesting is used, it is impossible to reject rot bunches. In this case, the only measure is to control analytically the health quality of grapes in order to process the poor quality grapes separately.

Despite the fact that South African and Australian red and white wines did not show different OTA incidence (Leong et al. 2006a), most studies indicate that red wines, in the same region, contain higher OTA concentrations than rosés and white wines (for references see review of Varga and Kozakiewicz 2006). This may be explained by OTA release from grape skins (Otteneder and Majerus 2000) or by fungal spore release (Atoui et al. 2007) during red grape maceration. Therefore, winemaking processes should significantly influence its content. Several studies showed that OTA content in wines increased with the maceration time and decreased with solid-liquid separation steps, such as red wine racking or clarification of white juices (Fernandes et al. 2003; Leong et al. 2006b). The reported reductions may range from 50% to 80% of initial OTA concentrations (Hocking et al. 2007). In recent work performed with grapes from Northwest Portugal, Fernandes et al. (2007) showed that, in different red vinification trials, the mean carry-over of OTA from artificially infected grapes to wine was 8.1 wt% after malolactic fermentation, even without use of enological adjuvants (fining agents), corroborating the findings of Leong et al. (2006b). Reduction of OTA was associated with removal of spent fractions during winemaking, such as wine lees after fermentation or sediment after racking, in which OTA contents were higher than in the original grapes. Some studies were performed to assess the effect of enological adjuvants, such as bentonite, gelatin, charcoal, and yeast cell wall preparations on the removal of OTA from wines (Leong et al. 2006b; Mateo et al. 2007). Most of them reduce the OTA content of wines, but the necessary concentrations have a strong effect on the wine quality. In addition, up to 29% of OTA spontaneous reduction has been observed during wine storage over 10–14 months (Hocking et al. 2007).

Little is known about the eventual degradation or binding of OTA by yeasts or lactic acid bacteria during the fermenting process, though this has been demonstrated as possible (Angioni et al. 2007; Hocking et al. 2007). However, when compared with the physical removal of OTA during the vinification, this practice is quantitatively irrelevant. So, the most important measures for a Code of Good Enological Practices to prevent or reduce the OTA content in wines are:

- To train the harvesting staff to reject rotten bunches, particularly those affected by dark brown or black moulds;
- To vinify separately the mechanical harvested grapes, when the sanitary quality of the crop is poor;
- To monitor, in large wineries and cooperatives, the sanitary quality of the grapes by FTIR instruments in order to (i) favour with fair pricing the sanitary status of grapes, and (ii) process the grapes accordingly;
- To avoid long periods of maceration and to use enological adsorbents, such as activated charcoal or yeast hulls, in red wine, and bentonite, in white wine, when the crop has a relevant percentage of rotten grapes;
- To perform a rapid grape drying and avoid water condensation overnight for grapes used in the vinification of dessert wines;
- To implement a complete HACCP plan, from the vine to the bottled wine or raisin, in the wine regions where the OTA occurrence is higher.

11.3 Volatile Phenols

Volatiles phenols (VP) are secondary metabolites produced by yeasts, moulds and bacteria which affect the flavour of several fermented food commodities (Loureiro and Malfeito-Ferreira 2006). These molecules have been under study since the first detection in fermented grains (Steinke and Paulson, 1964). Later, Dubois and Brulé (1970) reported their presence in wines and presently the importance of VP is mainly due to their role in the mediatic "horse sweat" taint in red wines. Available toxicological data suggest that VPs do not warrant concerns about acute or long-term effects (Rayne and Eggers 2007b).

11.3.1 Chemical Structure and Occurrence

Volatile phenols are a group of molecules included in the non-anthocyanin phenolic compounds of white and red wines. The chemical structure of VP are characterised by a phenolic ring and a radical with different compositions (Fig. 11.4). The main volatile phenols of wines are 4-vinylphenol (4-VP), 4-vinylguaiacol (4-VG), 4-vinylcathecol (4-EC), 4-vinylsyringol (4-VS) and their respective reduced forms 4-ethylphenol (4-EP), 4-ethylguaiacol (4EG), 4-ethylcathecol (4-EC) and 4-ethylsyringol (4-ES). These molecules occur in wines with different concentrations, being vinylphenols typically associated with white wines and ethylphenols with red wines. Table 11.3 shows recent results on the average concentration range of these molecules in wines. In addition, vinylphenols may be also present in red wines as anthocyanin-vinylphenol adducts, like malvidin 3-glucoside-4-vinylphenol (Cameira-dos-Santos et al. 1996) and similar 4-VP, 4-VC and 4-VS derivatives (Schwarz et al. 2003; Suárez et al. 2007).

11.3.2 Origin

11.3.2.1 Availability of Precursors

The precursors of VPs are hydroxycinnamic acids which are enzymatically decarboxylated by a cinnamate decarboxylase, leading to vinylphenol derivatives, and

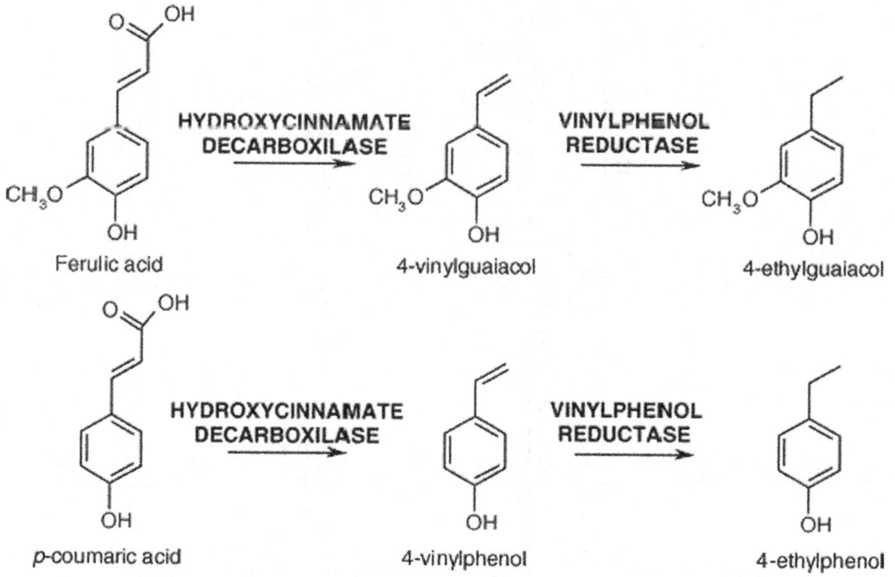

Fig. 11.4 Conversion of hydroxycinnamic acids to vinylphenols and ethylphenols (reprinted from Suárez et al. 2007, permission to be obtained)

Table 11.3 Range of volatile phenol concentrations in wines (μg/L)

Wine type	Volatile phenols					Observations	Reference
	4-Vinylguaiacol	4-Vinylphenol	4-Ethylguaiacol	4-Ethylphenol	4-Ethylcathecol		
White	200–324	1341–2802	Nd[a]-238	Nd-228	–	5 samples	Mejías et al. (2003)
Red	Nd-880	Nd-2174	72–255	97–782	–	5 samples	
	5.4–236	8.1–98	0.53–420	8.6–1500	–	57 aged red wines	López et al. (2002)
	–	–	46.6–169.3	399.5–2231.6	–	18 red wines	Tat et al. (2007)
	–	–	4.3–410.5	<0.5–586.2	–	54 red barrel aged wines	Rayne and Eggers (2007)
	–	–	Nd-60	127–494	42–160	9 tainted Pinot Noir wines	Hesford and Schneider (2004a)
	–	–	Nd-329	298–3780	49–427	11 tainted red wines	
Sherry	820–1170	800–2000	1000–2000	300–550	–	4 wines	Domínguez et al. (2002)

[a] Not detected

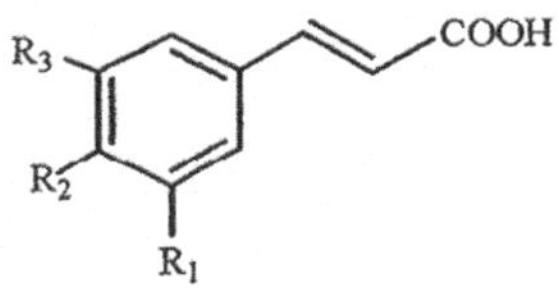

Hydroxycinnamic acid	R_1	R_2	R_3
p-Coumaric	H	OH	H
Caffeic	OH	OH	H
Ferulic	OCH_3	OH	H
Sinapic	OCH_3	OH	OCH_3

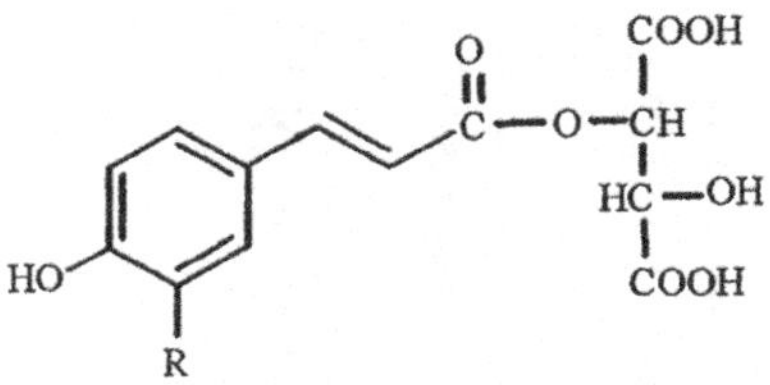

Hydroxycinnamic ester	R
Trans- caffeoyltartaric acid (caftaric acid)	OH
Trans-p-coumaroyltartaric acid (cutaric acid)	H
Trans- feruloyltartaric acid (fertaric acid)	OCH_3

Fig. 11.5 Hydroxycinnamic acids and their esters (obtained from Monagas et al. 2005, permission to be obtained)

reduced by a vinylphenol reductase, originating the ethylphenol derivatives (Heresztyn 1986a and Fig. 11.4). In grape juices, hydroxycinnamic acids are esterified, mainly to tartaric acid (Fig. 11.5). In wines they may be present in the free or esterified form, either with tartaric acids or other polyphenols (Table 11.4). Recent work has shown that most of the free acids from tartaric esters appear after the malolactic fermentation (Hernández et al. 2006, 2007; Cabrita et al. 2007). The other esterified forms of hydroxycinnamic acids are cinnamoyl-glucoside anthocyanins (Romero and Bakker 2000; Monagas et al. 2005; Oliveira et al. 2007) and *trans-p*-coumaric acid hexoses (Monagas et al. 2005; Hernández et al. 2006, 2007). Boselli (2006) also mentioned the existence of hydroxycinnamic acid esters with ethanol, like ethyl caffeoate, in white wines.

The above-mentioned release of hydroxycinnamic acids from anthocyanin esters during wine maturation may be only due to chemical reactions but conversions of acid precursors to volatile phenols are typically dependent on enzyme or microbial activity.

In grapes or grape juices, the tartaric esters may be hydrolysed by enzymes from contaminant fungi or from commercial pectolytic preparations, both with cinnamoyl decarboxilase activity, releasing free hydroxycinnamic acid forms (Dugelay et al. 1993; Gerbaux et al. 2002). However, the tartaric esters are mostly hydrolysed after malolactic fermentation (Hernández et al. 2006, 2007), it being hypothesised that the hydrolytic activity of lactic acid bacteria follows the completion of malic conversion to lactic acid (Cabrita et al. 2007) (see Table 11.4).

It is accepted that yeasts only metabolise the free acid forms, although brewing *S. cerevisiae* was supposed to possess feruloyl esterasic activity (Coghe et al. 2004). Then the availability of free hydroxycinnamic acids appears to be crucial for the production of VPs either by yeasts or bacteria.

Table 11.4 Range of concentrations of hydroxycinnamic acids and their esters (mg/L)

Hydroxycinnamic acids and their esters						Observations	Reference
trans-Caftaric	*trans*-Caffeic	*trans*-Coutaric		2-Glutathionylcaftaric	Malvidin-(6-coumaroyl)-3-glucoside		Ibern-Gómez et al. (2002)
12.9–34.3	5.5–11.8	6.1–22.1		3.1–6.6	6.7–43.1	6 red wines	
Caftaric	Caffeic	Coutaric	Coumaric	Fertaric	Syringic		Pérez-Margariño and González-San José (2005)
8.89–26.06	2.36—9.95	5.35–10.04	2.87–11.21	0.40–2.14	7.08–10.61	6 red wines not aged	
Caftaric	Caffeic acid	Coutaric	Coumaric acid				Makris et al. (2006)
4.3–122.3	0—31.2	2.8–39.0	0–7.1			41 young red wines	
Caftaric	Caffeic	*p*-Coutaric	*m*-Coutaric	Fertaric	Ferulic		Gómez-Míguez et al. (2007)
13.43–25.29	0.04—0.98	0.76–3.59	0.12–0.14	0.03–0.08	0.13–0.04	White musts	
27.13–57.14	0.78—4.11	3.08–4.94	1.03–0.48	1.34–4.20	0.58–1.44	White wines	
trans-Caftaric	*trans*-Caffeic	*trans*-Coutaric	*trans*-*p*-Coumaric	*trans*-Ferulic	*trans*-*p*-coumaric hexose		Gómez-Míguez et al. (2007)
14.98	2.68	13.75	2.06	-	1.23	Red wine before MLF[a]	
6.85	26.24	8.68	16.01	0.84	1.84	Red wine after MLF	
cis-Caftaric		*cis*-Coutaric	*cis*-*p*-Coumaric				
2.11		4.33	1.91			Red wine before MLF	
0.18		2.17	0.52			Red wine after MLF	
Caftaric	Caffeic	Coutaric	*p*-Coumaric	Fertaric	Ferulic		Cabrita et al. (2007)
4.10	0.47	0.85	0.24	0.40	–	Red wine before MLF	
0.25	6.90	0.12	2.31	0.06	0.95	Red wine after MLF	

[a]Malolactic fermentation

11.3.2.2 Conversion of Hydroxycinnamic Acids

Once hydroxycinnamic acids are released in the grape juice or in the wine, they may be converted into vinylphenol and ethylphenol derivatives depending on the presence of specific growing microbial populations.

During fermentation, *S. cerevisiae* may produce vinylphenol derivatives due to the presence of cinnamate decarboxylase enzymes (Chatonnet et al. 1992, 1993) which are inactive in red juices due to the polyphenol components of red wine (Chatonnet et al. 1997). Several grape juice contamination yeast species also have the ability to form vinylphenols (Dias et al. 2003a) but their contribution to the vinylphenol content of wines may only be relevant when are not inhibited by *S. cerevisisiae* (Barata et al. 2006).

After the decarboxylation step, vinylphenols may be reduced to ethylphenols but the sequential decarboxylase and reductase activities, regarding wine yeasts, have only been demonstrated in *D. bruxellensis* and in *P. guilliermondii* (Barata et al. 2006). The former species may also convert 4-VP into 4-EP in the absence of hydroxycinnamic acids (Dias et al. 2003b).

The production of 4-ethyphenol in wines is dependent on the presence of growing yeast populations. As *P. guilliermondii* does not grow in wines with average ethanol of 12 vol.% it is not likely to produce significant levels of 4-EP during wine maturation. Concerning *D. bruxellensis* it does not grow in white wines explaining the absence of phenolic taint in this type of wines (Malfeito-Ferreira et al. 2001). Several lactic acid bacteria (*Lactobacillus* spp., *Pediococcus* spp.) have also been characterised concerning the production of ethylphenols in synthetic media (Table 11.5) but in wines they are not regarded as significant 4-EP producers (Chatonnet et al. 1995, 1997). The main starter used in wines for malolactic conversion, *Oenococcus oeni*, does not seem to produce vinyl or ethylphenols even in synthetic media (Couto et al. 2006).

The conversion of *p*-coumaric acid into 4-EP only occurs when *D. bruxellensis* is growing on a carbon and energy source, the conversion rate being dependent on the substrate (Dias et al. 2003b). The conversion of the other hydroxycinnamic acids by yeasts has not been deeply studied. Most studies are related with *p*-coumaric acid metabolism but the conversion of ferulic, caffeic acids or sinapic acids may not be equally efficient, as demonstrated, in synthetic medium, for *D. bruxellensis* (Heresztyn 1986a), *S. cerevisiae* (Chatonnet et al. 1989) and *D. anomala* (Edlin et al. 1995). Knowing that caffeic acid is more concentrated than *p*-coumaric acid in wines (see Table 11.5) it would be expected that 4-EC would be present in higher concentration that 4-EP, but the few results published do not corroborate this hypothesis (see Table 11.4). Then, *D. bruxellensis*, although utilising caffeic acid (Heresztyn 1986a), may not produce 4-EC with the same efficiency as 4-EP in wines.

11.3.2.3 Changes in Wine Composition

The above-mentioned metabolic activities of microorganisms should be taken into account when studying wine compositional alterations. In fact, the effect of

Table 11.5 Metabolic activity of microorganisms related with production of volatile phenols in wine industry

Species	Function	Metabolic activity	Reference
Aspergillus niger	Mould for commercial enzyme production (pectinase, hemicellulase)	Active cinnamoyl esterase releasing free hydroxycinnamic acids in juices	Dugelay et al. (1993)
Saccharomyces cerevisiae	Fermenting yeast	Active hydroxycinnamate decarboxilase producing vinylphenols in fermenting white juices	Dugelay et al. (1993); Chatonnet et al. (1989); Shinohara et al. (2000); Barata et al. (2006)
Dekkera bruxellensis	Spoilage yeast	Active hydroxycinnamate decarboxylase and vinylphenol reductase producing ethylphenols in synthetic media, juices and wines	Heresztyn (1986); Chatonnet et al. (1995); Shinohara et al. (2000); Rodrigues et al. (2001); Dias et al. (2003a, 2003b)
Pichia guilliermondi	Contamination yeast	Active hydroxycinnamate decarboxylase and vinylphenol reductase producing ethylphenols in synthetic media and grape juices	Barata et al. (2006)
C. albidus, C. laurentii, C. stellata, C. wickerhamii, D. hanseniiI, H. anomala, H. uvarum, K. apiculata, K. thermotolerans, M. pulcherrima, P. guilliermondii, P. membranifaciens, R. rubra, S. pombe, Z. bailii	Contamination yeasts	Active hydroxycinnamate decarboxilase activity producing vinylphenols in synthetic media and grape juices	Chatonnet et al. (1992); Shinohara et al. (2000); Dias et al. (2003a)
Lactobacillus spp., *Pediococcus* spp.	Fermenting and spoilage lactic acid bacteria	Active hydroxycinnamate decarboxylase and vinylphenol reductase producing ethylphenols in synthetic media	Cavin et al. (1993); Couto et al. (2006)
Oenococcus oenii, *Lactobacillus plantarum*	Fermenting lactic acid bacteria	Active cinnamoyl esterase releasing free hydroxycinnamic acids in red wines	Hernández et al. (2007); Cabrita et al. (2007)

microorganisms on the polyphenolic composition does not seem to have been considered when establishing differentiation between grape varieties (Makris et al. 2006), between ripening stages (Pérez-Magariño and González-San José 2005) or bottle aging (Monagas et al. 2005). However, as observed in Table 11.4, changes in hydroxycinnamic acid compositions may be explained by microbial activity, which are higher than changes between grape varieties or ripening stages. These variations may explain, at least partially, the controversy on the evolution of hydroxycinnamates during wine processing (Monagas et al. 2005 and references cited therein).

The balance of VP and precursors is also influenced by non-microbial reactions. The esterified forms of hydroxycinnamic acids or vinylphenols form a pool of molecules which release or combine the acids during wine maturation, apparently without the influence of microorganisms (Hernández et al. 2006, 2007; Suárez et al. 2007). In addition, hydroxycinnamic acids may suffer oxidative condensation and browning during aging (Yokotsuka and Singleton 2001). Oak chips may also release 4-EG up to 0.15 μg/g, or 4-VG up to 7.76 μg/g, as influenced by higher toasting intensity (Natali et al. 2006). Overall, sources of VPs other than microbial should not account by more than 100 μg/L (Rayne and Eggers 2007b).

11.3.3 Effect of Volatile Phenols on Product Quality

The sensorial effect of a volatile compound may be positive or negative to wine depending on its smell and concentration. In wines it is not easy to define beneficial or detrimental effect because the odours of mixtures of different compounds are perceived differently than those of single compounds and there is also a matrix effect on the perception. In addition, the rejection of an odour occurs at higher concentrations than the detection, leading to different detection and preference thresholds. The value of a detection/preference threshold may measure the spoilage effect of molecules with sensorial activity. These may be defined as the minimum concentration under which 50% of the tasters, in a 70 person jury, statistically detected/rejected the sample (Chatonnet et al. 1992). For instance, in Bordeaux red wines, the preference threshold for 4-EP is about 620 μg/L, and for the mixture (10:1) of 4-EP and 4-EG is 426 μg/L (Chatonnet et al. 1992). Below these concentrations, volatile phenols may contribute favourably to the complexity of wine aroma by imparting aromatic notes of spices, leather, smoke or game, appreciated by most consumers. Above those levels, wines are clearly substandard for some consumers but remain pleasant for others. To increase the difficulty in the definition of spoiling concentrations, these thresholds are dependent on grapevine variety and on the style of wine (Gato et al. 2001; Coulter et al. 2003).

In the case of vinylphenols, they contribute to the spicy, floral and pharmaceutic character of white wines. The 4-VG has been detected in high levels in the variety Gewürztraminer (Grando et al. 1993). The depreciation due to high levels of 4-VG plus 4-VP in white wines of the German variety Kerner was associated with hot regions (e.g. South Africa) or exposure of grapes to sunlight, but no explanation was given to the fact (Rapp 1998). The overall incidence of vinylphenols in white wines is not known and but it seems to have decreased after the improvement in the purity of commercial pectolytic enzymes used in juice clarification.

In the case of ethylphenols, concentrations of 4-EP and 4-EG above the preference thresholds dominate the flavour contributing to the phenolic character. The mouthfeel sensations are also altered by increasing the metallic notes (Coulter et al. 2003). Volatile phenols are currently determined by gas-liquid chromatography after wine extraction with organic solvents (Loureiro and Malfeito-Ferreira 2006). However, the main flaw of this technique is the absence of 4-EC quantification, which requires derivatisation (Hesford and Schneider 2004). Despite this fact, numerous recent improvements in volatile phenol analysis were only directed to extraction procedures (López et al. 2002; Mejías et al. 2003; Díez et al. 2004; Fariña et al. 2007; Pizarro et al. 2007; Rayne and Eggers 2007a).In contrast, Carrillo and Tena (2007) presented an HS-SPME extraction followed by GC-MS of derivatised samples accounting for 4-EC. In addition, liquid chromatography has been present as an alternative to GC, having the advantage of avoiding sample extraction (Vanbeneden et al. 2006; Caboni et al. 2007; Larcher et al. 2007; Nicolini et al. 2007) but the proposed methods do not account for 4-EC.

The real incidence of VP in wines world wide is not known perhaps due to the difficulty in performing routine instrumental analysis in wineries. Normally, winemakers analyse samples suspected to have problems and so the reported proportion of affected wines is most probably biased. Some data from analytical laboratories have shown that from 6% to 74% of analysed samples may bear levels of 4-EP plus 4-EG higher than the preference threshold (426 μg/L) of a 10:1 mixture (Loureiro and Malfeito-Ferreira 2006). Although precise numbers are not available, we believe that the phenolic taint is the main microbiological problem leading to higher economical losses in winemaking industry. Moreover, as already mentioned, the sensory detection of VP depends on the type of wine (Gato et al. 2001; Coulter et al. 2003) and so a higher proportion of wines may be badly affected by these compounds.

The 4-EG is present in about one tenth of the 4-EP concentration (Chatonnet et al. 1992) but this rate is not always observed (Rodrigues et al. 2001; Coulter et al. 2003). The other odour active VP, 4-EC, has just begun to be studied. The fact that the precursors of 4-EC, caffeic acid and its esters, are present in relatively high concentrations in wines (see Table 11.4) and that the detection threshold for 4-EC (described as having a phenolic smell similar to that of 4-EP) is about 50 μg/L (Porret et al. 2004) suggests that its influence in phenolic taint should be not be neglected. Occasional discrepancies between sensorial detection and concentration of 4-EP and 4-EG may be explained by the hidden presence of 4-EC. Sinapic acid gives 4-VS and 4-ES but the syringols do not affect medium odour (Heresztyn 1986a).

The wines affected by ethylphenols are practically only red wines. All types are susceptible to the phenolic taint depending on the growth of *D. bruxellensis*. However, red wines matured in oak barriques are the typical wine product affected by this taint because these containers provide a highly favourable ecological niche for *D. bruxellensis* (Loureiro and Malfeito-Ferreira 2006).

11.3.4 Control Measures

When wines are affected by volatile phenols there is, at present, no effective curative process. In this situation, oenologists always weigh the possibility of blending tainted wine with "clean" wine. Although this measure may attenuate the defect of the tainted wine by dilution it cannot be seen as a curative measure. In fact, mixtures of wines with null or low levels of 4-ethylphenol are only effective for small proportions of tainted wines because large volumes of "clean" wine must be used to obtain a blend with 4-EP levels lower than the preference threshold. Then, effective curative measures would depend either (i) on the reduction or elimination of the sensorial effect or (ii) on the extraction of odour active molecules from the wine. These strategies have not yet been effectively tested in practical conditions. Guilloux-Benatier et al. (2001) hypothesised that yeast lees have the property to adsorb volatile phenol, which was later shown, at least partially, in laboratory conditions, by Chassagne et al. (2005). A reverse osmosis procedure to reduce volatile phenols is commercially available in New Zealand (http://www.armourtech.co.nz/memstarreverseos.html). Salameh et al. (2007) showed that *p*-coumaric acid may be adsorbed on *Brettanomyces* cells, decreasing its availability as substrate. Commonly, when adsorbents are added to wine, favourable aroma compounds are also removed and a balance must be drawn between benefits and losses of wine attributes.

If a curative approach is not effective, then prevention is, at present, the most reasonable way to deal with the problem. Bearing in mind that, to produce volatile phenols microorganisms needs the substrate to be available and active, the preventive measures may be directed either to minimise the release of free acid or to avoid microbial activity.

In juices, prevention should be based on (i) decreasing the release of free acids that is favoured by mould infections of grapes and by the decarboxylase activity of commercial enzyme preparations and (ii) avoiding the production of volatile phenols that is favoured by the uncontrolled activity of contamination yeasts growing in damaged grapes or in juices. Then, the main measures to be adopted are:

- Separation of sound grapes from damaged grapes
- Use of sulphur dioxide to prevent yeast contaminations
- Use of pure commercial enzymes, if necessary
- Initiatiation of active fermentation with *S. cerevisiae* as soon as possible

These measures are good manufacturing practices of winemaking, irrespective of the risk of phenolic taint. The main preventive measures should be performed during wine storage, aging (mainly in oak barrels) and bottling. As the release of precursors is unavoidable, the main preventive measures are directed to reduce the activity of contaminating populations of yeasts and bacteria, especially towards *D. bruxellensis* (see review of Loureiro and Malfeito-Ferreira 2006):

- Use adequate hygienic practices and respective efficiency assessment
- Use adequate levels of sulphur dioxide or DMDC (if legally authorised)
- Minimise residual nutrient contents (sugar or nitrogen)
- Minimise oxygen dissolution
- Handle oak barrel aging properly (disinfection, toppings, rackings, cellar temperature)
- Perform microbiological monitoring especially when wines from external sources are processed and oak aging is used
- Thermal treatment or filter sterilisation is advised when risk of bottle infection is high

11.4 Tetrahydropyridines

Tetrahydropyridines (THP) are secondary metabolites produced by *D. bruxellensis* and lactic acid bacteria in wines and are responsible for a taint described as mousy off-flavour or mousiness. This problem has been known since late nineteenth century (see review of Snowdon et al. 2006) but, in spite of its obnoxious flavours, has been only vaguely studied perhaps due to its low frequency of occurrence.

11.4.1 Chemical Structure and Origin

Tetrahydropyridines (THP) include 2-ethyl-tetrahydropyridine (ETHP), 2-acetyl-tetrahydropyridine (ATHP) and 2-acetylpyrroline (APY) (Fig. 11.6). ETHP is present in tautomeric forms, but the second tautomer is minor. ATHP also occurs in two tautomeric forms, of which the distribution is pH dependent. These molecules are uncommon components of wines and are not currently analysed.

Tucknott (1977) first identified ETHP and other unknown compounds as the molecules imparting mousiness, showing that they were not present in the absence of microorganisms. The origin of THP is related with activity of *D. bruxellensis* and of lactic bacteria, mainly heterofermentative strains, but the possible role of acetic bacteria should not be discarded (Heresztyn 1986b; Snowdon et al. 2006). APY is not produced by *Dekkera* spp. but by lactic acid bacteria, being an indicator of bacterial spoilage. The pathways for production of THP by heterofermentative bacteria or *D. bruxellensis* have been proposed (Costello and Henschke 2002; Snowdon et al. 2006) but require further confirmation. However, it is established that both pathways require L-lysine and ethanol to THP synthesis. L-Lysine and L-ornithine are responsible for ring formation of the heterocycles whereas ethanol and acetaldehyde are responsible for the acetyl side chain (Snowdon et al. 2006;

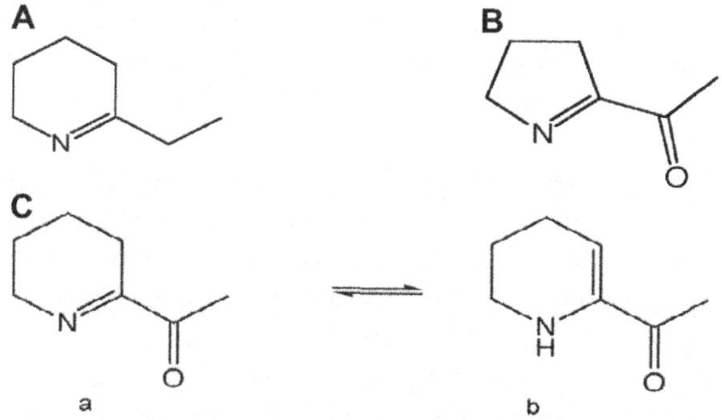

Fig. 11.6 Chemical structure of (A) 2-ethyltetrahydropyridine (ETHP), (B) 2-acetylpyrroline (APY) and tautomers of (Ca) 2-acetyl-3,4,5-tetrahydropyridine (ATHP) and (C) 2-acetyl-1,4,5,6-tetrahydropyridine (adapted from Snowdon et al. 2006, permission to be obtained)

Grbin et al. 2007). ATHP reduction may lead to EHTP. As ethanol is a precursor, mousy off-flavour occurs after alcoholic fermentation, preferably after lactic acid bacteria activity. It seems that the formation of mousiness may be induced by oxidation but it is not clear if the effect is on the microorganisms or in any chemical reaction stimulated by the redox potential. Other agents claimed to affect its production (high pH, low sulphite, residual sugar content) (Lay 2004; Snowdon et al. 2006; Romano et al. 2007) are also stimulators of microbial activity and so the true mechanisms are not yet clarified, but the non-enzymatic chemical synthesis has been ruled out in *D. anomala* (Grbin et al. 2007).

11.4.2 Effect on Wine Quality and Occurrence

The "mousy off-flavour" is described as resembling the smell of mice urine and once tasted becomes unforgettable. The taint is mainly perceived by the after-mouth sensation and has a long persistence (may exceed 10 min). The compounds responsible are not volatile at wine pH and so are only perceived by the increase in pH due to saliva. These features justify the use of practical sensorial detection methods without the need to swallow the wine, like rubbing the wine in the palm of the hand and sniffing the skin, or dipping an alkaline paper strip in the wine and smelling. ATHP is the main molecule responsible for the fault (Strauss and Heresztyn 1984), being present in levels up to 108 μg/L (Snowdon et al. 2006) and having a detection threshold of 1.6 μg/L in water (Colagrande et al. 1988). The imino tautomeric form provides the mousy perception and its prevalence at high pH explains the detection after increasing sample pH. Its off-flavour also resembles cracker biscuit, and this molecule is also present in some cereal based products (Snowdon et al. 2006). EHTP has been detected in wines only recently and has a detection threshold of 150 μg/L in wines (Snowdon et al. 2006). APY is also a major contributor to mousy off-flavour, with detection threshold of 0.1 μg/L in water and being detected in trace levels up to 7.8 μg/L (Snowdon et al. 2006). In other food products it has been described as "roasty" and "popcorn-like". Acetamide, although occasionally linked to mousiness, is not the cause of mousy off-flavour because it is odourless (Snowdon et al. 2006).

The incidence of mousiness in wines is not known. In our experience we have only tasted it a handful of times. Earlier reports and classical enology treatises already mention this problem (Grbin and Henschke 2000 and references cited therein). The activity of *Dekkera/Brettanomyces* spp. has been linked to this fault since early studies leading to some confusion between mousiness and phenolic taint. The development of each other is independent (Romano et al. 2007) and we are not aware of the factors stimulating one instead of the other. Although it is not a common taint, it is known to affect wines all over the wine countries (Grbin and Henschke 2000). As it depends on *Dekkera/Brettanomyces* activity it is likely that red wines are more affected. Also, white wines are usually not subjected to malolactic fermentation and so the activity of lactic bacteria may preferentially affect red wines. These yeasts also produce THP in grape juices (Grbin and Henschke 2000) but the real incidence in this product is not known.

11.4.3 Control Measures

There is no available method to remove this taint effectively (Lay 2004). The removal of precursors (L-lysine and ethanol) is not feasible. As it depends on microbial activity, the preventive measures are similar to those suggested for volatile phenols when there is the risk of *D. bruxellensis* infection. The prevention of spoilage by heterofermentative lactic bacteria usually advised, like decreasing wine pH values and rapid inactivation by sulphur dioxide, once malolactic conversion is finished, should also be effective against bacterial mousiness.

11.5 Final Remarks

The wine spoilage effects described in this chapter illustrate two different approaches undertaken to solve the problems. One, OTA production, is associated with food safety and so profits from the allocation of relatively large funds under specific EU research frameworks. The other problems, associated mainly with VP, are of technological nature and are not specifically supported by research frameworks and so the respective scientific outputs are much less relevant. The present awareness of OTA incidence and knowledge of preventive measures make this a relatively minor problem to the wine industry, which can easily be kept under control by using an adequate HACCP system from vineyard to the bottle. In contrast, concerning phenolic taint, this continues to be a major problem of wine microbial spoilage which involves significant economical losses. The primary sources and routes of contamination of *D. bruxellensis*, the role of esterified VP precursors, the role of lactic acid bacteria, the white wine enigmatic resistance to VP production, the reasons explaining unexpected blooms, the choice of effective prevention measures, and the different wine susceptibility to spoilage are only some of the issues not fully understood by scientists and wine technologists. From a practical point of view, the slow response of the production sector to the problem is surprising, particularly concerning *D. bruxellensis* monitoring in wines and equipment surfaces. The utilisation of microbiological control and the adoption of guidelines by attributes for bulk and bottled wines would certainly avoid most problems, given that the essential of the knowledge on *D. bruxellensis* spoiling activity is already available.

References

Angioni, A., Caboni, P., Garau, A., Farris, A., Orro, D., Budroni, M., & Cabras, P. (2007). In Vitro Interaction Ochratoxin A and Different Strains of *Saccharomyces cerevisiae* and *Kloeckera apiculata*. *J. Agric. Food Chem., 55*, 2043–2048.

Astoreca, A., Magnoli, C., Barberis, C., Chiacchiera, S.M., Combina, M., & Dalcero, A. (2007). Ochratoxin A production in relation to ecophysiological factors by *Aspergillus* section *Nigri* strains isolated from different in Argentina. *Sci. Total Environm., 388*, 16–23.

Atoui, A., Mitchell, D., Mathieu, F., Magan, N., & Lebrihi, A. (2007). Partitioning of ochratoxin A in mycelium and conidia of *Aspergillus carbonarius* and the impact on toxin contamination of grapes and wine. *J. Appl. Microbiol., 103*, 961–968.

Bakker, M., & Pieters, M. (2002). Risk assessment of Occhratoxin A in the Netherlands. RIVM rport 388802025/2002.

Barata, A., Correia, P., Nobre, A., Malfeito-Ferreira, M., & Loureiro, V. (2006). Growth and 4-ethylphenol production by the yeast *Pichia guilliermondii* in grape juices. *Am. J. Enol. Vitic., 57*, 133–138.

Bellí, N., Marín, S., Coronas, I., Sanchis, V., & Ramos, A.J. (2007). Skin damage, high temperature and relative humidity as detrimental factors for *Aspergillus carbonarius* infection and ochratoxin A production in grapes. *Food Control, 18*, 1343–1349.

Bennett, J.W., & Klich, M. (2003). Mycotoxins. *Clinic. Microbiol. Rev., 16*, 497–516.

Blesa, J., Soriano, J., Molto, J., & Mañes, J. (2006). Factors affecting the presence of ochratoxin A in Wines. *Crit. Rev. Food Sci. Nutr., 46*, 473–478.

Boselli, E., Minardi, M., Giomo, A., & Frega, N. (2006). Phenolic composition and quality of white D.O.C. wines from Marche (Italy). *Anal. Chim. Acta, 563*, 93–100.

Burdaspal, P., & Legarda, T. (2007). Occurrence of ochratoxin A in sweet wines produced in Spain and other countries. *Food Add. Contam., 24*, 976–986.

Caboni, P., Sarais, G., Cabras, M., & Angioni, A. (2007). Determination of 4-Ethylphenol and 4-Ethylguaiacol in Wines by LC-MS-MS and HPLC-DAD-Fluorescence. *J. Agric. Food Chem., 55*, 7288–7293.

Cabrita, M.J., Torres, M., Palma, V., Alves, E., Patão, R., & Freitas, A.M.C. (2007). Impact of malolactic fermentation on low molecular weight phenolic compounds. *Talanta*, doi: 10.1016/j.talanta.2007.08.045.

Cameira-dos-Santos, P., Fulcrand, H., Sarni-Manchado, P., Cheynier, V., & Favre-Bonvin, J. (1996). Structure of new anthocyanin-derived wine pigments. *J. Chem. Soc. Perkin Trans., 1*, 735–739

Carrillo, J.D., & Tena, M.T. (2007). Determination of ethylphenols in wine by in situ derivatisation and headspace solid-phase microextraction–gas chromatography–mass spectrometry. *Anal. Bioanal. Chem.*, Doi: 10.1007/s00216-006-1086-x.

CAST (2003). Mycotoxins: Risks in Plant, Animal and Human Systems. Ames, IA, USA: Council for Agricultural Science and Tecnology Task Force report n° 139.

Cavin, J., Andioc, P., Etievant, P., & Divies, C. (1993). Ability of wine lactic bacteria to metabolize phenol carboxylic acids. *Am. J. Enol. Vitic.*, *1*, 76–80.

Chassagne, D., Guilloux-Benatier, M., Alexandre, H., & Voilley, A. (2005). Sorption of wine volatile phenols by yeast lees. *Food Chem.*, *91*, 39–44.

Chatonnet, P., Dubourdieu, D., & Boidron, J.N. (1989). Incidence de certains facteurs sur la decarboxylation des acides phenols par la levure. *Conn. Vigne Vin*, *23*, 59–62.

Chatonnet, P., Dubourdieu, D., Boidron, J.N., & Pons, M. (1992). The origin of ethylphenols in wines. *J. Sci. Food Agric.*, *60*, 165–178.

Chatonnet, P., Boidron, J., & Dubourdieu, D. (1993). Influence des conditions d'élevage et de sulfitage des vins rouges en barriques sur le teneur en ácide acétique et en ethyl-phenols. *J. Int. Vigne Vin*, *27*, 277–298.

Chatonnet, P., Dubourdieu, D., & Boidron, J.N. (1995). The influence of *Brettanomyces/ Dekkera* sp. yeasts and lactic acid bacteria on the ethylphenol content of red wines. *Am. J. Enol. Vitic.*, *46*, 463–468.

Chatonnet, P., Viala, C., & Dubourdieu, D. (1997). Influence of polyphenolic components of red wines on the microbial synthesis of volatile phenols. *Am. J. Enol. Vitic.*, *48*, 443–448.

Cigič, I.K., Strlič, M., Schreiber, A., Kocjančič, M., & Pihlar, B. (2006). Ochratoxin A in wine: Its determination and photostability. *Anal. Lett.*, *39*, 1475–1488.

Coghe, S., Benoot, K., Delvaux, F., Vanderhaegen, B., & Delvaux, F.R. (2004). Ferulic acid Release and 4-Vinylguaiacol formation during brewing and fermentation: Indications for feruloyl esterase activity in *Saccharomyces cerevisiae*. *J. Agric. Food Chem.*, *52*, 602–608.

Colagrande, O., Mazzoleni, V., & Silva, A. (1988). Genesi degli odori e sapori anomali dei vini. *Vignevini*, 7–8, 23–30.

Commission Regulation (EC) (2006). Laying down the methods of sampling and analysis for the official control of the levels of mycotoxins in foodstuffs. No 401/2006 of 23 February 2006. Assessed at http://eur-lex.europa.eu/LexUriServ/LexUriServ.do?uri=OJ:L:2006: 070:0012:01:EN:HTML.

CONTAM (2006). Mechanism of ochratoxin A induced carcinogenicity as a basis for an improved risk assessment. Ochratoxin A – Risk assessment Project No. QLK-2001-01614. Assessed at: http://www.uni.wuerzburg.de/toxikologie/EU-OTA /ochratoxinA. html

Costello, P., & Henschke, P. (2002). Mousy off-flavor of wine: Precursors and biosynthesis of the causative N-Heterocycles 2-ethyltetrahydropyridine, 2-acetyltetrahydropyridine, and 2-acetyl-1-pyrroline by Lactobacillus hilgardii DSM 20176.*J. Agric. Food Chem.*, *50*, 7079–7087.

Coulter, A., Robinson, E., Cowey, G., Francis, I.L., Lattey, K., Capone, D., Gishen, M., & Godden, P. (2003). *Dekkera/Brettanomyces* yeast: An overview of recent AWRI investigations and some recommendations for its control. In: Bell, S.M., de Garis, K.A., Dundon, C.G., Hamilton, R.P., Partridge, S.J. & Wall, G.S. (eds.), Grapegrowing at the edge; Managing the wine business; Impacts on wine flavour: proceedings of a seminar; 10–11 July 2003, Barossa Convention Centre, Tanunda, S.A. Adelaide, S.A: Australian Society of Viticulture and Oenology, 41–50.

Couto, J., Campos, F., Figueiredo, A., & Hogg, T. (2006). Ability of lactic bactéria to produce volatile phenols. *Am. J. Enol. Vitic.*, *57*, 166–171.

Dias, L., Dias, S., Sancho, T., Stender, H., Querol, A., Malfeito-Ferreira, M., & Loureiro, V. (2003a). Identification of yeasts isolated from wine related environments and capable of producing 4-ethylphenol. *Food Microbiol.*, 20:567–574.

Dias, L., Pereira-da-Silva, S., Tavares, M., Malfeito-Ferreira, M., & Loureiro, V. (2003b). Factors affecting the production of 4-ethylphenol by the yeast *Dekkera bruxellensis* in enological conditions. *Food Microbiol.*, *20*, 377–384.

Díez, J., Domínguez, C., Guillén, D.A., Veas, R., & Barroso, C.G. (2004). Optimisation of stir bar sorptive extraction for the analysis of volatile phenols in wines. *J. Chromatogr. A*, *1025*, 263–267.

Domínguez, C., Guillén, D. A., & Barroso, C.G. (2002). Determination of volatile phenols in fino sherry wines. *Anal. Chim. Acta*, *458*, 95–102.

Dubois, P., & Brulé, G. (1970). Etude des phénols volatiles des vins rouges. *C.R. Acad. Sci. Paris*, *271*, 1597–1598.

Dugelay, I., Gunata, Z., Sapis, J.C., Baumes, R., & Bayonove, C. (1993). Role of cinnamoyl esterase activities from enzyme preparations on the formation of volatile phenols during wine-making. *J. Agric. Food Chem., 41*, 2092–2096.

Edlin, D., Narbad, A., Dickinson, J., & Lloyd, D. (1995). The biotransformation of phenolic compounds by *Brettanomyces anomalus. FEMS Microbiol. Lett., 5*, 311–315.

European Commission (1997). SCOOP. Task 3.2.2. Assessment of dietary intake of ochratoxin A by the population of EU Member States, Reports on tasks for scientific cooperation. Report EUR 17523.

European Commission (2002). SCOOP Task 3.2.7. Assessment of dietary intake of ochratoxin A by the population in EU Member States. European Commission, January 2002.

Fariña, L., Boido, E., Carrau, F., & Delacassa, E. (2007). Determination of volatile phenols in red wines by dispersive liquid-liquid microextraction and gas chromatography-mass spectrometry detection. *J. Chromatogr. A, 1157*, 46–50.

Fernandes, A., Venâncio, A., Moura, J., Garrido, J., & Cerdeira, A. (2003). Fate of ochratoxin A during a vinification trial. *Aspects Appl. Biol., 68*, 73–80.

Fernandes, A., Ratola, N., Cerdeira, A., Alves, A., & Venâncio, A. (2007). Changes in ochratoxin A concentration during winemaking. *Am. J. Enol. Vitic., 58*, 92–96.

Frisvad, J.C., & Samson, R.A. (2000). *Neopetromyces* gen. nov. and an overview of teleomorphs of *Aspergillus* subgenus *Circundati. Stu. Mycol., 45*, 201–207.

Frisvad, J., Frank, J., Houbraken, J., Kuijpers, A., & Samson, R. (2004). New ochratoxin A producing species of *Aspergillus* section *Circumdati. Stud. Mycol., 50*, 23–43.

Frisvad, J., Thrane, U., Samson, R., & Pitt, J. (2006). Important mycotoxins and the fungi which produce them. In: Hocking, A., Pitt, J., Samson, R., & Thrane, U. (eds.), *Advances in Food Mycology* (pp. 3–31). Springer, New York, USA.

Gato, O., Cota, M., Borbinha, S., & Laureano, P. (2001). Perception thresholds of 4-ethylphenol in Portuguese wines (in Portuguese). Proceedings of the 5th Vitiviniculture Symposium of Alentejo, 23–25th May, Évora, Portugal.

Gerbaux, V., Vincent, B., & Bertrand, A. (2002). Influence of maceration temperature and enzymes on the content of volatile phenols in Pinot Noir wines. *Am. J. Enol. Vitic., 53*, 131–137.

Gómez-Míguez, M., González-Miret, M.L., Hernanz, D., Fernández, M.A., Vicario, I.M., & Heredia, F.J. (2007). Effects of prefermentative skin contact conditions on colour and phenolic content of white wines. *J. Food Eng., 78*, 238–245.

Grando, M.S., Versini, G., Nicolini, G., & Mattivi, F. (1993). Selective use of wine yeast strains having different volatile phenols production. *Vitis, 32*, 43–50.

Grbin, P.R., & Henschke, P.A. (2000). Mousy off-flavour production in grape juice and wine by *Dekkera* and *Brettanomyces* yeasts. *Aust. J. Grape Wine Res., 6*, 255–262.

Grbin, P.R., Herderich, M., Markides, A., Lee, T., & Henschke, P.A. (2007). The role of lysine amino nitrogen in the biosynthesis of mousy off-flavor compounds by *Dekkera anomala. J. Agric. Food Chem., 55*, 10872–10879.

Guilloux-Benatier, M., Chassagne, D., Alexandre, H., Charpentier, C., & Feuillat, M. (2001). Influence of yeast autolysis after alcoholic fermentation on the development of "*Brettanomyces/Dekkera*" in wine. *J. Int. Sci. Vigne Vin., 35*, 157–164.

Heresztyn, T. (1986a). Metabolism of volatile phenolic compounds from hidroxycinnamic acids by *Brettanomyces* yeast. *Arch. Microbiol., 146*, 96–98.

Heresztyn, T. (1986b). Formation of substituted tetrahydropyridines by species of *Brettanomyces* and *Lactobacillus* isolated from mousy wines. *Am. J. Enol. Vitic.*, 3, 127–132.

Hernández, T., Estrella, I, Carlavilla, D., Martín-Álvarez, P.J., & Moreno-Arribas, M.V. (2006). Phenolic compounds in red wine subjected to industrial maloliactic fermentation and ageing on lees. *Anal. Chim. Acta, 563*, 116–125.

Hernández, T., Estrella, I., Pérez-Gordo, M., Alegría, E.G., Tenorio, C., Ruiz-Larrrea, F., & Moreno-Arribas, M.V. (2007). Contribution of malolactic fermentation by *Oenococcus oeni* and *Lactobacillus plantarum* to the changes in the nonanthocyanin polyphenolic composition of red wine. *J. Agric. Food Chem., 55*, 5260–5266.

Hesford, F., & Schneider, K. (2004). Die Entdeckung eines dritten Ethylphenols als Mitverursacher des *Brettanomyces*-Fehltons. *Schweiz Z. Obst-Weinbau, 13*, 11–13.

Hocking, A., Leong, S., Kazi, B., Emmett, R., & Scott, E. (2007). Fungi and mycotoxins in vineyards and grape products. *Int. J. Food Microbiol., 119*, 84–88.

Iamanaka, B., Taniwaki, M., Vicente, E., & Menzes, H. (2006). Fungi producing ochratoxin in dried fruits. In: Hocking, A., Pitt, J., Samson, R., & Thrane, U. (eds.), *Advances in Food Mycology* (pp. 181–188). Springer, New York, USA.

IARC (International Agency for Research on Cancer) (1993). Some naturally occurring substances: food items and constituents, heterocyclic aromatic amines and mycotoxins (vol. 56). Lyon, France: World Health Organization.

Ibern-Gómez, M., Andrés-Lacueva, C., Lamuela-Raventós, R.M., & Waterhouse, A.L. (2002). Rapid HPLC analysis of phenolic compounds in red wines. *Am. J. Enol. Vitic., 53*, 218–221.

JECFA (Joint FAO/WHO Expert Committee of Food Additives) (2001). Ochratoxin A. In: *Safety Evaluation of Certain Mycotoxins in Food.* Prepared by the Fifty-sixth meeting of the JECFA. FAO Food and Nutrition Paper 74, Food and Agriculture Organization of the United Nations, Rome, Italy.

Jorgensen, K. (2005). Occurrence of ochratoxin A in commodities and processes food – A review of EU occurrence data. *Food Add. Cont. Suppl., 1*, 26–30.

Kiessling, K.H., Pettersson, H., Sandholm, K., & Olsen, M. (1984). Metabolism of aflatoxin, ochratoxin, zearalenone, and three tricothecenes by intact rumen fluid, rumen protozoa, and rumen bacteria. *Appl. Environ. Microbiol., 47*, 1070–1073.

Larcher, R., Nicolini, G., Puecher, C., Bertoldi, D., Moser, S., & Favaro, G. (2007). Determination of volatile phenols in wine using high-performance liquid chromatography with a coulometric array detector. *Anal. Chim. Acta, 582*, 55–60.

Lay, H. (2004). Research on the development of the "Mousy taint" in wine and model solutions. *Mitt. Klosterbeuburg Rebe Wein Obstb. Früchteverwert, 54*, 243–250.

Leong, L.S., Hocking, A.D., Pitt, J.I., Kazi, B.A., Emmett, R.W., & Scott, E.S. (2006a). Black *Aspergillus* Species in Australian Vineyards: From Soil to Ochratoxin A in Wine. In: Hocking, A.D., Pitt, J.I., Samson, R.A., & Thrane, U. (eds), *Adv. Exp. Med. Biol., 571*, 153–171.

Leong, L.S., Hocking, A.D., & Scott, E.S. (2006b). The effect of juice clarification, static or rotary fermentation and fining on ochratoxin A in wine. *Aust. J. Grape Wine Res., 12*, 245–252.

López, R., Aznar, M., Cacho, J., & Ferreira, V. (2002). Determination of minor and trace volatile compounds in wine by solid-phase extraction and gas chromatography with mass spectrometric detection. *J. Chromatogr. A, 966*, 167–177.

Loureiro, V., Malfeito-Ferreira, M. (2006). Spoilage activities of *Dekkera/Brettanomyces* spp. In: Blackburn, C. (ed.) *Food spoilage microorganisms* (pp. 354–398). Woodhead Pubs, Cambridge.

Magan, N. (2006). Mycotoxin contamination of food in Europe: Early detection and prevention strategies. *Mycopathologia, 162*, 245–253.

Makris, D.P., Kallithraka, S., & Mamalos, A. (2006). Differentiation of young red wines based on cultivar and geographical origin with application of chemometrics of principal polyphenolic constituents. *Talanta, 70*, 1143–1152.

Malfeito-Ferreira, M., Rodrigues, N., & Loureiro, V. (2001). The influence of oxygen on the "horse sweat taint" in red wines. *Italian Food Bev. Technol., 24*, 34–38.

Mateo, R., Medina, A., Mateo, E.M., Mateo, F., & Jiménez, M. (2007). An overview of ochratoxin A in beer and wine. *Int. J. Food Microbiol., 119*, 79–83.

Medina, A., Mateo, R., Valle-Algarra, F., Mateo, E., & Jiménez, M. (2007). Effect of carbendazim and physicochemical factors on the growth and ochratoxin A production of *Aspergillus carbonarius* isolated from grapes. *Int. J. Food Microbiol., 119*, 230–235.

Mejías, R.C., Marín, R.N., Moreno, M.V.G., & Barroso, C.G. (2003). Optimisation of headspace solid-phase microextraction for the analysis of volatile phenols in wine. *J. Chromatogr. A, 995*, 11–20.

Mínguez, S., Cantus, J.M., Pons, A., Margot, P., Cabanes, F.X., Masque, C., Accensi, F., Elorduy, X., Giralt, L.L., Vilavella, M., Rico, S., Domingo, C., Blasco, M., & Capdevila, J. (2004).

Influence of the fungus control strategy in the vineyard on the presence of ochrotoxin A in the wine. *Bull l'OIV,* 885–886, 821–831.

Monagas., M., Bartolomé., B., & Gómez-Cordovés, C. (2005). Updated knowledge about the presence of phenolic compounds in wine. *Critic. Rev. Food Sci. Nutr., 45*, 85–118.

Murphy, P.A., Hendrich, S., Landgren, C., & Bryant, C.M. (2006). Food Mycotoxins: An Update. *J. Food Sci., 71*, R51–R65.

Natali, N., Chinnici, F., & Riponi, C. (2006). Characterization of volatiles in extracts from oak chips obtained by accelerated solvent extraction (ASE). *J. Agric. Food Chem., 54*, 8190–8198.

Ng, W., Mankotia, M., Pantazopoulos, P., Neil, R.J., & Scott, P.M. (2004). Ochratoxin A in wine and grape juice sold in Canada. *Food Addit. Contam., 21*, 971–981.

Nicolini, G., Larcher, R., Bertoldi, D., Puecher, C., & Magno, F. (2007). Rapid quantification of 4-ethylphenol in wine using high-performance liquid chromatography with a fluorimetric detector. *Vitis, 46*, 202–206.

OIV (2005). Code of sound vitivinicultural practices in order to mimise levels of Ochratoxin A in vine-based products. Resolution Viti-Oeno 1/2005. Assessed at http://news.reseau-concept.net/images/oiv_uk/Client/VITI-OENO_1-2005_EN.pdf.

Oliveira, J., Freitas, V., Silva, A.M.S., & Mateus, N. (2007). Reaction between hydroxycinnamic acids and anthocyanin-pyruvic acid adducts yielding new portisins. *J. Agric. Food Chem., 55*, 6349–6356.

Otteneder, H., & Majerus, P. (2000). Occurrence of ochratoxin A (OTA) in wines: Influence of the type of wine and its geographical origin. *Food Addit. Contam., 17*, 793–798.

Pérez-Magariño, S., & González-San José, M.L. (2005). Effect of ripening stage of grapes on the low molecular weight phenolic compounds of red wines. *Eur. Food Res. Technol., 220*, 597–606.

Perrone, G., Mulè, G., Susca, A., Battilani, P., Pietri, A., & Logrieco, A. (2006a). Ochratoxin A production and amplified fragment length polymorphism analysis of *Aspergillus carbonarius*, *Aspergillus tubingensis*, and *Aspergillus niger* strains isolated from grapes in Italy. *Appl. Environm. Microbiol., 72*, 680–685.

Perrone, G., Susca, A., Epifani, F., & Mulè, G. (2006b). AFLP characterization of Southern Europe population of *Aspergillus* section *Nigri* from grapes. *Int. J. Food Microbiol., 111*, S22–S27.

Perrone, G., Nicoletti, I., Pascale, M., De Rossi, A., De Girolamo, A., & Visconti, A. (2007). Positive correlation between high levels of ochratoxin A and resveratrol-related compounds in red wines. *J. Agric. Food Chem., 55*, 6807–6812.

Pizarro, C., Pérez-del-Notario, N., & González-Sáiz, J.M. (2007). Determination of Brett character responsible compounds in wines by using multiple headspace solid-phase microextraction. *J. Chromatogr. A, 1143*, 176–181.

Porret, N.A., Schneider, K., Hesford, F., & Gafner, J. (2004). Früherkennung unerwünschter Mikroorganismen im wein: *Brettanomyces bruxellensis. Schweiz. Z. Obst-Weinbau, 6*, 13–15.

Rapp, A. (1998). Volatile flavour of wine: correlation between instrumental analysis and sensory perception. *Nahrung, 42*, 351–363.

Rayne, S., & Eggers, N.J. (2007a). Quantitative determination of 4-ethylphenol and 4-ethyl-2-methoxyphenol in wines by a stable isotope dilution assay. *J. Chromatogr. A, 1167*, 195–201.

Rayne, S., & Eggers, N.J. (2007b). 4-Ethylphenol and 4-ethylguaiacol in wines: estimating non-microbial sourced contributions and toxicological considerations. *J. Environm. Sci. Health. Part B, 42*, 887–897.

Ribéreau-Gayon, P., Dubourdieu, D., Donèche, B., & Lonvaud, A. (2006). Handbook of Enology (vols. 1 and 2). The Microbiology of Wine and Vinification. John Wiley and Sons, Ltd, Chichester, UK.

Richard, J.L. (2007). Some major mycotoxins and their mycotoxicoses – An overview. *Int. J. Food Microbiol., 119*, 3–10.

Ringot, D., Chango, A., Schneider, Y.J., & Larondelle, Y. (2006). Toxicokinetics and toxicodynamics of ochratoxin A, an update. *Chem. Biol., 159*, 18–46.

Rodrigues, N., Gonçalves, G., Pereira-da-Silva, S., Malfeito-Ferreira, M., & Loureiro, V. (2001). Development and use of a new medium to detect yeasts of the genera *Dekkera/Brettanomyces* spp. *J. Appl. Microbiol., 90*, 588–599.

Romano, A., Perello, M.C., De Revel, G., & Lonvaud-Funel, A. (2007). Growth and Volatile Compound production by *Brettanomyces/Dekkara* bruxellensis in red wine. *J. Appl. Microbiol., 104*, 1577–1585.

Romero, C., & Bakker, J. (2000). Effect of storage temperature and pyruvate on kinetics of anthocyanin degradation, vitisin A derivattive formation, and color characteristics of model solutions. *J. Agric. Food Chem., 48*, 2135–2141.

Salameh, D., Brandam, C., Medawar, W., Lteif, R., & Strehaiano, P. (2007). Highlight on the problems generated by p-coumaric acid analysis in wine fermentations. *Food Chem.*, (2007) doi: 10.1016/j.foodchem.2007.09.052.

Samson, R.A., Houbraken, J.A., Kuijpers, A.F., Frank, J.M., & Frisvad, J.C. (2004). New ochratoxin A or sclerotium producing species in *Aspergillus* section *Nigri*. *Stud. Mycol., 50*, 45–61.

Schwarz, M., Wabnitz, T.C., & Winterhalter, P. (2003). Pathway leading to the formation of anthocyanin-vinylphenol adducts and related pigments in red wines. *J. Agric. Food Chem., 51*, 3682–3687.

Serra, R., Cabañes, J.F., Perrone, G., Castellá, G., Venâncio, A., Mulè, G., & Kozakiewicz, Z. (2006). *Aspergillus ibericus*: A new species of section *Nigri* isolated from grapes. *Mycologia, 98*, 295–306.

Shinohara, T., Kubodera, S., & Yanagida, F. (2000). Distribution of phenolic yeasts and production of phenolic off-flavors in wine fermentation. *J. Biosci. Bioeng., 90*, 90–97.

Snowdon, E.M., Bowyer, M.C., Grbin, C.B., & Owyer, P.K. (2006). Mousy Off-Flavor: A review. *J. Agric. Food Chem., 54*, 6465–6474.

Steinke, R.D., & Paulson, M.C. (1964). The production of steam-volatile phenols during the cooking and alcoholic fermentation of grain. *J. Agric Food Chem., 12*, 381–387.

Strauss, C.R., & Heresztyn, T. (1984). 2-Acetyltetrahydropyridines a cause of "mousy" taint in wine. Chem. Ind. (London) 1984, 109–110.

Suárez, R., Suárez-Lepe, J.A., Morata, A., & Calderón, F. (2007). The production of ethylphenols in wine by yeasts of the genera *Brettanomyces* and *Dekkera*: A review. *Food Chem., 102*, 10–21.

Taniwaki, M. (2006). An update on ochratoxigenic fungi and ochratoxin A in coffee. In: Hocking, A., Pitt, J., Samson, R., & Thrane, U. (eds.), *Advances in Food Mycology* (pp. 189–202). Springer, New York, USA.

Tat, L., Comuzzo, P., Battitutta, F., & Zironi, R. (2007). Sweet-like off-flavor in Aglianico del Vulture wine: Ethyl phenylacetate as the mainly involved compound. *J. Agric. Food Chem., 55*, 5205–5212.

Tucknott, O.G. (1977). Taints in fermented juice products: mousy taint in cider. Long Ashton Research Station Annual Report 129–130.

Valero, A., Marín, S., Ramos, A.J., & Sanchis, V. (2007a). Effect of preharvest fungicides and interacting fungi on *Aspergillus carbonarius* growth and ochratoxin A synthesis in dehydrating grapes. *Lett. App. Microbiol., 45*, 194–199.

Valero, A., Marín, S., Ramos, A.J., & Sanchis, V. (2007b). Survey: Ochratoxin A in European speciality wines. *Food Chem.*, doi: 10.1016/j.foodchem.2007.11.040.

Van der Merwe, K.J., Steyn, P.S., Fourie, L., Scott, D.B., & Theron, J.J. (1965). Ochratoxin A, a toxic metabolite produced by *Aspergillus ochraceus* Wilh. *Nature, 205*, 1112–1113.

Van Egmond, H.P., & Schothorst, R. (2007). Regulations relating to mycotoxins in food. *Anal. Bioanal. Chem., 389*, 147–157

Vanbeneden, N., Delvaux, F., Delvaux, F.R. (2006). Determination of hydroxycinnamic acids volatile phenols in wort and beer by isocratic high-performance liquid chromatography using electrochemical detection. *J. Chromatogr. A, 1136*, 237–242.

Varga, J., & Kozakiewicz, Z. (2006). Ochratoxin A in grapes and grape-derived products. *Trends Food Sci. Technol., 17*, 72–81.

Yokotsuka, K., & Singleton, V. (2001). Effects of seed tannins on enzymatic decolorization of wine pigments in the presence of oxidizable phenols. *Am. J. Enol. Vitic., 52*, 93–100.

Zimmerli, B., & Dick, R. (1996). Ochratoxin A in table wine and grape-juice: Occurrence and risk assessment. *Food Add. Contam., 13*, 665–668.

Part IV
Automatic Analysers and Data Processing

Chapter 12
Automatic Analysers in Oenology

Marc Dubernet

Contents

12.1 Introduction

Automatic analysers are nowadays universally used in oenological laboratories, and this includes control laboratories who often employ these techniques for screening purposes before applying reference methods when limit values have been identified.

M. Dubernet (✉)
Laboratoires Dubernet, 9 quai d'Alsace, F-11100 Narbonne
e-mail: marc.dubernet@dubernet.com

M.V. Moreno-Arribas, M.C. Polo (eds.), *Wine Chemistry and Biochemistry*,
DOI 10.1007/978-0-387-74118-5_24, © Springer Science+Business Media, LLC 2009

Automated methods frequently exhibit remarkable performance not only in terms of sample throughput and cost, but in relation to the quality of results, especially in relation to repeatability and reproducibility. Several systems are currently available that enable simple automation of manual gestures; these include automatic titrators, pH-meters with a circulating cell, etc., and involve approaches that are not fundamentally different to the corresponding manual method. These techniques are not, however, described in this chapter, nor are gas and liquid chromatography and capillary electrophoresis, automated techniques, that may be present but are not routinely used in oenological laboratories.

Four major families of automated analysers are routinely used today in oenology:

- Flow injection analysers
- Sequential analysers
- NIR (near infra-red) analysers
- FTIR (Fourier transform interferometer) analysers

For each of these families, the general principles, the particularities and limits of their applications will be presented.

Note: The methods presented below are described in the OIV green document N° 1276–2008 (Dubernet 2008).

12.2 Flow Injection (FIA) Analysers

The earliest studies on the automation of chemical analysis in oenology by FIA were carried out by Sarris et al. (1970). However, it was only from 1974 onwards that this technique started to become widely used in routine oenological determinations with the development of reliable and repeatable analytical methods.

12.2.1 Composition and Principle of an FIA Analyser

FIA analysers are composed of separate modules, each of which has a specific function.

12.2.1.1 The Sample Distributor

This module consists of a tray that can generally hold between 20 and 60 sample vials whose volume is usually between 1 and 3ml. A pivoting arm to which is a needle is attached aspirates an aliquot of each sample with a rinse step between each sampling step. The needle is linked to the analytical circuit via a PVC tube. The rinsing liquid, the sample, the rinsing liquid and the next sample are successively aspirated by the needle. The rate of analysis is determined by the sample distributor.

12.2.1.2 The Peristaltic Pump

The peristaltic pump is at the heart of the FIA analyser; containing 20–30 channels in general, it regulates all liquid movements in the analytical circuit. Each channel is constituted of a tube, composed generally of PVC which is pressed by a series of rotating rollers against a circular surface to create pressurised flow. The flow rate within each tube depends on its internal diameter which varies from 0.005 to 0.110cm. In this way the rhythm of introduction of the sample and different reagents may be controlled with a high degree of precision.

12.2.1.3 The Analytical Cassette

This component contains all the elements of the analytical circuit. It is here that the sample may be subjected to different treatments according to the requirements of the analysis, i.e.

Dilution
Dialysis
Addition of reagent
Mixing
Distillation
Passage through a water bath (variable temperature and duration)
Part of the circuit may be thermostatted

From the time it exits the pump the liquid flow in the analytical circuit is segmented regularly by air bubbles every 1 or 2cm; these bubbles are formed when the needle passes from the rinsing vial to the sample vial. Due to surface tension forces, the air bubbles prevent the diffusion of compounds in solution from one section of the liquid column formed between two air boluses and the preceding or succeeding column. In addition, the air bubble wipes the internal wall of the glass or PVC tube, thus minimising the problem sample-to-sample cross contamination. A second generation of FIA technology has been proposed, based on micro-flow techniques which obviates the need for segmentation. This technology has remained rather marginal due to problems with clogging of the tubes which can be difficult to clear. Otherwise, the analytical principles remain the same as for segmentation-based instruments.

12.2.1.4 The Detector

The most widely-used detectors in FIA methods are colorimetric.

12.2.1.5 The Recorder

A property of an analyte in the sample is converted by the detector into a signal whose intensity is proportional to the analyte's concentration. The signal is recorded in the form of peaks whose height is proportional to the concentration.

The reliability of the FIA method reposes on the recording the signal according to the following principles:

1. The creation of an equilibrium in the analytical circuit such that return to baseline is always obtained by the rinsing step and that a given analyte concentration gives a peak of constant height
2. The measurement method is comparative. All the measurements are made by comparison with a calibration curve constructed from standards of accurately known concentration

If these principles are applied, any malfunctioning of the circuit translates into a disruption in the recording (baseline drift, variation in peak shape or heights of the calibration standard peaks etc.).

12.2.1.6 The Digital Interface

This component enables the analogue signal from the colorimeter to be converted into a numerical value which may be directly expressed as analyte concentration. It may often include a control module by which the functioning of the system may be verified.

12.2.2 Commonly FIA Analytical Methods

According to Bouvier (1993) and Bouvier et al. (1995).

12.2.2.1 Determination of Volatile Acidity in Wine and Must (Grape Juice)

According to Pilone (1967), Sarris et al. (1970) and Dubernet (1976)

Principle

Volatile acidity is derived from the acids of the acetic series present in wine in the free state and combined as salts.

Acids of the acetic series are isolated from wine by micro-distillation at 98°C under a stream of nitrogen. Sulphur dioxide is oxidised to sulphuric acid using hydrogen peroxide and eliminated before distillation. Lactic acid is removed by rectification during the distillation, and carbon dioxide does not interfere with the determination. The distillate is mixed with a redox reagent (potassium iodide or bromphenol blue) whose variations in colour intensity are proportional to the level of volatile acidity.

Characteristics of the Method

Intra-laboratory reproducibility is 0.06g/L in acetic acid
Inter-laboratory reproducibility is 0.12g/L in acetic acid

12.2.2.2 Determination of L-Malic and L-Lactic Acids in Wines and Musts

According to Trossais and Asselin (1985), Battle et al. (1987a), Battle and Bouvier (1986), and Curvelo-Garcia (1993).

Principle

The acids are oxidised, in the presence of nicotinamide-adenine-dinucleotide (NAD), in a reaction catalysed by a specific enzyme:

Malate dehydrogenase (MDH) in the case of malic acid
Lactate dehydrogenase (LDH) in the case of malic acid

The equilibria of these reactions are forced in the direction of the products by the addition of an alkaline hydrazine buffer and an excess of NAD.

$$\text{L-malate} + NAD^{+} \overset{\text{MDH}}{<==>} \text{oxaloacetate hydrazone} + NADH + H^{+}$$

$$\text{L-lactate} + NAD^{+} \overset{\text{LDH}}{<==>} \text{pyruvate hydrazone} + NADH + H^{+}$$

The quantity of NADH produced is proportional to the concentration of acid in the sample, and its absorbance is measured at 340 nm.

Characteristics of the Method

Inter-laboratory reproducibility: 0.15g/L

12.2.2.3 Determination of Tartaric Acid in Wines and Musts

According to Battle et al. (1978b).

Principle

This involves the Rebelein method. A sample of dialysed wine is reacted with metavanadate to produce a red colour at 37°C in acid medium. The absorbance is measured at 520 nm. High concentrations of malic acid and sugars can interfere with the analysis.

Characteristics of the Method

Inter-laboratory reproducibility: 0.7g/L

12.2.2.4 Determination of Citric Acid in Wines and Musts

According to Battle and Bouvier (1986).

Principle

Citric acid is transformed into oxaloacetate and acetate in a reaction catalysed by citrate-lyase (CL).

$$\text{citrate} \overset{\text{CL}}{\longleftarrow\!=\!=\!=\!=\!>} \text{oxaloacetate} + \text{acetate} \qquad (1)$$

In the presence of malate deshydrogenase (MDH), oxaloacetate and pyruvate, its decarboxylation derivative, are reduced to L-malate by reduced NADH. The equilibrium of the reaction is forced in the direction of the products by the judicious selection of operating conditions (buffer pH7.8 and an excess of NADH):

$$\text{oxaloacetate} + \text{NADH} + \text{H}^+ \overset{\text{MDH}}{\longleftarrow\!=\!=\!=\!>} \text{L-malate} + \text{NAD} \qquad (2)$$

The quantity of NADH oxidised to NAD is proportional to the quantity of citrate present and the oxidation of NADH is tracked by the reduction in its absorbance at 340 nm.

12.2.2.5 Determination of Reducing Sugars in Wines and Musts

According to Curvelo-Garcia (1993) and Curvelo-Garcia and Godinho (1988).

Principle

A copperIII -neocuproine chelate is reduced by reducing sugars in alkaline medium at approximately 85°C. A dialysis step eliminates macromolecules that can interfere with the colour reaction, the absorbance of which is measured at 460 nm. Must analysis requires a 40-fold dilution prior to analysis.

Characteristics of the Method

Inter-laboratory reproducibility: 20%

12.2.2.6 Determination of Glucose and Fructose in Wines and Musts

According to Battle and Bouvier (1986).

Principle

This method enables the concentrations of glucose and fructose to be determined separately or the sum of their concentrations to be determined in wines and musts.

Phosphorylation of glucose and fructose.

This reaction is catalysed by hexokinase (HK):

$$\text{glucose} + \text{ATP} \overset{\text{HK}}{\longleftrightarrow} \text{glucose-6-phosphate} + \text{ADP} \quad (1)$$

$$\text{fructose} + \text{ATP} \overset{\text{HK}}{\leftarrow\!=\!>} \text{fructose-6-phosphate} + \text{ADP}$$

ATP: adenosine triphosphate.
ADP: adenosine diphosphate.

Oxidation glucose-6-phosphate (G-6-P).

This reaction involves nicotinamide adenine dinucleotide phosphate (NADP), in the presence of its specific enzyme, glucose-6-phosphate dehydrogenase (G6PDH).

$$\text{G-6-P} + \text{NADP+} \overset{\text{G6PDH}}{\longleftrightarrow} \text{gluconate-6-phosphate} + \text{NADPH} + \text{H+} \quad (2)$$

The equilibrium of the reaction is forced in the direction of the products by the appropriate of operating conditions (buffer pH 7.6 and excess NADP). The amount of NADPH formed, which is proportional to the concentration of glucose in the sample is measured by its absorbance at 340 nm.

This reaction, which enables the sum of glucose + fructose to be determined, is catalysed by the enzyme phosphoglucose-isomerase (PGI):

$$\text{fructose-6-phosphate} \overset{\text{PGI}}{\leftarrow\!=\!=\!>} \text{glucose-6-phosphate} \quad (3)$$

The glucose-6-phosphate formed reacts with NADP as shown in reaction (2), producing NADPH, which is measured by its absorbance at 340 nm.

The determination of fructose requires that the standards and samples are injected twice, thus doubling the analysis time by comparison with the determination of glucose alone.

12.2.2.7 Determination of Free and Total Sulphur Dioxide in Wines and Musts by Dialysis

According to Scholten (1982) and Dubernet et al. (1997).

Principle

After acidification, free sulphur dioxide in the sample is allowed to diffuse across a dialysis membrane. Combined sulphur dioxide is liberated by alkaline hydrolysis.

Two determinations are made at 560 nm in the presence of para-rosaniline (or basic fuschine) and segmentation of the stream is with nitrogen to avoid oxidation of sulphur dioxide into sulphuric acid.

$$SO_2 + \text{rosaniline} \longrightarrow \text{colourless complex} + \text{formaldehyde} \longrightarrow \text{coloured complex}.$$

The method is applicable over the concentration range 0–200mg/L.

Characteristics of the Method

Inter-laboratory reproducibility free sulphur dioxide: 7mg/L
Inter-laboratory reproducibility total sulphur dioxide: 27mg/L

12.2.2.8 Determination of Free and Total Sulphur Dioxide in Wines and Musts by Distillation

Principle

An alternative to the above method is also largely applied. Dialysis is replaced by a distillation step using a micro distillation column. In the case of total sulphur dioxide, alkaline hydrolysis is replaced by strong acidification and a higher distillation temperature.

Characteristics of the Method

Inter-laboratory reproducibility free sulphur dioxide: 7mg/L
Inter-laboratory reproducibility total sulphur dioxide: 27mg/L

12.3 Sequential Analysers

12.3.1 Principle and Organisation of a Sequential Analyser

These are often multi-parameter analysers and enable several determinations to be carried out on the same sample; furthermore, the number of determinations can be programmed for each individual sample. An aliquot of the sample is placed in a transparent measurement cuvette; following addition of the reagents, a colorimetric measurement is then carried out directly on the cuvette. The analysis rate varies from 50 to 1000 determinations per hour, and the principal applications are carried out either by chemical or enzymatic analysis. The volumes of reagent required for sequential analysis are small by comparison with FIA which substantially reduces the cost per analysis, particularly for enzymatic determinations.

12.3.2 Composition of a Sequential Analyser

12.3.2.1 The Sample Tray

This contains a variable number of vials into which the samples to be analysed are placed.

12.3.2.2 The Reaction Tray

This is composed of transparent cuvettes into which are placed the sample and the reagent(s). The cuvettes are housed in a thermostatted bath to ensure constant temperature for the various colour reactions. The cuvettes may be single use or may be washed in a cleaning ramp before being re-used.

12.3.2.3 The Reagent Tray

This tray contains reservoirs of the reagents to carry out the desired analyses. They are placed in refrigerated wells which enables the use of fragile, temperature-sensitive reagents in the case of enzymatic methods.

12.3.2.4 The Sampling Arm(s)

These robotic arms equipped with an automatic syringe system are used to take pre-programmed aliquots from the sample and reagent tray and place them into the reaction cuvettes.

12.3.2.5 The Optical Sector

This component carries out colorimetric measurements of the different reaction cuvettes, either at specific moments (at the start and end of the reaction, for example) or in a continuous fashion in order to establish a reaction curve for each reaction cuvette.

12.3.2.6 The Electronic-Computing Module

This component drives and controls the system and carries out the calculations required to transform the signal into a numerical result.

12.3.3 Analytical Scheme

A wide variety of analytical schemes may be used and the example described below corresponds to the most classical cases. It consists of the following steps.

12.3.3.1 Sampling

A pre-determined volume of sample is removed by the robotic arm from the vial in the sampling tray and placed in the reaction cuvette.

12.3.3.2 Addition of the First Reagent

A pre-determined volume of the first reagent (R1) is taken from the reagent tray and placed in the reaction cuvette containing the sample. The sample and reagent are generally mixed by a stirring mechanism.

12.3.3.3 Delay 1

The sample and reagent are allowed to react for a specified period – usually from 4 to 6 min.

12.3.3.4 Absorbance Measurement

The absorbance is first measured at the selected wavelength. This is the point zero of the reaction and may be used to take the colour of the sample into account.

12.3.3.5 Addition of the Second Reagent

The second reagent (R2) is added to the reaction cuvette and this starts the reaction leading to the formation of the coloured product.

12.3.3.6 Delay 2

The sample and reagent are allowed to react for a specified period – usually from 4 to 6 min.

12.3.3.7 Absorbance Measurement

A second absorbance reading measures the variation in absorbance due to the reaction itself. This basic scheme may vary and thus with more sophisticated machines readings are taken more frequently, for example every 12s. Thus the reaction curve may be established and the initial reaction rate measured which, under precise analytical conditions, may be proportional to the concentration of the compound being measured.

12.3.3.8 Calculations

For oenological applications, calibrations always involve the preparation of synthetic wine standards containing known concentrations of the analyte. The number of calibration standards may be variable and a greater number of calibration points is required if the reaction curve is not linear.

Certain instruments are only capable of adding one reagent and this precludes reactions for which two reagents are required, though of course the reverse is not the case.

12.3.4 Usual Methods Based in Sequential Analysis

12.3.4.1 Determination of Acetic Acid

According to McCloskey (1980), Doneche and Sanchez, (1985) and Dubernet et al. (1997).

Principle

Enzymatic Method

In the presence of ATP, acetic acid is transformed into acetyl-phosphate in a reaction catalysed by acetate kinase:

$$\text{Acetate} + \text{ATP} \xrightarrow{\text{AcetateKinase}} \text{Acetyl-phosphate} + \text{ADP} \qquad (1)$$

The ADP formed by this reaction is reconverted into ATP by reaction with phosphoenolpyruvate in the presence of pyruvate kinase:

$$\text{ADP} + \text{Phosphoenolpyruvate} \xrightarrow{\text{PyruvateKinase}} \text{Pyruvate} + \text{ATP} \qquad (2)$$

Pyruvate is reduced to l-lactate by reduced nicotinamide-adenine-nucleotide (NADH) in the presence of lactate-deshydrogenase:

$$\text{Pyruvate} + \text{NADH} + \text{H}^+ \xrightarrow{\text{Lactatedeshydrogenase}} \text{Lactate} + \text{NAD}^+ + \text{H}_2\text{O} \qquad (3)$$

The quantity of NADH oxidised in reaction (3) is determined by measuring its absorbance at 340 nm, which is proportional to the concentration of acetic acid in the wine.

A fourth reaction, the elimination of acetylphosphate is used to force the equilibrium of reaction (1) in favour of the products:

$$\text{Acetyl-phosphate} + \text{CoA} \xrightarrow{\text{Phosphotransacetylase}} \text{Acetyl-CoA} + \text{inorganic phosphate} \qquad (4)$$

Polyvinylpyrrolidone (PVP) is added to the reaction medium in order eliminate interference due to wine phenolic compounds.

Characteristics of the Method

Intra-laboratory reproducibility: 0.04g/L
Inter-laboratory reproducibility: 0.10g/L

12.3.4.2 Determination of Total Sulphur Dioxide in Wines and Musts

According to Ellman (1959), Dubernet et al. (1995a) and Dubernet et al. (1998).

Principle

Chemical Method

The sample to be analysed is diluted in a solution of phosphate buffer, pH8. After stabilisation, a buffered solution of DTNB (5,5′-dithio-bis(2-nitrobenzoic acid) (3,3′-6)) or 3-carboxy-4-nitrophenyl disulphide (otherwise known as Ellman's reagent) is added to the reaction mixture. This reagent is used for specific detection of disulphide bonds and results in a yellow colour which is measured at 405 nm.

Characteristics of the Method

Intra-laboratory reproducibility: 12 mg/L
Inter-laboratory reproducibility: 20 mg/L

12.3.4.3 Determination of Free Sulphur Dioxide in Wines and Musts

According to Dubernet et al. (1995b).

Principle

Chemical Method

Free sulphur dioxide is stabilised in acid medium and reacted in the presence of formaldehyde with Fuschine which is de-coloured by phosphoric acid. The reaction results in a pink colour which is measured by colorimetry. A blank reaction is carried out to compensate for parasite reactions with wine phenolic compounds.

Characteristics of the Method

Intra-laboratory reproducibility: 6mg/L
Inter-laboratory reproducibility: 12mg/L

12.3.4.4 Determination of Glucose and Fructose in Wines and Musts

Principle

Enzymatic Method

Glucose and fructose are phosphorylated in the presence of adenosine triphosphate (ATP) in a reaction is catalysed by hexokinase (HK) producing respectively, glucose-6-phosphate (G6P) and fructose-6-phosphate (F6P):

$$\text{Glucose} + \text{ATP} \xrightarrow{\text{HK}} \text{G6P} + \text{ADP}$$

$$\text{Fructose} + \text{ATP} \xrightarrow{\text{HK}} \text{F6P} + \text{ADP}$$

Glucose-6-phosphate is oxidised to form gluconate-6-phosphate in the presence of NAD in a reaction catalysed by glucose-6-phosphate deshydrogenase (G6PDH):

$$\text{G6P} + \text{NAD} \xrightarrow{\text{G6PDH}} \text{gluconate-6-phosphate} + \text{NADH} + \text{H}^+$$

Fructose-6-phosphate is transformed into glucose-6-phosphate by phosphoglucose isomerase (PGI):

$$\text{F6P} \xrightarrow{\text{PGI}} \text{G6P}$$

The increase in the absorbance at 340 nm is proportional to the quantity of d-glucose and d-fructose.

Characteristics of the Method

Intra-laboratory reproducibility: 0.3g/L from 0 to 5g/L and 0.6g/L above 5g/L.
Inter-laboratory reproducibility: 0.35g/L from 0 to 2g/L

12.3.4.5 Determination of L-Malic Acid in Wines and Musts

Principle

In the presence of NAD, L-malic acid is oxidised to oxaloacetate in a reaction catalysed by L-malate deshydrogenase (L-MDH). The reaction equilibrium is forced in the direction of the products by the elimination of oxaloacetate, via its reaction with l-glutamate, resulting in the production of L-aspartate. This reaction is catalysed by glutamate-oxaloacetate-transaminase (GOT):

$$\text{L-malate} + \text{NAD}^+ \;{->}\; \text{oxaloacetate} + \text{NADH} + \text{H}^+$$
(in the presence of L-malate deshydrogenase)

$$\text{Oxaloacetate} + \text{L-glutamate} \;{->}\; \text{L-aspartate} + \alpha - \text{cetoglutarate}$$
(in the presence of GOT)

The formation of NADH, measured by the increase in its absorbance at 340 nm, is proportional to the quantity of L-malate initially present.

Characteristics of the Method

Intra-laboratory reproducibility: 0.4g/L
Inter-laboratory reproducibility: 0.6g/L

12.3.4.6 Determination of L-Lactic Acid in Wines and Musts

Principle

Enzymatic Method

L-Lactic acid, in the presence of NAD, is oxidised to pyruvate in a reaction catalysed L-lactate deshydrogenase (L-LDH). The equilibrium of the reaction is forced in the direction of the products by the elimination of pyruvate by reacting it with L-glutamate, resulting in the formation of L-alanine. This reaction is catalysed by glutamate-pyruvate-transaminase (GPT):

$$\text{L-lactate} + NAD^+ \; -> \; \text{pyruvate} + NADH + H^+$$
(in the presence of L-lactate deshydrogenase)

$$\text{Pyruvate} + \text{L-glutamate} \; -> \; \text{L-alanine} + \alpha - \text{cetoglutarate}$$
(in the presence of GPT)

The formation of NADH, measured by the increase in its absorbance at 340 nm, is proportional to the quantity of lactate initially present.

Characteristics of the Method

Intra-laboratory reproducibility: 0.2g/L
Inter-laboratory reproducibility: 0.5g/L

12.3.4.7 Determination of Total Phenolic Compounds in by the Folin-Ciocalteu Index

Principle

Chemical Method

Total wine polyphenols are oxidised by the Folin-Ciocalteu reagent – composed of a mixture of phosphotungstic and phosphomolybdic acids which are reduced by the oxidation of the phenols, forming a mixture of blue oxides of tungsten and molybdenum. The blue coloration has an absorption maximum at approximately 750 nm, the intensity of which is proportional to the level of phenolic compounds present in the wine. The sequential analyser method is a direct automation of the manual method and results are expressed as a unit-less index.

Characteristics of the Method

Intra-laboratory reproducibility: 4.5
Inter-laboratory reproducibility: 8.5

12.3.4.8 Determination of Iron in Wines and Musts

Principle

Chemical Method

The method is based on the Aquaquant® kit for iron commercialised by Merck. Ferrous ions react with the reagent, Ferrospectral (the sodium salt off (pyridyl-2)-3 bis (phenyl-4-sulfonique)-5,6 triazine-1,2,4,acid) forming a violet complex. Only the ferrous form of iron reacts with this reagent, so the inclusion of an additional reagent containing ammonium thioglycolate digests weak iron complexes and reduces trivalent iron to the bivalent form.

Characteristics of the Method

Intra-laboratory reproducibility: 0.5mg/L

12.3.4.9 Determination of α-Amino Nitrogen in Wines and Musts

Principle

Chemical Method

Total α-amino nitrogen is determined by its reaction with *o*-phthaldialdehyde. The intensity of the absorbance 340 nm is compared to that obtained for standards of known concentrations of isoleucine.

12.3.4.10 Determination of Ammoniacal Nitrogen in Wines and Musts

Principle

Enzymatic Method

In the presence of reduced nicotinamide-adenosine-dinucleotide (NADH), NH_4^+ ions are reacted with 2-oxoglutarate resulting in the formation of L-glutamate, in a reaction catalysed by glutamate deshydrogenase (GLDH):

$$NH_4^+ + \text{2-oxoglutarate} + \text{NADH} \;->\; \text{L-glutamate} + \text{NAD}^+ + H_2O$$
$$(\text{ in the presence of GLDH})$$

The quantity of NADH oxidised into NAD^+ is proportional to the quantity of ammoniacal nitrogen present in the sample.

12.4 Near Infra-Red Analyser

According to Cabanis et al. (1983), Bouvier (1986) and OIV (1994).

12.4.1 Principle

Most organic compounds when subjected to infra-red radiation present characteristic absorption bands. The spectral data are compared to those obtained for standard reference wines used to calibrate the instrument. The concentration of the analytes is calculated using multiple linear regression; the apparatus is computerised and can be linked to an automatic sampler. The analyte should exhibit strong absorption bands for the method to be exploitable, and it is thus only suitable for major wine or must constituents, essentially ethanol and sugars. The major qualities of this technique are simplicity of operation, high sample throughput and the lack of necessity for sample preparation – the only required sample re-treatment is the removal of carbon dioxide from musts in fermentation.

12.4.2 Material

The IR equipment used in oenology operates by reflectance. The circulation cell containing the sample is equipped with a reflector at its base which reflects the incident IR radiation; this essentially means that the radiation traverses the sample twice. The various components of the system are described in the following sections.

12.4.2.1 Sample Pumping System

In general this is composed of a peristaltic pump which is responsible for filling the cells. The instrument is usually equipped with a thermostatted bath to control the temperature of analysis.

12.4.2.2 Light Source

This is a tungsten polychromatic light source producing light in the spectral range 320–2500 nm. It is essential that the power supply is perfectly stable to ensure constant light intensity.

12.4.2.3 Wavelength Selector

The instruments used in oenological analysis are equipped with interference filters of fixed wavelength or monochromators to select the characteristic wavelengths of the analytes.

12.4.2.4 Measurement Cell

The part of the cell traversed by the incident and reflected rays is made of quartz, and the base of the cell may be composed of ceramic covered with a film of gold to provide reflectance. The cell is maintained at a constant measurement temperature, generally using a system based on the Peltier effect.

12.4.2.5 Detectors

The reflected rays are directed towards two lead sulphide photoelectric cells.

12.4.2.6 Computer

This component carries out the mathematical and statistical operations which enable comparisons with the calibrated standards and determination of the analyte concentration.

12.4.3 Operation

The installation and operation of a near infra-red analyser involves several steps.

12.4.3.1 Initial Calibration

This step involves calibration of the apparatus which will serve as a reference. It consists of analysing the greatest number possible (minimum 50) wines or must samples containing different and accurately known concentrations of each analyte. The concentration points should be uniformly distributed over the probable scale of measure for each analyte. The matrices should mimic as accurately as possible the wines and musts destined for analysis using that particular instrument. For each calibration sample, a measurement is carried out at a maximum number of wavelengths in the infra-red. Multi-linear regression is then carried out on the results which enables the following relationship to be established:

$$C = K_0 + K_1R_1 + K_2R_2 + K_3R_3 + \ldots + K_iR_i$$

where

C is the concentration
K_0 is a constant characteristic of the individual instrument and varies according to the analyte, independent of the wavelength
K_i is a constant of the instrument, a given compound at a specific wavelength
R_i is an expression of the spectral measure at wavelength L_i

For each analyte, 2–10 characteristic wavelengths are selected. The quality of the calibration is then tested using a separate set of reference wine and must samples.

12.4.3.2 Routine Calibrations

It is necessary to re-calibrate the instrument periodically, notably when routine control indicates an instrument-related drift due to factors such as aging of the electronic components, reparations, changing of spare parts, etc. Re-calibrating the instrument does not modify the measurement wavelengths but does involve recalculation of the K_0 and K_i constants.

12.4.3.3 Routine Correction of Bias

Each time the equipment is used, control sample(s) containing a known concentration of the analyte are analysed. If the there is bias in the result by comparison with the expected value, a correction may be carried out.

12.4.4 Usual NIR Methods

12.4.4.1 Determination of Ethanol in Wines and Musts During Fermentation

This is a very widely used determination, but suffers from the drawback of matrix effects, for instance if the wine contains atypical concentrations of glycerol, as was first reported by Bouvier (1986).

Characteristics of the Method

Inter-laboratory reproducibility: 0.19vol.%

12.4.4.2 Determination of Reducing Sugars in Musts

This is a less widely-used application. No inter-laboratory studies have been carried out to describe characteristics of the method.

12.5 Fourier-Transform Infra-Red Analysers (FTIR)

According to Dubernet et al. (1999, 2000, 2001, 2006), Dubernet and Dubernet, (2000), Bertrand and Dufour (2000), Patz et al. (1999), and Boulet et al. (2007a, b).

12.5.1 Material and Principle of the Method

The apparatus consists of an interferometer using Fourier transform infra-red spectrophotometry covering the range 2000–10000 nm which corresponds to part of the mid infra-red range of the electromagnetic spectrum. After calibrating the instrument for different organic compounds, spectral analysis of the data enables the simultaneous determination of several analytes in must or wine.

12.5.2 Interferometry and Fourier Transform

Interferometry is an alternative method to classic techniques of spectral acquisition, which tend to be onerous and time-consuming if fine resolution is required. It enables all wavelengths emitted by a single infrared source to be treated simultaneously in real time. As preliminary wavelength selection is not required, complete spectra may be obtained in less than 1 s. The first step (Fig. 12.1) consists of producing an interferogram of the sample to be analysed. The interferogram is based on the separation of polychromatic infra-red light (emitted by an incandescent filament) on a blade. Before arriving at the detector the two components of the split signal follow different pathways – one part traverses the sample directly while the other part is reflected against a mobile mirror before passing through the sample.

Each elementary wavelength of the infra-red radiation arrives at the detector as two component parts with a phase difference of **p**. By the action of the mobile mirror, this phase difference will vary continually during measurement. The final signal

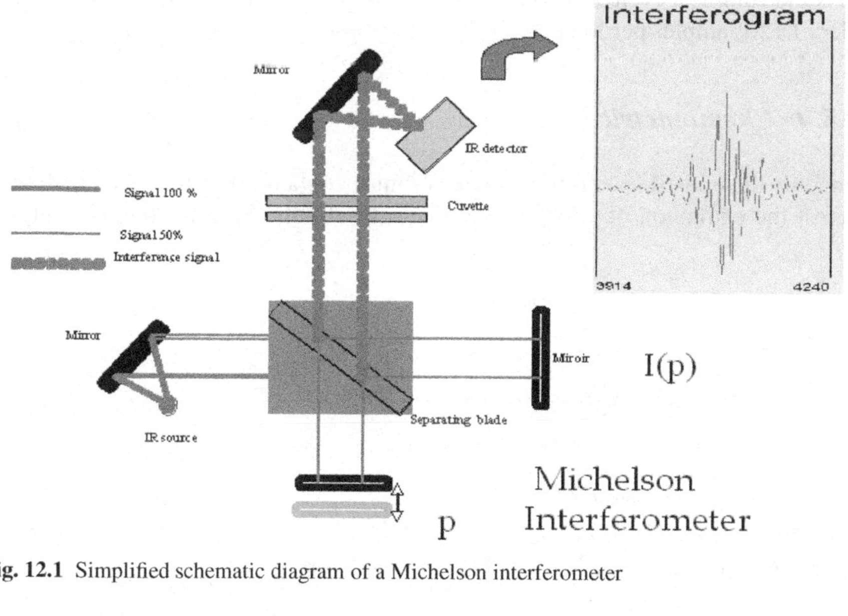

Fig. 12.1 Simplified schematic diagram of a Michelson interferometer

obtained is thus is two light beams of the same wavelengths with a phase interference of **p**. According to the phase difference recombination of the interference signals is constructive or destructive. In other words, the intensity of the interference signal will vary according to the value of **p**. The plot of the variation in intensity as a function of the phase is called an interferogram and its mathematical model is an integral. Fourier transform is a mathematical procedure which converts the interferogram into a signal the intensity of which varies according to wavelength and which in turn enables the infra-red spectrum to be constructed. This is an intricate calculation which can only be carried out using powerful computing techniques.

The spectrum thus obtained enables the technique to be used as a classical method of spectrophotometric analysis, in which specific absorption wavelengths of organic compounds are employed to determine concentrations of the target analytes.

12.5.3 Installation and Acquisition of Infra-Red Spectra

The sample to be analysed does not require any particular sample pre-treatment. However, in the case of musts or cloudy wines, a preliminary clarification by centrifugation or filtration is recommended in order to prevent system blockage. Carbon dioxide levels exceeding 750mg/L should be reduced or eliminated before analysis to avoid degassing of the sample in the analytical circuit.

The circuit (Fig. 12.2) starts with a sampling needle. A peristaltic pump transfers the sample into a heating chamber where it is heated to 40°C. After passing through a filter, the sample is transferred into the measurement cuvette which is constructed of fluorosilicate as glass or quartz absorb strongly in the mid infra-red. After analysis, the sample is ejected to waste.

As the complete cycle lasts only 30s, with automation, the analysis rate is of the order of 120 samples per hour.

12.5.4 Chemometrics

Generally the mid-IR spectrum of a wine or must contains information of analytical interest the extraction of which in the majority of cases requires highly complex

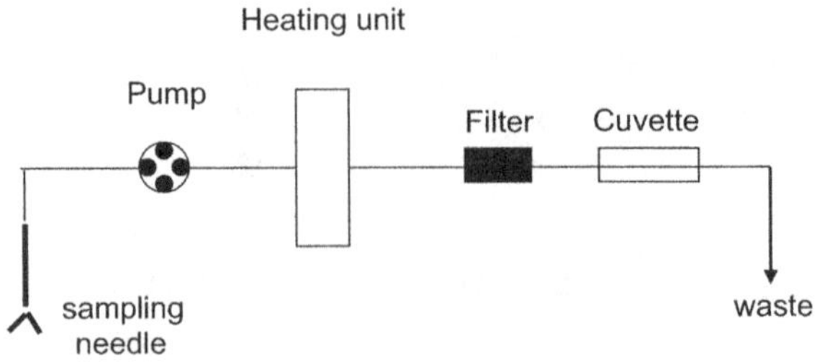

Fig. 12.2 Simplified schematic diagram of the analytical circuit

mathematical treatment. The methods used to extract this information form part of a relatively recent scientific technique, chemometrics. This has been defined by GELADI and DABAKK, (1995) as "the science of the use of mathematical, statistical and computing methods which aim to extract useful information from data obtained by chemical measurements".

Analysts usually have two principal objectives, the first of these being an exploratory method which enables spectra to be classified into pre-defined families of compounds. Such a tool enables the sample to be identified as a must, a must in fermentation, a dry wine, a liqueur wine or a naturally sweet wine. The second objective involves a quantitative approach which enables the attribution of analytical values or indices particular to the wine or must on the basis of the previously acquired reference data (calibration). It is above all this second approach which is used by analytical laboratories where is possible to replace classical analytical techniques by FTIR.

There are several chemometric methods applicable which provide numerical analytical data and their implementation remains a specialist affair. They vary from simple statistical tools such as principal component analysis (PCA) to highly complex methods such as neural networks.

All manufacturers offer optical measurement tools, the performance of which are in general satisfactory, though methods of standardisation of the measurement, tools for the treatment of spectral data and the quality of the chemometric tools available are usually more variable. The quality of the chemometrics used to obtain spectral data of a wine or a must depend on the reliability and the precision of the analytical data.

12.5.5 Analysis of Near Infra-Red Spectra

It is useful to take into consideration the great complexity of the information contained in a mid-IR spectrum of a wine or a must. This complexity arises from several factors:

1. The organic and mineral composition of these products which is one of the richest found in the agri-food industry
2. The absorption bands of organic molecules are multiple, particularly those arising from significant resonance effects
3. There are several types interactions between the different organic compounds resulting largely from matrix effects which are impossible to model

The procedure applied in the experimental calibration approach consists of acquiring spectral data for the maximum of sample varieties of precisely known values for the compound or indices to be measured, before applying mathematical statistical methods for quantification It is essential to understand that this step is fundamental for the construction of a high-performance analytical tool. Not only must the operator have at his or her disposal the greatest number of representative samples possible, but he/she must apply the most appropriate chemometric methods and if possible, carry out critical comparative analyses of the results obtained by these methods. Such an approach should always be validated by a rigorous statistical comparison of the results obtained using the selected model with measurements obtained by classical analytical techniques. Several thousand determinations may be required for this last step. Calibration of an FTIR instrument is therefore an onerous undertaking which demands significant time and investment. To understand better the difficulty of the approach, it is appropriate to specify the following critical points.

12.5.5.1 Analyte Behaviour

Not all the organic compounds of wine behave in the same manner analytically for several reasons.

First, it is clear that compounds at elevated concentrations (above 1g/L, for instance) are likely to be more easily measured by virtue of the fact that they will exhibit strong absorbance due to their concentration.

Second, the absorbance capacity of a compound depends on its molecular structure. Certain compounds, such as carbon dioxide absorb strongly in the mid-IR, whereas other compounds, even those of substantial molecular complexity, have less intensive absorbance.

Third, the influence of the other wine compounds (matrix effects) is far from negligible. It is obvious that if the spectral bands of a target compound overlap with those of water or ethanol, present in high proportions, the measurement sensitivity for the compound in question will be severely compromised. Similarly, interactions between constituents may lead to displacement of absorption bands.

It is easily acknowledged that here, again, theoretical knowledge is not of great use and only an experimental approach can establish whether a compound can readily be determined by FTIR. For instance, the determination of sugars is less reliable for values less than 1g/L; it is evident that the more inaccessible the analyte, the more complex the calibration and the more it needs to be executed with attention and rigour.

12.5.5.2 Extrapolation

This is an essential point. In executing the calibration, it is not possible to assume that extrapolation of the system is possible. Therefore, if the initial sampling does not cover the entire analytical domain equally, the results obtained are less rich in information and will thus be qualitatively deficient. By the same token, if there is a significant matrix effect which arises in cases of matrices rich in polyphenols, it is essential that the instrument is calibrated for wines of all probable polyphenol concentrations at each concentration point for each analyte. If this is not the case that part of the curve which is not calibrated for all possible phenol concentrations will be more sensitive to matrix effects resulting in less reliable measurements.

12.5.5.3 Matrix Effects

It has been outlined above that matrix effects are due to other compounds present in the analytical medium which absorb in the mid-IR. It is possible, however, to circumvent matrix effects. As in expert systems, all that is required is the availability of the necessary information so that the instrument can, by experience, measure the impact of the matrix effect on the determination of the target compound. Matrix effects may thus be eliminated by reference to the calibration value whenever the analyte occurs in an unknown sample type. At this stage, it is obvious that the greater the number of different matrices presented to the instrument during calibration, the more robust the method will be, and less subject to errors arising from matrix effects. This is one of the strong points of the FTIR method, but nonetheless there in an inherent disadvantage. The weakness lies in the fact that the more elaborate of the instrument's capacity to manage matrix effects, the lower its performance in terms of sensitivity and precision. Therefore, the operator developing the calibration is obliged to seek the optimum compromise between robustness, sensitivity and precision. Here again, the approach rapidly becomes extremely complex and its implementation can only be envisaged if significant means are available for that purpose. In oenological practice, and the level of understanding permitting, wines and musts are grouped into five basic classes of matrices:

Unfermented musts
Musts in fermentation
Dry wines or wines containing low sugar concentrations
Liqueur wines
Naturally sweet wines

It is important to point out that matrix effects engendered by different grape varieties, wine colour or geographical origin are in general of no consequence in well-executed calibrations.

The complexity of matrix effects has another important consequence: it is impossible to use synthetic matrices or even samples spiked with the analytes, even for the controls – only natural wines and musts can be applied for this purpose.

12.5.5.4 Reference Values

The quality of a calibration also depends on the quality of the reference samples used. Glucuronic acid is a particular example; the enzymatic method for this analysis is not reliable in that it lacks robustness. This problem has not been identified in the literature and is probably linked to non-identified matrix effects. The first FTIR calibrations were carried out using the enzymatic method as a reference and, even though gluconic acid has highly qualitative absorption characteristics, the results obtained by FTIR were unreliable. Subsequent calibrations carried out using more reliable methods such as capillary electrophoresis as a reference technique have resulted in FTIR analyses of high quality in terms of precision and accuracy.

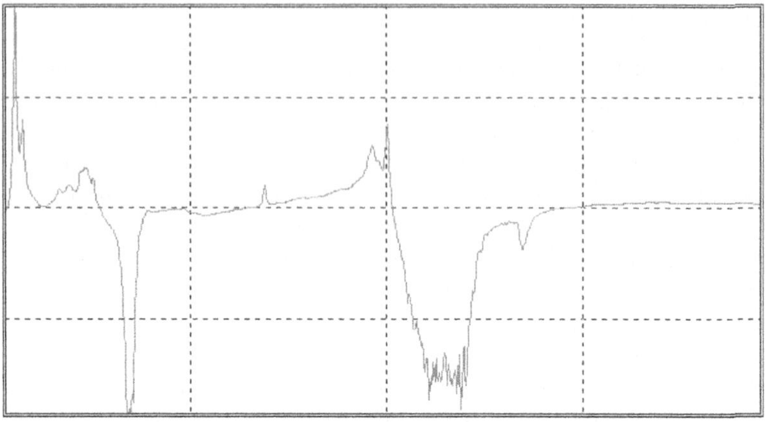

Fig. 12.3 Example of a wine mid-IR spectrum

12.5.6 Principal Steps in Spectral Analysis

An example of a wine IR spectrum is presented in Fig. 12.3.

Based on the data base constructed with reference samples of known values for the target analytes, chemometric methods enable the most appropriate spectral zones (usually about 10) to be selected and then a regression model is calculated which is used to determine the concentration of unknown samples.

12.5.7 The Importance of Quality Control of the Results

It is important to point out that the use of FTIR requires that the laboratory implements a system to control the quality of the results, and this system ought to be based on a specific and original strategy. The highly powerful automation of the system allied to the simplicity of instrument operation constitute a risk in terms of drift or of results that are either aberrant of unidentified. The quality of results can only be assured by a very thorough system with strict identification of matrices, the elimination of results of samples that do not comply with precise definition, the use of a sufficient number of control samples and perfect data management. Each laboratory should be aware that the time spent in controlling the quality of results in FTIR is significantly greater than the analysis time, which is not the case with other methods used in oenological laboratories.

It is now accepted that laboratories which have implemented the necessary measures to ensure quality control of results obtain, above and beyond an unequalled facility of operation, considerable optimisation of laboratory performance, and an overall quality of analysis that is vastly superior to that offered by other automated methods. It is strongly recommended that laboratories employing FTIR in routine analysis retain alternative methods for each significant wine parameter to be analysed.

12.5.8 Usual IR Methods

The number of parameters that may be determined by FRIR is constantly increasing due to the continuous development of new calibrations. The most important methods currently available may be divided into three distinct categories.

12.5.8.1 "Direct" Methods

These are methods involving the determination of organic compounds for which there is a direct relationship between molecular absorbance and concentration. Routinely used methods are presented in Table 12.1.

12.5.8.2 "Indirect" Methods

In oenology two different cases may arise:

1. The target compound has little or no absorbance in the mid-IR either because of its native structure or because it is present at low concentrations
2. The target compound itself is not determined; rather an index reflecting a group of organic compounds is measured

The determination of potassium is an example of the first case. As it obviously does not absorb in the IR, compounds to which its concentration is correlated are detected and the correspondence obtained by chemometric approaches.

An example of the second case is the determination of pH or titratable acidity, which are reliably determined by the measurement of the complex equilibrium of the organic compounds that determine their values. Routinely used methods are presented in Table 12.2.

12.5.8.3 "Biological" Methods

These methods involve the measurement of indicative values based on biological data identifiable in the mid IR.

Routine application are as follows.

Table 12.1 Usual methods of determination in oenology by near IR analysers

Analyte	Measurement scale	Intra-laboratory reproducibility	Inter-laboratory reproducibility
Glucose and fructose	0–400 g/L^{--}	1.3g/L	
Malic acid	0.3–6 g/L	0.15g/L^{--}	0.6g/L^{--}
Lactic acid	0.3–6 g/L^{--}	0.15g/L	0.6g/L
Carbon dioxide	60–1300 mg/L^{--}	95mg/L^{--}	230mg/L
Glycerol	0.2–10 g/L	10%	
Tartaric acid	1–15 g/L^{--}	10%	
Gluconic acid	0.1–15 g/L	10%	

Table 12.2 Indirect methods of determination in oenology by near IR analysers

Method	Measurement scale	Intra-laboratory reproducibility	Inter-laboratory reproducibility
Density	0.0012–1.40g/cm^{--3}	0.00015 g/cm^3	0.0010 g/cm^3
Alcoholic volume	8–16vol.%	0.10vol.%	0.21vol.%
Titratable acidity	3–33g/L as tartaric acid	0.11 g/L as tartaric acid	0.15 g/L as tartaric acid
Volatile acidity	0.30–0.96g/L as acetic acid	0.036 g/L as acetic acid	0.12 g/L as acetic acid
PH	2.6–4.7	0.06	0.12
Potassium	600–8000 mg/L	10%	
Total polyphenol index	15–100	14	
Ammoniacal nitrogen	5–250 mg/L	16%	
α-Amino nitrogen	5–250 mg/L	10%	

Botrytis Cinerea Index of Musts

When this parasite invades the grape-berry, it leaves a characteristic metabolic footprint which has been shown to be closely correlated with the degree of infection and the associated technological problems. Although the correlation is not linear it may nonetheless be modelled using neural networks and thus it is quite possible to express the degree of Botrytis infection in the form of an index.

The Fermentation Index of Musts

The fermentation activity of grapes or musts, which for example may start during transport from the vineyard to the cellar, may be determined by an index that measures the organic compounds metabolised during fermentation.

Acetobacter Index

Acetobacter attack of grapes results in acid rot which has serious consequences in terms of harvest quality. The level of *Acetobacter* attack is determined by indices that measure the compounds produced by this micro-organism.

Aspergillus Carbonarius in Musts

This fungal parasite is responsible for the production of ochratoxin A (OTA), the presence of which is identified by the determination of related compounds.

The Determination of Ochratoxin A (OTA) in Wines

OTA is an organic compound that does have IR absorbance; however, it is present in concentrations (some μg/L) that are too low to permit its direct determination. In this particular case it has been demonstrated that *Aspergillus carbonarius*, the organism responsible for the production of OTA, also produces other organic compounds such as citric and gluconic acids and glycerol in correlated and measurable concentrations. The determination of these compounds in the IR produces a value indicative of the concentration of OTA in wine.

References

Battle, J.L., & Bouvier, J.C. (1986). Automatisation of determinations enzymatiques pour analyse oenologique. *Rev. Fr. Oenol.*, 101, 38–43.

Battle, J.L., Joubert, R., Colion, Y., & Jouret, C. (1978a). Determination enzymatique en flux continuo of the L-malate et of the L-lactate in musts of raisin and wines. *Ann. Fals. Exp. Chim.*, 71(766), 223–228.

Battle, J.L., Joubert, R., Colion, Y., & Jouret, C. (1978b). Utilisation of theflux continu pour the determination colorimetrique of acide tartrique inmusts of raisin and wines by method of Rebelein. *Ann. Fals. Exp. Chim.*, 71(764), 155–158.

Bertrand, D., & Dufour, E. (2000). The spectroscopie infrarouge and ses applications analytiques. Technique and documentation Ed. Technique & Documentation.

Boulet, J.C., Williams, P., & Doco, T. (2007a). A Fourier transform infrared spectroscopy study of wine polysaccharides, *Carbohyd. Polym.*, 69, 79–85.

Boulet, J.C., Doco, T., & Roger, J.M. (2007b). Improvement of calibration models using two successive orthogonal projection methods. Application to quantification of wine polysaccharides, *Chemometr. Intell. Lab.*, 87, 295–302.

Bouvier, J.C. (1986). Influence of theglycerol sur the mesureof the degre alcoolique by infrarouge.*Rev. Fr. Oenol.*, 102, 21–22.

Bouvier, J.C. (1993). Analyse Automatique en Œnologie. *FV OIV* 951

Bouvier, J.C., Barros, P., Boulanger, P., Curvelo-Garcia, A.S., & Dubernet, M. (1995). Methods d'analyses automatiques des wines and des musts.*FV OIV* 993.

Cabanis, M.T., Cabanis, J.C., Raffy, J., Peyronnenche, M., Viotte, O., & Leboeuf, J.P. (1983). Infralyseur 400: Une determination automatique of the titre alcoometrique des wines. *Rev. Fr. Oenol.*, 89, 75–79.

Curvelo-Garcia, A.S. (1993). The fiabilite des methods of flux continu segmente appliquees to analyse des wines and des musts. *FV OIV* 941.

Curvelo-Garcia, A.S., & Godinho, M.C. (1988). the determination colorimetrique en flux continu des sucres reducteurs in wines. *Sciencia Tec. Vitiv.*, 7, 5–10.

Doneche, B., & Sanchez, P.J. (1985). Automatisation en flux continu of the determination enzymatique of acide acetique in wines.*Conn. Vigne Vin.* n°3, 161–169.

Dubernet, M. (1976). Determination automatique of acidite volatile in wines. *Conn. Vigne Vin*, 10(3), 297–309.

Dubernet, M. (1997). automatisation of analyse chimique en œnologie. *Rev. Fr. Oenol*, 66, 45–61.

Dubernet, M. (2008). Analyser automatiques en œnologie. *FV OIV* 1276.

Dubernet, M., & Dubernet, M. (2000). Utilisation of analyse infrarouge to transformee of Fourier pour analyse œnologique of routine. *Rev. Fr. Oenol*, N°181, 10–13.

Dubernet, M., Pennequin, F., & Grasset, F. (1995a). Adaptation of the determination of sulphur dioxide total in wines sur sequential analyser Hitachi 717.*FV OIV* 997.

Dubernet, M., Pennequin, F., & Grasset, F. (1995b). Determination of sulphur dioxide libre in wines sur sequential analyser Hitachi 717. *FV OIV* 998.

Dubernet, M., Pennequin, F., & Grasset, F. (1997). Adaptationof the determination of acide acetique in wines sur sequential analyser Hitachi 717, *FV OIV* 1001.

Dubernet, M., Leboeuf, J.P., & Grasset, F. (1998). Method automatisee of determination colorimetriqueof sulphur dioxide total in wines.*FV OIV* 1068.

Dubernet, M., Dubernet, M., & Grasset, F. (1999). Utilisation of analyse infrarouge multiparametrique to transformee of Fourier en œnologie. *FV OIV* 1089.

Dubernet, M., Dubernet, M., Dubernet, V., Coulomb, S., Lerch, M., & Traineau, I. (2000b). Analyse objective of the qualite des vendanges by spectrometrie infra-rouge to transformee of Fourier (IRTF). *Congrès OIV 2000, Paris.*

Dubernet, M., Dubernet, M., Grasset, F., & Garcia, A. (2001). Analyse of nitrogen assimilable inmusts by Interferometrie Infrarouge to Transformee of Fourier, *Rev. Fr. Oenol.*, N°187, 9–13.

Dubernet, M., Dubernet, M., & Sage, L. (2006). Modification of the composition des musts and des wines issus of raisins parasites by *Aspergillus carbonarius*. Evaluation rapide of the contamination of the raisin and of the teneur en Ochratoxine A in musts and wines by infrarouge to transformee of Fourier. *Rev. Fr. Oenol.*, N°216.

Ellman, G.L. (1959). For the modification and quantitative detection of sulfhydryl groups. *Arch. Biochem. Biophys.*, 82, 70.

Geladi, P., & Dabakk, E. (1995). An overview of chemometrics application in NIR spectrometry. *J. NIR Spectrosc.*, 3, 119–132.

McCloskey, L. P. (1980).An improved enzymatic assay for acetate in juice and wine. *Am. J. Enol. Vitic.*, 31, 170–173.

OIV, (1994). Recueil des methods d'analyse des boissons spiritueuses. O.I.V ed Paris

Patz, C.-D., David, A., Thente, K., Kurbel, P., & Dietrich, H. (1999). Wine analysis with FTIR spectrometry. *Wein Wissenschaft*, 54, 80–87.

Pilone, G.J. (1967). Effect of lactic acid on volatile acid determination of wine. *J. Oenol.*, 18, 149–156.

Sarris, J., Morfaux, J.N., & Derwine, L. (1970). Determination automatique of acidite volatileof thewine. *Ind. Alim. Agric.*, 87, 115–121.

Scholten, G., Woller, R., & Holbach, B. (1982). Automatisierte weinanalyse, 2. *Mittteilung. Die Wein. Wiss.*, 38, 397–426.

Trossais, J., & Asselin, C. (1985). Influence des teneurs en malic acid des musts sur the determination of acide tartrique by analyse en flux continu by colorimetrie with the metavanadate. *Conn. Vigne Vin.*, 19(4), 249–259.

Chapter 13
Statistical Techniques for the Interpretation of Analytical Data

Pedro J. Martín-Álvarez

Contents

In this chapter, I will try to explain some of the statistical methods (univariate, bivariate and multivariate) most used by our group, in the hope that it will be useful and accessible for the majority of readers, given that the emphasis is on comprehension of the principles of the methods, their applications and interpretation of the results obtained.

13.1 Univariate Statistical Techniques

In this first section, we will consider the statistical methods to process data, originating in the observation of *a single continuous random variable*. We will distinguish three possible situations, with one, two or more than two data sets of the observed variable. In the last case, we will present the Analysis of Variance (ANOVA) for one or more factors.

P.J. Martín-Álvarez (✉)
Instituto de Fermentaciones Industriales, Consejo Superior de Investigaciones Científicas (CSIC), C/Juan de la Cierva 3, 28006 Madrid, Spain
e-mail: pjmartin@ifi.csic.es

M.V. Moreno-Arribas, M.C. Polo (eds.), *Wine Chemistry and Biochemistry*,
DOI 10.1007/978-0-387-74118-5_25, © Springer Science+Business Media, LLC 2009

13.1.1 Statistical Treatment for Only One Sample

Let $\{x_1, x_2, x_3, \ldots, x_n\}$ be a random sample of n observations of a continuous random variable X, from a population where X has a mean μ and a standard deviation σ. We have the *sample mean* $(\bar{x} = \sum x_i/n)$ and the *sample standard deviation* $\left(s = \sqrt{\sum (x_i - \bar{x})^2/(n-1)}\right)$. The median, the lower (Q_1) and upper (Q_3) quartiles, the relative standard deviation $\left(RSD(\%) = CV(\%) = 100\frac{s}{\bar{x}}\right)$, and the standard error $(s/\sqrt{n})$ can also be calculated. For the graphical processing we can use the histogram or box plot (with *min*, Q_1, *median*, Q_3, and *max* values, or with $Q_1 - 1.5(Q_3 - Q_1)$, Q_1 , *median*, Q_3, $Q_3 + 1.5(Q_3 - Q_1)$ values to show potential outliers). Accepting X is a normally distributed random variable with mean μ and standard deviation σ $(X \sim N(\mu, \sigma))$, that can be verified with the *normal probability plot* or with the *normality tests* (Shapiro-Wilks, Kolmogoroff-Smirnov-Lilliefors, etc.), and since $\bar{x}$ and s are estimators of μ and σ, respectively, with a fixed *significance level* α (e.g. $\alpha = 0.05$), we can calculate the corresponding *confidence intervals* at $100(1 - \alpha)\%$ for μ : $\left[\bar{x} - t_{1-\alpha/2,n-1}s/\sqrt{n}, \bar{x} + t_{1-\alpha/2,n-1}s/\sqrt{n}\right]$ and for σ^2 : $\left[(n-1)s^2/\chi_{1-\alpha/2,n-1}, (n-1)s^2/\chi_{\alpha/2,n-1}\right]$, where $t_{1-\alpha/2,n-1}$ is the critical value of the t-Student distribution with $n - 1$ degrees of freedom (df) such that $prob(t_{n-1} \leq t_{1-\alpha/2,n-1}) = 1 - \alpha/2$, $\chi^2_{\alpha/2,n-1}$ and $\chi^2_{1-\alpha/2,n-1}$ are the critical values of the χ^2 -distribution with $n - 1$ df, such that $prob(\chi^2_{n-1} \leq \chi^2_{\alpha/2,n-1}) = \alpha/2$ and$prob(\chi^2_{n-1} \leq \chi^2_{1-\alpha/2,n-1}) = 1 - \alpha/2$.

13.1.1.1 Hypothesis Test for a Mean or One-Sample T Test

To test the null hypothesis $H_0 \equiv \mu = \mu_0$ against the *two-sided* alternative hypothesis $H_1 \equiv \mu \neq \mu_0$ (H_0 can be rejected equally by $\mu < \mu_0$ or by $\mu > \mu_0$), we can use the statistic: $t_{cal} = \dfrac{\bar{x} - \mu_0}{s/\sqrt{n}}$ that has a t-Student distribution with $n - 1$ df, if H_0 is true. For a fixed value of α, if $|t_{cal}| > t_{1-\alpha/2,n-1}$, $H_0 \equiv \mu = \mu_0$ is rejected and $H_1 \equiv \mu \neq \mu_0$ is accepted (the test is *statistically significant at level* α); otherwise ($|t_{cal}| \leq t_{1-\alpha/2,n-1}$), there is no reason to reject H_0 (the test is *statistically nonsignificant*). With the associated probability ($P = 2prob(t_{n-1} > |t_{cal}|)$), facilitated by the statistical programs, if the P value is less than α then the null hypothesis ($H_0 \equiv \mu = \mu_0$) is rejected, otherwise ($P > \alpha$) $\mu = \mu_0$ is not rejected. In the case of the *one-sided* alternative hypothesis (e.g. $H_1 \equiv \mu > \mu_0$ or $H_1 \equiv \mu < \mu_0$), if $|t_{cal}| > t_{1-\alpha,n-1}$, or if $P = prob(t_{n-1} > |t_{cal}|) < \alpha$ then H_0 should be rejected.

13.1.1.2 Example of Application

The one-sample t test can be used to test for systematic errors in the analytical method for a standard material with a knownconcentration μ_0 ($H_0 \equiv \mu = \mu_0$ *vs* $H_1 \equiv \mu \neq \mu_0$), or to verify if the changes in the elaboration process of a certain product affect the previous concentration μ_0 of a compound ($H_0 \equiv \mu = \mu_0$ *vs* $H_1 \equiv \mu > \mu_0$) (Massart et al. 1990; Miller and Miller 2000; Martín-Álvarez 2000, 2006). Table 13.1 shows the results of the one-sample t test

Table 13.1 Results of the one-sample t test for 15 replicate values from a new analytical method for a standard material with known concentration of 0.34 μg (Massart et al. 1990)

Mean	SD	N	Reference constant (μ_0)	t-value	df	P-value
0.316	0.056	15	0.34	−1.67	14	0.117*

t-value= value of the statistic t_{cal}, df= degrees of freedom, P= associated probability
*The two means μ and μ_0 are not different (P>0.05).

for 15 replicate values from a new analytical method for a standard material with known concentration of 0.34 μg (Massart et al. 1990), obtained with *t-test for single mean* procedure in the *Basic Statistics and Tables* module of STATISTICA program version 7.1 (StatSoft, Inc., http://www.statsoft.com/). Since the *P*-value associated with the t-value is greater than 0.05, we cannot reject the null hypothesis at the 95% confidence level, and there are no systematic errors in this analytical method for this sample. The same conclusion is obtained from the 95% confidence interval for μ (0.285, 0.347) that includes the reference value. Using the *Descriptive statistics* procedure in the same module, the value of the W statistic in Shapiro-Wilk's test is nonsignificant, (W = 0.970, P = 0.86 > 0.05), and the normality could be accepted (also from Fig. 13.1).

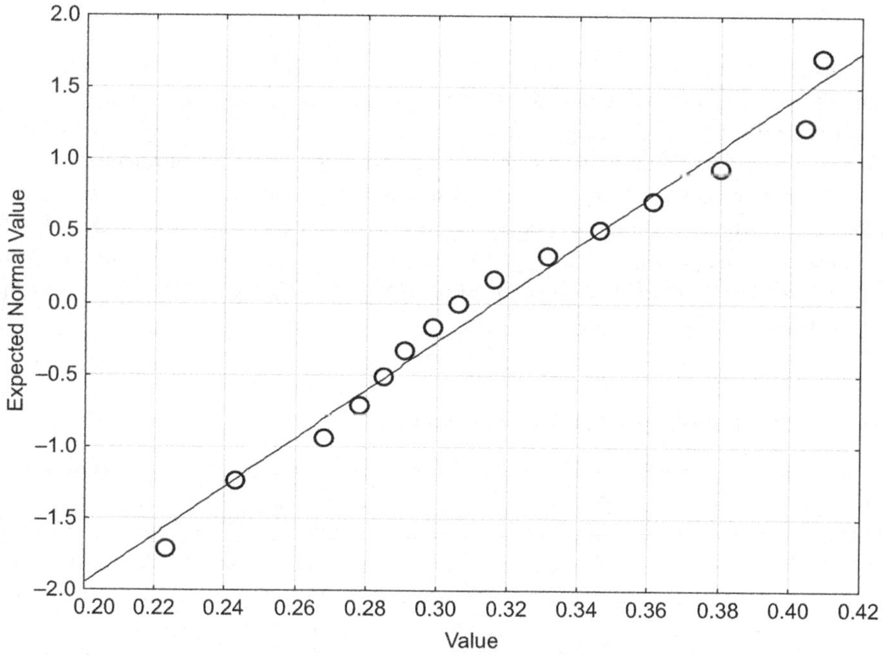

Fig. 13.1 Normal probability plot for data

13.1.2 Statistical Treatment to Compare Two Independent Samples

Let $\{x_{1,1}, x_{2,1}, x_{3,1}, \ldots, x_{n_1,1}\}$ and $\{x_{1,2}, x_{2,2}, x_{3,2}, \ldots, x_{n_2,2}\}$ be two independent random samples of n_1 and n_2 observations of a continuous random variable X, from two populations W_1 and W_2 where X has mean values μ_1, μ_2 and standard deviation values σ_1, σ_2, respectively. We have the descriptive values $\bar{x}_j, s_j\ j = 1, 2$. For the graphic processing we can use the histograms or the box plots. Accepting normal distributions ($X \sim N(\mu_j, \sigma_j)\ \ in\ \ W_j$), and for a *significance level* α, we can calculate the corresponding confidence intervals for the parameters μ_j, σ_j, or to test hypotheses about them.

13.1.2.1 Hypothesis Test for Two Means or Two-Sample T Test

To test the null hypothesis $H_0 \equiv \mu_1 = \mu_2$ against the alternative hypothesis $H_1 \equiv \mu_1 \neq \mu_2$, assuming normal distributions and equality of variances ($\sigma_1^2 = \sigma_2^2$), the test statistic is $t_{cal} = \dfrac{\bar{x}_1 - \bar{x}_2}{s_p\sqrt{\frac{1}{n_1} + \frac{1}{n_2}}}$ which, under H_0, has a t-Student distribution with $\nu = n_1 + n_2 - 2$ df, and where $s_p^2 = \frac{(n_1-1)s_1^2+(n_2-1)s_2^2}{n_1+n_2-2}$ is the pooled sample variance. When the variances are not equal, the test statistic is: $t_{cal} = \dfrac{\bar{x}_1 - \bar{x}_2}{\sqrt{\frac{s_1^2}{n_1} + \frac{s_2^2}{n_2}}}$ which, under H_0, has a t-Student distribution with $\nu = \dfrac{(s_1^2/n_1 + s_2^2/n_2)^2}{(s_1^2/n_1)^2/(n_1+1) + (s_2^2/n_2)^2/(n_2+1)}$ $- 2$ df. For a fixed value of α, if $|t_{cal}| > t_{1-\alpha/2,\nu}$ then $H_0 \equiv \mu_1 = \mu_2$ is rejected and $H_1 \equiv \mu_1 \neq \mu_2$ is accepted, otherwise ($|t_{cal}| \leq t_{1-\alpha/2,\nu}$), H_0 should not be rejected. With the associated probability ($P = 2prob(t_\nu > |t_{cal}|)$), if $P < \alpha$ then $H_0 \equiv \mu_1 = \mu_2$ is rejected, otherwise ($P > \alpha$) $H_0 \equiv \mu_1 = \mu_2$ is not rejected.

The nonparametric Mann-Whitney U-test for the comparison of two independent sets of data, without accepting normal distributions, uses the ranks of all the observations, previously arranged from the lowest to the highest value, to test $H_0 \equiv median_1 = median_2$ (Massart et al. 1990; O'Mahony 1986). If $P < \alpha$, H_0 is rejected.

13.1.2.2 Hypothesis Test for Two Variances

To test the null hypothesis $H_0 \equiv \sigma_1^2 = \sigma_2^2$ against $H_1 \equiv \sigma_1^2 > \sigma_2^2$ (one-tail), and assuming normal distributions and $s_1^2 > s_2^2$, the test statistic is $F_{cal} = s_1^2/s_2^2$ which, under H_0, has an F-distribution with $n_1 - 1$ and $n_2 - 1$ df. For a fixed value of α, if $F_{cal} < F_{1-\alpha, n_1-1, n_2-1}$, $H_0 \equiv \sigma_1^2 = \sigma_2^2$ should not be rejected, otherwise ($F_{cal} > F_{1-\alpha, n_1-1, n_2-1}$) $H_1 \equiv \sigma_1^2 > \sigma_2^2$ is accepted. With the associated probability ($P = prob(F_{n_1-1,n_2-1} > F_{cal})$), if $P < \alpha$ then the variances are significantly different. This test can be used to compare the precision of two analytical methods for the same sample.

Table 13.2 Mean and the standard deviation (SD) values for octanoic acid in white and rosé wines and the results of the two-sample t test

White wines (n=8)		Rosé wines (n=8)		Assuming equal variances			No assuming equal variances			Test of variances	
Mean	SD	Mean	SD	t-value	df	P	t-value	df	P	F-value	P
2.54	0.92	8.26	3.01	–5.14	14	.0002	–5.14	8.31	.0008**	10.61	.0059*

t-value= value of the statistic t_{cal}, df= degrees of freedom, P= associated probability
*the two variances are different (P<0.05).
**the two means are different (P<0.05).

13.1.2.3 Example of Application

The two-sample t test can be used to compare the results obtained by two laboratories for the same sample of reference, to compare the concentrations of a certain compound in wines elaborated with grapes of two varieties, etc., and in a general way, to compare the mean values of two groups of independent observations (Miller and Miller 2000; Massart et al. 1990; Martín-Álvarez 2000, 2006). Table 13.2 shows the mean and the standard deviation values for octanoic acid in white and rosé wines and the results of the two-sample t test, obtained with the STATISTICA program (procedure *T-Test for Independent samples*, in the *Basic Statistics and Tables* module). The results of the test of variances are also included in the table, and since the *P*-value is less than 0.05, the variances are significantly different. The rosé wines have a higher octanoic acid content than the white wines ($P < 0.05$).

13.1.3 Statistical Treatment to Compare Two Related Samples

Let $\{(x_{1,1}, x_{1,2}), (x_{2,1}, x_{2,2}), \ldots, (x_{n,1}, x_{n,2})\}$ be a random sample of n paired observations of two continuous random variables X_1 and X_2, from a population W where the variables have mean values μ_1, μ_2. From these n paired observations, n differences may be calculated $\{d_i = x_{i,1} - x_{i,2}\}$, with descriptive values $\bar{d}$ and s_d. Accepting normality of the differences, to test the null hypothesis $H_0 \equiv \mu_d = 0$ (or $H_0 \equiv \mu_1 = \mu_2$), against $H_1 \equiv \mu_d \neq 0$ (or $H_1 \equiv \mu_1 \neq \mu_2$), the test statistic is $t_{cal} = \dfrac{\bar{d}}{s_d/\sqrt{n}}$, which has a t-distribution with $n - 1$ df. For a fixed value of α, if $|t_{cal}| > t_{1-\alpha/2,n-1}$, or if $P < \alpha$, $H_0 \equiv \mu_d = 0$ should be rejected and $H_1 \equiv \mu_d \neq 0$ accepted, otherwise, $|t_{cal}| < t_{1-\alpha/2,n-1}$ or $P > \alpha$, $H_0 \equiv \mu_d = 0$ is not rejected.

If a normal distribution is not assumed, the Wilcoxon matched-pairs signed-ranks test can be used to test $H_0 \equiv \mu_d = 0$.

13.1.3.1 Applications

This t test for related samples can be used to compare a new analytical method and the reference method (Massart et al. 1990), to compare the amino acid content between different stages of the elaboration processes for wines from the same batches (Marcobal et al. 2006b), and more generally, to compare paired samples.

Table 13.3 Mean and the standard deviation values of the histamine concentrations in 30 commercial wines analyzed by a direct ELISA and an HPLC method, and the results of t test for related samples

	Mean	SD	n	Diff. ($\bar{d}$)	SD. Diff. (s_d)	t-value	df	P
HPLC	4.51	3.90						
ELISA	5.49	3.91	30	–0.978	1.722	–3.110	29	0.0042*

t-value= value of the statistic t_{cal}, df= degrees of freedom, P= associated probability
*the two means are different (P<0.05).

Table 13.3 shows the mean and the standard deviation values of the histamine concentrations in 30 commercial wines analyzed by a direct ELISA and HPLC methods (Marcobal et al. 2005) and the results of the t test obtained with the STATISTICA program (procedure *T-Test for dependent samples*, in the *Basic Statistics and Tables* module). The results revealed slightly higher results for ELISA (P < 0.05).

13.1.4 Statistical Treatment to Compare More than Two Independent Samples

Let $\{x_{1,1}, x_{2,1}, x_{3,1}, \ldots, x_{n_1,1}\}$, $\{x_{1,2}, x_{2,2}, x_{3,2}, \ldots, x_{n_2,2}\}, \ldots,$ $\{x_{1,k}, x_{2,k}, x_{3,k}, \ldots, x_{n_k,k}\}$ be k ($k > 2$) independent random samples of n_j observations of a continuous random variable X, from k populations W_j, and let μ_j and σ_j be the mean and standard deviation of X in population W_j, $j = 1, \ldots, k$. For each sample the corresponding estimates of μ_j and σ_j ($\bar{x}_j$ and s_j) can be calculated. Accepting normal distributions, and for a fixed *significance level* α, we can calculate the corresponding confidence intervals for the parameters μ_j , σ_j, or to test hypotheses about them.

13.1.4.1 Hypothesis Test for k Means or One-Way ANOVA

Assuming normal distributions and equality variances ($X \sim N(\mu_j, \sigma) \quad in \quad W_j$), the null hypothesis $H_0 \equiv \mu_1 = \mu_2 = \ldots = \mu_k$ may be tested using the statistic $F_{cal} = \frac{\sum_{j=1}^{k} n_j(\bar{x}_j - \bar{x})^2/(k-1)}{\sum_{j=1}^{k}\sum_{i=1}^{n_j}(x_{i,j} - \bar{x}_j)^2/(n-k)}$, which has an F-distribution with $k - 1$ and $n - k$ df, under H_0, and where $n = \sum_{j=1}^{k} n_j$, and $\bar{x}$ is the sample general mean. For a fixed value of α, if $F_{cal} > F_{1-\alpha, k-1, n-k}$, or if $P < \alpha$, $H_0 \equiv \mu_1 = \mu_2 = \ldots = \mu_k$ should be rejected and we conclude that there are some differences among the k means ($H_1 \equiv not\ all\ \mu_i (i = 1, \ldots, k) are\ equal$). If the null hypothesis H_0 is rejected, the posterior tests for means comparisons, Least Significant Difference (LSD), Scheffé, Tukey, Bonferroni or Student-Newman-Keuls (S-N-K) tests can be used to characterize the differences. The error plots, with the 95% confidence interval for the means in the k groups, can be used for a graphical comparison; if the mean value of a group remains inside the confidence interval for the mean of another, it is accepted that the two groups have similar means.

The F_{cal} ratio is also the result of a statistical method known as the *one-way Analysis of Variance* (ANOVA), which assumes the model $x_{i,j} = \mu_j + \varepsilon_{i,j}$ for each observation $x_{i,j}$, where $\varepsilon_{i,j}$ are the independent and normally distributed random errors ($\varepsilon_{i,j} \sim N(0, \sigma)$), and which has as objective to test $H_0 \equiv \mu_1 = \mu_2 = \ldots = \mu_k$. The results of the ANOVA procedure, which include the decomposition of the total sum of squares of the deviation of all observations around the general mean ($SS_{total} = \sum_{j=1}^{k} \sum_{i=1}^{n_j} (x_{i,j} - \bar{x})^2$) in two parts: the sum of the deviations between the groups ($SS_{between} = \sum_{j=1}^{k} n_j(\bar{x}_j - \bar{x})^2$) and the sum of the deviations within the groups ($SS_{within} = \sum_{j=1}^{k} \sum_{i=1}^{n_j} (x_{i,j} - \bar{x}_j)^2$), with their df, $(n-1) = (k-1)+(n-k)$, their mean squares, $MSS_{between} = SS_{between}/(k-1)$ and $MSS_{within} = SS_{within}/(n-k)$, and the value of the statistic $F_{cal} = \frac{MSS_{between}}{MSS_{within}}$, with $k - 1$ *and* $n - k$ df, along with their associated probability values (P), are summarized in the corresponding ANOVA table.

To check the assumptions of the model, Bartlett's or Levene's tests can be used to assess the assumption of equality of variance, and the normal probability plot of the residuals ($e_{i,j} = x_{i,j} - \bar{x}_j$) to assess the assumption of normality. If either equality or normality are inappropriate, we can transform the data, or we can use the nonparametric Kruskal-Wallis test to compare the k groups. In any case, the ANOVA procedure is insensitive to moderate departures from the assumptions (Massart et al. 1990)

13.1.4.2 Example of Application of One-Way ANOVA

One-way ANOVA has been used to determine if there are significant differences between varietal wines (Pozo-Bayón et al. 2001), to compare the concentrations from seven laboratories for a same reference sample (Massart et al. 1990), to compare the biogenic amine contents in the wines grouped according to the type of aging (Marcobal et al. 2005), or from five wineries (Martín-Álvarez et al. 2006), and more generally, to compare the mean values of k groups of independent observations. Tables 13.4 to 13.6 show the results of the application of one-way ANOVA for octanoic acid content in four varietal wines, obtained with the STATISTICA program (*One-way ANOVA* procedure in the *ANOVA* module).

Table 13.4 Results of the application of one-way ANOVA for octanoic acid content in four varietal wines. ANOVA table:

Source of variation:	Degrees of freedom (df)	Sums of Squares (SS)	Mean square (MSS)	F-value	P-Value
Variety	3	176.7010	58.9003	29.6947	0.000008*
Error	12	23.8024	1.9835		
Total	15	200.5034			

*There is a statistically significant difference between the four means.

Table 13.5 Results of S-N-K test for means comparison:

		Homogenous Groups ($\alpha = 0.05$):		
Variety	Mean	1	2	3
Malvar	2.49	* * * *		
Airén	2.58	* * * *		
Monastrell	5.87		* * * *	
Trepat	10.65			* * * *

Means values with * * * * in the same homogenous group indicate that there are no significant differences between them (P > 0.05).

Table 13.6 Means and standard error (Std. Err.) values:

Variety	Mean	Std. Err.	95% Confidence interval		N
Malvar	2.49[a]	0.32	1.46	3.52	4
Airén	2.58[a]	0.63	0.58	4.58	4
Trepat	10.65[c]	1.02	7.40	13.90	4
Monastrell	5.87[b]	0.67	3.75	7.99	4

[a–c]From S-N-K test mean values with the same letter indicate that there are no significant differences between them (P > 0.05).

They include: the ANOVA table (since the P-value associated with the F-value is less than 0.05, the four means are significantly different, $P < 0.05$), the results of S-N-K test for means comparisons (there are three homogenous groups of means), and the table with the mean and the standard error values of octanoic acid in the wines of the four varieties. All means are different except for the means of the Malvar and Airén varieties, and the higher concentration is that of the Trepat variety (Pozo-Bayón et al. 2001).

13.1.4.3 One Factor Experimental Design or One-Way ANOVA

Accepting that previous values $\{x_{1,1}, x_{2,1}, \ldots, x_{n_1,1}\}, \{x_{1,2}, x_{2,2}, \ldots, x_{n_2,2}\}, \ldots, \{x_{1,k}, x_{2,k}, \ldots, x_{n_k,k}\}$, correspond to the data from k fixed levels (*treatments*) of a factor, obtained from a *completely randomized design*, and with normal distribution ($x_{i,j} \sim N(\mu_j, \sigma)$, $j = 1, \ldots, k$), we can consider for each observation $x_{i,j}$, the *fixed effects model:* $x_{i,j} = \mu + \alpha_j + \varepsilon_{i,j}$, $j = 1, \ldots, k$, $i = 1, \ldots, n_j$, where μ is the overall mean ($\mu = \sum n_j \mu_j / n$), α_j is a fixed quantity representing the *effect of treatment j* on the overall mean ($\alpha_j = (\mu_j - \mu)$, $\sum_j n_j \alpha_j = 0$), and $\varepsilon_{i,j}$ the random errors, independent and with normal distribution ($\varepsilon_{i,j} \sim N(0, \sigma)$). For this model, we are interested in testing the hypothesis $H_0 \equiv \alpha_1 = \ldots = \alpha_k = 0$ using the previous statistic $F_{cal} = \frac{SS_{factor}/(k-1)}{SS_{error}/(n-k)}$ with a $F_{k-1,n-k}$ distribution, except that now the sources of variation between and within groups are assigned to the factor and to the error ($H_0 \equiv \mu_1 = \ldots = \mu_k$ implies $H_0 \equiv \alpha_1 = \ldots = \alpha_k = 0$). After fixing the value of α, if the null hypothesis $H_0 \equiv \alpha_1 = \ldots = \alpha_k = 0$ is rejected ($F_{cal} > F_{1-\alpha,k-1,n-k}$ or $P < \alpha$), we conclude that the factor influences the analyzed variable, and the posterior tests for means comparisons can be used to characterize the differences (Afifi and Azen 1979; Jobson 1991).

Accepting that previous values $\{x_{1,1}, x_{2,1}, \ldots, x_{n_1,1}\}, \{x_{1,2}, x_{2,2}, \ldots, x_{n_2,2}\}, \ldots, \{x_{1,k}, x_{2,k}, \ldots, x_{n_k,k}\}$ correspond to the data from k random subpopulations, with normal distribution ($x_{i,j} \sim N(m_j, \sigma)$, $j = 1, \ldots, k$), and assuming that $m_1, \ldots, m_j$ represents a random sample from a population with distribution $N(\mu, \sigma_a)$, we can consider for each observation $x_{i,j}$ the *random effects model (components of variance model)* $x_{i,j} = \mu + a_j + \varepsilon_{i,j}$ where $a_j = m_j - \mu$ are independent $N(0, \sigma_a)$ and $\varepsilon_{i,j}$ are independent $N(0, \sigma)$. For this model, we are interested in estimating the two components of variance σ^2 and σ_a^2, and in testing the hypothesis $H_0 \equiv \sigma_a^2 = 0$ using the same statistic $F_{cal} = \frac{SS_{factor}/(k-1)}{SS_{error}/(n-k)}$ with a $F_{k-1,n-k}$ distribution as in the fixed effects case. The components of variance are estimated as $\sigma^2 = MSS_{within}$ and $\sigma_a^2 = (MSS_{between} - MSS_{within})/q$, where $q = \frac{1}{k-1}(\sum n_j - \frac{\sum n_j^2}{\sum n_j})$ (Afifi and Azen 1979; Massart et al. 1990; Gardiner 1997). This *random effects model* is often used in analytical chemistry to breakdown a total precision into its components such as between-days and within-days, or between-laboratories and within-laboratories in collaborative trials, to validate an analytical method using reference material.

13.1.4.4 Two Factor Experimental Design or Two-Way ANOVA

In the case of two factors A and B, with a and b fixed levels respectively, and m replicate measurements of response variable X in each of $a \cdot b$ treatment combinations, we accept for every observation $x_{i,j,k}$, the k-th replicate of experimental units receiving factor A level i and B level j, the following model: $x_{i,j,k} = \mu + \alpha_i + \beta_j + \gamma_{i,j} + \varepsilon_{i,j,k}$, with $i = 1, \ldots, a$, $j = 1, \ldots, b$, y $k = 1, \ldots, m$, and where μ is a global mean, α_i representing the effect on the response variable X of factor A at level i, β_j representing the effect on the response variable of factor B at level j,$\gamma_{i,j}$ representing the combined effect on the response variable X of factor A at level i with factor B at level j, and $\varepsilon_{i,j,k}$ is the error with normal distribution $N(0, \sigma)$. The two-way ANOVA breaks down the total variation of the observations into four sources associated with both factors, their interaction and with other factors not controlled, in agreement with the terms in the model ($SS_{total} = SS_A + SS_B + SS_{AxB} + SS_{error}$). There are three null hypotheses associated with two factors: factor A does not influence the response variable X, $H_0^1 \equiv \alpha_i = 0, \ \forall i$, factor B does not influence the response, $H_0^2 \equiv \beta_j = 0, \ \forall j$, and no interaction effect occurs, $H_0^3 \equiv \gamma_{i,j} = 0, \forall i, j$, that can be confirmed with the three following F-statistics: $F_{cal}^1 = \text{MSS}_A/\text{MSS}_{error}$ with $(a-1)$ and $ab(m-1)$df, $F_{cal}^2 = \text{MSS}_B/\text{MSS}_{error}$ with $(b-1)$ and $ab(m-1)$ df and $F_{cal}^3 = \text{MSS}_{AB}/\text{MSS}_{error}$ with $(a-1)(b-1)$ and $ab(m-1)$ df, respectively. The sums of squares associated with the source of variation, their df and mean squares, and the F-ratios, along with their associated probabilities, are summarized in the two-way ANOVA table (see Table 13.7). After fixing the value for α, if $F_{cal}^3 < F_{1-\alpha,(a-1)(b-1),ab(m-1)}$ or $P_{AxB} > \alpha$, then H_0^3 is accepted, the interaction does not exist, and the influence of one of the factors will not depend on the levels of another factor. If $F_{cal}^1 > F_{1-\alpha,(a-1),ab(m-1)}$ or $P_A < \alpha$, then H_0^2 is rejected, factor A influences the analyzed variable, and if

Table 13.7 Results of two-way ANOVA for 1-propanol content in wines resulting from the fermentation of a same must with or without skins and in the presence or absence of sulfur dioxide (SO_2). ANOVA table:

Source of variation:	Degrees of freedom (df)	Sums of Squares (SS)	Mean square (MSS)	F-values	P-Values
SO_2 (Factor A)	1	41.95	41.95	7.007	0.057157
Skins (Factor B)	1	1011.15	1011.15	168.882	0.000202*
SO_2*Skins (Interaction)	1	1.14	1.14	0.190	0.685095
Error	4	23.95	5.99		
Total	7	1078.19			

*The skins factor has a significant influence (P<0.05)

Table 13.8 Means and standard error values for 1-propanol content in wines:

Factors: SO_2 (A)	Skins (B)	Mean	Std. Err.	95% Confidence interval		N
Without	Without	38.60	0.10	37.33	39.87	2
Without	With	61.84	0.46	56.00	67.68	2
With	Without	34.78	3.36	–7.85	77.40	2
With	With	56.50	0.71	47.55	65.46	2

Table 13.9 S-N-K test for comparison of means values for 1-propanol content in wines according to skins' factor:

		Homogenous Groups ($\alpha = 0.05$):	
Skins	Mean	1	2
Without	36.68750	****	
With	59.17250		****

$F^2_{cal} < F_{1-\alpha,(b-1),ab(m-1)}$ or, $P_B > \alpha$,then H^3_0 is accepted, factor B does not influence the analyzed variable. When the interaction exists, the influence of one of the factors will depend on the levels of another factor.

13.1.4.5 Example of Application of Two-Way ANOVA

Tables 13.5 to 13.9 show the results of applying two-way ANOVA to the 1-propanol content in wines resulting from fermentation of the same must with or without skins and in the presence or absence of sulfur dioxide, that corresponds to a 2^2 factorial design (Herraiz et al. 1990; Martín-Álvarez et al. 2006), obtained with the STATISTICA program (*Factorial ANOVA* procedure, in the *ANOVA* module). They include the two-way ANOVA table (only the skins factor has a significant influence at the 95% confidence level, $P < 0.05$), the mean and the standard error values of 1-propanol content for each combination of levels of factors (greater values in wines elaborated with the addition of skins), and the results of the S-N-K test for comparison of means of the skins' factor. Figure 13.2 shows plots of the means of 1-propanol content for each level of factors.

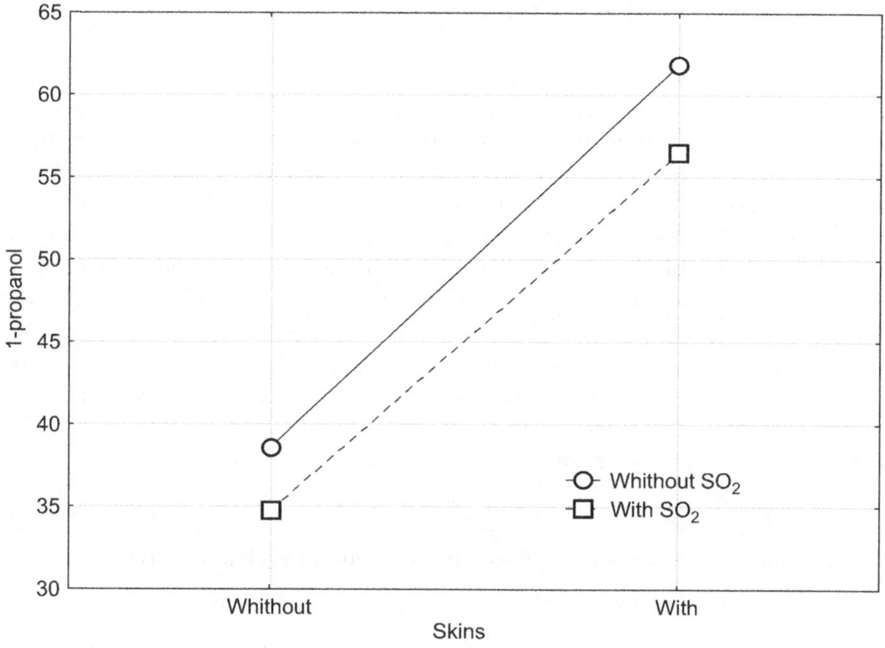

Fig. 13.2 Means plot of 1-propanol content for each levels of factors

13.1.4.6 Two-way ANOVA with no Replication

When there is only one observation for each combination of the levels of the two factors ($m = 1$), the model is $x_{i,j} = \mu + \alpha_i + \beta_j + \varepsilon_{i,j}$, without the interaction term. The purpose of this factorial design is only to test the main effects because interaction and error are confounded. This model is also valid for the case of one factor *randomized block design* or repeated measures design. Two-way ANOVA with no replication has been used: to test the influence of the yeast strain and aging time factors on different compounds in sparkling wines (Martínez-Rodríguez et al. 2002; Hidalgo et al. 2004), to test differences between the quality scores of three wines from ten tasters (O'Mahony 1986), to compare the concentrations reported by five analytical methods for six samples (Sharaf et al. 1986), to test the effect of vineyard and aging time factors on the phenolic, volatile and nitrogen compounds of the wines (Pozo-Bayón et al. 2004), to test the effect of time and blend factors in wines (Monagas et al. 2007), to test the effect of technological and time factors on the nitrogen compounds in wines (Alcaide-Hidalgo et al. 2007) and on the phenolic compounds in wines (Hernández et al. 2006).

Multifactorial ANOVA can be used to test the effect of more than two factors (Pozo-Bayón 2003a; Martín-Álvarez et al. 2006).

13.2 Bivariate Statistical Techniques

In this section we consider the statistical techniques, correlation and regression analysis, to *study the interrelationship between two continuous random variables* (X_1, X_2), from the information supplied by a sample of n pairs of observations $\{(x_{1,1}, x_{1,2}), (x_{2,1}, x_{2,2}), \ldots, (x_{n,1}, x_{n,2})\}$, from a population W. In the *correlation analysis* we accept that the sample has been obtained of random form, and in the *regression analysis* (linear or not linear) we accept that the values of one of the variables are not subject to error (independent variable $X = X_1$), and the dependent variable ($Y = X_2$) is related to the independent variable by means of a mathematical model ($Y = f(X) + \varepsilon$).

Mean and standard deviation values $(\bar{x}_1, s_1, \bar{x}_2, s_2)$ for every variable can be calculated, and the scatterplot with the n points can be used to see the form of the association between the two variables. In the case of random samples and assuming a bivariate normal distribution, the 95% confidence ellipse: $(x_1 - \bar{x}_1, x_2 - \bar{x}_2)\begin{pmatrix} s_1^2 & s_{12} \\ s_{12} & s_2^2 \end{pmatrix}^{-1}\begin{pmatrix} x_1 - \bar{x}_1 \\ x_2 - \bar{x}_2 \end{pmatrix}\frac{n(n-2)}{2(n^2-1)} = F_{1-\alpha,2,n-2}$, that can be used to detect outliers, can also be included in the scatterplot. The covariance ($s_{12} = \sum_{i=1}^{n}(x_{i,1} - \bar{x}_1)(x_{i,2} - \bar{x}_2)/(n-1)$) and correlation coefficient ($r = s_{12}/(s_1 s_2)$) values, which take into account the joint variation of both variables, can also be calculated (Afifi and Azen 1979; Jobson 1991).

13.2.1 Correlation Analysis

Accepting normal bivariate distribution, *Pearson's correlation coefficient*, defined by $r = \frac{\sum_{i=1}^{n}(x_{i,1}-\bar{x}_1)(x_{i,2}-\bar{x}_2)}{\sqrt{\sum_{i=1}^{n}(x_{i,1}-\bar{x}_1)^2\sum_{i=1}^{n}(x_{i,2}-\bar{x}_2)^2}}$, that is the estimator of the population's correlation coefficient ρ, measures the intensity of the linear relation between both variables X_1, X_2. It is possible to calculate the $100(1-\alpha)\%$ confidence interval for ρ, and/or test the null hypothesis $H_0 \equiv \rho = 0$ vs the alternative $H_1 \equiv \rho \neq 0$, by means of the statistic $t_{cal} = \frac{r\sqrt{n-2}}{\sqrt{1-r^2}}$ which has a t-distribution with $n-2$ df; and if $|t_{cal}| > t_{1-\alpha/2,n-2}$, or if the associated probability is less than α, H_0 is rejected and $\rho \neq 0$ is accepted.

If normality of the data cannot be accepted, *Spearman's correlation coefficient* and its corresponding nonparametric test can be used for the null hypothesis $H_0 \equiv \rho = 0$.

13.2.1.1 Applications

As an example, correlation analysis has been applied: to confirm the correlation between biogenic amine formation and disappearance of their corresponding amino

Table 13.10 Pearson's correlation coefficients between the color parameters and phenolic components

Phenolic components	CI	%yellow	%red	%blue	%dA	Tint
Anthocyanin glucosides	–0.89*	–0.96*	0.95*	–0.78*	0.95*	–0.95*
Acetyl glucoside anthocyanins	–0.94*	–0.94*	0.91*	–0.72	0.91*	–0.93*
Cinnamoyl-glucoside anthocyanins	–0.89*	–0.97*	0.94*	–0.75	0.94*	–0.96*
Anthocyanin-piruvic acid adducts	–0.89*	–0.99*	0.98*	–0.82*	0.98*	–0.99*
Anthocyanin-vinylflavanol adducts	0.79*	0.80*	–0.77*	0.58	–0.78*	0.79*
Hydroxybenzoic acids	–0.80*	–0.91*	0.88*	–0.69*	0.89*	–0.90*
Hydroxycinnamic acids	0.87*	0.90*	–0.86*	–0.62	–0.86*	0.88*
Flavanols	–0.80*	–0.92*	0.91*	–0.76*	0.91*	–0.92*
Flavonols	–0.56	–0.74	0.77*	–0.74	0.77*	–0.77*

CI=Color intensity
*Correlation significantly different from zero (p < 0.05).

acids (Marcobal et al. 2006b), to correlate the autolytic capacity of the strains and isobutanol production levels (Barcenilla et al. 2003), to examine the linear relationships between chemical composition and foam characteristics in sparkling wines (Moreno-Arribas et al. 2000), or to confirm the relationship between the wine color and the phenolic composition during aging time in bottle (Monagas et al. 2006a). In Table 13.10, Pearson's correlation coefficients between the color parameters and phenolic components during aging time in bottle in Tempranillo wines are shown. Color parameters were significantly correlated ($p < 0.05$) with the majority of components.

13.2.2 Simple Linear Regression Analysis

The simple linear regression accepts that the variables X, Y are related by the *mathematical model* $Y = \beta_0 + \beta_1 X + \varepsilon$, where β_0 is the intercept, β_1 the slope, ε the error term, which means summarising the dependence of X (independent variable) on Y (dependent variable) by a straight line ($\hat{Y} = \beta_0 + \beta_1 X$). The following hypotheses are accepted (Draper and Smith 1981): the X values are fixed and measured without error, and the errors are independent and have a normal distribution with a common standard deviation for each X value ($\varepsilon \sim N(0, \sigma)$), which means that the response Y can be assumed to be normally distributed with a common standard deviation for each X value ($Y \sim N(\beta_0 + \beta_1 X, \sigma)$). From the experimental data, $\{(x_i, y_i)\}_{i=1,\ldots,n}$, the parameters β_0 *and* β_1 can be *estimated* as b_0 *and* b_1, by applying the *technique of least-squares* ($min \sum_{i=1}^{n}(y_i - \beta_0 - \beta_1 x_i)^2$), according to $b_1 = \sum_{i=1}^{n}(x_i - \bar{x})(y_i - \bar{y}) / \sum_{i=1}^{n}(x_i - \bar{x})^2$ and $b_0 = \bar{y} - b_1\bar{x}$. With these regression coefficients b_0 *and* b_1 it is possible to obtain the fitted value $\hat{y}_i$ for a fixed x_i value of X ($\hat{y}_i = b_0 + b_1 x_i$), and the residual $\hat{y}_i - y_i$.

To assess the statistical validity of the fitted regression equation, the ANOVA principle of splitting the variation in the Ys about their mean, $SS_{tot} = \sum_{i=1}^{n} (y_i - \bar{y})^2$, into an explainable component due to the fitted regression model, $SS_{reg} = \sum_{i=1}^{n} (\hat{y}_i - \bar{y})^2$, and another unexplainable one due to the error, $SS_{res} = \sum_{i=1}^{n} (\hat{y}_i - y_i)^2$, can be applied ($SS_{tot} = SS_{reg} + SS_{res}$). A measure of the precision of fit is the *coefficient of determination*, $R^2 = SS_{reg}/SS_{tot}$, that measures the proportion of total variation about the mean $\bar{y}$ explained by the regression ($0 \leq R^2 \leq 1$), which is often expressed as a percentage ($0 \leq R^2(\%) \leq 100$), and that should not be too far from 1. Another measure of the precision is the *standard deviation of residuals, or standard error of estimate*, $s = \sqrt{SS_{res}/(n-2)}$, that is the estimate of standard deviation σ, and should be as small as possible. These sums of squares, their corresponding df, $(n-1) = (1) + (n-2)$, and the mean squares, MSS_{reg} and MSS_{res}, are presented in the *ANOVA table for regression*. The statistic $F_{cal} = MSS_{reg}/MSS_{res}$ has an F-distribution with 1 and $n-2$ df, and can be used to test the null hypothesis $H_0 \equiv \beta_1 = 0$ vs $H_1 \equiv \beta_1 \neq 0$ (*F test of linear model*). When the value of α is fixed, if $P < \alpha$ the null hypothesis is rejected and the fitted equation appears statistically valid, and if $P > \alpha$, H_0 is accepted and a constant model would be accepted for Y ($Y = \beta_0 + \varepsilon$). It is also possible to test the null hypotheses $H_0 \equiv \beta_1 = 0$ and $H_0 \equiv \beta_0 = 0$ using the statistics $t_{cal} = \frac{b_1}{s_{b_1}}$ and $t_{cal} = \frac{b_0}{s_{b_0}}$ that follow a t-distribution with $n-2$ df, and where s_{b_1} and s_{b_0} are the standard errors of b_1 and b_0, respectively ($s_{b_1} = \frac{s}{\sqrt{\sum (x_i - \bar{x})^2}}$, $s_{b_0} = s\sqrt{\frac{\sum x_i^2}{n \sum (x_i - \bar{x})^2}}$) or calculate the *confidence intervals for the parameters* $\beta_1 \in \left(b_1 \pm t_{1-\alpha/2,n-2}\, s_{b_1}\right)$ and $\beta_0 \in \left(b_0 \pm t_{1-\alpha/2,n-2}\, s_{b_0}\right)$.

When there are genuine repeated runs in the data, we can use the *lack of fit F-test*, $F_{cal} = \frac{SS_{lackf}/(n-2-g)}{SS_{pureer}/g}$ with $n-2-g$ and g df, to check whether or not the model is correct ($H_0 \equiv$*the model has no lack-of-fit, or there are no reasons to doubt the adequacy of the model*, and $H_1 \equiv$*the model appears to be inadequate*). The statistic $SS_{pureer} = \sum_j (n_j - 1)\, s_j^2$ is the sum of squares of pure error of repeated points, with $g = \sum (n_j - 1)$ df, and $SS_{lackf} = SS_{res} - SS_{pureer}$ is the sum of squares of lack of fit with $n-2-g$ df. If $F_{cal} > F_{1-\alpha,n-2-g,g}$ this indicates that the model appears to be inadequate and we should use another model, such as the quadratic polynomial model, $Y = \beta_0 + \beta_1 X + \beta_2 X^2 + \varepsilon$, that is a second-order (in X) linear (in the $\beta's$) regression model, or diminishing the interval of X values.

Diagnostic checking of the residuals can be used to assess the validity of model assumptions, and to check the practical validity of the predictions.

If the model is accepted as valid, it can be used to *predict* the Y value for a given value x_0 of X ($\hat{y}_0 = b_0 + b_1 x_0$), and to calculate the confidence interval for the true mean value of Y for the x_0 value, $\mu_{Y|X=x_0} \in \left(\hat{y}_0 \pm t_{1-\alpha/2,n-2}\, s\sqrt{\frac{1}{n} + \frac{(x_0 - \bar{x})^2}{\sum (x_i - \bar{x})^2}}\right)$, that defines the *confidence bands* for any value x_0.

13.2.2.1 Applications

Linear calibration is the most important application of this technique in chemical experimentation (Miller and Miller 2000; Cela 1994; Massart et al. 1990). The fitted equation can be used to estimate the value of X (concentration, $\hat{x}_0$) corresponding to some measurement $\hat{y}_0$ (mean value of m replicates) of the response Y on an unknown sample ($\hat{x}_0 = (\hat{y}_0 - b_0)/b_1$), and its confidence interval $\left(\hat{x}_0 \pm t_{1-\alpha/2,n-2} \frac{s}{b_1} \sqrt{\frac{1}{m} + \frac{1}{n} + \frac{(\hat{y}_0 - \bar{y})^2}{b_1^2 \sum (x_i - \bar{x})^2}}\right)$ (*inverse regression*). As an example, Tables 13.11 and 13.12 show the results of the linear regression for relative area (Y) vs concentration of the analyte (X) in the standard solution using two replicates at five points, obtained with the STATISTICA program (procedure *Multiple Regression*, in the *Multiple Linear Regression* module), that include estimation of the model parameters ($Y = 0.005889 + 0.030635{*}X$) and the statistics for the fit: correlation coefficient ($r = 0.9798$), determination coefficient ($R^2 = 0.960$), and standard error of estimation ($s = 0.0124$) and the ANOVA table (there is a statistically significant relationship between Y and X at the 99% confidence level). Table 13.13 shows the ANOVA

Table 13.11 Results of linear regression for relative area (Y) vs concentration of the analyte (X) in the standard solution. Estimation of the model parameters and statistics for the fit:

n = 10	Beta	Std. error of Beta	B (*regression coefficients*)	Std. error of B	t-value	P-level
Intercept			0.005889	0.007092	0.83033	0.430433
Octanoic	0.979809	0.070687	0.030635	0.002210	13.86117	0.000001*

R=.97981 R^2=.96003 Adjusted R^2=.95503
Std. Error of estimate:.01244
*Parameter $\beta_1 \neq 0$ (P<0.01).

Table 13.12 ANOVA table for the regression:

Source of Variation	Sums of Squares (SS)	Degrees of freedom (df)	Mean Squares (MSS)	F-value	P-level
Regress.	0.029736	1	0.029736	192.1322	0.000001*
Residual	0.001238	8	0.000155		
Total	0.030974				

*There is a statistically significant relationship between Y and X at the 99% confidence level.

Table 13.13 ANOVA table for the regression with the lack of fit test

Source of Variation	Sums of Squares (SS)	Degrees of freedom	Mean Squares (MSS)	F-value	P-level
Regress.	0.0297356	1	0.0297356	192.13	0.00000[a]
Residual	0.00123813	8	0.00015477		
Lack-of-Fit	0.00022487	3	0.000074956	0.37	0.779[b]
Pure Error	0.00101326	5	0.000202653		
Total	0.0309737	9			

[a]There is a statistically significant relationship between Y and X at the 99% confidence level
[b]the model appears to be adequate for the observed data (P-value for lack-of-fit > 0.10).

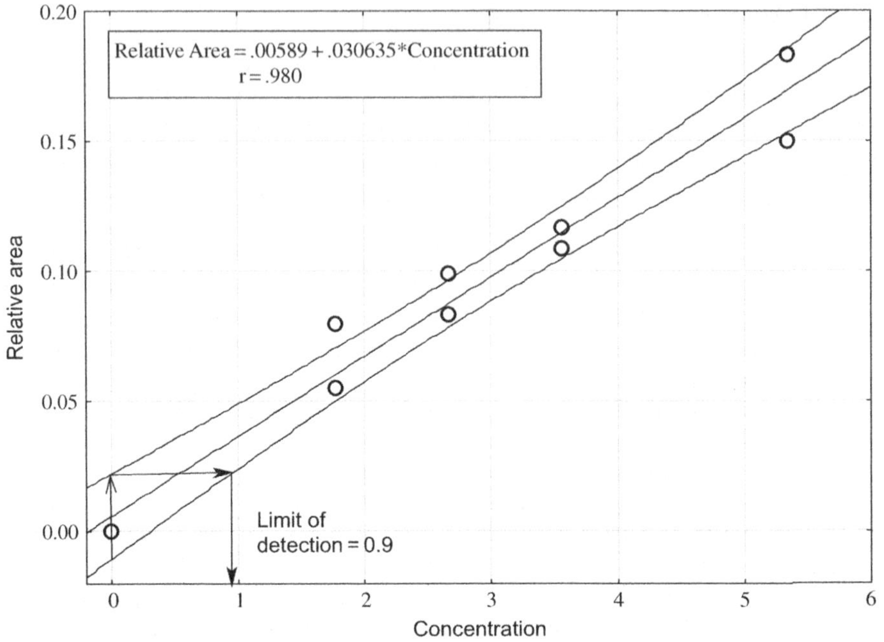

Fig. 13.3 Plot of the observed relative area versus the concentration of analyte in the standard solution, the estimated calibration line with its 95% confidence interval, and a possible way of calculating the limit of detection

table with the *lack of fit test* obtained with the STATGRAPHICS program (Statistical Graphics Corp., http://www.statgraphics.com/) using the *Calibration Models* procedure in the *Special Advanced Regression* module (the model appears to be adequate for the observed data, because $P>0.10$). Figure 13.3 displays plots of the observed relative area to the internal standard versus the concentration of analyte in the standard solution, the estimated calibration line with its 95% confidence interval, and a possible way of calculating the *detection limit* (the upper value of the 95% confidence interval for the estimated X concentration corresponding to the upper value of the 95% confidence interval for the true mean value of Y at $X = 0$ (Sharaf et al. 1986). The calibration range for the analyte and the results of recovery experiments to evaluate the occurrence of proportional systematic errors are often included in the calibration results.

13.3 Multivariate Statistical Techniques

13.3.1 Introduction

Multivariate analysis can be defined as the set of *statistical and mathematical methods* to analyse, describe and interpret *multidimensional observations*. Application of these techniques can have the following objectives: (1) *to reduce the dimensions of the data* without losing important information, (2) *to look for clustering of observations or variables* based on some similar measure, (3) *to define the rules to decide on the assignation of an observation* to a given group, (4) *to study measures of dependence* between variables, (5) *to predict the values* of the variables from others by applying a mathematical model, and (6) *to develop and compare hypotheses* about some population parameters. As working tools these methods use algebraic geometry, matricial calculus and numerical calculus.

13.3.1.1 Data Matrix

To apply these multivariate techniques, we require a data matrix with the information corresponding to n observations of p quantitative variables $(X_1, X_2, \ldots, X_p)$. We could, also, have some qualitative variables, coded numerically, to classify the observations into groups. From a geometric perspective, the n observations of the data matrix would correspond to n points of the Euclidean space of the p variables, and the Euclidean distance between observations would correspond to a measure of proximity (similarity).

13.3.1.2 Graphical Representation of the Data

The bidimensional methods of representation most used by multivariate techniques are: *direct methods*, such as matricial dispersion diagrams, and icon plots based on histograms, profiles or stars; *projection approach techniques*, that represent observations in the new variables obtained, and which fulfil a specific objective (principal components, canonical variables, etc.) and *dendrograms* that inform about the similarity of observations or variables (Krzanowski 1988).

13.3.1.3 Classification of Methods

Taking into account the groups of variables studied and the origin of the observations of the data matrix, we can have *non-supervised techniques* for matrices with n observations derived from one population and a single group of variables, *supervised methods* for k matrices with n_k observations derived from k populations and a single group of variables, and methods for the *dependence study* for matrices with n observations from a single population and two groups of variables.

13.3.2 Multivariate Statistical Non Supervised Techniques

To apply these techniques, we have a data matrix with n observations in p variables $(X_1, X_2, \ldots, X_p)$:

$$\textit{Observations}\quad \begin{array}{c} \\ 1 \\ 2 \\ \dots \\ n \end{array} \begin{array}{c} X_1\ X_2\ \dots\ X_p \\ \begin{pmatrix} x_{1,1} & x_{1,2} & \dots & x_{1,p} \\ x_{2,1} & x_{2,2} & \dots & x_{2,p} \\ \dots & \dots & \dots & \dots \\ x_{n,1} & x_{n,2} & \dots & x_{n,p} \end{pmatrix} \end{array}$$

The vectors of means $\vec{x}\,' = (\bar{x}_1, \bar{x}_2, \ldots, \bar{x}_p)$ and deviations $\vec{s}\,' = (s_1, s_2, \ldots, s_p)$, and matrices of covariances $S = (s_{i,j})$ and correlations $R = (r_{i,j})$ can be calculated. For this data matrix, the most used non-supervised methods are *Principal Components Analysis* (PCA), and/or *Factorial Analysis* (FA) in an attempt to reduce the dimensions of the data and study the interrelation between variables and observations, and *Cluster Analysis* (CA) to search for clusters of observations or variables (Krzanowski 1988; Cela 1994; Afifi and Clark 1996). Before applying these techniques, variables are usually first standardised ($X_i \rightarrow X_i^*$) to achieve a mean of 0 and unit variance.

13.3.2.1 Principal Components Analysis (PCA)

The main objective of this technique is to *reduce the dimensions of data without losing important information, starting with the correlation between variables, to explore the relationship between variables and between observations.* The aim is to obtain p new variables ($Y_1, Y_2, \ldots, Y_p$), that we will call *principal components*, which are (1) a normalised linear combination of the original variables ($Y_i = a_{1,i}X_1 + a_{2,i}X_2 + \ldots + a_{p,i}X_p; \sum_k a_{k,i}^2 = 1$), (2) uncorrelated ones ($\mathrm{cov}(Y_i, Y_j) = 0 \forall i \neq j$), (3) with progressively diminishing variances ($var(Y_1) \geq var(Y_2) \geq \ldots \geq var(Y_p)$), and (4) the total variance (VT) coincident with that of the original variables ($\sum_{i=1}^{p} var(Y_i) = \sum_{i=1}^{p} var(X_i) = VT$) (Afifi and Azen 1979). The aim is, therefore, to obtain the matricial transformation $Y_{(n,p)} = X_{(n,p)}\,A_{(p,p)}$.

If the original variables are previously standardised (most programmes do this), the coefficients ($a_{i,j}$) will be determined from the eigenvalues and eigenvectors of the correlations' matrix $R_{(p,p)} = (r_{i,j})$, where the *j*-th column ($\vec{a}_j$) of the matrix $A_{(p,p)} = (a_{i,j})$ is the eigenvector associated with the *j*-th greatest eigenvalue λ_j, complying with $Var(Y_j) = \lambda_j$, $\sum_{i=1}^{p} var(Y_i) = \sum_{i=1}^{p} var(X_i^*) = VT = p$, with the $A_{(p,p)}$ matrix being orthogonal ($A_{(p,p)}^{-1} = A_{(p,p)}^t$). In many applications, if the first q components ($Y_1, \ldots, Y_q$) can, together, explain a high percentage of the total variance, e.g. $\frac{\lambda_1+\lambda_2+\ldots+\lambda_q}{VT}100\% > 80\%$, and if q is very much minor that p, the dimensions of the original data will have been reduced, without having lost more than a small proportion of non-essential data. This number q of components usually corresponds to the number of $\lambda_i > 1$. From a geometric perspective, the transformation $Y_{(n,q)} = X_{(n,p)}\,A_{(p,q)}$ corresponds to an orthogonal rotation of the coordinate axes in the directions of maximum variance, and since the $A_{(p,p)}$ matrix is orthogonal, it is also possible to consider the equation $X_{(n,p)} = Y_{(n,q)}A'_{(q,p)}$ that corresponds to the *factorial model of principal components*.

The results of PCA are (1) the number q of principal components, (2) the matrix $Y_{(n,q)} = X_{(n,p)}A_{(p,q)} = (y_{i,j})$, with the scores of n observations in the q new variables $(Y_1, Y_2, \ldots, Y_q)$, and (3) the matrix $A_{(p,q)} = (a_{i,j})$ with information about the contribution of the original variables $(X_1, X_2, \ldots, X_p)$ in the definition of the q components. The coefficients $(a_{i,j})$ are usually transformed so that they correspond with the correlations between the principal components and the original variables and to help to know the variables that define the principal components. The bidimensional representation of the n observations in the plane defined by the first two principal components is usually used *to explore or confirm possible clusters of observations* and to *detect possible outliers.*

13.3.2.2 Factor Analysis (FA)

FA is a more general *method for describing the dependence structure of the p variables* $(X_1, X_2, \ldots, X_p)$ *from other q non-observed variables, called factors*, that we accept to be responsible for the original ones, and manages to reduce the dimension of the data (if q very much minor that p). The *orthogonal factorial model* accepts for each original standardised variable (X_i^*), the following model: $X_i^* = b_{1,i}F_1 + b_{2,i}F_2 + \ldots + b_{q,i}F_q + \varepsilon_i$, where $(F_1, F_2, \ldots, F_q)$ are the q common factors, that are incorrelated; $(\varepsilon_1, \varepsilon_2, \ldots, \varepsilon_p)$ are the specific factors of each variable X_i, that are also incorrelated; and $b_{i,j}$ the *loadings* of the factors. It is also accepted that the common and specific factors are independent and have a mean of 0 and a variance of one. The *factorial model, in matricial form, is* $X^*_{(n,p)} = F_{(n,q)}B_{(q,p)} + E_{(n,p)}$, where $X^*_{(n,p)}$ is the matrix with standardised observations, $F_{(n,q)}$ is the matrix with the coordinates of the observations in q factors, $B_{(q,p)}$ is the matrix with the loadings of the factors in the p original variables, and $E_{(n,p)}$ is the matrix of the model errors. The number of factors (q), the scores' matrix ($F_{(n,q)}$), and the loadings' matrix ($B_{(q,p)}$), are the results of FA. The graphical representation of the observations in the plane defined by the first two factors informs of the possible clusters of observations and of the presence or not of outliers. From a geometrical perspective, the aim is to find the subspace that best fits the n points in space of the variables $(X_1, X_2, \ldots, X_p)$, to minimise the sum of the modules of the n vectors row of the matrix $E_{(n,p)}$.

The *factorial model in principal components* $X^*_{(n,p)} = Y_{(n,q)}A^t_{(q,p)} + E_{(n,p)}$, which is frequently used, considers *principal components as factors* ($F_{(n,q)} = Y_{(n,q)}$), and the transposed components matrix as the saturations matrix ($B_{(q,p)} = A^t_{(q,p)}$), and this fulfils all the previous requirements since $A_{(p,p)}$ is an orthogonal matrix. Calculation of the number of factors (components) can be carried out from the eigenvalues or by means of a cross-validation procedure (Brereton 1990; Cela 1994).

Sometimes, for a better definition of the contribution of factors in the variables, it is possible to rotate the q factors extracted ($B^*_{(q,p)} = B_{(q,p)}Q_{(p,p)}$), and the *Varimax rotation* is the most frequently used (Afifi and Azen 1979).

Table 13.14 Results of applying principal components analysis to the 10 volatile compounds analyzed in 16 varietal wines. Factor loadings matrix:

	PC1	PC2
Methanol	–0.707*	0.375
1-Propanol	0.870*	0.193
Isobutanol	–0.634	0.334
Isoamylic alcohols	–0.590	0.720*
1-Hexanol	–0.885*	–0.126
cis-3-Hexen-1-ol	–0.888*	–0.228
Hexanoic acid	0.589	–0.569
Octanoic acid	–0.861*	–0.438
Decanoic acid	–0.507	–0.793*
Ethyl octanoate	0.103	–0.242
*Loadings >0.7		
Explain variance	4.935	2.078
Proportion total	0.493	0.208

Table 13.15 Results of applying principal components analysis to the 10 volatile compounds analyzed in 16 varietal wines. Scores for the samples:

	PC1	PC2
Ma98	1.543	–0.539
Ma98	1.462	–0.537
Ma99	0.891	0.453
Ma99	0.695	0.358
Ai97	0.669	0.639
Ai97	0.422	0.940
Ai98	0.526	0.160
Ai98	0.408	0.493
Tr97	–1.644	–0.003
Tr97	–1.807	–0.163
Tr98	–0.581	–1.808
Tr98	–0.735	1.328
Mo97	–0.419	–2.188
Mo97	–1.015	1.560
Mo98	–0.221	–0.409
Mo98	–0.195	–0.286

Applications

As an example, Tables 13.14 and 13.15 show the results of applying principal components analysis to the 10 volatile compounds (methanol, 1-propanol, isobutanol, 2- and 3-methyl-1-butanol, 1-hexanol, *cis*-3-hexen-1-ol, hexanoic acid, octanoic acid, decanoic acid and ethyl octanoate) analyzed in 16 varietal wines (Pozo-Bayón et al. 2001), obtained with the STATISTICA program (*Factor Analysis* procedure in *Multivariate Exploratory Techniques* module, and using Principal Components as Extraction method). The results include: the factor loadings matrix for the two first principal components selected ($q = 2$), which explains 70.1% of the total variance (Table 13.14). The first principal component is strongly correlated with *cis*-3-hexen-1-ol (–0.888), 1-hexanol (–0.885), 1-propanol (0.870), and

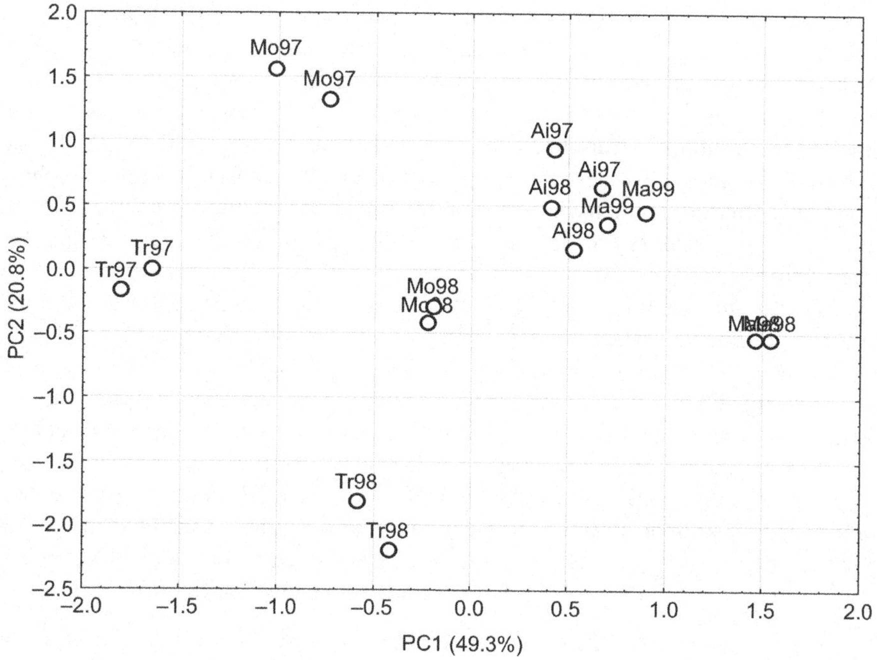

Fig. 13.4 Plot of the 16 varietal wines in the plane defined by the first two principal components

octanoic acid (–0.861), and explains 49.3% of the total variance, while decanoic acid (–0.793) and isoamylic alcohols (0.720) contribute more to the second principal component, which explains 20.8% of the total variance; the scores for the 16 samples of wine in the first two principal components (Table 13.15). In Fig. 13.4, the 16 wines are plotted on the plane defined by the first two principal components. From the figure, wines of the red varieties (Trepat and Monastrell) appear on the left side of the plane, grouped by year of harvest, with lower values for PC1, while wines of white varieties (Malvar and Airén) are found on the right side of the plane, that is to say, they have greater values for PC1 and are grouped by year of harvest (the red varieties essentially have lower concentrations of 1-propanol and higher concentrations of *cis*-3-hexen-1-ol, 1-hexanol and octanoic acid than the white varieties). The second principal component mainly differentiates between wines from the two harvests. It can be observed that the greatest cause of variation among the samples is due to the factor variety, followed by harvest.

13.3.2.3 Cluster Analysis (CA)

The *objective of this technique is to look for natural clusters among the n observations* (sometimes between the p variables) of the data matrix $X_{(n,p)}$. Considering these observations as points of space for the variables $(X_1, X_2, \ldots, X_p)$, there are two techniques to search for groups: *hierarchical ones* that reveal similarities among observations, which are generally agglomerative (initially starting with the same number of observations as groups), and *non-hierachical ones* that only indicate if an observation belongs to one cluster or another. The following considerations must be taken into account to apply this technique: (1) a measure of the similarity among observations (or variables) must be chosen, depending on the type of data analysed (Krzanowski 1988); (2) an algorithm must be chosen to unite the clusters; (3) the number of clusters to be formed must be established in the case of non-hierarchical and, (4) if the variables are different in nature they must first be standardized.

After deciding on the number k of clusters (C_i) that we want to form, the *non-hierachical* techniques can be used to obtain a division of the order k $\{C_1, C_2, \ldots, C_k\}$ of the set of n observations $W = \{1, 2, 3, \ldots, n\}$, such that $W = C_1 \cup C_2 \cup \ldots \cup C_k$ and $C_i \cap C_j = \phi$, $i \neq j$. Each cluster C_i will be comprised of n_i observations, and will have a *centroid* (c^i) the coordinates of which will correspond to the mean values of p variables in the n_i observations, in other words, $c^i = (\bar{x}_1^i, \bar{x}_2^i, \ldots, \bar{x}_p^i)$. For each cluster, we can define its dispersion given by the sum of the squares of distances between the n_i observations and the centroid, in other words, ($E_i = \sum_j d^2(\vec{w}_j, c^p)$, $\forall \vec{w}_j \in C_i$). We can, thus, define for a given division $\{C_1, C_2, \ldots, C_k\}$, the *total dispersion*, defined by: $D_T(C_1, C_2, \ldots, C_k) = \sum_{i=1}^{k} E_i$. The aim of these techniques is to find the division $\{C_1, C_2, \ldots, C_k\}$ of order k of $W = \{1, 2, 3, \ldots, n\}$, that minimizes this total dispersion $D_T(C_1, C_2, \ldots, C_k)$. One of the most frequently used algorithms is *McQueen's k-means algorithm* that consists in (1) assigning the n observations randomly to the k groups, (2) calculating the centroids of each group, (3) assigning each individual to the group with the nearest centroid, and (4) repeating steps (2) and (3) to achieve stability.

Although stability can be guaranteed after a finite number of steps, this number can be reduced if step (3) is modified and the centroids are recalculated after each assignation of the observations. As a result of applying this technique, as well as a description of k clusters, computer programs usually provide the mean values of the variables in each of these, and the comparison of these mean values.

With the application of *agglomerative hierarchical techniques*, valid for grouping together observations (or variables), the interrelation between observations (or variables) can be established by a bidimensional graph called a *dendrogram*. The algorithms to apply these techniques, in the case of grouping together observations, have the following steps in common: (1) they start with as many clusters as observations ($C_1 = \{1\}$, $C_2 = \{2\}$, $C_n = \{n\}$), the matrix of the distances between them is calculated, $D = (d_{i,j})$; (2) the two clusters (C_p *and* C_q) with the smallest distance ($d(C_p, C_q) = min_{i,j}\ d(C_i, C_j)$) are looked for; (3) the clusters C_p *and* C_q are combined to form a new group and the new matrix of distances between the groups is calculated, and (4) steps (2) and (3) are repeated until a single cluster is formed for all n observations $\{1, 2, 3, \ldots, n\}$. In general, the distances' matrix from the first step usually corresponds to the Euclidean distance. The different ways of defining the distance $d(C_i, C_j)$ between the two clusters C_i and C_j, in step (3), give rise to different linkage, or amalgamation, rules:

- *Single link (or nearest neighbour) method*, $d(C_i, C_j)$ is the smallest distance between the observations for both clusters ($d(C_i, C_j) = min\, d(\vec{w}_i, \vec{w}_j)$, $\forall \vec{w}_i \in C_i, \forall \vec{w}_j \in C_j$).
- *Complete link (or furthest neighbour) method*, $d(C_i, C_j)$ is the largest distance between the observations for both clusters ($d(C_i, C_j) = max\, d(\vec{w}_i, \vec{w}_j)$, $\forall \vec{w}_i \in C_i, \forall \vec{w}_j \in C_j$).
- *Centroid method*, $d(C_i, C_j)$ is the distance between the centroids of both clusters ($d(C_i, C_j) = d(c^i, c^j)$).
- *Average link (or unweighted pair-group average) method*, $d(C_i, C_j)$ is calculated as the average distance between all pairs of observations in the two clusters ($d(C_i, C_j) = mean \left\{d(\vec{w}_i, \vec{w}_j)\right\}, \forall \vec{w}_i \in C_i, \forall \vec{w}_j \in C_j$).
- *Ward method*, that takes into account, within each group, the dispersion of the observations in relation to the centroid ($E_p = \sum_i d^2(\vec{w}_i, c^p), \forall \vec{w}_i \in C_p$). The clusters ($C_p$ *and* C_q) are joined, from step (3), if $E_{(p,q)} - E_p - E_q$ is minimum. In general, this method is regarded as very efficient, although it tends to create small clusters.

Before applying these hierarchical methods, the data matrix is usually standardised to give equal importance to all variables. The sequence of steps of the algorithm, is illustrated graphically in the *dendrogram*, in which the groups obtained will be observed.

In the case of clustering of variables, the algorithm is similar, using one minus the correlation coefficient to measure the distance between variables.

Applications

We have used CA to discover natural groupings of the wine samples and to obtain a preliminary view of the greatest cause of variation among them (Moreno-Arribas et al. 1998, 1999; Pozo-Bayón et al. 2003b, 2005; Marcobal et al. 2005; Hernández et al. 2006; Alcaide-Hidalgo et al. 2007). As an example, the application of cluster analysis to the 10 volatile compounds analyzed in 16 varietal wines (Pozo-Bayón et al. 2001) produces the dendrogram shown in Fig. 13.5, obtained with the STATISTICA program (*Cluster Analysis* procedure in *Multivariate Exploratory Techniques* module). The squared Euclidean distance was taken as a measure of the proximity between two samples, and Ward's method was used as a linkage rule. The variables were previously standardized. Two groups are observed, one comprised of wines of the red Monastrell and Trepat varieties, and the other formed by the white varieties Airén and Malvar. The wines of the four varieties are grouped according to variety. In turn, each of these groups are grouped by year of harvest. It can be observed that the greatest cause of variation among the samples is due to type of variety, followed by harvest.

13.3.3 Multivariate Statistical Supervised Techniques

To apply these techniques, we have k groups with observations in the same p variables ($X_1, X_2, \ldots, X_p$), from k W_i populations, with mean vectors $\vec{\mu}_i$ and covariance matrices Σ_i:

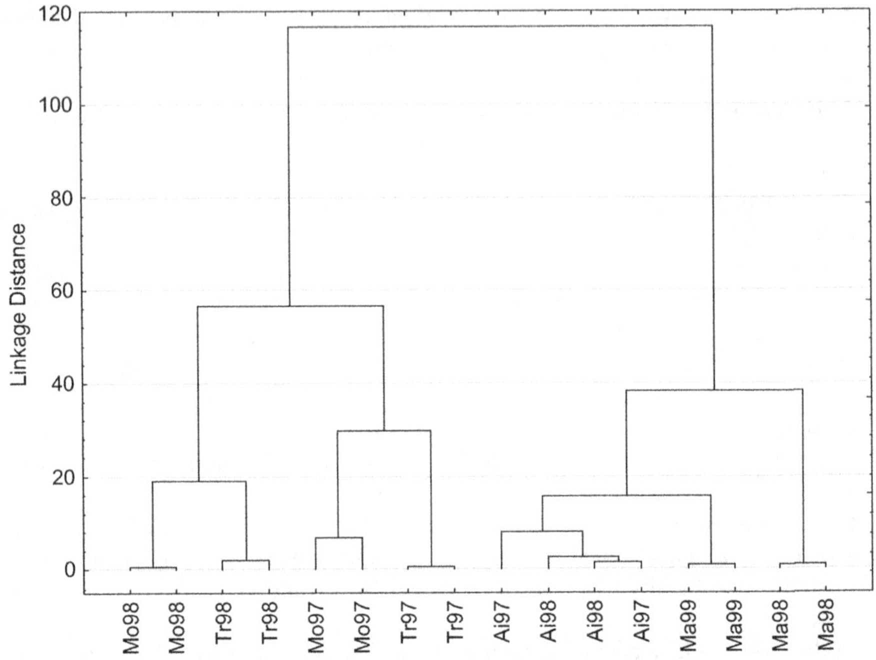

Fig. 13.5 Dendrogram for the 16 wines according to their volatile compounds content

$$\begin{array}{ccc} & & \begin{array}{cccc} X_1 & X_2 & \ldots & X_p \end{array} \\ \begin{array}{c} Observ. \\ W_1 \end{array} & \begin{array}{c} 1 \\ 2 \\ \ldots \\ n_1 \end{array} & \left(\begin{array}{cccc} x_{1,1} & x_{1,2} & \ldots & x_{1,p} \\ x_{2,1} & x_{2,2} & \ldots & x_{2,p} \\ \ldots & \ldots & \ldots & \ldots \\ x_{n_1,1} & x_{n_1,2} & \ldots & x_{n_1,p} \end{array} \right) \end{array}$$

$$\begin{array}{cccc} & & \begin{array}{cccc} X_1 & X_2 & \ldots & X_p \end{array} & \\ \begin{array}{c} Observ. \\ W_2 \end{array} & \begin{array}{c} 1 \\ 2 \\ \ldots \\ n_1 \end{array} & \left(\begin{array}{cccc} x_{1,1} & x_{1,2} & \ldots & x_{1,p} \\ x_{2,1} & x_{2,2} & \ldots & x_{2,p} \\ \ldots & \ldots & \ldots & \ldots \\ x_{n_2,1} & x_{n_2,2} & \ldots & x_{n_2,p} \end{array} \right) & \ldots \end{array}$$

$$\begin{array}{ccc} & & \begin{array}{cccc} X_1 & X_2 & \ldots & X_p \end{array} \\ \begin{array}{c} Observ. \\ W_k \end{array} & \begin{array}{c} 1 \\ 2 \\ \ldots \\ n_k \end{array} & \left(\begin{array}{cccc} x_{1,1} & x_{1,2} & \ldots & x_{1,p} \\ x_{2,1} & x_{2,2} & \ldots & x_{2,p} \\ \ldots & \ldots & \ldots & \ldots \\ x_{n_k,1} & x_{n_k,2} & \ldots & x_{n_k,p} \end{array} \right) \end{array}$$

The main objective of these techniques is to obtain, from the information of $n = \sum n_i$, genuine observations (*training set*), *classification rules* that can be used to differentiate the k groups and to use these rules to assign new samples (*test set*) into one of k groups. The most frequently used supervised classification methods are *Linear Discriminant Analysis* (LDA), *Quadratic Discriminant Analysis* (QDA), SIMCA (Soft Independent Modelling Class Analogy) and kNN (k-Nearest Neighbour) methods. Neural Networks correspond to another recent method to classify samples into one of the known k groups. Also, to achieve *maximum graphical differentiation* of the k groups, *Canonical Variate Analysis* (CVA) can be used, and *Multivariate Analysis of Variance* (MANOVA) to test for differences between the groups. Descriptive values of the variables in the groups will permit their characterization.

13.3.3.1 Discriminant Analysis

This supervised classification method, which is the most used, accepts a normal multivariate distribution for the variables in each population ($(X_1, \ldots, X_p) \sim N(\vec{\mu}_i, \Sigma_i)$), and calculates the classification functions minimising the possibility of incorrect classification of the observations of the training group (Bayesian type rule). If multivariate normality is accepted and equality of the k covariance matrices ($(X_1, \ldots, X_p) \sim N(\vec{\mu}_i, \Sigma)$), *Linear Discriminant Analysis* (LDA) calculates *klinear classification functions*, one for each group, $\left\{d_i = c_i + \sum_{j=1}^{p} a_{i,j} X_j\right\}_{i=1,\ldots,k}$, that would permit samples of the training group to be classified according to the *assignation rule*: the sample is assigned to the group with the highest score ("$(x_1, x_2, \ldots, x_p) \in W_i$ *if* $d_i(x_1, x_2, \ldots, x_p) = \max \left\{d_j(x_1, x_2, \ldots, x_p)\right\}_{j=1,\ldots,k}$"). If the p variables have a high discriminant power the percentage of correct classification will be high, and the assignation rule can be applied to new samples. The most important results are: classification functions, classification of n samples, the posterior probabilities ($\{e^{d_i} / \sum_j e^{d_j}\}_{i=1,2,\ldots,k}$), the classification matrix with the correct percentage of assignment of the samples for validation purposes, and classification of the samples in the test set. The leave-one-out cross-validation procedure can also be used to validate the classification process. The *Stepwise Linear Discriminant Analysis* (SLDA) provides these same results, but using less variables, selecting in each step the variable that most favours discrimination of the k groups. If the covariance matrices are unequal ($(X_1, \ldots, X_p) \sim N(\vec{\mu}_i, \Sigma_i)$), *Quadratic Discriminant Analysis* (QDA) can be used to obtain *quadratic functions to classify the samples.*

13.3.3.2 SIMCA Method

The SIMCA method defines a factorial model with a_i principal components for each of the k groups, starting with the corresponding matrix of standardised data, $\left\{X^{*(i)}_{(n_i,p)} = F^{(i)}_{(n_i,a_i)} B^{(i)}_{(a_i,p)} + E^{(i)}_{(n_i,p)}\right\}_{i=1,\ldots,k}$, and using these k models to assign the samples to each of the groups. The observation $\vec{w}$ *is assigned* in relation to its degree of fit to each model, comparing the error of fit to each class with the mean fit error of the observations of the class. The results include the table for classification of observations and the graphical representation of the *degree of fit* of the samples to each pair of classes, known as *Coomans plot.*

13.3.3.3 kNN Method

This is a *non-parametric* method that calculates the *distances matrix* between all n observations and uses the following assignation rule: "sample $\vec{w}$ is assigned to the group most represented among the nearest k observations". Generally k is odd, and the size of the groups is also taken into account.

13.3.3.4 Canonical Variate Analysis(CVA)

To obtain the best $\boldsymbol{q}$-dimensional ($q = min(p, k - 1)$) graphical representation of the observations of the training group, by maximizing the differences between k groups, q new variables are obtained ($Y_1, \ldots, Y_q$), called *canonical variables*, which are a linear combination of the original ones, incorrelated and maximizing the differences between the groups by the transformation $Y_{(n,q)} = X_{(n,p)}A_{(p,q)}$. The following results are obtained by applying this technique: the transformation matrix ($A_{(p,q)}$), the coordinates of the observations in the canonical variables ($Y_{(n,q)}$) for their graphical representation, the coordinates of the centroids for each group, and the structure matrix with the correlations between the canonical and original variables. The distances to the centroids could be used to assign new samples.

13.3.3.5 MANOVA

Assuming a normal multivariate distribution, with the same covariance matrices, in each of the populations, $(X_1, X_2, \ldots, X_p)\ N(\vec{\mu}_i, \Sigma)$, the *multivariate analysis of variance* (*MANOVA*) for a single factor with k levels (extension of the single factor ANOVA to the case of p variables), permits the equality of the k mean vectors in p variables to be tested ($H_0 \equiv \vec{\mu}_1 = \vec{\mu}_2 = \ldots = \vec{\mu}_k$), where $\vec{\mu}'_i = (\bar{\mu}^i_1, \bar{\mu}^i_2, \ldots, \bar{\mu}^i_p)$ is the mean vector of p variables in population W_i. The statistic used in the comparison is the Λ *of Wilks*, the value of which can be estimated by another statistic with F-distribution. If the calculated value is greater than the tabulated value, the null hypothesis for equality of the k mean vectors must be rejected. To establish whether the variables can distinguish each pair of groups a statistic is used with the F-distribution with p and $n - p - k + 1$ df, based on the square of Mahalanobis' distance between the centroids, that permits the equality of the pairs of mean vectors to be compared ($H_0 \equiv \vec{\mu}_i = \vec{\mu}_j$) (Afifi and Azen 1979; Martín-Álvarez 2000).

We have used SLDA to select the variables most useful to differentiate groups of wine samples (Cabezudo et al. 1986; Martín-Álvarez et al. 1987, 1988; Pozo-Bayón et al. 2001, 2003a,b; Monagas et al. 2007) to distinguish between grape, apple and pineapple juice (Dizy et al. 1992), to classify commercial orange juices (Simó et al. 2004), and to differentiate whisky brands (Herranz et al. 1989). SIMCA and kNN were used to differentiate Spanish wines (Martín-Álvarez et al. 1987) and whisky brands (Martín-Álvarez et al. 1988). LDA, QDA, CVA and kNN were used to differentiate gin brands (Martín-Alvarez and Herranz 1991) and rum brands (Herranz et al. 1990). As an example, Tables 13.16 to 13.22 show the results of

Table 13.16 Results of applying LDA to the data of 10 volatile compounds in order to discriminate the four groups of varietal wines. Classification functions:

	Malvar	Airén	Trepat	Monastrell
Methanol	10.799	10.257	0.256	5.117
1-Propanol	22.060	19.429	–10.130	4.492
Isobutanol	7.578	10.969	9.830	3.557
Isoamylic alcohols	2.758	2.107	–2.919	0.675
1-Hexanol	–426.708	–326.405	357.505	–62.965
cis-3-Hexen-1-ol	–689.479	–641.758	442.618	–86.874
Hexanoic acid	18.915	19.051	5.176	11.013
Octanoic acid	–25.394	–27.983	11.366	–8.608
Decanoic acid	7.380	40.821	153.799	61.847
Ethyl octanoate	–32.454	–7.918	81.870	6.050
Constant	–818.041	–813.798	–761.745	–259.392

Table 13.17 Results of applying LDA to the data of 10 volatile compounds in order to discriminate the four groups of varietal wines. Posterior probabilities:

Wine:	Observed classif.	Malvar	Airén	Trepat	Monastrell
Ma98	Malvar	1.00	0.00	0.00	0.00
Ma98	Malvar	1.00	0.00	0.00	0.00
Ma99	Malvar	1.00	0.00	0.00	0.00
Ma99	Malvar	1.00	0.00	0.00	0.00
Ai97	Airén	0.00	1.00	0.00	0.00
Ai97	Airén	0.00	1.00	0.00	0.00
Ai98	Airén	0.00	1.00	0.00	0.00
Ai98	Airén	0.00	1.00	0.00	0.00
Tr97	Trepat	0.00	0.00	1.00	0.00
Tr97	Trepat	0.00	0.00	1.00	0.00
Tr98	Trepat	0.00	0.00	1.00	0.00
Tr98	Trepat	0.00	0.00	1.00	0.00
Mo97	Monastrell	0.00	0.00	0.00	1.00
Mo97	Monastrell	0.00	0.00	0.00	1.00
Mo98	Monastrell	0.00	0.00	0.00	1.00
Mo98	Monastrell	0.00	0.00	0.00	1.00

Table 13.18 Results of applying LDA to the data of 10 volatile compounds in order to discriminate the four groups of varietal wines. Classification matrix:

		Predicted classifications:			
Observed classifications	Percent correct	Malvar	Airén	Trepat	Monastrell
Malvar	100.0	4	0	0	0
Airén	100.0	0	4	0	0
Trepat	100.0	0	0	4	0
Monastrell	100.0	0	0	0	4
Total	100.0	4	4	4	4

Table 13.19 Results of applying CVA to the data of 10 volatile compounds in order to discriminate the four groups of varietal wines. Standardized Coefficients for Canonical (Root) Variables:

	Root 1	Root 2	Root 3
Methanol	−1.1987	−0.3426	0.0429
1-Propanol	−1.9686	−0.4525	0.6012
Isobutanol	−0.0210	−5.3379	0.5914
Isoamylic alcohols	−4.140	2.5574	−1.7933
1-Hexanol	3.8768	−0.9482	−0.8557
cis-3-Hexen-1-ol	4.2968	1.2231	0.2976
Hexanoic acid	−0.6603	−0.3863	−0.1481
Octanoic acid	0.9565	0.2778	1.2102
Decanoic acid	1.2183	−1.2147	−0.6512
Ethyl octanoate	1.5857	−1.7593	−0.3254
Eigenvalues	703.5202	38.9719	5.6440
Cum. Proportion	0.9404	0.9925	1.0000

Table 13.20 Results of applying CVA to the data of 10 volatile compounds in order to discriminate the four groups of varietal wines. Unstandardized Canonical (root) Scores:

	Group	Root 1	Root 2	Root 3
Ma98	Malvar	−24.068	0.426	4.769
Ma98	Malvar	−22.078	1.826	2.500
Ma99	Malvar	−22.447	2.930	2.205
Ma99	Malvar	−22.428	−0.445	2.079
Ai97	Airén	−17.647	−7.432	−2.442
Ai97	Airén	−18.098	−5.638	−3.028
Ai98	Airén	−17.348	−4.570	−0.604
Ai98	Airén	−18.936	−5.900	−2.943
Tr97	Trepat	33.058	−3.201	0.758
Tr97	Trepat	37.019	−4.041	1.288
Tr98	Trepat	35.529	−4.025	1.372
Tr98	Trepat	35.093	−2.968	0.458
Mo97	Monastrell	5.687	8.399	−0.419
Mo97	Monastrell	5.669	8.537	−2.631
Mo98	Monastrell	5.758	8.144	−1.279
Mo98	Monastrell	5.238	7.957	−2.083

Table 13.21 Results of applying CVA to the data of 10 volatile compounds in order to discriminate the four groups of varietal wines. Means of Canonical (root) Variables:

	Root 1	Root 2	Root 3
Malvar	−22.76	1.18	2.89
Airén	−18.01	−5.89	−2.25
Trepat	35.17	−3.56	0.97
Monastrell	5.59	8.26	−1.60

Table 13.22 Results of applying CVA to the data of 10 volatile compounds in order to discriminate the four groups of varietal wines. Factor structure matrix:

	Root 1	Root 2	Root 3
Methanol	0.0265	0.0682	–0.3696
1-Propanol	–0.1184	–0.0282	0.6616
Isobutanol	0.0175	–0.0707	–0.0332
Isoamylic alcohols	0.0061	0.0201	–0.2463
1-Hexanol	0.0579	0.0541	–0.0282
cis-3-Hexen-1-ol	0.0994	0.0792	0.1281
Hexanoic acid	–0.0129	–0.0597	0.3269
Octanoic acid	0.1023	–0.0101	0.0959
Decanoic acid	0.0535	–0.0530	0.0800
Ethyl octanoate	0.0009	–0.2568	–0.3631

applying LDA and CVA to the data of 10 volatile compounds in order to discriminate the four groups of varietal wines (Pozo-Bayón et al. 2001), obtained with the STATISTICA program (*Discriminant Analysis* procedure in *Multivariate Exploratory Techniques* module). The results of LDA include: the classification functions (one for each group) for assigning the samples(Table 13.16), the posterior probabilities for the classification of wines(Table 13.17), and the classification matrix in Table 13.18(100% correct assignment of the wines was obtained). The results from CVA include: standardized coefficients for canonical (roots) variables (Table 13.19), unstandardized canonical scores (Table 13.20), means of canonical variables in the four groups (Table 13.21), and in the Table 13.22 the factor structure matrix (correlations between original and canonical variables). In Fig. 13.6, the wines are plotted on the plane defined by the first two canonical variables. The red varieties essentially have lower concentrations of 1-propanol and higher concentrations of *cis*-3-hexen-1-ol and octanoic acid than the white varieties. Ethyl octanoate differentiates the two red varieties and also the two white varieties. The wines of the Malvar and Airén varieties are close together in the figure. The results of hypotheses tests about the mean vectors for each pair of groups are shown in Table 13.23. The value of F-statistic (19.450 with 30 and 9 df), corresponding to the approximation of Λ-Wilks (0.00001) revealed the existence of significant differences between the four mean vectors (P<0.01), in other words, the variables did differentiate the groups. The value of F-statistic with 10 and 3 df to establish the differences between mean vectors in each pair of groups reveals significant differences (P<0.01) for all pairs except for the Malvar and Airén groups ($F_{cal} = 4.948 < F_{.95,\ 10,3} = 8.79$).

Table 13.23 F values (df = 10,3) from results of hypotheses tests about the mean vectors for each pair of groups

	Malvar	Airén	Trepat	Monastrell
Malvar		4.948	169.103*	43.679*
Airén	4.948		142.207*	37.862*
Trepat	169.103*	142.207*		51.082*
Monastrell	43.679*	37.862*	51.082*	

*Significant differences between mean vectors in each pairs of groups (P<0.01).

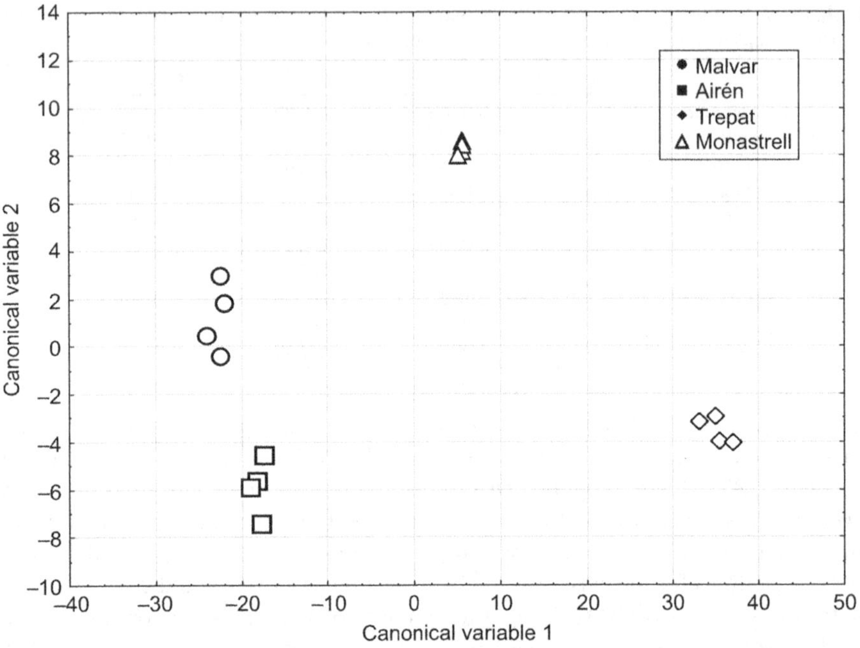

Fig. 13.6 Plot of the wines on the plane defined by the first two canonical variables

13.3.4 Multivariate Statistical Techniques to Study Dependence

The principal objective of these techniques is to study the dependence between two groups of variables, $(X_1, X_2, \ldots, X_p)$ and $(Y_1, \ldots, Y_q)$, from n observations in these $p+q$ variables:

$$
\begin{array}{cc}
 & \begin{array}{ccc} \overline{X_1 \; X_2 \ldots X_p} & & \overline{Y_1 \;\; Y_2 \ldots \; Y_q} \end{array} \\
Observ. \begin{array}{c} 1 \\ 2 \\ \ldots \\ n \end{array} & \begin{pmatrix} x_{1,1} \; x_{1,2} \ldots x_{1,p} & y_{1,1} \; y_{1,2} \ldots y_{1,q} \\ x_{2,1} \; x_{2,2} \ldots x_{2,p} & y_{2,1} \; y_{2,2} \ldots y_{2,q} \\ \ldots \; \ldots \; \ldots \; \ldots & \ldots \; \ldots \; \ldots \; \ldots \\ x_{n,1} \; x_{n,2} \ldots x_{n,p} & y_{n,1} \; y_{n,2} \ldots y_{n,q} \end{pmatrix}
\end{array}
$$

Besides the descriptive values, it is also interesting to know the correlations between the two groups of variables (r_{X_i, Y_j}). The multivariate statistical methods for this data matrix are *Canonical Correlation Analysis* (CCA) to investigate the relationship between both sets of variables, and *Multivariate Regression* with a view to predicting the values of the response variables in the Y-block in function of the variables in the X-block, using a mathematical model.

13.3.4.1 Canonical Correlation Analysis (CCA)

In order to measure the association between the two sets of variables, CCA calculates m new variables ($m = min(p, q)$) in each block ($F_1, \ldots, F_m, S_1, \ldots, S_m$), called canonical variables, that are linear combinations of the original variables ($F_{(n,m)} = X_{(n,p)}A_{(p,m)}$ and $S_{(n,m)} = Y_{(n,q)}B_{(q,m)}$), and with the largest possible correlation ($corr(F_1, S_1) \geq \ldots \geq corr(F_m, S_m)$). The results obtained with this statistical technique are: the transformation matrices ($A_{(p,m)}$, $B_{(q,m)}$), the score matrices ($F_{(n,m)}$, $S_{(n,m)}$), and the canonical correlation values ($R_i = corr(F_i, S_i)$) and their statistical significances. Observation of the successive columns of the matrices $A_{(p,m)}$ and $B_{(q,m)}$ will enable us to establish the variables most correlated with every canonical variable. It is, also, possible to obtain the graph of dispersion of F_1 vs S_1. However, this method cannot be used to predict values of the variables of the Y-block, and requires $n > p + q$. CCA was used to examine the linear relationship between chemical composition and foam characteristics of wine and cava samples (Pueyo et al. 1995).

13.3.4.2 Multivariate Regression

The aim of this technique is to predict values of the response or dependent variables ($Y_1, \ldots, Y_q$), as a function of the predictive, or independent variables ($X_1, X_2, \ldots, X_p$), by applying a mathematical model $Y_j = f(X_1, X_2, \ldots, X_p)$, that will be estimated using n observations of the calibration set, $\{(x_{i,1}, x_{i,2}, \ldots, x_{i,p}, y_{i,1}, \ldots, y_{i,q})\}_{i=1,\ldots,n}$. These observations may have been selected by a fixed or randomised experimental design.

Multiple Linear Regression (MLR)

MLR accepts for the observed value of each random dependent variable the following linear model: $y_i = \beta_0 + \beta_1 x_{i,1} + \beta_2 x_{i,2} + \ldots + \beta_p x_{i,p} + \varepsilon_i$, where β_i are the unknown parameters, and ε_i the independent error variables with normal distribution ($\varepsilon_i \sim N(0, \sigma)$). If we assume that ($x_{i,1}, x_{i,2}, \ldots, x_{i,p}$) are fixed values of the independent variables ($X_1, X_2, \ldots, X_p$), then the y_i values will have a normal distribution with a common standard deviation ($y_i \sim N(\beta_0 + \sum_{j=1}^{p} \beta_j x_{i,j}, \sigma)$). Using the *ordinary least squares* (OLS) procedure, which minimizes the sum of squares of errors ($\sum_{i=1}^{n} \varepsilon_i^2 = \sum_{i=1}^{n} (y_i - \beta_0 - \beta_1 x_{i,1} - \beta_2 x_{i,2} - \ldots - \beta_p x_{i,p})^2$), the estimated linear model (*regression equation*) is $\hat{y}_i = b_0 + b_1 x_{i,1} + b_2 x_{i,2} + \ldots + b_p x_{i,p}$. The *regression coefficients* b_i, estimators of the parameters β_i, can be calculated according to $\vec{b} = (X^t X)^{-1} X^t \vec{y}$, where $\vec{b} = \begin{pmatrix} b_0 \\ b_1 \\ \ldots \\ b_p \end{pmatrix}$, $\vec{y} = \begin{pmatrix} y_1 \\ y_2 \\ \ldots \\ y_n \end{pmatrix}$, $X = \begin{pmatrix} 1 & x_{1,1} & \ldots & x_{1,p} \\ 1 & x_{2,1} & \ldots & x_{2,p} \\ \ldots & \ldots & \ldots & \ldots \\ 1 & x_{n,1} & \ldots & x_{n,p} \end{pmatrix}$, providing that the matrix X is not singular. Among the estimators of β_i which are both unbiased and linear functions of the y_i values, the OLS estimators have the smallest variance.

The most important results obtained using MLR are:

- The *regression coefficients* (b_i) and their standard errors, the confidence interval for β_i, and values of the t-statistic to test the null hypothesis $H_0 \equiv \beta_i = 0$ and their associated probabilities
- The *ANOVA table*, with the decomposition of the total variation of y_i values with respect of their mean (SS_{tot}), in two parts ($SS_{tot} = SS_{reg} + SS_{res}$): the variation explained by the regression model (SS_{reg}) and the non-explained variation (SS_{res}), and the F-value ($F_{cal} = \dfrac{SS_{reg}/p}{SS_{res}/(n-p-1)}$) with p and $n-p-1$ df, to test the global hypothesis $H_0 \equiv \beta_1 = \ldots = \beta_p = 0$ vs $H_1 \equiv$ *not all of the coefficients are equal to zero* (*F-Test for overall significance*)
- The calculated values ($\hat{y}_i = b_0 + b_1 x_{i,1} + b_2 x_{i,2} + \ldots + b_p x_{i,p}$), the residuals ($e_i = y_i - \hat{y}_i$) and the graphical representation of the calculated ($\hat{y}_i$) vs. observed (y_i) values
- The statistics about the *goodness of fit*: the *coefficient of determination* $R^2 = \dfrac{SS_{reg}}{SS_{tot}} = 1 - \dfrac{SS_{res}}{SS_{tot}}$, the *adjusted coefficient of determination* $R^2_{adj} = 1 - \dfrac{SS_{res}/(n-p-1)}{SS_{tot}/(n-1)}$, the *multiple correlation coefficient* ($R = \sqrt{R^2}$), which is the correlation coefficient between observed and calculated values of Y, and the *standard error of estimation*, $s = \sqrt{SS_{res}/(n-p-1)} = \sqrt{\sum_{i=1}^{n} (y_i - \hat{y}_i)^2/(n-p-1)}$, that is an unbiased estimator of σ

To validate the estimated model, the hypotheses established on the errors (independence, equality of variances and normal distribution) must be confirmed. When there are genuine repeated runs in the data, the *lack of fit F-test* can be used.

Although the OLS estimation provides the smallest variance, the presence of *multicolinearity* among the X variables can give rise to unreliable predictions of Y, and in this case, the parameters β_i can also be estimated using other procedures such as:

- *Principal Components Regression* (PCR), which first carries out PCA on the ($X_1, X_2, \ldots, X_p$) variables, and then uses the scores of the q-first uncorrelated PCs, as independent variables ($\hat{y}_i = b^*_0 + b^*_1 PC_{i,1} + b^*_2 PC_{i,2} + \ldots + b^*_q PC_{i,q}$), or
- *Partial Least Squares (PLS) regression*, which is an alternative to PCR, using other q PCs components, that are also combinations of the variables but calculated assuming that it has a high correlation with the response variable Y

The results with PCR and PLS regression include the number of PCs obtained by leave-one-out cross-validation procedure, the values of regression coefficients for X variables, the value of R^2, and the *root mean square error of calibration* ($RMSEC$) and the *root mean square error of prediction* by cross-validation procedure ($RMSEP$), which are defined by the equations $RMSEC = \sqrt{\sum_{i=1}^{n} (y_i - \hat{y}_i)^2/n}$ and $RMSEP = \sqrt{\sum_{i=1}^{n} (y_i - \hat{y}_{(i)})^2/n}$, where y_i is the observed value of Y, $\hat{y}_i$, is the predicted value, and $\hat{y}_{(i)}$ is the predicted value when the regression model is constructed without the sample i, and n is the number of samples. $RMSEP$ is a measure of the model's ability to predict the values of the response variables in new samples.

In many cases, it is possible to use only a subset of the p variables X without a serious loss predictive ability with the *forward stepwise regression* (SWMLR) procedure, which, in each step, selects the predictor that more increases the variation explained and verifies if a previously selected predictor can be removed (values for F-statistics to enter and to remove variables should be fixed); or with the *best subsets regression* procedure.

Applications

MLR was used to investigate the factors influencing growth and tyramine production (Marcobal et al. 2006a); SWMLR, PCR and PLS regression have been used to predict the aging time of the wine samples from the volatile composition (Pérez-Coello et al. 1999), and from the nitrogenous fraction (Moreno-Arribas et al. 1998). MLR was applied to detect the proportion of each fruit present in mixtures of grape and apple juices (Dizy et al. 1992). SWMLR has been used: to predict the CIELAB variables using the colorimetric indices as possible predictor variables (Monagas et al. 2006b), to predict the foam characteristics of sparkling wines (Moreno-Arribas et al. 2000), and to find out the phenolic compound that provided the best predictive model of the antioxidant capacity (Monagas et al. 2005). PLS was used to model quantitative relationships between foam characteristics and chemical composition of cava samples (Pueyo et al. 1995). Table 13.24 shows the results of the application of PLS regression for the prediction of aging times in wines from five amino acids in peptides <700-Da fraction (Moreno-Arribas et al. 1998), obtained with The Unscrambler program version 7.8 (CAMO PROCESS AS, http://www.camo.no/). Figure 13.7 shows the predicted values obtained using the fitted regression equation

Table 13.24 Results of PLS regression for prediction of the aging times in wines from five amino acids in peptides <700 Da fraction

Regresión coefficients:						Statistics:			
Intercept	Aspartic acid + asparagine	Threonine	α-Alanine	γ-Amino butyric acid	Ornithine	NC	R^2	RMSEC	RMSEP
4.13513	1.14990	–.77765	–.89973	–.69257	1.07405	2	.968	1.67	1.94

NC = number of components selected by cross-validation, R^2 = determination coefficient, RMSEC = Root Mean Square Error of Calibration, RMSEP = Root Mean Square Error of Prediction

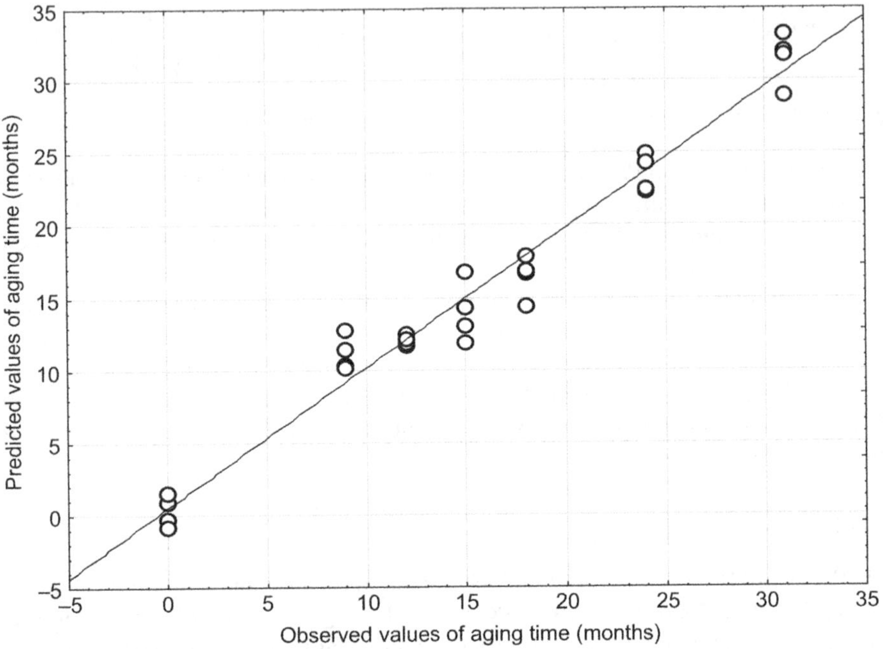

Fig. 13.7 Predicted values obtained using PLS regression vs observed values for aging time

vs the observed values for time. As can be seen, the fit for the predictions of aging time can be considered to be appropriate.

We have also used these techniques in the area of dairy products, more specifically to determine the percentage composition of milk blends (Molina et al. 1995, 1999; Recio et al. 2004), to predict the ripening time of cheeses (Poveda et al. 2004; García-Ruíz 1998), and to predict sensory attributes of cheeses (Cabezas et al. 2006; González-Viñas et al. 2007).

Acknowledgments This work was supported by grants AGL2006-04514 and AGL2005-03381, from the Spanish Ministerio de Educación y Ciencia.

References

Afifi, A.A., & Azen, S.P. (1979). Statistical analysis. A computer oriented. Academic Press, Inc. New York.

Afifi A.A., & Clark V. (1996). Computer-Aided Multivariate Analysis. Chapman & Hall, London.

Alcaide-Hidalgo, J.M., Moreno-Arribas, M.V., Martín-Álvarez, P.J., & Polo, M.C. (2007). Influence of malolactic fermentation, postfermentative treatments and ageing with lees on nitrogen compounds of red wines. *Food Chem., 103*, 572–581.

Barcenilla, J.M., Martin-Alvarez, P.J., Vian, A., & González, R. (2003). Characterization of yeast strains on the basis of autolytic activity and volatile compounds. *Food Sci. Technol. Int., 9*,

95–100.

Brereton, R.G. (1990). Chemometrics: applications of mathematics and statistics to laboratory system. Ellis Horwood Limited, Chichester, England.

Cabezas, L., González-Viñas M.A., Ballesteros C., & Martín-Álvarez P.J. (2006). Application of partial least squares regression to predict sensory attributes of artisanal and industrial Manchego cheeses. *Eur. Food Res. Technol., 222*, 223–228.

Cabezudo M.D., Polo, M.C., Herráiz, M., Reglero, G., González-Raurich, M., Cáceres, I., & Martín-Álvarez, P.J. (1986). Using discriminant analysis to characterize Spanish variety white wines. In: The Shelf Life of Foods and Beverages, pp. 186–204. Ed. by G. Charalambous. Elsevier Science Publishers B.V., Amsterdam.

Cela, R. (1994). Avances en Quimiometría Práctica. Servicio de Publicaciones de la Universidad de Santiago de Compostela. Santiago de Compostela.

Dizy, M., Martín-Álvarez, P.J., Cabezudo, M.D., & Polo, M.C. (1992). Grape, apple and pineapple juice characterisation and detection of mixtures. *J. Sci. Food Agric., 60*, 47–53.

Draper, N.R., & Smith, H. (1981). Applied regression analysis. John Wiley and Sons, Inc., New York.

García-Ruíz, A., Cabezas L., Martín-Álvarez P.J., & Cabezudo M.D. (1998). Prediction of the ripening times of Manchego cheese using multivariate statistical analyses: A preliminary study. *Z. Lebensm. Unters. Forsch., 206*, 382–386.

Gardiner, W.P. (1997). Statistical Analysis Methods for Chemists. A Software Based Approach. The Royal Society of Chemistry, Cambridge, UK.

González-Viñas, M.A., Ballesteros, C., Martín-Álvarez, P.J., & Cabezas, L. (2007). Relationship between sensory and instrumental measurements of texture for artisanal and industrial manchego cheeses. *J. Sens. Stud., 22*, 462–476.

Hernández, T., Estrella, I., Carlavilla, D., Martín-Álvarez, P.J., & Moreno-Arribas, M.V. (2006). Phenolic compounds in red wine subjected to industrial malolactic fermentation and ageing on lees. *Anal. Chim. Acta, 563*, 116–125.

Herraiz, T., Martín-Álvarez, P.J., Reglero, G., Herraiz, M., & Cabezudo, M.D. (1990). Effects of the presence of skins during alcoholic fermentation on the composition of wine volatiles. *Vitis, 29*, 239–249.

Herranz, A., Serna, P., Barro, C., Martín-Álvarez, P.J., & Cabezudo, M.D. (1989). Application of the statistical multivariate analysis to the differentiation of whiskies of different brands. *Food Chem., 31*, 73–81.

Herranz, A., Serna, P., Barro, C., & Martín-Álvarez, P.J. (1990). Multivariate statistical methods applied to the differentiation of rum brands. *J. Sci. Food Agric., 51*, 555–560.

Hidalgo, P., Pueyo, E., Pozo-Bayón, M.A., Martínez-Rodríguez, A.J., Martín-Álvarez, P.J., & Polo, M.C. (2004). Sensory and analytical study of rosé sparkling wines manufactured by second fermentation in the bottle. *J Agric. Food Chem., 52*, 6640–6645.

Jobson J.D. (1991). Applied Multivariate Data Analysis. Volume I: Regression and Experiemtnal Design. Springer-Verlag. New York, Inc.

Krzanowski, W.J. (1988). Principles of multivariate analysis. A User's Perspective. Oxford University Press Inc., New York.

Marcobal, A., Polo, M.C., Martín-Álvarez, P.J., & Moreno-Arribas, M.V. (2005). Biogenic amine content of red Spanish wines: Comparison of a direct ELISA and an HPLC method for the determination of histamine in wines. *Food Res. Int., 38*, 387–394.

Marcobal, A., Martín-Álvarez, P.J., Moreno-Arribas M.V., & Muñoz, R. (2006a). A multifactorial design for studying factors influencing growth and tyramine production of the lactic acid bacteria *Lactobacillus brevis* CECT 4669 and *Enterococcus faecium* BIFI-58. *Res. Microbiol., 157*, 417–424.

Marcobal, A., Martín-Álvarez, P.J., Polo, M.C., Muñoz, R., & Moreno-Arribas, M.V. (2006b). Formation of biogenic amines throughout the industrial manufacture of red wine. *J. Food Protect., 69*, 397–404.

Martín-Álvarez, P.J. (2000). Quimiometría Alimentaria. Universidad Autónoma de Madrid, Madrid.

Martín-Álvarez, P.J. (2006). Prácticas de tratamiento estadístico de datos con el programa SPSS para Windows. Aplicaciones en el área de Ciencia y Tecnología de Alimentos. Ed. CSIC, Madrid.

Martín-Álvarez, P.J., & Herranz, A. (1991). Application of multivariate statistical methods to the differentiation of gin brands. *J. Sci. Food Agric., 57*, 263–272.

Martín-Álvarez, P.J., Polo, M.C., Herráiz, M., Cáceres, I., González Raurich, M., Herráiz, T., Reglero, G., Cabezudo M.D., & Sanz, J. (1987). Application of chemometrics to the characterization of Spanish wines. In: Flavour Science and Technology, pp. 489–499. Ed. by M. Martens, G.A. Dalen & H. Russwurm Jr. John Wiley & Sons Ltd., Oslo, Norway.

Martín-Álvarez, P.J., Cabezudo M.D., Sanz, J.. Herranz, A., de la Serna. P., & Barro, C. (1988). Application of several statistical classification techniques to the differentiation of whisky brands. *J. Sci. Food Agric., 45*, 347–358.

Martín-Álvarez, P.J., Marcobal, A., Polo, C., & Moreno-Arribas, M.V. (2006). Influence of technological practices on biogenic amine contents in red wines. *Eur. Food Res. Technol., 222*, 420–424.

Martínez-Rodríguez, A.J., Carrascosa, A.V., Martín-Álvarez, P.J., Moreno-Arribas, M.V., & Polo, M.C. (2002). Influence of the yeast strain on the changes of the amino acids, peptides and proteins during sparkling wine production by the traditional method. *J. Ind. Microbiol. Biotecht., 29*, 314–322.

Massart, D.L., Vandeginste, B.G.M., Deming, S.N., Michotte, Y., & Kaufman, L. (1990). Chemometrics: A textbook. Elsevier Science Publishing Company Inc., New York.

Miller, J.C., & Miller, J.N. (2000). Statistics for analytical chemistry. Pearson Education Limited. Harlow, England.

Molina, E., Ramos M., & Martín-Álvarez P. J. (1995). Prediction of the percentages of cows', goats', and ewes' milk in Ibérico cheese by electrophoretic analysis of whey proteins. *Z. Lebensm. Unters. Forsch., 201*, 331–335.

Molina, E., Martín-Álvarez P. J., & Ramos M. (1999). Analysis of cows', ewes' and goats' milk mixtures by capillary electrophoresis: Quantification by multivariate regression analysis. *Int. Dairy J., 9*, 99–105.

Monagas, M., Hernández-Ledesma, B., Garrido, I., Martín-Álvarez, P.J., Gómez-Cordovés, C., & Bartolomé, B. (2005). Quality assessment of commercial dietary antioxidant products from *Vitis vinifera L.* grape seeds. *Nutr. Cancer, 53*, 244–254.

Monagas, M., Martín-Álvarez, P.J., Bartolomé, B., & Gómez-Cordovés, C. (2006a). Statistical interpretation of the color parameters of red wines in function of their phenolic composition during aging in bottle. *Eur. Food Res. Technol., 222*, 702–709.

Monagas, M., Martín-Álvarez, P.J., Gómez-Cordovés, C., & Bartolomé, B. (2006b). Time course of the colour of young red wines from *Vitis vinifera L.* during ageing in bottle. *Int. J. Food Sci. Technol., 41*, 892–899.

Monagas, M., Martín-Álvarez, P.J., Gómez-Cordovés, C., & Bartolomé, B. (2007). Effect of the modifier (Graciano vs. Cabernet sauvignon) on blends of Tempranillo wine during ageing in the bottle. II. Colour and overall appreciation. *LWT-Food Sci. Technol., 40*, 107–115.

Moreno-Arribas, M.V., Pueyo, E., Polo, M.C., & Martín-Álvarez, P.J. (1998). Changes in the amino acid composition of the different nitrogenous fractions during the aging of wine with yeasts. *J. Agric. Food Chem., 46*, 4042–4051.

Moreno-Arribas, M.V., Cabello, F., Polo, M.C., Martín-Álvarez, P.J., & Pueyo, E. (1999). Assessment of the native electrophoretic analysis of total grape must proteins for the characterization of *Vitis vinifera* L. cultivars. *J. Agric. Food Chem., 47*, 114–120.

Moreno-Arribas, V., Pueyo, E., Nieto, F.J., Martín-Álvarez, P.J., & Polo, M.C. (2000). Influence of the polysaccharides and the nitrogen compounds on foaming properties of sparkling wines. *Food Chem., 70, 309–317.*

O'Mahony, M. (1986). Sensory evaluation of food. Statistical methods and procedures, Marcel Dekker, Inc., New York.

Pérez-Coello, M.S., Martín-Álvarez, P.J., & Cabezudo M.D. (1999). Prediction of the storage time in bottles of Spanish white wines using multivariate statistical analysis. *Z. Lebensm. Unters. Forsch. A, 208*, 408–412.

Poveda, J.M., García, A., Martín-Álvarez P.J., & Cabezas L. (2004). Application of partial least squares (PLS) regression to predict the ripening time of Manchego cheese. *Food Chem., 84*, 29–33.

Pozo-Bayón, M.A., Pueyo, E., Martín-Álvarez, P.J., & Polo, M.C. (2001). Polydimethylsiloxane solid-phase microextraction–gas chromatography method for the analysis of volatile compounds in wines. Its application to the characterization of varietal wines. *J. Chromatogr. A, 922*, 267–275.

Pozo-Bayón, M.A., Hernández, M.T., Martín-Álvarez, P.J., & Polo, C. (2003a). Study of low molecular weight phenolic compounds during the aging of sparkling wines manufactured with red and white grape varieties. *J. Agric. Food Chem., 51*, 2089–2095.

Pozo-Bayón, M.A., Pueyo, E., Martín-Álvarez, P.J., Martinez-Rodriguez, A.J., & Polo, M.C. (2003b). Influence of yeast strain, bentonite addition, and aging time on volatile compounds of sparkling wines. *Am. J. Enol. Viticult., 54*, 273–278.

Pozo-Bayón, M.A., Polo, M.C., Martín-Álvarez, P.J., & Pueyo, E. (2004). Effect of vineyard yield on the composition of sparkling wines produced from the grape cultivar Parellada. *Food Chem., 86*, 413–419.

Pozo-Bayón, M.A., García-Alegría, E., Polo, M.C., Tenorio, C., Martín-Álvarez, P.J., Calvo de la Banda, M.T., Ruiz-Larrea, F., & Moreno-Arribas, M.V. (2005). Wine volatile and amino acid composition after malolactic fermentation: Effect of *Oenococcus oeni* and *Lactobacillus plantarum* starter cultures. *J. Agric. Food Chem., 53*, 8729–8735.

Pueyo, E. Martín-Álvarez, P.J., & Polo, M.C. (1995). Relationship between foam characteristics and chemical composition in wines and cavas (sparkling wines). *Am. J. Enol. Viticult., 46*, 518–524.

Recio, I., García-Risco, M.R., Amigo, L., Molina, E., Ramos, M., & Martín-Álvarez, P. J. (2004). Detection of milk mixtures in Halloumi cheese. *J. Dairy Sci., 87*, 1595–1600.

Sharaf, M.A., Illman, D.L., & Kowalski, B.R. (1986). Chemometrics. John Wiley & Sons, Inc., New York.

Simó, C., Martín-Álvarez, P.J., Barbas, C., & Cifuentes, A. (2004). Application of stepwise discriminant analysis to classify commercial orange juices using chiral micellar electrokinetic chromatography-laser induced fluorescence data of amino acids. *Electrophoresis, 25*, 2885–2891.

Index

M

N

Z

Made in United States
North Haven, CT
04 November 2026

21743222R00411